Units Used in the SI System

Quantity	Unit	Symbol	Formula
Base Unit			
Length	meter	m	
Mass	kilogram	kg	
Time	second	s	
Electric current	ampere	A	
Thermodynamic temperature	kelvin	K	
Amount of substance	mole	mol	
Luminous intensity	candela	cd	
Supplementary Units			
Plane angle	radian	rad	
Solid angle	steradian	sr	
Derived Units			
Acceleration	meter per second squared		m/s^2
Angular acceleration	radian per second squared		rad/s^2
Angular velocity	radian per second		rad/s
Area	square meter		m^2
Density	kilogram per cubic meter		kg/m^3
Electric capacitance	farad	F	$A \cdot s/V$
Electric inductance	henry	H	$V \cdot s/A$
Electric potential difference	volt	V	W/A
Electric resistance	ohm	Ω	V/A
Electromotive force	volt	V	W/A
Energy	joule	J	$N \cdot m$
Entropy	joule per kelvin		J/K
Force	newton	N	$kg \cdot m/s^2$
Frequency	hertz	Hz	cps
Luminous flux	lumen	lm	$cd \cdot sr$
Magnetic flux	weber	Wb	$V \cdot s$
Magnetomotive force	ampere	A	
Power	watt	W	J/s
Pressure	pascal	Pa	N/m^2
Quantity of electricity	coulomb	C	$A \cdot s$
Quantity of heat	joule	J	$N \cdot m$
Specific heat	joule per kilogram-kelvin		$J/kg \cdot K$
Stress	pascal	Pa	N/m^2
Thermal conductivity	watt per meter-kelvin		$W/m \cdot K$
Velocity	meter per second		m/s
Voltage	volt	V	W/A
Volume	cubic meter		m^3
Weight (force)	newton	N	$kg \cdot m/s^2$
Work	joule	J	$N \cdot m$

MANUFACTURING PROCESSES AND SYSTEMS

NINTH EDITION
MANUFACTURING
PROCESSES AND SYSTEMS

PHILLIP F. OSTWALD
University of Colorado
Boulder, Colorado

JAIRO MUÑOZ
Iowa Precision Industries, Inc.
Cedar Rapids, Iowa

JOHN WILEY & SONS
New York   Chichester   Brisbane   Toronto   Singapore   Weinheim

ACQUISITIONS EDITOR	Sharon Smith
PRODUCTION SERVICE	Spectrum Publisher Services
MARKETING MANAGER	Harper Mooy
MANUFACTURING MANAGER	Dorothy Sinclair
ILLUSTRATION COORDINATOR	Jaime Perea

This book was set in 10/12 pt Times Roman by Bi-Comp, Inc., and printed and bound by Hamilton Printing. The cover was printed by Lehigh Press.

Recognizing the importance of preserving what has been written, it is a policy of John Wiley & Sons, Inc. to have books of enduring value published in the United States printed on acid-free paper, and we exert our best efforts to that end.

The paper in this book was manufactured by a mill whose forest management programs include sustained yield harvesting of its timberlands. Sustained yield harvesting principles ensure that the number of trees cut each year does not exceed the amount of new growth.

Library of Congress Cataloging-in-Publication Data
Ostwald, Phillip F., 1931–
    Manufacturing processes and systems / Phillip F. Ostwald, Jairo
Muñoz.—9th ed.
        p.   cm.
    Includes bibliographical references.
    ISBN 0-471-04741-4 (alk. paper)
    1. Manufacturing processes.   I. Muñoz, Jairo.   II. Title.
TS183.088   1996
670—DC21   96-47300
                                                           CIP

Printed in the United States of America

10 9 8 7 6 5 4

PREFACE

The business of manufacturing is no longer a "processes only" activity. Traditionally, manufacturing processes had a limited perspective of dealing with transformation of material. Manufacturing leaders now proclaim that the process is *no more or no less* important than the system. For manufacturing to excel, it must include dramatic improvements in *systems.* In acknowledging this new reality, we have changed the name of the book Manufacturing Processes to *Manufacturing Processes and Systems* for the ninth edition.

Student learning and instructor teaching are two different sides of the coin. Figure P.1 shows how our gears of manufacturing learning and teaching works. Processes and support systems can only abide in this new world of manufacturing, and the instructor is the key to this paradigm.

In this edition, the instructor will notice several new chapters: Electronic Fabrication (Chapter 19), Operations Planning (Chapter 23), Quality Systems (Chapter 27), Process Automation (Chapter 29), Operator-Machine Systems (Chapter 30), and Cost Estimating (Chapter 31). These chapters focus on the systems side of manufacturing. Throughout the book, much more stress is given to this aspect of manufacturing. Integration is now the means to successful practice. Learning about systems is essential to manufacturing education, and the field is incomplete without it.

With increasing emphasis of design over manual skills in modern manufacturing courses, the text is introducing chapters on The Manufacturing System (Chapter 1), Geometric Dimensioning and Tolerancing (Chapter 24), Tool Design (Chapter 25), and Quality Systems (Chapter 27). This integration is coupled to the processes/systems context of manufacturing.

This text can be used for a variety of teaching situations: for lecture only, for lecture with a laboratory menu, or with professional mentoring with business, and developed field trips. A companion use of video and television learning through the lecture mode is possible. The instructor will notice Internet requirements that search for information. We provide Internet addresses throughout the book for various assignments. (Regrettably, these addresses may change from time to time.) In the interactive environment of teaching, this book is a part of modern courseware.

The instructor also will appreciate the versatility of this book. The book has more material than can be covered in one semester or quarter, and thus independent chapters can be selected to meet the various objectives of each class. For example, if the students already have an understanding of materials, Chapter 2, Nature and Properties of Materials, can be excluded from the course curriculum. Other chapters also can be excluded depending on student preparation and course objectives. However, there is enough material to cover two semesters or three quarters, if a longer teaching opportunity is available.

Some colleges or universities will use the book with a laboratory, which can be coordinated with the chapter selection. Other colleges or universities that simultaneously employ video or television assistance may adopt the book to give academic content to the instruction.

Practicing professionals, who have climbed from the ranks in industry, will find this

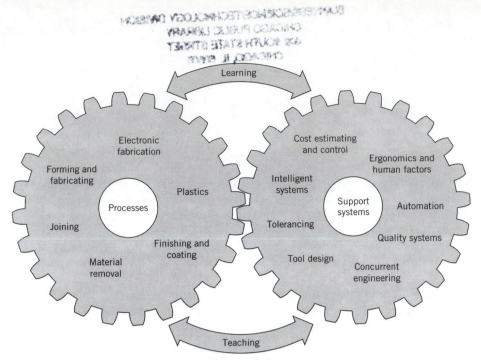

Figure P.1
Gears of learning and teaching for the manufacturing enterprise.

text useful for self-study. Even the apprentice needs a companion, and this text is a friend to many motivated young people. Many companies that provide in-plant courses to their employees also will see a need for this text.

The ninth edition has an increased number of Questions, Problems, More Difficult Problems, Practical Applications, and Case Studies. There is a range of difficulty for Questions. Some only require looking back in the chapter and responding with a few sentences. Other Questions seek a more thoughtful response. The Problems also have levels of difficulty. We stress manufacturing as a design activity; therefore, Problems and More Difficult Problems request sketches, CAD drawings, or computations. The student can provide this design using CAD, board and T-square, or sketches. Whenever the student is asked to prepare designs, much learning occurs.

For the instructor, a comprehensive Solution's Manual is available. This manual can be requested from the John Wiley college representative or from John Wiley and Sons directly.

There is a new end-of-chapter addition for the ninth edition, which we call Practical Applications. For example, it encourages field trips and communication through the Internet to technologists, engineers, and other manufacturing professionals. It asks for observations of ongoing operations, of operation of equipment to make parts, of visiting machine tool shows, and so on. The instructor will appreciate this experiential approach, allowing him or her to use Practical Applications in exciting ways.

The end-of-chapter Case Studies are open ended, perhaps having several solutions. Students are often disturbed by this anomaly, but instructors recognize that manufacturing courses are unlike calculus courses with their singularity of answers.

Academic requirements for this book/course can vary, but we believe that the text is widely suitable for a number of teaching approaches. The book has been written to appeal to a range of college/university/technical school settings. The student need only have a mathematics level of algebra and a high school or college courses in chemistry and physics. Word-processing computer skills are assumed, and some CAD ability is always helpful. A previous machine shop course is not necessary, nor is a concurrent laboratory course mandatory. Various academic levels and backgrounds are appropriate and the instructor will find that this text is suitable to a variety of teaching styles.

We have added information on the latest technology, keeping this text current with the rapidly changing field of manufacturing. For example, chapters on Electronic Fabrication (Chapter 19), Finish Processes (Chapter 22), Metrology and Testing (Chapter 26), and Computer Numerical Control Systems (Chapter 28) have been added or strengthened. When one realizes that electronic fabrication is an annual trillion dollar business, it is folly to ignore this pivotal manufacturing field of study. The book has been thoroughly rewritten, and now encourages the *design* pedagogy of learning.

The ninth edition of *Manufacturing Processes and Systems* is the longest running active book in this field. There are many international translations. If this book is considered exemplary, it is due to the late Professor Myron Begeman, who was the original author. He added Professor B.H. Amstead, now retired, and Bill's imprimatur is unmistakable and vitally important to the current authors. For the seventh and eighth editions, Professor Phillip F. Ostwald contributed as a co-author. Now with the ninth edition, a new co-author is continuing the book's grand heritage, Dr. Jairo Muñoz. He brings real-world experience as a practicing professional manufacturing engineer/manager/researcher to this book.

The authors are grateful to many people. Their advice and information has made this a much better text. Our special gratitude is given to helpful people such as William W. Wainwright, the late Franklin Essenburg, Chung ha Suh, Sander Friedman, Thomas Carson, Gordon Nordyke, Steve Hirsch, and Bill Kennedy. Students have graciously consented to field test the problems and have assisted in numerous other ways. The names used in the case studies are of real people, and they are mentioned because of our sincere regard for their contribution and friendship. For in writing a book of this magnitude, the authors are always aware that friends and colleagues are the hidden but important advisers.

Phillip F. Ostwald
Boulder, Colorado

Jairo Muñoz
Iowa City, Iowa

CONTENTS

CHAPTER 1
THE MANUFACTURING SYSTEM

Manufacturing is an industrial activity that changes the form of raw materials to create products. To be profitable, an enterprise establishes and nurtures a *manufacturing system* that facilitates the flow of information to coordinate inputs, processes, and outputs. Development of modern manufacturing, for example, is dependent on research in materials that may require a variety of new production processes. Success demands implementation of robust *manufacturing processes* and systems. This chapter gives students a good understanding of the *manufacturing enterprise* as a system, a view that is needed to excel in today's competitive markets.

The inventions of power-transforming machines, such as the water wheel, steam engine, and electric motor, substituted machine energy for human labor. The development of ferrous and nonferrous materials, and plastics and their shaping into useful products, moved humanity forward in technology with a giant step. The use of flyball governors, cams, electricity, electronics, and computers enabled people to control machine processes. These inventions came slowly, however, and along with them was the necessity for manufacturing. Computer-aided design and manufacturing, robotics, new alloys, electronic and aerospace designs, and safety and antipollution laws have stimulated development. In turn, improvements in manufacturing have enhanced the products that consumers use. Unfortunately, we overlook the history of manufacturing, which is an interesting field of study and culture. The authors, convinced that nations that wish to improve their standard of living always develop and strengthen their manufacturing systems, want the student to appreciate that manufacturing does have a glorious past and a promising future.

1.1 EVOLUTION OF THE ENTERPRISE

Manufacture is derived from the Latin, *manu factus*, or literally "made by hand." But the power of the hand tool is limited. Domesticated animals and water power from the water wheel were steps in the evolution. Harnessing water power provided an opportunity for the development of power-driven machine tools and manufacturing. In the United States the small shops were often located near a pond that had a waterfall. In fact, water-driven machine tools were used to build the first practical steam engine. Wilkinson's boring machine was used for machining the cylinders of James Watt's steam engine, which was invented in 1776. This is generally accepted to be the beginning of the Industrial Revolution. This power source could be located wherever it was needed for work.

Motive Power

With the availability of power and its transmission by overhead power belts, other machine tool developments followed. Note Figure 1.1, which is a woodcut illustration of

Figure 1.1
Interior of machine shop, 100 × 300 ft (30 × 90 m) using overhead belt-driven
power, circa 1885.

a machine shop interior. Overhead line shafts were driven by steam engines. The screw
cutting lathe, invented in 1797 by Henry Maudsley, was one of the first important machines. By the early 1900s the kinematics of the basic machine tools and structures for
mechanical cutting methods had been developed. The electrical motor, developed in the
early 1900s, improved the efficiency of power transformation, and eventually power line
shafts disappeared by the 1940s.

The Manufacturing System

An interesting aspect of the history of manufacturing is the relationship between people
and technology. The beginning of modern manufacturing processes, for example, may
be credited to **Eli Whitney** and his cotton gin, to *interchangeable production*, or to the
first milling machine (shown in Figure 1.2) of the early 1800s, or to any number of other
developments that took place in the world at that time. All these events, however, have
one thing in common: The skills of the operator were the dominant reason for product
quality. A method to transmit knowledge about the use of materials, tools, and machines
was very rudimental.

The origin of experimentation and analysis in manufacturing processes, credited in
large measure to Fred W. Taylor, marked a second stage in the manufacturing enterprise.
The use of machines gave people the roles of supervisors and monitors of semiautomatic
processes. Taylor's experiments determined that many functions were based more on
rules and knowledge than just on physical skills. Furthermore, it was clear that people
had developed *mental models* of the processes and machines under their control. Such

Figure 1.2
Eli Whitney's milling machine that is displayed by the New Haven Colony Historical Society. The belt drive was attached to the left spindle and the cutter to the right spindle, 1818.

models became operator's manuals, which are symbolic representation of knowledge about tools, machines, and materials, and their interactions with the environment.

Taylor's published papers on the art of cutting metals gave a scientific basis to production and allowed for a systems approach to the analysis of manufacturing. This is crucial for a society undergoing another revolution in which more and more skill-based labor is being replaced by automated knowledge-based equipment; that is, equipment that has "intelligent" capabilities such as speech recognition, picture understanding, and self-adapting control among others.

To understand the manufacturing system, we can describe it as having three essential parts: *input, process*, and *output*. Figure 1.3 shows the broad details of a manufacturing system.

1. *Input.* In a free enterprise system, *consumer demand* serves as the stimulant to encourage business to provide products. *Materials*, which are the minerals of nature, such as coal, ores, hydrocarbons, and many more, are converted into these products. It takes financing and money, gained from bank loans, from capital investments from stockholders, or from plowback of profit into the business, to sustain this activity. *Working capital* is

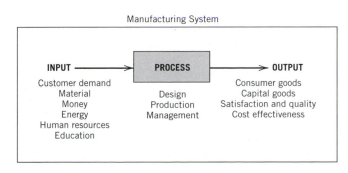

Figure 1.3
Diagram of a manufacturing system.

money used to buy materials and pay employees. *Fixed capital* is the money for tools, machines, and factory buildings. A manufacturing enterprise needs money for these and other requirements. *Energy* is an important input to manufacturing because it exists in many different forms, such as electricity, compressed air, steam, gas, or coal.

2. *Process.* Processes are the next step of the enterprise. *Management* provides planning, organization, direction, control, and leadership of the business enterprise to make it productive and profitable. Managers have responsibilities to the owners, employees, customers, general public, and the enterprise itself. It is essential for the business enterprise to make a profit or it will fail.

The *design* element consists of creating plans for products so that they are attractive, perform well, and give service at low cost. Manufactured products are designed before they are made. They may be designed by workers in the shop, draftspersons, or engineers, but design is usually handled by trained engineering specialists.

The third element is *production*. The processes needed to manufacture a product must be designed and engineered in great detail. General plans for the processes are recognized during the design stage, and now the techniques of manufacturing engineering are used. The best combination of machines, processes, and people are selected to satisfy the objectives of the firm, shareholders, employees, and customer.

3. *Output.* The output of a manufacturing system is a product. Look around you; products of manufacturing are everywhere. Classes of goods can be divided into consumer or capital goods. *Consumer goods* are those products that people buy for their personal consumption or use, such as food or cars. *Capital goods* are products purchased by manufacturing firms to make the consumer products. Machine tools, computer-controlled robots, and plants are examples of capital goods.

It used to be that processes were critical to efficient operation. Although they still are, they are not the only factor. International competitive markets, changes in consumer quality values, concern for the environment, and new ergonomics and safety standards demand a more global view to manufacturing practices that only a systems approach (the study of all the components and their relationships) can give.

1.2 CLASSIFICATION OF BASIC MANUFACTURING

There are a number of ways to classify manufacturing. Broadly speaking, they are mass production, moderate production, and job-lot production. A part is said to be mass produced if produced continuously or intermittently at high volume for a considerable period of time. Some authorities say that more than 100,000 parts per year must be produced to qualify as a mass-produced part, but this is a restrictive definition. In *mass production,* industry sales volume is well established and production rates are independent of individual orders. Machines producing these parts are usually incapable of performing operations on other work. Unit costs must be kept to an absolute minimum. Common examples of mass-produced items are matches, bottle caps, pencils, automobiles, nuts, bolts, washers, light globes, and wire.

Parts made in *moderate production* operations are produced in relatively large quantities, but the output may be more variable than for mass produced parts and more dependent on sales orders. The machines likely will be multipurpose ones, although this is not true in plants producing specialty items with less demand or sales than is the case of mass-produced parts. The number of parts may vary from 2500 to 100,000 per year depending on complexity. Examples of this type of industry are more descriptive: printing of books, aircraft compasses, and radio transmitters.

The *job-lot* industries are more flexible and their production is usually limited to lots closely attuned to sales orders or expected sales. Production equipment is multipurpose, and employees may be more highly skilled, performing various tasks depending on the part or assembly being made. Lot sizes, customarily varying from 10 to 500 parts per lot, are moved through the various processes from raw material to finished product. The company usually has three or more products and may produce them in any order and quantity depending on demand. In some cases, the plant may not have its own product and, if so, it then "contracts" work as a subcontractor or vendor. Product changes are rather frequent and in some the percent profit per item exceeds other types of manufacture. For example, these products may be produced in job-lot industries: airplanes, antique automobile replacement parts, oil field valves, special electrical meters, and artificial hands.

Mass, moderate, and job-lot production require different equipment and systems. The specification of the equipment identifies what it is and what it can do. A manufacturing system design may specify a robotic loading station for a mass-produced product. Figure 1.4 shows an isometric of a robotic system inserting a relay into a product presented to it on a pallet. The pallet is synchronized to the robot. Parts are picked up by the gripper from a feeder unit. Special orientation of the part in the gripper and egg-shell forces are features of the system. The loading rate is 438 units per hour, which is considered moderate to mass production volume.

Throughout this text, pictures, diagrams, and sketches are shown of processes, tools, and machines. The student will want to study these figures for details and overall impressions. In addition, performance statements are listed along with their special purpose.

Organizing for Manufacturing

Manufacturing depends on *organization* to achieve its purpose. Resources such as people, power, materials, machines, and money are necessities. For efficient and economical

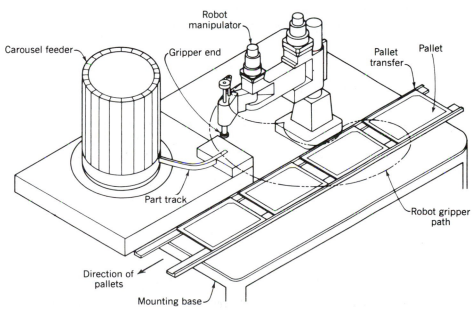

Figure 1.4
Elements of robot loader for a mass-produced electronic product.

Table 1.1   **Overview of the Evolution of Strategic Manufacturing Systems**

	1960s	1970s	1980s	1990s	2000s
Driving Force	Cost	Market	Product quality	Time to market	Service and value
Manufacturing Strategies	High volume Cost minimization Stabilize Product focus	Functional integration Closed loop Automation Diversification	Process control Material velocity World class manufacturing Overhead cost reduction	New product introduction Responsiveness Manufacturing metrics Reengineering Rapid prototyping	Customer-centered mission Information sharing Global integration Environment safety Virtual enterprise
Systems to Support the Strategy	Production and inventory control systems Numerical control	Material requirements planning Master production scheduling Computer numerical control Push systems	Manufacturing resource planning Just in time Statistical quality control Computer-aided design and manufacturing Simulation Pull systems	Computer-integrated manufacturing Decentralization Simplification Total quality management Self-directed workforce Activity-based costing	"Intelligent" manufacturing systems Flexible and agile automated systems Continuous benchmarking systems Community involvement Continuous infrastructure improvement Paperless systems Ergonomics and safety systems

production, these resources must be organized and coordinated. In other words, a well-defined manufacturing strategy is needed. Strategic planning is a dynamic, iterative, and living process that can be a powerful advantage in a competitive marketplace. Table 1.1 shows the evolution of strategic manufacturing since the 1960s.

Concurrent Engineering

Concurrent engineering (CE) is a philosophy that promotes interactive design and manufacturing efforts to develop product and processes simultaneously, thus optimizing the use of company resources and reducing time to market cycles. Although CE carries the word "engineering" in its title, CE is not an engineering discipline. Concurrent engineering is an organizational, human resources, and communications discipline, which requires on-going checks and balances among design, manufacturing, purchasing, quality and business development representatives. Table 1.2 represents a typical CE team.

General CE Phases

For a new project, four general phases are followed.

Phase 1: Technology and Concept Development.   This phase includes

a. Definition of customer requirements implementation (i.e., preparation of preliminary product definition).

Table 1.2 **Functions Represented in a Concurrent Engineering Team**

Supplier's Representatives	Customer's Representatives
● Manufacturing engineering	● Product engineering
● Product/design engineering	● Manufacturing engineering
● Business development	● Materials
● Quality engineering	● Purchasing
● Purchasing	● Marketing
● Scheduling	● Quality assurance
● Assemblers	● Production experts

   b. Assessment of primary alternatives and the feasibility of each alternative, product design, prototyping, manufacturing, and procurement strategy.

   c. Development of required resources and schedule.

   d. Preparation of product plan.

Phase 2: Product and Process Development and Prototype Validation. This is the phase in which the CE team will

   a. Develop product improvements for assembly and manufacturing.

   b. Develop/adapt the manufacturing and assembly processes.

   c. Establish baseline process technology.

   d. Establish program goals and risk-reduction plans.

Generally, other project elements are addressed in this phase, such as financial business plans, product concept demonstration, freezing of engineering content, and final project approval.

Phase 3: Process Validation and Product Design Confirmation. This is the execution phase where

   a. The production process and methods are finalized.

   b. Simulations and pilot runs are conducted to test critical steps.

   c. Machine and equipment are validated for production readiness, quality, and performance.

Phase 4: Production and Continuous Improvement. This phase is the acceleration of the production floor into *"full-rated capacity"* to attain the quality, cost, and productivity goals of the program. An endless journey to better equipment in the factory floor, ergonomics practices across departments, waste elimination in all functions, and, above all, incorporation of the customer's voice in the enterprise's everyday functions is the overriding goal of this phase.

1.3 DESIGN, MATERIALS, AND PRODUCTION

The cost of a product depends on raw materials, production costs for machines and labor, management and sales, warehousing and logistics, and overhead. Machine and labor costs are inexorably related and comprise, along with raw materials expenditures, the bulk of production costs. When a material is chosen, the process, including the machine, is subsequently specified. Or, if a machine is available, the raw material that can be processed on that machine may be utilized. One could say that the purpose of economical production is to produce a product at a *profit*. This infers that the cost must be acceptable and competitive; also, a demand for the product must exist or must be created.

Efficiency in Production

Since the first use of machine tools, there has been a gradual, but steady, trend toward *automation*; that is, making machines more efficient by combining operations and by transferring more skill to the machine, thus reducing time and labor. To meet these needs, machine tools have become complex both in design and in control. Automatic features have been built into many machines, and some are completely automatic. This technical development has made it possible to attain a high production rate with a low labor cost, which is essential for any society wishing to enjoy high living standards. Computer-aided design and manufacturing are a significant step of progress. This text devotes much attention to this development.

Along with the development of production machines, the quality in manufacturing must be maintained. Quality and accuracy in manufacturing operations demand that dimensional control be maintained to provide parts that are interchangeable and give the best operating service. For mass production, any one of a quantity of parts must fit in a given assembly. A product made of *interchangeable parts* is quickly assembled, lower in cost, and easily serviced. To maintain this dimensional control, appropriate inspection facilities must be provided.

Three criteria that determine economical production are

1. A functional but simple design that has appropriate aesthetic quality.
2. A material choice that represents the best compromise among physical properties, appearance, cost, and workability or machinability.
3. The selection of the manufacturing processes that will yield a product with no more accuracy or better surface finish than necessary and at the lowest possible unit cost.

Product Engineering and Design

It is important that the product be designed with material, manufacturing, and engineering to be competitive. For any manufactured product it is possible to specify a stronger, more corrosive-resistant or longer-life material, for example. However, it is the engineer's obligation not to overlook the opportunity of economical production. This leads to *value engineering*, which is the substitution of cheaper materials or the elimination of costly materials or unnecessary operations.

To produce parts of greater accuracy, more expensive machine tools and operations are necessary, more highly skilled labor is required, and rejected parts may be more numerous. Products should not be designed with greater accuracy than the service requirements demand. A good design includes consideration of a finishing or coating operation because a product is often judged for appearance as well as for function and operation. Many products, such as those made from colored plastics or other special materials, are more salable because of appearance. In most cases the function of the part is the deciding factor, especially where great strength, wear, corrosion resistance, or weight limitations are encountered.

For *mass-produced parts,* the design should be adaptable to mass-production-type machines with a minimum of different setups. Whenever a part is loaded, stored, and reloaded into another machine, costs are involved that may not add value to the product.

Engineering Materials

In the design and manufacture of a product, it is essential that the material and the process be understood. *Engineering materials* differ widely in physical properties, machinability characteristics, methods of forming, and possible service life. The designer should consider these facts in selecting an economical material and a process that is best suited to the product. Materials are of two basic types, *metallic* or *nonmetallic.* Nonmetallic materials are further classified as *organic* or *inorganic* substances. Since there is an infinite number of nonmetallic materials, as well as pure and alloyed metals, considerable study is necessary to choose the appropriate one.

Few commercial materials exist as elements in nature. For example, the natural compounds of metals, such as oxides, sulfides, or carbonates, must undergo a separating or refining operation before they can be processed further. Once separated, they must have an atomic structure that is stable at ordinary temperatures over a prolonged period. In metal working, iron is the most important natural element. Iron has little commercial use in its pure state, but when combined with other elements into various alloys it becomes the leading engineering metal. The nonferrous metals, including copper, tin, zinc, nickel, magnesium, aluminum, lead, and others, all play an important part in our economy; each has specific properties and uses.

Manufacturing requires tools and machines that can produce economically and accurately. Economy depends on the proper selection of the machine or process that will give a satisfactory finished product, its optimum operation, and maximum performance of labor and support facilities. The selection is influenced by the quantity of items to be produced. Usually, there is one machine best suited for a certain output. In small lot or job shop manufacturing, *general-purpose machines* such as the lathe, drill press, and milling machine may prove to be best since they are adaptable, have lower initial cost, require less maintenance, and possess the flexibility to meet changing conditions. However, *special-purpose machines* should be considered for producing large quantities of a standardized product. A machine built for one type of work or operation, such as the grinding of a piston or the surfacing of a cylinder head, will do the job well, quickly, and at low cost with a semiskilled operator.

Many special-purpose machines or tools differ from the standard type in that they have built into them some of the skill of the operator. A simple bolt may be produced on either a lathe or an automatic screw machine. The lathe operator must know not only how to make the bolt but also must be sufficiently skilled to operate the machine. On the automatic machine the sequence of operations and movements of tools are controlled by motors and electronics, cams and stops, and each item produced is identical to the previous one. This "transfer of skill" into the machine allows less skillful operators but it does require greater skill in supervision and maintenance. Often it is uneconomical to make a machine completely automatic as the cost may become prohibitive.

The selection of the best machine or process for a given product requires knowledge of production methods. Factors that must be considered are volume of production, quality of the finished product, and the advantages and limitations of the equipment capable of doing the work. Most parts can be produced by several methods, but usually there is one way that is most economical.

1.4 ENGLISH METRIC PRACTICE

Système International d'Unités, officially known worldwide as *SI*, is a modernized metric system and incorporates many advanced concepts. Adopting SI is more than converting from inch, pound, gallon, and degree Fahrenheit units to meter, kilogram, ampere, and degree Celsius units. It is a chance to introduce a new, simplified, coherent, decimalized, and absolute system of measuring units. The practice frequently followed in this book is to provide the English or customary units followed by SI in parentheses.

Base Units of SI

Built on seven base units (meter, kilogram, second, ampere, Kelvin, candela, and mole), SI provides a coherent array of units obtained on a direct one-to-one relationship without intermediate factors or duplication of units for any quantity. Thus, *1* newton (N) is the force required to accelerate a mass of *1* kilogram (kg) at the rate of *1* meter per second squared (m/s^2). *One* joule (J) is the energy involved when a force of *1* newton moves a distance *1* meter along its line of action. *One* watt (W) is the power that in *1* second gives rise to the energy of *1* joule. The SI units for force, energy, and power are the same regardless of whether the manufacturing process is mechanical, electrical, or chemical. Some new names receive more usage, such as newton, pascal (Pa), joule, and hertz (Hz). Eventually SI will replace such terms and abbreviations as poundals, horsepower, BTU, psi, weight, feet per minute, and thousandths.

Rules for Conversion and Rounding

The front and back endpapers of the book contain conversion factors that give exact or six-figure accuracy and are used for converting *English units* throughout this book. The conversion of quantities should be such that accuracy is neither sacrificed nor exaggerated.

The proper conversion procedure is to multiply the specified quantity by the conversion factor exactly as given in the endpapers and then *round* to the appropriate number of significant digits. For example, to convert 11.4 ft to meters: $11.4 \times 0.3048 = 3.474$, which rounds to 3.47 m.

The practical aspects of measuring must be considered when using SI equivalents. If a scale divided into sixteenths of an inch is suitable for making the measurement, a metric scale having divisions of 1 mm (millimeter) is suitable for measuring in SI units, and the equivalents should not be closer than the nearest 1 mm. Similarly, a caliper graduated in division of 0.02 mm is comparable to one graduated in divisions of 0.001 in. A measurement of 1.1875 in. may be an accurate decimalization of a noncritical 1¾₆ in., which should have been expressed as 1.19 in. However, the value 3 may mean "about 3" or it may mean "3.0000."

It is necessary to determine the intended precision of a quantity before converting. This estimate of intended precision should never be smaller than the accuracy of measurement, and as a rule of thumb, should usually be smaller than one-tenth the tolerance. After estimating the precision of the dimension, the converted dimension should be rounded to a minimum number of significant digits such that a unit of the last place is equal to or smaller than the converted precision. For example, a piece of stock is 6 in. long. Precision is estimated to be about ½ in. ($\pm$¼ in.). This converted precision is 12.7 mm. The converted dimension 152.4 mm should be rounded to the nearest 10 mm as 150 mm.

QUESTIONS

1. Give an explanation of the following terms:

Manufacturing system   Profit
Manufacturing processes   Automation
Manufacturing enterprise   Mass-produced parts
Eli Whitney   Engineering materials
Interchangeable production   Special-purpose machines
Job-lot production   SI
Organization   English units
Concurrent engineering   Round

2. Prepare a list of career opportunities in manufacturing processes from the classified want ads of a newspaper.

3. Indicate the criteria for economical production.

4. Why are materials and design important to manufacturing processes?

5. What is the purpose of a strategic manufacturing plan? Discuss.

6. Contrast the advantages and disadvantages for mass, moderate, and job-lot production.

7. Define automation. Is it more suitable for mass, moderate, or job-lot production? Discuss.

8. Show the advantages of SI units. Indicate the base units of SI. Why is it important to know English and SI units, conversions, and practices?

9. List the rules for converting and rounding English units to SI.

10. Describe the fundamentals of concurrent engineering. What do you think about this philosophy?

PROBLEMS

Convert the following from customary (English) units to SI, or from SI to English units. Show correct abbreviations. Use conversion factors as found in the book endpapers.

1.1. Area

a. 15 ft², 2.4 ft² to square meter (m²)

b. 0.15 in.², 0.035 in.² to square millimeter (mm²)

c. 0.51 mm², 1.060 mm² to in.²

1.2. Energy

a. 28,000 British thermal units (BTUs) to joule (J)

b. 8,500,000 kilowatt hour (kWh) to J

1.3. Force

a. 19 pound-force (lbf) to newton (N)

b. 2400 ton-force to meganewton (MN)

c. 0.6 MN to lbf

1.4. Length

a. 1.9 ft, 2.0 ft to meter (m)

b. 11 in., 18.1 in., 250 in. to m

c. 10 in., 3.5 in., ⅛ in., 0.500 in., 0.0005 in. to mm

d. 9 microinch (μin.), 35 μin. to nanometer (nm)

e. 300 nm, 1250 nm to μin.

f. 6 mil, 32 mil to mm

1.5. Mass

a. 16 ounce-mass (ozm) to kilogram (kg)

b. 15 gram (g), 1700 kg to oz

c. 15 pound-mass (lbm), 2.3 lbm, 1.3 lbm to kg

d. 35 ton (short) to megagram (Mg)

1.6. Density

a. 14 pound-mass per cubic foot (lbm/ft³) to kilogram per cubic meter (kg/m³)

b. 63 lbm/ft³ to kg/m³

c. 0.2 ounce-mass per cubic inch (ozm/in.³) to kg/m³

d. 0.33 kg/m³ to lbm/ft³

1.7. Pressure or stress (force/area)

a. 2600 pound-force per square foot (lbf/ft²) to megaPascal (MPa; 1 Pa = 1 N/m²)

b. 185 pounds per square inch (psi), 1750 psi to Pa

c. 2500 psi, 1.8×10^6 psi to MPa

1.8. Temperature

a. -40 degree Fahrenheit (°F), 180°F, 250°F, 1750°F to degree Celsius (°C)

b. 1300°C, 200°C, 1000°C to °F

1.9. Velocity

a. 150 feet per minute (fpm), 500 fpm, 750 fpm to meter per second (m/s)

b. 400 m/s, 150 m/s, 1000 m/s, 2000 mm/s to fpm

1.10. Volume

a. 0.35 ft³, 120 ft³, 750 ft³ to m³

b. 0.02 in.³, 12 in.³, 150 in.³ to mm³

c. 155 in.³, 179 in.³, 1900 in.³ to m³

d. 1 ounce (oz), 173 oz, 210 oz to m³

e. 25 gallon (gal), 55 gal to m³

f. 1000 yard³ to m³

g. 350 mm³, 80 mm³, 1500 mm³ to in.³

1.11. Volume/time

a. 90 ft³/min, 64 ft³/s to m³

b. 24 yard³/min, 44 yard³/min to m³

c. 11 in.³/min, 67 in.³/min, 0.1 in.³/min to mm³/s

d. 1.3 mm³/s, 0.005 mm³/s to in.³/min

MORE DIFFICULT PROBLEMS

1.12. In this problem units are mixed intentionally. Figure P1.12 shows a cylinder C that weighs 3500 lb and has a cross section area of 1858 square centimeters. The piston P has a cross section area of 5 square inches and negligible weight. If the apparatus is filled with oil of specific gravity 0.78, find the force F in the SI system that is required for equilibrium.

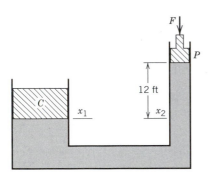

Figure P1.12

1.13. Determine the horsepower and the wattage required to pump 200 cubic feet of water per minute to an outlet 7.62 meters above water level. Neglect frictional effects, assume that the velocity of the water at the outlet is negligible, and that the pump is 60% efficient.

1.14. A plate of an unknown metal is 0.158 in. thick and has a temperature difference of 32°C between its faces. It transmits 2×10^5 calories per hour through an area of 5 cm². Calculate the thermal conductivity (k) of the metal in the English system. (Hint: The quantity of heat, H, transmitted per second from one face to the opposite one is proportional to the area of cross section, A, and to the temperature gradient; i.e., $H = k \times$ area $\times$ temperature gradient.) Bonus: What metal is it?

1.15. Study Table 1.1, then call a professional in the manufacturing field and ask questions about expected changes and future trends in manufacturing processes and in the systems that are needed to support them. Write a one page report on the interview.

PRACTICAL APPLICATION

As a new manufacturing professional you are interested in identifying career opportunities that fit your educational background. Access the Internet and contact one of the following: American Society of Mechanical Engineers, Society of Manufacturing Engineers, Institute of Industrial Engineers, Institute of Electrical and Electronic Engineers, or the National Association of Industrial Technology to ask questions about the future of manufacturing. (Hints: Society forums can be accessed through the Internet by checking www\nist.gov, www\iienet.org, and others. Your instructor can give you additional advice.)

CASE STUDY: PROFESSOR SMITH

"Good morning, Professor Smith. I'm Rusty Green, and I'm in your eight o'clock Manufacturing Processes class."

"Uh huh," replied the professor without looking up from the desk. "What can I do for you, Rusty?"

"Well, it's like this. I'm not sure that I belong in your class." Rusty smiles, and continues as the professor looks up. "Oh it's not you, Professor, but I really want to be an engineer and it's unclear what this course can do for me later on."

"Yes, go on," and Professor Smith, leaning forward in his chair, continues by saying, "What are you wanting?"

"I do want a career oriented program, but is manufacturing processes going to aim in that direction?" Rusty now looks expectantly at the professor.

"That's an important question, Rusty. There are many, many considerations."

Help the professor answer Rusty. Prepare a report on the opportunities in manufacturing for you in your field by taking the concurrent engineering framework into consideration. In addition, you may consider the following:

Call and talk to a manufacturing practitioner. Check the newspaper want ads, trade journals, technical magazines, and your college career office for employment opportunities. Speculate on the training or educational requirements and experience that are necessary for these careers. Consider tangential situations that can develop. List the titles and their relationship to manufacturing. Do these situations change after 1 year, 5 years, or 10 years?

CHAPTER 2

NATURE AND PROPERTIES

OF MATERIALS

Materials are one of the key elements of a manufacturing system. Understanding their properties, reliability, and manufacturability is critical for their effective and efficient application. *Composites*, polymers, and plastics are being used extensively in manufacturing processes that range from automotive and aerospace applications to medical and electrical applications. The commonly used metals, however, continue to provide the major load-carrying and heat-resistant structures. These metals are plentiful, relatively cheap, have known properties, and are easily machined or formed.

2.1 CLASSIFICATION OF MATERIALS

Materials for manufactured parts or machines have such diversified properties that even when performance and cost are considered, it is often difficult to decide the proper material for a given purpose. One material may have higher strength, another may have better corrosion properties, and yet another may be more economical; hence, most choices are a compromise among a number of materials using the best engineering data and judgment available. Copper, for example, may be alloyed in hundreds of ways to produce materials with special properties.

In general, materials may be classified as *metallic* and *nonmetallic*. The former are subdivided as *ferrous* and *nonferrous,* while the latter are arranged as *organic* and *inorganic*.

Table 2.1 shows the principal metals used in current processes and a few of their important properties. The values indicate the approximate range of properties that can be expected depending on alloying and heat treatment. Other important commercial metals include tin, silver, platinum, manganese, vanadium, and titanium. Because special properties can be obtained by alloying, few pure metals are used extensively.

The nonferrous metals are generally inferior in strength but superior in corrosion resistance as compared to ferrous materials, and most are more expensive.

Nonmetallic materials are considered organic if they contain animal or vegetable cells (dead or alive) or carbon compounds. Leather and wood are examples. Nonmetallic materials are classified as inorganic if they are other than animal, vegetable, or carbon compounds. Table 2.2 shows some common nonmetals used and a few of their important properties.

There are fundamental differences between organic and inorganic materials. Organic materials will usually dissolve in organic liquids such as alcohol or carbon tetrachloride, but they will not dissolve in water. Inorganic materials tend to dissolve in water. In general, inorganic materials resist heat more effectively than organic substances.

A number of materials are necessary in manufacturing finished products. Figure 2.1 shows the materials used in making an automobile. The choice between metallic and nonmetallic and between organic and inorganic materials is a compromise between service

Table 2.1 **Approximate[a] Properties of Common Metals**

Metal	Tensile Strength lb/in.$^2 \times 10^3$	Tensile Strength MPa	Ductility (%)	Melting Point °F	Melting Point °C	Brinell Hardness	Density lb/ft^3	Density kg/m^3
Ferrous								
Gray cast iron	16–30	110–207	0–1	2500	1370	100–150	450	7209
Malleable iron	40–50	276–345	1–20	2475	1360	100–145	480	7689
Steel	40–300	276–2070	15–22	2700	1480	110–500	485	7769
White cast iron	45	310	0–1	2500	1370	450	480	7689
Wrought iron	35–47	242–324	30–35	2800	1540	90–110	493	7897
Nonferrous								
Aluminum	12–45	83–310	10–35	1220	660	30–100	165	2643
Copper	50–100	345–689	5–50	1977	1080	50–100	556	8906
Magnesium	12–50	83–345	9–15	1200	650	30–60	109	1746
Nickel	60–160	414–1103	15–40	2650	1450	90–250	545	8730
Lead	2–33	18–23	25–40	620	325	3.2–4.5	706	11,309
Titanium	80–150	552–1034	0–12	3270	1800	158–266	282	4517
Zinc, cast	7–13	48–90	2–10	792	422	80–100	446	7144

[a] Depending on the alloy.

Table 2.2 **Approximate Properties of Common Nonmetals**

Material	Specific Gravity	Hardness	Specific Heat BTU/lb °F	Tensile Strength (10^3 psi)	Thermal Expansion (10^{-5}) in./in. °F
Resins					
Acetal	1.42	—	0.35	8.8	4.5
Nylon 6	1.14	119–121[a]	0.3–0.5	11.8	4.6
Polycarbonate	1.20	—	—	9.0	3.7
Polyester (PBT)	1.31	112–120[a]	0.3–0.55	8.5	5.3
Polypropylene	0.91	80–100[a]	0.45–0.48	5.0	4.0
Polystyrene	1.07	—	0.26	7.0	3.6
Cellulose acetate	1.33	40–120[a]	0.3–0.4	8.0	4.4–9.0
Ceramics					
Beryla	2.7–3	9[b]	0.25	15	0.528
Alumina	3.3–3.9	9[b]	0.19	20–30	0.43
Calcia	3.3–5	6[b]	0.08–0.21	—	0.7
Rubber	0.92	30–100[c]	—	>0.4	6–37
Graphite	1.88	—	—	340	0.67
Glass	2.2–2.8	—	—	—	0.18–0.58

[a] Rockwell number.

[b] Mohs' scale.

[c] Shore A scale.

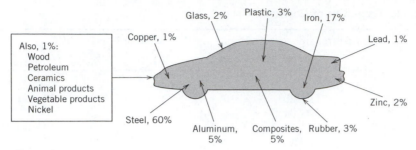

Figure 2.1
Representative materials in a full-size automobile. Percentage of weight distribution.

characteristics and cost. To reduce fuel consumption and weight, or to improve some specific properties, *advanced composites* are gaining favor. *Advanced composite materials*, sometimes referred to as *fiber reinforced plastics*, are defined as those materials composed of two or more constituent materials, of which one is a *fiber* (e.g., carbon, fiberglass, Kevlar) and the other is a *binder or matrix* (e.g., epoxy, thermoplastic, polyester). The controllable orientation of the fibers in the matrix set advanced composites apart from other random-oriented fiber composites.

2.2 STRUCTURE OF MATERIALS

Materials are all composed of the same basic components: protons, neutrons, and electrons. Their properties—electrical, chemical, thermal, and mechanical—are ultimately related to their atomic structure and to the energy level or forces that exist between the atoms. The energy level, driven by temperature and pressure, determines the phase—gas, liquid, or solid—of the material.

Grain Formation

Consider the three-phase diagram in Figure 2.2. When temperature and pressure conditions fall under the gas-solid line, a material solidifies and atoms arrange themselves geometrically. The initial formations in a solidifying liquid create the nuclei for crystals

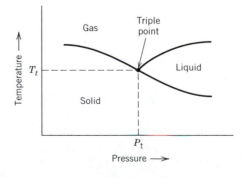

Figure 2.2
Three-phase diagram.

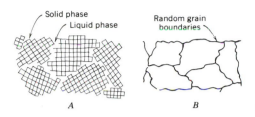

Figure 2.3
Growth of crystals forming grains. *A*, Crystal growth. *B*, Grains.

that grow in orderly form. Figure 2.3*A* schematically illustrates the way enlarging crystals form. Growth continues randomly, and when one crystal comes in contact with another one, growth of both crystals ceases, and the surfaces where they meet form the *grain* boundaries.

Individual grains are the structures that determine many of the physical properties of the material. A fine grain structure, for example, impedes the sliding and relative displacement of crystals, which makes the material less prone to deformation. Similarly, a material with a fine grain structure is harder and less ductile than a material with a coarse grain structure.

Most crystals do not develop uniformly but progress more rapidly in one direction than another. As the crystal growth advances, the crystal front branches out in a treelike fashion. Such growth is called *dendritic*, and the crystal formation is called a *dendrite*. The growth is always uneven with branches of the dendrite thickening or new branches forming as solidification progresses. Figure 2.3*B* shows the completely developed grain boundaries of several crystals. The grains of a metal may be studied with a microscope after the material has been etched to make the boundaries stand out.

The grain size of a metal depends on the rate at which it was cooled and the extent and nature of the hot- or cold-working process. A metal with fine or small grains will have superior strength and toughness as compared to the same metal with large grains. This results because with the atoms closer together there is more "slip interference" in the lattice structure when a deforming force is applied. The larger-grained materials are characterized by easier machining, better ability to harden through heat treatment, and superior electrical and thermal conductivity. Although the larger-grained metals will harden more uniformly during heat treatment, the fine-grained materials are less apt to crack when quenched. Additives can be made to a molten metal to ensure a predetermined grain size: Aluminum, for example, may be added to steel to promote fine grains. The desired grain size is usually a compromise depending on the properties sought. For *brass,* which is used to make cartridge cases, a large grain enables the case to be formed more easily, but surface finish and strength are improved with a finer grain.

The hardness as well as the grain size is affected by the temperature history of a metal. *Quenching* steel from an elevated temperature will usually harden it, and cooling it slowly will bring out its maximum softness. *Annealing*, the slow cooling of a metal from an elevated temperature, is used to soften, add toughness, remove stresses, and increase ductility of metals.

Microscopic Examination

Using a metallurgical microscope to examine a polished specimen reveals the constituents of some metals as well as surface deformities. The polishing operation creates a mirror

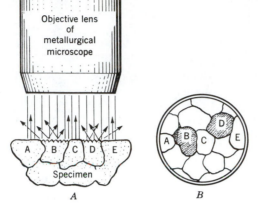

Figure 2.4
Metallurgical examination of an etched specimen. *A*, Etched specimen under examination. *B*, The specimen as viewed through the microscope.

like surface, but a coating of thin metal known as *smear metal*, which prevents critical analysis, is left on the specimen. If the specimen is *etched* with a suitable chemical solution, the smear metal is removed, the surface becomes slightly dull, grain boundaries are partially dissolved, and certain constituents are revealed because of the selective action of the etching solution. A common etching reagent for steel is a mixture of 3 parts nitric acid and 97 parts alcohol.

Figure 2.4*A* illustrates the effect of etching on the reflection of light from the specimen. Grains B and D have been attacked by the etching reagent and do not reflect the light in the same way as the other grains. When light is reflected into the lens of the microscope, the surface appears light; when the light is scattered, that area appears dark. The grain boundaries are like small canyons surrounding each grain, and the light is not reflected back to the observer. Thus the grain boundaries are well-defined black lines. Figure 2.4*B* illustrates the appearance of the specimen in Figure 2.4*A* when viewed through a microscope. Metal grains are usually examined at a magnification of more than 1000. Selective etching reagents are used to reveal particular constituents.

2.3 SOLIDIFICATION OF METALS AND ALLOYS

Pure metals solidify in a unique manner, as indicated by Figure 2.5. The liquid cools to the point at which the first nucleus forms. From the time solidification begins until it is complete, the temperature of the solid liquid mixture does not change. Once solidification is complete, the temperature drops with respect to time. During the freezing of the metal, the latent heat of solidification just balances the heat lost by the metal, thus preserving the constant-temperature condition.

Only a few metals are used commercially in their pure state; copper for electrical wiring and zinc for galvanizing are examples. When other elements are added to a pure metal to enhance its properties, the combination is called an *alloy*. *Brass* is an alloy of copper and zinc, *bronze* an alloy of copper and tin, and *steel* an alloy of iron and carbon. Since the number of alloys is very large, the prediction of their properties and characteristics is impossible.

Although pure metals solidify at a constant temperature, alloys do not, as shown in Figure 2.5. The first nuclei form at a higher temperature than that at which complete solidification occurs. This change in temperature as solidification progresses causes the

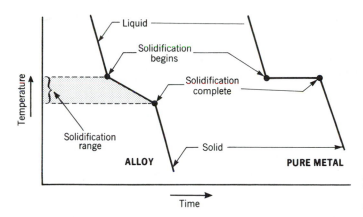

Figure 2.5
Time–temperature curve for an alloy and a pure metal.

solid being formed to change in chemical composition, because each element in an alloy has its own peculiarities relative to temperature.

An *equilibrium diagram* shows the way an alloy forms what is called a *solid solution*, that is, a solid that is, in effect, a solution of two or more materials. Many types of equilibrium diagrams exist depending on the alloys involved, but one of the simplest is the one for a copper–nickel alloy (Figure 2.6).

Monel is a metal composed of 67% nickel and 33% copper, which resists saltwater corrosion and is used in packaging beverages and foods. It has a range of working temperatures from $-100°$ to $400°F$ ($-75°$ to $205°C$). At the dotted line in Figure 2.6, monel begins to solidify when cooled to temperature ℓ_1, but at that point the first material to solidify will have a composition of 23% copper and 77% nickel as indicated by s_1. The liquid at ℓ_1 will be 67% nickel and 33% copper composition when freezing begins. When the temperature falls to the $\ell_2 s_2$ line, the liquid composition will be that indicated by ℓ_2

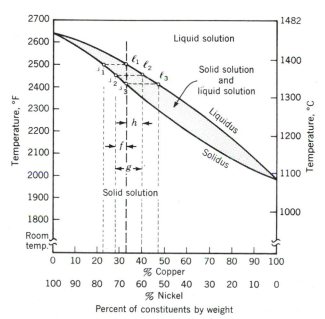

Figure 2.6
Equilibrium diagram for copper–nickel alloys.

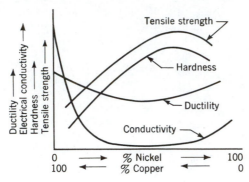

Figure 2.7
Effects of alloy composition on properties.

or 41% copper and 59% nickel, whereas the composition of the solid is indicated by point s_2. At the temperature $l_3 s_3$, the last liquid to solidify is of composition l_3 but the solid solution is monel with the 67% nickel and 33% copper content.

This type of diagram enables the engineer to determine the constituents of an alloy as well as certain other properties of the resulting solid solution.

The percentage of solid (S) at any temperature, s, can be calculated for a given composition. For example, for a 67% nickel and 33% copper alloy corresponding to the $s_2 - l_2$ line, at about 2455°F, S is expressed as

$$S = \frac{l_2 - s_3}{l_2 - s_2} \times 100 \tag{2.1}$$

where s_3, s_2, and l_2 represent lengths in the horizontal; that is, letting $s_3 - s_2 = f$, $l_2 - s_3 = h$, and $l_2 - s_2 = g$; then the percentage solid is

$$S = \frac{h}{g} \times 100 \tag{2.2}$$

Similarly, the percentage liquid (l) is calculated

$$l = \frac{s_3 - s_2}{l_2 - s_2} \times 100 = \frac{f}{g} \times 100 \tag{2.3}$$

Figure 2.7 shows how physical and mechanical properties of the copper–nickel alloys vary with respect to the two metals. Interestingly enough, the 5-cent American coin or "nickel" is 75% copper and 25% nickel.

Iron-carbon alloys are very important in manufacturing. A discussion of these alloys as well as the equilibrium diagram is shown in Chapter 17. It should be noted that one of the differences among wrought iron, steel, and cast iron is the carbon content. The approximate carbon limits for each are as follows:

Wrought iron   C <0.08% + 1 to 3% slag
Steel   C >0.08% but <2.0%
Cast iron   C >2.0%

2.4 PHYSICAL CHARACTERISTICS OF MATERIALS

The physical properties of materials include density, vapor pressure, thermal expansion, thermal conductivity, and electric and magnetic properties. Students are encouraged to consult materials handbooks to become familiar with values of such properties and to pay close attention to the conditions in which values are taken. For example, cast metals and plastics are *isotropic*, that is, their characteristics are the same measured in any direction, while rolled metals and extruded plastics are *anisotropic*, that is, their properties vary when measured in the longitudinal and transverse directions. It is also important to know if the material is *homogeneous* (uniform) or *heterogeneous* (nonuniform). In a completely uniform material (e.g., an unfilled thermoplastics) the smallest sample has the same physical properties as the entire body, which is not the case for nonuniform materials. In a heterogeneous body (e.g., a glass-reinforced composite) the composition varies from point to point. For most practical purposes, however, these materials are treated as homogeneous, but safety factors must be added.

A brief definition of some important physical characteristics follows.

The *density, ρ,* of a material is a measure of the mass per unit volume, usually expressed in lb/in.^{3} (g/cm^3) at a temperature of 73.4°F (23°C).

$$\rho = \frac{\text{mass of body}}{\text{volume of body}} = \frac{m}{V} \qquad (2.4)$$

Once the design of a part is complete, the drawings allow one to calculate the volume of the part, then density information is used to calculate the amount of material required to make the part.

A related property is the *specific gravity*, which is the ratio of the density of the material, ρ, to the density of a substance taken as a standard. Solids and liquids are referred to water as standard, while gases are often referred to air as standard. Since the ratio is dimensionless, it is useful to evaluate the cost of different materials in the design process.

A similar and sometimes helpful property is the *weight-density, $g\rho$,* defined as weight per unit volume,

$$g\rho = \frac{\text{weight of body}}{\text{volume of body}} = \frac{mg}{V} \qquad (2.5)$$

where g is the normal acceleration due to gravity 32.2 ft/s^2 (9.807 m/s^2). This characteristic is of great importance in finishing, cleaning, and electronic manufacturing as described later in Chapters 19 and 20.

Another important physical characteristic, *water absorption*, is the amount of increase in weight of a material because of the absorption of water. Water absorption serves as a guide to stability of electrical or mechanical properties, dimensions, or appearance. Parts made from materials with low water absorption rates tend to have greater dimensional stability. To calculate water absorption, specimens of about 3 in. $\times$ 1 in. are dried for 24 hours, then weighed and immersed in water at 73.4°F (23°C) for a specified time. Absorption is expressed as a ratio of weight increase to initial weight.

When working with castings, plastics, and composites, *shrinkage* becomes a critical characteristic to evaluate. Shrinkage, expressed in in./in. (cm/cm), is the ratio of the dimension of the die or the plastics molding to the corresponding dimension of the mold. Measurements are taken at room temperature to estimate the mold cavity dimensions needed to produce a part of the required size.

Plasticity and elasticity, as seen in Section 2.5, are properties of critical importance when evaluating a material for manufacturing. If the material returns to its original size and shape after being deformed, it is said that it has high *elasticity*. If, however, it tends to stay deformed, it is said that the material has high *plasticity*.

The significance of physical properties is very high during the design function. For any structural analysis, for example, if materials are isotropic and elastic, only two constants are needed for a homogeneous structure, the Young's modulus (E) and the Poisson's ratio (ν). These two constants are defined in Section 2.5. Assumptions of linear elasticity, isotropy, and homogeneity, however, lead to significant errors when advanced materials, composites, and plastics are used. As the degree of anisotropy increases, instead of two constants, up to 21 moduli are required to describe the material. Such a situation is outside the scope of this textbook and is addressed in part by finite-element analysis methods.

2.5 ENGINEERING PROPERTIES OF MATERIALS

Engineering properties include tensile strength, compressive strength, torsional strength, modulus of elasticity, and hardness. Other properties such as machinability or ease of forming are discussed in subsequent chapters.

Tensile Strength

Tensile strength is the ability of a material to support or carry an applied axial load. The property enables a material to resist being pulled apart. One test to measure tensile strength is conducted by pulling on the two ends of a specimen machined like that shown in Figure 2.8. The specimen, about 8 in. (205 mm) in total length, is machined from ¾-in. (19 mm) material. The reduced section is 2¼ in. (57.2 mm) long and ½ in. (12.7 mm) in diameter and contains center punch or "gage marks" 2 in. (50.8 mm) apart. When the specimen is pulled, the smaller diameter section necks down from an area A to an area A_1 and the gage length increases from L to L_1. For most engineering purposes, the area A is used in all calculations because A_1 is difficult to measure. The results are described by determining the changes that take place in length as the force is increased to the breaking point. From the data collected, while pulling the specimen, a curve can be plotted for two values, the *stress* (σ) and *strain* (ε, dimensionless), where

$$\text{Stress} = \frac{\text{force}, F}{\text{area}, A} = \sigma \tag{2.6}$$

and

$$\text{Strain} = \frac{L_1 - L}{L} = \varepsilon \tag{2.7}$$

Although such curves, known as *engineering stress–strain curves*, differ with material and heat treatment, the results generally describe a curve similar to those shown in the insert in Figure 2.9. A family of stress–strain curves for several metals also is shown in Figure 2.9

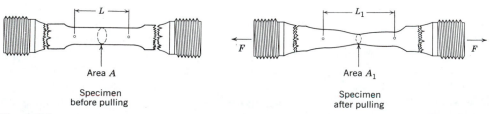

Figure 2.8
Tensile test specimen.

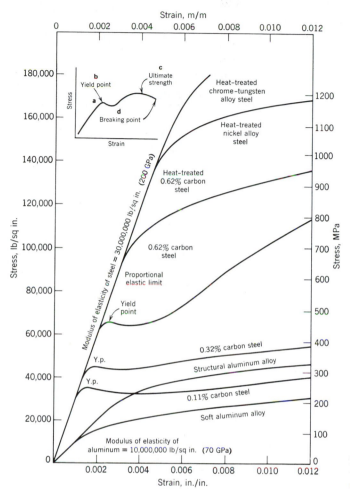

Figure 2.9
Engineering stress–strain
curves for various materials.

Referring to Figure 2.9, the point **a** represents the *elastic limit*. Any force causing a greater stress or strain would result in a permanent deformation of this material; that is, a tensile specimen would not return to its original length when unloaded. This point **a** also represents the *proportional limit*, and from that point on the stress–strain curve is no longer a straight line. The slope of this straight line is known as the *modulus of elasticity* or *Young's modulus*, *E* (named after Thomas Young who lived from 1773 to 1829), which is an indication of the stiffness of the material.

$$E = \sigma/\varepsilon \tag{2.8}$$

At some point **b** the material undergoes more rapid deformation, and it is at this point that its *yield strength* is determined. At point **c** the *ultimate strength* is determined, which is often referred to simply as the tensile strength of the material. In the case of a tensile specimen, this is the stress when necking begins. Point **d** notes the point on a stress–strain curve where fracture has occurred after maximum deformation.

Table 2.3 **Poisson's Ratio and Modulus of Elasticity Values for Some Common Engineering Materials**

Material	Young's Modulus of Elasticity[a]		Poisson's Ratio
	lb/in.$^2 \times 10^6$	GPa	
Metals (slab, bar, or block form)			
Aluminum alloys	10–12	69–83	0.33
Brass	14–16	96–110	0.34
Copper, pure	17.0	117.20	0.33
Lead, pure	2.0	14.79	0.40–0.45
Nickel, pure	32.0	220.61	0.31
Steel (hot-worked)	29–30	200–207	0.33
Titanium B 120 VCA (aged)	14.8	102	0.30
Nonmetals			
Concrete (short-term tests)	3–5	21–35	n/a
Commercial glasses	5–9	35–62	0.23
Construction lumber	1.2–2.1	8–14	n/a
Composites			
Boron epoxy[b]	10–12	70–79	n/a
Graphite epoxy[b]	15	103	n/a
DuPont's Kevlar epoxy	12.5	86.18	n/a

[a] Room-temperature properties are given.

[b] Unidirectionally reinforced in direction of loading.

The modulus of elasticity, E, may be determined as the slope of the linear portion of the stress–strain curve. This relationship was first noted in 1678 by Sir Robert Hooke (1635–1703) and is called *Hooke's law*, which states that the change in length per unit length of a bar is proportional to the force intensity in the bar. Typical values of E are given in Table 2.3.

When a bar is subject to a simple tensile load, there is an increase in length of the bar in the same direction of the load, but at the same time there is a contraction of the lateral dimension perpendicular to the load's direction. The relationship of these changes is the dimensionless *elastic constant* of the material, ν,

$$\nu = \frac{-\text{ normal strain in transverse direction}}{\text{normal strain in loaded direction}} \qquad (2.9)$$

also known as *Poisson's ratio* (after S. D. Poisson who lived from 1781 to 1840). Values for ν are given in Table 2.3. For most metals, ν is about 0.33, and for cork it is close to zero, which makes cork so useful for stoppers on bottles.

EXAMPLE 2.1

A 9-ft hinged beam is supported at its free end by a steel wire. The wire makes a 45° angle with the beam and experiences a 2500-lb tension. What is the elongation of the wire?

The equation of the straight line that passes through the origin ($y = mx$) in Figure 2.9 is given by

$$\sigma = E\varepsilon$$

replacing σ and ε by Equations (2.6) and (2.7), respectively, we obtain

$$F/A = E\,(L_1 - L)/L$$

then the elongation is

$$(L_1 - L) = L\,F/(A\,E)$$

The modulus of elasticity for steel is 29×10^6 (Table 2.3). The required area, A, of the wire is obtained by knowing the steel's yield strength. Using the minimum value given in Table 2.1 (40,000 psi), we get

$$A = F/\sigma = 2500/40{,}000 = 0.063 \text{ in.}^2$$

The length, L, of the wire is found using the right angle relationship

$$L = (\text{length of beam})/\cos 45 = 9 \times 12/0.707 = 152.75 \text{ in.}$$

Finally, the elongation of the wire is

$$(L_1 - L) = 2500 \times 152.75/(0.063 \times 29 \times 10^6) = 0.209 \text{ in.}$$

Normally, a factor of safety is added. If a factor of two is used, the required cross section of the wire would double and the elongation would be half of the one found in this example.

Shear, Compressive, and Torsional Strength

There is no universally used standard test for determining shear or torsion characteristics, but the relative differences in the properties between materials can be found in handbooks. The shear strength of a material is generally about 50% of its tensile strength and the torsional strength about 75%.

Shear stress, τ, results from a parallel force (or shear load), F, applied to a cross-sectional area, A, of a body and is proportional to the displacement angle γ, known as the shear angle or *shear strain* (see Figure 2.10A). The relationship is given by

$$\tau = G\gamma \tag{2.10}$$

where G is defined as the *shear modulus* or the *modulus of rigidity* of the material. For many metals G is about $(3/8)E$, but it can be calculated as

$$G = \frac{E}{2(1 + \nu)} \tag{2.11}$$

Compressive strength (Figure 2.10B) is easily determined for brittle materials that will fracture when a sufficient load is applied; but for ductile materials a strength in compression is valid only when the amount of deformation is specified. The compression strength of cast iron, a relatively brittle material, is three to four times its tensile strength, whereas for a more ductile material such as steel, predictions cannot be made so easily.

For design purposes it is necessary to define a *factor of safety*, FS. This factor is a number by which we reduce the maximum stress to obtain an allowable working stress. Factor of safety is defined as the ratio of the actual stress to the allowable strength of the material, that is,

$$\text{FS} = \frac{\sigma_{\text{actual}}}{\sigma_{\text{allowable}}} \tag{2.12}$$

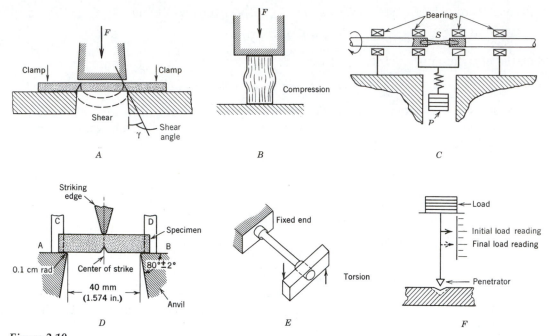

Figure 2.10
Schematic drawings of methods used in determining certain engineering properties of materials. *A*, Shear strength. *B*, Compressive strength. *C*, Fatigue strength. *D*, Impact strength. *E*, Torsional strength. *F*, Penetration hardness.

Recommended values of FS range from 1 to 3. High values are used for applications in which loading is not severe but material properties are not reliable, or when loading is severe but reliable materials are used. Low FS values are used for highly reliable materials used under benign conditions and where weight is an important consideration.

Ductility

Ductility is a property that enables a material to be bent, drawn, stretched, formed, or permanently distorted without rupture. A material that has high ductility has a relatively large tensile strain up to the point of rupture (e.g., aluminum and structural steel). Hard materials are usually brittle, lack ductility, and have a relatively small strain (e.g., cast iron, concrete). The tensile test can be used as a measure of ductility by calculating the *percentage of elongation* of the specimen upon fracture. Hence,

$$\text{Percentage elongation} = \frac{L_f - L}{L} \times 100 \qquad (2.13)$$

where

L = Original gage length
L_f = Separation of gage marks measured on the reassembled bar after fracture

An arbitrary strain of 0.05 in./in. (mm/mm) is frequently taken as the dividing line between brittle and ductile materials.

Creep and Notch Sensitivity

Creep is a permanent deformation resulting from the loading of members over a long period of time. High temperature creep may lead to the failure of loaded units such as high-pressure steam piping. Elongating caused by creep will occur below the yield strength of the material. Heat treatment, grain size, and chemical composition appreciably affect creep strength. *Notch sensitivity*, on the other hand, is a measure of the ease with which a crack progresses through a material from an existing notch, crack, or sharp corner.

2.6 OTHER METHODS FOR MATERIAL EVALUATION

In addition to the methods described to evaluate physical and engineering properties of materials, there are other characteristics and methods with which a manufacturing practitioner should be familiar. Those methods are briefly summarized in Table 2.4.

2.7 IMPACT AND ENDURANCE TESTING

A metal may be very hard and have a high tensile strength yet be totally unacceptable for a use that requires the metal to withstand impact or sudden load. A number of tests can be used to determine the impact capability of a metal, but the test most generally used is the *Charpy* test. Figure 2.10D shows a notched specimen that is struck by an anvil. The energy in foot-pounds required to break the specimen is an indication of the impact resistance of the metal.

The yield strength of metals can be used in designing parts that will withstand a static load, but for cyclic or repetitive loading, the endurance or *fatigue strength* is useful. An endurance test is made by loading the part and subjecting it to repetitive stress. Figure 2.10C shows one way that the endurance or fatigue strength of a material may be found. Generally, a number of specimens of a metal are tested at various loadings and the number of cycles to failure are noted. A curve of stress in pounds per square inch versus the number of cycles to failure is plotted, and these data can be used for designs involving repetitive loading.

Hardness

Hardness is a characteristic of the material, not a fundamental physical property. It is defined as the resistance to indentation and is determined by measuring the permanent depth of the indentation. In other words, when using a test force and a given indenter, the smaller the indentation, the harder the material. Indentation hardness value is obtained by measuring the depth or the area of the indentation using one of more than 12 different test methods (see Table 2.5). Figure 2.10F shows a penetration hardness tester.

The first step in obtaining a reading is to force the penetrator and material into contact with the specimen with a predetermined initial load. Next, an increased load is applied to the penetrator, and the hardness reading is obtained by noting the difference in penetration caused by the final load as compared to the initial load. Different scales of hardness depend on the shape and type of the penetrator and the loads applied (see Table 2.5).

The *Rockwell* hardness tester is the most flexible of the many types, because through a variation of different types of penetrators and loads, hardnesses can be measured on

Table 2.4 Selected Methods for Material Evaluation

Property	Test Procedure	Importance
Thermal Conductivity, k The specimen is a circular disk with thickness no more than ⅓ the diameter of the heater, and diameter larger than the heater's.	Two identical specimens are sandwiched with a heating plate in the center and cooling plates on the outside. Insulation and auxiliary heaters are used to simulate infinite expanse. Power required to maintain steady-state temperature is measured.	Measures heat transfer rate from hot to cold portions of material. Environmental conditions may affect value.
Specific Heat, C_s The specimen's size varies depending on the apparatus.	Specimen heated to uniform temperature is immersed in insulated container of water at known temperature and mass. When specimen is completely cooled, final temperature is measured. Amount of heat released to water can be calculated from water temperature change.	Measures the amount of heat energy absorbed or released when temperature changes. Can be used to calculate heating rates, power requirements, etc.
Coefficient of Thermal Expansion, α Specimen: 2–5 in. (50–127 mm) long.	Specimen must fit in tube without excessive play or friction. Specimen is heated in a quartz tube. Dimension changes are indicated by motion of rod resting on specimen.	Measures dimension changes with varying temperatures. Since temperature response is not linear, it should be measured in similar conditions as service environment.
Kinetic Coefficient of Friction, μ One specimen is round or a strip mounted on a wheel. The other, 2-mm thick, 20-mm diameter, must weigh 5 g.	5-g specimen is mounted on a counter-weighted pendulum and presses surface under spring pressure. Other specimen slides on rotating wheel of mating material. Specimen holder pivots until counterweight balances friction force.	Measures friction force between two materials. Actual force in service may vary depending on mating materials and surface conditions. Kinetic friction is usually lower than static friction.
Fatigue Strength or Life Size and configuration of the specimen is chosen to relate to intended use.	Load is repeatedly applied until specimen fails or predetermined number of cycles is attained. Tests may include tensile, compressive, or bending loads.	Fatigue strength is the maximum stress that can be applied without failure for a number of cycles. Fatigue life is the number of cycles before failure at a given load or strain.
Salt Spray Size and configuration of the specimen varies.	5% NaCl, 95% water solution is sprayed in chamber for specified time. Spray does not impinge on part. Rate is set such that 1.0 to 2.0 mL/hr is collected in 80 cm^2 horizontal drain.	Can serve to compare corrosion resistance of metals. However, test is not indicative of corrosion resistance in other environments, since other chemical reactions may be involved.

(continued)

Table 2.4 (continued)

Property	Test Procedure	Importance
Izod Impact (For Plastics)		
Specimen size is usually ⅛ × ½ × 2 in. ¹⁄₁₀ notch in center of specimen face.	A notched specimen is held on one end with notch facing pendulum. Notch is flush with top of clamp jaws. Results are more closely correlated to notch sensitivity than to brittleness.	Energy absorbed during fracture is calculated from height of pendulum swing after specimen is struck.
Sliding Wear Resistance		
A 0.62 × 0.40 × 0.25-in. block and a 1.377-in. diameter ring.	Test block is pressed with specified force against rotating ring. Volumes of material lost from block and ring are measured after given number of revolutions at given speed.	Evaluates wear resistance of pairs of materials. Service conditions can be simulated with use of lubricants or contaminants.

Table 2.5 **Hardness Tests and How to Read Hardness Numbers**

Test Principle	Test Method	Test Force Range	Indenter Types
Rockwell	Regular	Scale A, 60 kg; Scales B and D, 100 kg; Scale C, 150 kg	Cone diamond and small ball
	Superficial	15, 30, 45 kg	Cone diamond and small ball
	Light load	3, 5, 7, kg	Small cone diamond
	Micro	0.5, 0.1 kg	Small cone diamond
	Macro	0.5 to 3000 kg	5-, 10-mm ball
Microhardness	Vickers	5 to 2000 g	136° pyramid diamond
	Knoop	5 to 2000 g	130° × 172° diamond
	Rockwell	500, 1000 g	Small cone diamond
	Dynamic	0.01 to 200 g	Triangular diamond
Brinell	Optical	500 to 3000 kg	For both methods:
	Depth	500 to 3000 kg	5-, 10-mm ball
Shore	Regular	Scale A, 822 g; Scale D, 4550 g	For both methods: Scale A,
	Micro	Scale A, 257 g; Scale D, 1135 g	35° cone; Scale D, 30° cone
IRHD	Regular	597 g	2.5-mm ball
	Micro	15.7 g	0.395-mm ball

Example of Hardness Number	How to Read It			Test Principle and Method That It Represents	Tensile Strength[a] (10^3 psi)
	Value	Hardness	Scale		
60 HRC	60	Rockwell	Scale: C	Regular, light load, and micro Rockwell hardness	364
72 HR30N	72	Rockwell	Scale 30N	Superficial Rockwell hardness	292
350 HMR 10/3000	350	Rockwell Macro	10-mm ball Scale: 3000 kg	Macro Rockwell hardness (depth Brinell)	170
221 HBS 10/3000	221	Brinell Steel ball	10-mm ball Scale: 3000 kg	Brinell hardness (optical)	104
550 HV$_{500}$	550	Vickers	Test force: 500 g	Vickers hardness, 500 g load	270
350 HK$_{100}$	350	Knoop	Test force: 100 g	Knoop hardness, 100 g load	155
80 Shore A	80	Shore	Scale: A	Shore hardness of scale A	—
50 IRHD	50	Int'l Rubber	Scale: Degree	IRH hardness, degree scale	—

[a] From ASTM Standard E 140.

a range of materials from thin films to the hardest steel. The "C" scale, using a diamond penetrator and a 331-lb load, is used for hard steel; the "B" scale with a 1/16-in. diameter ball and 220-lb load is used for softer steels and nonferrous metals.

The *Shore scleroscope* method of measuring hardness utilizes a diamond-tipped hammer that drops on the specimen. The rebound height is a measure of hardness. The *Vickers* hardness test utilizes a diamond pyramid-shaped penetrator whose apex angle is 136° and is loaded with a force (F) from 1 to 120 kg, depending on the hardness and thickness of the specimen to be checked. Specimens of any size with thickness greater than 1.5 times length of the diagonal of the impression made by the penetrator can be used as long as the surface is smooth enough to allow edge of impression to be discernible. Measurement of the impression after load is removed leads to Vickers hardness numbers. The following equation may be used to obtain the Vickers hardness (VH or HV).

$$\text{VH} = 1.8544 \frac{F\,(\text{kg})}{\text{diagonal}^2\,(\text{mm}^2)} \tag{2.14}$$

The *Brinell* method is widely used on casting and forgings. Brinell hardness (BH or HB) is determined by applying a predetermined test force (F), ranging from 500 to 3000 kg, to a hard steel or carbide ball of fixed diameter (D), which is held for a predetermined time and then removed. The resulting indentation is measured across at least 2 diameters (d). Thus, the BH can be calculated as

$$\text{BH} = \frac{2F}{\pi D\,(D - \pi\sqrt{(D^2 - d^2)})} \tag{2.15}$$

where force is given in kilograms and the diameters are in millimeters. In practice, sometimes, the BH Scale C value is used to estimate the tensile strength of plain carbon steels as

$$\text{Tensile strength (psi)} = 500\,\text{BHC} \tag{2.16}$$

An approach to determine the hardness of rubber is the *International Rubber Hardness Degree* (IRHD) method. It provides a repeatable result on rubber parts of all sizes and shapes. To test by this method, a preliminary test force, f, ranging from 8.46 to 295.74 gram, is applied to a ball indenter. The test is zeroed at this position. Then the total test force, F, ranging from 15.7 to 597 gram, is applied. The distance between the positions of f and F is then measured and converted to an IRHD hardness number.

Hardness of natural materials can be compared by referring to the *Mohs' scale*, which has a range of 1 to 10 as follows: talc = 1, gypsum = 2, calcite = 3, fluorite = 4, apatite = 5, feldspar = 6, quartz = 7, topaz = 8, titanium nitride = 9, and diamond = 10. The different systems for determining hardness are related. ASTM standard E140 and handbooks provide tables that show the equivalent values of hardness for the various methods and scales. Table 2.5 shows how to read a hardness number.

Measurements of hardness have value in the nondestructive inspection of many industrial parts, because hardness and tensile strength are often proportional to each other and wear is inversely proportional to hardness. Figure 2.11 indicates the general manner in which some properties of ferrous material are related.

Because hardness measurements are relatively easy to make and seldom are destructive to the workpiece, such tests are used to control heat-treating processes and hot- and cold-working processes. A uniform hardness in parts made of the same ferrous material usually indicates that the engineering properties of the parts are identical.

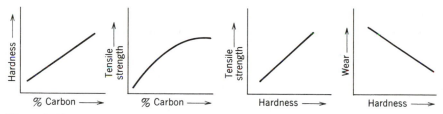

Figure 2.11
General relationships involving tensile strength and hardness for steels.

2.8 COMPOSITES AND NEW MATERIALS

The concept of composites is not really new. Egyptians reinforced mud with straw to make bricks. Wood is considered to be a composite of cellulose fibers bonded by a matrix of natural polymers. Concrete can be classified as a ceramic composite in which stones are dispersed among cement. In addition, in the mid-1900s short glass fibers impregnated with thermosetting resins, known as fiberglass, became the first composite with a plastic matrix.

Designers now have the ability to tailor materials and their reinforcements to meet specific requirements. Composites also offer a chance to build in only as much strength as needed and when needed. Figure 2.12, for example, illustrates the ply and loading terminology of laminate plastics. *Laminate plastics* are a special form of polymer-matrix composite consisting of layers of reinforcing materials that have been impregnated with thermosetting resins, bonded together, and cured under heat and pressure.

Laminates are identified by the number of plies and fiber lay angle. In Figure 2.12, identification for the laminate is $[90_1/-45_1/+45_1/0_1]$. Plies are identified from the outside to inside and subscripts label the number of fiber layers for each angle. Often, a laminate repeats layer and angle sequence in reverse order; that is, the lays are symmetrical about a center plane. In such a case a subscript s is added to the notation and would read $[90_1/-45_1/+45_1/0_1]_s$.

The strength of a laminated plastic is higher than that of a machined shape, because the reinforcing plies are not cut, as they are in a machined part. Furthermore, complex shapes can be laminated at a fraction of the cost of comparable machined parts. Table 2.6 shows the effect of reinforcing a resin to change its properties. Values in the table

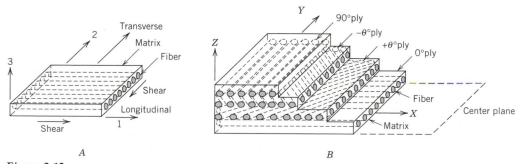

Figure 2.12
A, Ply and loading terminology. B, Laminated plastics.

Table 2.6   **Average Property Values for Selected Fiber Composites***

Base Resin	Specific Gravity		Mold Shrinkage (in./in.)		Water Absorption 24-hr (%)		Tensile Strength (10^3 psi)		Thermal Expansion ($10^{-5} \times$ in./in. °F)	
Acetal	1.63	**1.42**	0.003	**0.020**	0.30	**0.22**	19.5	**8.8**	2.2	**4.5**
ABS[a]	1.28	**1.05**	0.001	**0.006**	0.14	**0.30**	14.5	**5.0**	1.6	**5.3**
ETFE[b]	1.89	**1.70**	0.003	**0.018**	0.02	**0.02**	14.0	**6.5**	1.6	**4.0**
Nylon 6	1.37	**1.14**	0.004	**0.016**	1.1	**1.8**	23.0	**11.8**	1.7	**4.6**
Polycarbonate	1.43	**1.20**	0.001	**0.006**	0.07	**0.15**	18.5	**9.0**	1.3	**3.7**
Polyester (PBT)	1.52	**1.31**	0.003	**0.020**	0.06	**0.08**	19.5	**8.5**	1.2	**5.3**
Polypropylene	1.13	**0.91**	0.004	—	0.03	**0.03**	9.7	**5.0**	2.0	**4.0**
PPS[c]	1.56	**1.34**	0.002	**0.010**	0.04	**0.05**	20.0	**10.8**	1.1	**3.0**
Polystyrene	1.28	**1.07**	0.001	**0.004**	0.05	**0.10**	13.5	**7.0**	1.9	**3.6**
SAN[d]	1.31	**1.08**	0.001	**0.005**	0.10	**0.25**	17.4	**9.8**	1.8	**3.4**

* Lightface, values for glass-reinforced resins. **Boldface**, values for unreinforced resins for comparison.

[a] Acrylonitrile butadiene styrene.

[b] Ethylene tetrafluoroethylene.

[c] Polyphenylene sulfide.

[d] Styrene-acrylonitrile copolymer.

are representative since both higher and lower values may be obtained in commercially available resins and compounds. Values for the reinforced compounds are typical of 30% glass-reinforced formulations.

Although polymer-matrix composites are very popular, new components are challenging them. Metal-matrix, and ceramic-matrix composites are now a reality. *Metal-matrix* composites use a hardened material that is either precipitated or dispersed within a tougher matrix. One such material is an aluminum-lithium alloy developed for the space shuttle fuel tanks. It has 10% higher modulus, is 5% lighter, and is 30% stronger than the material used in the first vehicles, saving nearly 4 tons on the shuttle. *Ceramic-matrix* composites, however, show considerable promise for providing fracture toughness values similar to those for metals such as cast iron. SiC fibers of 15 μm can withstand temperatures up to 1200°C, SiC filaments of 140 μm can be used to reinforce aluminum and titanium matrices, SiC whiskers of diameters as low as 0.5 μm can stand up to 15.9 GPa and are stable at temperatures to 1800°C.

All these developments reveal a larger and more complicated materials-choice menu. This diversity has made the job of selecting the best materials from such a huge array of candidates quite challenging.

QUESTIONS

1. Give an explanation of the following terms:

Composites
Grain
Dendrite
Brass

Strain
Yield strength
Elastic constant
Shear modulus

Etched
Alloy
Solid solution
Tensile strength
Stress

Percentage elongation
Creep
Hardness
Mohs' scale

2. What are some reasons for not using inorganic materials in outdoor applications?

3. If a manufacturing application needs a metal very resistant to corrosion, would you use a ferrous or a nonferrous metal? What if the key requirement is strength? Explain.

4. Classify the principal materials used in the manufacture of an automobile. Indicate their principal ores or raw materials.

5. Classify the following materials as organic or inorganic: cardboard, telephone case, rubber cement, polyethylene bottle, copper tubing, eraser, alcohol, peanut butter, salt, scissors, gasoline, cotton cloth, and polyester cloth.

6. Name the desired properties for a metal if the application is an anvil. A draglike bucket?

7. Describe the solidification of the following materials by using a time–temperature curve: cast iron, magnesium, steel, pure iron, and gold.

8. Which characteristics of wrought-iron water pipe enable it to be recognized after 5000 years?

9. Study Table 2.1. Recall the relationship between strength and grain size and then arrange the nonferrous metals from smaller grain size to larger grain size.

10. Same as question 9, but for ferrous metals.

11. What can be said about a fine-grained metal as compared to large- or coarse-grained metal?

12. Give the purpose of an etching reagent. Why is it applied after polishing?

13. What general statement can be made about the relationship between hardness and tensile strength in steels?

14. What is the difference, if any, among ductility, plasticity, and elasticity?

15. How are metal properties affected by quenching? By annealing?

16. A piece of steel is hardened by heat treating. What is the effect on tensile strength and wear resistance?

17. How could one build a crude hardness tester using a hand press, a spring, and an industrial diamond?

18. Describe the problems in measuring the hardness of very thin, extremely hard, and very soft metals.

19. For ferrous metals, what alloying element is the most important in determining properties?

20. What are the components of an advanced composite material? Give examples.

PROBLEMS

2.1. Sketch a general time–temperature curve for steel, magnesium, and brass.

2.2. Sketch a golf club driver and identify the materials used in its manufacture. Classify the materials.

2.3. A stress–strain test is performed on a steel specimen with 0.32% carbon. At what value of stress and strain does necking of the specimen first occur? If the bar brakes and $L = 2.0$ in. and $L_f = 2.3$ in., what is the percentage elongation and what is its significance?

2.4. A lead cylinder is subjected to a simple tensile load. If its diameter decreases 0.002 mm while its length increases 0.005 mm, what is the elastic constant of the material?

2.5. The harder a steel is the more difficult it is to weld. What is the relationship between weldability and percentage carbon; between weldability and tensile strength?

2.6. The diameter of a 6-in. long brass cylinder is 1.2 in. If it increases 0.004 in. when the cylinder is compressed, what is the length change? (Hint: Poisson ratios given in Table 2.3.)

2.7. The maximum shear stress that can be placed on a titanium shaft is 8500 psi. Use a factor of safety of 2 and estimate the shear strain. (Hint: E and ν values given in Table 2.3.)

2.8. White cast iron is used to make a 0.5-in. diameter eyebolt. If it is to support a 1600-lb load, what is the factor of safety? (Hint: The maximum allowable stress is typically the yield stress of the material. This value is given in Table 2.1.)

2.9. Using Figure 2.6, determine the composition of the material at ℓ_2, ℓ_3, and s_3.

2.10. A common alloy used in the past to make nickels had what composition at the liquidus and solidus lines? (Hint: Use Figure 2.6.)

2.11. Give the composition at the liquidus and solidus lines for an alloy of 40% nickel and 60% copper.

2.12. If 588 lb of an alloy are needed to make a test part and the material occupies 3.50 ft^3, find the density, weight-density, and specific gravity of the alloy. If a pure metal with density $\rho = 7.20$ g/cm^3 were used and it cost 2.5 times less than the alloy, which material would you recommend? Why?

2.13. In the automobile shown in Figure 2.1, assume you will replace a steel part that weighs 130 lb with one made of aluminum of the same size. How much weight reduction will result if steel weighs 2.9 times as much as aluminum? Suppose the new part is plastic. Steel weighs 4.8 times the weight of plastic. How much weight is saved?

2.14. The Brinell hardness number of a plain carbon steel is 275. If a 10-mm carbide ball is used and a test force of 2500 kg is applied for the test, calculate the indentation made and estimate the strength of the material.

MORE DIFFICULT PROBLEMS

2.15. In Figure P2.15, determine the total elongation of an initially straight bar of length L, cross-sectional area A, and modulus of elasticity E if a tensile load P acts on the ends of the bar.

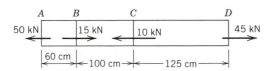

Figure P2.15

2.16. An aluminum alloy bar of cross section 350 mm^2 is acted on by the forces shown in Figure P2.16. Determine the total elongation of the bar. (Hint: E values are given in Table 2.3. Use average values.)

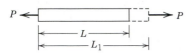

Figure P2.16

2.17. Two prismatic bars are rigidly fastened together and support a vertical load of 8000 lb, as shown in Figure P2.17. The upper bar is steel having specific weight 0.283 lb/in.^{3}, length 240 in., and cross-sectional area 10 in.^{2}. The lower bar is brass having specific weight 0.300 lb/in.^{3}, length 180 in., and cross-sectional area 8 in.^{2}. Determine the maximum stress in each material and the allowable stress if a safety factor of 60% is required. (Hint: E values are given in Table 2.3. Use average values.)

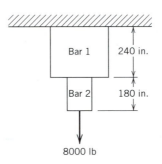

Figure P2.17

2.18. A block of aluminum alloy is 400 mm long and of rectangular cross section 25 by 30 mm. A compressive force $P = 60$ kN is applied in the direction of the 400-mm dimension and lateral contraction is completely restrained in each of the other two directions. Find the effective modulus of elasticity as well as the change of the 400-mm length. (Hint: E and ν values are given in Table 2.3. Use average values.)

PRACTICAL APPLICATION

Some government programs promote economic growth by introducing advanced materials technologies into machinery and equipment products. Contact a local materials laboratory, or the National Institute of Standards and Technology (NIST, Internet: "www\nist.gov") to identify current or research materials that can be utilized by the automotive industry to reduce the cost of engine, powertrain or chassis components by approximately 5% or $3 billion per year.

CASE STUDY: THE UNKNOWN MATERIAL

The Wilsan Machinery Company was asked to repair a foreign-made ski lift that had failed. A 1.13-in. OD steel support pin failed by necking down in tension, and subsequent tests showed the tensile strength of the material used in the original lift was about 50,000 psi. A discussion of the failure is held and you are asked to make general recommendations for a replacement part.

The expected load on the pin is 41,000 psi, but it is pragmatic to have a built-in factor of safety of 2.0 (2 × 41,000).

Some information about the pin is known and recommendations can be made. When steel is requisitioned for the pin, the inspection department checks it by making a standard steel tensile specimen of the material available. It is pulled by a local laboratory, and the results are as follows:

Force		$L_1 - L$
lb	N	in.
32,000	142,336	0.002
62,000	275,776	0.004
91,000	404,768	0.006
105,000	467,040	0.008

From the analysis an estimate of the percentage of carbon in the steel and its suitability for the pin can be made. What conclusions can be made about the carbon content of the failed pin and the material chosen to replace it? What comments can be made about other physical characteristics when the original pin and the proposed pin are compared?

CHAPTER 3

PRODUCTION OF FERROUS METALS

Iron and steel comprise about 95% of the tonnage of metal produced annually worldwide. On the average, they are the least expensive of the metals. In some applications, no other materials will serve. Yet iron and steel have competition from other metals and nonmetals.

Early ironmaking used ore removed from bogs, charcoal made from nearby forests, clam and oyster shells for flux, and a waterwheel provided power to pump bellows for the air necessary for the combustion. Later, a *Bessemer process* forced air through molten iron, which was contained in a pear-shaped vessel called a converter. As the air moved upward through the iron, it "burned out" many impurities contained in the molten metal. Even later, the addition of manganese caused chemical changes, which converted the iron into steel during the Bessemer conversion.

3.1 PRODUCTION OF PIG IRON

The principal raw material for all ferrous products is pig iron and to a much lesser extent, direct iron. *Pig iron* is the product of the blast furnace and is obtained by smelting iron ore with coke and limestone.

The final analysis of pig iron depends primarily on the characteristics of the ores. The principal *iron ores* used in the production of pig iron are Hematite (Fe_2O_3, 70% iron), Magnetite (Fe_3O_4, 72.4% iron), Siderite ($FeCO_3$, 48.3% iron), and Limonite ($Fe_2O_3 \times H_2O$, 60–65% iron). The percent iron is for pure ores. Most ores contain impurities, however, and the actual percentage of iron is less. *Hematite* is the most important iron ore used in the United States. Vast quantities of iron pyrite (FeS_2) are available, but are not used because of the sulfur content. If used, the sulfur must be eliminated by an additional roasting process.

Blast Furnace

The traditional method of making iron starts with the *blast furnace*, a vessel lined with insulating brick. The blast furnace is the most widely used method of iron production in the world. Figure 3.1 is a schematic of the blast furnace. The materials added to the blast furnace are coke, iron ores, and limestone. These coke-burning, fire-belching furnaces are efficient producers of pig iron.

A metallurgical grade of *bituminous coal* is cleaned and baked in coking ovens into *coke*, which burns inside as well as outside and does not fuse into a sticky mass, or cause the smoke of the original coal. The coal is dumped into coking ovens, where it is heated to 2400°F (1315°C) and most of the coal's gases are removed. Coke is used in the blast furnace because it burns with intense heat.

Iron bearing materials, or ore and limestone, are brought to the top of the furnace; for example, 28-ft diameter, 120-ft tall, with a skip hoist and then dumped into the

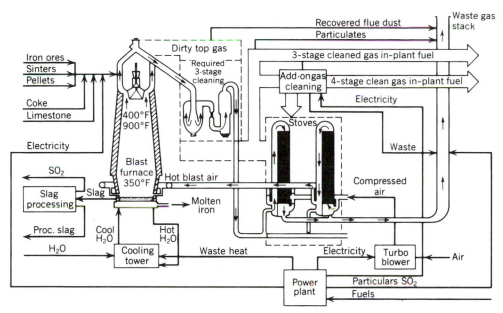

Figure 3.1
Simplified schematic of the blast furnace process.

double-bell hopper. These materials are sprinkled in alternating layers into the hopper. As the charge is released into the stove, the double *bell* confines the gaseous discharge and aids in the prevention of pollution releases. The ore is heated to high temperatures to produce iron. To produce 1000 tons of iron, a plant uses 2000 tons of ore, 800 tons of coke, 500 tons of limestone, and 4000 tons of hot air.

The use of *heated air* instead of cold air enables the coke to burn more effectively; therefore, less coke is used. The hot air is heated in tall cylindrical towers or stoves to about 1000°F. This hot air is "blasted" in from the bottom to fuel the burning coke, furthering the production of carbon monoxide, which reacts with ore to produce iron and carbon dioxide. The hot air and ash are separated to be used again in the furnace and as an in-plant fuel.

Limestone is added to the blast furnace to combine with impurities in the ore. These impurities rise to the top of the molten iron, and the *slag* is skimmed off the top of the molten iron mass and drained into a slag bucket. The slag is either recycled within the mill or, for example, sold as railroad ballast, landfill, or for road and highway use.

The blast furnace is tapped after slag removal, and the molten iron flows into *ladles, or hot metal cars*. Pig iron can be sold as an approximate 40-lb solid product, used in molten form for steelmaking, or solidified into large ingots. Many different grades of iron are produced from the blast furnace, depending on the charged raw materials. Various grades of pig iron are given in Table 3.1.

Direct Reduction

The *direct reduction process* employs either a solid or gaseous reducing agent reacting with ground ore fines, pellets, or lump ore, usually Hematite, Fe_2O_3, to produce a spongelike iron. This heating in the reactor is in the presence of hydrogen and methane,

Table 3.1   **Classification of Pig Iron**

Grade of Iron	Chemical Content (%)			
	Silicon	Sulfur	Phosphorus	Manganese
No. 1 foundry	2.5–3.0	<0.035	0.05–1.0	<1.0
No. 2 foundry	2.0–2.5	<0.045	0.05–1.0	<1.0
No. 3 foundry	1.5–2.0	<0.055	0.05–1.0	<1.0
Malleable	0.75–1.5	<0.050	<0.2	<1.0
Bessemer	1.0–2.0	<0.050	<0.1	<1.0
Basic	<1.0	<0.050	<1.0	<1.0

and the hydrogen combines with the oxygen in the ore to form water, while the carbon in the methane combines with the iron. The processes consist of crushing the iron ore and reacting it, usually at an elevated temperature (700° to 1000°C) with the reducing agent. This reducing agent may be coke, natural gas, fuel oil, carbon monoxide, hydrogen, or graphite. The product of the process is a spongy granular, clinkerlike material that either is used for making metal powders or is converted in an electric arc furnace to iron or steel.

Figure 3.2 is a diagram of the process. The reaction in which 100 tons (90 Mg) of iron ore are reduced to about 63 tons (60 Mg) of sponge iron takes from 10 to 14 h. Approximately 25,000 ft³ (700 m³) of methane are required per ton (0.9 Mg) of sponge iron. Because this process is effective in reducing oxygen and sulfur, relatively low-grade ores may be employed.

Although the direct reduction process of making iron was known before the invention of the blast furnace, less than 2% of the world's pig iron is made by this method. Those steel mills, which depend on the electric furnace more or less exclusively, may be users of direct reduction iron. As electric furnaces do not depend on molten pig iron, sponge iron is another source in addition to scrap.

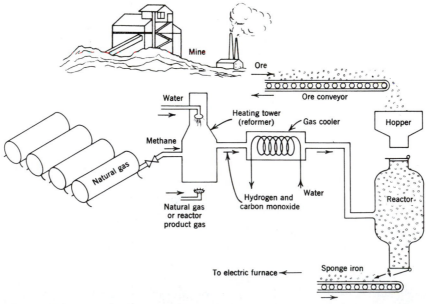

Figure 3.2
Diagram of direct reduction process for producing sponge iron from iron ore.

3.2 FURNACES FOR STEELMAKING AND IRONMAKING

Blast furnace pig iron is either cast into permanent iron molds or is transferred liquid in hot-ladle cars to a furnace, where it may be refined for manufacturing wrought iron, steel, cast iron, and malleable or ductile iron. The capacity of hot-metal cars may exceed 100 tons (90 Mg), and more than 90% of all pig iron is transported to steelmaking furnaces by this method.

The principal processes used for converting pig iron and remelting ferrous metals are described in Figure 3.3. Table 3.2 is a general table for furnaces used for ferrous materials. Three furnaces—basic oxygen, electric arc, and open hearth—account for approximately 45%, 40%, and 15%, respectively, of steel produced in the United States. These furnaces are not interchangeable, because each furnace system requires different materials and energy sources. Steelmaking by all processes refines the pig iron and aims at precise chemical control.

Basic Oxygen Furnace

The basic oxygen furnace (BOF) process uses molten pig iron as its principal raw material (65%–80%) from a blast furnace. In addition, scrap and lime are added. As the name implies, heat is generated with oxygen.

Scrap, about 30% of the total charge, is loaded into a basic refractory lined vessel as shown in Figure 3.4. Hot metal is poured into the mouth of the tilted vessel. A water-cooled, oxygen-carrying *lance* is lowered 4 to 8 ft (1.2 to 2.4 m) above the bath in the vertical vessel. With oxygen blowing over the surface of up to 6000 cfm at 160 psi, ignition starts immediately and the temperature climbs close to iron's boiling point of about 3000°F. Carbon, manganese, and silicon are oxidized. Lime and fluorspar are added to collect impurities such as phosphorus and sulfur in slag form. When the refining process is over, the vessel is tilted for tapping. The tap-to-tap time to produce 300 tons (270 Mg) is approximately 45 min. The oxygen needed to produce 1 ton (0.9 Mg) of steel is about 1600 ft³ (45 m³).

A BOF plant (Figure 3.5) is designed to conduct the gases to air treatment facilities. The gas cleaning equipment is so complex that it may account for one-third or more of

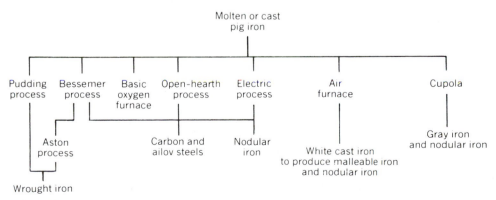

Figure 3.3
Principal processes for remelting or refining pig iron.

Table 3.2 Furnaces for Ferrous Metals

Type of Furnace	Primary Fuel	Predominant Metal Charge	Special Atmosphere Available	Product
Air or reverberatory	Pulverized coal, oil	Molten or solid pig iron, scrap		Gray cast iron, white cast iron
Basic oxygen	Oxygen	Molten pig iron and scrap		Steel
Converter	Air	Molten pig iron or molten cupola iron		Raw material for wrought iron and steel
Crucible	Gas, coke, oil	Select scrap		Small quantities of steel and cast irons
Cupola	Coke	Solid pig iron and scrap		Gray cast iron, nodular iron
Electric furnace	Electricity	Scrap	Vacuum or inert gas	Steel, gray iron
Induction	Electricity	Select scrap	Vacuum or inert gas	Steel
Open-hearth furnace	Natural gas, coke oven gas, pulverized coal, oil	Molten pig iron		Steel

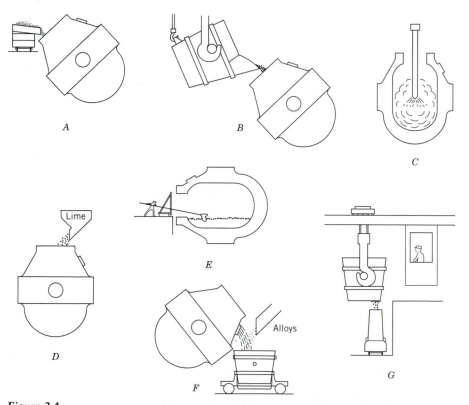

Figure 3.4
Ladle positions in the production of steel by the basic oxygen furnace (BOF). *A*, Charging scrap. *B*, Charging hot metal. *C*, Blowing with oxygen. *D*, Charging fluxes. *E*, Testing. *F*, Tapping and addition of alloys. *G*, Teeming steel.

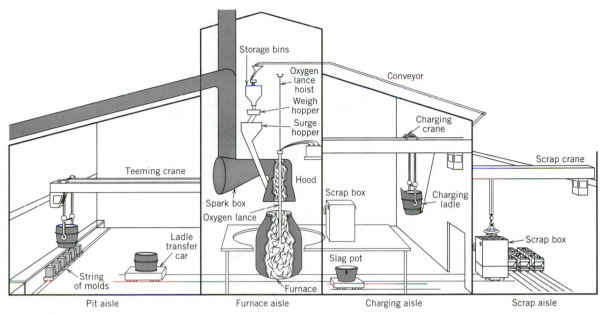

Figure 3.5
Basic oxygen steel plant.

the overall structure of a BOF plant. Experienced BOF operators identify the refining progress by a decrease in the flame and change in the sound level and flame color. When a batch of steel is complete, the oxygen is shut off and the lance is retracted through the hood. The furnace is tilted to remove the slag. Following testing for acceptable temperature of the bath, which may be about 3000°F, and carbon, the furnace is tilted to the opposite side for pouring to the ladle car.

Electric Furnace

The electric furnace (Figure 3.6) is charged with carefully selected steel scrap rather than molten iron. In fewer cases, pig iron or prereduced iron ore, which may contain up to 98% of the element iron, is the charge of material. The load is charged through the top of the furnace. By close control of the charge, including small amounts of burned lime and mill scale, and by adding alloying materials, ingot and castings of stainless steel, heat-resistant steel, tool steel, and many general-purpose alloy steels are poured from the electric furnace. In addition to steel, high-test gray iron is produced, because electric furnaces do not pollute the atmosphere as much as other gray iron furnaces.

The roof of the furnace is designed to allow entrance of three carbon or graphite electrodes, which are lowered into the furnace. The three graphite electrodes sit above the scrap heap. Three-phase current arcs from one electrode to the charge and then arcs back to another electrode. These electrodes give the electric arc furnace its name because, in operation, the current arcs from electrode to the metallic charge and then from the charge to the next electrode.

The *direct arc furnace* may have the hearth either basic lined or acidic lined. The acid-lined furnace, with a hearth of ground ganister and side walls of silica brick, is used in a limited way to produce low-carbon, low-alloy steels provided the scrap is low in phospho-

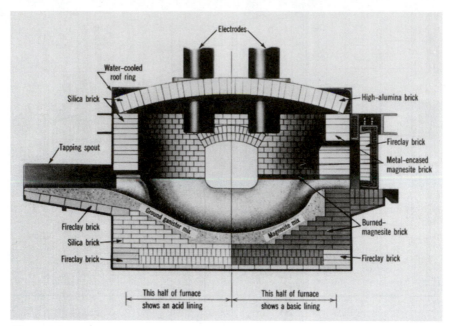

Figure 3.6
Cutaway drawing of electric furnace showing both the acidic- and basic-type linings.

rus and sulfur. The *basic-lined furnace*, with a hearth of magnesite and side walls of magnesite and alumina brick, produces any grade of steel or steel alloy. The basic-lined furnace controls phosphorus, reduces the sulfur, and maintains close control of temperature and composition. The molten metal can be sampled to determine the precise chemical composition of the melt, and before the metal is poured into ingots, the composition may be adjusted.

Internal diameters of furnaces range from 7 to 30 ft and will process 4.5 to 325 tons of steel per heat. For a 125-ton (115 Mg) tap-to-tap time, 3 h are required and 50,000 kWh of power. A refining step in the process is to inject high-purity oxygen, which reduces the tap-to-tap time and oxidizes phosphorus, silicon, manganese, carbon, and iron.

Graphite electrodes up to 30 in. (760 mm) in diameter and more than 80 ft (24 m) long may be used in a large furnace. All furnaces operate at approximately 40 V and at currents that may exceed 12,000 A.

Open-Hearth Furnace

The open-hearth furnace was once the most popular process for making steels. Figure 3.7 is typical. These furnaces hold from 10 to 600 tons (9.9 to 540 Mg) of metal in a shallow pool that is heated by action of a gas, tar, or oil flame passing over the charge. The furnace is said to be *reverberatory*, because the low roof of the furnace reflects the heat onto the long, shallow hearth. It is *regenerative*, because the chambers on either side of the furnace are capable of being heated by combustion gases, which in turn allow the air and fuel entering the furnace to be increased in temperature, thus ensuring an increased combustion efficiency and temperature. The chambers, left and right, are alternately heated so that as one set is being used to raise the temperature of air and fuel, the checker work of bricks in the opposite chamber is being heated.

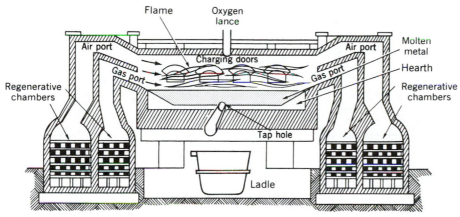

Figure 3.7
Sectional view of open-hearth furnace.

The *open-hearth furnace* may have a hearth of basic or acidic material, although in practice more than 90% are basic open hearths. In the basic unit, which can remove phosphorus, sulfur, silicon, manganese, and carbon, the hearth is lined with magnesite. The acidic open hearth, which can remove only silicon, manganese, and carbon, has a hearth of acid brick or sand whose principal ingredient is silica.

The time may be reduced 25% by inserting an *oxygen lance* through the roof of the furnace into the combustion area after the molten iron is introduced. When using oxygen, the fuel cost may be reduced by 30%, but this is offset by the cost of supplying up to 500 ft³ (14 m³) of oxygen. The charge must be controlled carefully and frequent samples of the molten metal are tested to determine the extent to which additions must be made.

Cupola

A cupola does not produce steel. Instead, iron castings are poured by remelting scrap along with pig iron in a furnace called a *cupola*. The cupola is simple in construction, economical to operate, and melts iron continuously with a minimum of maintenance. As the metal is melted in contact with the fuel, some elements are picked up while others are lost. This affects the final analysis of the metal and necessitates close regulation of the cupola. A water-cooled, unlined cupola is shown in Figure 3.8.

The openings for introducing air to the coke bed are called *tuyères*. Small windows covered with mica are located at each tuyère, so that conditions in the cupola can be observed. Surrounding the cupola at the tuyères is an insulated *wind box* or jacket for the air supply. The *air blast,* furnished by a positive displacement or centrifugal-type blower, enters the side of the wind box.

The opening through which the metal flows to the spout is called the *tap hole*. The overflow or slag spout is slightly below the tuyères.

In operation, a coke bed is ignited and alternate charges of coke and iron are made in a ratio of 1 part coke to 8 or 10 parts iron, measured by weight. A fluxing material, usually limestone ($CaCO_3$), fluorspar (CaF_2), or soda ash (Na_2CO_3), is added to protect the iron from oxidation and render the slag more fluid. For limestone about 75 lb (35 kg) are added for each ton (0.9 Mg) of iron. The amount of air required to melt a ton (0.9 Mg) of iron depends on the quality of coke and the coke-iron ratio. Theoretically,

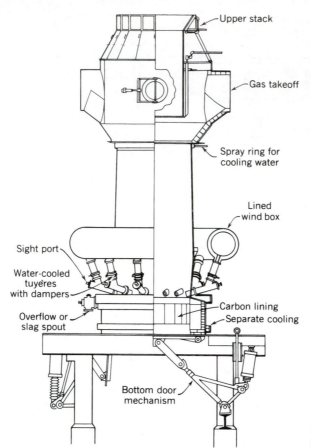

Upper stack

Gas takeoff

Spray ring for
cooling water

Lined
wind box

Sight port

Water-cooled
tuyeres
with dampers

Overflow or
slag spout

Carbon lining

Separate cooling

Bottom door
mechanism

Figure 3.8
Sectional drawing of a cupola.

113 ft³ (3.19 m³) of air at 14.7 psi (100 kPa) and 60°F (15.5°C) is required to melt 1 lb (0.5 kg) of carbon.

Combustion is improved by preheating the air as in regenerative-type melting furnaces. Such a furnace is called a *hot-blast cupola.* Alternatively, another method of introducing the charge into the furnace is profitable. Pellets of iron ore, coke dust, lime, and silica are cold bonded and fed into the cupola.

Pollution control costs are significant in cupola operation, causing some foundries to use horizontal, oil-fired, rotary furnaces (rotary cupolas) for making gray iron. Furnaces of this type use compressed air to atomize the fuel and can melt up to 25 tons (22.7 Mg) per day in a temperature environment of up to 2800°F (1540°C). The units have relatively low initial and maintenance costs and can utilize scrap as small as machine borings.

3.3 STEEL INGOTS AND STRAND CASTING

In each of the steelmaking processes described so far—BOF, electric arc, and open hearth—the process ended with molten steel in a ladle. For steel to be useful, it must solidify into forms that are suitable for later *hot working,* which is discussed in Chapter 14. Two methods for solidifying the molten steel are *ingots* and strand casting. The traditional method of handling molten steel from a furnace is to "teem" it from the ladle into ingot molds of various sizes and shapes.

The cast-iron ingot mold may be rectangular, square, or round in cross section, and final casting varying in size from a few hundred pounds to 350 tons (317 Mg). The kind of metal cast and the product determine the ingot size. Both rectangular and square-section ingots have rounded corners, and the sides are corrugated. Rounding the corners reduces the tendency for columnar grains to meet and form a plane of weakness. Cooling is accelerated by corrugating the sides, a process that reduces the size of the columnar grains.

The two types of ingot molds in Figure 3.9 are used for top pouring. The big-end-down type shown in Figure 3.9A is easy to strip from the ingot, but the loss in metal is high as a result of the shrinkage cavity or *pipe* that is formed during the cooling operation. This loss is lower when the big-end-up type shown in Figure 3.9B is used.

Solidification is progressive, starting at the mold surface and progressing toward the center and from the bottom upward. During this stage there is considerable shrinkage of the metal. As layer after layer solidifies, the volume of metal decreases, resulting in the formation of a pipe when solidification is complete. The rate of cooling is an important factor in the production of a sound ingot.

Ingots made in *big-end-up molds* have a large volume of hot metal available at the top of the mold during the cooling operation and when solidified show little loss of metal resulting from piping. Losses from pipe formation in ingots can be reduced either by adding metal during cooling or by using refractory risers. Metal in the riser remains molten until the ingot has solidified, and during the solidification period supplies the ingot with needed metal to compensate for shrinkage.

Several types of ingot structures are obtained by controlling or eliminating a gas evolution in the metal during solidification. *Killed steel* has been deoxidized and it evolves no gas during solidification. The top surface of such ingots solidifies immediately as do the walls, and because of the shrinkage of the metal on solidification, a larger cavity or pipe is formed within the ingot. The process of producing killed steel is complex and depends on starting with a higher carbon steel than finally desired; then, when the carbon is reduced to the exact amount, the steel is deoxidized by furnace or ladle additions of high-silicon pig iron or an alloy high in silicon. All steels having more than 0.30% carbon are killed. Such ingots have a minimum of segregation, good structure, and a large cavity in the center.

Another ingot structure is known as *rimmed steel*, which is either not or only lightly deoxidized in the furnace or ladle. This type of steel is characterized by a semiboiling action in the ingot after it has been poured resulting from rapid evolution of carbon monoxide gas during the solidification period. This causes the formulation of a honeycomb structure that if controlled compensates for most shrinkage loss. These small blow holes do not constitute a defect if they have not had contact with the outside atmosphere and are closed by pressure welding in the hot-working processes. Rimmed-steel ingots have

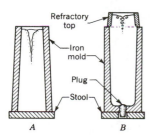

Figure 3.9
Type of ingot molds. *A*, Big end down. *B*, Big end up.

a good surface and there is little or no opportunity for pipes to form. Semikilled and other ingot structures are obtained by controlling the formation of gas during solidification.

Longitudinal and cross sections of a medium-carbon, killed-steel ingot are shown in Figure 3.10. The coarse dendritic crystalline structure, indicated in the cross section, is eliminated by the effect of hot working. The impurities in ingots tend to segregate in the shrink head during the process of solidification. Cutting off the end of the ingot, either before the rolling starts or after, largely eliminates this defect but decreases overall efficiency.

Much more modern than the traditional pouring metal in ingots, a process that converts molten steel into semifinished solid products, such as a slab or billet continuously, is called *strand casting* (shown in Figure 3.11), the procedure that bypasses ingot *teeming*, stripping, soaking, and initial rolling of the ingot.

Molten steel from a furnace is carried in a ladle to the top of the strand caster. A stopper in the bottom of the furnace ladle is lifted allowing molten metal to flow into the tundish of the caster. This transfer to the tundish provides an even pool of molten metal for feeding the casting machine. Molten steel drops in a controlled flow to the mold section where cooling water chills the outside to form a solid skin. The skin thickens as the column of steel descends through the cooling system. Breakout of molten steel from the skin is a potential problem, but the drastic cooling and the controlled rate of withdrawal from the tundish are in equilibrium within the surface skin. The solidified steel curves into a straight section; the steel is flame cut by an oxyacetylene torch into lengths by a traveling frame that moves at the same rate as the steel.

The caster uses brass or copper or graphite molds of a thickness that permits a heat flow rate that is sufficient to prevent the mold from being damaged by the metal being cast. In addition, rapeseed oil is often poured over the surface of the mold to prevent unnecessary friction between metal and the walls of the mold, and to aid in the stripping process. The brass or copper mold has a high heat conductivity and is not easily wetted by molten steel. Rapid mold cooling is essential for the success of this process, which results in improved mold life, less segregation, smaller grain structure, and better surface. Actually, the metal next to the mold wall solidifies only a few inches below the top surface and shrinks slightly from the mold walls. As the cast section leaves the cooled

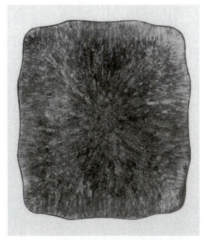

Figure 3.10
Ingot macrographs of longitudinal and cross sections of a medium-carbon, killed-steel ingot.

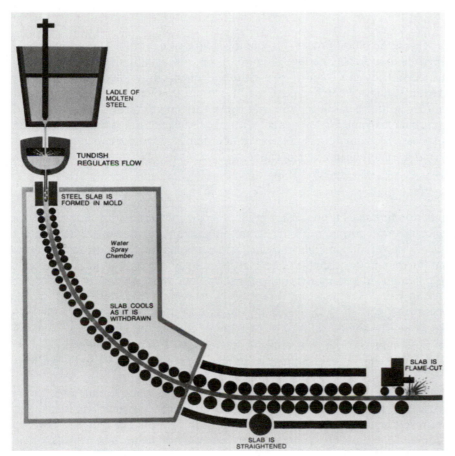

Figure 3.11
Semihorizontal method for continuously casting steel.

mold, it passes through a section that controls the rate of cooling where roller guides prevent expansion of the casting. The steel emerges from the curving mold as continuous rectangular ribbons—either slabs, blooms, or square billets depending on the proportions of the end product. The "strands" then pass through a straightner prior to torch cutting to length.

This *continuous-slab method* can convert 300 tons (270 Mg) of molten metal to solid slabs in 45 min instead of 12 h for ingot processing. Strand casting may be up to 15 fpm for some billets and slabs. Casting machines can produce about 1.5 million tons/year/strand and operate with nearly 80% uptime. Slab casters produce slabs with thicknesses anywhere from 2 to 12 in. thick. Development of strip casters give products as thin as 0.1 in. thick.

3.4 REFINING FURNACES AND VESSELS

Raw steel is sometimes enhanced in furnaces, hearths, ladles, and vessels for certain applications. Small but growing tonnages are involved. Special products for aerospace, nuclear engineering, medical tools, and computers are requiring advanced melting and refining vessels.

Crucible

The crucible process is the oldest process for making steel castings, but it is used predominantly in nonferrous foundries. *Crucibles*, Figure 3.12*A*, are usually made of a mixture of graphite and clay. They are fragile when cold and must be handled with care, but they possess greater strength when heated. Crucibles are heated with coke, oil, or natural gas and are handled with special fitting tongs to prevent damage. Wrought iron, washed metal, steel scrap, charcoal, and ferroalloys constitute the raw materials for steel manufactured by this process. These materials are placed in crucibles having a capacity of about 100 lb (45 kg) and are melted in a regenerative furnace.

Induction

The *electric induction furnace*, Figure 3.12*B*, induces a current to melt the charge. Essentially, the furnace is a transformer with the scrap metal acting as the core. The coreless *induction furnace* is powered by a high-frequency current supplied to the primary water-cooled coil that surrounds the crucible. The high-frequency current, about 1000 cycles per second (Hz), is supplied by a motor generator set or a mercury arc frequency system. The scrap or pig iron charge is induced with a much heavier secondary current, which is heated by resistance, and gives an eventual molten pool of steel or iron.

The crucible is charged with solid pieces of metal, scraps, or chips, and a heavy secondary current is induced. The resistance of this induced current in the charge in 50 to 90 min will melt the charge in even large crucibles containing up to 4 tons (3.6 Mg) of steel.

Induction furnaces, available in crucible sizes holding from a few pounds to 4 tons (3.6 Mg), are relatively low in cost, almost noise free, and generate very little heat. The temperature required is no higher than that required for melting the charge, so scrap alloys can be remelted without "burning out" the valuable alloying material. Because the furnace is closed, little or no slag is encountered. After the melting is complete, small additions of alloys or deoxidizers bring the steel or iron up to the specified chemical composition. For these reasons, induction furnaces are often found in experimental laboratories or foundries.

Melting in a Vacuum and Special Atmospheres

Molten metals tend to absorb gases because of moisture in the furnace, ladle, and atmosphere, or because of entrapped hydrocarbons in the charge. The oxidation rate of

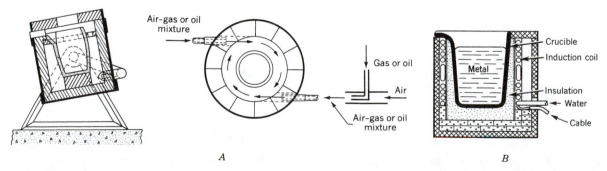

Figure 3.12
A, Oil- or gas-fired tilting crucible furnace. *B*, Induction furnace.

metals exposed to the atmosphere increases as the temperature increases. The lowest melting temperature consistent with sufficient fluidity will minimize gas absorption and oxidation, which are both detrimental.

For some metals a *slag coating* or *dross* is allowed to accumulate over the molten metal to protect it from excessive oxidation. Slag inducements are often added. In aluminum this dross is troublesome during pouring, and care must be exercised to prevent its entrapment in the casting.

If a vacuum or special atmosphere is to be employed, the furnace and mold must be enclosed in a chamber. The furnace, an induction or electric arc type, is seldom larger than 4 tons (3.6 Mg) in capacity. The steam ejector-type vacuum pumps can reduce the pressure to the equivalent of that supporting 0.000394 in. (0.01 mm) of mercury, as compared to atmospheric conditions of about 29.92 in. (760 mm) of mercury. Vacuum melting and pouring improve the tensile strength and fatigue life of most metals. Some metals such as titanium must be melted in a vacuum.

A high-vacuum, *consumable-electrode furnace*, Figure 3.13, in which a steel ingot acts as the electrode, is used to remelt and purify steel. The arc between the electrode and the furnace melts the ingot in a progressive manner, and solidification takes place almost as rapidly as the material is melted. Furnaces of this type are capable of making ingots weighing up to 20 tons (18 Mg).

Magnesium is melted and cast under an atmosphere of sulfur dioxide. Castings that must be absolutely free of absorbed gases, pin holes, and entrapped slag are melted and cast under a vacuum.

Some metals, particularly aluminum and magnesium, are often degassed after melting by passing an inert gas such as argon through the molten metal. Steel may be degassed with carbon monoxide.

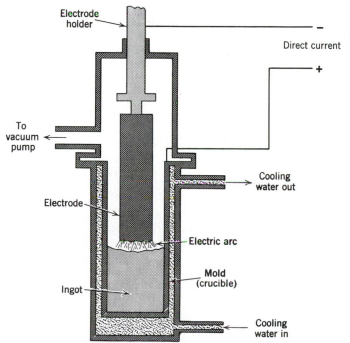

Figure 3.13
Consumable-electrode furnace.

AOD Process

AOD is an acronym for *argon oxygen decarburization*. Argon is a gas that is recovered from the air separation for oxygen production plants. Argon is used in steel refining vessels that have separate tuyères to permit both argon and oxygen to enter the bath of molten steel simultaneously. Argon dilutes carbon-oxygen atmosphere in the melt. With stainless steel, this increases the affinity of carbon for oxygen, thus minimizing the oxidation of chromium. Since argon is inert, it also stirs the molten steel, promoting rapid equilibrium, slag, and metal.

About 80% of the stainless steel production now employs this process, which has reduced the cost of manufacture by enabling utilization of less expensive feed metals while producing a steel of greater toughness, strength, and ductility, as well as a material that can be more easily machined. Instead of using the oxygen lance as in the BOF process, the gases are introduced to decarburize and reduce sulfur and dissolved gas in the steel. By diluting oxygen with argon, less oxidation takes place. The time for sulfur removal is about one-fifteenth as long as in an electric furnace. When the steel treatment is complete, it is transferred to a pouring ladle.

3.5 ENERGY REQUIRED FOR MELTING

The energy required for melting a ton of any metal is dependent on the *specific heat* of the material, which is defined as the ratio of the heat capacity of a material to that of water. Knowing that it takes 1 BTU of heat energy to raise 1 lb of water 1°F, aluminum with a specific heat of 0.21 would require roughly only one-fourth as much energy to raise the temperature of 1 lb a single degree. Specific heat of a metal varies with temperature, particularly around its melting point. The specific heat of iron varies from about 0.1228 at 500°F to 0.1666 at 2700°F. Figure 3.14 illustrates the approximate heat content of several metals as a function of temperature. This graph gives the melting characteristics of metals and illustrates why lead is so easily melted in comparison to brass or steel. Moreover, the metals with higher melting temperatures have much lower efficiencies in employing BTUs and furnace efficiencies are lower. The energy required to raise the temperature of a metal can be expressed:

$$H_t = h_t \times W/f \tag{3.1}$$

where

H_t = Energy required to reach a temperature of t (BTU/lb)
f = Furnace efficiency, %
h_t = Heat content of material, BTU/lb, at a temperature of t
W = Weight of material

Referring to Figure 3.14, it is possible to estimate the energy required to melt 1 lb of metal if you know the furnace efficiency. If one assumes it is 50%, then to melt 1 ton of aluminum requires approximately 560 BTU/lb $\times$ 2000 lb/ton = 11.2 $\times$ 10^5 BTU/ton. If this material is melted in 1 h, the kilowatt hours (kWh) per ton can be calculated.

$$\text{kWh} = (11.2 \times 10^5 \text{ BTU/ton})(2.928 \times 10^{-4} \text{ kW/BTU})$$
$$= 327.9 \text{ kWh/ton}$$

The 2.928 $\times$ 10^{-4} constant converts British thermal units to kilowatts. Gaseous and liquified fuels are sold by their BTU content, and electric furnaces are rated in kilowatts.

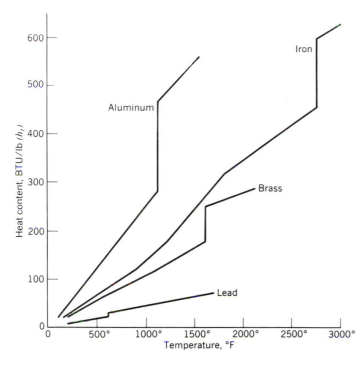

Figure 3.14
Specific heat as a function of temperature.

3.6 FERROUS METALS

Ferrous metals differ in physical properties according to their carbon content. Alloying elements can alter the properties of ferrous metals.

Wrought Iron

Wrought iron is a ferrous metal containing less than 0.1% carbon with 1% to 3% finely divided slag distributed uniformly throughout the metal. It has been produced for many centuries by a variety of processes; the two in use today are the puddling process and the Aston process. The latter accounts for the greatest tonnage of wrought iron.

In the *puddling process* pig iron and iron scrap are melted in a small, reverberatory, 500-lb (230-kg) capacity puddling furnace fired with coal, oil, or gas. Most of the elements are removed by oxidation, because they come in contact with the basic refractory lining of the furnace. This process uses mechanical puddling furnaces and eliminates the hand labor of stirring and gathering the metal into large balls. Freed from impurities, the product is removed from the furnace as a pasty mass of iron and slag, then mechanically worked to squeeze out the slag and to form it into some commercial shape.

In the *Aston process* the pig iron is melted in a cupola and refined in a Bessemer converter. The blown metal is then poured into a ladle (shotting) containing a required amount of slag previously prepared in a small open-hearth furnace. Because the slag is at a lower temperature, the mass solidifies rapidly, liberating the dissolved gases with sufficient force to blow the metal into small pieces. These pieces settle to the bottom of the ladle and weld together as sponge iron. Each ball of iron collected in the ladle weighs 3 to 4 tons (2.7 to 3.6 Mg), and the rate of production is about one ball every 5 min. A press squeezes out the surplus slag and welds the mass of slag-coated particles of iron into a solid bloom that can be rolled into various shapes.

Wrought iron produced by this method usually has a carbon content less than 0.03%, silicon about 0.13%, sulfur less than 0.02%, phosphorus about 0.18%, and manganese less than 0.1%. A photomicrograph of a piece of polished and etched wrought iron is shown in Figure 3.15. The particles of slag, which constitute about 2.5% by weight, are seen as dark streaks running through the metal. The direction of rolling is visible in the direction of the slag distribution.

Wrought iron is used principally in the production of pipe and other products subjected to deterioration by rusting, metals used around shipyards, railroads, farms, and oil companies. Advantages, other than its resistance to corrosion, include ease of welding, high ductility, and ability to hold protective coatings.

Steel

Steel is a crystalline alloy of iron, carbon, and several other elements. It hardens when quenched above its critical temperature. It contains no slag and may be cast, rolled, or forged. *Carbon* is an important constituent because of its ability to increase the hardness and strength of the steel. More tons of steel are used than all other metals combined. Although steel may be cast into molds to conform to a definite and complex shape and size, it is cast most often into ingots for use in making pipe, bar stock, sheet steel, or structural shapes.

Steel is classified according to the alloying elements it contains. Carbon is the most important element; therefore, all steels are classified according to carbon content. *Plain carbon steels* contain primarily iron and carbon, and they are classified as 10XX steels. The first two digits refer to plain carbon steel. The third and fourth digits refer to the carbon content in hundredths of a percent. Thus, a 1035 steel is a plain carbon steel with 0.35% carbon. There are varying amounts of other materials in carbon steel, but their content is so small that they do not affect physical properties.

Alloy steels are classified by the Society of Automotive Engineers (SAE) and by the American Iron and Steel Institute (AISI). Some of the designations accepted by them as standard are shown in Table 3.3. Often as many as five or more alloying elements may be present, and it is impractical to describe the alloy correctly by a simple numbering system. Steels may be more broadly classified as follows:

A. Carbon steel
  1. Low carbon—less than 0.30%
  2. Medium carbon—0.30% to 0.70%
  3. High carbon—0.70% to 2.0% (nominally the upper limit is 1.40%)

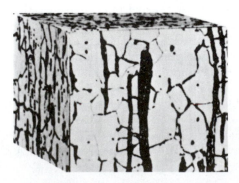

Figure 3.15
Photomicrograph of wrought iron showing slag distribution resulting from rolling. Magnification ×200.

Table 3.3 **Classification of Steels**

Classification	Number	Range of Numbers
A. Carbon Steels		
Carbon steel SAE–AISI	1XXX	
Plain carbon	10XX	1006–1095
Free machining (resulfurized)	11XX	1108–1151
Resulfurized, rephosphorized	12XX	1211–1214
B. Alloy Steels		
Manganese (1.5%–2.0%)	13XX	1320–1340
Molybdenum	4XXX	
C–Mo (0.25% Mo)	40XX	4024–4068
Cr–Mo (0.70% Cr, 0.15% Mo)	41XX	4130–4150
Ni–Cr–Mo (1.8% Ni, 0.65% Cr)	43XX	4317–4340
Ni–Mo (1.75% Ni)	46XX	4608–4640
Ni–Cr (0.45% Ni, 0.2% Mo)	47XX	
Ni–Mo (3.5% Ni, 0.25% Mo)	48XX	4812–4820
Chromium	5XXX	
0.5% Cr	50XX	
1.0% Cr	51XX	5120–5152
1.5% Cr	52XXX	52095–52101
Corrosion-heat resistant	514XX	(AISI 400 series)
Chromium–vanadium	6XXX	
1% Cr, 0.12% V	61XX	6120–6152
Silicon manganese		
0.85% Mn, 2% Si	92XX	9255–9262
Triple-alloy steels		
0.55% Ni, 0.50% Cr, 0.20% Mo	85XX	8615–8660
0.55% Ni, 0.50% Cr, 0.25% Mo	87XX	8720–8750
3.25% Ni, 1.20% Cr, 0.12% Mo	93XX	9310–9317
0.45% Ni, 0.40% Cr, 0.12% Mo	94XX	9437–9445
0.45% Ni, 0.15% Cr, 0.20% Mo	97XX	9747–9763
1.00% Ni, 0.80% Cr, 0.25% Mo	98XX	9840–9850
Boron (~0.005% Mn)	XXBXX	

Boron is denoted by addition of B. Boron–vanadium is denoted by addition of BV. Examples: 14BXX, 50BXX, 80BXX, 43BV14. The letters appearing before the number indicate the following: A, alloy-basic open hearth; B, carbon-acid Bessemer; C, carbon-basic open hearth; D, carbon-acid open hearth; E, electric furnace.

Stainless and heat-resisting steels:
   2XX Chromium–nickel–manganese types
   3XX Chromium–nickel types
   4XX Straight chromium types
   5XX Low chromium types
All stainless steels are produced in the electric furnace.

  B. Alloy steel
    1. Low alloys—special alloying elements totaling less than 8.0%
    2. High alloys—special alloying elements totaling more than 8.0%

Low-carbon steel is used for wire, structural shapes, and screw machine parts such as screws, nuts, and bolts. *Medium-carbon steels* are used for rails, axles, gears, and parts requiring high strength and moderate to great hardness. *High-carbon steels* find use in cutting tools such as knives, drills, taps, and for abrasion-resisting properties.

The microstructure of a medium-carbon cast steel is shown in Figure 3.16. The light areas are ferrite and the dark areas are pearlite. The grain structure of most cast steels is large because of the high casting temperature of the metal combined with relatively slow cooling.

More than 50% of all steel castings are of medium-carbon steel. Steel castings are used in the transportation industry, industrial machinery field, and the construction field. They have ductility and good tensile strength in a normalized condition ranging from 60,000 to 100,000 psi (400 to 690 MPa). The chemical composition of medium-carbon castings is approximately as follows: carbon, 0.21% to 0.46%; manganese, 0.55% to 0.73%; silicon, 0.28% to 0.45%; phosphorus and sulfur together, less than 0.1%; and ferrite the remainder.

The alloy steels, which account for only about 15% of the steel produced, are selected for many uses because they have characteristics that are superior to those of plain carbon steel. Although every alloy steel does not include each characteristic, the advantages of steel alloys should be considered.

1. Improvement in ductility without a lowering of tensile strength.
2. Ability to be hardened by quenching in oil or air instead of water, thus lowering the chance of cracking or warping.
3. Ability to retain physical properties at extremes of temperature.
4. Lower susceptibility to corrosion and wear, depending on the alloy.
5. Promotion of desirable metallurgical properties such as fine grain size.

Stainless Steels

Three types of *stainless steels* are listed here:

1. *Ferritic*. These steels, designated as 405, 430, 430F, and 446, cannot be hardened by heat treatment because their ratio of carbon to chromium is low. They also have good resistance to corrosion.

2. *Martensitic*. These steels, some of which are designated as 410, 414, 416, 431, and 440A, 440B, and 440C, are hardenable because their ratio of carbon to chromium is high, and they have high strength and are moderately corrosion resistant. Martensitic steels can be employed as forgings, can be hot worked, and are used for machine parts and cutlery.

Figure 3.16
Structure of medium-carbon cast steel. Magnification ×200.

3. *Austenitic. Austenitic steels*, some of which are designated as 301, 302, 303, 304, 310, 310S, and 384 (wire), cannot be hardened by heat treatment and are nonmagnetic. They are also highly corrosion resistant and can be cold formed. Austenitic steels harden and become stronger when cold worked and can be welded. *Type 303 stainless steel* is the easiest of all stainless steels to machine and is widely used for machine parts and fasteners.

Cast Iron

Cast iron is a general term applied to a wide range of *iron–carbon–silicon alloys* in combination with smaller percentages of several other elements. Cast iron contains so much carbon or its equivalent that it is not malleable. Cast iron has a wide range of properties, because small percentage variations of its elements may cause considerable change. Cast iron should not be thought of as a metal containing a single element but rather as one with at least six elements in its composition. All cast irons contain iron, carbon, silicon, manganese, phosphorus, and sulfur. Alloy cast iron has other elements that have important effects on the physical properties.

Gray iron is ordinary commercial iron with a grayish-colored fracture. The gray color is caused by flake graphite, the principal form of carbon present. Gray iron is easily machined and has a high compression strength. The tensile strength varies from 20,000 to 60,000 psi (140 to 415 MPa), but the ductility is usually low. The percentages of the several elements may vary considerably but are usually within the following limits: carbon, 3.00% to 3.50%; silicon, 1.00% to 2.75%; manganese, 0.40% to 1.00%; phosphorus, 0.15% to 1.00%; sulfur, 0.02% to 0.15%; and iron, the remainder.

Figure 3.17*A* and *B* are photomicrographs showing the structure of gray cast iron. The dark lines are small flakes of graphite that reduce the strength of the iron. Strength is greater if these flakes are small and uniformly distributed throughout the metal. The light-colored constituent in the etched specimen is *steadite*, a structural component in cast iron that contains phosphorus. The other constituent is known as *pearlite*. Steadite is identified by its white dendritic formation. It is a eutectic structure of alpha iron and iron phosphide. *Ferrite*, or pure iron, also appears as a constituent of gray iron having a high silicon content or irons that have been slowly cooled. Pearlite, composed of

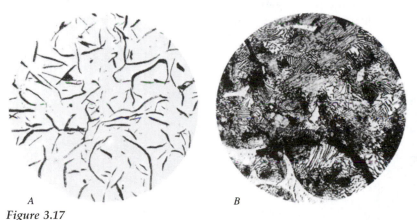

A   B

Figure 3.17
Structure of gray cast iron (ASTM Class 40). *A*, Graphite flakes in un-etched matrix. Magnification ×125. *B*, Etched in 5% nital acid showing graphite, pearlite, and steadite. Magnification ×562.

alternate lamellae of ferrite and iron carbide, is found in most irons and is similar to the pearlite iron. The dark graphite flakes may also be seen in Figure 3.17. White cast iron shows a white fracture because the carbon is in the form of a carbide, Fe_3C. The carbide known as *cementite* is the hardest constituent of iron.

White iron with a high percentage of carbide can be grounded but not machined. The principal constituents visible in the micrograph in Figure 3.18 are cementite and pearlite. The dark area is the pearlite and the light area cementite.

White cast iron can be produced by casting against metal chills or by regulating the analysis. Chills are used when a hard, wear-resisting surface is wanted for such products as rail wheels, rolls for crushing grain, and jaw crusher plates. The first step in the production of malleable iron is to produce a white iron casting by controlling the analysis of the metal. One specification for the production of these castings is as follows: carbon, 1.75% to 2.30%; silicon, 0.85% to 1.20%; manganese, less than 0.40%; phosphorus, less than 0.20%; sulfur, less than 0.12%; and iron, the remainder. *Mottled cast iron* is an intermediary product between gray and white cast iron. The name is derived from the appearance of the fracture. It is obtained in castings in which surfaces subjected to wear have been chilled.

Malleable cast iron or *malleable iron* is made from white cast iron. White-heart castings are packed in pots and placed in an annealing oven arranged to allow free circulation of heat around each unit. The annealing time lasts three to four days at temperatures varying from 1500° to 1850°F (815° to 1010°C). In this process the hard iron carbides are changed into nodules of temper or graphitic carbon in a matrix of comparatively pure iron, as shown in the micrograph in Figure 3.19A. Such iron has a tensile strength of around 55,000 psi (380 MPa) and an elongation of 18%.

Malleable castings, which have considerable shock resistance and good machinability, are used principally by the railroad, automotive, plumbing, and agricultural implement industries. When the castings are placed in pots with an oxidizing material, the resulting material is called black-heart malleable iron. Because of the difficulty in casting white cast iron in large sections, malleable iron's use is predominantly for small castings.

Ductile iron or nodular iron is the high-strength, high-ductility iron shown in Figure 3.19B. Ductile iron is classified by three sets of numbers (XX-YY-ZZ), where the first set represents the tensile strength, the second set the yield strength, and the last set the percentage elongation. For example, a 60-50-15 would have 60,000 psi tensile strength, 50,000 psi yield strength, and 15% elongation. Ductile iron has carbon in the form of graphite nodules and is produced by adding a small amount of a magnesium-containing

Figure 3.18
Structure of white cast iron as cast. Etched in 5% nital acid showing pearlite and cementite. Magnification ×125.

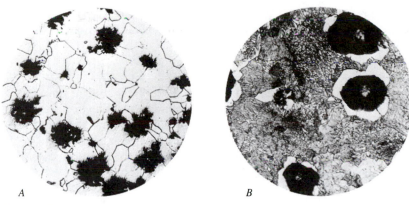

Figure 3.19
A, Structure of malleable iron. Etched in 5% nital acid showing ferrite matrix. Magnification ×125. *B*, Structure of nodular or ductile cast iron. Magnification ×250.

agent such as magnesium–nickel or magnesium–copper–ferrosilicon alloy to gray iron. The amount of magnesium required to produce graphite depends on the amount of sulfur present. Sulfur is first eliminated by being converted to magnesium sulfide. Additional magnesium present changes the graphite to the nodular form. This type of iron is normally used in the as-cast condition; however, casting followed by a short annealing period is often employed to obtain certain required properties. In this process, the time for annealing is much shorter than that used in the manufacture of malleable iron. The improved physical properties of this iron allow it to be competitive for casting crankshafts and for miscellaneous parts in a variety of machines, even though it is more costly than gray cast iron.

3.7 EFFECTS OF CHEMICAL ELEMENTS ON CAST IRON

Carbon

Although any iron containing more than 2.0% carbon is in the cast iron range, gray cast iron has a carbon content of 3% to 4%. The final properties of the iron depend not only on the amount of *carbon,* but also on the form in which it exists. The formation of graphitic carbon depends on slow cooling and silicon content. High silicon promotes the formation of graphitic carbon, which reduces the shrinkage and improves machinability. The strength and hardness of iron increases with the percentage of carbon in the combined form. The properties of cast iron may be changed by heat treatment.

Silicon

Silicon up to 3.25% is a softener in iron and is the dominating element in determining the amounts of combined and graphitic carbon. It combines with iron that otherwise would combine with carbon, thus allowing the carbon to change to the graphitic state. After an equilibrium is reached, additional silicon unites with the ferrite to form a hard compound. Silicon above 3.25% acts as a hardener. In melting the average loss of silicon is about 10%. High silicon content is recommended for small castings and low for large castings.

When it is used in amounts from 13% to 17%, an alloy having acid and corrosion resistance is formed. Gray irons that are low in silicon respond best to heat treatment.

Manganese

Manganese in small amounts does not have an appreciable effect, but in amounts of more than 0.5% it combines with sulfur to form a manganese sulfide that has a low specific gravity and is eliminated from the metal with the slag. It acts as a deoxidizer as well as a purifier and increases the fluidity, strength, and hardness of the iron. If the percentage is increased appreciably, manganese will promote the formation of combined carbon and rapidly increase the hardness of the iron.

Sulfur

Nothing good can be said for sulfur in cast iron. It promotes the formation of combined carbon with accompanying hardness and causes the iron to lose fluidity with resultant blow holes. Each time the iron is remelted there is a slight pickup in sulfur, frequently as much as 0.03%. To counteract this increase, manganese should be added to the charge in the form of ferromanganese briquettes or *spiegeleisen*.

Phosphorus

Phosphorus increases the fluidity of the molten metal and lowers the melting temperature. For this reason phosphorus up to 1% is used in small castings and in those having thin sections. There is a slight increase (about 0.02%) in the phosphorus content during the melting process. Phosphorus also forms a constituent known as *steadite*, a mixture of iron and phosphide that is hard, brittle, and of rather low melting point. It contains about 10% phosphorus, so that an iron with 0.50% phosphorus would have 5% steadite by volume. Steadite appears as a light, structureless area under the microscope but may appear as a network if sufficient phosphorus is present.

QUESTIONS

1. Give an explanation of the following terms:

Blast furnace	Continuous casting
Bell	Crucibles
Slag	Induction furnace
BOF	Specific heat
Indirect arc furnace	Wrought iron
Basic-lined hearth	Aston process
Oxygen lance	Plain carbon steels
Tuyères	Austenitic steels
Windbox	Type 303 stainless steel
Ingot	Types of cast iron
Killed steel	Cementite
Teeming	Malleable cast iron

2. How is bituminous coal used in the blast furnace?

3. What type of iron ore is predominantly used in the United States?

4. What is an essential ingredient in producing direct reduction pig iron as compared to that in a blast furnace?

5. What are the kinds of cast iron and their distinguishing characteristics?

6. How is heat generated in a converter?

7. How can combustion be improved in a blast furnace?

8. Referring to Figure 3.14, what generalizations can be made about the number of BTUs necessary to melt aluminum, iron, brass, and lead?

9. Give reasons that the graph of Figure 3.14 may not tell the whole story. For example, why does melting brass require more BTUs of fuel than melting aluminum?

10. Describe the process of making steel in the BOF.

11. What is the difference between the process of degassing and controlled-atmosphere melting?

12. What is the general relationship between carbon content and hardenability in ferrous materials?

13. If you were trying to make a high-grade steel chain, would a high or low Bhn be sought?

14. Give the range of carbon content that might be used for the following parts: lawnmower blade, bridge railing, padlock, screwdriver, screw for a computer, gasoline tank, pliers, and drive shaft.

15. From what type of cast iron is malleable iron produced? What happens to the carbon?

16. Why does an ingot solidify on the bottom first?

17. How is steel "killed"?

18. How can the sulfur content be reduced in the remelting process?

19. Name the metallurgical constituents found in white cast iron, gray cast iron, and malleable iron.

20. How is wrought iron totally unlike the irons described in question 19?

21. What kinds of steels would be used for the following: knife, car fender, spring, restaurant sink, paper clip?

22. How can a magnet be used to determine if a piece of stainless steel is easy to machine?

23. What is the classification number of a stainless-steel restaurant sink?

24. Why is it important to have phosphorus in a cast iron? What happens to a cast iron that has too much phosphorus?

25. What can you state about the ability of a machinist to machine a cast iron that is almost 100% cementite? What type of cast iron would this be? What are some uses for this type of cast iron?

26. How are malleable iron castings made?

27. What is the effect of both the basic and acidic linings in an open-hearth furnace?

28. Zinc-coated wrought iron is used for "tin" roofs. Is there tin in this type of iron? Why will wrought iron stand corrosion so well even if it has no zinc coating?

29. Describe how slag is made and used in the manufacture of wrought iron.

30. Oxygen is an expensive gas. How can its cost be justified in the melting process?

31. What is the purpose of argon gas in the AOD process?

32. Describe the following steels: 1030, 14B30, E4130, 446, 1040.

33. Assign the general classifications of steel, wrought iron, or cast iron to ferrous materials having the following carbon contents: 0.15%, 0.005%, 0.45%, 2.9%, 1.2%, 0.90%, and 1.17%.

34. Describe the basic differences among the three types of stainless steels.

35. What are the properties of a modular iron classified as 80–60-18?

PROBLEMS

3.1. If the heating value of methane is 1000 BTU/ft³ and it costs $3 per 1000 ft³, what is the cost of the methane to produce 1 ton of iron? How many BTUs are required?

3.2. How many tons of coke are used to melt 113 tons of iron in a cupola? How many tons of limestone are required? If the heating value of coke is 13,000 BTU/lb and it burns at 50% efficiency, how many pounds of coke are required? How many BTUs are gainfully employed?

3.3. It cost 9 cents per kilowatt-hour of electricity—that is, per volt × amperes × hours = 1 Wh. How many dollars per heat are spent on electricity to operate the electric furnace described in the text?

3.4. Iron that is used for remelting has 0.11% sulfur. What is its sulfur content after remelting? What can be said about the cast iron before and after remelting?

3.5. Convert the following to metric units: 60,000 psi tensile strength; 3000 BTU/lb; 1064 tons; 42 ft; 50 lb/ft^3.

3.6. What is the approximate electrical cost per pound of electric-furnace produced steel? How does this compare with electrical cost per pound in the production of aluminum? Assume electrical cost is $0.08 per kWh.

MORE DIFFICULT PROBLEMS

3.7. A 10,000-kg charge is loaded into an induction furnace and melting occurs at 1500°C. If the furnace efficiency is 75%, how much energy is necessary for the operation? (Hint: Assume that it takes 350 kWh to melt 1000 kg.)

3.8. An examination of electric furnace operation finds that 20% of total energy is lost to waste-gas air-pollution control system. It is suggested that the waste gas be used to preheat the scrap, but it takes 10 kWh/ton to divert the gas through the scrap, with an increased labor and maintenance cost of $1/ton. Would the suggested design be cost effective? (Hints: Neglect capital costs, and cost of electricity = $0.10/kWh.)

PRACTICAL APPLICATION

An ingot is rolled into various rail sections. For Rail Section 902, the rolling practice reduces an ingot into two blooms, which are further finished rolled into three standard 39-ft length rails, providing six rails from each ingot. This practice has a yield of less than 81% from ingot to finished rail. Larger ingots and longer mill run-outs from ingot to rail suggests possible improved yield in Section 902 by cutting one or both blooms into four rails, giving either seven or eight rails.

Section 902 rail has a cross-sectional area of 8.8207 in.^{2}, and at a density of 0.2833 lb/in.^{3} for rolled high-carbon steel, the end product weight is 1169.493 lb per rail. Design information for this improved process is as follows.

Type of Loss	%	Remarks
Heating and scale	2.50	Of poured ingot weight
Bottom crop discarded	3.75	Of poured ingot weight
Top crop discarded	7.40	Minimum per average ingot
Rail ends hotcropping	2.20	For seven-rail ingot
Rail ends hotcropping	1.85	For eight-rail ingot
Nick and break tests	0.64	Of poured ingot weight
Allowances for cutback	0.10	Of poured ingot weight

Constraints on size for the large ingots are 59¼ in. minimum height and 8586.5 lb poured weight to 80½ in. maximum height and 11,432.7 lb.

Possible ingot height–weight pouring tables are given as:

Height, in.	Weight, lb
78	11,100.9
78¼	11,134.1
78½	11,167.3
78¾	11,200.5
79	11,233.7

Optimize for the best ingot height. (Hints: Consider a seven- or eight-rail choice. Base your choice on maximum yield, which is 100% − losses. This is a real-life problem faced by a company that rolls rails.)

CASE STUDY: MELTING COST ESTIMATE FOR A FOUNDRY

Your consulting firm has been engaged to estimate fuel costs for a new product by Farwell Company, a foundry that builds robot castings. These robots hug the floor as they move about and thus need a weight for ballast. The chief engineer wants to design the ballast using gray cast iron, because it is relatively easy to machine. The ballast design is 35 in. in diameter and 9.5 in. thick. The production schedule calls for 30 units per day, for 22 working days per month.

The electrical energy for a melting furnace costs $0.11 per kilowatt-hour, and the manufacturer of the furnace promises a furnace efficiency of 49.5%. If electrical cost is one-half the cost of producing a casting, is it practical for Farwell to operate a foundry if they can buy the castings for $0.38 per pound? Report your findings and point out the "break-even" cost of electrical energy.

If the chief engineer suggests using malleable iron instead of gray cast iron, what is your advice?

CHAPTER 4

PRODUCTION OF
NONFERROUS METALS

Nonferrous metals are employed widely in architectural, transportation, chemical, and electrical applications. For example, aluminum is surpassed only by steel in its use as a structural material.

On the basis of weight, less than 25% of metals used for industrial products are nonferrous. Although nonferrous metals in the pure state possess some useful properties, they are seldom employed for industrial products because they lack structural strength. For this reason they are blended with one or more elements to form an alloy having improved properties. Yet, the popular aluminum 12-ounce beverage container is nearly pure aluminum. This range of application demonstrates the importance of nonferrous materials. Selection of an alloy is a compromise among strength, ease of fabrication, weight, cost of material and labor, and the appearance of the product.

4.1 PROPERTIES

Important properties of nonferrous alloys are resistance to corrosion, electrical conductivity, and ease of fabrication. One essential difference in metals is density or unit weight, as noted in Table 2.1. Although few applications of materials are dependent on mass alone, the range of densities of the metals listed in Table 2.1 does offer opportunities. Zinc-coated steel roofing, known as corrugated iron, is, for example, three times heavier than aluminum roofing. An airplane made of steel would weigh about three and a half times as much.

Most nonferrous metals are more resistant to water or moisture and may be weathered without paint or coating. The *corrosion resistance* of nonferrous alloy varies, because each is selective in its qualities. Although magnesium is resistant to ordinary atmospheres, it corrodes more rapidly than steel in ocean water. The green oxide on ancient copper artifacts is considered beautiful, but the iron oxide or rust on steel is a sign of a deteriorating part. In general, the nonferrous alloys with the highest densities are the more resistant to corrosion. The exception is aluminum, however, because it forms an impermeable oxide film on its surface that prevents or slows down attack by most corrosive materials other than strong alkalies.

The *natural color* of aluminum, copper, bronze, brass, tin, and other nonferrous metals allows the engineer to select a material that aesthetically will enhance the product. Special coatings on some materials, such as the anodized organic coatings on aluminum, add flexibility to color design.

The *electrical properties* of nonferrous materials are usually superior to those made of iron. Copper has 5.3 times and aluminum has 3.2 times the electrical conductivity of iron.

As shown in Table 2.1, the melting points of the principal cast nonferrous materials vary from about 620° to 3270°F (327° to 1799°C). The pouring temperature is usually about 400° to 600°F (200° to 315°C) above the melting point. Nonferrous materials are more difficult to weld than ferrous ones, and the lower density ones are the most difficult. Nonferrous materials can be cast, formed, or machined with varying degrees of ease. The difficulty of fabricating the nonferrous alloys varies with the material and the process. There are many different compositions among aluminum alloys, some of which can be cold formed. In metal-cutting processes the light nonferrous materials are easier to machine than steel, but a heavier one such as nickel is difficult.

4.2 NONFERROUS METALS

Smelting

Nonferrous metal ores are seldom found in the pure state in commercial quantities. Because they must be separated from the *gangue* before the ore can be reduced, a process known as *ore dressing* is performed. One method of concentrating, or "dressing the ore," is familiar to those who have panned for gold. Metals and metal compounds are heavier than the gangue, so they settle to the bottom quicker than if a mixture is agitated in water. Methods using this principle have been developed to accelerate the accumulation of metal compounds.

In another method of ore dressing, the ore and gangue are finely powdered and mixed with water. An oil is added and violent mixing is induced. At this stage frothing occurs, and the metallic compounds suspend in the froth, which is drawn off for processing.

The affinity that most metals have for oxygen increases the difficulty of separation and recovery. The close association of many nonferrous metals with one another leads to difficulties in the *smelting process*. The ores of copper, lead, and zinc are often found in the same mine. It is reported that at least 21 elements can be recovered in useful quantities during their processing, with possibly another 9 that may be discarded and 3 that may be utilized as catalysts in the smelting process. The complexity of a smelting process involving so many elements cannot be overemphasized. The cost associated with recovering some elements exceeds their value, but they must be removed to ensure the purity of the desired product. Although the processes discussed in this chapter are simplified, they illustrate production methods for ideal ore conditions.

Furnaces for Nonferrous Smelting

The *blast furnace* used in the manufacture of pig iron is also used for smelting copper, tin, and other nonferrous metals. It has the same proportions but is smaller than the furnace used for iron ore reduction. The fuel, usually coke, is mixed with the ore, and a cold-air blast supports combustion. The coke or ore pellets or chunks should not be less than ½ in. (12.7 mm) in diameter, or else the updraft will carry it out of the flue or the furnace. Fluxes are added to the charge to have a purer metal and a more fluid slag.

The *reverberatory furnace* is most often used in nonferrous smelting. *Slag inducers* or fluxes are added to reduce oxidation. Furnaces have fume and dust collectors arranged to trap not only harmful but also valuable by-products.

In addition to furnaces, nonferrous refining operations require *roasting ovens* where sulfide ores are oxidized. In these ovens, oxidizing gases are forced through grates on which the ore lies. Roasting ovens are used for copper and zinc.

4.3 PRODUCTION OF ALUMINUM

Although aluminum ores are widely distributed in the earth's crust, only *bauxite* is economical as an ore source. Bauxite is usually mined by the open-pit method, crushed, sometimes washed to remove clay, and dried. It is then refined into aluminum oxide, or alumina.

The Bayer process, named for the German chemist Karl Josef Bayer, is the most widely used method for producing pure alumina. Dried, finely ground bauxite is charged into a digester where it is treated with caustic (NaOH) solution under pressure and at a temperature above the boiling point. This caustic solution reacts with the bauxite to form sodium aluminate, which is soluble in the liquor.

The pressure is reduced following digestion, and the residue, which contains insoluble oxides or iron, silicon, titanium, and other impurities, is forced out of the digester through filter presses and discarded. The liquor, which contains extracted alumina in the form of sodium aluminate, is pumped to tanks called precipitators. In the precipitators, fine crystals of aluminum hydroxide from a previous cycle are added to the liquor. These crystals are continuously circulated through the liquor and serve as seed crystals that grow in size as the aluminum hydroxide separates from the solution.

The aluminum hydroxide that settles out from the liquor is filtered, then calcined in kilns at temperatures above 1800°F (980°C). This converts the alumina to a form suitable for smelting.

Metallic aluminum is produced by an *electrolytic process* that reduces the alumina into oxygen and aluminum. In this process pure alumina is dissolved in a bath of molten cryolite (sodium aluminum fluoride) in large electrolytic furnaces called *reduction cells* or "pots." By means of a carbon anode suspended in the bath, electric current is passed through the bath mixture causing metallic aluminum to be deposited on the carbon cathode at the bottom of the cell. The heat generated by passage of this electric current keeps the bath molten, so that alumina can be added as necessary to make the process continuous. At intervals aluminum is siphoned from the pots, and the molten metal is transferred to holding furnaces for either alloying or impurity removal. It is then cast into ingots of various sizes for further fabrication.

At Aluminum Company of America's Rockdale (Texas) Works, each cell (Figure 4.1)

Figure 4.1
Electrolytic cells in pot room of aluminum reduction plant.

will produce approximately ½ ton (0.5 Mg) of aluminum per day. The production of 1 lb (0.45 kg) of aluminum requires 2 lb (0.9 kg) of alumina [obtained from about 4 lb (3.8 kg) of bauxite], 0.6 lb (0.27 kg) of carbon, small amounts of cryolite and other materials, and approximately 8 kWh of electricity. The cost of a pound of aluminum is closely dependent on the electricity. Aluminum is second only to steel in tonnage produced and used each year.

4.4 PRODUCTION OF MAGNESIUM

The largest tonnage of magnesium produced in the United States utilizes seawater as the "ore." Figure 4.2 is a sketch of the Dow Chemical Company process and illustrates the key steps in the production of magnesium ingots. Seawater containing approximately 1300 parts per million (ppm) of magnesium is treated with milk of lime. The lime is made from oyster shells in a kiln that operates at approximately 2400°F (1320°C). When the lime and seawater react, magnesium hydrate settles to the bottom of the settling tank and is drawn off as a thin slurry containing about 12% magnesium hydrate. The slurry is filtered and a more concentrated hydrate is obtained. This is converted to magnesium chloride by the addition of hydrochloric acid. The chloride solution is evaporated to remove the water. After subsequent filtration and drying, the magnesium chloride has a 68% concentration.

The magnesium chloride that is converted to granular form is then transferred to the electrolytic cells. The cells are approximately 25,000 gal (95 m³) in capacity and operate at about 1300°F (700°C). The graphite electrodes serve as anodes and the pots are the cathodes. A direct current of 60,000 A decomposes magnesium chloride and the magnesium metal floats to the top. Each pot produces about 1200 lb (540 kg) of magnesium per day, which is cast into 18-lb (8.1-kg) ingots. Approximately 90% of the magnesium in seawater is recovered. The process generates chlorine gas, which converts the magnesium hydrate to magnesium chloride.

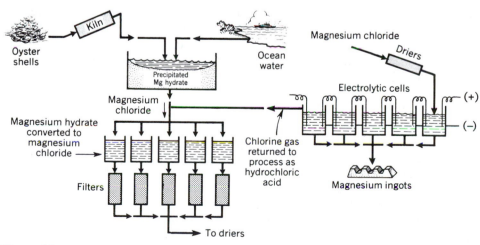

Figure 4.2
Manufacture flow process chart for magnesium from seawater.

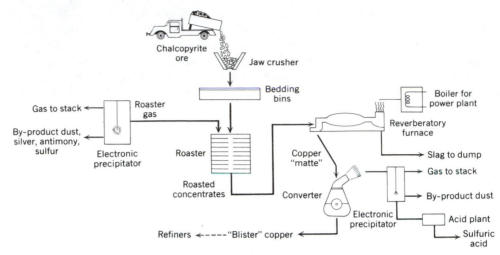

Figure 4.3
Manufacture flow process chart for copper smelting by the reverberatory converter
process.

4.5 PRODUCTION OF COPPER

The United States produces one-fourth of the world's copper from ores, known as *chalcopyrite*. These ores contain primarily Cu_2S and $CuFeS_2$, and usually lie deep underground. The process of producing copper is shown in Figure 4.3. The ore is crushed and mixed with lime and siliceous flux material. To concentrate the copper, flotation tanks or bedding bins are employed. The ores are partially roasted to form a mixture of FeS, FeO, SiO_2, and CuS. This mixture, called calcine, is fused with limestone as a flux in a reverberatory furnace. Most of the iron is removed as slag and the remaining iron and copper—or matte, as it is commonly known—is poured into a *converter*, much like the Bessemer converter used in steelmaking.

The air supplied to the converter for about 4 or 5 h oxidizes the impurities, and many pass off as volatile oxides. The iron forms a slag that is occasionally skimmed off. The heat of oxidation keeps the charge molten and the copper sulfide eventually becomes copper oxide or sulfate. When the air is shut off, the cuprous oxide reacts with the cuprous sulfide to form blister copper and sulfur dioxide. Blister copper, between 98% and 99% pure, is electrolytically refined to higher purity.

4.6 PRODUCTION OF LEAD

In the production of lead a number of other elements are recovered. Figure 4.4 shows the complexity of the operations. The concentrate is 65% to 80% lead and is roasted to remove the sulfides. Limestone, iron ore, sand, and granulated slag are mixed with the lead concentrate before *sintering*. The sulfur dioxide driven off by the sintering is made into sulfuric acid, and the sintered material is fed into a coke-fired blast furnace. The gases and dust given off contain chloride of cadmium, which can be processed into cadmium. The bullion is drossed, the floating copper dross combines with sulfur to induce separation of copper from the dross, and the liquid lead mixture is oxidized in a furnace known as a softening furnace.

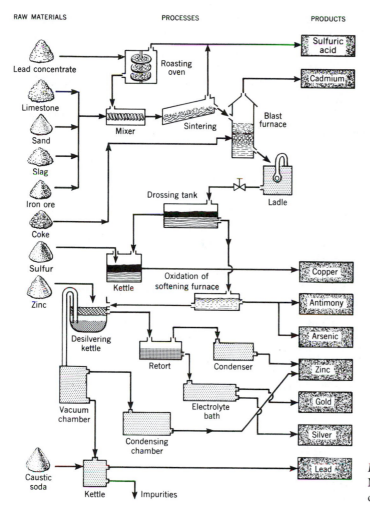

RAW MATERIALS   PROCESSES   PRODUCTS

Figure 4.4
Manufacture flow process
chart for producing lead.

The slag that is skimmed off in the softening furnace contains antimony and arsenic. Zinc is added to the lead mixture in a desilvering kettle, and any gold and/or silver present becomes soluble with zinc. The zinc alloy is skimmed off and fed into a retort. The zinc vapor is condensed to produce solid zinc, and any residual liquid is electrolytically separated into gold and silver. The lead from the kettle is cleaned further of zinc before being combined with caustic soda. This process is done by shooting a small stream of hot lead into a vacuum chamber causing the zinc to vaporize. The impurities are chemically removed in the final kettle operation, and the lead is cast in 56-lb (25-kg) bars or 2000-lb (900-kg) blocks. The process is limited by the blast furnace production, which seldom exceeds 300 tons (270 Mg) of lead bullion per day.

4.7 CASTING NONFERROUS MATERIALS

The common elements used in *nonferrous castings* are copper, aluminum, zinc, tin, and lead. Many alloys, however, have small amounts of other elements such as antimony, phosphorous, manganese, nickel, and silicon.

The foundry methods for making nonferrous casting in sand differs little from that used for iron castings. Molds are made in the same way and with the same kind of tools and equipment. The molding sand is usually a finer grain size, because most castings are small, a smooth surface is desired, and the melting temperatures are lower for nonferrous metals. The sand need not be so refractory as for iron and steel castings because of the lower melting temperatures. Alloying is accomplished by the precise addition of the alloying elements to the base charge. The addition of an *alloy* to a pure metal always results in a material with a lower melting point. The pouring temperature is generally about 20% above the melting temperature of the metal, but the type of mold and the section thickness have a bearing on this value.

Crucible furnaces are frequently used for this type of work. They may be either the stationary or the tilting type. Coke is commonly used as the fuel for the stationary pit furnaces, although oil or gas can be used equally well, if available. Oil and gas have the advantage of heating more quickly than coke. Electrical resistance, indirect arc, and induction furnaces, having accurate temperature control and low melting losses, can be used under certain conditions. Electric furnaces are widely used for laboratory work as well as for installations requiring mass production.

Two of the most common alloys using copper are brass and bronze. *Brass* is essentially an alloy of copper and zinc. The percentages of each element may vary considerably, but in most cases the zinc content ranges from 10% to 40%. The strength, hardness, and ductility of the alloy are increased as the proportion of zinc is increased to 40%. Zinc contents more than 40% are not desirable because there is a consequent rapid decrease in strength and a tendency to volatilize in melting. Addition of a small percentage of lead (0.5% to 5%) increases machinability. Brass is used extensively in industry because of its strength, appearance, resistance to corrosion, and the ability to be rolled, cast, or extruded.

Table 4.1   **Typical Copper-Based Casting Alloys**

Material	UNS Designation[a]	Melting Temp °F	Melting Temp °C	Tensile Strength (psi)	Machinability Rating[b]
Manganese bronze	C67500	1650	899	65,000–84,000	30
Leaded red brass	C83600	1850	1010	37,000	84
Leaded yellow brass	C85200	1725	941	38,000	80
Manganese bronze	C86200	1725	941	95,000	30
Navy bronze	C92200	1810	988	40,000	42
Aluminum bronze	C95400	1900	1038	85,000–105,000	60
Nickel silver	C97300	1904	1040	35,000	70

[a] The Unified Numbering System (UNS) for metals and alloys is an extension of the system formerly used by the copper and brass industry and has been accepted by the American Society for Testing and Materials (ASTM).

[b] Based on free-cutting brass, C36000 = 100.

Table 4.2   **Typical Aluminum-Based Casting Alloys**[a]

Alloy	Melting Point °F	Melting Point °C	Tensile Strength (psi)	Type Casting	Uses
201-T4	1135	613	31,000	Sand	Aircraft parts
356-T6	1135	613	33,000	Sand and PM	Manifolds, wheels
A390-5	1140	616	24,000	Sand and PM	Automotive housings

[a] Casting alloys carry Aluminum Association designation, and permanent-mold (PM) alloys use chemical content or nearest Aluminum Association designation.

Bronze is usually a copper base alloy containing tin with manganese and several other elements. Most elements used as alloys with copper add to the hardness, strength, or corrosion resistance of the metal. Table 4.1 lists properties of some copper base alloys.

Because of light weight and ability to resist many forms of corrosion, aluminum alloys have many applications. Many respond to heat treatment and are suitable where high strength is needed. Copper is one of the principal alloying elements and is used in amounts up to 8%, which adds to strength and hardness. Aluminum alloys containing silicon have excellent casting characteristics and increased resistance to corrosion. Magnesium as an alloying element improves machining, makes the castings lighter, and assists in resisting corrosion. Table 4.2 shows typical properties of aluminum alloys suitable for sand casting.

Magnesium alloys are useful where light weight is essential, because they are about two-thirds the weight of aluminum and one-fourth the weight of cast ferrous metals. Magnesium alloys have excellent machinability and respond to treatments that improve their physical properties. Aluminum, the principal alloying element, increases hardness and strength. Manganese in small amounts increases the resistance of the metal to salt water. Sand castings made from magnesium alloys find use in portable tools, aircraft, and other construction where weight savings is important.

4.8 WROUGHT ALLOYS

A *wrought material* is one that has been or can be formed to a desired shape by hot or cold working. A wrought part may be more or less costly than a casting, but as a rule superior properties are obtained. Chapters 14 and 15, covering hot and cold working, describe the processes used to fabricate a part by rolling, hammering, or forging. The physical properties of several special wrought alloys are shown in Table 4.3.

Aluminum Alloys

Although aluminum has a strength as shown in Table 2.1, vigorous *cold or hot working* can double its tensile strength. By judicious use of one or more alloying materials, the application of hot or cold work and heat treating, it is possible to have an alloy strength of more than 100,000 psi (690 MPa). Aluminum alloys are available for forgings, extrusions, bending, drawing, spinning, coining, embossing, roll forming, and wire. These alloys are

Table 4.3   **Typical Nonferrous Wrought Alloys and Approximate Properties**

Material	Designation or Name	Tensile Strength	Elongation (%)	Uses and Remarks
Aluminum[a]	1100-0	13,000	45	Spinning, drawn parts, cooking utensils
	6061-0	18,000	30	Boats
	6061-T6	45,000	17	Rail cars
Copper[b]	C34000	52,000	50	Nuts, rivets, screws, dials
	C36000	55,000	18–53	Gears, screw machine parts; easiest copper alloy to machine
	C65500	58,000–108,000	60	Hydraulic lines, marine hardware
Magnesium[c]	AZ318-B	34,000	10	Excellent machinability, nonmagnetic, often easily corroded
Lead	(0.07% Ca)	4700	30	Sheet lead; age hardness

[a] Aluminum Association designation.

[b] International Annealed Copper Standard designation.

[c] American Society for Testing and Materials (ASTM) designation.

available in commercial forms such as wire, foil, sheets, plates, and bar shapes. All wrought aluminum alloys can be machined, welded, and brazed.

Copper Alloys

Although there are hundreds of copper alloys, most of them can be broadly classified as coppers, brasses, bronzes, and cupronickels and nickel-silvers. The tensile strength of copper alloys varies from approximately 30,000 psi (200 MPa) for almost pure copper to 200,000 psi (1380 MPa) for beryllium copper, and yet the amount of beryllium in such an alloy will be less than 2%. Copper with less than 5% alloying materials is used for low-resistance electrical wiring and conductors, refrigeration, and water tubing. It is relatively soft but becomes harder and brittle when cold worked. The copper–zinc alloys, commonly known as brasses, are used in heat exchangers and for a variety of parts where corrosion resistance and strength must be balanced against a desired ductility. Nickel, silver, and the bronzes (alloys of copper and tin) are more expensive than brasses. They are used in springs, bells, and in corrosive atmospheres, particularly where high tensile strength is a factor. Lead can be added to any copper alloy to improve its machining characteristics.

The decision to use copper is based on its color, easy forming, machining capabilities, resistance to corrosion, and versatile properties. Copper can be purchased in all forms, as is the case with aluminum.

Magnesium Alloys

Magnesium is the lightest structural metal. Two-thirds the weight of aluminum, magnesium is alloyed to produce a metal of good strength, excellent machinability, good weldability, and formability. It is easily extruded if sharp corners are not involved. Both strong and weak acids attack magnesium alloys, and aqueous salt solutions or brines corrode them rapidly. There are few uses for magnesium near the ocean in unprotected environments. Magnesium alloy must be used at temperatures below about 300°F (150°C), because it does not hold its strength at higher temperatures. At cryogenic temperatures magnesium performs well. Magnesium has a very high coefficient of expansion, so care must be taken in delicate assemblies. More costly than aluminum or steel, magnesium is normally used where savings in weight and machinability result in overall product advantages.

Magnesium alloys are used extensively in aircraft, cameras, binoculars, low-temperature engine parts, portable tools, vacuum cleaners, and for high-speed rotating equipment where it is desirable to minimize inertia. Magnesium can be purchased in bar shapes, plates, and sheets.

4.9 DIE-CASTING ALLOYS

A relatively wide range of nonferrous alloys can be die cast. The principal base metals used in order of commercial importance are zinc, aluminum, magnesium, copper, lead, and tin. The alloys may be further classified as low-temperature alloys and high-temperature alloys; those having a casting temperature below 1000°F (540°C) such as zinc, tin, and lead are in the low-temperature class. The low-temperature alloys have the advantages of lower cost of production and lower die maintenance costs. As the casting temperature increases, ferrous alloy dies in the best treated condition are required to resist the erosion and heat checking of die surfaces. The destructive effect of high temperatures on the dies is the principal factor in retarding the development of high-temperature die castings.

Other considerations that influence alloy selection are mechanical properties, weight, machinability, resistance to corrosion, surface finish, and, of course, cost. Obviously the least expensive alloy that will give satisfactory service should be selected.

Zinc Base Alloys

More than 75% of *die castings* are zinc based. These alloys cast easily with a good finish at fairly low temperatures, have considerable strength, and are the lowest cost of the nonferrous metals. Die cast wall thickness compare favorably with sheet metal, such as 0.015 in. for small parts. Holes of 0.125-in. ID can be cast, and threads of 32 pitch are possible.

The purest grades of commercial zinc, 99.99+% zinc, known as Special High Grade, should be used because such elements as lead, cadmium, and tin are impurities that cause serious casting and aging defects. The usual elements allowed with zinc are aluminum, copper, and magnesium; all are held within close limits.

Nominal compositions of the two standard zinc die-casting alloys are indicated in Table 4.4. These alloys are similar in composition except for the copper content, and in most cases they can be used interchangeably. Aluminum in amounts around 4% greatly improves the mechanical properties of the alloys and reduces the tendency of the metal to dissolve iron in the molding process. Copper increases the tensile strength, ductility, and hardness. Magnesium, which is usually held to an optimum of 0.04%, is used because of the beneficial effect it has in making the castings permanently stable.

Zinc alloys are widely used in the automotive industry and for other high-production markets such as washing machines, oil burners, refrigerators, radios, television, business

Table 4.4   **Typical Zinc Die-Casting Alloys**

Alloy[a]	Alloy Number	SAE Description	As-Cast Tensile Strength (psi)	As-Cast Elongation (%)
AG40A	3	903	41,000	10
AC41A	5	925	47,600	7
—	7	903	41,000	14

[a] These alloys have multiple designations; Zinc Institute, Society of Automotive Engineers (SAE), and American Society for Testing and Materials (ASTM).

machines, parking meters, and small machine tools. The United States automotive industry uses more than 300,000 tons annually.

Aluminum Base Alloys

Many die castings are made of aluminum alloys because of their light weight and resistance to corrosion, but they are more difficult to die cast than zinc. Because molten alloys of aluminum will attack steel pots and dies if kept in continuous contact with it, the cold-chamber process of die casting is generally used. (See Chapter 6.) The melting temperature of aluminum alloys is around 1100°F (540°C).

The principal elements used as alloys with aluminum are copper, manganese, silicon, magnesium, and zinc. Silicon increases the hardness and corrosion-resisting properties, copper improves the mechanical properties slightly, and magnesium increases the lightness and resistance to impact. Table 4.2 shows the nominal composition of the principal aluminum die-casting alloys.

The specific gravity of these alloys is 2.7 times that of water, and they have a wide range of helpful properties including resistance to corrosion, high electrical conductivity, ease of applying surface finishes, and good machinability. The two principal aluminum die-casting alloys, as based on the Aluminum Association designation, are 360.0F and 380.0F. They each have a tensile strength of about 47,000 psi.

Copper Base Alloys

Die castings of brass and bronze presents a problem in pressure casting because of their high casting temperatures. These temperatures range from 1600° to 1900°F (870° to 1040°C) and make it necessary to use heat-resisting alloy steel for the dies to reduce their rapid deterioration.

Most of the casting alloys listed in Table 4.1 can be die cast, and there are many others. Copper base alloys have extensive use in miscellaneous hardware; electric machinery parts; small gears; golf putters; marine, aircraft, and automotive fittings; chemical apparatus; and numerous other small parts. Because the high casting temperatures and pressures shorten the die life, the cost of brass die castings is higher than that of other metals. However, these alloys are useful where high strength, resistance to corrosion, or wear resistance is important. Since thinner wall sections may be produced, the savings in metal cost, coupled with high production rates, help offset the disadvantage of a shortened die life.

Magnesium Base Alloys

Magnesium is alloyed principally with aluminum but may contain small amounts of silicon, manganese, zinc, copper, and nickel. Its alloys, the lightest in weight of all die-cast metals, are about two-thirds the weight of aluminum alloys. Although the price is slightly higher than for aluminum, the extra cost may be compensated by light weight and improved machinability.

Corrosion resistance of magnesium alloys is inferior to other die-casting alloys, especially in moist or sea atmospheres, and usually necessitates a chemical treatment as well as the subsequent application of a special priming coat shortly after the casting is produced. These treatments render the casting suitable for a range of applications.

ASTM Specification B 94, alloy AZ91B, is the principal die-casting alloy. It has good casting characteristics and fairly high mechanical properties. This alloy contains 9% aluminum, 0.5% zinc, 0.13% manganese, 0.5% maximum silicon, 0.3% copper, 0.03% nickel, and the remainder magnesium. It is desirable that copper and nickel content be kept very low to minimize corrosion.

Magnesium alloys are cast in much the same manner as aluminum alloys and require a casting temperature between 1200° and 1300°F (650° and 700°C). Best results are obtained in so-called cold-chamber machines, and it is necessary to ladle the alloy from a crucible that is hooded, keeping the metal covered by a nonoxidizing atmosphere. The lightness of these alloys, combined with good mechanical properties and excellent machinability, merits their use for aircraft, motor and instrument parts, portable tools, textile machinery, household appliances, and many other similar applications.

Lead Base Alloys

Pure lead, which melts at 621.3°F (327°C), will melt at around 470°F (240°C) when alloyed with about 16% antimony. This element is the principal one used with lead and its antimony ranges from 9.25% to 16%. Antimony hardens lead and reduces its shrinkage value. Lead alloys have poor mechanical properties but are inexpensive and easily cast. They are used principally for light-duty bearings, weights, battery parts, X-ray shields, and applications requiring a noncorrosive metal.

Tin Base Alloys

Die-casting alloys based on tin are in about the same category as the lead alloys as far as mechanical properties are concerned, but are high in price. Tin alloys are high in corrosion resistance, and some of them are well suited for contact with foods and beverages. Also, tin alloys have excellent bearing properties and can be cast within remarkably close dimensional tolerances. This fact, together with high corrosion resistance, accounts for their use in small parts such as number wheels, especially where contact with corrosive inks may be involved. Tin alloys also can be used for low-cost jewelry, and certain grades are classed as *pewter*.

4.10 CONTINUOUS CASTING OF ALUMINUM

Recycling of aluminum is an accepted practice. Aluminum cans, structural components, and pressworking waste from the production of containers are significant sources of

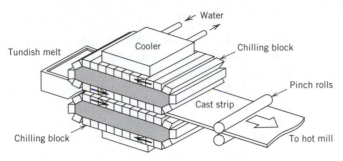

Figure 4.5
Continuous casting of molten aluminum.

aluminum. Molten aluminum from an open hearth furnace can be further processed by either an ingot or continuous casting procedure.

Continuous casting is a process that cools the metal and casts it into a thin continuous sheet. Figure 4.5 illustrates the process. Metal at a temperature of 1550°F (843°C) is poured into a *tundish*, which is a ceramic tub that feeds the aluminum through ceramic nozzles into the caster. With dams on either side of the caster, metal is forced along the chilling blocks, as it begins to cool and form a mixture of solid and liquid metal until it is solidified completely. The chilling blocks, moving at 16.5 fpm, are steel blocks with a thick copper surface. At the side opposite the cooling metal, the blocks are cooled by water. Stock leaving the caster has a thickness of 0.75 in. and a width of 32 in.

Pinch rollers, just outside the caster, compensate for shrinkage in the metal and regulate speed variances between the caster and subsequent rolling operations. A rolling mill reduces the thickness and widens the stock ready for coils, which are recycled for use in aluminum container production.

QUESTIONS

1. Give an explanation of the following terms:

Gangue	Alloy
Slag inducers	Brass
Roasting ovens	Wrought material
Bauxite	Die casting
Reduction cells	Pewter
Chalcopyrite	Tundish

2. What generalizations can be made concerning nonferrous materials with regard to cost, strength, electrical properties, and resistance to corrosion? Explain.

3. Explain how the corrosion resistance of magnesium would compare to that of wrought iron.

4. As tensile strength of an alloy increases, what can be stated generally about its ductility?

5. For what purpose might a large lime plant use its surplus in the production of a nonferrous metal?

6. How and why does the cost of aluminum depend to some extent on the price of electricity?

7. What is the general rule for the metal temperature if it is to be used in a sand casting? To what temperature would aluminum be raised before pouring?

8. For what purposes are the electrostatic precipitators used in the manufacture of blister copper?

9. What are precipitators used for in the manufacture of aluminum? How do they differ from the electrostatic type?

10. Examine your local newspaper or *The Wall Street Journal* and tabulate the prices for ferrous and nonferrous materials.

11. Why would you think brass costs less per pound than bronze?

12. Explain the difference in properties between an aluminum alloy that is used for wrought products instead of cast ones.

13. List 10 products made from wrought aluminum alloys.

14. List 10 products made from wrought copper-based alloys.

15. What is the essential difference in a blast furnace used for making pig iron as compared to one used in smelting tin? Describe the method of heating in a blast furnace.

16. State the advantages of continuous casting as compared to ingot casting for aluminum.

PROBLEMS

4.1. Compare the weight of an 8-in. sphere made of steel, aluminum, and magnesium. (Hints: The volume V of a sphere is $V = \frac{4}{3} \pi r^3$, where r = radius in feet and V is expressed in cubic feet. Use Table 2.1.)

4.2. How much weight reduction is possible for a 14.6-lb aluminum auto part if it is magnesium? What about the use of magnesium in a high-temperature application?

4.3. A country is preparing for its anniversary, and a politician wants 2500 busts of the president for gifts. Copper alloy will be used, but the weight of the bust depends on the copper alloy. When the master of the bust is submerged in water, it displaces 0.00098 m^3 of water. An 80%–20% copper zinc alloy requires $2.50 per kg for copper and $2.07 per kg for zinc. Find the cost for the lot of 2500 busts. (Hint: Densities of copper and zinc are 8906 and 7144 kg/m^3.)

4.4. An airplane design requires a shaft of 2.34 cm in diameter and a length of 6 cm. Calculate the mass of the shaft for aluminum and magnesium. If the cost of aluminum and magnesium is $0.96/lb and $1.63/lb, what is the cost difference? (Hint: The density of aluminum and magnesium is 2.70 and 2.34 g/cm^3.)

PRACTICAL APPLICATION

Call and ask for a tour of a material service supply center (a business that sells ferrous and nonferrous products, such as barstock, plate, sheet, etc.) and interview the customer service representative. These centers are warehouses, and the representative is knowledgeable about materials. These warehouses are surprising and interesting. In your telephone arrangements prior to the visit, be sure to inquire about safety requirements.

The visit needs a purpose, so in the previsit planning, discuss among the class or team the specifics of the plant tour. For example, you or your team can concentrate on fabrication services and how that differs for ferrous or nonferrous materials. You can check on the nonferrous specifications, those materials that have the greatest local demand, and so on. The questions above can give ideas. The outcome of the visit will be a team report by two to four members. Your instructor will give the specifications for the report.

CASE STUDY: PROFIT ANALYSIS

As president of Rempe Die Castings, Inc., you have been asked to bid on a new Olympic symbol, the unit cross (Figure C4.1). Promoters want the solid cross in a nonferrous metal. The bid is for aluminum, brass, zinc, magnesium, and lead. They want a smooth finish without machining, so the units are die cast to final shape. Sprues necessary in the die-casting process are cut off. The sprue is the pipe through which metal feeds to the casting cavity. You are authorized to charge five times the cost of the metal in a single cross.

Overhead and production costs per unit
= 2.5 (metal cost per unit + $0.30/lb)

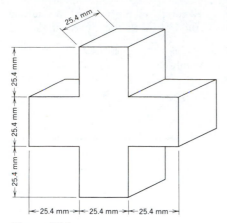

Figure C4.1
Case study.

The $0.30/lb is for handling costs. The bid quantities are 50,000, 100,000, and 300,000 units. Prepare a short discussion on the advantage and disadvantage of the use of each material for the supplier's edification.

Tabulate for each quantity and each material the essential figures including metal cost, handling cost, production and overhead, selling price, and profit. Let your last column be the profit per unit. If you have a computer available, the problem can be programmed and the printout submitted.

(Hints: Figure your cost for each metal. Then calculate your production and overhead cost. Cost analysis reveals the following: aluminum die cast alloy, $1.10/lb; brass die cast alloy, $1.44/lb; zinc die cast alloy, $0.59/lb; magnesium die cast alloy, $1.43/lb; and lead, $0.29/lb. Densities are aluminum, 165 lb/ft^3; brass 535 lb/ft^3; zinc 446 lb/ft^3; magnesium 109 lb/ft^3; and lead, 706 lb/ft^3.)

CHAPTER 5

FOUNDRY PROCESSES

Foundry processes consist of making molds, preparing and melting the metal, pouring liquid metal into the molds, cleaning the castings, and reclaiming the sand for reuse. Machining a block of metal to an intricate shape can be very expensive, which encourages a foundry worker to say, "Why whittle when you can cast?" High rates of production, good surface finish, small dimensional tolerances, and improved material properties allow the casting of small and large intricate parts of almost all metals and their alloys.

Molds may be made of metal, plaster, ceramics, and other refractive substances. But this chapter is concerned with the preparation of *sand molds*. The principles used in sand casting of metals are important because of their economic advantage over some manufacturing methods.

5.1 SAND CASTINGS AND MOLDING PROCEDURES

Two different methods by which sand castings can be produced are classified according to the type of pattern used: *removable pattern* and *disposable pattern*.

In the methods employing a removable pattern, sand is packed around the pattern. Later the pattern is removed and the cavity is filled with molten metal. Disposable patterns are made from *polystyrene* and, instead of being removed from the sand, are vaporized when the molten metal is poured into the mold.

To understand the foundry process, it is necessary to know how a mold is made and the important factors in producing a good casting. These factors are molding procedure, patterns, sand, cores, mechanical equipment, metal, and pouring and cleaning the casting. Molds are classified according to the materials.

Green-Sand Molds

This most common method, consisting of forming the mold from damp molding sand, is used in both of the processes previously described. The term *green sand* does not refer to the color of the sand, which is a dark brown or black, but rather to the sand that is uncured and uses clay and water for bonding. Figure 5.1 illustrates the procedure for making green-sand molds.

Loam Molds

Loam molds are used for *large castings*. The mold is first built up with bricks or large iron parts. These parts are plastered over with a thick loam mortar, the shape of the mold being obtained with sweeps or skeleton patterns. The mold is allowed to dry thoroughly, so it can resist the heavy rush of molten metal. Because of the production time, loam molds are seldom used.

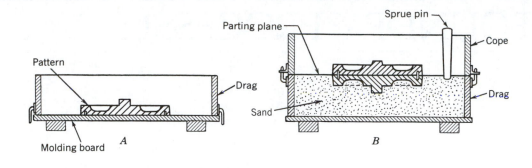

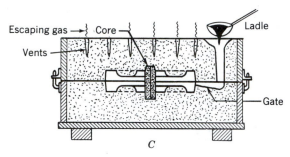

Figure 5.1
Procedure for making green-sand molds. *A*, Pattern on molding board ready to ram up drag. *B*, Drag rolled over and pattern assembled ready to ram cope. *C*, Mold complete with dry-sand core in place.

Furfuryl Alcohol Binder Molds

The process is also known by its trademark, Furan. This process is good for making molds using disposable patterns and cores. Dry *sharp sand* is mixed with phosphoric and sulfuric acid, which acts as an accelerator. The resin is added and mixing is continued only long enough to distribute the resin. The sand material begins to air-harden almost immediately but the time delay is sufficient to allow molding. In use with disposable patterns, furan resin sand is employed as a wall or shell around the pattern supported by green or sharp sand, or it can be used as the complete molding material.

CO$_2$ Molds

In this process, clean sand is mixed with *sodium silicate* and the mixture is rammed about a pattern. When CO$_2$ gas is pressure-fed to the mold, the sand mixture hardens. Very smooth and intricate castings are obtained by this method, although the process was originally developed for making cores. Shakeout and core removal are difficult.

Skin-Dried Molds

Two general methods are used in preparing skin-dried molds. First, the sand around the pattern is mixed with a binder to a depth of about ½ in. (12.7 mm). After drying there is a hard surface on the mold. The remainder of the mold is ordinary green sand. The other method makes the mold of green sand and then coats its surface with a *spray* or *wash* that hardens when heat is applied. Sprays used for this purpose include linseed oil, molasses water, gelatinized starch, and similar liquid solutions. In both methods the mold is dried either by air or by a torch to harden the surface and drive out excess moisture.

Dry-Sand Molds

These molds are made entirely from fairly coarse molding sand mixed with a binding material similar to those already discussed. Since dry-sand molds must be *oven-baked* before being used, the flasks are of metal. A dry-sand mold holds its shape when poured and is free from gas problems caused by moisture. Both the skin-dried and dry-sand molds are not used widely in foundries.

Metal and Special Molds

Metal molds are used mainly in *die-casting* low-melting-temperature alloys. Castings are accurately shaped and have a smooth finish, thus eliminating final machining. Plastics, cement, plaster, paper, wood, and rubber are mold materials for particular applications. Metal and special molds are discussed in Chapter 6.

Molding processes in the conventional foundry are classified as follows:

1. *Bench molding.* Bench-type molding is for small work done on a bench at a height convenient to the molder.

2. *Floor molding.* When castings increase in size and are difficult to handle, the work is done on the foundry floor. This type of molding is used for medium-size and large castings.

3. *Pit molding.* Extremely large castings are frequently molded in a pit instead of a flask. The pit acts as the drag part of the flask, and a separate cope is used above it. The sides of the pit are brick-lined, and on the bottom is a thick layer of cinders with connecting vent pipes to the floor level.

4. *Machine molding.* Machines do a number of the operations that the molder ordinarily does by hand. Ramming the sand, rolling the mold over, forming the gate, and drawing the pattern are performed by machines much better and more efficiently than by hand.

5. *Removable pattern molds.* A simple procedure for molding a cast-iron gear blank is illustrated in Figure 5.1. The mold for this blank is made in a *flask* that has two parts. The top part is called the *cope* and the lower part is called the *drag*. If the flask is made in three parts, the center is called a *cheek*. The parts of the flask are held in a definite relation to one another by pins.

The first step in making a mold is to place the pattern on a molding board that fits the flask being used. Next, the drag is placed on the board with the pins down (Figure 5.1*A*). Molding sand is then riddled in to cover the pattern. *Riddling* is the use of a sieve and a vibrating motion that encourages the sand to fall gently on the surface of the pattern. The sand should be pressed around the pattern with the fingers or by mechanical means and the drag should be completely filled. Small molds are packed by a hand rammer, and mechanical ramming is used for large molds and in high-production molding. If the mold is not sufficiently rammed, it will fall apart when handled or when the molten iron strikes it; if it is rammed too hard, steam and gas cannot escape when the molten metal enters.

After *ramming* is completed, the excess sand is leveled off with a straight bar called a strike rod. To ensure the escape of gases when the casting is poured, small vent holes are forced through the sand to within a fraction of an inch of the pattern.

The lower half of the mold is then turned over, so that the cope may be placed in position and the mold finished. Before turning, a little sand is sprinkled over the mold and a bottom board placed on top. The drag is then rolled over and the molding board removed, exposing the pattern. The surface of the sand is smoothed over with a trowel and covered with a fine coating of dry parting sand. *Parting sand* is a fine-grained, dry silica sand without strength that prevents bonding of sand in the cope with sand in the drag.

At this point the mold is complete except for removal of the pattern and sprue pin. The *sprue pin* is withdrawn first and a funnel-shaped opening is scooped out at the top so that there will be a fairly large opening into which to pour the metal. The cope half of the flask is then carefully lifted off and set to one side. Before the pattern is withdrawn, the sand around the edge of the pattern is usually moistened with a swab so that the edges of the mold hold firmly together when the pattern is removed. To loosen the pattern a draw spike is driven into it and rapped lightly in all directions. The pattern can then be withdrawn by lifting the draw spike.

Before the mold is closed, a small passage known as a gate is cut between the cavity made by the pattern and the sprue opening. This passage is shallow at the part entrance, so that after the metal is poured the gate is broken off close to the casting. To allow for metal shrinkage, a cavity is cut into the cope, which provides a supply of hot metal as the casting cools. This opening is called a *riser*.

The mold surfaces may be sprayed, swabbed, or dusted with a prepared coating material. These coatings often contain silica flour and graphite for cast iron, and their composition varies depending on the material being cast. Steel uses refractory coatings, such as zircon or alumina. A mold coating improves the surface finish of the casting and reduces possible surface defects. The completed mold is shown in Figure 5.1*C*.

Before the mold is poured, a weight is placed on top to prevent the *hydrostatic pressure* of the liquid metal from floating the cope and allowing the metal to run out of the mold at the parting line. The net force tending to float the cope is

$$F_n = A(\rho \times h) - (W_c + W_{sc}) \tag{5.1}$$

where

F_n = Net force or the weight that must be added, lb (kg)
A = Area of metal surface at the parting line, ft^2 (m^2)
ρ = Density of the metal, lb/ft^3 (kg/m^3)
h = Height of cope, ft (m)
W_c = Weight of cope, lb (kg)
W_{sc} = Weight of sand in cope, lb (kg)

If $W_c + W_{sc} > A(\rho \times h)$, additional weight is unnecessary.

6. *Disposable pattern molds.* In making a disposable pattern mold, the pattern—usually one piece including the sprue, gate, and riser—is placed on a follow board and the drag is molded in the conventional way. Study the process given by Figure 5.2. Vent holes are added and the drag is turned over for molding the cope. Although green sand is the most common material, other special-purpose sands can be used, particularly as facing immediately around the pattern. No parting sand is applied, because the cope and drag will not be separated until the casting is removed. Instead, the cope is filled with sand and rammed. Either the sprue is cut into the gating system or else, as usually happens, it is part of the disposable pattern. Vent holes are added and a hold-down weight is placed upon the cope. The pattern, usually polystyrene or Styrofoam, including the gating and pouring system, is left in the mold.

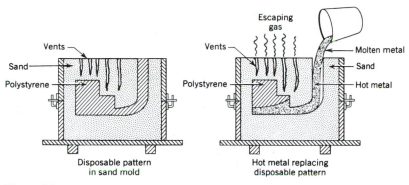

Figure 5.2
Disposable pattern mold.

Molten metal is poured rapidly into the sprue, and the pattern vaporizes. The mold is poured quickly to prevent combustion of the polystyrene with the resulting carbonaceous residue. Instead, the gases caused by vaporization of the material are driven out through the permeable sand and vent holes. A refractory coating is usually applied to the pattern to ensure a better surface finish for the casting and added pattern strength. Considerable hold-down weight and at times even side binding are necessary to accommodate the relatively high pressures within the mold.

Disposable patterns can be made in one of two ways. First, they may be glued and hot-wire or hot-knife carved. Sprues, runners, and gates may be glued on or attached with wire or nails. Second, they may be made from expandable or foam polystyrene in the form of small beads that contain penthane as a blowing agent. The polystyrene thus transformed into foam is injected under heat and pressure into a metal die. The die is then cooled. Complex shapes can be formed at high production rates. Because no allowance is necessary for removing the pattern from the mold, the only allowances necessary are for shrinkage, finish, and distortion.

The advantages of this disposable pattern process include the following.

1. For one-of-a-kind castings the process requires less production time.
2. Allowances are unnecessary in removing the pattern from the sand.
3. Finish is uniform and reasonably smooth.
4. A complex wooden pattern with loose pieces is unnecessary.
5. Cores are seldom required.
6. Molding is greatly simplified.

The disadvantages include the following.

1. The pattern is destroyed in the process.
2. Patterns are more delicate to handle.
3. There is no opportunity to inspect the finished cavity.
4. Environmental concerns with collection of toxic gases.

Casting Weight

The weight of a complex casting is not easily determined but can be estimated by weighing the pattern. The following relationship is used.

$$W_p/\rho_p = W_c/\rho_c \tag{5.2}$$

where

$$W_p \text{ and } W_c = \text{Weight of pattern and casting, lb (kg)}$$
$$\rho_p \text{ and } \rho_c = \text{Density of pattern and casting, lb/ft}^3 \text{ (kg/m}^3)$$

The density of polystyrene foam is about 21 lb/ft³ (336 kg/m³) and soft pattern pine is 25 lb/ft³ (400 kg/m³).

5.2 GATING SYSTEM AND SOLIDIFICATION CHARACTERISTICS

The passageway for bringing the molten metal to the mold cavity, known as the *gating* system, is constructed of pouring basin, a down gate or vertical channel known as a sprue, and a gate through which the metal flows from the sprue base to the mold cavity.

Aspiration of mold gases is likely to occur in the sprue. By tapering the sprue to proper proportions, the metal does not pull away from the mold as it accelerates downward. For freely falling metal, the velocity increases with the square root of the falling distance. A conservative estimate of the taper, which neglects frictional effects of the sprue, entrance and exit effect, and back pressure of the sprue occurring as the mold fills, is given as

$$A_1/A_2 \leq (h_a/h_b)^{1/2} \tag{5.3}$$

where

$$A = \text{Cross-sectional area ft}^2 \text{ (m}^2)$$
$$h_a = \text{Head of metal of pouring basin height, ft (m)}$$
$$h_b = \text{Head of metal of pouring basin plus sprue, ft (m)}$$

Subscripts 1 and 2 are entrance and exit locations of the sprue.

Abrupt changes in direction of flow channels causes aspiration or bubbling. One approach is to make the horizontal channels so large that flow velocity in these portions are low, which is achieved by making the smallest cross section at the sprue base. These proportions are described by gating ratios. Typically, aluminum has a gating ratio of $1:3:3$, meaning that the ratio of the area at the runner to that of the sprue base is 3, and the sum of the cross-sectional areas of all gates is equal to that of the runner. Gating ratios vary from $1:1.5:1.5$ to $1:5:5$, according to the material cast with the higher ratios being for alloys that easily form oxides or those that have slag or oxide, such as ductile iron. Large castings have several runners to distribute the metal from the sprue base to several gate passageways around the cavity.

The gating system is the highway for delivery of molten metal to the mold and involves a number of factors.

1. Metal should enter the cavity with as little turbulence as possible at or near the bottom of the mold cavity when pouring small castings.
2. Erosion of the passageway or cavity surfaces should be avoided by regulating the flow of metal properly or by the use of dry-sand cores. Formed gates and runners resist erosion better than those that are cut.
3. Metal should enter the cavity so that directional solidification is provided if possible. The solidification should progress from the mold surfaces to the hottest metal so that there is always molten metal available to compensate for shrinkage.

4. Slag and foreign particles should be prevented from entering the mold cavity. A pouring basin next to the top of the sprue hole often is provided on large molds to simplify pouring and to keep slag from entering the mold. Skimming gates such as the one shown in Figure 5.3 trap slag and other light particles into the second sprue hole. The gate to the mold is restricted somewhat to allow time for floating particles to rise into the skimmer. A strainer made of baked dry sand or ceramic material also can be used at the pouring basin to allow entry of only clean metal.

Risers are often provided in molds to feed molten metal into the main casting cavity to compensate for the shrinkage. They should be large in section so as to remain molten as long as possible and should be located near heavy sections subject to large shrinkage. If they are placed at the top of the section (Figure 5.3), gravity will assist in supplying metal into the casting.

Blind risers are domelike risers found in the cope half of the flask that are not the complete height of the cope. They are normally placed directly over the gate where the metal feeds into the mold cavity and thus supply the hottest metal when pouring is completed. Risers should be the last to solidify.

Volumetric *shrinkage* usually occurs when metal solidifies. A shrinkage cavity results if the solidification is not directed. Voids caused by shrinkage should occur in the gate, risers, or sprue. The shrinkage occurs in the location where the metal stays molten the longest. Shrinkage for aluminum, zinc, and iron is 6%, 4.2%, and 3%, respectively.

Figure 5.4 illustrates temperature gradient or isotherm lines in a casting and the directions of heat flow from the solidifying metal to the sand. Shrinkage-caused voids occur in the region of highest temperature, and mold design must be modified so as to remove this tendency if such a void is detrimental to a casting.

Metal inserts called *chills* are sometimes used to control solidification by carrying heat away from the solidifying metal at a more rapid rate. *Exothermic chemical* compounds may be packed next to a part of the casting so that heat is retained in that area. This exothermic material burns at or close to the melting point of the metal so that it does not add heat to the metal. It does, however, act as an adiabatic surface to prevent heat from leaving the casting zone.

5.3 PATTERNS

Figure 5.5 shows various pattern construction. The simplest form is the solid or *single-piece* pattern shown in Figure 5.5*A*. Many patterns cannot be made in a single piece because they cannot be removed from the sand. To eliminate this difficulty some patterns

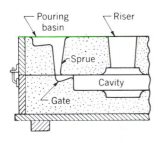
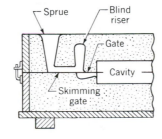

Figure 5.3
Methods used in introducing metal to mold cavity.

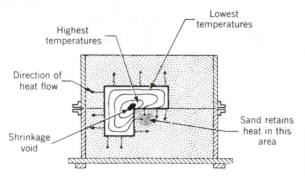

Figure 5.4
Isotherms indicating the area of shrinkage void.

are made in two parts (Figure 5.5B) so that half of the pattern will rest in the lower part of the mold and half in the upper part. The split in the pattern occurs at the parting line of the mold. Figure 5.5C shows a pattern with two loose pieces that are necessary to facilitate withdrawing it from the mold.

In production work where many castings are required, gated patterns, as shown in Figure 5.5D, may be used. These patterns are made of metal to give them strength and to eliminate warping. *Match plates* provide a substantial mounting for patterns and are

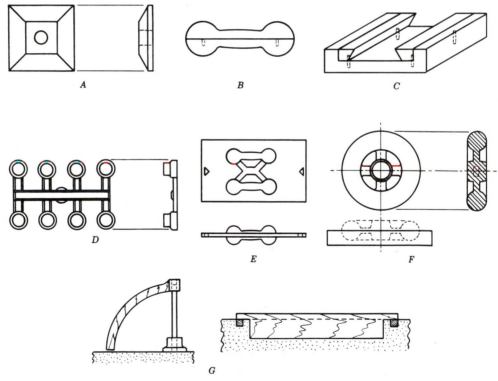

Figure 5.5
Types of patterns. *A*, Solid pattern. *B*, Split pattern. *C*, Loose-piece pattern. *D*, Gated pattern. *E*, Match plate. *F*, Follow board for wheel pattern. *G*, Sweep patterns: curved sweep for shaping large green-sand core, and straight sweep.

widely used with machine molding. Figure 5.5*E* shows a match plate, upon which are mounted the patterns for two small dumbbells. The follow board, shown in Figure 5.5*F*, may be used with either single- or multiple-gated patterns. Practically all high-production work on molding machines uses the match-plate pattern. Patterns requiring follow boards are usually somewhat more difficult to make as a split pattern. The board is routed out, so that the pattern rests in it up to the parting line. This board acts as a molding board for the first molding operation. Many molds of regular shape may be constructed by the use of *sweep* patterns as illustrated in Figure 5.5*G*. The principal advantage of this pattern is that it eliminates expensive pattern construction.

Allowances

In pattern work the question asked is why a finished part could not be used for making molds where there would not be the expense of making a pattern. In some cases they might be used, but in general this procedure is impractical because allowances are added to the pattern. These allowances are shrinkage, draft, finish, distortion, and shake.

Shrinkage. When any pure metal and most alloy metals cool, they shrink. To compensate for shrinkage, a *shrink rule* is used in laying out the measurements for the pattern. A shrink rule for cast iron is $\frac{1}{8}$ in. longer per foot than a standard rule. If a gear blank has an outside diameter of 6 in. (152 mm) when finished, the *shrink rule* would actually make it $6\frac{1}{16}$ in. (154 mm) in diameter, thus compensating for the shrinkage. The shrinkage for brass varies with its composition, but is close to $\frac{3}{16}$ in./ft, steel $\frac{1}{4}$ in./ft, and aluminum and magnesium $\frac{5}{32}$ in./ft. When metal patterns are cast from original patterns, double shrinkage is allowed.

Draft. When a removable pattern is drawn from a mold, the tendency to tear away the edges of the mold in contact with the pattern is decreased if the surfaces of the pattern, parallel to the direction it is being withdrawn, are slightly tapered. This tapering of the sides of the pattern, known as *draft*, provides a slight clearance for the pattern as it is lifted. Draft is added to the exterior dimensions of a pattern and is usually $\frac{1}{8}$ to $\frac{1}{4}$ in./ft. Interior holes may need draft as large as $\frac{3}{4}$ in./ft.

Finish. Each surface that is machined is indicated by a finish mark such as ... $\overset{64}{\nabla}$. This mark shows the pattern maker where additional metal stock is provided for metal machining, hence, a finish allowance. The amount that is to be added to the pattern depends on the size and shape of the gating, but in general the allowance for small and average-size castings is $\frac{1}{8}$ in. (3.2 mm). When patterns are large, this allowance is increased.

Distortion. *Distortion allowance* applies to castings of irregular shape that are distorted in the process of cooling because of metal shrinkage. A horseshoe-shaped piece is an irregular designed component.

Shake. When a removable pattern is rapped in the mold before it is withdrawn, the cavity in the mold increases slightly. In an average-size casting, this increase in size can be ignored. In large castings, or in ones that must fit together without machining, a shake allowance is occasionally considered by making the pattern slightly smaller.

5.4 REMOVABLE PATTERNS

Most patterns are made of wood, which is inexpensive and can be easily worked. *Wood patterns* are usually given at least three coats of shellac or a synthetic varnish that will not redissolve on contact with moisture. This finish fills the pores of the wood, creates a moisture seal, and gives the pattern a smooth surface.

If the production quantity is very high, the pattern is constructed of metal. Metal patterns do not change their shape when subjected to moist conditions, and they require minimum maintenance to keep them in operating condition. Metals used for patterns include brass, white metal, cast iron, and aluminum. Aluminum is probably the best all-around metal because it is easy to work, lightweight, and resistant to corrosion. Metal patterns are usually cast from a master pattern constructed of wood. The plastics are especially adapted for pattern materials because they do not absorb moisture, are strong and dimensionally stable, and have a smooth surface.

The details for laying out a cast-iron V block are shown in Figure 5.6, where the end view is drawn first using a shrink rule. Because the detail calls for "finish" all over, more metal must be provided. This is indicated by the second outline of the V block on the layout. In providing for the draft, the method of molding the pattern must be considered. The final outline on the layout board represents the actual size and shape that are used for constructing the pattern.

Sharp interior corners are filleted to eliminate the development of metal shrinkage cracks, or *stress concentrations*. A fillet is a concave-connecting surface or the rounding out of a corner at two intersecting planes. Rounded corners and fillets assist materially in molding, in that the sand is not likely to break out when the pattern is withdrawn. Fillets are usually designed with a radius of ⅛ to ¼ in./ft.

Loose-piece patterns are required when projections or overhanging parts occur, making it impossible to remove them from the sand even though they are parted. In such patterns the projections are fastened loosely to the main pattern by wooden or wire dowel pins. The loose pieces remain in the mold until after the pattern is withdrawn. They are then withdrawn separately through the cavity formed by the main pattern. The use of loose pieces is illustrated in the pattern for a *gib block* casting that fits over a dovetailed slide; a detail is shown in Figure 5.7.

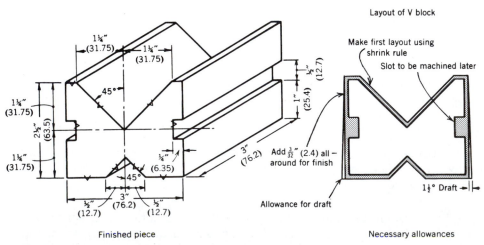

Figure 5.6
Method of making a pattern layout for a cast-iron block.

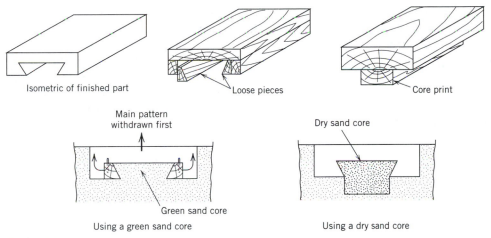

Figure 5.7
Two methods of molding a gib block.

Loose pieces may be eliminated by a dry-sand core. If this construction is used, the pattern is made using a core print. In addition, a *core box* is necessary. This latter method is less economical because of the expense in making the *core box and core*.

To improve the surface finish of the completed castings, the mold is often brushed, swabbed, or sprayed with a zirconite wash. The facing sand placed next to the pattern can be green sand, sodium silicate-bonded sand, and furan sand.

5.5 SAND TECHNOLOGY

Silica sand (SiO_2), found in many natural deposits, is suited for molding because it can withstand high temperatures without breakdown. Sand is low in cost, has long life, and is available in a wide range of grain sizes and shapes. The disadvantages are that sand has a high expansion rate when subjected to heat and has a tendency to fuse with the metal. If sand contains a high percentage of fine dust, it may constitute a health hazard.

The pouring temperature of iron ranges from 2400° to 2800°F, while for steel it is 2750° to 3000°F. These temperatures require a coarser sand for iron, while a highly refractory sand is necessary for steel.

Pure silica sand is not suitable for molding because it lacks binding qualities. The binding qualities may be obtained by adding 8% to 15% clay. The three clays commonly used are kaolinite, illite, and bentonite. The latter, used most often, is weathered volcanic ash.

Some natural molding sands are bonded with clay when quarried, and only water is added to have an adequate molding sand for nonferrous castings. The large amount of organic material found in natural sands prevents them from being sufficiently refractory for high-temperature applications such as in the molding of higher melting point metals and alloys. Synthetic molding sands are composed of washed, sharp-grained silica (sharp sand) to which 3% to 5% clay is added. Less gas is generated with synthetic sands because less than 5% moisture is necessary to develop adequate strength.

The size of the sand grains depends on the molded work. For small and intricate castings a fine sand is desirable to allow all the details of the mold to be brought out sharply. As the casting size increases, the sand particles should be coarser to permit the escape of gases. Sharp, irregular-shaped grains are usually preferred because they interlock and add strength to the mold.

Periodic tests are necessary for *foundry sand*. Sand properties change by contamination from foreign materials, washing action in pouring, change and distribution of grain size, and by high metal temperatures. Tests may be either chemical or mechanical, but, aside from determining undesirable elements in the sand, chemical tests are seldom used. Various tests determine the following properties of a molding sand.

1. *Permeability*. Porosity of the sand enables the escape of gas and steam formed in the mold.
2. *Strength*. Sand must be cohesive to the extent that it has sufficient bond. Both water and clay content affect the cohesive properties.
3. *Refractoriness*. Sand must resist high temperatures without fusing.
4. *Grain size and shape*. Sand must have a grain size suitable with the surface to be produced, and grain shape must be irregular to permit interlocking.

Mold and Core Hardness Test

Penetration of a steel ball into the sand surface is a measure of the *hardness* or firmness of the sand. A spring-loaded steel ball 0.2 in. in diameter is pressed into the surface of the mold, and the depth of penetration is indicated on the dial in thousandths of an inch. Medium-rammed molds give a value around 75. A hardness tester is shown by Figure 5.8.

Fineness Test

The fineness test finds the percentage distribution of grain sizes in sand and is performed on a clay-free, dried-sand sample. A set of standard testing sieves is used. The U.S. National Bureau of Standards meshes are 6, 12, 20, 30, 40, 50, 70, 100, 140, 200, and 270. The stacked sieves have motor-driven shakers. The sand is placed on the coarsest sieve at the top, and after 15 min of vibration, the weight of the sand retained on each sieve is obtained and converted to a percentage basis.

To obtain the American Foundrymen Society *fineness number* (see Table 5.1), each percentage is multiplied by a factor as given in an example. The fineness number is obtained by adding all the resulting products and dividing the total by the percentage of sand retained. This number allows the comparisons of different foundry sands.

Figure 5.8
Mold hardness tester for measuring the surface hardness of green-sand molds.

Table 5.1   **Example of AFA Fineness Calculation**

Mesh	Percentage Retained	Multiplier	Product
6	0	3	0
12	0	5	0
20	0	10	0
30	2.0	20	40.0
40	2.5	30	75.0
50	3.0	40	120.0
70	6.0	50	300.0
100	20.0	70	1400.0
140	32.0	100	3200.0
200	12.0	140	1680.0
270	9.0	200	1800.0
Pan	4.0	300	1200.0
Totals	90.5		9815.0

$$\text{Grain fineness number} = \frac{9815}{90.5} = 108$$

Test for Moisture Content

The moisture content of foundry sands varies according to the type of molds being made and the kind of metal being poured. For a given condition, there is a close range within which the moisture percentage should be held to produce satisfactory results.

The moisture teller contains electric heating units and a blower for forcing warm air through the filter pan containing the sand sample. By weighing the sand after it is dried and noting the difference in the initial and final readings, the percentage of moisture can be determined. The moisture content should vary from 2% to 8%, depending on the type of molding being done.

Clay Content Test

The equipment to determine the percentage of clay in molding sands consists of a drying oven, a balance and weights, and a sand washer. A quantity of sand is dried and a water-based caustic soda solution is added. Following a timed mixing the caustic solution, which has absorbed the clay, is siphoned off. This process is repeated three times. The sample is dried, weighted, and compared to the original sample weight to determine the loss in clay.

Permeability Test

An essential quality in molding sand is *porosity* to permit the escape of gases. This depends on the shape of sand grains, fineness, degree of packing, moisture content, and amount of binder. Permeability is measured by the quantity of air that passes through a sample of sand in a prescribed time and under standard conditions. Coarse-grained sands are more permeable, but when coarse grains of sand are added to a fine-grained sand, the permeability initially decreases and then increases. Permeability increases with moisture content up to approximately 5% moisture. Figure 5.9 illustrates a procedure for determining the permeability of a specimen.

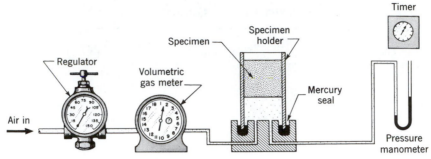

Figure 5.9
Equipment used to measure sand permeability.

Sand Strength Test

Strength tests find the holding power of various bonding materials in green and dry sand. *Compression tests* are common, although tension, shear, and transverse tests are sometimes used. The principle of the sand strength machine is shown in Figure 5.10.

Sand Conditioning

Properly conditioned sand is an important factor in obtaining good castings. New and used sand accomplishes the following results.

1. Binder is distributed more uniformly around the sand grains.
2. Moisture content is controlled and particle surfaces are moistened.
3. Foreign particles are eliminated.
4. Sand is aerated.
5. Sand is cooled to about room temperature.

The *muller* for preparing the sand (Figure 5.11) has two circular pans in which are mounted a combination of plows and mullers driven by a vertical shaft. The two mullers are arranged to process the sand continuously and to provide an intensive kneading and rubbing action.

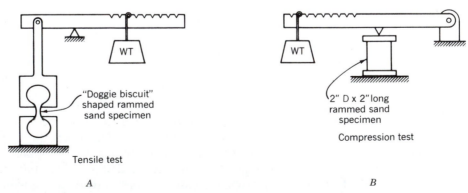

Figure 5.10
Principle of sand tester. *A*, Chemical bonded sand. *B*, Green sand.

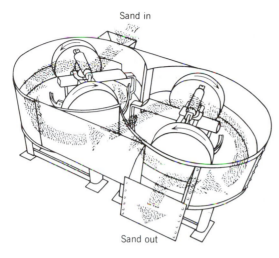

Sand in

Sand out

Figure 5.11
Continuous multimull for renewing used
foundry sand.

A sand reclamation and conditioning installation is illustrated in Figure 5.12. After
the metal in a mold is solidified and is shaken out at the ends of the roller conveyor line,
the sand falls through a grate on a belt conveyor as shown in the figure. This conveyor
carries the warm used sand to a smaller belt conveyor equipped with a magnetic separator.
The sand is then discharged onto a bucket elevator from which it passes through an
enclosed revolving screen into the storage bin. The sand is delivered from the storage
bin to one or more mullers and conditioned for reuse. From here it is discharged to an
overhead belt conveyor through an aerator that separates the sand grains and improves
their flowability for molding. The cycle is completed when the sand is discharged into
the several hoppers serving the molding stations.

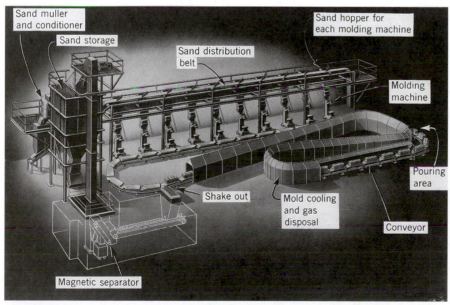

Sand muller
and conditioner

Sand storage

Sand distribution
belt

Sand hopper for
each molding machine

Molding
machine

Pouring
area

Shake out

Mold cooling
and gas
disposal

Conveyor

Magnetic separator

Figure 5.12
Progressive foundry unit for reclaiming foundry sand.

5.6 CORES

When a casting is to have a cavity or recess, a *core* is introduced into the mold. A core is sometimes defined as "any projection of the sand into the mold." The projection is formed by the pattern itself or made separately from the mold and inserted into the mold after the pattern is withdrawn. Internal or external surfaces of a casting can be formed by a core.

Green-Sand Cores

Cores are either *green sand* or *dry sand*. Figure 5.13 shows various cores. Green-sand cores (Figure 5.13*A*) are formed by the pattern and made with the same sand as the mold. It shows how a flanged casting is molded with the hole through the center, or "cored out" with green sand.

Dry-sand cores are formed separately and inserted after the pattern is withdrawn, but before the mold is closed. They are usually made of clean river sand, which is mixed with a binder and then baked to increase strength. The box in which they are formed to shape is called a core box.

Several types of dry-sand cores are also illustrated. The design for supporting a core when molding a cylindrical bushing is shown in Figure 5.13*B*. The projections on each end of the cylindrical pattern are known as core prints and form the seats that support and hold the core in place. A vertical core is shown in Figure 5.13*C*. The upper end requires considerable taper so as not to tear the sand in the cope when the flask is assembled. Cores that are supported only at one end must have the core print of sufficient length to prevent the core from falling into the mold. Such a core (Figure 5.13*D*) is known as balanced core. A core supported above and hanging into the mold is shown in Figure 5.13*E*. This type of core usually requires a hole through the upper part to permit the metal to reach the mold. A drop core (Figure 5.13*F*) is required when a hole is not in line with the parting surface and must be formed at a lower level.

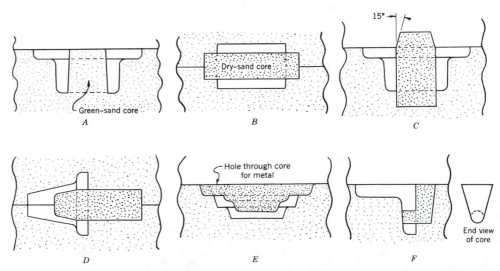

Figure 5.13
Typical cores. *A*, Solid pattern with green-sand core. *B*, Dry-sand core supported on both ends. *C*, Vertical dry-sand core. *D*, Balanced dry-sand core. *E*, Hanging dry-sand core. *F*, Drop core.

Green-sand cores are preferred to minimize pattern and casting cost. Separate cores naturally increase production cost.

Dry-Sand Cores

Core boxes are required for dry-sand cores. The cores must be formed separately, baked, and placed properly in the molds. More accurate holes can be made with dry-sand cores, because they give a better casting surface and are less likely to be washed away by the molten metal.

In setting dry-sand cores into molds, adequate supports must be provided. Ordinarily these supports are formed into the mold by the pattern; however, for large or intricate cores, *chaplets* (small metal shapes made of low-melting-point alloy) are placed in the mold to give additional support to the core until the molten metal enters the mold and fuses the chaplets into the casting. The use of chaplets should be limited because of the difficulty in fusing the chaplet with the metal.

A core must have strength to support itself. Porosity or permeability is an important consideration in making cores. As the hot metal surrounds the core, gases are generated by the heat in contact with binding material. Provision is required to vent these gases. The core must have a smooth surface to ensure a smooth casting. Cores require refractoriness to resist the action of the heat until the hot metal has stabilized.

Core Making

The core is formed by being rammed into a core box or by the use of *sweeps*. Fragile and medium-size cores are reinforced with wires for added strength to withstand deflection and the hydrostatic action of the metal. In large cores, perforated pipes or arbors are used. In addition to giving the core strength, they also serve as a large vent. Cores having round sections are often made in halves and glued together after baking.

Cores are produced not only in hand-filled boxes but also on a variety of molding machines, including many of the conventional types such as the jolt, squeeze, rollover, jolt-squeeze, and sandslinger machines.

Pneumatic *core-blowing machines* also offer a rapid means of producing small and medium-size cores in quantity production work. In this method sand is blown under pressure and at high velocity from a sand magazine into the core box. Figure 5.14 illustrates the process. Suitable vents are built into the core box or blow plate to permit the air to escape. These vents are small enough to prevent sand seepage and to allow air venting. This equipment is adapted to production work in which the cost of metal core boxes is justified.

Stock cores of uniform cross section may be produced continuously by an extrusion process. The machine consists of a hopper in which the sand is mixed. Below it, in a horizontal position, is a spiral screw conveyor that forces the prepared sand through a die tube at uniform speed and pressure.

Binders and Core Mixtures

Oil binders are used in making cores. Linseed oil is frequently used in small cores. The oil forms a film around the sand grain that hardens when oxidized by heat. Such cores should be baked for 2 h at a temperature between 350° and 425°F (180° and 220°C). A common mixture uses 40 parts river sand and 1 part linseed oil. An advantage to this proportion is its strength and resistance to water absorption.

Another group of water-soluble binders includes wheat flour, dextrin, gelatinized starch, and other commercial preparations. The ratio of binder to sand in these mixtures

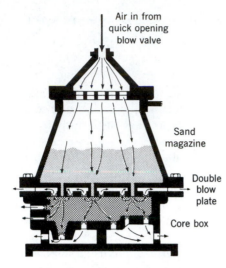

Figure 5.14
Diagram of core-blowing operation.

is rather high, 1 : 8 or more parts of sand. Frequently a small percentage of old sand is used in place of new sand. In addition, pulverized pitch or rosin may be used.

Several types of thermosetting plastics, including urea and phenol formaldehyde, are used as core binders. They are made in both liquid and powder form and are mixed with silica flour, cereal binder, water, kerosene, and a parting liquid. Urea *resin binders* are baked at 325° to 375°F (165° to 190°C), and the phenolic binders at 400° to 450°F (200° to 230°C). Both respond to dielectric heating and are combustible under the heat of the metal. Their success as core binders is based on their high adhesive strength, moisture resistance, burnout characteristic, and ability to provide a smooth surface to the core.

Furfuryl alcohol resin binders with sand are replacing many of the core binders that require baking. These resin binders either are air-dried or are blown or tamped into a hot core box maintained at about 452°F (220°C). The hot box-made cores can be removed from the mold in 10 to 20 s. If hot boxes are not used, the furfuryl alcohol resin is mixed with formaldehyde or urea formaldehyde resins from which air-dried cores can be made. They are known as furan or no-bake cores.

Cores are also made from a mixture of sand and sodium silicate. After being rammed into a core box, they are hardened by the application of carbon dioxide gas, which is known as the CO_2 process. Because cores made this way do not need to be baked, they can be produced rapidly at low cost in an air-conditioned environment.

5.7 MOLDING EQUIPMENT

Machines eliminate much of the labor in molding and produce better molds. Molding machines, varying considerably in design and method of operation, are named according to the way the ramming operation is performed. Figure 5.15 illustrates the principles used in packing a mold. The shading indicates the density of sand packing for each process.

Jolt Machine

The plain jolt-molding machine is equipped with adjustable flask-lifting pins to permit various flask sizes within the capacity of the machine. Molds weighing up to 13,000 lb (5850 kg) can be made on the larger machines. The table is pneumatically raised a short

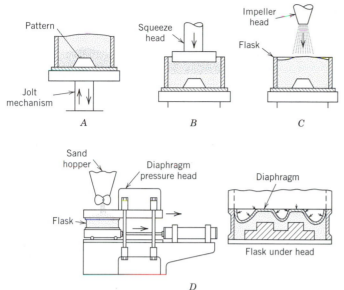

Figure 5.15
Machine molding principles.
A, Jolt. *B*, Squeeze. *C*, Sandslinger. *D*, Contour diaphragm molding machine.

distance and then dropped. This action causes the sand to be packed evenly about the pattern. The density of the sand is greatest around the pattern and at the parting line, and varies according to the height of the drop and depth of the sand in the flask. The uniform ramming about the pattern gives added strength to the mold and reduces the possibility of swells, scabs, or runouts. Castings produced under such conditions vary little in size or weight. Jolt machines handle the cope or drag separately, and are suitable for production quantities.

Squeezer Machine

Squeezer machines press the sand in the flask between the machine table and an overhead platen. Greatest mold density is obtained at the side of the mold from which the pressure is applied. Because it is impossible to obtain uniform mold density by this method, squeezing machines are limited to molds only a few inches in thickness.

Jolt-Squeeze Machine

Many machines use both jolt and squeeze. To produce a mold in this machine the flask is assembled with the match plate between the cope and drag, and the assembly is placed upside down on the machine table. Sand is added to the drag and leveled off and a bottom board is placed on top. The jolting action then rams the sand in the drag. The assembly is turned over and the cope filled with sand and leveled off. A pressure board is placed on top of the flask and the top platen of the machine is brought into position. By application of pressure the flask is squeezed between the platen and table, packing the sand in the cope to the proper density. After the pressure is released, the platen is swung out of the way. The cope is then lifted from the match plate while the plate is vibrated, after which the plate is removed from the drag. This machine eliminates six separate hand operations: ramming, smoothing the parting surface, applying parting sand, swabbing around the pattern, rapping the pattern, and cutting the gate.

Sandslinger

Uniform packing of sand in molds is an important part in producing molds. For large molds a mechanical device known as the *sandslinger* has been developed. Figure 5.15*C* illustrates the principle. These units are mobile and some may be a self-propelled unit operating on a narrow-gage track. Portable wheel units are available. The supply of sand is carried in a large tank with a capacity of about 300 ft³ (8.5 m³), which is refilled at intervals by overhead-handling equipment.

A delivery belt feeding out of a hopper conveys the sand to the rotating impeller head. The impeller enclosed head contains a single rotating cup that slings the sand into the mold. This cup, rotating at high speed, slings more than 1000 small buckets of sand a minute. The ramming capacity of this machine is 7 to 10 ft³ (0.2 to 0.28 m³), or 1000 lb (450 kg) sand/min. The density of the packing can be controlled by the speed of the impeller head. For high production, machines of this type are available having a capacity of 4000 lb sand/min (30 kg/s).

Diaphragm Molding Machines

The *diaphragm molding* machine utilizes a pure gum rubber diaphragm for packing the sand over the pattern contour, as illustrated in Figure 5.15D. The process uses a uniform air pressure to force the rubber diaphragm over the surface of the mold regardless of the pattern contour.

In Figure 5.15*D* the flask is shown at the filling position. The flask and sand chute are then moved to under the diaphragm pressure head. Air is admitted to the pressure head and the diaphragm is forced against the molding sand in the flask. The flask is returned to original position, striking off any sand above the flask. A pin lift removes the match plate from the flask. The entire process is very rapid, and close tolerances can be maintained because of the uniform packing of the sand.

5.8 POURING AND CLEANING CASTINGS

In low-technology foundries the molds are lined up on the floor as they are made and the metal is taken to them in hand ladles. When more metal is required or if heavier metal is poured, ladle tongs designed for two workers are used. In modern foundries engaged in mass-producing castings, molds are placed on conveyors and passed slowly by a pouring station. The pouring station may be located permanently next to the furnace, or metal may be brought to certain points by overhead-handling equipment. The conveyor serves as a storage location for the molds while they are being transported to the cleaning room.

After a casting is solidified and cooled to a suitable temperature for handling, it is shaken from the mold. Very often shaking is done at a ventilated mold shakeout. The dust is collected by a cyclone dust collector while the sand is collected underneath and transported to the conditioning station. Castings are retained on the movable grate bars of the shakeout.

Iron and steel castings are covered with a layer of sand and scale that is somewhat difficult to remove. The gates and risers on iron castings may be broken off, but to remove them from steel castings a cutting torch or a high-speed cutting-off wheel is necessary. The process of removing gates, runners, risers, flash, and cleaning the casting

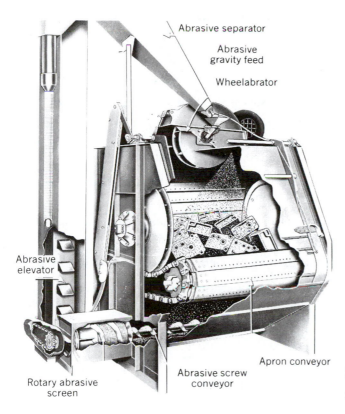

Abrasive separator

Abrasive
gravity feed

Wheelabrator

Abrasive
elevator

Rotary abrasive
screen

Abrasive screw
conveyor

Apron conveyor

Figure 5.16
Tumbler and blasting machine
for cleaning castings.

is termed *fettling*. The fettling operation accounts for 15% to 25% of the labor cost of manufacturing a casting.

Several methods of cleaning castings are used depending on the size, kind, and shape of the castings. The most common piece of equipment used is the rotating cylindrical tumbling mill. It cleans by the tumbling action of the castings upon one another as the mill rotates. A similar piece of equipment is shown in Figure 5.16. It cleans 65 to 100 lb (30 to 45 kg) of gray iron or malleable castings in 5 to 8 min. Larger machines of this type have capacities of more than 1 ton (0.9 Mg) per charge. The machine consists of cleaning barrel formed by an endless apron conveyor. The work is tumbled beneath a blasting unit located just above the load, and a metallic shot is blasted onto the castings. After striking the load, the shot falls through holes in the conveyor and is carried overhead to a separator and storage hopper. From there it is fed by gravity to the blasting unit. The unit is unloaded by reversing the apron conveyor. A dust collector is installed with the machine to eliminate dust hazards.

Sandblasting units may be used separately for cleaning castings. Sharp sand is blown against castings inside a blasting cabinet, removing scale and cleaning the surface. Plated or galvanized castings are frequently pickled in a weak acid solution and rinsed in hot water. Large castings that are difficult to handle are often cleaned hydraulically. The casting is placed on a rotating table, and streams of pressurized water wash away the sand.

In addition to these cleaning processes, castings may require chipping or grinding to remove surface and edge defects. Stand, portable, and swing frame grinders are used for this work. Fast, free-cutting abrasive wheels, operating at a cutting speed of around 9500 ft/min (48 m/s), are recommended.

QUESTIONS

1. Give an explanation of the following terms:

Green sand	Gib block
Sharp sand	Permeability
Sodium silicate	Fineness number
Parting sand	Muller
Blind risers	Chaplets
Sweep	Core blowing machines
Shrink rule	Resin binders
Draft	Diaphragm molding

2. How would the gear blank shown in Figure 5.1 be molded by the CO_2 technique and by a Styrofoam pattern?

3. Describe the step-by-step process to mold the V block shown in Figure 5.6 by the disposable-pattern method and the solid-pattern method.

4. How can a sweep-type pattern cast a solid vase-shaped casting?

5. Explain the problems in slowly pouring a mold with a polystyrene pattern.

6. What happens to conventional molds if they are poured too slowly?

7. Why is a blind riser more apt to "force feed" a shrinkage void?

8. Where and how can exothermic materials be used in casting a dumbbell? Describe the isothermal lines.

9. What are the disadvantages and advantages of a skimming gate?

10. Outline the steps to mold the split pattern shown in Figure 5.5B.

11. Is an allowance made for draft in a disposable mold casting? Why?

12. For a steel casting, state the differences in allowances between a wood and a polystyrene pattern.

13. Describe the "full mold" process.

14. Which surfaces on the gib block patterns (Figure 5.7) require draft?

15. Describe a method of producing 5000 polystyrene patterns for an electric motor base.

16. What happens to gates, risers, and sprues for polystyrene patterns?

17. Would one use a match plate with polystyrene patterns?

18. What are the reasons for reconditioning used sand?

19. Discuss the factors relating to permeability.

20. How does grain size of molding sand affect permeability?

21. How does the grain size distribution of molding sand affect strength and permeability?

22. How is permeability artificially introduced into the cope of a mold?

23. How does clay content affect molding sand strength, permeability, and surface finish?

24. List ways to improve the surface finish of a mold.

25. Why are weights placed on top of the cope before pouring? Why are weights more necessary when pouring cast iron as compared to aluminum?

26. Discuss the effects of moisture on molding sand.

27. How does moisture in a molding sand affect a wood pattern, core, polystyrene, permeability, strength?

28. Why are cores seldom necessary in disposable-pattern molding?

29. What are the choices in selecting a core binder?

30. State the principal use for a furan binder.

31. What is the purpose and advantage of dielectric heating ability for a core binder?

32. Why is it best to use green sand in making a core, if the mold permits?

33. Explain the differences in permeability that come from jolting, squeezing, or sandslinging a mold. What differences can be expected in the casting surface?

34. What is fettling?

35. List the comparative advantages and disadvantages of sandblasting.

36. Why is coarser sand better for steel castings than fine grained sand? Why is it that as castings increase in size, it is often better to use increasingly coarser sand?

37. What is a core print? Give its purpose.

38. If an aluminum casting is used as a pattern for a steel casting, what allowances are necessary in the wood pattern to make the aluminum casting?

PROBLEMS

5.1. If approximately 53% of metal poured as castings actually end up as the casting, what makes up the remainder of the metal poured? Sketch a casting with the sprue, runner, gates, etc., in the mold and indicate their approximate proportions. (Hints: A sketch is not an engineering drawing using instruments. Rather, a 0.5-mm pencil and engineering computation paper are used. The sketch can be a multiview or an isometric drawing.)

5.2. Sketch the isothermal lines for a dumbbell-shaped casting. Explain the use of risers, and redraw the isothermal lines with risers employed.

5.3. Sketch a pattern requiring a cheek. Label the parts. (Hints: Use engineering computation paper, a 0.5-mm pencil, and be neat and careful. A sketch is not an engineering drawing, nor is a CAD software package required.)

5.4. What is the actual length measurement for a cast iron shrink rule used for laying off 10 in. and 12 in.?

5.5. Sketch a part that requires chaplets. (Hint: Point out the location of the chaplets with arrow leaders.)

5.6. Sand weighs 100 lb/ft^3 and a cope contains 1.4 ft^3 of sand and 8 lb of wood. The cross section at the parting line of the casting is 14 in.^{2} If the cope is 10 in. high, how much weight is required to keep the cope from "floating" if aluminum is being poured? Determine in conventional and SI units. If steel is being cast, how much weight on the cope is required?

5.7. For 1 ton of sharp clean sand, how many pounds of bentonite are added for a nonferrous molding sand? How many pounds of water? How many gallons?

5.8. If 40% of a casting is sprues, gates, and risers, how much aluminum is melted per day to produce 350 finished castings weighing 1.9 kg each? Determine in SI and English units.

5.9. What is the actual length measurement for a cast iron shrink rule used for laying off the following lengths: 14 in. (355 mm) and 3 in. (76 mm)? Determine in English and SI units.

5.10. Calculate the grain fineness number of sand passed through a set of standard testing sieves. The following percentages are retained on each sieve: No. 6, 0%; No. 12, 0%; No. 20, 4.0%; No. 30, 4.0%; No. 40, 2.0%; No. 50, 3.5%; No. 70, 5.0%; No. 100, 18%; No. 140, 25.0%; No. 200, 17.0%; No. 270, 9.0%; and pan, 5.0%.

5.11. If an abrasive grinder used to snag off casting irregularities rotates at 1260 rpm, what is the approximate diameter of the wheel, if it operates at the optimum cutting speed? Determine in conventional and SI units.

5.12. Make a graph showing sandslinger impeller speed against impeller diameter.

5.13. If a 10-in. diameter sandslinger delivers sand at 60 ft/s velocity to a mold, how fast is the impeller turning?

5.14. Sketch the orthographic views of a pattern for a 4-in. cube where the poured material is gray iron. (Hints: An orthographic view is top, front, and right-hand side. This wood pattern is one piece. Provide allowances for draft, shrinkage, and machine finishing. Provide essential dimensions. Remember that a sketch is not a CAD or instrument drawing.)

5.15. Sketch the views of a pattern for a 4-in. open cube where the poured material is gray iron. The internal opening is 3 by 3 in. (Hints: The pattern is two pieces, and the opening is provided by a dry-sand core. Consider the core to be horizontal in the cope and drag. Core prints are necessary, and a core box is required for the core. Evaluate the application of chaplets. Sketch the core box also.) What other methods are available for this casting?

5.16. A 4-in. cube is cast of gray iron. It is top-gated through a tapered sprue. The cross-sectional area at the base is 2 in.^{2} The pouring basin height is 4 in., while the sprue plus pouring basin height is 16 in.

a. Neglecting frictional losses, what is the area at the top of the sprue to prevent aspiration or bubbling?

b. If there is one gate, what is its cross-sectional area? Repeat for two gates.

c. If there is no riser, what volumetric shrinkage is expected in the cylinder casting? (Hint: Let iron shrinkage be 3%.)

MORE DIFFICULT PROBLEMS

5.17. Calculate the grain fineness number of sand passed through a set of standard testing sieves. The following percentages are retained on each sieve: No. 6, 0%; No. 12, 0%; No. 20, 2%; No. 30, 3%; No. 40, 8%; No. 50, 12%; No. 70, 14%; No. 100, 18%; No. 140, 20%; No. 200, 12%; No. 270, 3%; and pan, 1%.

5.18. A cylinder, 5 cm in diameter and 10 cm high, is cast of aluminum. It is top-gated through a tapered sprue. The cross-sectional area at the base is 1.5 cm². The pouring basin is 4 cm height, while the sprue plus pouring basin height is 15 cm.

a. Neglecting frictional losses, what is the area at the top of the sprue to prevent aspiration?

b. If there is one gate, what is its cross-sectional area? Repeat for four gates.

c. If there is no riser, what volumetric shrinkage is expected in the cylinder casting?

PRACTICAL APPLICATION

Arrange for a plant tour of a foundry where liquid metal is poured into green-sand molds. In the planning for the tour, as your instructor suggests, determine goals for the tour. Plant tours can be poor learning opportunities unless there is guided inquiry and responsibilities for the student. In addition to an overall impression of the foundry, each student should concentrate on a particular subject that will be seen. Prepare a list of questions, make notes, observe details of the operation, and especially important, read Chapter 5 before the tour. Immediately after the tour, prepare a report.

CASE STUDY: FOUNDRY BUSINESS

The Jesaitis Castings Company is asked by Store Display Company to bid on the manufacture of a casting in the form of a soft drink can, except for a 1.5-in. hole down through the center and along the longitudinal axis. The can is 2.5 in. in diameter and 4.75 in. long. The required quantity for store displays is 5000 units in steel and 100 units in aluminum.

Recommend a foundry process from pattern to delivery, including fettling for each item. Make a step-by-step analysis and give your reason for each step. Organize your presentation to Store Display for maximum clarity, brevity, and understanding. Because the aluminum cans are for display, they need to be machined. Sketch the pattern for this unit and the method of molding to help Store Display purchasing agents understand the added costs per unit of the aluminum units. Estimate the cost of steel and aluminum units. Explain to Store Display why the same molds and molding sand cannot be used for both castings. Steel costs $0.50/lb and aluminum costs $1.50/lb.

CHAPTER 6
CONTEMPORARY
CASTING PROCESSES

The system for producing a casting depends on quantity, the metal to be cast, and the intricacy of the part. Most commercial metals are cast in sand molds and sizes vary from small to large. However, sand molds are used once and then destroyed after metal solidification.

Permanent molds offer cost savings for big production quantities and whenever the size of the casting is not large. This chapter is titled "contemporary," that being relative to foundry processes, which are centuries old.

6.1 METAL MOLDS

Permanent molds are constructed of metals capable of withstanding high temperatures. Because of their high cost they are recommended when many castings are produced. Although permanent molds are impractical for large castings and alloys of high melting temperatures, they are advantageous for small- and medium-size castings.

Die Casting

In die casting, molten metal is forced by pressure into a metal mold known as a *die*. Because the metal solidifies under a pressure from 80 to 40,000 psi (0.6 to 275 MPa), the casting conforms to the die cavity in shape and surface finish. The usual pressure is from 1500 to 2000 psi (10.3 to 14 MPa).

Die casting is the most popular of the *permanent-mold processes*. Two methods are employed: (1) *hot chamber* and (2) *cold chamber*. The principal distinction between the two methods is determined by the location of the melting pot. In the hot-chamber method, a melting pot is included within the machine and the injection cylinder is immersed in the molten metal at all times. The injection cylinder is actuated by either air or hydraulic pressure, which forces the metal into the dies to complete the casting.

Machines using the cold-chamber process have a separate melting furnace, and metal is introduced into the injection cylinder by mechanical means or by hand. Hydraulic pressure then forces the metal into the die.

The process is rapid because the die and core are permanent. A smooth surface not only improves the appearance but also minimizes the work to prepare the castings for plating or other finishing operations. The wall thickness is more uniform than in sand castings and, consequently, less metal is required. The optimum production quantity ranges from 1000 to 200,000 pieces. The maximum weight of a brass die casting is about 5 lb (2.3 kg), but aluminum die castings of more than 100 lb (45 kg) are common. Small- to medium-size castings are produced at a cycle of 100 to 300 die fillings/h. The size is so accurately controlled that little or no machining is necessary. Waste is low because the sprue, runners, and gates are remelted. There is a stack loss of about 4% of the hot liquid metal, which is later captured in a pollution recovery system, but the air suspended metal fines are largely unusable.

The process largely eliminates secondary operations such as drilling and even certain types of threading. *Die-casting tolerances* vary according to the size of the casting and metal. For small castings, the tolerance ranges from ± 0.001 to 0.010 in. (± 0.03 to 0.25 mm). The closest tolerances are obtained for zinc alloys.

One of the limitations of die casting is the cost of equipment and dies. However, this is not an important factor in mass production, although cost is limiting in short-run jobs. Also, there is undesirable chilling of the metal unless high temperatures are sustained before the cast is removed. Metals having a high coefficient of contraction are removed from the mold as soon as possible because of the inability of the mold to contract with the casting.

Die castings were once limited to low-melting alloys, but with a gradual improvement of heat-resisting metals for dies, this process is used for numerous alloys. Although gray cast iron and low-carbon and alloy steels have been produced in dies made of unalloyed sintered molybdenum, die casting is commercially limited to nonferrous alloys.

Dies

Dies for hot- and cold-chamber machines are similar in construction because there is little difference in design and operation. They are made in two sections to allow casting removal and are usually equipped with heavy dowel pins to maintain the halves in proper alignment. Metal enters the stationary side when the die is locked in the closed position. As the die opens, the ejector plate in the movable half of the die is advanced, so that pins project through the die half and force the casting from the cavity and fixed cores. The dies are provided with a separate mechanism for moving the ejector plate or movable cores. The life of these molds depends on the metal cast and may range from 10,000 fillings if brass castings are made, to several million if zinc is used.

For large or complex castings a single-cavity mold is used. The casting, *gate,* and *runners* from such a mold are shown in Figure 6.1. The magnesium part is cast with a steel insert in a 600-ton (540-Mg) machine. If the quantity of castings is large and they are relatively small in size, a multiple-cavity die is used.

Vents and small overflow wells on one side of a die facilitate the escape of air and catch surplus metal that passes through the die mating surface. *Flash* metal is found at

Figure 6.1
Gate and casting of a chain saw crankcase from a single-cavity die.

the die half mating surface that must be trimmed off in the finishing operation. Notice the flashing around the part in Figure 6.1. A trim press is used to remove flashing.

A combination die has two or more cavities that are different; they are frequently made of insert blocks, which are removed to allow the substitution of other die blocks. Most dies are provided with channels for water cooling to keep the die at the correct temperature for rapid production.

Hot-Chamber Die Casting

Low-melting alloys of zinc, tin, and lead are the materials cast in hot-chamber machines. Chapter 4 discusses these alloys. Most other materials either have too high a melting point, an affinity for iron, or cause other problems that reduce the life of the machine and the die. Hot-chamber castings vary in size from 1 oz to 90 lb (0.03 to 40 kg), although very small castings are usually cast in multiple-mold dies.

Metal is forced into the mold, and pressure is maintained during solidification either by a plunger or by compressed air. The plunger-type machine, illustrated in Figure 6.2, is hydraulically operated for both the metal plunger and the mechanism for opening and closing the die. The plunger operates in one end of a gooseneck casting that is submerged in the molten metal. With the plunger in the upper position, metal flows by gravity into this casting through several holes, just below the plunger. On the downstroke these holes are closed by the plunger and pressure is applied on the entrapped metal, causing it to be forced into the die cavity. Pressures more than 5000 psi (35 MPa) are used in some machines of this type, resulting in castings of dense structure. As soon as the casting is solid, the pressure is relieved, the dies are forced open, and the casting is ejected by knockout pins. The sprue is removed with the runner and castings.

Air-operated machines have a gooseneck operated by a lifting mechanism. In the starting position, the casting is submerged in the molten metal and is filled by gravity. It is then raised so that the nozzle is in contact with the die opening and locked in position. Compressed air at pressures ranging from 80 to 600 psi (0.5 to 4.0 MPa) is applied directly on the metal, thus forcing it into the die. When solidification is nearly complete, the air pressure is turned off and the gooseneck lowered into position to receive more metal. The operation of opening the dies, withdrawing cores, and ejecting the castings is the same as for the plunger-type machine.

Cold-Chamber Die Casting

Die casting brass, aluminum, and magnesium requires higher pressures and melting temperatures and necessitates a change in the melting procedure from that previously

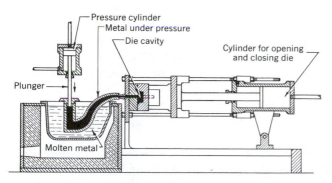

Figure 6.2
Plunger-activated actuated hot-chamber die-casting machine.

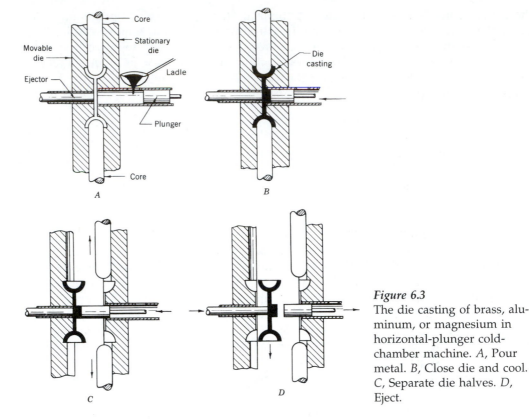

Figure 6.3
The die casting of brass, aluminum, or magnesium in horizontal-plunger cold-chamber machine. *A*, Pour metal. *B*, Close die and cool. *C*, Separate die halves. *D*, Eject.

described. Chapter 4 discusses the alloys used in cold-chamber casting. These metals are not melted in a self-contained pot, because the life of the pot would be very short. The usual procedure heats the metal in an auxiliary furnace and ladles it to the plunger cavity next to the dies. It is then forced into the dies under hydraulic pressure. These machines are built to withstand the heavy pressures exerted on the metal as it is forced into the dies. Of the two machines in general use, one has the plunger in a vertical position, the other in a horizontal position.

A sketch illustrating the operation of horizontal-plunger cold-chamber machines is shown in Figure 6.3. In Figure 6.3*A*, the dies are shown closed with cores in position and the molten metal ready to be poured. As soon as the ladle is emptied, the plunger moves to the left and forces the metal into the mold (Figure 6.3*B*). After the metal solidifies, the cores are withdrawn and the dies are opened. In Figure 6.3*C*, the dies are opened and the casting is ejected from the stationary half. To complete the process of opening, an *ejector rod* comes into operation and ejects the casting from the movable half of the die (Figure 6.3*D*). This operating cycle is used in a variety of machines that operate at pressures ranging from 5600 to 22,000 psi (39 to 150 MPa). These machines are fully hydraulic and semiautomatic. After the metal is poured, the rest of the operations are automatic.

The 2500-ton (22-MN), hydraulically operated, cold-chamber die-casting machine shown in Figure 6.4 is for making die castings up to 84 lb (38 kg) of aluminum, brass, or magnesium.

The manufacture of brass die castings is an important achievement. The difficulties of the high temperatures and the rapid oxidation of steel dies have been overcome largely

Figure 6.4
A 2500-ton (22-MN) cold-chamber die-casting machine.

by improvements in die metals and casting at as low a temperature as possible. One machine is designed to use metal in a semiliquid or plastic state to permit operation at lower temperatures than those used for liquid metal. To protect the dies from overheating further, water is circulated through plates adjacent to the dies. Metal is maintained under close temperature control and is ladled by machine or hand to the compression chamber. The pressure used in this machine is 9800 psi (68 MPa); 100 to 200 shots/h can be made depending on the size of the machine.

Two variations of this process, each with the injection plunger in a vertical position, are illustrated in Figure 6.5. In Figure 6.5*B*, the compression chamber into which the

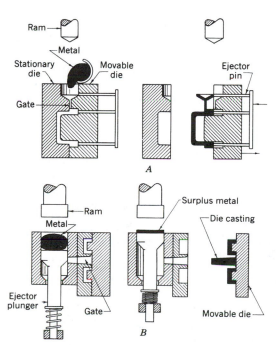

Figure 6.5
Two methods of die construction for pressing brass castings. *A*, Compression chamber in dies. *B*, Compression chamber separate from dies.

metal is ladled is separate from the dies. The metal is poured into this cavity onto a spring-backed plunger. As the ram descends, this plunger is forced down until the gate opening is exposed, forcing the metal into the die cavity. As the ram returns to its upper position, the ejector plunger also moves upward, carrying with it any surplus metal. As the die opens, the casting is ejected.

A variation of this machine with the compression chamber a part of the die is illustrated in Figure 6.5A. Metal is poured into this chamber at the upper part of the die and forced by pressure into the die cavity as the ram descends. As soon as the ram moves up, the dies open and the casting is ejected by ejector pins. The sprue and excess metal are later trimmed off in the finishing operation.

Low-Pressure Permanent-Mold Casting

In the *low-pressure permanent-mold process*, a metal mold is mounted over an induction furnace as shown in Figure 6.6. The furnace is sealed and an inert pressurized gas forces the molten metal in the furnace up through a heated refractory "*stalk*" into the cavity. Vacuum pumps are sometimes utilized to remove entrapped air from the mold and to ensure a more dense structure and faster filling. Small castings may be left to cool in the mold for a minute or less, but castings up to 65 lb (29 kg) in weight are reported to have a cycle time of only 3 min. The process is most economical if the production rate is at least 5000 to 50,000 parts per year. Castings produced by this method are dense, free from inclusions, have good dimensional accuracy, and the scrap loss is usually less than 10% and can be as low as 2%.

Gravity or Permanent-Mold Casting

This method utilizes a *permanent metal* or *graphite mold*. The molds are usually coated with a refractory wash and then lampblack, which reduces the chilling effect on the metal and facilitates the removal of the casting. No pressure is used except that obtained from the head of metal in the mold. The process is used successfully for both ferrous and nonferrous castings, although the latter type does not present as many problems as ferrous castings because of the lower pouring temperatures.

The simplest type of permanent mold hinges at one end of the mold with provision for clamping the halves together at the other. Some production machines, as illustrated

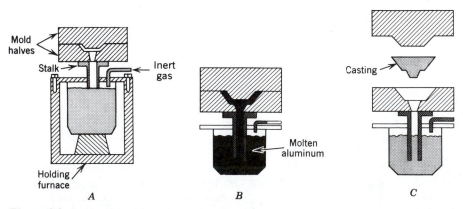

Figure 6.6
Low-pressure permanent mold. *A*, Closed. *B*, Inject. *C*, Open.

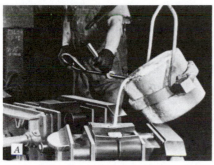

Figure 6.7
Multistation machine for permanent mold casting. *A*, Pouring into metal mold. *B*, Mold opening for removal of casting.

in Figure 6.7, are circular in arrangement and have molds placed at a number of stations. The cycle consists of pouring, cooling, ejecting the casting, blowing out the hot molds, coating them, and in some cases setting the cores. Both metal and dry-sand cores are used. If metal cores are used, they are withdrawn as soon as the metal starts to solidify.

Permanent molds produce castings free from embedded sand and with good finish and surface detail. They are especially adapted to the quantity production of small- and medium-size castings and are capable of maintaining tolerances ranging from 0.0025 to 0.010 in. (0.064 to 0.25 mm). The high initial cost of equipment and the cost of mold maintenance is a disadvantage of this process. This process turns out products such as aluminum pistons, cooking utensils, refrigerator parts, electric irons, and small gear blanks.

Cast iron is being increasingly cast by the gravity method. Often the dies are lined with sand about ¼ in. (6 mm) thick by utilizing a core blower. In this process, only a small amount of sand is used, compared to sand casting, the cleaning costs are lower, and the accuracy is better.

Slush Casting

Slush casting produces hollow casting in metal molds without the use of cores. Molten metal is poured into the mold, which is turned over immediately so that the metal remaining as liquid runs out. A thin-walled casting results, the thickness depending on the chilling effect from the mold and the time of the operation. The casting is removed by opening the halves of the mold. This method is used for ornamental objects, statuettes, toys, and other novelties. The metals used for these objects are lead, zinc, and various low-melting alloys. Parts cast in this fashion are either painted or finished to represent bronze, silver, or other more expensive metals.

Pressed or Corthias Casting

This method of casting resembles both the gravity and the slush processes but differs somewhat in procedure. A definite amount of metal is poured into an open-ended mold, and a close-fitting core is forced into the cavity pressurizing the metal into the mold cavities. The core is removed as soon as the metal sets, leaving a hollow thin-walled casting. This process, developed in France by *Corthias*, is limited to ornamental casting of open design. Nonmetallic molds are used with both high- and low-temperature alloys.

Electroslag Casting

The *electroslag casting process* is unusual because it does not employ a furnace. Instead, consumable electrodes melting or striking beneath a slag layer furnish molten metal to fill a water-cooled permanent mold. The molten metal continually drips or runs into the mold. It does not contact the atmosphere because of the slag layer. Gates or risers are unnecessary, and usually the electrodes withdraw from the mold in concert with its filling from bottom to top. Studies indicate metal cast in this way may be superior to forging. One interesting application comes about when the electrode material is changed in carbon content to effect a varying property in the casting.

Centrifugal Casting

Centrifugal casting is the process of rotating a mold while the metal solidifies, so as to utilize centrifugal force to position the metal in the mold. Greater detail on the surface of the casting is obtained, and the dense metal structure has superior physical properties. Castings of symmetrical shape lend themselves to this method, although other types of castings are produced.

Centrifugal casting is often more economical than other methods. Cores in cylindrical shapes and risers or feed heads are eliminated. The castings have a dense metal structure, with all impurities forced back to the center where they can be machined out. Because of the pressure exerted on the metal, thinner sections are cast than would be possible in static casting.

Although there are limitations on the size and shape of centrifugally cast parts, piston rings weighing a few ounces and paper mill rolls weighing more than 42 tons (38 Mg) have been cast in this manner. Aluminum engine blocks utilize centrifugally cast-iron liners. If a metal can be melted, it can be cast centrifugally, but for a few alloys the heavier elements tend to be separated from the base metal. This separation is known as *gravity segregation*.

The methods of centrifugal casting may be classified as follows: (1) true centrifugal casting, (2) semicentrifugal casting, and (3) centrifuging.

True Centrifugal Casting. True centrifugal casting is used for pipe, liners, and symmetrical objects that are cast by rotating the mold about its horizontal or vertical axis. The metal is held against the wall of the mold by centrifugal force, and core is not required to form a cylindrical cavity on the inside. There are two types of horizontal-axis molds used for producing cast-iron pipe. Massive, thick metal molds with a thick refractory coating allow the molten metal to solidify faster and the solidification proceeds from the wall of the mold toward the inside of the cast pipe. Such a mold encourages a preferred solidification that ensures a more solid casting with any impurities on the inside wall. Figure 6.8 illustrates such a casting machine. The mold is spinning at the time the molten metal is introduced, and the spinning action is not stopped until solidification is complete. The wall thickness of the pipe is controlled by the amount of metal poured into the mold.

Another type of *horizontal centrifugal casting* uses a thick, highly insulating sand interface between the mold and the casting. Such a sand lining is spun into the mold. When metal is introduced, the insulating nature of the sand prevents directional solidification, and, hence, the metal solidifies from the wall and from the inside pipe face at the same time. This can cause a spongy, less dense midsection that has entrapped inclusions.

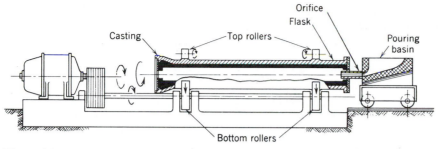

Figure 6.8
Centrifugal casting machine for casting steel or cast-iron pipe.

Another example of true centrifugal casting is shown in Figure 6.9, which illustrates two methods that cast radial engine cylinder barrels. The horizontal method of casting is similar to the process followed in casting pipe lengths, and the inside diameter is a true cylinder requiring a minimum amount of machining. In vertical castings, the inside cavity forms a paraboloid as illustrated by Figure 6.9. The slope of the sides of the paraboloid depends on the speed of rotation, the dotted lines at *A* representing a higher rotational speed than shown by the paraboloid *B*. To reduce the inside-diameter differences between the top and the bottom of the cylinder, spinning speeds are higher for vertical than for horizontal casting.

If the centrifugal force is too low, the metal will slip, slide, or rain. If the force is too great, the surface will develop detrimental abnormalities. Most horizontal castings are spun so that a force of about 65 *g*'s or 65 × (force of gravity) is developed. Vertically cast parts are usually spun at 90 to 100 *g*'s. Vertical castings are smaller in size and weight because of the instability of a spinning vertical cylinder, the higher *g* force necessary to

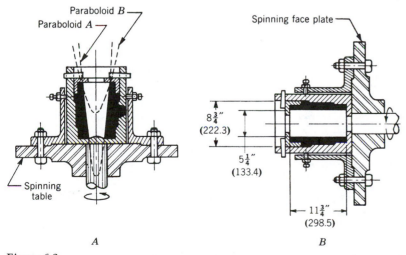

Figure 6.9
True centrifugal method of casting radial engine cylinder barrels. *A*,
Vertical. *B*, Horizontal.

overcome the paraboloidal shape and the increased pressures on the mold. The revolutions per minute necessary to produce a given number of g's are independent of the density of the metal or the total weight being cast. The centrifugal force, CF, is expressed as

$$CF = \frac{mv^2}{r} \tag{6.1}$$

where

$$
\begin{aligned}
CF &= \text{Centrifugal force, lb} \\
m &= \text{Mass} = \frac{W}{g} = \frac{\text{weight (lb)}}{\text{acceleration of gravity (ft/s)}^2} \\
&= \frac{W}{32.2} \\
v &= \text{Velocity, ft/s} = r \times w \\
r &= \text{Radius, ft} = \tfrac{1}{2}D \\
w &= \text{Angular velocity, rad/s} \\
&= 2\pi/60 \times \text{rpm} \\
D &= \text{Inside diameter, ft}
\end{aligned}
\tag{6.2}
$$

The number of g's is

$$g\text{'s} = CF/W$$

Hence

$$g\text{'s} = \frac{1}{W}\left[\frac{W}{32.2 \times r}\left(\frac{r \times 2\pi}{60}\right)^2\right] \tag{6.3}$$

$$= r \times 3.41 \times 10^{-4}\,\text{rpm}^2$$

$$= 1.7 \times 10^{-4} \times D \times (\text{rpm})^2 \tag{6.4}$$

The spinning speed for horizontal-axis molds may be found in English units from the equation

$$N = \sqrt{(\text{Number of } g\text{'s}) \times \frac{70{,}500}{D}} \tag{6.5}$$

where

$$
\begin{aligned}
N &= \text{rpm} \\
D &= \text{Inside diameter of mold, ft}
\end{aligned}
$$

Semicentrifugal Casting. In semicentrifugal casting the mold is completely full of metal as it is spun about its vertical axis, and risers and cores may be employed. The center of the casting is usually solid, but because the pressure is less there, the structure is not as dense, and inclusions and entrapped air are often present. This method is normally used for parts in which the center of the casting will be removed by machining. The *stack mold* shown in Figure 6.10 can produce five semicentrifugally cast track wheels. The number of castings made in a mold depends on the size of the casting and the convenience in handling and assembling the molds. Rotational speeds for this form of centrifugal casting are not as great as for the true centrifugal process. The process produces a dense structure at the outer circumference, whereas the center metal is usually removed.

Centrifuging. In the *centrifuge method* several casting cavities are located around the outer portion of a mold, and metal is fed to these cavities by radial gates from the center. Either single or stack molds can be used. The mold cavities are filled under pressure

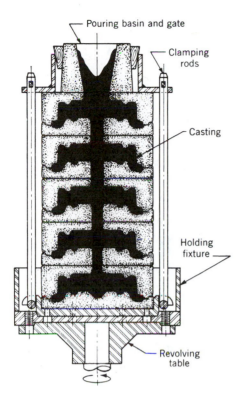

Figure 6.10
Semicentrifugal stack molding of track wheels.

from the centrifugal force of the metal as the mold is rotated. Figure 6.11 shows five castings made in one mold by this process. The internal cavities are irregular in shape and are formed by dry-sand cores. The centrifuge method, not limited to symmetrical objects, produces castings of irregular shape such as bearing caps or small brackets. The dental profession uses this process for casting gold inlays.

6.2 PRECISION OR INVESTMENT CASTING

Precision or *investment casting* employs techniques that enable very smooth, highly accurate castings to be made from both ferrous and nonferrous alloys. Figure 6.12 shows a small investment casting made from a chrome–molybdenum–steel alloy. No other casting method other than die casting can ensure production of so intricate a part. The process is useful in casting unmachinable alloys and radioactive metals. There are a number of

Figure 6.11
Centrifuged castings with internal cavities of irregular shape.

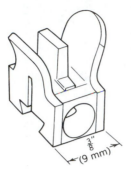

Figure 6.12
Investment casting of a front rifle sight.

processes employed, but all incorporate a sand, ceramic, plaster, or plastic shell made from an accurate pattern into which metal is poured. Although most castings are small, the investment process produces castings weighing more than 100 lb (45 kg).

The advantages of precision or investment techniques are as follows:

1. Intricate forms with undercuts can be cast.
2. A very smooth surface is obtained without a parting line.
3. Dimensional accuracy is good.
4. Certain unmachinable parts can be cast to preplanned shape.
5. It may be used to replace die casting where short runs are involved.

The investment process is expensive. It is usually limited to small castings and presents some difficulties when cores are involved. Holes cannot be smaller than 1/16 in. (1.6 mm) and should be no deeper than about 1.5 times the diameter.

"Lost Wax" Precision Casting Process

The *lost wax* process derives its name from the *wax pattern* used in the process, which is subsequently melted from the mold, leaving a cavity having all the details of the original pattern. The process, as originally practiced by artisans in the sixteenth century, consisted of forming the object in wax by hand. The wax object or pattern is then covered by a plaster investment. When this plaster becomes hard, the mold is heated in an oven, melting the wax and at the same time further drying and hardening the mold. The remaining cavity, having all the intricate details of the original wax form, is then filled with metal. Upon cooling the plaster investment is broken away, leaving the casting.

In large casting such as statuary, plaster cores were used to provide relatively thin walls in the casting. Benvenuto Cellini's famous chapter on the casting of his bronze statue, Perseue, contains information on methods of molding used during the Renaissance. Cellini used a form of the "lost wax" process.

Current practice requires that a replica of the part be made from steel or brass. From this replica a bismuth or lead alloy split mold is made. After wax is poured into the mold and solidification takes place, the mold is opened and the wax pattern removed. In the forming operation the mold is held in a water-cooled vise and the heated wax is injected into it under pressure. Thermoplastic polystyrene resin is sometimes used in place of wax.

Several patterns are usually assembled with necessary gates and risers by heating the contact surfaces (wax welding) with a hot wire. This cluster is then supported in a metal flask. A finely ground refractory material, thinned by a mixing agent such as alcohol or water, is poured into the flask after first spraying the pattern with a fine silica-flour mixture. After the plaster sets, the mold is placed upside down and heated in an oven for several hours to melt out the wax and to dry the mold. The casting is produced by gravity, vacuum, pressure, or centrifugal casting. Pressures ranging from 3 to 30 psi (0.02 to 0.2 MPa) are generally used in the casting operation. After the mold cools, the plaster is broken away. After gates and feeders are cut off, the castings are cleaned by grinding, sandblasting, or other finishing operations.

Ceramic Shell Process

This process, similar to the "lost wax process," involves the removal of a heat-disposable pattern from a refractory investment. The pattern is made from wax or a low-melting-point plastic. Often a number of them are joined by "wax welding" into a cluster, as shown in Figure 6.13. The cost of producing plastic patterns is less than that for wax. A handle is usually mounted in one end of the wax sprue that forms the pouring basin. The pattern cluster is repeatedly dipped into a ceramic slurry and dusted with refractory material. This process, called stuccoing, is repeated until the shell is ³⁄₁₆ to ½ in. (4.8 to 12.7 mm) thick. The pattern is then melted out of the mold, which is first dried and then fired at 1800° to 2000°F (980° to 1095°C) to remove all moisture and organic material. The mold, free of any parting lines, is usually poured immediately after it is removed from the furnace. The shell breaks away from the casting as cooling takes place. Good accuracy and surface finish are obtained with both ferrous and nonferrous metals. Tolerances of ± 0.005 in. (± 0.13 mm) are common, and as-cast tolerances can be improved by coining or sizing, but the cost is increased.

Frozen mercury is sometimes used in the place of wax or plastic patterns. A metal mold or die is made of the part to be cast with the necessary gates and sprue hole. When assembled and ready for pouring, it is partially immersed in a cold bath and filled with acetone, which acts as a lubricant. As the mercury is poured into the mold, the acetone is displaced. Freezing takes place in a liquid bath held at around −76°F (−60°C) and is completed in 10 min.

The patterns are removed from the mold and invested in a cold ceramic slurry by

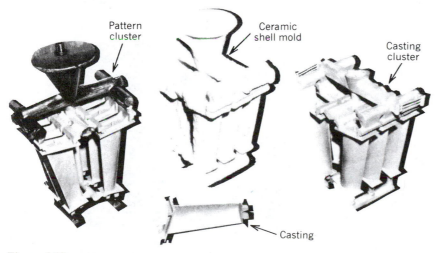

Figure 6.13
Ceramic shell process.

repeated dippings until a shell about ⅛ in. (3.1 mm) thick is built up. Mercury is melted and removed from the shell at room temperature and, after a short drying period, is fired at a high temperature, resulting in a hard permeable form. The shell is placed in a flask, surrounded by sand, preheated, and filled with metal. Casting is usually done by the centrifugal method. Although the process yields castings of high accuracy, it is limited because of high production costs and the hazards associated with mercury.

Plaster Mold Casting

The gypsum-based plasters used as a casting investment dry quickly with good porosity but are not permanent, because they are destroyed when the casting is removed.

Patterns are made of a free-machining brass and are held to a close tolerance. They are assembled on bottom plates of standard-size flasks, as shown in Figure 6.14*A*. Before receiving the plaster, they are sprayed with a parting compound. The plaster, which is made of gypsum with added strengtheners and setting agents, is dry-mixed and water is added. The plaster is then poured over the patterns and the mold is vibrated slightly to ensure that the plaster fills all small cavities. The plaster sets in a few minutes and is removed from the flask by a vacuum head. All moisture is driven from the molds after baking in an oven conveyor at a temperature around 1500°F (815°C). These molds are shown in Figure 6.14*B* as they are emerging from the drying oven. After pouring, the castings are removed by breaking up the mold. Any surplus plaster is removed by a washing operation.

Mold porosity for the removal of any gases developed in the mold is controlled by the water content of the plaster. When the mold is dried, the water driven out leaves numerous fine passageways that act as vents. In addition to having adequate porosity, plaster molds have the structural strength for casting plus enough elasticity to allow some contraction of the metal during cooling.

Plaster molds are suitable for nonferrous alloys only. The wide variety of small castings made by this process includes miscellaneous airplane parts, small gears, cams, handles, pump parts, small housings, and numerous other intricate castings.

One of the principal advantages of plaster mold casting is the resulting high degree of dimensional accuracy. This, coupled with the smooth surface obtained, enables the process to compete with sand castings requiring machining to obtain a smooth surface.

 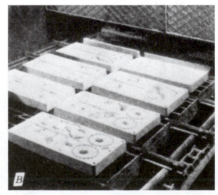

Figure 6.14
A, Assembling metal patterns in flask for plaster mold. *B*, Finished plaster molds emerging from drying oven.

Figure 6.15
Shapes of vacuum-cast, plaster-cored aluminum impellers.

Because of the low thermal conductivity of plaster, the metal does not chill rapidly and very thin sections are cast. There is little tendency toward internal porosity in plaster mold castings, and there is no difficulty with sand or other inclusions. In general, the process competes more successfully with die casting using high-temperature alloys such as brass rather than metals such as zinc and aluminum. At high temperatures metal molds have a relatively short life; with plaster molds, which are used only once, the temperature is not a problem.

A tolerance of $\pm$ 0.005 in. ($\pm$ 0.13 mm) is maintained for simple castings; slightly more is required if the dimension crosses the parting line. The process is used for both small and quantity production runs.

The complex aluminum impellers shown in Figure 6.15 may be cast by vacuum pouring into a cast-iron mold containing a plaster core. Vacuum melting and pouring techniques along with a moistureless mold and a low pouring temperature yield high strength and good surface finish.

The two principal variations of this process are (1) foam plaster molds and (2) the Antioch process. In the foam process a foaming agent is added to the water, then the plaster is added and the materials mixed until the desired slurry and density is obtained. After oven-drying, the more permeable molds are poured. Material is saved in this process and permeability is increased.

In the Antioch process a typical plaster mold is steam-autoclaved for about 10 h to cause the plaster to granulate and become more permeable. Then the mold is dipped in water and allowed to granulate further at room temperature, then oven-dried. Even though the mold granulates internally, the surface is only moderately roughened.

Shell Molding Process

The mold is made up of a mixture of *dried silica sand* and *phenolic resin* formed into thin, *half-mold shells* that are clamped together for pouring, as illustrated by the series of sketches in Figure 6.16. The sand, free from clay, is first mixed with either urea or phenol formaldehyde resin. The mix is put into a *dump box* or blowing machine. A metal pattern must be used, because it is preheated to a temperature around 450°F (230°C) and sprayed with a silicon release agent before being placed on top of the dump box. The box is then inverted, causing the sand mix to drop on the pattern, and is held there for 15 to 30 s before it is returned to its original position. The pattern, with a thin shell of sand ⅛ to ³⁄₁₆ in. (3.1 to 4.7 mm) thick adhering to it, is then placed in an oven and the shell cured ½ to 1 min. until it is rigid. Finally the shell is removed from the pattern by ejector pins and the mold halves assembled with clamps, resin adhesives, or other devices. They are placed in a flask supported against one another or by some backing

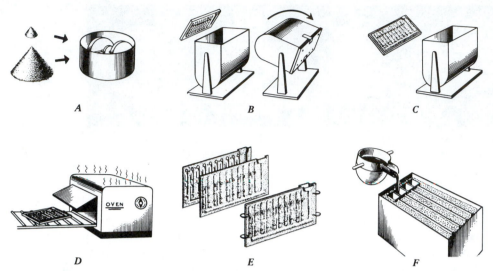

Figure 6.16
Diagram of shell molding process. *A*, Mulling sand and resin. *B*, Resin sand mixture applied to pattern. *C*, Excess resin sand mixture falls back into dump box. *D*, Curing shells on pattern. *E*, Mold halves are aligned and joined. *F*, Molds supported and poured.

material such as shot or gravel. Some are poured while they are resting flat on the floor with a weight on top.

The advantages of the shell molding process include fine tolerances of 0.002 to 0.005 in./in. (0.2% to 0.5%), low cleaning costs, and smooth surfaces. Little molding skill is necessary and the sand requirements are low. Shell molding is readily adapted to automation. The disadvantages are that the process requires metal patterns and fairly expensive equipment for making and heating the molds.

CO$_2$ Mold Hardening Process

The process of hardening molds and cores using CO$_2$ and a sodium silicate liquid base binder is widely used. Because of inherent advantages and rapid sand hardening, this process is used in many foundries. The process consists of thoroughly mixing clean, dry silica or other dry conventional sand (fineness number around 75) with 3.5% to 5% sodium silicate liquid base binder in a muller. It is then ready for use and is packed in flasks and core boxes by standard molding machines, by core blowers, or by hand. The sand should be free from moisture and clay, but other ingredients such as coal dust, pitch, graphite, or wood flour may be added to improve certain properties such as collapsibility.

When the packing is complete, CO$_2$ is blown into the mold or core at a pressure of around 20 psi (0.1 MPa). The reaction is complex but is usually represented by the following chemical equation.

$$Na_2SiO_3 + CO_2 = SiO_2(aq) + Na_2CO_3$$

The *silica gel* that is formed hardens and acts as a cement to bond the sand grains together. The method of introducing the gas, important to the success of the process, must

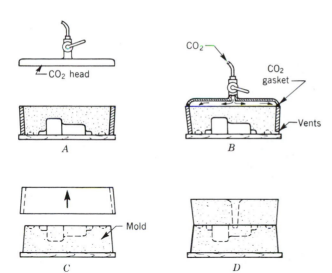

Figure 6.17
Diagram of CO_2 mold hardening process. *A*, Rammed mold. *B*, Hardening mold with CO_2. *C*, Mold jacket removed. *D*, Assembled mold.

be simple, rapid, and uniform throughout the sand body, and should not be cumbersome to apply. The time to harden a small or medium-size body of sand is 15 to 30 s. For small cores a gasketed, funnel-shaped head is placed over the core box. Larger molds may be hardened by placing a hood over the mold, by running small tubes into the mold as illustrated in Figure 6.17, or by introducing the gas into a hollow, vented pattern. The figure shows the steps in preparing a mold. In some cases the sprue is at the ends of the cope and drag, and a number of molds may be book-staked between end braces for pouring.

A CO_2 mold or core can be made quickly, and baking is not required. Semiskilled help can be used. The surface finish of the casting is good, and the same sand is used for both cores and molds.

Some sodium silicate-bonding agents permit self-setting action without CO_2 and permit shear strengths of 90 psi (0.6 MPa) and compression strengths of 300 psi (2 MPa). Seacoal, coal in the amount of 2%, is the most used additive to improve shakeout, sand removal from the casting, and to reduce pollution. Special shakeout sands having carbonaceous material also are used. Most silicate-bonded sands can be shaken out relatively easily if organic esters are used in the mix. Sugar up to 12% is added as a breakdown agent so the core will deteriorate more fully after solidification of the mold. The sand is reused if about 30% new sand is added, although the economics of reclamation are questioned.

The molds have a relatively short storage life, and sometimes poor collapsibility causes problems. The process is used in both the ferrous and nonferrous casting industry.

Molds of Other Materials

Various materials such as rubber, paper, and wood can be used for molds for low-melting-temperature metals. Costume jewelry and similar small items are cast successfully in rubber molds. An alloy of 98% tin, 1% copper, and 1% antimony is frequently used in this type of work.

Figure 6.18 illustrates a mold made of a Dow Corning silicone rubber product known as Silastic. This material is used to cast wax patterns, plastics, or low-melting-point alloys. The molds withstand 500°F (260°C) and reproduce as fine a detail as found on a high-

Figure 6.18
Silastic RTV silicone rubber mold.

fidelity record. The material is so flexible that it can be removed from intricate shapes without difficulty.

The Shaw process uses a mixture of sand, hydrolyzed ethyl silicate, and other ingredients that permit the investment mold to be "peeled" from the pattern. The pattern need not be wax or mercury, because in the "as-poured" state the mold material is rubberlike. Once removed from the pattern, the mold is ignited and later baked to provide a rigid,

Table 6.1 **Characteristics of Casting Processes**

Type Process	Surface Finish (μin.)	Dimensional Accuracy for Small Casting[a]	Dimensional Accuracy for Large Casting[b]	Intricacy[c]	Tooling Cost[d]	Cost per Unit[d]
High-pressure die casting	30	1	1	3	1	8
Low-pressure die casting	50	4	5	5	2	7
Plaster casting	32	3	2	2	7	1–5
Investment casting	60	2	4	1	6	4
Gravity casting	70	5	7	7	3	6
Shell	125	6	3	4	5	3
Sand	500	7	6	6	4	2

[a] Order, 1 being best.

[b] Order, 1 being able to produce most accurate.

[c] Order, 1 the most intricacy.

[d] Order, 1 being lowest.

permeable, high-quality surface finish mold. The Shaw process, adaptable to complex shapes and reusable patterns, is adapted to automatic operation but is relatively time consuming and costly except for certain castings.

Characteristics of Molding Processes

Various methods of casting have been discussed in this chapter. Characteristics have been described but they have not been compared to one another. Table 6.1 provides comparisons among the processes.

6.3 CONTINUOUS CASTING

There are many opportunities for cost economies in the *continuous casting* of metals. In addition, metals starting as continuous castings have a degree of soundness and uniformity not possessed by ingots, bars, and billets. Briefly, the process consists of continuously pouring molten metal into a mold that has the capability for rapidly chilling the metal to the point of solidification, and then withdrawing it from the mold. The following processes are typical.

Reciprocating Mold Process

A reciprocating, water-cooled copper mold is used, the down stroke being synchronized with the discharge rate of the slab. Molten metal is poured into the holding furnace and is discharged to the mold after being metered through a ⅞-in. (22-mm) orifice at the needle valve. The down spout tube is 1⅛ in. (28 mm) in diameter and delivers metal at the rate of 30,000 lb/h (225 kg/s).

The molten metal is distributed across the mold from a submerged horizontal cross-piece. The level of the metal is held constant at all times. The pouring rate of the molten metal is controlled by a needle valve through the top of the holding furnace. As the metal becomes chilled in the lower part of the mold, it discharges at a constant rate and enters the withdrawing rolls. These are synchronized with the downward movement of the mold and are mounted just above a circular saw that cuts the slab to required lengths. Brass slabs are further processed by cold rolling into sheets and strips. Large quantities of 7- to 10-in. (178- to 254-mm) round billets for hot-extrusion processes are also produced in this manner.

Asarco Process

This process (Figure 6.19) differs from other continuous processes in that the forming die or mold is integral with the furnace, and there is no problem of controlling the flow of metal. The metal is fed by gravity into the mold from the furnace as it is solidified continuously and withdrawn by the rolls below. An important feature of this process is the water-cooled graphite forming die, which is self-lubricating, is resistant to thermal shock, and is not attacked by copper-base alloys. The upper end in molten metal acts as a riser and compensates for any shrinkage that might take place during solidification, while simultaneously acting as an effective path for the dissipation of evolved gases. These dies are easily machined to shape, and products may be produced ranging from ⁷⁄₁₆ to 9 in. (11 to 229 mm) in diameter. Multiple production from a single die permits casting small, cross-section rods.

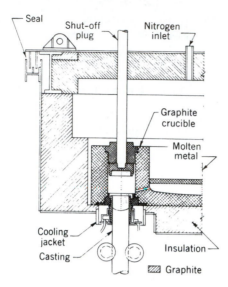

Figure 6.19
Process for continuous-cast shapes.

In starting the process a rod of the same shape as that to be cast is placed between the drawing rolls and inserted into the die. This rod is tipped with a short length of the alloy to be cast. As the molten metal enters the die, it melts the end surface of the rod, forming a joint. The casting cycle is then started by the drawing rolls, and the molten metal is solidified continuously as it is chilled and withdrawn from the die. When the casting leaves the furnace, it ultimately reaches the sawing floor where it is cut to desired length while still in motion. A tilting receiver takes the work and drops it to a horizontal position, and from there it goes to inspecting and straightening operations.

The process has proved successful for phosphorized copper and many of the standard bronzes. The alloy composition may be produced with satisfactory commercial finish as rounds, tubes, squares, or special shapes. Physical properties are superior to those obtained from permanent-mold and sand casting.

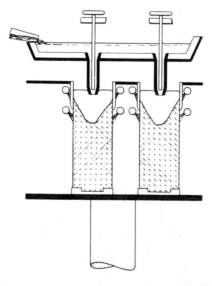

Figure 6.20
Casting aluminum ingots by the direct-chill process.

Direct-Chill Process

In the direct-chill process aluminum and aluminum alloy ingots are continuously cast by forming a shell in a vertical, stationary, water-cooled mold. Solidification is completed by direct water application beneath the mold. The mold is closed at the start by a block on an elevator or by a dummy ingot. Molten metal is fed from a furnace through troughs and spouts with flow regulated manually or automatically by float control to coincide with the casting rate controlled by the elevator or driven rolls. The process is shown in Figure 6.20. Cross sections up to 1500 in.^{2} (0.97 m^2) are produced. Lengths 100 to 150 in. (2.54 to 3.81 m) long and limited by the stroke are cast where an elevator is employed, or ingots are sawed to finish length where rolls are used. Surface quality is adequate as cast for certain alloys and products, or it may require scalping for other alloys or more critical applications.

QUESTIONS

1. Give an explanation of the following terms:

Hot chamber

Flash

Ejector rod

"Stalk"

Corthias

Stack mold

Investment casting

Lost wax

Dump box

Continuous casting

2. Describe the effect of melting temperature on die life.

3. How is pressure applied to a cold-chamber die-cast part?

4. What special casting processes are used to produce a cast-iron dumbbell?

5. Write a paragraph on the characteristics of zinc. (Refer to earlier chapters.)

6. Give the principal reasons that account for 75% of die castings being made from zinc.

7. Compare the characteristics of a small zinc casting made by the processes listed in Table 6.1.

8. List the metals and alloys normally cast in a hot-chamber machine.

9. Why is an inert gas used in low-pressure, permanent-mold castings?

10. List the metals and alloys normally cast in a cold-chamber machine.

11. What is done with gates and risers that are trimmed from die castings?

12. What is the difference between a multiple cavity die and a combination die?

13. Why are ejector pins or rods required in a die casting? Do they mar the surface in any way? Why?

14. Describe three products that are made by the slush casting process.

15. Why are brass die castings difficult and expensive to produce?

16. How is air removed from a die so that that metal may completely fill it?

17. What is the function of the "stalk" in a low-pressure permanent mold? What is done with sprues and gates in a low-pressure permanent molding?

18. Explain the difficulties of making large castings with the vertical true centrifugal casting method.

19. Compare the advantages between gravity-type permanent molds and die casting. How are molds protected in making a gravity-type permanent-mold casting?

20. Would alloys or pure metals be most adaptable to slush casting? Why? (Refer to earlier chapters.)

21. Describe the steps to make a bronze bust from a clay carving.

22. Why are ferrous metals unsuitable for plaster molds?

23. Rank the horizontal true centrifugal casting process. (Refer to Table 6.1.)

24. Rank the vertical true centrifugal casting process. (Refer to Table 6.1.)

25. Describe the manufacture of a hollow brass decorative head for a walking cane using the lost wax process.

26. How are small wax patterns attached to the gates in a multiple-cavity investment pattern?

27. What is meant by "vacuum pouring"? Would the use of an inert gas instead of a vacuum do the same thing?

28. Using the principles of shrinkage examined in Chapter 5, discuss continuous casting as influenced by shrinkage.

29. Comparing the shell mold and CO_2 processes, which is more adaptable to mass production? Why?

30. What special expenses are incurred in a shell molding operation? Give reasons why it cannot be employed in job-lot or small production runs.

31. What process is used for the following: small zinc castings, statuettes, aluminum pistons, aluminum ingot, small brass gears, carburetors, automobile door handles, automobile grills, and aluminum alloy ingots?

32. Describe the method of making a Silastic mold from a decorative brass duck.

PROBLEMS

6.1. Sketch a die that is used for casting a small dumbbell out of aluminum and label the principal parts. (Hints: A sketch is not a mechanical drawing made with instruments or a CAD drawing. Use engineering computation paper and a 0.5-mm pencil. Be neat.)

6.2. Sketch a design for a small vase that is cast from lead by the slush casting process.

6.3. A small aircraft cylinder liner 9-in. OD and 7-in. ID is spun in a vertical true centrifugal mold. How many revolutions per minute (rpm) are necessary? The same casting described is spun horizontally. How many rpm's are necessary? If too low a speed is used, what difficulties can be encountered? Describe the mold and coatings for these processes.

6.4. Describe the methods for manufacturing 24-in. diameter, 20-ft lengths of cast-iron pipe. How many pounds of cast iron are required? (Hint: Density = 450 lb/ft^3.)

6.5. Estimate the time to make a shell mold half. Write down these process cycle times for each step and multiply your total time by 2.5 to allow for breakdowns, bad molds, and coffee breaks. (Hint: Make telephone inquiries and ask others about this process.)

6.6. If 35% of a die casting is trimmed away and remelted with a 5% loss during melting, how many metric tonnes of metal are required per day in a plant making and shipping 3500 nutcrackers per day that weigh 200 g each? Discuss the advantages and disadvantages of the various ways of casting the nutcracker. (Hint: Refer to Table 6.1.) Estimate total monthly costs involved in producing the 3500 nutcrackers.

6.7. If a 48 in. diameter metal pipe 10 ft long is centrifugally cast with a force of 60 g's, what rotational speed in revolutions per minute should be used? What is the effect of length of pipe on the rotational speed? Why?

6.8. Plot a curve of revolutions per minute versus inside diameter for a centrifugal process. Use diameters of 1 in. to 20 in.

PRACTICAL APPLICATION

Invite a specialist to speak to the class about one of the topics described in this chapter. Consult with the instructor about protocol and the class schedule, but select one member of the class to make arrangements on the details and topic. (Hints: There are professional groups that can give assistance, such as the American Foundrymen's Association or the Society of Manufacturing Engineers. Look in the telephone book for the classification of casting industries in your area. Businesses are usually honored by the opportunity to speak to a class.)

CASE STUDY: WATER PIPE

The LaGrande Company in South America has been casting 15 ft (4.6 m) long, 8 in. (203 mm) in diameter gray cast iron pipe in sand molds for 75 years using conventional techniques employing a long core supported by chaplets. As a consultant, you are asked to compare the cost with a proposed centrifugal casting process. The Bolivar is the monetary unit and it is worth U.S. $0.133. The iron costs 1.2 Bolivars/kg, and labor is paid, on the average, 10.7 Bolivars/h. Advise LaGrande of the type of centrifugal process to use and the revolutions per minute it must maintain during pouring and solidification. Can you make an estimate of the savings that would benefit LaGrande if they wish to produce 1000 units/month? What about the space requirements for the two types of operations and any logistic problems? (Hints: Tabulate your cost estimates. Prepare your report on a computer.)

CHAPTER 7

BASIC MACHINE TOOL ELEMENTS

The study of manufacturing processes must pay attention to machine tools. While manufacturing processes are clearly more than *traditional machine tools*, their importance to the broad field of manufacturing processes and systems is critical. This chapter concentrates on basic machines, recalling that their role is seen as a fundamental building block to manufacturing systems.

Most machine tools are constructed by using two or more elements, such as a base and *headstock*. These elements, although they have different functions in the lathe, mill, or drill press, have common characteristics.

Important requirements for machine-tool structures include *rigidity, shape, operator and part accessibility, ease of chip removal,* and *safety.* In terms of machine tool performance, static and dynamic stiffness is necessary for accuracy and precision. Stability of the machine structure is required to prevent machining chatter.

The purpose of this chapter is understanding basic machine tool elements necessary to appreciate the breadth of modern machining methods in the manufacture of products.

7.1 MACHINE TOOLS

Machine tools differ not only in the number of cutting edges they employ, but also in the way the tool and workpiece are moved in relation to each other. Refer to Figure 7.1 for metal-cutting processes as found with *traditional machine tools.*

In vertical machining centers, drill presses, boring machines, milling machines, shapers, and grinders, the workpiece remains virtually motionless and the tool moves. In others (planers, lathes, and boring mills) the tool is virtually fixed and the workpiece moves. Seldom are these simple elements applied without modification. The word *virtually* suggests a small but very important movement, and that will become clearer later. The words *rotation, feed, reciprocating, revolving, stationary, translational,* and *curvilinear* are important descriptions. These descriptions have meaning in the context of specific machining operations.

The single-point tool-shaping machines are the easiest to visualize. The lathe and the boring machine are *kinematic inversions* of each other. They both employ the *single-point tool.* In Figure 7.2A, the work rotates in the lathe, but the cutting tool is nearly stationary. Lathe operations do turning, the tool moves in a carriage, and the longitudinal movement is called feed, where the velocity of the feed is relatively slow and is measured in inches per revolution (*ipr*). In the boring machine, Figure 7.2B, the tool rotates while the work is stationary. Rotation is about a fixed center. *Translational* means straight-line movement, while curvilinear means motion not along a straight line and not circular rotation. Although the lathe tool and the boring machine worktable are not truly stationary, this is overlooked for the moment. To feed a tool carriage past rotating work is

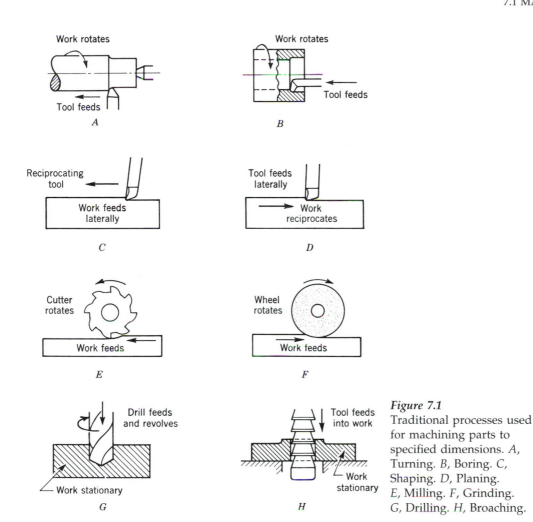

Figure 7.1
Traditional processes used for machining parts to specified dimensions. *A*, Turning. *B*, Boring. *C*, Shaping. *D*, Planing. *E*, Milling. *F*, Grinding. *G*, Drilling. *H*, Broaching.

usually easier than to feed rotating work with headstock and supports past a stationary tool post.

The *shaper* and *planer* use single-point cutting tools. Figures *7.2C* and *D* point out that size of the workpiece is a factor in machine structural design. A smaller workpiece is more efficiently machined on the shaper than on the planer. The general appearance of the machine is changed by reversing the kinematic relationship of work and tool. However, the cutting action principle is identical. While shaping and planing are *seldom* used in production, it is important for the student to know their characteristics, as these tool movements are found in prominent and current equipment.

Before the introduction of the milling cutter by Eli Whitney in the early 1800s, the rotating tool was used only as a boring tool. However, Whitney gave it a new application. The milling cutter was no longer used for cutting circular bores exclusively, but was used for cutting keyways, slitting, sawing, slab and face milling, gear cutting, and shaping irregularly formed pieces. The rotating tool combined with *traversing* work was introduced in the milling machine as shown in Figure 7.2*E* and Figure 7.2*F*. The kinematic inversion of the standard milling machine is the floor-type horizontal boring, drilling, and milling machine illustrated in Figure 7.2*G*.

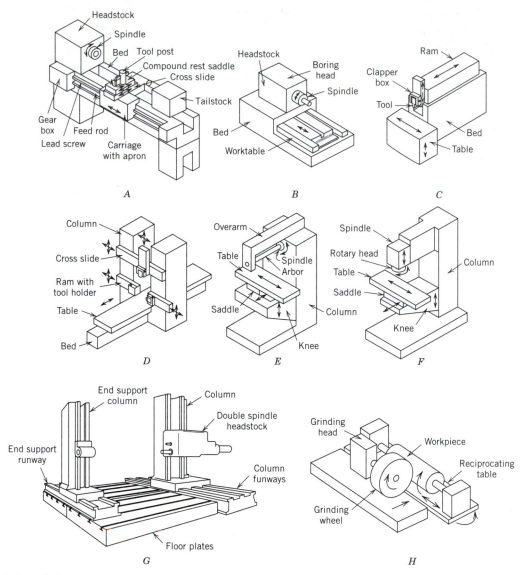

Figure 7.2
Basic structural elements in conventional machine tools. *A*, Lathe. *B*, Horizontal boring machine. *C*, Shaper. *D*, Planer. *E*, Horizontal milling machine. *F*, Vertical milling machine. *G*, Horizontal boring machine. *H*, Grinding machine.

The *cylindrical grinder* adopts motions of the lathe and boring machine except for the substitution of rotating tools (the grinding wheel) for single-point tools. The work and cutting tool rotate in the grinder as shown in Figure 7.2*H*. The characteristics of these basic cutting machines are tabulated in Table 7.1. The student should note the heading of the columns, pay attention to the description, and relate the table comments to Figures 7.1 and 7.2. These machine tools are discussed further in later chapters.

Table 7.1 **Cutting and Feed Movement for Conventional Machines**

Machine	Cutting Movement	Feed Movement	Types of Operation
Lathe	Workpiece rotates	Tool and carriage	Cylindrical surfaces, drilling, boring, reaming, and facing
Boring machine	Tool rotates	Table	Drilling, boring, reaming, and facing
Planer	Table traverses	Tool	Flat surfaces (planing)
Shaper	Tool traverses	Table	Flat surfaces (shaping)
Horizontal milling machine	Tool rotates	Table	Flat surfaces gears, cams, drilling, boring, reaming, and facing
Horizontal boring	Tool rotates	Tool traverses	Flat surfaces
Cylindrical grinder	Tool (grinding wheel) rotates	Table and/or tool	Cylindrical surfaces (grinding)
Drill press	Tool rotates	Tool	Drilling, boring, facing, and threading
Saw	Tool	Tool and/or work- piece	Cut off
Broaching	Tool	Tool	External and internal surfaces

7.2 ELEMENTS

Metal-cutting machines are composed of separate elements, each having a special function. The elements consist of machine frame, headstock, column, table, carriage or saddle, bed, base or runways, and cross or slide rails.

Machines identify with combinations of these elements. For example, there are four distinct types of horizontal boring, drilling, and milling machines: table, floor, planer, and multiple head. The table machine has a table and a saddle, and the workpiece is placed on the table. The floor machine combines floor plate sections and runways. The planer machine derives its name from a reciprocating table. The multiple-head machine incorporates additional headstocks, a cross rail, and column supports. These basic units can be employed in a number of ways to machine a part.

An important element of the machine tool is the *machine frame*, which is either cast from gray cast iron or steel, or manufactured by welding steel plate. The frame is usually cast if it is too intricate for welding fabrication, or if the weight of the frame is relatively important. Gray cast iron is preferred for most moderate-size machines in which vibration forces are a problem, because cast iron has a high damping capacity. Large, heavy-duty frames that must withstand impact loads are often made of cast steel.

Welded or fabricated frames as compared to cast frames have the following advantages.

1. Saving in weight may amount to 25%.
2. Repairs are relatively easy for damaged frames.
3. Various grades of steel may be used in the same frame depending on the design requirements of a machine member.
4. Design changes are less costly because there is no investment in patterns or cores.
5. Errors in machining or design are easier to correct.

6. Additional material can be located near the stress zones to control vibration and deflection.

The disadvantages of the fabricated frame are as follows.

1. Gray cast iron absorbs vibrations better than welded steel frames.
2. Cast material is homogeneous and, hence, chemical reactions are negligible.
3. Casting process may be more adaptable to high production rates.
4. Frames for heavy-duty machines may need considerable weight to absorb the loads.

The *headstock* drives and feeds the cutting tool or rotates the part. Figure 7.3 is a cutaway view of a headstock where complicated gearing is shown. This headstock is used on boring, drilling, and milling machines. As an example of the performance of the machine tool, the spindle speed is infinitely variable from 25 to 1120 rpm, and it is driven by a 25 to 40 hp (18 to 30 kW) DC drive.

Spindle *rotation* may be reversed to accommodate threading and tapping operations. For certain machines spindles are provided with power feed and adjustments in several directions. The *rpm* or the *feed*, in inches per revolution (mm per revolution), is usually variable unless the machine is designed for one specific production operation only, as in a turning operation of railroad car wheels.

A cutaway of a grinding wheel spindle is shown in Figure 7.4. This spindle is 1½ hp (1100 W), 3600 rpm, and is direct motorized.

The *bed* or *base* supports other basic machine tool elements as shown in Figure 7.5*A*. In this case, the bed allows for linear motion of the table, and the table is guided by double *V*-ways. In the lathe the bed supports the headstock, tailstock, cross slide, and carriage. In a boring machine the bed supports a rotating table, while in the planer the bed supports a reciprocating table.

Figure 7.3
Open view of a single-spindle headstock showing gearing.

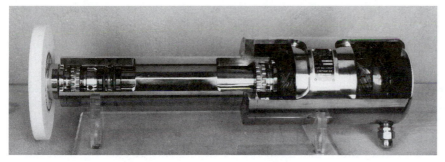

Figure 7.4
Grinding spindle. Cut away view.

The mass of the structure affects its performance. One example is the planer machine bed, which is "held on" to its base by gravity alone. As the effect of the planer's mass is greater than the upward component of the cutting forces, it is not lifted from the sliding ways during metal removal.

The fixed and movable elements, which form a machine-tool structure, locate and guide each other in accordance with the relative position between workpiece and cutting tool. Whenever basic elements slide over each other, as a *carriage* over a bed, it is necessary to provide accurate movements. This is provided with either *flat* or *V-ways* or a combination of the two. The V-shape acts as a guide in two directions. The ways are oiled by force-feed lubrication to avoid scoring or seizing. A V-way has the advantage of not becoming loose as wear occurs, although the sliding unit may ride up one surface when the side thrust is large. Ways may be constructed with roller or ball bearings. Some

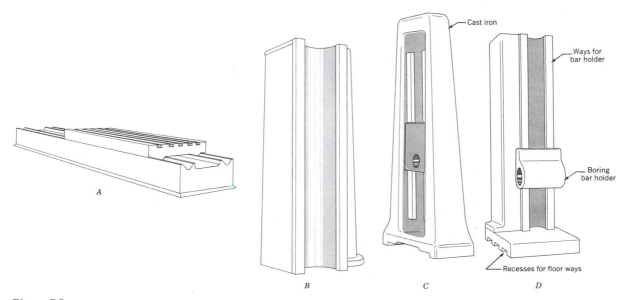

Figure 7.5
A, Planer-type table with T-slots on cast-iron, double-V way bed. *B*, Cast-iron column with scraped flat ways. *C*, End-support column, enclosed. *D*, End-support column, open.

machines use hard plastic inserts in the ways to reduce friction, and wipers that prevent entry of chips between the ways become necessary.

The *column*, Figure 7.5*B*, provides vertical support and guides the headstock for a class of machines. For large machines the column is ribbed and heavily constructed, and has provision for counterweighting heavy headstocks to reduce lifting horsepower. The counterweight may be a chain over a pulley, and the counterweight is hidden inside the column, as shown in Figure 7.5*B*.

When cutting forces or loads are not severe, ways and contacting members have scraped cast-iron sliding members. *Scraping* is performed manually with a sharp chisel-shaped carbide tool point much like a flat file. A rectangular carbide tip is brazed on the end. The surface is scraped leaving small 0.0002-in. (0.005-mm) oil pockets randomly arranged. During scraping, the ways are compared to a true flat master surface plate. The plate is coated with red lead or Prussian blue, and by rubbing the master on the ways, the high spots show dye. These high spots are removed by hand scraping away minute amounts of metal. The process of scraping and checking with a master and compound is repeated until a true plane is obtained. Figure 7.5*A* and *B* illustrate flat and V-ways that may have scraped or ground surfaces.

End supports or *tailstocks*, as shown by Figure 7.5*C* and *D*, serve as an outboard support for cutting tools, such as a long and heavy boring bar or the workpiece.

The function of a table is to support the workpiece and to provide for locating and clamping. In some machines the table is provided with power feeds in one to three directions. Figure 7.6*A* illustrates a table with a saddle that has two-axis movement in the horizontal plane. Carriages are found in lathes and provide movement along the axis of the bed.

Runways carry a floor planer-type machine column and rotary tables, which are used primarily with larger machines. When a column base with a column and headstock traverse as a complete unit and are supported on an element, the element is called a runway and not a base. When a column base, column, and headstock are an integral unit, the supporting element is called a bed and not a runway, although both are similar in principle, as shown in Figure 7.6*B*.

The power capacity and the performance of the machine determine *static* and *dynamic stiffness*. The size and shape of the workpiece, together with the cutting processes and operating and loading conditions, affect the shape and design of the structure. Load-carrying capacity is limited by the allowable stresses in the material and the shapes and

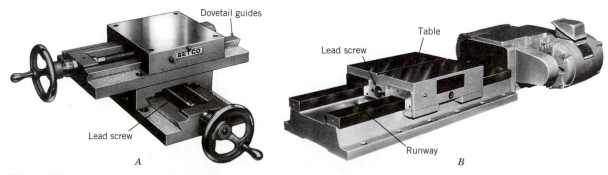

Figure 7.6
A, Feed table and saddle for compound movement using hand feed. *B*, Motorized feed system using lead screws on runway.

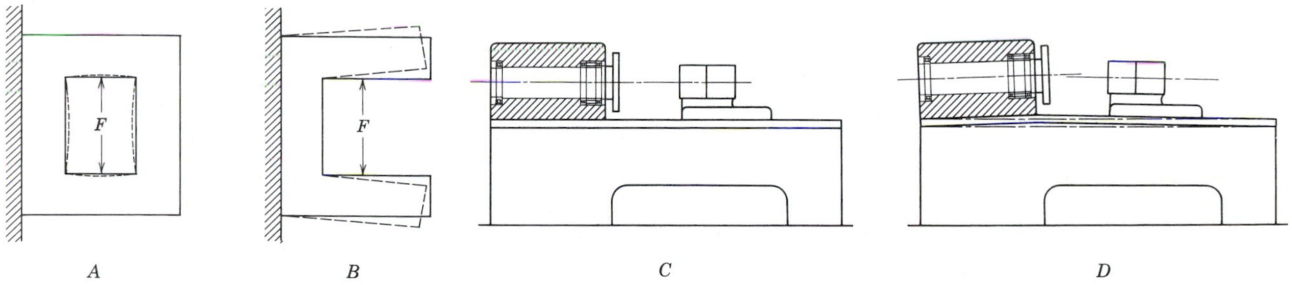

Figure 7.7
A, B, Deformations of open and closed frame under axial load. C, D, Misalignment effect of temperature changes on the bearings of lathe headstock.

sizes of the various cross sections of the members. In a structure subjected to impact loads, excessive stiffness may be unwanted and elasticity desired. An open C-shaped frame has lower stiffness than a closed frame, as illustrated in Figure 7.7A and B. Most machine tool structures can be resolved into elements, which may be modeled as beams subject to transverse bending and torsion.

When a machine is idle, deviations from defined straightness and flatness are caused by inaccuracies in the machine manufacture. During machining, static and dynamic deformations and changes in oil film thickness can cause conditions similar to those shown in Figure 7.7C and D. Temperature misalignment effects are caused by heat distortion of lathe headstock bearings if there is insufficient cooling or overloading.

The machining of metals is often accompanied by a violent relative vibration between work and tool, which is called *chatter*. Chatter is undesirable because of its adverse effects on surface finish, machining accuracy, and tool life. Vibration and chatter are complex phenomena.

Two types of vibration are forced and self-induced vibration. *Forced vibration* is generally caused by some periodic applied force present in the machine tool. Imbalance of machine tool components, misalignment, gear drives, or motors and pumps may be the source of forced vibrations. A basic solution to forced vibration is to isolate or remove the forcing element once it is found. If the forcing frequency is at or near the natural frequency of an element of the machine-tool system, one of the frequencies may be raised or lowered.

Self-induced vibration, or chatter, is caused by the interaction of the chip removal process and the structure of the machine tool. Chatter typically begins with a disturbance in the cutting zone, which is caused by a nonhomogeneous material, its surface condition, changes in the type of chip, or changes in the surface friction between the chip and workpiece interface. Chatter can be stopped by changing the tool or redesigning the work-piece.

7.3 MOTORS

Power for machine tools is electric. The individual motor, either internal to or mounted on the machine tool, was generally introduced to each machine tool in the 1920s. This signaled a tremendous gain in productivity. Before that period the power to machine tools was by overhead line shafts, with each line shaft powered by a central prime mover, perhaps a steam engine or another electric motor. Each machine tool received its power by belts from this central line shaft. The "engine" lathe was coined because of its dependence on the steam engine, which was coupled directly to the lathe. Therefore, motors are important to machine tools.

An electric motor converts electrical energy to mechanical energy and is composed of many elements. An *induction motor* contains an electromagnet, armature, and a commutator with its brushes. Motors are designed for various applications, such as high starting torque and speed control. A motor and its application depend on whether the power source is AC (alternating current) or DC (direct current), frequency of starting, starting torque magnitude, one or infinite speeds, load and variability, overload capacity, temperature regime, dust or explosive or corrosive gases, horizontal or vertical position, and electrical load. Most motors are classified based on their power source, either AC or DC.

DC motors have variable speed control and speed/torque characteristics that find extensive industrial applications in machine tools, steel works, and cranes, for example. Direct-current motors are often used with numerically controlled equipment. Although industrial plants seldom have an extensive DC supply, motor generator sets and static converters provide the power. There are two basic types of DC motors that deal with the type of windings employed: the series motor and the *shunt motor*.

The *series-wound motor* is controlled by a variable resistance in series with the field coils. It has a high starting torque, but speed is decreased with an increased load. Because of this speed dropoff characteristic and the possibility of dangerously high speeds at light load, this motor is not suitable for any load where there is a significant drop in torque. The shunt motor can maintain a more constant speed when the load is increased, but it has average starting torque applications.

The *alternating-current motors* are generally of the single-phase or induction type. The series motor has the motor field winding and the armature connected in series. The connections to the armature are made through brushes to the segments of a commutator. The speed of the motor is controlled by a variable series resistance, and it has a high starting torque. Arcing between the brushes and the commutator, and between the commutator segments themselves, is a disadvantage of this motor. It must be totally enclosed if it is used around inflammable materials or gases.

The *AC induction motor* can be operated from single-, two-, or three-phase current. Polyphase motors consist of *squirrel cage* induction, wound rotor induction, and synchronous types. Single-phase motors consist of either capacitor-start or repulsion-start induction motors. The windings on the armature form closed circuits known as a "squirrel cage." This motor derives its name from the similarity to the wheel of a squirrel cage. Its popularity is due to its mechanical simplicity, ruggedness, and its suitability to industrial applications. This motor is primarily constant speed. Low or average starting torque is a disadvantage of this motor.

The motors are either directly connected through gearing to work and tool-holding devices or else by one or more "V" belts. High-production machines are usually "geared head," whereas small laboratory-type machine tools are often belt-driven. Fluid drives employing hydraulic motors are used when power requirements cause sharp power surges.

7.4 HOLDING WORKPIECES

Methods for holding a workpiece depend on part size and shape, machine tool, and quantity. If the quantity is high, there is a necessity for rapid production. Holding devices may be actuated mechanically, hydraulically, pneumatically, electrically, magnetically, or by cam action. Control systems are integrated into workpiece holding, as with auto-mated or numerically controlled machines, and holding devices are programmed to release the workpiece at the conclusion of the machining cycle and to clamp automatically the next part.

Supporting Work between Centers

One common way to support a rotating workpiece for a turning operation is to mount the part *between centers*. This method supports heavy cuts and is convenient for long parts. Since the work is mounted between two tapered and conical centers, the part will not rotate once heavy cutting starts, unless the part is required to turn and is attached to a *faceplate*. Two tapered centers are insufficient to cause turning. Attachment is made through a pear-shaped forging, called a dog, which consists of an opening to accommodate the diameter of the stock being turned, a setscrew to fasten the work securely, and an elongated portion at the top (known as the tail), which is bent at a right angle—parallel with the stock—so that it may engage a slot in the faceplate. After the setscrew is tightened, the tail of the dog is fitted into the faceplate and the work is ready to be turned. The center in the headstock turns with the work; consequently, no lubrication of that center is necessary. The tailstock center or *dead center* acts as a conical bearing and is kept clean and lubricated, or else the center is a ball-bearing type. Allowance for expansion of the workpiece because of the heat generated by the machining operation and the rotation of the stock is necessary. Figure 7.8 shows a dog and a ball-bearing center rest.

In turning long slender shafts or boring and threading the ends of spindles, a *center rest* gives additional support to the work and prevents deflection of the part during cutting loads. The center steady rest is attached to the bed of the lathe and supports the work with three jaws or rollers. Another similar rest is known as a follower rest. It is attached to the saddle of the carriage and supports work of small diameter that is likely to spring away from the cutting tool. The follower rest moves with the tool, whereas the center rest is stationary.

In machines employing high rotational speeds for carbide tool turning, the hydrostatic steady rest is used. This steady rest has three jaws, which are held against the workpiece with a constant pressure.

Mandrel

Cylindrical work that is bored or reamed is held between centers by a *mandrel*, shown in Figure 7.9. Solid mandrels have hardened, ground surfaces and are available in standard sizes. The surface is ground with a taper of about 0.0006 in./in. in length, the small end being 0.0005 in. (0.013 mm), undersized to facilitate insertion into the work. Expanding mandrels hold work with holes that are not accurate to size or where several parts are to be machined in one setup.

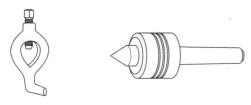

Figure 7.8
A, "Dog." B, Ball-bearing center rest.

A   B

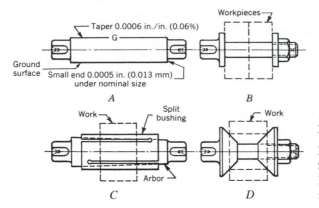

Figure 7.9
Various types of mandrels used for holding stock between centers. *A*, Solid mandrel. *B*, Gang mandrel. *C*, Expanding mandrel. *D*, Cone mandrel.

Faceplate

The workpiece is held to the *faceplate* by clamps, bolts, or in a fixture or special holding device attached to it. Figure 7.10 illustrates work bolted to a faceplate. Mounting is suitable for flat plates and parts of irregular shape. Figure 7.10 illustrates a boring operation of an eccentric hole in a part, and that part is mounted off-center to a faceplate for boring.

Chuck

Chucks hold small or large and regular or irregularly shaped parts, and are either bolted or screwed to the spindle making a rigid mounting. Chucks are made in several designs.

1. *Universal chuck.* Jaws maintain a concentric relationship when the chuck wrench is turned or power is used to adjust the diameter of the jaws.
2. *Independent chuck.* Each jaw has an independent adjusting screw as shown in Figure 7.11.
3. *Combination chuck.* Each jaw has an independent adjustment. In addition, it has a separate wrench opening that controls all jaws simultaneously.

Figure 7.10
Boring an eccentric hole on the faceplate of a lathe.

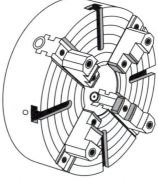

Figure 7.11
Independent jaw lathe chuck.

4. *Drill chuck.* A small universal screw chuck is used principally on drill presses, but also used on lathes for drilling and centering.

Power chucks operated pneumatically, hydraulically, or electrically relieve the operator of the effort involved in tightening and loosening the work. Power chucks are quick acting, the chucking pressure can be regulated, and chucks are used for both bar and chucking workpieces. "Chucking" refers to an irregular-shaped casting, forging, or previously machined part. "Bar" refers to *bar stock* that is usually round, square, or hexagonal in cross section.

Because it is difficult to mount all types of work on standard equipment, special chuck jaws or holding fixtures are designed. Standard faceplates mount such fixtures. The holding device is held to the faceplate either by bolting or by T-slots on the face of the plate.

Collet

Collets, commonly used for bar stock material, are made with jaws of standard sizes to accommodate round, square, and hexagonal stock. Collets of the parallel-closing type are sometimes used for large stock, but ordinarily collets of the spring type are more common. These collets are solid at one end and split on the tapered end. The tapered end contacts a similarly tapered hood or bushing and when forced into the hood, the jaws of the collet tighten around the stock. Spring collets are made in three types: push out, draw back, and stationary.

The *push-out* and *draw-back collets* operate in a similar way. The push-out collet shown in Figure 7-12*A* operates in the following manner. When the plunger is moved to the right, the tapered split end of the collet is forced into the taper of the head, which

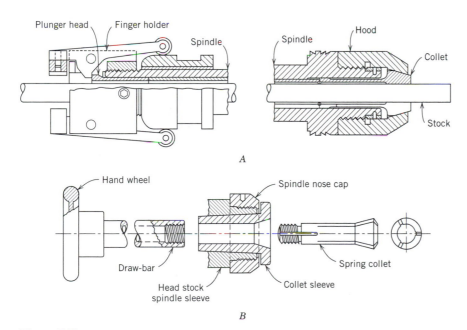

Figure 7.12
A, Push-out type of collet. *B,* Section showing construction of draw-in collet attachment.

causes the collet to tighten about the stock. The hood is screwed on a threaded spindle. The draw-back collet operates in the same way except that the collet is drawn back against a tapered hood for tightening action. Push-out collets are recommended for bar work, since a resulting slight movement of the stock pushes it against the bar stop. The bar stop ensures that all material protrudes from the collet to the same length. Draw-back collets are not widely employed for bar stock, but are useful when the collets are of extra capacity size and are utilized for holding short pieces. The slight back motion in closing forces the work against the locating stops.

A simple draw-in collet used on engine lathes is shown in Figure 7.12*B*. The proper-size collet is placed in the sleeve and screwed to the *draw bar* extending through the spindle. Work can be placed in the collet and held by turning the hand wheel on the end of the draw bar. This forces the collet back against the tapered surface of the sleeve and causes the collet jaws to grip the work.

With both the push and draw collets there is a slight movement of the workpiece because the collet moves as it is tightened. In most cases this is not a disadvantage, but when it is, the stationary collet can be used. A shoulder on the collet comes to rest against the head to provide this endwise accuracy. Because there are more sliding surfaces in the stationary collet, the rotational concentricity is not as accurate as for other types of collets.

Arbor

Expanding or threaded *arbors* hold short pieces of stock that have a previously machined accurate hole. The action in holding the work is controlled by a mechanism similar to collets. An expanding plug type arbor is shown in Figure 7.13. The work is placed on the arbor against the stop plate and as the draw rod is pulled, the tapered pin expands the partially split plug and grips the work. The *threaded arbor* operates in a similar fashion except that the work is screwed on the arbor by hand until it is forced back against a stop or flange. Collets and arbors are powered pneumatically, hydraulically, or electrically.

T-Slot and Vise

Worktables on mills, planers, and shapers are constructed with *T-slots* on their surfaces for holding and clamping parts. Most work is held by clamping directly to the table; a variety of clamps, stop pins, and holding devices are shown in Figure 7.14.

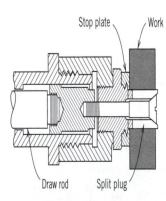

Figure 7.13
Expanding plug-type arbor.

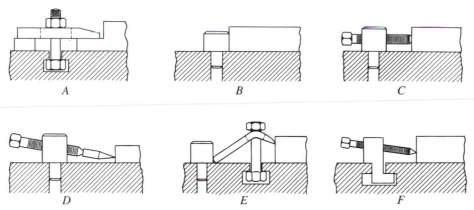

Figure 7.14
Methods of holding work on worktable. *A*, Holding strap. *B*, End stop pin. *C*, Adjustable screw stop. *D*, Inclined screw with toe dog. *E*, Compressive clamp. *F*, Inclined screw for T-slot.

Vises can be clamped to a worktable provided with T-slots. Delicate or complex-shaped parts are often held in a pocketed chuck by casting them in place with a low melting-point lead alloy or a hard tooling wax.

Magnetic Chuck

Ferrous work is held on surface grinders and other machine tools by *magnetic chucks*. This method is both simple and rapid, but is adaptable only when forces required by the operation are low.

Two types of chucks are the *permanent magnet chuck*, those magnetized by direct current, and the direct-current chuck, which is made in both rectangular and circular shapes. The rectangular style is suitable for reciprocating grinders or for light milling machine work. Rotary chucks, designed for lathes and rotating table grinders, are shown in Figure 7.15. The problem of energizing the current to these chucks is overcome by collector rings and a brush unit mounted at the back end of the chuck or spindle. The

Figure 7.15
Concentric-gap and radial-pole rotary chucks for holding work by magnetism.

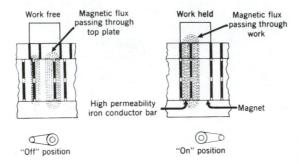

Figure 7.16
Schematic showing how work is held on permanent magnet chuck.

pulling power varies according to the type of winding used and may be as high as 165 psi (1.1 MPa). The equipment for furnishing the direct current consists of a motor generator set and demagnetizing switch or silicon-controlled rectifiers.

Parts held on a magnetic chuck should be demagnetized after the work is finished. Demagnetizers are available, operating on either AC or DC, which removes the residual magnetism.

Permanent magnet chucks do not require any electric power. Figure 7.16 shows the operation when the operating lever is shifted. In the "off" position the conductor bars and separator are shifted in a way that the magnetic flux passes through the top plate and is short-circuited from the work. When the handle is turned to the "on" position, the conductor bars and nonmagnetic separators line up so that magnetic flux in following the line of least resistance goes through the work in completing the circuit. The holding power, obtained by the magnetic flux, is able to withstand the force of grinding wheels and other light machining operations. This chuck and the DC chucks may be used for either wet or dry operations.

7.5 HANDLING WORKPIECES

For small quantities the customary method of handling a workpiece is manual, if the weight is less than 30 to 50 lb (10 to 25 kg), or by crane or conveyor if heavier. The Occupational Safety and Health Act (OSHA) and management practice encourage operator safety and welfare, which provides for robots, automatic equipment, and other handling devices if the quantity is large or if the weight is prohibitive.

If the quantity of production is sufficient, mechanical loaders have an economic advantage over manual loading. A variety of mechanisms are available to load, position, control the cycle, and unload the workpiece. Systems are available that can completely process and assemble the item. (See Chapter 29.)

Mechanical loading reduces operator fatigue. If a workpiece has a weight of 35 lb (16 kg) at a production rate of 60 pieces per hour, the operator will load and unload about 32,000 lb (7600 kg) in a working day.

Figure 7.17 is a loader–unloader that can pick up and place small parts or pieces weighing several hundred pounds. It loads and unloads vertical chucking machines from conveyors. These robots include press feeders, mechanical hands, vibrating positions, and assembly systems.

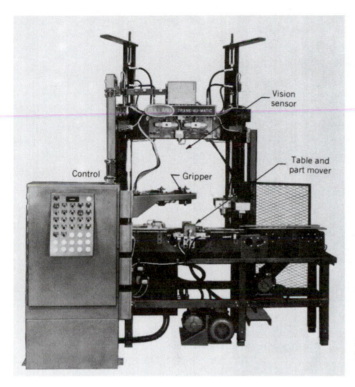

Control

Vision
sensor

Table and
part mover

Gripper

Figure 7.17
Semiautomatic loading–
unloading "robot" machine.

7.6 HANDLING TOOLS

Many computer-controlled machines employ tool changers. Selection and changing of cutting tools is employed with numerical control and high-speed mechanisms. These tool changers move dimensioned and preset tools from their storage bin to the spindle.

In the more basic approach the spindle moves to a tool storage rack or carousel located on the machining envelope and swaps the tool directly. After the spindle descends into contact with the new tool and inserts it into its spindle taper, a power-driven drawbar actuates a split collet that closes around the tool-retention knob opposite its working end. If the *tool magazine* is some distance from the spindle, then the new and existing tools are conveyed to the spindle mechanically.

The most popular method for tool changing is the two-hand arm. The desired tool is collected from a drum in one end of the special arm. At the moment of the tool change the empty end of the arm grips the tool in the spindle, removes it, indexes 180°, and inserts the new tool into the spindle taper.

If the magazine is adjacent to the spindle, the *changer arm* can stay open, remove the working tool, and replace it with the next tool. This is the arrangement shown by Figure 7.18. In the photograph, the gripper is moving a drill from the magazine to the spindle. Note the other tools on the drum that are garrisoned ready for work. The gripper takes a tool from the spindle, rotates 180° to swing down, around, and up to deposit the tool in an injection-molded pocket in a 20-tool rotary magazine. The magazine indexes to the predetermined position according to the program, ready for the next tool for the arm to grip and carry back to the spindle.

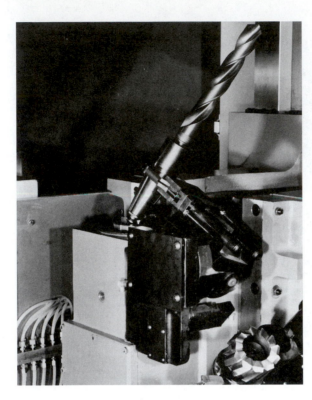

Figure 7.18
Magazine in lower-right corner and
tool gripper with tapered shank drill.

7.7 CONTROL SYSTEMS

Few production machines are controlled entirely by human intervention, although some
small general-purpose machines are. Machines having only hand feed, which implies that
muscle power replaces the motors that give power to these components, move the table,
spindle, or carriage and are known as *sensitive machines*. "Sensitive" implies that the
operator is able to feel the response of the tool while it is cutting, and the operator is
able to "control" the action. A different muscle response is required, for example, if the
drill tool is dull than if it is sharp. Sensitive machines are simple, usually found only in
school or home shops.

The first machine tools relied on operators to turn the wheels and pull the levers that
moved the cutting tools and the workpiece. For high-volume parts, the machinist installed
mechanical stops on the ways. Machine tool builders, however, designed the first automatic
controls, mechanical systems of *cams*, gears, and levers that caused repetitive and sequential actions. The operator continued to supervise overall control. Later mechanical controls adopted hydraulics and used piston actuators, which moved the workpiece and
positioned the table or the spindle. These hydraulic devices were powerful and relatively
cheap, and are accurate in simple positioning applications. They are still found with
transfer machines, or the closing of vises in simple holding applications, but have become
overshadowed with the more versatile electronic controls.

The heart of the first electric control is the *limit switch*, which is a simple on–off device
that is normally open or closed, and alternates state when something bumps it. A milling
machine, for example, would continue to mill until the table hits a limit switch, and that
switch turns off the power, or reverses the table to trip another switch at the other end
of the bed.

Control components have changed from electric on–off or reversing switches to *electronic transducers*, to integrated circuits, and finally to *microprocessors*. The purpose of these controls is much the same, however: to turn components on or off, to reverse direction, and to provide information on the status of the operation.

In earlier years, electric relays were grouped for a variety of actions. The controls were "hard wired"; that is, reprogramming meant manually reconnecting the wires linking the electric components. In the 1950s, relay networks developed into *"ladder-logic" systems* that could be mechanically programmed. This would later develop into *numerical control* (NC). Ladder logic later progressed into the *programmable logic controller* (PLC). The PLC holds sufficient data to perform simple sequential machining programs. Programmable logic controllers regulate transfer lines or coordinate subroutines for computer numerical control; they are an important element in automatic systems. However, they now rely on "soft wired" electronic circuits, which allow for adjustment of instructions through reprogramming of the software code. More than any other device, PLCs are responsible for converting stand-alone equipment into integrated systems.

A control system is a connected group of components forming a configuration that provides a desired system response. There have been input–output relationships, and a process or component to be controlled can be represented by a block, as shown by Figure 7.19*A*. A simple open-loop system is shown in Figure 7.19*B*. An open-loop system is without feedback. A closed-loop system may have feedback, as illustrated in Figure 7.19*C*. A feedback-control system tends to maintain a desired relationship of one system variable to another by comparing functions of these variables and using the difference as a means of control. It may be that this difference is amplified so that the difference between the desired and the actual is reduced. The *feedback* concept is the foundation for control systems analysis and design.

Chapter 28 is devoted to *computer numerically controlled (CNC) equipment*. Simply stated, this is a system where electronic and computer commands that control the machine are stored on a mylar 1-in. wide tape or the more popular hard computer drive. The tape is uniquely pierced with holes, which imply the presence of a signal. General-purpose equipment, such as milling machines, is provided with numerical control and may be used for a "one of a kind" job or in semi-mass production by simply changing the programming and the tooling. These machines are not automatic since they may not have loading and unloading capabilities. Numerical programming equipment can be both open and closed loop.

The *lathe tracer* system of Figure 7.20*A* is an example of an open-loop mechanical control system. The template or cam is mounted to the bed of the lathe, and a tracer head and stylus contact the surface of the template. The tracer head is mounted to the carriage of the lathe and maintains a fixed position relative to the cutter. In and out

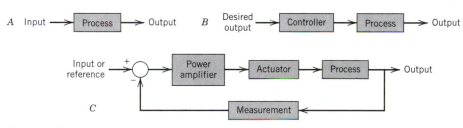

Figure 7.19
A, Simple block diagram. *B*, Open-loop control system. *C*, Basic closed-loop control system.

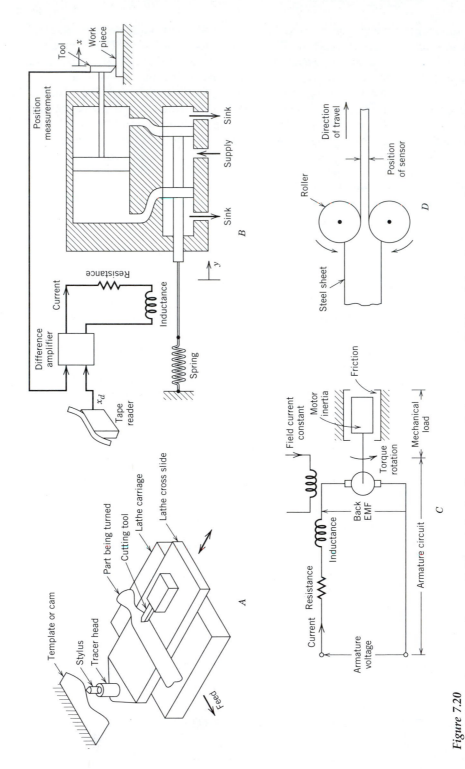

Figure 7.20

A, Mechanical tracing of taper. *B,* Electric–hydraulic system. *C,* Servomotor. *D,* Time-delayed response for control of thickness of rolling of steel sheet.

motion of the carriage is controlled by the tracer, whereas the axial motion follows the axial feed. The tracer head compares the desired position and the actual position because of the fixed distance. Control also is accomplished because the tracer head is mounted on the carriage.

Older machine tools may be automatically controlled by a *punch tape reader*, as shown in Figure 7.20*B*. These automatic systems are called numerically controlled machine tools. Considering one axis, the desired position of the tool *xd* is compared with the actual position *x*. This difference is used to actuate a solenoid coil and the shaft of a hydraulic control piston. The displacement of the piston is *y*. A hydraulic motor forces fluid into the supply, and the smaller piston is able to drive the larger piston. The piston is able to move back and forth in its guiding bearings and open and close ports to have the tool respond to changes in feedback signals.

Servomechanisms are important for controlling mechanical position or rotation of a motor. The servomotor, as illustrated in Figure 7.20*C*, is a DC motor designed for a closed-loop control system. The armature voltage is considered to be the system input; the armature circuit has a resistance and inductance. There is a voltage generated in the armature coil because of the coil in the motor's magnetic field, and is sometimes called back EMF or back electromotive force. There is a moment of inertia that includes all inertia connected to the motor shaft. Air and bearing friction of the motor and the mechanical load are influencing factors for the loads. Torque and motor shaft angle are the desired output.

A control system may have a time delay in the measurement of the output variable. Notice Figure 7.20*D*, which shows a schematic to control the thickness of the rolled steel. The rollers are controlled hydraulically. Important variables are the speed of the sheet, roller pressure and rotation, thickness entering, and the desired output thickness. The sensor is downstream from the rollers, and measurement occurs, not at the current moment of the rolling, but seconds later. Timing cycles are used frequently in automatic machine tools. Besides controlling product variation, timers may actuate microswitches or solenoids that, in turn, control machine movements. When there is a time delay, stability problems can occur and oscillations of the control system become possible.

Tapers are examples of surfaces that can be machined by several methods. One way, other than CNC, is with taper guides, which are essentially linear cams. Figure 7.20*A* shows a mechanical guiding component for the carriage. These mechanical additions to lathes may be found in your student shop. Another way to visualize surfaces that control the motion of a carriage with a tool is in the design and construction of cams.

Machines using *cam action* are semiautomatic. After material is loaded and started, each successive operation in the cycle begins when the previous operation is completed. The cams actuate changes in speeds or feeds, or in the cutting tool itself. Many automatic screw machines are cam-controlled. Although cam control is not elegant, for its purpose, it is highly efficient and economical. Certainly maintenance is minimized, although each new design will require new cams, and thus its disadvantage is loss in flexibility.

A cam is a plate or cylinder having a curved outline rotating about a fixed axis, and thus imparting motion to a tool, or the part itself. Cams produce motion in a single plane, usually up or down, or in or out. The curved outline of the cam produces a rise or fall in a follower, that is in contact with the cam. Cams utilize the principle of the inclined wedge, with the surface of the cam causing a change in the slope of the plane, thereby producing the desired motion in the machine tool.

The object in contact with the cam is called a follower. Dwell is that part of the cam that imparts no motion to the follower. In a plate cam, a dwell period has the contour of the circumference of a circle concentric with the axis of the cam.

Figure 7.21 describes a graphic method to design cams. The four steps start out with the construction of a semicircle whose diameter is the rise of the cam follower. The semicircle is divided into the same number of divisions as there are between 0° and 180° on the horizontal axis of the displacement diagram. In Figure 7.21*B*, measurements are removed from the displacement diagram. In Figure 7.21*C*, a base circle is formed to a specified diameter. The base circle intersects the center of the roller follower. The circle is divided into the same number of segments as given with the displacement diagram. The measurements are from the base circle. Circles are drawn to represent the roller as the cam revolves. Then the cam profile is drawn tangent to all the rollers.

This *plate cam* is used to cutoff a bar stock, and the diameter is equal to two times the rise of the displacement diagram; however, the motion of the cutter would be "harmonic." Cams can be designed for other motion patterns, such as uniform, gravity, or a combination of several patterns. Control systems can involve more than electronic and mechanical components. Hydraulic drives and controls are used primarily when a machine has a reciprocating member, on presses, and when constant velocity is difficult to maintain by mechanical means, especially during drawing operations of sheet metal stock.

Control is also thought of as *adaptive*. This notion gives the machine tool the opportunity to report and correct errors in its own operation. For example, the tool is able to

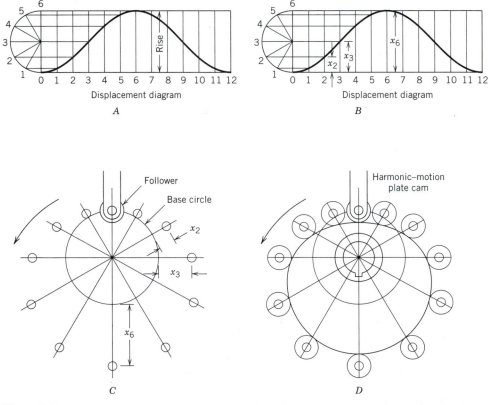

Figure 7.21
Cam design. *A*, Displacement diagram for 360°. *B*, Measurements of rise for sectors of cam rotation. *C*, Layout of roller follower with base circle. *D*, Harmonic-motion plate cam.

sense the wearing of a tool that may affect the surface roughness. The *adaptive control*, perhaps sensing the increased torque or force, will cause an alarm or even change the tool automatically.

QUESTIONS

1. Give an explanation of the following terms:

Machine tools	Mandrel
Headstock	Chucks
Kinematic inversions	Collets
Saddle	Draw bar
Scraping	Arbors
Tailstocks	T-slots
Chatter	Magnetic chucks
Shunt motor	Tool magazine
Squirrel cage	Sensitive machines
Faceplate	Cams
Dead center	Feedback
Center rest	Adaptive control

2. In a paragraph, state the advantages and disadvantages for a welded machine tool frame as compared to a cast iron machine tool frame.

3. What is a multiple head machine?

4. Identify two ways for a spindle or table to traverse a distance.

5. Discuss design features that encourage machine table versatility.

6. Describe methods for supporting long slender shafts that are to be machined.

7. Describe the difference between a bed and a runway.

8. Why are mandrels tapered?

9. What is dynamic stiffness? What is its purpose for basic machine tool elements?

10. What are the advantages of a universal chuck over an independent chuck and vice versa?

11. Discuss the differences between an open or C-frame and a closed frame.

12. Why is a stationary collet less accurate in concentricity but more accurate for stock location so far as endwise dimensions are concerned?

13. What is the purpose of a tailstock?

14. List the advantages of flexible systems from an operator's point of view. From a management point of view.

15. What are the principal advantages of V-type ways?

16. What is the fundamental difference between conventional and flexible machining systems?

17. Trace the history of controls for machine tools.

PROBLEMS

7.1. Determine the total force holding a 4-in. diameter plate on a direct current chuck.

7.2. What is the maximum force holding a 4-in. diameter plate to a rotary magnetic chuck?

7.3. What is the customary taper on a mandrel when expressed in mm/m?

7.4. If a workpiece weighs 20 lb and a part is loaded every 6 min, how many pounds must a worker handle each 8-h shift? Calculate also in SI units.

7.5. Sketch a simple machine using basic machine tool elements that will simultaneously drill the holes holding a cylinder head to an engine block. (Hint: Assume that there are four holes on each side and one per end, and that the holes are on a level and top plane.)

7.6 Sketch the components for a single machine to turn and thread a hexagonal bolt and slot it for a screwdriver. Label the parts.

7.7. Sketch a cam that will give a tool rise of 0.50 in. Standard machine tool dimensions for cams are base circle 3.50 in., roller follower 0.50 in. diameter, shaft 0.75 in. diameter, hub 1.25 in. diameter, and direction of rotation is clockwise. (Hints: Use engineering computation paper, 0.5 mm lead pencil, and keep the sketch roughly to scale. Provide one frontal view.)

7.8. Design a plate cam to move a tool into stock for plunge metal cutting a depth of 2 in. (Hints: Set up a semicircle that is 2 in. in diameter and sector the semicircle into units that correspond to horizontal displacement. Make necessary assumptions.)

7.9. Repeat problem 7.7 if the tool rise is 1 in.

7.10. Figure P7.10 shows a bent T-bolt holding a part on a machine table. Sketch an improved layout considering the height of the fulcrum block and workpiece, force of the T-nut, and consider the number of T-bolts for a block that is 12 in. wide by 30 in. long with a 1-in. high by 2-in. wide step completely around the block top. Are there problems with too much force in tightening an object?

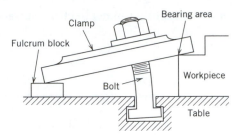

Figure P7.10

What about equal spacing of the T-nuts around the part? If the object is to be machined on the top 12 by 30-in. surface, is the location of the T-bolts important?

MORE DIFFICULT PROBLEMS

7.11. As assigned by your instructor, do one of the following:

a. Discuss the lathe tracer system of Figure 7.20*A*.

b. Draw a block diagram of input-output and feedback analysis for Figure 7.20*B*.

c. Draw a block diagram of input-output and feedback analysis for Figure 7.20*C*.

d. Draw a block diagram of input-output and feedback analysis for Figure 7.20*D*.

7.12. As assigned by your instructor, sketch and dimension a faceplate for Figure P7.12 where specific dimensions are given for each subproblem. The operation is to bore the eccentric hole. Assume that the hole has been rough drilled and that a boring bar will enter. The part is uniform in thickness. (Hints: Remember that the center line for the eccentric is "concentrically" machined. Size the faceplate allowing for clamping, positioning, and mounting to the spindle of an engine lathe. Indicate standard components for the faceplate with arrow leaders. Devise a simple bill of material that further describes the faceplate. Give some overall dimensions, although it is not necessary to be complete. Use engineering computation paper, and provide a top view.)

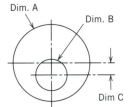

Figure P7.12

	A	B	C	Thickness
a.	1.2783 ± 0.00005	0.875 ± 0.003	0.156 ± 0.002	1.125
b.	8.75 ± 0.05	1.200 ± 0.005	0.156 ± 0.002	2.00
c.	5.275 ± 0.010	1.200 ± 0.005	0.557 ± 0.005	1.50

7.13. Reconsider problem 7.12, but now use a three jaw chuck instead of a faceplate. Each jaw of the chuck is separately positioned. Assume the rotational locations of the jaws at 0°, 120°, and 240°. (Hints: Sketch the eccentric component and then lay out the jaws at the appropriate positions. Locate the jaw surfaces from the lathe spindle center and find the dimension for the location.)

PRACTICAL APPLICATION

Machine tool shows are frequent occurrences in metropolitan areas. These commercial events are intended for practitioners who are interested in state-of-the-art equipment, but manufacturing classes are welcome. These exhibitions are excellent opportunities for classes to learn about machine tools and to see demonstrations. A tour of this kind is appropriate for this chapter. Whereas these exhibitions are not an annual affair in many cities, nevertheless, they are scattered in many areas and occur frequently. Ask machine tool distributors about this possibility.

CASE STUDY: TURBINE WHEEL

An impeller turbine blade, Figure C7.1, is attached to a water turbine wheel. The turbine is located at a remote hydroelectric plant. Each turbine blade is bolted to the turbine wheel through three holes. After many years of on-line service, the holes in the wheel and blade have elongated. The maintenance department believes that reboring and rebolting the blades to the wheel is necessary. New and larger diameter steel bolts will be used to fasten the turbine blades to the water wheel. The maintenance supervisor, Carlos Ramirez, says, "We will unbolt the 'buckets' from the wheel. The buckets can be rebored in a machining center in Central City. No need to do that here, but the wheel is a different matter." Down time of the unit, disassembly of the wheel, and transportation to a repair facility are costly, because the hydroelectric plant is remote.

Devise a system of machine components to allow reboring the wheel to the dimensions given by Figure C7.1. Assume that the wheel remains mounted on the turbine generator shaft and that it can be rotated and braked after the shroud is removed. Specify and sketch the machine component building blocks, accessories, platform, and machine that will do the work. (Hints: Sketch an elevation

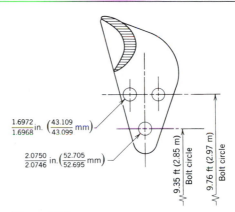

Figure C7.1

and top view of the new machine tool. A crane is available for assembly of the machine components at the site. Give your special machine tool a name and identify the major components by arrows and notes. Indicate the direction of the travel for the machine tool elements.)

CHAPTER 8
METAL CUTTING

After basic metals are refined, they are changed by primary processes to shapes suitable for metal machining. Final products are obtained by machining materials to design dimensions.

For this fabrication it is important that metal-cutting principles be understood. These first fundamentals are found in the mechanics of the forces acting between the tool and the workpiece. A model, called orthogonal cutting, allows understanding of machining processes. This simple model gives intriguing considerations to guide the machining operation.

An appreciation of tool geometry is instructive, and it indicates the efficiency of operation and surface appearance. Parts are produced by removing metal in the form of small chips. One can study the chip, or the waste of the operation, and gain insight into the optimization of the cutting process. The cutting tool, which removes these chips, is the focus of many important principles.

Metal-cutting principles are useful for turning, machining centers, milling and drilling operations, and many other processes performed by machine tools.

8.1 METAL-CUTTING THEORY

The simplest form of the cutting tool is the single-point tool, which is used in a lathe cutoff operation, a planer, or a shaper machine tool. Multiple-point cutting tools are merely two or more single-point tools ganged together as a unit. The milling cutter and broaching tool are examples of multiple-point cutters. Discussion in this chapter is limited to orthogonal or two-dimensional cutting rather than the more common oblique, or three-dimensional cutting. A sketch of oblique cutting is given later.

In *orthogonal cutting*, the cutting tool edge is perpendicular to the direction of the cut, and there is no lateral flow of metal; nor is there chip curvature in these idealized forms. All parts of the chip have the same velocity. This form of cutting is illustrated in Figure 8.1.

In analyzing the cutting process it is assumed that the chip is severed from the workpiece by a shearing action across the plane, as shown in Figure 8.1, although other theories exist on chip formation. Since the deformed chip is in compression against the face of the tool, a frictional force is developed. The work of making the chip must overcome both the shearing force and the friction force.

In our version of the orthogonal cutting model, the tool is considered stationary and the workpiece moving. An opposite motion pattern does not change the concept. The region of the workpiece in the vicinity of the shear plane undergoes a plastic deformation. A shear plane, ϕ, is determined by the rake angle of the tool and by the friction between the chip and the tool face.

The forces acting on the chip are a stress field at the shear plane and a distributed force field at the face of the cutting tool. The resultant of the stress field is a force R

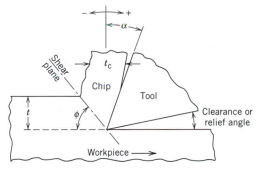

Figure 8.1
Orthogonal cutting tool model.

and the resultant of the distributed force is a force R'. The components of R and R' are shown in Figure 8.2A and are listed as follows:

F_s = Resistance to *shear* of metal in forming the chip; this force acts along the shear line.

F_n = Force *normal* to the shear plane; it is the resistance offered by the workpiece.

N = Force acting on the chip *normal* to the cutting face of the tool; it is provided by the tool.

F = *Frictional resistance* of the tool acting on the chip; it acts against the chip as it moves along the face of the tool.

Figure 8.2B is a *free-body diagram* showing the forces acting on the chip. Forces F_s and F_n are replaced by their resultant R and forces F and N by their resultant R'. This reduces the two combined forces to act as one on the chip. It is assumed that the dynamic effects are sufficiently small that they may be neglected; that is, that the forces acting on the chip are an equilibrium system. Consequently, R and R' are colinear, equal in magnitude, and opposite in direction.

The components of the force of the workpiece on the chip are the shear force F_s and the normal compressive force F_n. Conversely, the forces of the chip on the workpiece are $-F_s$ and $-F_n$.

The *shearing force* and the *angle* of the shear plane are affected by the *frictional force* of the chip against the tool face. The frictional force, in turn, depends on smoothness

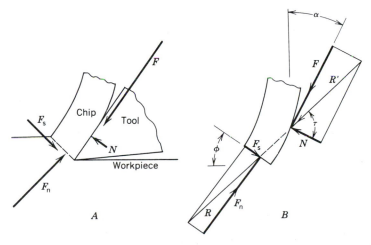

Figure 8.2
A, Forces on chip. B, Two force triangles on free-body diagram.

and keenness of the tool, whether or not a coolant is used, materials in the tool and workpiece, cutting speed, and the shape of the tool. A large frictional force results in a thick chip having a low shear angle, whereas the reverse is true if the frictional force is low. The efficiency of metal removal is higher when the friction force is minimized.

The two-force triangles of Figure 8.2B can be superimposed by placing the two equal forces R and R' together. The angle between F_s and F_n is a right angle, and together with F and N, a circle is shown in Figure 8.3. From this convention two more forces are drawn, F_c and F_t, which are the *horizontal* cutting force and the vertical, or *tangential*, force. These two forces are found during machining with a force *dynamometer*. F_c is the horizontal cutting force of the tool on the workpiece, and F_t is the *vertical cutting force* to hold the tool against the work. The rake angle α is determined by measurements of the tool in its holder. Even ϕ, the *shear angle*, can be approximated from photomicrographs. Another method results from measurement of the thickness of the chip, t_c, and the depth of uncut chip thickness, or by measurement of the length of chip.

Various quantities can be determined using the force diagram. For example, the *coefficient of friction* μ can be found in terms of the magnitudes of the forces or

$$F = F_t \cos \alpha + F_c \sin \alpha \tag{8.1}$$

and

$$N = F_c \cos \alpha - F_t \sin \alpha \tag{8.2}$$

Therefore,

$$\mu = \tan \beta = \frac{F}{N} \tag{8.3}$$

or

$$\mu = \frac{F_t + F_c \tan \alpha}{F_c - F_t \tan \alpha} \tag{8.4}$$

where

$$\beta = \text{Friction angle}$$

Customarily the chip is thicker than the depth of cut or, noticing Figure 8.1, $t_c > t$. The ratio of the uncut chip thickness t and the chip thickness t_c is called the *chip thickness*

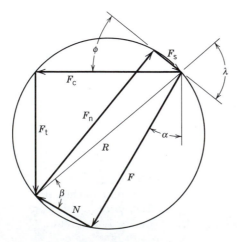

Figure 8.3
Relationship between forces in orthogonal metal cutting.

ratio r, and may be expressed as

$$r = \frac{t}{t_c} = \frac{\sin \phi}{\cos(\phi - \alpha)} \tag{8.5}$$

and solving for tan ϕ,

$$\tan \phi = \frac{r \cos \alpha}{1 - r \sin \alpha} \tag{8.6}$$

The ratio r can be found by knowing the depth of cut and measurement of chip thickness using a *point* micrometer. However, the back of the chip is rough, making an accurate measurement for the chip thickness unlikely when using a point micrometer. An alternate way uses length rather than thickness. This method assumes that the density does not change during cutting, and the volume of the chip is equal to the volume of the metal removed. It has already been assumed that there is no lateral flow of the chip during cutting; this is especially true for orthogonal cutting. This correspondence gives equal volume before and after cutting or

$$wtL = w_c t_c L_c \tag{8.7}$$

where

L_c = Length of chip
L = Corresponding length of material removed from workpiece
w = width of orthogonal cut

and

$$r = \frac{t}{t_c} = \frac{L_c}{L} \tag{8.8}$$

The velocity of the chip as it moves along the face of the tool is less than the cutting velocity. This holds since the chip is thicker than the depth of cut. Velocities can be found; the velocity relationships are shown in Figure 8.4, where the velocity of the workpiece is given as V_c. The chip moves along the cutting face of the tool with a velocity V_f. The velocity of the chip relative to the workpiece, denoted by V_s, is in the direction of the shear plane. Thus,

$$V_c + V_s = V_f \tag{8.9}$$

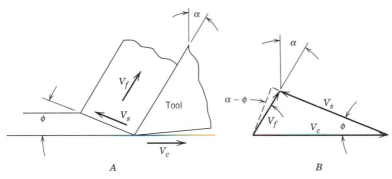

Figure 8.4
A, Tool and chip velocity relationship. B, Velocity polygon.

The velocity of the chip sliding along the cutting face of the tool is therefore

$$V_f = \frac{V_c \sin \phi}{\cos(\phi - \alpha)} \tag{8.10}$$

or

$$V_f = rV_c \tag{8.11}$$

From Figure 8.4, the velocity of the chip sliding along the shear plane is

$$V_s = \frac{V_c \cos \alpha}{\cos(\phi - \alpha)} \tag{8.12}$$

Consider the following example. A shaper is machining a steel block where the length of cut is 3 in. The length of the chip is measured after annealing and straightening as $L_c = 2.4$ in. The tool is known to have a rake angle $\alpha = 40°$. Other known conditions are $V_c = 35$ ft/min, depth of cut $t = 0.008$ in., horizontal cutting force $F_c = 275$ lb, and vertical force $F_t = 95$ lb.

The *coefficient of friction* is found as

$$\mu = \frac{F_t + F_c \tan \alpha}{F_c - F_t \tan \alpha} = \frac{95 + 275 \times 0.839}{275 - 95 \times 0.839} = 1.7$$

The length of cut is the block length or 3 in. and the chip ratio r is

$$r = \frac{L_c}{L} = \frac{2.4}{3} = 0.8$$

The thickness of the chip now becomes

$$t_c = \frac{t}{r} = \frac{0.008}{0.8} = 0.010 \text{ in.}$$

The shear plane angle ϕ is found as

$$\tan \phi = \frac{r \cos \alpha}{1 - r \sin \alpha} = \frac{0.8 \times 0.766}{1 - 0.8 \times 0.643} = 1.26$$

or

$$\phi = 52°$$

The velocity of the chip along the tool face is found as

$$V_f = \frac{V_c \sin \phi}{\cos(\phi - \alpha)} = \frac{35 \times 0.642}{0.978} = 23 \text{ ft/min}$$

The velocity of shear along the shear plane is

$$V_s = \frac{V_c \cos \alpha}{\cos(\phi - \alpha)} = \frac{35 \times 0.766}{0.978} = 27.4 \text{ ft/min}$$

Measurement of forces acting on the tool are made with a dynamometer. Electronic load dynamometers are frequently used to measure these forces. As it is impossible to measure cutting forces at the point of the tool and workpiece, the reactions are measured away from the cutting point. Transducers and a platform are combined to measure 1, 2, or 3 forces and torques. A tool or workpiece is mounted on a platform. Figure 8.5 shows the workpiece mounted on a platform where transducers measure drilling feed, force, and torque.

Figure 8.5
Two-channel dynamometer measuring drill thrust and moment. Dynamometer transducers mounted in platform.

Transducers measure deformations using a change of inductance, capacitance, or resistance (strain gages). The transducers used in the load cell of Figure 8.6 are *piezoelectric*. A piezoelectric force measuring principle differs in that as a force acts on a quartz element, a proportional electric charge appears on the loaded surface, as shown by Figure 8.6. The piezoelectric properties of quartz are such that the crystals are sensitive either to pressure or shear in one axis. In this way, components of cutting force or torque are measured independently.

Typical forces acting on an oblique cutting tool, as measured by a dynamometer, are indicated by Figure 8.7*A*. These are cutting, thrust, and radial forces. Figure 8.7*B* indicates the approximate distribution of the forces. In most oblique tool-cutting operations, the cutting force is the most significant.

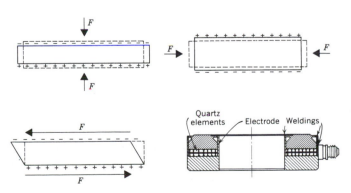

Figure 8.6
Diagram showing longitudinal, transverse, and shear effect on quartz element and construction of dynamometer load transducer.

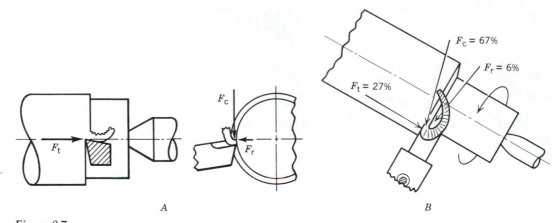

Figure 8.7
A, Forces acting on end of lathe tool: longitudinal thrust, F_t; tangential cutting, F_c; and radial, F_r. *B*, Distribution of forces on a single-point cutting tool.

For a given material the forces on a cutting tool depend on a number of considerations.

1. Tool forces are not changed significantly by a change in cutting speed.
2. The greater the feed of the tool, the larger the forces.
3. The greater the depth of cut, the larger the forces.
4. Cutting force increases with chip size.
5. Thrust force is decreased if the *cutting tool nose radius* is made larger or if the side cutting edge angle is increased.
6. Cutting force is reduced as the back rake angle is increased about 1% per degree.
7. Using a coolant reduces the forces on a tool slightly but greatly increases tool life.

A practical method to find the power for machining uses a wattmeter or ammeter at the drive motor. The wattmeter is considered more accurate when the motor is fully loaded. The horsepower at the cutter is found by subtracting the idle horsepower from the cutting horsepower.

Horsepower is calculated from measurement of the forces by a dynamometer and using F_c. This gives the horsepower at the spindle.

$$HP_s = \frac{F_c \times V_c}{33,000} \qquad (8.13)$$

where

$$F_c = \text{Cutting force, lb}$$
$$V_c = \text{Cutting speed, ft/min}$$

Sometimes this formula is seen with two additional terms that provide for the two additional force vectors, such as radial F_r and tangential F_t. Their contribution to the horsepower requirements are slight, because the force × distance for these two terms is minor.

Let

$$HP_m = \text{Horsepower at motor}$$
$$= \frac{HP_s}{E} \tag{8.14}$$

where

$$E = \text{Efficiency of spindle drive, \%}$$

Suppose that information indicates that $F_c = 275$ lb, $V_c = 35$ ft/min, and $E = 65\%$. Then $HP_s = 0.29$ at the spindle and $HP_m = 0.45$ at the motor.

The metal removal rate is approximately determined by the following:

$$Q = 12 \times t \times f_r \times V_c \tag{8.15}$$

where

$$Q = \text{Metal removal rate, in.}^3/\text{min}$$
$$t = \text{Depth of cut, in.}$$
$$f_r = \text{Feed, in. per revolution (ipr)}$$
$$V_c = \text{Cutting speed, ft/min}$$

Sometimes it is important to understand that unit power or unit horsepower equations can be calculated, given the information.

$$P = \frac{HP_m}{Q} \tag{8.16}$$

where

$$P = \text{Unit horsepower, hp/in.}^3/\text{min}$$

In metric units, for example, unit power has the dimension of (kW/cm³/s) as 1 hp = 0.746 kW, 1 in. = 2.54 cm, and 1 in.³ = 16.387 cm³. Some sample values are given by Table 8.1.

A digital wattmeter indicates 15 kW are required for a turning operation. The turning is at 500 ft/min, feed is 0.020 ipr, and the depth of cut is 0.150 in. The volumetric rate of machining Q is $12 \times 0.150 \times 0.020 \times 500 = 18$ in.³/min.

The power is 15 kW or 20.1 hp, and the unit horsepower is hp/in.³/min or 20.1/18 = 1.12 hp/in.³/min. For example, to remove material at a rate of 1 in.³, 1.12 hp is required.

Table 8.1 **Sample Values of Unit Horsepower**

Material	Hardness Bhn	Unit Horsepower (hp/in.³/min)	Unit Power (kW/cm³/s)
Steel 1020	150	0.58	1.58
1040	200	0.67	1.83
1330	260	0.92	2.51
3130	300	1.00	2.73
Cast iron	150	0.3	0.82
	250	0.5	1.36
	270	1.2	3.28
Leaded brass	35	0.23	0.63
	75	0.26	0.71
	130	0.30	0.82
Copper	40	0.90	2.46
Aluminum		0.30	0.82

Values of *unit horsepower* (unit power) vary with different materials and with the hardness. The sharpness of the cutting tool has an influence because a dull tool requires more energy to remove material as compared to a sharp one. Notice Table 8.1, which shows the range of values for unit horsepower. Values of the materials at the lower end are for freer-cutting materials, such as leaded steels.

8.2 METAL-CUTTING TOOLS

Successful application of metal-cutting tools depends on the tool geometry, material of the tool and workpiece, coolants, machine tools, and many other factors. Properties for effective tools include hot hardness, wear resistance, toughness, low friction, and relatively favorable cost. Later chapters amplify on cutting tools.

Cutting Tool Signatures

To understand the cutting action of a single-point tool as applied to a lathe, refer to Figure 8.8. This figure illustrates a brazed carbide tip tool, which is classified as *single point*, meaning that during cutting there is only one cutting edge in action. Tools are described by geometry and material, which vary significantly with the workpiece material and machine tool.

The *side-cutting edge angle* (SCEA) can vary from 0° to 90°, which influences the thickness and width of the chip. With a longer and thinner chip, the downward pressure is spread over greater length, increasing tool life because it is less likely that the cutting edge will be broken by the force of the chip. Tool life can be increased by decreasing the SCEA or increasing the feed when machining difficult materials. When the SCEA is 0°, the full length of the cutting edge is in contact with the workpiece at once and receives a strong initial shock. With a 30° SCEA, the tool receives a lower initial shock from the metal.

Side relief angle (SRA) eliminates tool breakage and increases tool life. If the angle is too small, the tool will rub against the workpiece, generate excessive heat, and cause

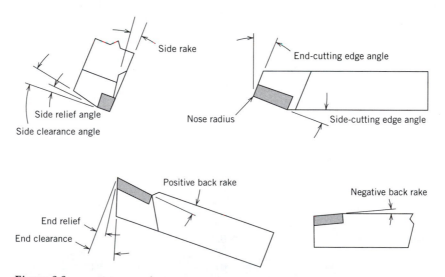

Figure 8.8
Terminology used to designate the surfaces, angles, and radii of single-point cutting tools. The tool shown here is the brazed-tip type, but the same definitions apply to indexable tools.

premature dulling of the tool. The SRA for chilled iron is $2° - 4°$, $4° - 6°$ for hard steels, and $7° - 12°$ for nonferrous materials.

The *end relief angle* allows the tool to cut without rubbing on the workpiece. If the back rake angle is downward toward the nose, it is negative back rake. If the slope is downward from the nose to the rear, it is positive. This angle controls the flow of the chip. A positive angle moves the chip away from the machine surface of the workpiece, and a negative back rake moves the chip toward the workpiece. A positive back rake of 5° moves the chip away and prevents the chip from marring the machine surface. If the angle is negative, it strengthens the tool, and provides longer tool life, especially under interrupted cutting.

Disposable Inserts

In addition to the solid single-point tool, a carbide tool material tip may be brazed to the body or shank of the tool base, or it may be inserted in a tool holder, such as illustrated in Figure 8.9. Many styles of tool holders and inserts are available. The *inserts* are disposable after wearout and are replaced by other inserts. The tool body is used over and over again. *Indexable tools* are more economical than brazed-on carbide tips. The indexable inserts provide a number of edges, while the brazed tip provides only one edge. If regrinding is necessary, the brazed shank unit is removed from the tool holder. Standard brazed tools are suitable for general-purpose machining operations, where maximum tool performance is of secondary importance. They are useful for short-run quantity.

The inserts vary in geometry from triangular, square, circular, and diamond to other special shapes. The insert in Figure 8.9 is triangular. There are six points that can be used for cutting with a triangular point. More heat is generated with carbide tooling, so an adequate coolant supply is necessary.

Tool Material

Current production practices make severe demands on machine tools. To accommodate the many conditions imposed on them, a variety of tool materials have been developed. The best material is the one that will produce the machined part at the lowest cost.

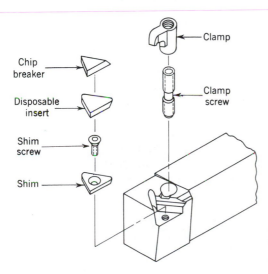

Figure 8.9
Exploded view of a tool with disposable, tri-angular, carbide cutting point insert.

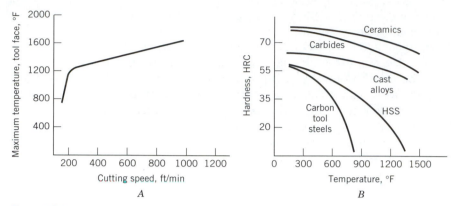

Figure 8.10
A, Maximum temperature attained on tool face in cutting steel. *B,* Approximate hardness of cutting tool materials as related to temperature.

Desirable properties for any tool material include the ability to resist softening at high temperature, low coefficient of friction, good abrasive resisting qualities, and sufficient toughness to resist fracture.

Figure 8.10 shows various properties of tools with respect to cutting velocity. In Figure 8.10*A,* the maximum temperature of the tool face is related to the cutting speed. In Figure 8.10*B,* an approximation of the decline of surface hardness is shown, as measured by Rockwell C, with increasing temperature of the surface of the tool. This property is known as *red or hot hardness.* Various materials respond differently with hot hardness.

Cemented Carbide

Carbide cutting tool inserts are made only by the powder metallurgy technique, as discussed in Chapter 20. Powders of tungsten carbide and cobalt are pressed to shape, semisintered to facilitate handling and formed to final shape, sintered in a hydrogen atmosphere furnace at 2800°F (1550°C), and finished by a grinding operation. The material tungsten carbide will not coalesce, but with the addition of cobalt, and the process of sintering, the tool materials are "cemented" and do not decompose.

Cobalt-bonded WC-based materials have been in use for at least 50 years. Their long-term success is due to their ability to combine high strength and hardness with excellent thermal shock resistance. Economically, carbide tools should always be used if possible. This material, along with a special coating on the cemented carbide tool, is dominant as a cutting tool commercially. Machines using carbide tools must be rigidly built, have ample power, and have a range of *cutting speeds and feeds* suitable to the material.

Those carbide tools that consist of tungsten carbide with a cobalt binder (WC-Co) are a class that differs from alloyed tungsten carbide and cobalt, which are augmented with varying quantities of titanium carbide (TiC) and/or tantalum carbide (TaC).

Carbide tools containing only tungsten carbide and cobalt (approximately 94% tungsten carbide and 6% cobalt) are suitable for machining cast iron and most other materials except steel. Steel cannot be machined satisfactorily by this composition because the chips stick or weld to the carbide surface and soon ruin the tool with excessive cratering. To eliminate this difficulty titanium and tantalum carbides are added, in addition to increasing the cobalt percentage. A typical analysis of a carbide suitable for steel machining is 82% tungsten carbide, 10% titanium carbide, and 8% cobalt. This composition has a low coefficient of friction and, as a result, has little tendency toward top wear or cratering. Since variation in composition alters the properties of carbide materials, various commercial grades are available for various applications.

The red hardness of carbide tool materials is superior to all others, with the exception of ceramic. It will maintain a cutting edge at temperatures of more than 2200°F (1200°C). Notice Figure 8.10B with a comparison of red hardness as measured by hardness number Rockwell C. It has high compressive strength, $450 - 850 \times 10^3$ psi. However, it is very brittle, has low resistance to shock, and must be supported rigidly to prevent cracking. Grinding is difficult and can be done only with silicon carbide or diamond wheels. Clearance angles should be held to a minimum. Carbide tools permit cutting speeds two to three times that of cast alloy tools, but at such speeds that a much smaller feed must be used.

A quality of cemented carbide, known as micrograin carbide, is a high strength, high hardness, fine grain size tungsten carbide that is used when cutting speeds are too low for regular carbides and conventional tools will not hold up for wear. A case in point is for milling where the high levels of thermal shock result from the continued heating and cooling of the insert's cutting edge as it enters and exits the workpiece. These WC-Co grades having very fine grain size run at similar conditions to their coarser-grained counterparts, but they display longer life.

Carbide tools may be coated with a 0.002- to 0.003-in. (0.05- to 0.08-mm) bonded layer of titanium carbide (TiC), aluminum oxide (Al_2O_3), or titanium nitride (TiN) to a tungsten carbide base. Coated carbide cutting tools have been in use since the early 1970s, and the majority of carbide cutting tools are now coated. These inserts have a golden color. Commercial application of uncoated carbide tool materials is declining.

The coatings are put down by a *chemical vapor deposition* (CVD) usually by reaction of the appropriate metal chloride with C, N, or O-containing gases at temperatures of the order of 1000°C. These are high temperatures and will influence the substrate material properties. The CVD techniques of high temperatures make it unsuitable for coating tool steels. Instead, physical vapor deposition (PVD) techniques are applying TiN at a much lower temperature, about 400°C. Physical vapor deposition also allows sharper corners and gives a lower coefficient of friction, both desirable properties.

With uncoated carbides there is a contradiction that occurs because of the need for high wear resistance on the one hand and high breakage resistance on the other hand. With coated carbide tools, however, the thin coating essentially provides the flank and crater wear resistance, permitting the cemented carbide substrate to prevent breakage and thermal deformation. Thus, coated carbides are able to trade off the material properties of the substrate and the coating to give optimized performance.

Combining two or three coatings in multiple layers can provide a broader performance base. Indeed, these coating combinations can be engineered to meet special requirements. Figure 8.11 illustrates both a PVD over a CVD multilayer coating on a cobalt-enriched substrate. The insert is designed to give resistance to edge chipping during interrupted cutting, as found with milling. This material is developed for 400 to 900 ft/min dry milling of steels.

High-Speed Steel

Fred W. Taylor, a famous engineer of the early 1900s, in his development of metal-cutting practices, and along with other engineers and metallurgists, created a steel for machining metals, which at that time greatly increased the speed of cutting over other hardened and tempered steels by as much as 90% for cast iron. The term *high speed* was

Figure 8.11
A steel milling material is based on insert coating technology having CVD-PVD multilayer coating on a cobalt-enriched substrate.

adopted then to reflect this extraordinary gain in the rotational velocity of bar stock. The term "high-speed steel" persists, and its use today reflects this early nickname. In comparison to today's materials, these materials are not high speed. Nowadays, the speed for this material is characterized as low speed in comparison to other more advanced tool materials.

High-speed steels (HSSs) are high in alloy content, have excellent hardenability, and retain a good cutting edge to temperatures of around 1200°F (650°C). The first tool steel that would hold its cutting edge to almost a red heat was developed by Taylor and M. White. This was accomplished by adding 18% tungsten and 5.5% chromium to steel as the principal alloying elements. Current practice in manufacturing high-speed steels still use these two elements in nearly the same percentage. If the HSS alloy is molybdenum-based, its identification is M, while tungsten-based HSS is called T. Other common alloying elements are vanadium, molybdenum, and cobalt. Although there are numerous high-speed steel compositions, they may be grouped in three classes.

1. *18-4-1 high-speed steel.* This steel, containing 18% tungsten, 4% chromium, and 1% vanadium, is one of the best all-purpose tool steels.

2. *Molybdenum high-speed steel.* Many high-speed steels use molybdenum as the principal alloying element, since one part will replace two parts of tungsten. Molybdenum steels such as 6-6-4-2 containing 6% tungsten, 6% molybdenum, 4% chromium, and 2% vanadium have excellent toughness and cutting ability.

3. *Super high-speed steels.* Some high-speed steels have cobalt added in amounts ranging from 2% to 15%, since this element increases the cutting efficiency, especially at high temperatures. One analysis of this steel contains 20% tungsten, 4% chromium, 2% vanadium, and 12% cobalt. The remainder is essentially iron. Because of the greater cost of this material, it is used principally for heavy cutting operations that impose high pressures and temperatures on the tool. These materials can be hardened to Rockwell C scale of 70, but for cutting tool applications, they are hardened to 67 to 68.

High-Carbon Steel

For many years, before the development of high-speed steel tools, high-carbon steels were used for all cutting tools. Their carbon content ranges from 0.80% to 1.30%. These steels have good hardening ability and with proper heat treatment, attain as great a hardness as any of the high-speed alloys. At maximum hardness the steel is brittle and if some toughness is desired, it must be obtained at the expense of hardness. Depth hardenability is poor, limiting the use of this steel to tools of small size.

Because these tools lose hardness at around 400° to 500°F (200° to 260°C), they are not suitable for metal cutting having high rotary speeds and heavy-duty work. Their usefulness is confined to work on soft materials, such as wood, aluminum, brass, or soft steels. When higher temperatures are exceeded, the tool edge wears quickly and softens during service. These tool materials should be operated at lower cutting speeds of one-third to one-half of high-speed steels.

The chief advantage of high-carbon steel for cutting tools is its low cost, but with very limited applications. Common examples include drills, reamers, hand taps, and threading dies. This material is not significantly used for production work.

Cast Nonferrous Alloy

A number of nonferrous alloys containing principally chromium, cobalt, and tungsten with smaller percentages of one or more carbide-forming elements, such as tantalum, molybdenum, or boron, are excellent materials for cutting tools. These alloys are cast to shape, have high red hardness, and are able to maintain good cutting edges on tools at temperatures up to 1700°F (925°C).

Compared with high-speed steels, these alloys can be used at twice the cutting speed and still maintain the same feed. However, they are more brittle, do not respond to heat treatment, and can be machined only by grinding.

Intricate tools can be formed by casting into ceramic or metal molds and finished to shape by grinding. Their properties are largely determined by the degree of chill given the material in casting. The range of elements in these alloys is 12% to 25% tungsten, 40% to 50% cobalt, and 15% to 35% chromium. In addition to one or more carbide-forming elements, carbon is added in amounts of 1% to 4%.

These alloys have good resistance to cratering and can resist shock loads better than carbide. As a tool material they rank midway between high-speed steels and carbides for cutting efficiency. However, like other cast materials, these alloys are relatively weak in tension and tend to shatter when subjected to shock load.

Diamond

Diamonds are the hardest known substance (Brinell hardness ≈ 7000). They have low thermal expansion (≈ 12% of steel), high heat conductivity (≈ twice of steel, ⅓ of aluminum), poor electrical conductivity, and very low coefficient of friction against metals (≈ friction of 0.05 on dry steel in air), but they have limited chemical stability. Research is progressing on vapor coating on diamonds.

Diamonds, obviously, are expensive, which limits their general application as cutting tools. Diamonds are not able to cut steels or ferrous material. At high temperatures, diamonds graphitize on these materials.

Diamonds used as single-point tools for light cuts and high speeds must be rigidly supported because of their high hardness and brittleness. They are used either for hard materials difficult to cut with other tool materials or for light, high-speed cuts on softer materials where accuracy and surface finish are important. Diamonds are used in machining plastics, hard rubber, pressed carbon, and aluminum with cutting speeds from 1000 to 5000 ft/min (5 to 25 m/s). Diamonds are also used for dressing grinding wheels, for small wire drawing dies, and in certain grinding and lapping operations.

Ceramic

Aluminum oxide powder, along with additives of titanium, magnesium, or chromium oxide, is mixed with a binder and processed into a cutting tool insert by powder metallurgy techniques. The insert is either clamped onto the tool holder or else bonded to it with an epoxy resin. The resulting material has an extremely high compressive strength but is quite brittle. Therefore, the inserts must be given a 5° to 7° negative rake to strengthen their cutting edge and be well supported by the toolholder.

Cermet

In metal cutting, the term *cermet* is usually reserved for metal-bonded materials containing a high proportion of cubic carbides (especially TiC) and that contain significant nitride levels. Cermet is a mixture of CERamic and METals. Cermets are primarily used for relatively light cuts at high speeds on cast iron and steel.

Cubic Boron Nitride

Cubic boron nitride (CBN) is a material that is next to diamond in hardness. The cutting tool uses a layer of 0.02- to 0.04-in. (0.5- to 1-mm) layer that is bonded to a tungsten carbide insert. *Cubic boron nitride tools* also are used for grinding wheels and abrasives; CBN is a brittle material, and bonding to the substrate gives stiffness and resistance to fracturing.

These tool materials have ranges of operations. For example, their rough performance may be classified as

Common Name	Important Component	Typical Grade	Maximum V_c (ft/min)
High-speed steel	18W + 4Cr + 1V	AISI T1	165
Tungsten carbide	WC + Co + TiC + Tac	C7, C8	500
Coated carbide	TiC, TiN, Al_2O_3		900
Cermet	TiN + TiC, N_0c		1200
CBN		Amborite	3000
Diamond	Synthetic		3000

8.3 CHIP SHAPE AND FORMATION

The mechanics and geometry of chip formation, and the relationship of the shape of the chip to tool life, surface finish, and machining performance are important to the effectiveness of metal cutting. Tool chips have been classified into three types as shown in Figure 8.12.

A *discontinuous* or *segmented chip* represents a condition in which the metal ahead of the cutting tool is fractured into fairly small pieces. This type of chip is obtained in machining most brittle materials, such as cast iron and bronze. As these chips are produced, the cutting edge smooths over the irregularities on the workpiece and a fairly good finish is obtained. Tool life is reasonably good, and failure usually occurs as a result of abrading action on the contact surface of the tool. *Discontinuous chips* also are formed on some ductile materials if the coefficient of friction is high. However, discontinuous chips from ductile materials are an indication of poor cutting conditions.

Figure 8.12
Basic chip types. *A*, Discontinuous. *B*, Continuous. *C*, Continuous with built-up edge.

An ideal type of chip from the standpoint of tool life and finish is the simple *continuous type*, which is obtained in cutting ductile materials having a low coefficient of friction. In this case, the metal is continuously deformed and slides up the face of the tool without being fractured. Chips of this type are obtained at high cutting speeds and are common when cutting with carbide tools. Because of their simplicity, *continuous chips* can be analyzed easily from the standpoint of the forces involved; they come off the bar stock as string from a ball and can be troublesome to handle and sometimes dangerous.

Another type chip is characteristic of those machined from ductile materials that have a fairly high coefficient of friction. As the tool starts the cut, some of the material, because of the high friction coefficient, builds up ahead of the cutting edge. Some of the workpiece may even weld on the tool point, and is thus known as a *built-up edge* (*BUE*) *chip*. As the cutting proceeds, the chips flow over this edge and up along the face of the tool. Periodically, a small amount of this BUE separates and leaves with the chip, or is embedded in the turned surface. Chatter may result. Because of this action the surface smoothness is not as good as the continuous-type chip. The BUE remains fairly constant during cutting and has the effect of slightly altering the rake angle. However, as the cutting speed is increased, the size of the BUE decreases and the surface finish is improved. This phenomenon also is decreased by either reducing the chip thickness or increasing the rake angle, but on many of the ductile materials it cannot be eliminated entirely.

An analog of the stress patterns involved in metal cutting can be seen in Figure 8.13. The photograph, which shows two motion picture frames of beeswax being cut at 50 ft/min (0.25 m/s), was taken at 7200 frames per second. The lines are 0.050 in. (1.27 mm)

Figure 8.13
High-speed photography of beeswax being machined.

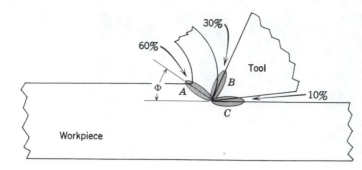

Figure 8.14
Approximate sources of heat in three zones. *A*, Shear plane. *B*, Friction plane. *C*, Surface plane.

apart. The vertical lines are compressed together as the chip is formed, but remain parallel, thus indicating the action of the compressive forces, causing the wax to expand in the direction perpendicular to the cutting direction. A shear angle of approximately 45° can be approximated.

Some investigators report that 97% of the work that goes into cutting is dissipated in the form of heat. Figure 8.14 shows the three *heat zones* for generating heat. As the shear angle, ϕ, is increased, the percent heat generated in the shear plane *A* will decrease, since the plastic flow of the metal takes place over a shorter distance. The shear angle can be increased by applying a coolant and reducing the friction between the chip and the tool as well as by properly grinding the tool. Of all the cutting variables, cutting speed has the most effect on temperature. To increase the rate of metal removal, an increase in feed is preferred over an increase in speed.

Chip Control

In high-rate production turning, control and disposal of chips are important for operator safety and protection of the part and tools. Long curling chips snarl about the workpiece and the machine. Their sharp edge and high tensile strength make their removal difficult and hazardous, particularly when the machine is in operation. A cutting tool will have a *chip breaker curl*, which curls and stresses the off-coming chips, and encourages their fracture into short lengths for easy removal. Short chips are desired because they handle well in the various chip disposal systems. Various chip breakers incorporated into the tool holder or tool are given in Figure 8.15.

Methods of chip control include:

1. Grinding a small flat to a depth of 0.015 to 0.030 in. (0.38 to 0.76 mm) on the face of the tool along the cutting edge is known as a *step-type* chip breaker. It may be either parallel with the edge or at a slight angle. The width varies according to the feed and depth of cut and may range from ⅟₁₆ to 0.25 in. (1.6 to 6.4 mm).

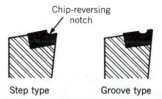

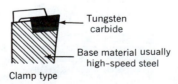

Figure 8.15
Typical chip breakers used on single-point tools.

2. Grinding a small groove about ¹⁄₃₂ in. (0.8 mm) or molding small *staggered cups* (by the powder metallurgy technique) in the insert behind the cutting edge to a depth of 0.010 to 0.020 in. (0.25 to 0.50 mm) is a popular technique. The correct dimension for the land distance and depth depends on the feed and should be increased slightly as the feed is increased.

3. Brazing or screwing a thin carbide *plate* or clamp on the face of the tool is used with insert-type tools. As the chip is formed, it hits the edge of the plate and is curled back to the extent that it breaks into short pieces.

4. Proper selection of *tool angles* controlling the direction of the curled chip is another way to cause breakage of the curling chip. The chip direction flows the chip into some obstruction and stresses the chip to its breaking point.

8.4 COOLANTS

Improvement in cutting action may be accomplished by using solids, liquids, emulsions, or gases in the cutting process. Interestingly, it was Fred Taylor who initially reported on the success in pouring a heavy stream of water (also supersaturated with carbonate of soda to prevent rusting) on the tool and chip interface, and this improvement increased the amount of work by 30% to 40%.

In forming and cutting operations, high temperatures develop as the result of friction. Unless temperatures and pressures are controlled, the metal surfaces tend to adhere to one another. Figure 8.14 indicates the principal locations of heat. In the empirical understanding of metal cutting, application of coolant to the cutting process is assumed to achieve the following advantages.

1. Reduce friction between chip, tool, and workpiece. (At high- and higher-cutting speeds, the entrance of coolant to the tool workpiece interface becomes more questionable.)
2. Reduce the temperature of the tool and work.
3. Wash away chips and control any dust.
4. Improve surface finish.
5. Reduce the power required.
6. Increase tool life.
7. Reduce possible corrosion on both the work and the machine.
8. Help prevent welding the chip to the tool.

A coolant should be nonobjectionable physiologically to the operator, harmless to the machine, and stable. It also should have good heat transfer characteristics, be nonvolatile, nonfoaming, and lubricating, and have a high flash temperature.

Solids that improve cutting ability, include materials added to the molten metal or inherent in the workpiece material, such as graphite in gray cast iron or lead in steel.

Gases include water vapor, carbon dioxide, and compressed air. Most coolants are liquid, however, because they are easily directed to the tool-chip interface and are easily recirculated. Liquids are applied principally in water- or oil-base solutions and with certain additives to increase their effectiveness.

Chemical coolants are blends of chemical components dissolved in water. Their purpose is to cool, but they may be used for both cooling and lubricating. The chemical agents used are

1. Amines and nitrites for rust prevention.
2. Nitrates for nitrite stabilization.
3. Phosphates and borates for water softening.
4. Soaps and wetting agents for lubrication and to reduce surface tension.
5. Compounds of phosphorus, chlorine, and sulfur for chemical lubrication.
6. Chlorine for lubrication.
7. Glycols as blending agents and humectants.
8. Germicides to control bacterial growth.

The advantages of a coolant result from cooling the tool and reducing friction, particularly between the chip and the tool. Because of the roughness of the chip as well as the machined workpiece, coolant is forced in small quantities to the cutting edge. *Capillary attraction* of the fluid between the rough chip and the tool may allow for converted vapor to reach the cutting point. It should be noted, however, that there is a divergent opinion on the capillary attraction of the coolant into the tool point workpiece interface. Simple flooding of the cutting area is not so effective as directing the coolant to the tool interface areas. The vibration of the tool and workpiece act to pump the coolant to the cutting edge.

Coolants are selected based on the material and the operation. The following are some nonchemical coolants for several common materials.

1. *Cast iron.* Compressed air, soluble oil, or worked dry. The use of compressed air necessitates an exhaust system to remove the dust. A reason for using a coolant is to control the dust.
2. *Aluminum.* Kerosene lubricant, soluble oil, or soda water. Soda water consists of water with a small percentage of some alkali, which acts as a rust preventive.
3. *Malleable iron.* Dry or water-soluble oil lubricant. The latter coolant consists of a light mineral oil suspended by caustic soda, sulfurized oil, soap, and other ingredients that form an emulsion when mixed with water.
4. *Brass.* Worked dry, paraffin oil, or lard oil compounds.
5. *Steel.* Water-soluble oil, sulfurized oil, or mineral oil.
6. *Wrought iron.* Lard oil or water-soluble oil.

8.5 MACHINABILITY

Machinability, or ease of machining, is influenced by the kind and shape of cutting tool. However, *machinability* is a loosely defined term; it is expressed as the time of tool life, power required for cutting, cost of removing a deformed amount of material, or surface condition obtained. The most important of these factors, so far as machining costs are concerned, is tool life, and values of machinability given in handbooks are based on this factor. Therefore, machinability is not a precise term, but a word that implies several concepts.

For many commercial applications, the engineering design strength of the part is not as important as economical machining. For this reason, materials are often selected because of the importance of tool life and machinability.

Plain-carbon steels have better machinability than alloy steels of the same hardness and carbon content. The addition of lead to steel, while teeming the ingots, adds to machinability, although it makes the steel more costly, softer, and more ductile. Materials have been developed, especially for certain applications. Automatic machine screw stock, which is very easily machined, is common, such as 12L14, a low-carbon leaded steel. The letter "L" in the specification number refers to lead. Yet lead, which contributes to the toxicity of the metal, is being replaced by graphite because of concerns of the toxicity of lead to human health.

A few hundredths of 1% of tellurium to steel will increase the machinability and cutting speed about 3½ times, but the cost of the element is comparable to that of gold. The addition of moderate amounts of phosphorus as well as sulfur adds to machinability. Phosphorus causes chips to be brittle and hence eliminates long, difficult chip formation. Even calcium and sulphur are being considered as additions to steel to improve machinability.

Machinability tests are conducted under standardized conditions, if results are to be comparable. These tests indicate the resistance of the material being cut, and the results are affected by its composition, hardness, grain size, microstructure, work-hardening characteristics, and size. Other influencing factors include type and rigidity of equipment, coolant properties, feed and depth of cut, and the kind of tool. It is important that machinability tests simulate actual machining operations.

Two factors that affect the machinability are ductility and hardness. Increasing the hardness of a metal makes penetration by the tool more difficult and machinability is decreased. More ductile materials do not lend themselves to the formation of discontinuous chips and, in cases where continuous chips are formed, the increased ductility contributes to the speed at which a built-up edge occurs. Low ductility, however, is an asset to good machinability.

Metal properties of metals influence its general machinability. For example, white cast iron has high hardness and low ductility. When malleabilized, it has higher ductility, but a large decrease in hardness. Hence, it can be machined with greater ease. A medium-carbon steel may be heat treated in such a way to have its machinability either increase or decrease. Predictably for most metals, hardness increases as ductility decreases, and the choice of a material or a heat treatment is a compromise between desired service conditions and machinability.

Good machinability does not necessarily mean a good surface finish or low forces or long tool life. Sometimes it refers to the costs associated with metal removal. Although these concepts are easy to state, a rigorous value consistent across many tool and part materials that is a *machinability rating* is difficult to find. Sometimes materials are given machinability ratings, such as 110% or 60%, and these ratings are referred to a standard material, such as B1118 having a rating of 100%, which is easily machined. The 60% material would be 60% of the standard material, for example, which directly lowers the cutting value V_c. However, this rating is sometimes discredited as a useful measure because of the uncertainty in its analysis. These ratings are sometimes useful as a starting point.

Material	Brinell Hardness	Machinability Rating
1020 steel	135	0.60
1040 steel	205	0.61
4140 leaded steel	187	0.70
Cast iron	160	0.60
Cast iron	195	0.40
Stainless steel	183	0.55
Stainless steel	207	0.45
Aluminum, cast		1.50

Machinability is an ideal feature that represents a variety of concepts. Accelerated tests are unable to provide accurately and consistently reliable machinability ratings that are comparable with each other. However, there are four tests that give broad machinability values.

1. A fixed shape tool is set to cut at a predetermined depth and feed. The cutting speed at which a tool can be run and still have a 60-min life is a measure of machinability.

2. The cutting tool wear rate is determined by inspection or by radioactive methods. Low wear rates indicate good workpiece machinability. This method is not suitable for carbide or ceramic tools.

3. A dynamometer records the tool forces for a given set of cutting conditions. The material that can be turned at the highest speed before causing an arbitrary force on the dynamometer will have the best machinability.

4. Operate the tool until an unsatisfactory finish is apparent.

A good surface finish is affected by many variables in single- or multiple-point turning. The factors improving surface finish are light cuts, small feeds, high-cutting speeds, cutting fluids, noise radii, and increased rake angles on well-ground tools.

8.6 TOOL LIFE

The life of a tool is important in metal cutting since considerable time is lost whenever a tool is replaced and reset. Cutting tools become dull as usage continues and their effectiveness drops. At some point in time it is necessary to replace, index, or resharpen and reset the tool. *Tool life* is a measure of the length of time a tool will cut satisfactorily and, like machinability, is measured in several ways. Sometimes tool life is expressed as the minutes between changes of the cutting tool.

There are five basic types of wear that affect a cutting tool.

1. *Abrasion wear.* This type of wear is caused by small particles of the workpiece "rubbing" against the tool surface. This gradual wear at the tool point is seen on the clearance face of the tool in the form of a wear land. Wear is also found on the cutting face of the tool, or cratering.

2. *Adhesion wear.* Because of plastic deformation and friction, high temperatures involved in the cutting process can cause a welding action on the surfaces of the tool and workpiece. The consequent stresses of the cutting process lead to fracture of the weld causing degeneration of the cutting tool.

3. *Diffusion wear.* Diffusion wear is caused by a displacement of atoms in the metallic crystal of the cutting element from one lattice point to another. This results in a gradual deformation of the tool surface.

4. *Chemical and electrolytic wear.* A chemical reaction between the tool and workpiece in the presence of a cutting fluid is the cause of *chemical wear. Electrolytic wear* is the result of possible galvanic corrosion between the tool and workpiece.

5. *Oxidation wear.* At high temperatures, oxidation of the carbide in the cutting tool decreases its strength and causes wear of the edge. The tip of the tool becomes too soft to function and failure is quick.

Wear is evident in two places on a tool, as shown in Figure 8.16. One is on the flank of the tool where a small land extends from the top to a distance below, which is abraded

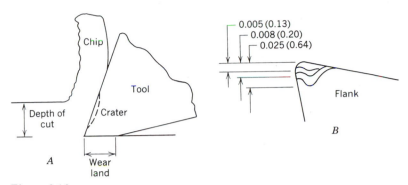

Figure 8.16
A, Wear of a cutting tool. *B,* Progressive flank wear measured during cutting tests.

away as cutting continues. On high-speed tools, a tool failure is arbitrarily defined to have taken place if this land extends 0.062 in. (1.58 mm), and for carbide tools when the wear land reaches 0.030 in. (0.76 mm). *Wear* also takes place on the face of the tool in the form of a small crater (Figure 8.16*A*) or depression behind the tip. This depression results from the abrading action of the chip as it passes over the tool face. Locations of wear are seen on the tip of the tool, thermal shock cracks, chipping resulting from local fractures at the cutting edge, and at the point of the depth of cut, which is called depth-of-cut notching.

In Figure 8.16*B*, the progressing amounts of *land wear* on the flank of the tool is shown, and it is this length distance that is plotted on Figure 8.17*A*. Some velocities may not result in wear that is noticeable, for example, turning aluminum bar stock with carbide. Note velocity V_1 in Figure 8.17*A*.

In 1906, Fred W. Taylor reported the relationship between tool life and cutting speed as follows

$$VT^n = C \tag{8.17}$$

where

V = Cutting speed, ft/min
T = Tool life, min
n = Exponent depending on cutting conditions, and empirically found by testing
C = Constant cutting speed for a tool life, min, and empirically found by testing

Since tool life decreases as the cutting speed is increased, *tool life curves* are plotted as tool life in minutes against cutting speed in ft/min (or in rare cases, in.³) of metal removed. In some cases, the life is determined by surface finish measurements and in others by an increase in force on a dynamometer.

Taylor tool life curves are shown in Figure 8.17*B*. One test might have a velocity, V_4, and the time is noted that it takes to reach a specific limit of land wear, say, 0.030 in. for carbide tools. Similarly, another velocity different from V_4 is tested, and the time it takes to reach the land wear limit of 0.030 in. is plotted, as shown in Figure 8.17*A*. The composite of these curves is shown in Figure 8.17*B*, the time it takes to reach the specified standard amount of land wear such as 0.030 in. limit for several velocities.

When the cutting speed is plotted as a function of the tool life on log-log paper, a straight line results, as shown in Figure 8.17*C*; this is the tool life curve. The value of n can be determined by using the formula shown on Figure 8.17*C*, or by statistical methods.

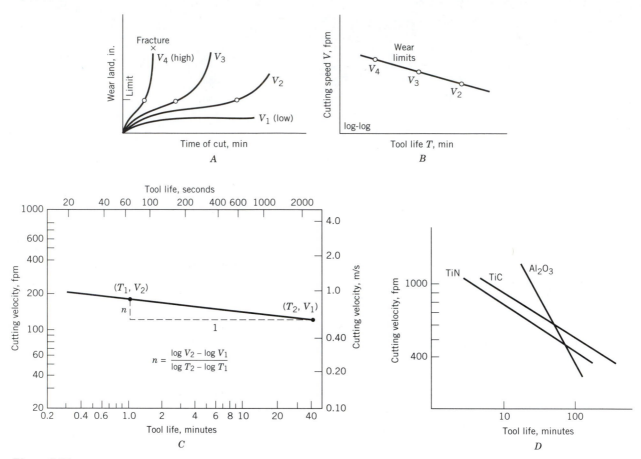

Figure 8.17
A, Several tool life curves for a constant velocity. B, Plotting time to reach wear land limit versus cutting velocity. C, Effect of cutting speed on tool life for a high-speed steel tool. D, Example of coated insert tool life.

The empirical tool life coefficients, n and C, are determined by statistical tests of cutting metal test logs under standard conditions and noting the time for flank wear to reach the specified limit against rotational velocity of the test log. These tests are usually undertaken with turning operations on lathes.

Since the data of Figure 8.17C depend on a given set of conditions, results may not be applicable if the depth of cut or feed is changed. If either of these operating factors is increased, a reduction must be made in cutting speed to obtain the same tool life.

Sometimes smaller flank wear limits are used, which results in *accelerated tool life* testing. Notice Figure 8.17D, where coated inserts are tested with a wear land of 0.010 in. flank wear. The material in this case is AISI 1045 steel, 190 Bhn hardness and the operating conditions were 0.016 ipr and 0.100 depth of cut. Tool life curves may not be directly comparable, as the material, tool, and workpiece, and operating conditions may differ from test to test.

With high velocities it is possible to fracture quickly or instantaneously the tool point and cause severe damage and safety problems. Note V_4 in Figure 8.17A, which resulted in an instantaneous fracture rather than progressive wear.

The exponent n depends on characteristics of the workpiece and tool material. Samples values are as follows.

Material	High-Speed Steel		Tungsten Carbide	
	C	n	C	n
Stainless steel	170	0.08	400	0.16
Medium-carbon steel	190	0.11	150	0.20
Gray cast iron	75	0.14	130	0.25

The intercept C in Figure 8.17C is found at the intersection of the line and tool life of 1 min. The slope is found using the equation

$$n = \frac{\log V_2 - \log V_1}{\log T_2 - \log T_1} \tag{8.18}$$

Note the locations of the points on the figure. Taking measurements from the figure $C = 190$ and $n = 0.1482$, the Taylor tool life equation for this curve is then $VT^{0.1482} = 190$. The equation finds either velocity or tool life given the other quantity. For example, if a life of 50 min is desired, the velocity is found as 106 ft/min. The HSS tool would be removed after 50 min, resharpened to *original signature*, and replaced in the lathe tool holder.

The correct choice of cutting velocity can enhance tool life, but at the same time, the tool should be used to its maximum capacity. Tool failures usually occur for the following reasons.

1. *Improper grinding of tool angles.* Cutting angles depend on the tool and workpiece. Values are given in handbooks and manufacturers' literature.

2. *Loss of tool hardness.* This is caused by excessive heat generated at the cutting edge. This situation is relieved by coolants or by reducing the cutting speed.

3. *Breaking or spalling of tool edge.* This may result by taking too heavy of a cut or by too small of a *tool lip* angle.

4. *Natural wear and abrasion.* Tools gradually become dull by abrasion. This process is accelerated by the development of a crater just back of the cutting edge. As the crater increases in size, the cutting edge becomes weaker and breaks off. This effect can be reduced by the proper selection of tool material.

5. *Fracture of tool by heavy load.* This condition is reduced if the cutting tool is rigidly supported with a minimum of overhang.

8.7 SURFACE FINISH

Machinability is often assessed by surface finish after the machining operation. A good surface finish is affected by many variables in single- or multiple-point machining. The factors improving surface finish are light cuts, small feeds, high-cutting speeds, cutting fluids, round-nose tools, and increased rake angles on well-ground tools. Many of these features, while highly desirable, may shorten tool life, increase forces, and may not be economical overall.

The outermost boundary of a body adjacent to the air is called a *surface*. When this surface is deformed by a sharp cutting edge, the term *surface finish* describes the boundary. When a rotating cylindrical body in contact with a tool has an existing surface removed, the substrate surface provides the new boundary. If the surface was magnified many times, it would have the appearance of successive jagged mountain ranges. Despite the

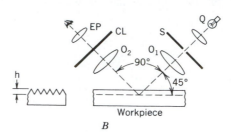

Figure 8.18
A, Light section on work-piece. B, Light principle through lens system.

seemingly random natural appearance of a mountainous profile, there is mathematical form and geometry to the topography. This analogy extends to the microscopic texture of grooves turned by a tool on a lathe.

The importance of surface finish in current technology is recognized by the consumer and engineer. Excessive cost can be attributed to overfinish. However, surface finish contributes to the precision of the dimension. Engineering properties such as fatigue, hardness, heat transfer, and the like, are affected by surface finish.

Surface roughness is subject to measurement electronically and visually. Traditional methods of measurements are discussed in Chapter 26. Another device that provides a nondestructive surface examination is based on the *light section principle*. A diagramatic sketch is given in Figure 8.18. According to this principle an optical cut is made through the surface without touching or scratching, which are two problems with traditional methods. In operation an incandescent lamp, Q, illuminates slit S. A razorlike beam projects through objective O_1 at a 45° angle to the workpiece surface. The band of light is observed through a microscope at the opposite 45° angle. The microscope objective, O_2, has the same magnification as O_1. This fine band of light can observe the peaks and valleys, and a cross line reticle, CL, in the eyepiece, EP, can be shifted within the field of vision. Two measurements are possible: roughness height, or the distance from the peak to the valley, and roughness width. The instrument can measure micron dimensions. With this approach it becomes possible to find surface finish effects that reasonably correspond to the tool-point profile.

In the ideal surface one presumes that the surface is assumed to be machined exactly as the geometrical factors of the tool suggest. Chatter or BUE effects or effects of tool wear are ignored. In turning, the tool point places a screw thread helix on the surface, but the penetration of the tool on the surface is so slight that appearance of a screw thread is not noticeable. Note Figure 8.19 for the single-point effect on a rotating bar stock.

A model of the ideal surface is possible using the geometric analog. Two tool geometries are shown in Figure 8.20 because of their popularity in finishing operations. For the tool

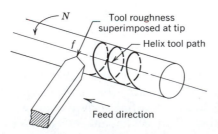

Figure 8.19
Helix effects caused by feeding a tool point along a bar stock.

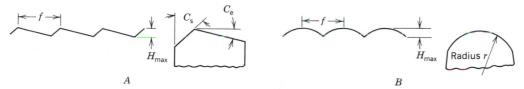

Figure 8.20
Surface geometry models.

plan shown in Figure 8.20A, the maximum surface roughness is given by

$$H_{max} = \frac{f}{\tan C_s + \cos C_e} \tag{8.19}$$

where

$$f = \text{Feed distance, in. (mm)}$$
$$C_s = \text{Side angle, degree (°)}$$
$$C_e = \text{End angle, degree (°)}$$

For the tool plan shown in Figure 8.20B, surface geometry results in the following:

$$H_{max} = \frac{f^2}{8r} \tag{8.20}$$

where

$$r = \text{Radius of the tool, in.}$$

H_{max} is not the same as the arithmetical average as shown by Figure 26.13 where the distance is measured by a midline roughness datum. Thus, those measures must be multiplied by two to compare to H_{max}.

For example, assume a feed dimension of 0.008 in. and $C_s = C_e = 15°$. The tool plan in Figure 8.20A will give a maximum roughness of 0.002 in., or about 100 μin., which is not a fine finish. However, a round-nose tool, say with a ⅟₃₂-in. radius, will result in a surface height maximum of 0.000256 in., or about eight times finer than the above tool. Now these values represent ideal figures, and in practice the operator will determine steady-state roughness using standard methods. If roughness exceeds these values for a predetermined tool geometry, feed, tool, and workpiece material, the operation has exceeded machinability standards.

8.8 CUTTING SPEEDS AND FEEDS

Cutting speed, V_c, is expressed in ft/min (m/s) and on a lathe it is the surface speed or rate at which the work passes the cutter. In turning the work rotates, whereas for milling, the tool rotates and the work is stationary. Thus, cutting speed measures the peripheral velocity of the lathe bar stock and the milling cutter. They are kinematic inversions of each other.

Cutting velocity is expressed by the formula

$$V_c = \frac{\pi DN}{12} \tag{8.21}$$

where

D = Diameter of the rotational workpiece, in.
N = Rotary speed of the workpiece, revolutions per minute (rpm) = $12V_c/\pi D$

The equivalent metric equation is given as

$$V_c = \pi DN/60{,}000 \qquad (8.22)$$

where

D = Diameter of the work, mm

The cutting speed in this expression is seldom unknown, since recommended cutting speeds for many tool and workpiece materials are found in handbooks, supplier recommendations, and textbooks. These recommendations are based, in a general way, on tool life, surface finish, machinability, depth of cut, and many other practical factors. Table 8.2 is an abbreviated table and is used for textbook problems given later. Most values of cutting speeds and feeds are given with a range, but our practice for problems in this book encourages the adoption of the midpoint value. Note that feed and inch per revolution (ipr), are listed in the footnotes of the table and are used with cutting velocity.

In lathe work the unknown factor is the rotary speed or the term N. Referring to Figure 8.21, it is seen that to maintain a recommended cutting speed of 90 ft/min, it is necessary to increase the work revolutions as the diameter is decreased from 5 in. (127 mm) to 1 in. (25.4 mm).

Table 8.2   Typical Cutting Speeds and Feeds for Turning

Material Being Cut	Cutting Material			
	High-Speed Steel		Carbide	
	Finish[a]	Rough[b]	Finish[a]	Rough[b]
Free cutting steels, 1112, 1315	250–350 (1.3–1.8)	80–150 (0.4–0.8)	600–750 (3.0–3.8)	350–450 (1.8–2.3)
Carbon steels, 1010, 1025	225–300 (1.1–1.5)	75–125 (0.4–0.6)	550–700 (2.8–3.5)	300–400 (1.5–2.0)
Medium steels, 1030, 1050	200–275 (1.0–1.4)	70–120 (0.4–0.6)	450–600 (2.3–3.0)	250–350 (1.3–1.8)
Nickel steels, 2330	200–275 (1.0–1.4)	70–110 (0.4–0.6)	425–550 (2.1–2.8)	225–325 (1.1–1.6)
Chromium nickel, 3120, 5140	150–200 (0.8–1.0)	50–75 (0.3–0.4)	325–425 (1.7–2.1)	175–260 (0.9–1.3)
Soft gray cast iron	120–150 (0.6–0.8)	75–90 (0.4–0.5)	350–450 (1.8–2.3)	200–250 (1.0–1.3)
Brass	275–350 (1.4–1.8)	150–225 (0.8–1.1)	600–700 (3.0–3.5)	400–500 (2.0–2.5)
Aluminum	225–350 (1.1–1.8)	100–150 (0.5–0.8)	450–700 (2.3–3.5)	200–300 (1.0–1.5)
Plastics	300–500 (1.5–2.5)	100–200 (0.5–1.0)	400–650 (2.0–3.3)	150–250 (0.8–1.3)

[a] Cut depth, 0.015–0.095 in. (0.38–2.39 mm). Feed, 0.005–0.015 ipr (0.13–0.38 mm/rev).
[b] Cut depth, 0.187–0.375 in. (4.75–9.53 mm). Feed, 0.030–0.050 ipr (0.75–1.27 mm/rev).

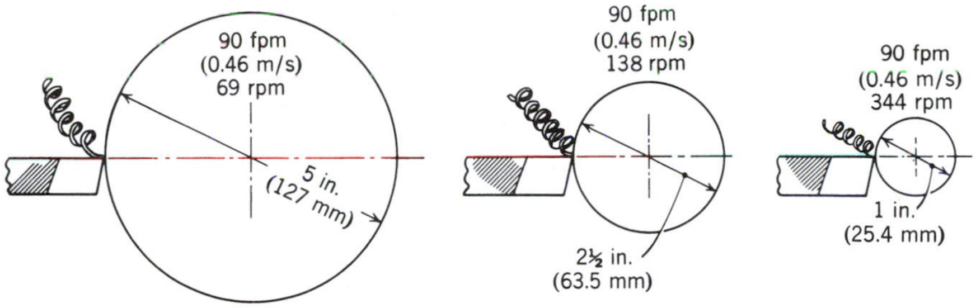

Figure 8.21

Relationship of revolutions per minute to surface velocity using $V_c = \dfrac{\pi D N}{12}$.

The term *feed* refers to the rate at which a cutting tool or grinding wheel advances along or into the surface of the workpiece. For machines in which the work rotates, feed is expressed in ipr (mm/rev). For machines in which the tool or work reciprocates, it is expressed in inches per stroke (mm/stroke); and for stationary work and rotating tools, it is expressed in ipr (mm/rev) of the tool. In milling work feed is expressed in in./tooth. Milling cutters will have from several to many teeth, which gives a different rate, or in./min, for milling.

The following factors require a reduced feed for a given tool: greatly increased cutting speed, harder workpiece, more ductile workpiece, less coolant, dulling of the tool, or a reduced rigidity in the workpiece or machine. Values of feeds corresponding to the speeds given in Table 8.2 are discussed further in later chapters.

QUESTIONS

1. Give an explanation of the following terms:

Orthogonal cutting
Dynamometer
Vertical cutting force
Coefficient of friction
Cutting tool nose radius
Side relief angle
Carbide cutting tool
Cutting speeds and feeds
High-speed steels
Cermet
Cubic boron nitride tools

Discontinuous chips
Continuous chips
BUE
Heat zones
Chip breaker curl
Machinability
Tool life
Accelerated tool life
Tool lip
Light section principle

2. What does the word *orthogonal* mean with regard to cutting tools?

3. List factors affecting the frictional force in single-point tools.

4. What are the reasons that account for the tangential force on a single-point cutting tool as used on a lathe being at least 10 times greater than the radially directed force?

5. Explain how tool forces are affected by changing cutting speed. What factors vary as a result?

6. How is the longitudinal force affected by an increase in side-cutting edge angle?

7. Describe the ideal chip. Why is it ideal?

8. What are effects of a large frictional force on a cutting tool and how can they be reduced?

9. What is a major disadvantage of long continuous chips?

10. Sketch a left-hand cutting tool and label the angles.

11. Why are liquid coolants usually preferred?

12. What are the advantages and limitations associated with using carbide tools?

13. What is the effect of shear angle on the heat generated in cutting?

14. Why are chip breakers used on some tools? How are they constructed?

15. Set up on paper an experiment for measuring machinability using a dynamometer. Explain the data needed and why.

16. List ideal characteristics for a coolant.

17. How can the vibration of a tool and workpiece increase coolant flow?

18. How does ductility affect machinability?

19. What factors cause the tool to lose hardness?

20. If tool life needs increasing for economic reasons, what are the most significant steps you can take?

21. List advantages of the materials used for cutting tools.

22. Categorize tool materials.

23. List three tool cutter materials.

24. Discuss the basic type chips and the materials that usually result in these types.

25. Discuss how tool shape influences surface integrity and surface finish.

PROBLEMS

8.1. A right-hand turning tool is shown as Figure P8.1. Identify the various angles and radius for this single-piece high-speed steel cutter. (Hint: Consider the nomenclature to be the following: side rake angle, end-cutting edge angle, nose radius, end relief angle, side-cutting edge angle, side relief angle, and back rake angle.)

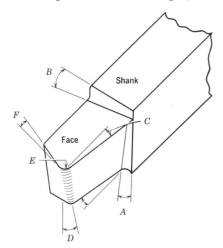

Figure P8.1

8.2. Calculate the rpm for a 2-in. OD gray cast-iron component, if the recommended cutting speed for this carbide insert cutter is 200 ft/min. Repeat for 300 ft/min. Repeat for 3 in. and 200 ft/min.

8.3. Estimate the difference in machinability of SAE 1020 and SAE 1050 steel. (Hint: Include cutting speeds given in Table 8.2.)

8.4. Calculate the revolutions per minute for a lathe-turning operation to cut the following: rough finish 4-in. (102

mm) diameter cast iron with high-speed steel; finish a 12-in. (300-mm) brass bar with high-speed steel; finish a ½-in. (12.7-mm) diameter bronze bar with carbide.

8.5. An orthogonal-type cut has a feed of 0.020 ipr and a tool rake angle of 10°. The chip is 0.035 in. thick.

a. Find the shear angle.

b. If the depth of cut is 0.030 in., what is the shear area of the cut? The yield shear stress of the material is 35,000 psi. Find the shear force.

8.6. An operation has the machining conditions of 0.020 ipr, tool rake angle of 5°, chip thickness after machining of 0.040 in., and depth of 0.030 in. Material yield shear stress is 25,000 psi. Find the shear angle, shear area, and shear force.

8.7. An orthogonal-type cut is made of a pipe 1½-in. diameter. Only one rotation is turned before the pipe is stopped suddenly. The cutter rake angle = 35°; cutting speed = 35 ft/min; depth of cut = 0.015 in.; length of continuous chip for one rotation of the pipe = 2.4 in.; cutting force = 425 lb; and vertical force = 170 lb. Find the coefficient of friction, chip ratio, shear plane angle, and velocity of the chip.

8.8. A block 6 in. long is shaped. Data from a test give a rake angle = 10°; cutting speed = 70 ft/min; depth of cut = 0.025 in.; length of chip = 5.2 in.; horizontal cutting force = 950 lb; and vertical force = 315 lb.

a. Find the coefficient of friction, chip ratio, shear plane angle, and velocity of the chip.

b. Repeat for a 20° rake angle.

c. Repeat for a 30° rake angle.

8.9. A machine tool is machining a steel block where the length of the cut is 3 in., and the chip is straightened to 2.4 in. The rake angle of the tool is 40°. Other known conditions are $V_c = 35$ ft/min, depth of cut = 0.008 in., and horizontal and vertical cutting force = 275 and 95 lb. Find the coefficient of friction, chip ratio, thickness of the chip, shear plane angle, and velocity of the chip along the tool face. Find the horsepower at the spindle and at the motor if the efficiency is 65%.

8.10. An AISI 1045 steel block 6 in. long is orthogonally machined. All testing conditions are consistent except that the rake angle of the tool is changed. Constant data for the test are cutting speed = 75 ft/min; depth of cut = 0.010 in.; length of chip is substantially identical over the tests and is found to be 5.2 in. Other factors vary as follows:

Rake Angle (°)	Horizontal Force (lb)	Vertical Force (lb)
10	925	320
20	950	330
30	974	331
50	1075	340

Determine the coefficient of friction. Discuss your results and provide conclusions for this test.

8.11. Cutting conditions are feed, 0.010 ipr; cutting speed, 175 ft/min; and depth of cut, 0.200 in. The stock diameter is 3.5 in. Find the rpm and the volumetric rate of machining.

8.12. Data from a machinability test indicate $F_s = 750$ lb and $V_c = 375$ ft/min. Find the spindle horsepower. If spindle efficiency = 85%, what is required motor power? Also convert to SI units.

8.13. A dynamometer for a turning operation indicates a cutting force of 1500 N and a circular cutting velocity of 80 m/min. Determine the power.

8.14. A turning operation of a bar stock requires a feed rate of 0.018 ipr, depth of cut = 0.25 in., and a cutting speed of 275 ft/min. Determine the metal removal rate. Repeat for 0.015 ipr and 300 ft/min.

8.15. A boring operation indicates a cutting force of 225 lb and cutting speed of 180 ft/min. Find the spindle horsepower if the efficiency of the spindle = 92%. Find the motor horsepower. Convert to SI units.

8.16. A machine tool is rigged with a wattmeter for input power and a dynamometer for cutting force during a cutting operation. There is a cutting velocity of 50 m/min, wattmeter reading of 2.5 kW, and a force from the dynamometer of 2000 N. Find the overall efficiency of the machine tool.

8.17. For a depth of cut = 0.015 in., cutting velocity = 425 ft/min, and a feed of 0.064 ipr, find the metal removal rate. Also convert to SI units.

8.18. If 3 hp drives a lathe and 70% of this power is used for cutting, how much energy is dissipated in the form of heat? Assume 760 watts equals 1 hp.

8.19. A digital wattmeter indicates 10 kW are required by a turning operation. Data for the operation are cutting speed = 400 ft/min, feed = 0.007 ipr, and depth of cut = 0.10 in. Find the unit horsepower. (Hints: Find the volumetric rate of machining, which is $12 \times V_c \times f \times d$, and has the units of in.³/min. Convert kW into horsepower by 1 hp = 0.746 kW. Unit horsepower has the units of hp/in.³/min.)

8.20. A turning operation on a lathe wattmeter shows 25 kW consumed. The operations process requires 600 ft/min, feed = 0.030 ipr, and a depth of cut of 0.25 in. Determine the unit horsepower.

8.21. Using Figure P8.21, find the tool life equation. What is the life if the high-speed steel tool runs at 200 ft/min? The material for the test is AISI 3140, and the operating conditions are feed = 0.013 ipr and depth of cut = 0.50 in. What is the life if the tool runs at 150 ft/min? What velocity may be expected for 10 minutes? (Hint: Find the velocity by the equation and confirm by the graph.)

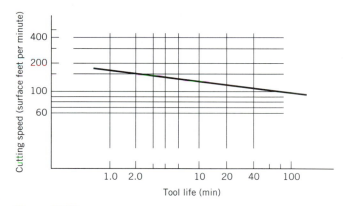

Figure P8.21

8.22. A gray cast-iron wheel is turned on a lathe. Find the cutting speed for 30 min of tool life with a tungsten carbide tool point $VT^{0.373} = 1022$. Repeat for 20 min. Repeat for 60 min.

8.23. Tool life equation parameters are given as $n = 0.15$ and $C = 150$. What circular velocity is found for a life of 60 minutes? If the diameter of the bar stock is 4 in., what rpm is expected?

8.24. Tool life equation parameters are $n = 0.15$ and $C = 190$. What velocity is found for a life of 60 minutes? If the diameter of the bar stock is 5 in., what rpms can be expected?

8.25. The Taylor tool life equation is $VT^{0.1} = 372$. What is the expected average tool life for $V = 275$ ft/min? For $V = 180$ ft/min? Find the velocity for 18 min of life.

8.26. A stainless steel material can be machined by either high-speed steel or tungsten carbide tools. What is the velocity for 30 min of cutting for these two materials? What percent advantage will a 400 ft/min velocity provide with tungsten carbide for tool life? (Hint: Establish your own Taylor tool life equation.)

8.27. A C-7 carbide tool was tested on AISI 8640 steel, 190 Bhn, and turned at several cutting velocities. Conditions were dry, depth of cut = 0.020 in., feed of 0.010 ipr, and a wear limit of 0.015 in. was selected. Plot the tool life curve on log-log scales for the following (T, V) data: (60, 340), (44, 400), (30, 500), (11, 600), and (8, 700). From the graph find the values of n and C. What is the expected life

for a velocity of 550 ft/min using the equation and the curve? Using the graph and equation, find the velocity for 25 min of life.

8.28. Let $C_s = 15°$ and $C_e = 25°$ for a sharp tool point. Find H_{max} for a feed of 0.005 in. Repeat for $C_e = 15°$ and 0.008-in. feed.

8.29. Find the peak to valley roughness with a $\frac{1}{32}$-in. nose radius and a 0.010-in. feed. For a 0.005-in. feed. Repeat for a $\frac{3}{64}$-in. radius tool nose.

8.30. The tool has a 60° included angle sharp point and is used for making V threads. Find the roughness for a 0.020-in. feed. What is the approximate arithmetic average for this roughness using ordinary standards?

8.31. A right-hand lathe-turning tool has a fine-finish depth-of-cut radius penetration of 0.015 in. and a feed of 0.008 ipr. The tool has a side angle of 15° and an end angle of 15°. In addition, the tool has a round nose $\frac{1}{32}$ in. radius. What is the minimum surface roughness expected?

MORE DIFFICULT PROBLEMS

8.32. A tool life graph of TiN is given by Figure 8.17D. For a cutting velocity of 500 ft/min, find rpm, expected tool life, and number of parts before replacement of the tool, if the material is AISI 1045, 190 Bhn, and the bar stock diameter is 8 in. and the length is 20 in. The depth of cut is 0.100 in. There are 250 units to be turned.

8.33. Consider the Taylor tool life model for the following tool materials and work materials

Tool	Work Material	n	C
High-speed steel	Cast iron	0.14	75
High-speed steel	Steel	0.125	47
Cemented carbide	Steel	0.20	150
Cemented carbide	Cast iron	0.25	130
Cast alloy	Stainless steel	0.40	145

What is the cutting velocity for a tool life of 10 min for each of these combinations?

8.34. A machine tool is machining a steel block where the length of the cut is 6 in., and the chip is straightened to 4.8 in. The rake angle of the tool is 40°. Other known conditions are $V_c = 35$ ft/min, depth of cut = 0.008 in., and horizontal and vertical cutting force = 300 and 100 lb. Find the coefficient of friction, chip ratio, thickness of the chip, shear plane angle, and velocity of the chip along the tool

face. Find the horsepower at the spindle and at the motor if the efficiency is 85%.

8.35. An AISI 1045 steel block 4 in. long is orthogonally machined. All testing conditions are consistent except that the rake angle of the tool is changed. Constant data for the test are cutting speed = 80 ft/min, depth of cut = 0.010 in; and length of chip is substantially identical over the tests and found to be 3.2 in. Other factors varied as follows

Rake Angle (°)	Horizontal Force (lb)	Vertical Force (lb)
10	925	320
20	950	330
30	974	331
50	1075	340

Make conclusions about this test and provide recommendations for cutting.

8.36. Note the tool life curve given in Figure 8.17C. Assume that the material is AISI 4140 steel, feed of 0.015 in., depth of cut 0.150 in., and a HSS tool material.

a. Use the curve and find the values for the Taylor's tool life equation, $VT^n = C$.

b. What is the tool life for 150 ft/min?

c. What are the rpms for 4-in. bar stock and $V = 100$ ft/min? For 200 ft/min?

d. What is the operating rpm for a 25-mm bar stock and $V = 0.50$ m/s?

8.37. A testing apparatus for a turning operation uses a wattmeter to measure input power and a dynamometer to measure the cutting force. The workpiece is 2.5 in. OD. The rotational rpm of the bar stock is 230 rpm, depth of cut is 0.125 in., and the feed is 0.006 ipr. The dynamometer gives a cutting force measurement 450 lb and the wattmeter for a concurrent value in time gives a measurement of 1.8 kW for the motor.

a. Find the input hp.

b. Calculate the hp at the cutting point.

c. Find the efficiency of the operation.

8.38. A metal-cutting operation has a feed of 0.010 ipr (0.254 mm/pr) and a tool rake angle of 5°. Measurement of the chip by a point micrometer gives a thickness of 0.043 in. (1.09 mm).

a. Determine the shear angle in customary units. Repeat in metric units.

b. If the tool rake angle is increased to 20°, and the chip thickness is found to be 0.025 in. (0.635 mm), what is the shear angle?

c. A graphic design of the geometry of the tool, workpiece, and chip is provided for sketches (C) and (D), as shown in Figure P8.38. (c) Using a scale determine the chip thickness and confirm your results analytically. (d) Using a protractor, determine the shear angle of (d), and confirm your results analytically.

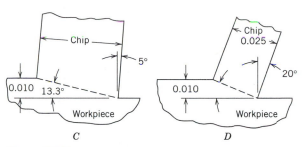

Figure P8.38

d. If the depth of cutting for problems a and b is 0.250 in., what are the areas of shear? (Hint: It is necessary to find the shear length.) If the material being cut flows plastically in shear with a yield shear stress of 27,000 psi, find the shear force for the two cases of cutting. Which is the preferred rake angle?

8.39. Aircraft alloy steel bar stock is turned on a lathe with a two-pass two-tool system using HSS tools. One tool is used for roughing and the second is used for finish machining. Initial diameter and length are 6 and 40 in. The roughing pass will remove 0.5 in. on the radius. The finish pass will remove 0.015 in. on the radius. Roughing and finishing machining circular speeds and feeds are (150, 0.015) and (200, 0.007). For rough turning the Taylor tool life coefficients are $n = 0.25$ and $C = 500$. Find the rpms, time to machine, and the recommended number of components before tool replacement is required for rough turning. (Hint: Time to machine is the length divided by the rate of cutting expressed in in./min.)

PRACTICAL APPLICATION

There are many opportunities for a practical experience in metal cutting. Your instructor may suggest several:

1. An interesting possibility is the collection of chips from several similar operations. In this study, one can tabulate the work material and its condition, tool material and its signature, cutting conditions of velocity, feed, depth of cut, and machine tool factors. The student can correlate the chips to operating conditions.

2. Another possibility is the actual machining of a part in the school machine shop, if there is one.

CASE STUDY: ADVANCED JOB SHOP

Bob Williams, manufacturing engineer for Advanced Job Shop, is considering the prospect of a large new order for turning AISI 8140 steel bar stock. He is discussing the order with the Vice President of Operations, Frank Essenburg.

"Frank, we are going to be making a lot of chips for this Acme job," Bob says, while looking at the purchase order. "It calls for turning diameters of at least 8 in., and we don't have much experience in that size and grade."

"This job doesn't have much profit in it, Bob," Frank comments while looking at the material specification. "We had to bid competitively to keep our shop labor working while we get through this slump. Perhaps, there is something that you can do to make a better profit on the job, Bob. After all, we have that new lathe."

"Umm, with the quantity that we are to deliver, it might be useful to conduct a machinability test before releasing the order to the shop."

"What is a machinability test, Bob?"

Bob explains the function of the test and at the end Frank remarks by saying, "Bob, it is important to AJS that we make money on this job because of the risk involved. Do your testing and get back to me with the results and recommendations."

Bob reasons that a tool life test with various velocities will economize on the machining part of the job. He selects bar stocks identical to the new job, and instructs the lathe operator to run a test with the following conditions: Depth of cut = 0.050 in. and a feed = 0.010 ipr. The lathe operator is instructed to vary the job with three different rpms based on selected cutting velocities; $V_1 \doteq 600$ ft/min, $V_2 = 525$ ft/min, and $V_3 = 480$ ft/min. The machinist is to examine the tool point for wear using a portable microscope especially made for measuring the length of wear land along the flank of the tool. From time to time, the machinist will stop the work, measure the wear, and note the time on a table that

Bob constructs. The machinist provides Bob with the following

\(V_1\)		\(V_2\)		\(V_3\)	
min	Wear (in.)	min	Wear (in.)	min	Wear (in.)
3	0.005	6	0.007	5	0.004
6	0.011	11	0.013	9	0.005
9	0.022	13	0.018	13	0.010
10	0.040	15	0.024	19	0.017
11	Fracture	17	0.036	23	0.027
				25	0.035
				35	0.041

Bob puzzles over the data and he begins to plot it similar to Figure 8.17A. He uses a graph to find the intersection of the point for 0.030 in. of wear for each velocity. Then he uses log-log graph paper and constructs the tool life line and finds n and C from the drawn line. At that point, he ponders his results, and says: "Each part should require about 25 min of machining, and it is better to replace the tool after the part is finished than during a run." What suggestions can you provide to help the optimization? Use log-log graph paper to plot the data points, or you may wish to do the analysis with a computer spreadsheet and plot the data on logarithm-scaled paper. Do the analysis and submit your recommendations for the machining operation for this material specification and diameter.

CHAPTER 9

TURNING, DRILLING, BORING, AND MILLING MACHINE TOOLS

This chapter discusses very common production processes. Their importance to manufacturing cannot be understated, and in terms of the dollar business nationally, turning, drilling, boring, and milling fabrication work rank number one. These machine tools are frequently found in school shops.

The oldest and most common machine tool is the *lathe*, which removes material by rotating the workpiece against a single-point cutter. Parts are held between centers, attached to a faceplate, supported in a jaw chuck, or gripped in a draw-in chuck or collet. Although this machine is suited to turning cylindrical work, it also is used for other operations. Plane surfaces can be machined by supporting the work in a faceplate or chuck. Work held in this manner can be faced, centered, drilled, bored, or reamed where the tool is approached from the end or "face" of the stock. In addition, a lathe can knurl, cut threads, or turn tapers.

The *drill press* is one of the simplest machine tools used in production and tool room work. Drilling produces a hole in an object by forcing a rotating drill against it. Holes can be accomplished by holding the drill stationary and rotating the work, such as drilling on a lathe with the work held and rotated by a chuck. Whereas the drill press is essentially a single-purpose machine, dissimilar operations are possible with other cutting tools on this machine tool. Although drilling is commonplace, it should be noted that "making holes" in manufacturing is an extraordinary activity, if only when one considers the billions of holes that are manufactured by the countless methods every day. "Hole making," in its broadest understanding, is a very frequent manufacturing operation.

Perhaps the *Power Age* was born through boring, as the *boring mill* made James Watt's steam engine a reality by machining round cylinders true throughout the cylinder's length. Boring is enlarging a hole that has already been drilled or cored. Principally, it is an operation of truing a hole that has been drilled previously with a single-point lathe-type tool. For this operation on a drill press, a special holder for the boring tool is necessary. However, the boring operation is found with machines other than the drill press.

A milling machine removes metal when the work is fed against a rotating cutter. Except for rotation, the circular-shaped cutter has no other motion. The milling cutter has a series of cutting edges on its circumference, each acting as an individual cutter in the cycle of rotation. The work is held on a table, which controls the feed against the cutter; this is called milling. In most machines there are three possible table movements, longitudinal, crosswise, and vertical, but in some the table also possesses a swivel or rotational movement.

The milling machine is the most versatile of all machine tools. Flat or formed surfaces may be machined with excellent finish and accuracy. Angles, slots, gear teeth, and recess

A   B

Figure 9.1
A, Diamond-shaped cutting insert, rough turning barstock. *B*, Machining center mounted with milling cutter having coated physical vapor deposition on cemented carbide inserts.

cuts can be made with various cutters. Drills, reamers, and boring tools can be held in the arbor socket. Since table movements have micrometer adjustments, holes and other cuts can be dimensioned accurately. Most operations performed on shapers, drill presses, gear-cutting machines, and broaching machines can be done on the milling machine. Heavy cuts can be taken with little appreciable sacrifice in finish or accuracy. Cutters are efficient in their action and tool life is excellent. In most cases the work is completed in one pass of the table. These advantages, plus the availability of a variety of cutters, make the milling machine indispensable in production, toolroom, and the school shop.

Turning is pictured in Figure 9.1*A*. A cemented carbide insert, with a chip breaker formed next to the edge, can be used in both stationary lathe and rotating (machining center/mill turn center) applications, enabling users to employ the same tools on a number of different machine tools. Figure 9.1*B* shows a milling cutter tooled with coated physical vapor deposition (PVD) on cemented carbide inserts. The tool is milling a steel workpiece. A milling cutter will have several inserts.

9.1 LATHE GROUP

Engine Lathe

The "engine" lathe derived its name from the steam "engine" of the eighteenth and nineteenth century where the lathe is connected to pulley power from overhead line shafts or connected directly to the steam engine. The principal parts of a modern lathe are labeled in Figure 9.2. Controls on the side of the headstock allow selection of any one of many speeds that are arranged in logical geometric progression. A combination electric chuck and brake is provided for starting, stopping, or jogging the work.

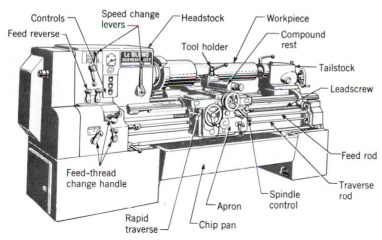

Figure 9.2
Heavy-duty engine lathe.

The *tailstock* can be moved along the bed of the lathe to accommodate different lengths of stock. It is commonly provided with a hardened ball-bearing center, which may be moved in and out by wheel adjustment. The tailstock can be adjusted in and out with respect to the center line of the bed. This allows for adjusting the alignment of the centers and for taper turning.

The *lead screw* is a long, carefully threaded shaft located slightly below and parallel to the bedways extending from the headstock to the tailstock. It is geared to the headstock and its rotation may be reversed; it is fitted to the *carriage assembly* and may be engaged or released from the carriage during cutting operations. The lead screw is for cutting threads and is disengaged when not in use to preserve its accuracy. Below the lead screw is a *feed rod*, which transmits power from the quick change box to drive the apron mechanism for cross and longitudinal power feed. Changing the speed of the lead screw or feed rod is done at the quick change gear box located at the headstock end of the lathe.

The carriage assembly includes the *compound rest, tool saddle*, and *apron*. Since it supports and guides the cutting tool, the carriage assembly must be rigid and constructed with accuracy. Two-hand feeds are provided to guide the tool on a crosswise motion. Obviously, these feeds can be power controlled as well. The upper-hand wheel or crank controls the motion of the compound rest, and since this rest is provided with a swiveling adjustment protractor, it can be adjusted for various angle positions for short taper turning. A third hand wheel moves the carriage along the bed, usually to pull it back to the starting position after the lead screw has carried it along the cut. The portion of the carriage that extends in front of the lathe is an apron. It is a double-walled casting, which contains the controls, gears, and other mechanisms for feeding the carriage, and cross slide by hand or power. On the face of the apron are mounted the various wheels and levers.

Lathe size is expressed in terms of the diameter of the work it will swing. A 16-in. (400-mm) lathe has sufficient clearance over the bed rails to handle work 16 in. (400 mm) in diameter. A second dimension is necessary to define capacity in terms of workpiece length. Some manufacturers use maximum work length between the lathe centers, whereas others express it in terms of bed length. The diameter that can be turned between centers is somewhat less than the swing because of the allowance for the carriage.

There are varieties of lathes, and their design depends on production or the nature of the workpiece. A *speed lathe* is the simplest of all lathes, and consists of a bed, headstock, tailstock, and an adjustable slide for supporting the tool. Usually, it is driven by a variable speed motor internal to the headstock, although the drive may be a belt to a step-cone pulley. Because hand tools are used and the cuts are light, the lathe is driven at high speed. The work is held between centers or attached to a faceplate on the headstock. The speed lathe is used in turning wood, centering metal cylinders, and in metal spinning.

The *engine lathe* differs from a speed lathe because of additional features for controlling the spindle speed and supporting and controlling the feed of the fixed cutting tool. There are several variations in headstock design for supplying power to the machine. Light- or medium-duty lathes receive their power through a short belt from the motor or from a small cone pulley countershaft driven by the motor. The headstock is equipped with a four-step cone pulley, which provides a range of four spindle speeds when connected directly from the motor countershaft. In addition, these lathes are equipped with back gears that, when connected with the cone pulley, provide four additional speeds. Figure 9.2 is a geared-head engine lathe. The spindle speeds are varied by a gear transmission, the different speeds being obtained by setting speed levers in the headstock. These lathes are usually driven by a constant speed motor mounted on the lathe, but in a few cases variable speed motors are used. A geared head lathe has the advantage of a positive drive and has a greater number of spindle speeds available than are found on a step-cone-driven lathe.

A *bench lathe* is a small lathe that is mounted on a workbench. It has the same features as engine lathes, but differs only in size and mounting. It is adapted to small work having a maximum swing capacity of 10 in. (250 mm) at the faceplate.

A toolroom engine lathe is equipped with all the accessories for accurate tool work, because it is an individually driven geared head lathe with a considerable range in spindle speeds. It is equipped with center steady rest, quick change gears, lead screw, feed rod, taper attachment, thread dial, chuck, indicator, draw-in collet attachment, and a pump for a coolant. Toolroom lathes are tested carefully for accuracy and, as the name implies, are especially adapted for making small tools, test gages, dies, and other precision parts.

Turret Lathe

Turret lathes are a major departure from engine or basic lathes. These machines possess special features that adapt them to production. The "skill of the worker" is built into these machines, making it possible for inexperienced operators to reproduce identical parts. In contrast, the engine lathe requires a skilled operator and requires more time to produce parts that are dimensionally the same. The principal characteristic of turret lathes is that the tools for consecutive operations are set up for use in the proper sequence. Although skill is required to set and adjust the tools properly, once they are correct, less skill is required to operate the turret lathe. Many parts can be produced before adjustments are necessary.

Horizontal turret lathes are made in two general designs and are known as the ram and saddle. In appearance they are similar, and both may be used for either bar or chucking work. The *ram-type turret lathe*, shown in Figure 9.3A, is so named because of the way the turret is mounted. The turret is placed on a slide or ram, which moves back and forth on a saddle clamped to the lathe bed. This arrangement permits quick movement of the turret and is recommended for bar and light-duty chucking work. The saddle, although capable of adjustment, does not move during the operation of the turret. Ram-

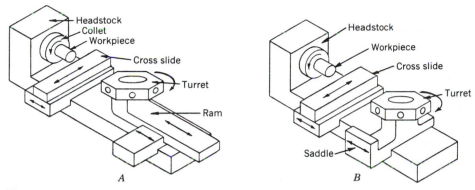

Figure 9.3
Horizontal turret lathes. *A*, Ram type. *B*, Saddle type.

type machines do not require the rigidity of chucking machines, because bar tools can be made to support the work.

A saddle-type lathe, shown in Figure 9.3*B*, is used for chucking work, and the turret is mounted directly on a saddle that moves back and forth with the turret. Since chucking tools overhang and are unconnected with the work through some sort of support, greater strain on both work and tool support results. Chucking tools must have rigidity. The stroke is longer, which is an advantage in long turning and boring cuts, and saddle mounts assist the rigidity.

Turret lathes are constructed similar to engine lathes. The headstock in most cases is geared with provision for 6 to 16 spindle speeds and double these with a two-speed motor. The spindle speeds, as well as forward and reverse movement, are controlled by levers extending from the head. The drive motor is sometimes located in or on the motor leg below the headstock and connected to the geared head sheave by V-belts, or a flange mounted motor may be connected directly to the drive shaft in the head. If a direct current motor is used, variable rpms are available.

The cross-slide unit, on which the tools are mounted for facing, forming, and cutting off, is somewhat different in construction from the tool post and carriage arrangement used on engine lathes. It is made up of four principal parts: cross slide, square turret, carriage, and the apron. These parts are discernable in the various turret lathe illustrations. Some of the cross slides are supported entirely on the front and lower front ways, permitting more swing clearance. This arrangement is frequently utilized on saddle-type machines, which are used for large diameter chucking jobs. The other arrangement for mounting has the cross slide riding on both upper bedways and is further supported by a lower way. This is used on machines engaged in bar work and in other processes where a large swing clearance is not necessary. An advantage of this type is the added tool post in the rear, frequently used in cutting off operations.

A *square turret* is mounted on the top of the cross slide and is capable of holding four tools. If several different tools are required, they are set up in sequence and can be quickly indexed and locked in correct working position. So that cuts may be duplicated, the slide is provided with positive stops or feed trips. Likewise, the longitudinal position of the entire assembly may be controlled by positive stops on the left side of the apron. Cuts may be taken with square turret tools simultaneously with tools mounted on the hexagon turret.

An outstanding feature is the *turret* in place of the tailstock. This turret mounted on either the sliding ram or the saddle, or on the back of the structure, carries anywhere from 4 to 18 tool stations. The tools are preset for the various operations. The tools are mounted in proper sequence on the various faces of the turret so that as the turret indexes between machining operations, the proper tools are engaged into position. For each tool there is a stop screw or electric/electronic transducer, which controls the distance the tool will feed and cut. When this distance is reached, an automatic trip lever stops further movement of the tool by disengaging the drive clutch.

The difference between the engine and turret lathes is that the turret lathe is adapted to quantity production work, whereas the engine lathe is used primarily for miscellaneous jobbing, toolroom, or single-operation work. The features of a turret lathe that make it a quantity production machine are

1. Tools may be set up in the turret in the proper sequence for the operation.
2. Each station is provided with a feed stop or feed trip so that each cut of a tool is the same as its previous cut.
3. Multiple cuts can be taken from the same station at the same time, such as two or more turning and/or boring cuts.
4. Combined cuts can be made; that is, tools on the cross slide can be used at the same time that tools on the turret are cutting.
5. Rigidity in holding work and tools is built into the machine to permit multiple and combined cuts.
6. Turret lathes may be fitted with attachments for taper turning, thread chasing, and duplicating, and can be tape controlled.

Numerical-Controlled Turret Lathe

A heavy duty two-axis turret lathe with numerical control is shown in Figure 28.1. It is designed especially for heavy-duty production. The control provides automatic functioning of spindle speed, slide movement, feeds, turret indexing, and other auxiliary functions. The slant bed, inclined rearward from the vertical, provides maximum rigidity and operator accessibility to the work area. This machine can be set up quickly for small-lot jobs, normally changing only jaw chucks, control tape, and possibly one or two cutters.

Vertical Turret Lathe

A vertical turret lathe resembles a vertical boring mill, but it has the characteristic turret arrangement for holding the tools. It consists of a rotating chuck or table in the horizontal position with the turret mounted above on a cross rail. In addition, there is at least one side head provided with a square turret for holding tools. All tools mounted on the turret or side head have their respective stops set so that the length of cuts can be the same in successive machining cycles. It is, in effect, the same as a turret lathe standing on the headstock end, and it has all the features necessary for the production of duplicate parts. This machine was developed to facilitate mounting, holding, and machining of large diameter heavy parts. Only chucking work is done on this kind of machine.

A vertical turret lathe, shown in Figure 9.4A, is provided with two cutter heads: the swiveling main turret head and the side head. Another side head is possible. The turret and side heads function in the same manner as the hexagonal and square turrets on a horizontal lathe. To provide for angle cuts both the ram and turret heads may be swiveled 30° right or left of center. The side head has rapid traverse and feed independent of the turret and, without interference, provides for simultaneous machining adjacent to

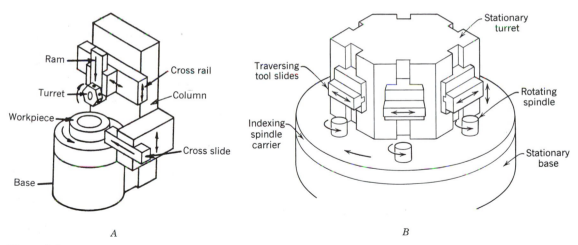

Figure 9.4
A, Vertical turret lathe. B, Multispindle vertical chucking lathe.

operations performed by the turret. The ram provides another tool station on the machine, which can be operated separately or in conjunction with the other two.

The machine can be provided with a control that permits automatic operation of each head including rate and direction of feed, change in spindle feed, indexing of turret, starting, and stopping. Once a cycle of operations is preset and tools are properly adjusted, the operator need only load, unload, and start the machine. Production rate is increased over those manually operated machines, because they operate almost continuously and make changes from one operation to another without hesitation or fatigue. By reducing the handling time and making the cycle automatic, an operator can attend more than one machine.

Automatic Vertical Multistation Lathe

Machines, as illustrated in Figure 9.4B, are designed for high production and are usually provided with either five or seven work stations and a loading position. In some machines two spindles are provided at each station, doubling the capacity for small diameter work. The work is mounted in chucks, the larger machines having chucking capacities up to 18 in. (460 mm) in diameter. Both plain and universal heads may be used, the latter providing for tool feed in any direction.

All varieties of machining operations can be performed, including milling, drilling, threading, tapping, reaming, and boring. The advantage of the automatic vertical multistation lathe is that all operations can be done simultaneously and in proper sequence. At each station, except the loading one, an operation is performed that leads to the completion of the part when it has been indexed through all the working stations. The work cannot be indexed to the succeeding positions until the operation requiring the longest time is complete. Each time when all operations are complete, the work spindles stop, the tools are retracted, and the worktable holding the spindles is indexed to the next position.

Automatic Lathe

Lathes that have their tools automatically fed to the work and withdrawn after the cycle is complete are known as automatic lathes. Since most lathes of this type require that the operator place the part to be machined in the lathe and remove it after the work is complete, they are perhaps incorrectly called automatic lathes. Lathes that are fully *automatic* are provided with a magazine feed so that a number of parts can be machined, one after the other, with little attention from the operator. Machines in this group differ principally in the manner of feeding the tools to the work. Most machines, especially those holding the work between centers, have front and rear tool slides. Others, adapted for chucking jobs, have an end tool slide located in the same position as the turret on the turret lathe. These machines also may have the two side tool slides.

Automatic Screw Machine

The automatic screw machine was invented by Christopher N. Spencer of the Billings and Spencer Company in 1873. The principal feature of the invention provides a controlling movement for the turret so that tools can be fed into the work at desired speeds, withdrawn, and indexed to the next position. This was accomplished by a cylindrical or drum cam located beneath the turret. Another feature, also cam controlled, is a mechanism for clamping the work in the collet, releasing it at the end of the cycle, and then feeding unworked bar stock against the stop. These features are still used in about the same way as originally designed.

The first machines of this type were used mainly for manufacturing bolts and screws; thus, the name "automatic screw machines." Since it can produce parts one after the other with little attention from the operator, it is called automatic. Most automatic screw machines not only feed in an entire bar of stock, but also are provided with a magazine so that bars can be fed through the machine automatically. Automatic screw machines may be classified according to the turret or the number of spindles.

Figure 9.5A illustrates an automatic screw machine designed for bar work of small diameter. This machine has a cross slide capable of carrying tools both front and rear and a turret mounted in a vertical position on a slide with longitudinal movement. The tools used in the machine are mounted around the turret in a vertical plane in line with the spindle. Usual machining operations such as turning, drilling, boring, and threading can be done on these machines. The bar stock, whether round, square, hexagonal, or of some special shape, is determined by the cross section preferred in the finished product. Collets are available for most commercial shapes.

Multispindle machines, however, are not usually spoken of as screw machines, but rather as *multispindle automatics*. The work that the two machines do is the same, although there is a difference in the design and production capacity.

Figure 9.5C and Figure 9.6 are the end view and schematic of a Swiss-type screw machine developed for precision turning of small parts. The single-point tools used on this machine are placed radially around the carbide-lined guide bushing through which the stock is advanced during machining operations. Most diameter turning is done by the two horizontal tool slides, whereas the other three are used principally for knurling, chamfering, cutting off, and recessing. The stock is held by a rotating collet in the headstock back of the tools, and all longitudinal feeds are accomplished by a cam that moves the headstock forward as a unit. This forward motion advances the stock through the guide bushing and to the single-point tools, which are controlled and positioned by cams. By coordinating their movement with the forward movement of the stock, small diameters on slender parts can be held to tolerances ranging from 0.0002 to 0.0005 in. (0.005 to 0.013 mm).

These machines are called "Swiss" because their development began in Switzerland, a land noted for the production of watches, and the necessity for producing very small shafts for the watch mechanism.

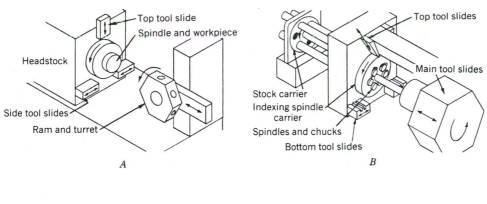

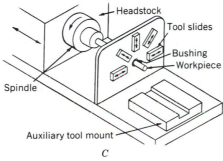

Figure 9.5
A, Single-spindle automatic screw machine. *B,* Multispindle automatic screw machine. *C,*
Swiss automatic screw machine.

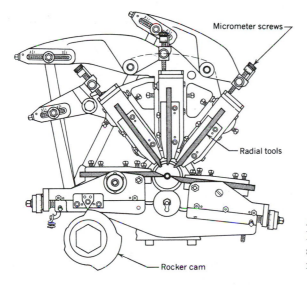

Figure 9.6
End view of Swiss-type screw machine
showing rocker cam and tool control
mechanism.

Multispindle Automatic

Multispindle automatic machines are the fastest type of production machines for bar work. They are fully automatic in their operations and are made in a variety of models with two, four, five, six, or eight spindles. In these machines the steps of the operation are divided so that a portion of it is performed at each of the several stations simultaneously, thereby shortening the time required to finish one part. One piece is completed each time the tools are withdrawn and the spindles indexed.

The construction of a multispindle automatic bar machine is shown in Figure 9.7. A schematic is shown in Figure 9.5*B*. The spindles carrying the bar stock are held and rotated in the stock reel. In front of the spindle is an end tool slide on which tools are placed in line with each of the spindles of the machine. The tool slide does not index or revolve with the spindle carrier, but moves forward and back to carry the end working tools to and from contact with the revolving bars of stock. The tooling section also includes one cross slide for each spindle position. Slides are independently operated and are used in connection with end slide tools for such operations as form turning, knurling, thread rolling, slotting, and cutting off. Tools for such operations as drilling and threading are mounted on the end tool slide.

Bars of stock are loaded into each spindle when it has been indexed to the first position. If automatic stock feeding is used, it is done in the lower spindle position at the rear of the machine. In operation the spindle carrier is indexed by steps to bring the bar of stock in each of the work spindles successively in line with the various tools held on the tool slides. All tools in the successive positions are at work on different bars at the same time. The time to complete one part is equal to the time of the longest operation, plus the time necessary for withdrawing the tools and indexing to the next position. This time can frequently be reduced to a minimum by dividing the long cuts between two or more operations. Multispindle automatics are not limited to bar stock but may be provided with hydraulic or air-operated chucks for holding individual pieces.

A variety of parts can be produced by a multispindle automatic, the only limiting factor being the capacity of the machine. Long-run jobs are necessary to offset the high initial investment, maintenance, and tooling costs. Both single-spindle automatics and

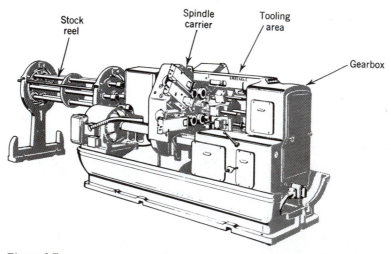

Figure 9.7
Multispindle automatic bar machine.

hand turret lathes have wide application, and in short- and medium-run work prove to be economical in operation. Each machine is good in its field, but care must be taken in making the initial selection.

Duplicating Lathe

Duplicating or *tracer lathes* reproduce a number of parts from either a master form or a sample of the workpiece. A schematic is shown by Figure 7.20A. Most any standard lathe can be modified for tracer work and special automatic tracer lathes are available. Reproduction is from a template, either round or flat, which is generally mounted at the rear of the lathe. It is engaged by a stylus that is actuated by air, hydraulic, or electric means. Many kinds of cuts can be made on the *duplicating lathe* using only a single-point tool. Other duplicating models are usually furnished with a point-to-point numerical control system having a direct reading decimal dial input.

9.2 DRILL PRESS GROUP

Drill Press

Drill presses are very common, useful, and inexpensive relative to other machine tools. Drill presses are often classified according to the maximum diameter drill that they will hold as in portable drilling units. The size of a sensitive or upright drilling machine is designated by the diameter of the largest workpiece that can be drilled. Thus, a 24-in. (600-mm) machine has at least 12-in. (300-mm) clearance between the center line of the drill and the machine frame. The size of radial drilling machines is based on the length of the arm in feet (meters). Usual sizes are 4 ft (1.2 m), 6 ft (1.8 m), and 8 ft (2.4 m). In some cases the diameter of the column in inches (millimeters) is also used in expressing size.

Portable and Sensitive Drill

Portable drills are small compact drilling machines used principally for drilling operations that cannot be done conveniently on a regular drill press. The simplest of these is the hand-operated drill. Most portable drills equipped with small electric motors operate at fairly high speeds and accommodate drills up to ½ in. (12.7 mm) in diameter. Similar drills using compressed air as a means of power are used where sparks from the motor may constitute a fire hazard.

The *sensitive drilling press* is a small high-speed machine of simple construction similar to the ordinary upright drill press. It consists of an upright, standard, horizontal table and a vertical spindle for holding and rotating the drill. Machines of this type are hand fed, usually by means of a rack and pinion drive on the sleeve holding the rotating spindle. These drills may be driven directly by a motor, belt, or *friction disk drive*. Sensitive drill presses are for light work and are seldom capable of rotating drills more than ⅝ in. (16 mm) in diameter. The term "sensitive" refers to the operator response when drilling hard or soft holes, as the operator's touch in forcing the drill responds appropriately to this resistance.

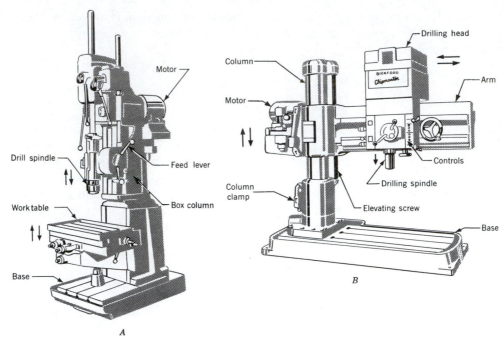

Figure 9.8
A, Thirty-nine inch (1 m) upright drill press. *B*, Radial drilling machine.

Upright Drill

Upright drills similar to sensitive drills have power-feeding mechanisms for the rotating drills and are designed for heavier work. Figure 9.8*A* shows a 39-in. (1-m) machine with a box-type column. A *box column* machine is more rigid than a round column machine and, consequently, is adapted to heavier work. These drilling machines tap as well as drill.

Radial Drilling Machine

The radial drilling machine is designed for large work when it is not feasible to move the work. Such a machine, shown in Figure 9.8*B*, consists of a vertical column supporting an arm that carries the drilling head. The *arm* may be swung around to any position over the work bed, and the drilling head has a radial adjustment along this arm. These adjustments permit the operator to locate the drill over any point on the work. Plain machines of this type will drill only in the vertical plane. On semiuniversal machines the head may be swiveled on the arm to drill holes at various angles in a vertical plane.

Gang Drilling Machine

When several drilling spindles are mounted on a single table, it is known as a *gang drill*. Note Figure 9.9*A* for a three-spindle schematic. This type is adapted to production work where several operations must be performed. The work is held in a jig, which can be moved on the table from one spindle to the next. If several operations must be performed, such as drilling two different-size holes and reaming, four spindles are set up. With

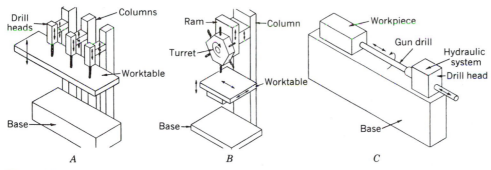

Figure 9.9
A, Gang drilling machine. B, Turret drilling machine. C, Gun drilling machine.

automatic feed control two or more of these operations may be simultaneous and attended by one operator. The arrangement is similar to operating several independent drill presses and is more convenient because of its compactness.

Turret Machine

A turret machine overcomes the floor space restriction caused by a gang drill press. A six-turret station NC drill press is shown in Figure 9.9B. The stations are set up with a variety of tools. Numerical control is also available. Two fixtures can be located side by side on the worktable, thus permitting loading and unloading of one part while the other part is being machined; this reduces the machine cycle.

Deep-Hole Drilling Machine

Several problems not encountered in ordinary drilling operations arise in drilling long holes in rifle barrels, long spindles, connecting rods, and certain oil well drilling equipment. As the hole length increases, it becomes more difficult to support the work and guide the drill. Rapid removal of chips from the drilling operation becomes necessary to ensure the operation and accuracy of the hole. Rotational speeds and feeds must be determined carefully, since there is greater possibility of deflection and wander of the drill than when a drill of shorter length is used.

Deep-hole drilling machines have been developed to overcome these problems. These machines are either horizontal or vertical. The work or the drill may revolve. Most machines are of horizontal construction using a center-cut gun drill, which has a single cutting edge with a straight flute running throughout its length. Observe Figure 9.9C. Oil under high pressure is forced to the cutting edge through a lengthwise hole in the drill. In gun drilling the feed must be light to avoid deflecting the drill and causing it to meander through its length.

Multispindle Drilling Machine

Multispindle or cluster drilling machines, as shown in Figure 9.10, drill several holes simultaneously. The holes may not be the same diameter. These production machines drill many parts with accuracy. Large quantity is necessary for this equipment. Usually, a jig or fixture provided with hardened bushings guides the drills into the work.

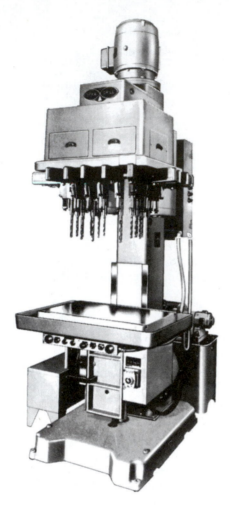

Figure 9.10
General-purpose, adjustable multispindle drilling
and tapping machines.

A common design of this machine has a head assembly with a number of fixed upper spindles driven from pinions surrounding a central gear. Corresponding spindles are located below this gear and are connected to the upper ones by a tubular drive shaft and two universal joints. Three lower spindles carrying the drills can be adjusted over a wide area.

Multispindle drilling machines frequently use a table feed eliminating the movement of the heavy-geared head mechanism that rotates the drills. This is done by rack and pinion drive, lead screw, or by a rotating plate cam. The last method provides varying motions, which give rapid approach, uniform feed, and quick return to starting position.

9.3 BORING MACHINE TOOL GROUP

Boring Machine

Several machines have been developed that are specially adapted to boring work. Boring is the operation that enlarges a preexisting internal hole. Figure 9.11 shows an indexable-insert boring bar with through-coolant.

Figure 9.11
Boring operation with carbide insert and through-bar coolant.

Jig-Boring Machines

The *jig borer* is constructed for precision work on jigs and fixtures. Similar in appearance to a drill press, a *jig-boring machine* will drill and end mill in addition to bore. The vertical boring mill and the horizontal boring machine are adapted to large work. Although the operations that these machines perform can be done on lathes and other machines, the costly and accurate construction of borer machines are justified by the ease and economy obtained in holding and machining the work.

Figure 9.12*A* is a schematic of a machine designed for locating and boring holes in jigs, fixtures, dies, gages, and other precision parts. Jig-boring machines resemble a vertical milling machine, but are constructed with greater precision and are equipped with accurate measuring devices for controlling table movements. On typical machines positioning to $\pm$ 0.0001 in. ($\pm$ 0.003 mm) can be dialed directly from a drawing. There are two sets of digital read-out panels, longitudinal and the other for transverse measure-

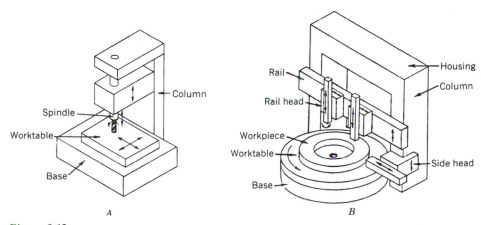

Figure 9.12
A, Jig-boring machine. *B*, Vertical boring mill.

ments. The operator enters the dimensions into the computer to correspond with the dimension on the drawing, and the workpiece is automatically and accurately positioned.

Jig-boring machines are also operated by numerical control. By putting part designs on tape, accurate repetition is ensured, jigs and fixtures are eliminated, and precision boring becomes practical for small-lot manufacturing.

Vertical Boring Mill

In the vertical boring mill the work rotates on a horizontal table in a fashion similar to the old potter's mill. The cutting tools are stationary except for feed movements and are mounted on the adjustable height cross rail. These tools are lathe and planer type, and are adapted to horizontal facing work, vertical turning, and boring. A vertical boring machine is sometimes called a rotary planer, and its cutting action on flat disks is identical with that of a planer. These machines, rated according to their table diameter, vary in size from 3 to 40 ft (0.9 to 12 m). A sketch is shown by Figure 19.12*B*.

Horizontal Boring Machine

The horizontal boring machine, nicknamed a "bar," differs from the vertical boring mill in that the work is stationary and the tool is revolved. It is adapted to the boring horizontal holes, and these holes and components are generally large. If these conditions of component size and hole size are not met, other smaller machines are used.

Figure 9.13 is an example. The horizontal spindle for holding the tool is supported in an assembly at one end, which is adjusted vertically within the limits of the machine. A worktable having longitudinal and crosswise movements is supported on ways on the bed of the machine. In some cases, the table is swiveled to permit indexing the work

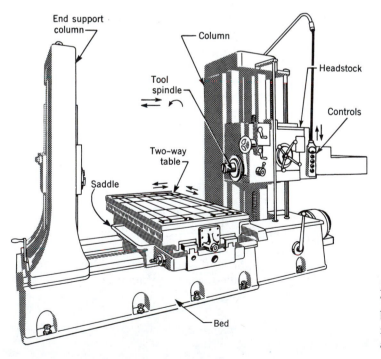

Figure 9.13
Table-type horizontal boring machine to perform boring, drilling, and milling operations.

and boring holes at various angles. At the other end of the machine there is an upright tailstock to support the outer end of a boring bar when boring in-line holes in large castings.

Tolerances for boring are ± 0.0002 in. (± 0.005 mm); positioning, ± 0.0001 in. (± 0.003 mm); repeatability, ± 0.0005 in. (± 0.013 mm); and depths to ± 0.0001 in. (± 0.003 mm).

9.4 MILLING MACHINE GROUP

Milling Machine

Milling machines are made in many types and sizes. The drive is either a cone pulley belt drive or an individual motor. The feed of the work is manual, mechanical, electric, or hydraulic. There are a variety of table movements.

Hand Milling Machine

The simplest milling machine is hand operated. It may have either the column and knee construction or the table is mounted on a fixed bed. Machines operated by hand are used principally in production work for light and simple milling operations such as cutting grooves, short keyways, and slotting. These machines have a horizontal arbor for holding the cutter and a worktable is usually provided with three movements. The work feeds to the rotating cutter either by manual movement of a lever or by a hand screw feed.

Plain Milling Machine

The plain milling machine is similar to the hand milling machine except that it is of much sturdier construction and is provided with a power-feeding mechanism to control the table movements. Plain milling machines of the column and knee type have three motions: longitudinal, transverse, and vertical. Figure 9.14A shows a knee and column plain milling machine.

Although it is a general-purpose machine, the plain milling machine is also used for quantity production work. Notice that it uses a horizontal arbor (not shown) and cutters having a center hole are mounted on this arbor. Other models are featured with universal or vertical head milling features. The machine uses stop dogs to control machine slide movements. Automatic table cycles are available. Cutters are mounted on a horizontal arbor rigidly supported by the overarm.

Vertical Milling Machine

A typical vertical machine, shown in Figure 9.14B, is so called because of the vertical position of the cutter spindle. The table movements are the same as in plain milling machines. Ordinarily no movement is given to the cutter other than the usual rotational motion. However, the spindle head may be swiveled, which permits setting the spindle in a vertical plane at any angle from vertical to horizontal. This machine has a short axial spindle travel to facilitate step milling. Some vertical milling machines are provided with rotary attachments or rotating worktables to permit milling circular grooves or continuous milling of small production parts. End mill cutters are used.

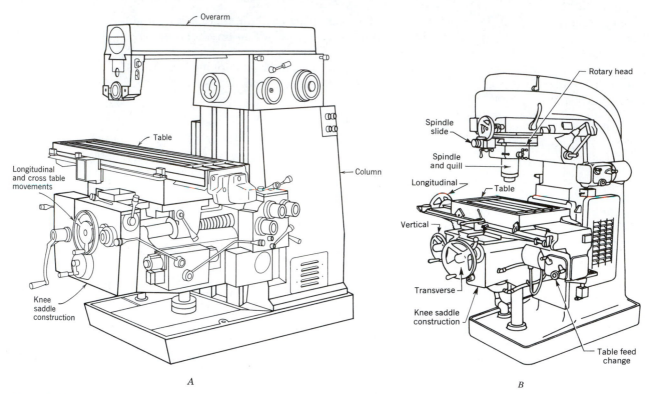

Figure 9.14
A, Knee and column milling machine. *B*, Rotary head vertical milling machine.

Planer Milling Machine

This milling machine receives its name from its resemblance to a planer. The work is carried on a long table having only a longitudinal movement and is fed against the rotating cutter at the proper speed. The variable table feeding movement and the rotating cutter are the principal features that distinguish this machine from a planer. Transverse and vertical movements are provided on the cutter spindle. Planer milling machines are designed for milling large work requiring heavy stock removal and for accurate duplication of contours and profiles. A hydraulically operated unit is shown in Figure 9.15. Work formerly machined on a planer has shifted to this machine.

Machining Center

These NC machines are designed for small- to medium-lot production. The term *machining center* was unknown before numerical control, which is covered in Chapter 28. Refer to Figure 28.2 for an illustration. A machining center may refer to one or more NC machines that have multipurpose capability machining.

It is incorrect to assume that these machines can only do milling. The machining center will mill, drill, bore, ream, tap, and contour, all in a single setup. Depending on the machine, machining centers will start and stop the work, select and change tools, perform two- or three-dimensional contouring using linear or other interpolation techniques, feed in any of two or three axes approximately 0.1 to 99 ipm (40 μm/s to 40 mm/s), position

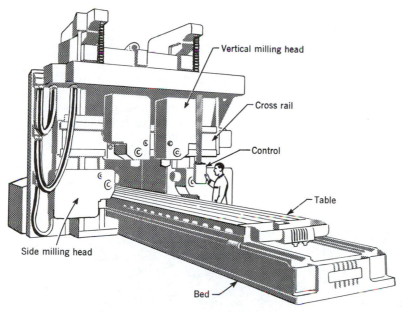

Figure 9.15
Planer milling machine.

any axis at a rapid traverse rate at 400 ipm (0.17 m/s), start or stop the spindle at programmed speed and direction of rotation, index the table to a programmed position, and turn the coolant on and off. Although these machines are versatile, the many features are optional depending on the cost of the machine.

These costly machines replace several other machines. Numerical control has made little improvement on the actual machining; however, its savings contribution depends on auxiliary and supporting functions. To maximize all practical savings from this versatile manufacturing machine tool, the parts should be completely machined with one setup without transfer to several machines.

A tool changer automatically changes tools in 4 s or more and cut-to-cut time is 8 to 10 s. A tool-changer magazine may have anywhere from 8 to 90 tools or more stored permanently or semipermanently. Accurate depth settings are possible by touching off tool points at work surfaces with a spindle hand wheel. This information is then recorded within the machine control unit. Numerical control machining centers have switchable inch or metric programming.

Planetary Milling Machine

Planetary milling machines are used for milling both internal and external short threads and surfaces. The work is held stationary and all movements necessary for the cutting are made by the milling cutters. At the start of a job, the rotating cutter is in center or neutral position. It is first fed radially to the proper depth and then given a planetary motion either inside or around the work. The relation between the work and the cutters is illustrated by the line diagram in Figure 9.16. Typical applications of this machine include milling internal and external threads on all kinds of tapered surfaces, bearing surfaces, rear axle end holes, and shell and bomb ends.

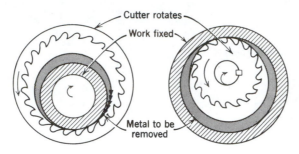

Figure 9.16
Planetary milling setup showing cutter action for both external and internal milling. Left, external milling. Right, internal milling.

Figure 9.17
A copy milling machine.

Duplicating Machine

The production of large-forming dies for automobile fenders, tops, and panels is an important use of duplicator machines. These machines reproduce a part from a model with no reduction or enlargement of size. A large-capacity machine is illustrated in Figure 9.17. This machine, known as a copy milling machine, has a true and mirror image copying range of 177 by 98 in. (4.5 by 2.5 m) with a table size of 216 by 126 in. (5.5 by 3.2 m). The 7-in. (178-mm) diameter spindle has an infinitely adjustable speed from 10 to 150 rpm.

The models or templates are made of hard wood, plaster of paris, or other easily worked materials because the only purpose they serve is to guide the tracer that controls the tool position.

The performance of milling machines in cutting is 7 in. 3/min (1900 mm 3/s) for steel and 14 in. 3/min (3800 mm 3/s) for cast iron.

9.5 TRANSFER-TYPE PRODUCTION MACHINE GROUP

Frequently designated as automated machines, transfer-type production machines complete a series of machining operations at successive stations and transfer the work from one station to the next. They are in effect a production line of connected machines that are synchronized in their operation so that the workpiece, after being loaded at the first station, progresses automatically through the various stations to its completion.

Figure 9.18
Thirty-five station palletized transfer machine for transmission cases; 75 parts per hour production.

Transfer machines perform a variety of machining, inspecting, and quality control functions. They drill, bore, mill, hone, and grind, as well as control and inspect the operations. Automatic machines are of the indexing table or the in-line transfer type.

1. *Indexing table.* Parts requiring only a few operations are adapted to indexing table machines, which are made with either vertical or horizontal spindles spaced around the periphery of the indexing table.

2. *Transfer.* The transfer machines are provided with suitable handling or transfer means between stations. A simple and economical method moves parts on a rail or conveyor between stations. When this is impractical because of the shape of the part, the part is clamped on a holding fixture or pallet, which then moves instead.

A 35-station automatic transfer machine performs a variety of operations on transmission cases, as shown in Figure 9.18. Palletized work-holding fixtures secure the transmissions during all operations. Transfer machines range from comparatively small units having only two or three stations to long straight line machines with more than 100 stations. They are used primarily in the automobile industry. Large production schedules offset their high initial cost by savings in labor. Products processed by these machines include cylinder blocks, cylinder heads, refrigeration compressor bodies, and similar parts.

QUESTIONS

1. Give an explanation of the following terms:

Tailstock	Box column
Lathe size	Gang drill
Turret lathes	Deep-hole drilling machines
Ram-type turret lathe	Jig-boring machine
Multispindle automatics	Machining center
Duplicating lathe	Planetary milling
Sensitive drill press	Transfer machines
Friction disk drive	

2. What functions do the carriage assembly perform in an engine lathe?

3. How are tools held for machining in an engine lathe? Also refer to earlier chapters.

4. Describe differences in horizontal turret lathes.

5. Distinguish between the cross slide on a turret lathe and an engine lathe.

6. What is the purpose of each of the following lathe parts: faceplate, compound rest, lead screw, back gears, and tailstock? Also refer to earlier chapters.

7. Discuss the accuracy that is expected in turning a part with a toolroom or production lathe.

8. Classify the differences in automatic screw machines that make small parts.

9. What are the problems of lathe with a worn lead screw?

10. How can a hole be bored in a short piece of bar stock with threading on the OD taking place at the same time? On an engine lathe? On a turret lathe?

11. What is the purpose of the feed rod? If trip dogs are worn, what problems might occur?

12. How many "working" stations are there on an eight-station vertical chucking lathe?

13. List the kinematic sequence of movements for a Swiss screw machine.

14. A large casting is to be drilled. The supervisor suggests that a sensitive drill press be used. Comment on her suggestion.

15. A sheet metal part is to be drilled and it has several diameters. What kind of a drill press do you select? Why?

16. Select the type of drill press for the following parts: small 3-lb casting with one hole diameter; 25 units of a gear blank requiring drilling and reaming of the center hole; thick metal with a schedule of 10 different size holes; milling machine base made of cast iron; and stainless steel tube 6 in. in diameter with 25 holes along a center line 72 in. long.

17. What type of machine do you recommend for making: tiny shafts for a watch; 10,000 precision ⅝-11 screws; 5 precision ⅝-11 screws; threading 10,000 ½-13 NC nipples?

18. Explain the terms multiple or single spindle when applied to stock feeders.

19. Describe the operation of a Swiss-type automatic screw machine.

20. What type of machine should be used for making ¾-10 by 1½ in. hexagon head cap screws when the quantities are 10, 200, and 10,000?

21. List various machines for making holes in heavy castings when the quantity is several; when the quantity is 2500.

22. What is the difference between a drilling and boring machine?

23. Radial drilling machines are used for what kind of parts?

24. Contrast the operation of a vertical boring machine with a horizontal boring machine. Give part examples for each type of machine.

25. Describe the operation of a tool changer. What machine has a tool changer?

26. What circumstances call for the use of a gun drilling machine?

27. What type of milling machine do you select for the following parts: cutting teeth on a spur gear, mill a pocket in a casting, recess a slot in bar stock, cut a keyway in a round bar, and surfacing a curving face of a die?

28. How does a vertical milling machine differ from a vertical drill press? What kind of parts would you run on each?

29. List similarities between a turret lathe and machining center. Also vertical boring mill and jig-boring machine.

PRACTICAL APPLICATION

Visit a machine shop having several machining centers. Your instructor will suggest businesses that are willing to provide a tour of the facilities. Prepare a survey of the machining centers, such as their size, structure (vertical, horizontal, multiple), ancillary equipment (pallets, conveyors, cranes), number of axis that can be machined (see later chapters), tool holders, and general methods of management (number of shifts, types of employees, and support personnel that assist the operator). Describe your impressions of the machining centers in a report typed on a computer.

CASE STUDY: AIRFRAME PART, PART NUMBER 50532

Advanced Machine Shop received an order for 1275 units of part number 50532, which is used on a new commercial passenger jet. The part material is ASTM 4340 alloy steel, which is purchased as annealed bar stock in 4-in. diameters. Each bar is 12 ft when received. The drawing requires that the part test to Rockwell C45 hardness and have a surface finish no rougher than 64 μin. Other than the tolerances shown by the simplified sketch, Figure C9.1, unstated tolerances are commercial level or ± 0.010 in. Other dimensions may be roughly scaled using the 1-in. dimension of the hub.

Jim Greene, the chief process planner, mentions to you, "Here is another job for that new jet. They want us to bid on this drawing. See what machines are necessary for the job and get back to me when you are finished. Oh yes, find out how many bars are necessary for the job. Don't forget to make allowances for the cutoff kerf and end facing stock."

As the new process planner your job requires you to select the possible machine tools that will transform the material from round 4-in. bar stock to finished dimension. (Hint: Form a rough operations sheet, such as the following:

Operation Number	Machine Tool Description	Descripbion of Operation

The columns are to be completed by listing successive operations 10, 20, . . . , choosing the machines, and writing operation descriptions. Follow that with sketches showing how each operation removes material. Use crosshatch conventions that correspond to the operations, such as 45° to

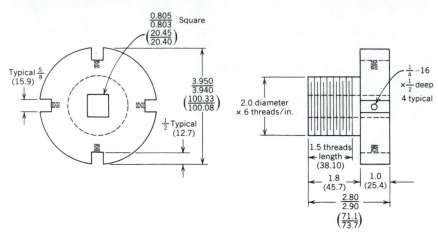

Figure C9.1
Case study.

represent a rough pass, another color pencil to represent a finish pass, and so on. Finally, determine the metal efficiency; that is, the ratio of the final metal in the part to original bar stock. Assume that only 12-ft lengths of bar stock can be purchased. Think about how the square hole will be machined. What various operations are able to machine a square hole? Consider broaching, found in Chapter 10.

SAWING, BROACHING, SHAPING, AND PLANING

The metal-cutting action of sawing, broaching, shaping, and planing is linear. Either the tool moves in a straight line across the workpiece, or the workpiece moves linearly past the tool. As we shall see, there are important variations to these movements, which distinguish the processes.

Although sawing might appear a totally different operation than shaping, which utilizes an orthogonal cutting tool, examination of a magnified chip from a sawing operation resembles one that is generated on a shaper or planer. The broaching operation produces a similar chip.

10.1 METAL SAWING MACHINES

Metal sawing is an important *first operation* in manufacturing. Although some machine tools can do *cutting-off operations* to a limited extent, special machines are necessary for mass production and work that requires cutting of nonsymmetrical shapes. Metal saws for power machines are circular, straight, or continuous.

Hacksaw

The *reciprocating power hacksaw*, which varies from light-duty crank-driven saws to large heavy-duty machines hydraulically driven, is simple in design and economical to operate. Machines vary in the method of feeding the saw into the work. The hacksaw is designed for manual, semiautomatic, or fully automatic operation.

Methods of feeding are positive or *uniform-pressure feed*. A positive feed has an exact depth of cut for each stroke, and the pressure on the blade varies directly with the number of teeth in contact with the work. In cutting a round bar, the pressure is light at the top of the bar and maximum at the center. A disadvantage of positive feed is that the saw is prevented from cutting fast at the start and the finish where contact is limited. With uniform-pressure feeds the pressure is constant regardless of the number of teeth in contact. This condition prevails in gravity or friction feeds. Here the depth of cut varies inversely with the number of teeth in contact, so that maximum pressure depends on the maximum load that a single tooth can stand. Some machines have incorporated both feed systems into their design. In all types, the pressure is released on the return stroke to eliminate wear on the teeth.

The simplest feed is *gravity*, in which the saw blade is forced into the work by the weight of the saw and frame. Uniform pressure is exerted on the work during the stroke, but some provision is usually made to control the depth of feed for a given stroke. Some machines have weights clamped on the frame for additional cutting pressure. Spring loading is another method for increasing cutting pressure. Positive-acting screw feeds with provision for overload provide a means of obtaining a definite depth of cut for each cutting stroke. Hydraulic feeds afford excellent control of cutting pressures.

The simplest drive for the saw frame employs a *crank* rotating at a uniform speed. With this arrangement, the cutting action takes place only 50% of the time because the time of the return stroke equals that of the cutting stroke. An improvement of this design provides a link mechanism that gives a quick-return action. Several link mechanisms are used, including the Whitworth mechanism found on some shapers.

Figure 10.1 shows a reciprocating bar feed hacksaw equipped with automatic bar feed and discharge track. Bars to be cut are loaded on a rolling dolly and vise and are moved forward by a chain arrangement. The usual cycle for automatic feed after the gage has been set is that a bar or bars move forward through an open vise, the vise is opened, and so on, until the length of bar has been cut up.

Power hacksaw blades are similar to those used for hand sawing. High-speed steel blades vary from 12 to 36 in. (300 to 900 mm) in length and are made in thicknesses from 0.050 to 0.125 in. (1.3 to 3.1 mm). The pitch is coarser than for hand sawing, ranging from 2½ to 14 teeth per in. (0.1 to 0.6 teeth per mm). The tooth construction of most hacksaw blades is indicated in Figure 10.2*A* and *B*. The most common type is the straight-tooth design having zero rake. The undercut tooth, which resembles a milling cutter tooth, is used for the larger blades. For efficient cutting of ordinary steel and cast iron, a pitch as coarse as possible should be used to provide ample chip space between teeth. However, two or more teeth should always be in contact with the stock.

High-carbon and alloy steels require a medium-pitch blade, whereas thin metal, tubing, and brass require a fine pitch. To provide ample clearance for the blade while cutting,

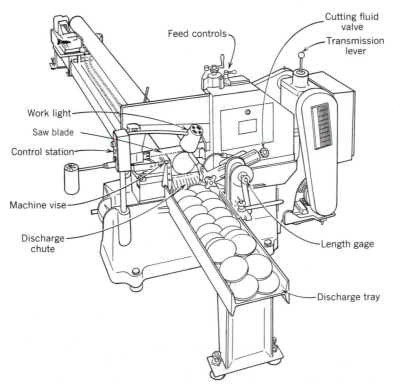

Figure 10.1
Automatic bar feed hacksaw machine.

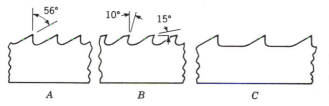

Figure 10.2
Tooth construction for metal saw blades. *A*, Straight tooth. *B*, Undercut tooth. *C*, Skip tooth.

the teeth are set to cut a slot or kerf slightly wider than the thickness of the blade. This is done by bending certain teeth slightly to the right or left, as seen in Figure 10.3. Set refers to the type of tooth construction on a saw. A straight-tooth saw has one tooth set to the right and the next tooth to the left. This type of saw is used for brass, copper, and plastic. On the *raker tooth saw* one straight tooth alternates with two teeth set in opposite directions. This tooth construction is used for most steel and iron cutting. A wave set consists of an alternate arrangement of several teeth set to the right and several teeth set to the left. This design is used in cutting tubes and light sheets of metal.

A lubricant is recommended for all power hacksaw cutting to lubricate the tool and to wash away the small chips accumulating between the teeth. Because there is little heat generated in most sawing operations, the problem is one of lubricating rather than cooling, and the cutting fluid should be chosen accordingly. Hacksaw machines cut between 40 and 160 spm, depending on the machinability of the metal. Surface speed is seldom specified, as it is not uniform throughout the stroke length.

Circular Saws

Machines using circular saws are commonly known as *cold sawing* machines. The saws are fairly large in diameter and operate at low rotational speeds. The cutting action is the same as that obtained with a milling cutter.

Figure 10.4 is a sketch of a cold sawing machine. This machine is hydraulically operated and saws round stock up to 10 in. (254 mm) in diameter with $\pm$ 1/64-in. ($\pm$ 0.4-mm) length tolerances. The saw is fed into the work, which is positively clamped by hydraulic, horizontal, and vertical vises. An automatic gripper-type feeder moves the stock when cutting to specified lengths.

Circular metal saws for rotating cutter machines are similar to the metal slitting saws used with milling machines. However, metal slitting saws are made only in diameters up to 8 in. (200 mm), which is not sufficient for large-size work. Solid blades with diameters up to 16 in. (400 mm) are used in circular sawing machines. Most large cutters have either replaceable inserted teeth or segmental-type blades (Figure 10.5). The segments, each having about four teeth, are grooved to fit over a tongue on the disk and are riveted in place. Both inserted teeth and segmental type blades are economical from the standpoint of cutter material cost and have the additional advantage that worn teeth can

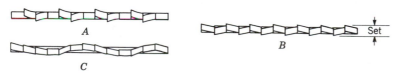

Figure 10.3
Types of set for metal saw blades. *A*, Raker set. *B*, Wave set. *C*, Straight set.

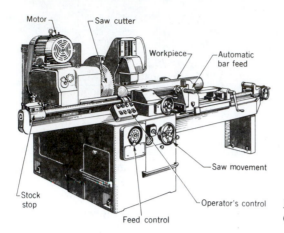

Figure 10.4
Cold sawing machine.

be replaced. The teeth are alternately ground so that one half of them are 0.010 to 0.020 in. (0.25 to 0.50 mm) higher than the rest.

Cutting speeds of circular blades range from 25 to 80 ft/min (0.1 to 0.4 m/s) for ferrous metals. For nonferrous metals, the cutting speed is from 200 to 4000 ft/min (1.0 to 20.3 m/s). The life of a saw is longer when the peripheral speed is not too high. The use of a lubricating fluid is recommended for all circular sawing work. Tolerances are $\pm$ 1/64 to $\pm$ 1/8 in. ($\pm$ 0.4 to 1/5 mm) and surface finish is 250 to 1000 μin. (6350 to 25,400 nm).

Operating at high-peripheral speeds, *steel friction disks* permit cutting through structural steel members and other steel sections. When the disk is rotating at rim speeds from 18,000 to 25,000 ft/min (90 to 125 m/s), the heat of friction quickly melts a path through the part being cut. About 30 s are required to cut through a 24-in. (600-mm) I-beam. Disks ranging in diameter from 24 to 60 in. (0.6 to 1.5 m) are available. They are usually furnished with small indentations on the circumference about 3/32 in. (2.4 mm) deep. The disks are ground slightly hollow to provide side clearance. Water cooling is recommended.

Friction cutting is not limited by the hardness of the material. Stainless steel and high-carbon steel are cut more easily than low-carbon steel. Cutting ability depends on the structure of the metal and its melting characteristics rather than metal hardness. During cutting the tensile strength of the steel decreases quickly as the temperature increases. Steel is weakened to the extent that the friction disk pulls it away from the colder metal. The separation temperature is below the melting point of steel. Nonferrous metals cannot be cut satisfactorily by friction sawing, because these metals tend to adhere to the disk and do not separate readily as a result of the disk action.

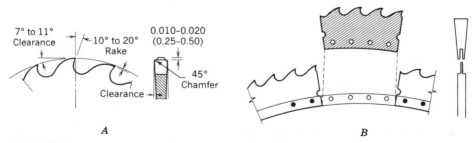

Figure 10.5
Tooth construction for circular metal saws. *A*, Regular teeth. *B*, Segmented teeth.

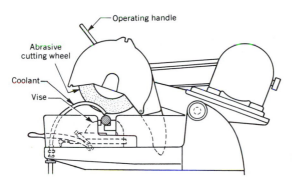

Figure 10.6
Abrasive-disk cutoff machine.

Abrasive Disk Sawing

An abrasive wheel machine suitable for either wet or dry cutting is shown in Figure 10.6. It will cut ferrous and nonferrous solids up to 2 in. (50 mm) in diameter or tubing up to 3½ in. (990 mm). Resinoid-bonded *abrasive wheels* should be used at speeds around 16,000 ft/min (80 m/s) for dry cutting. High peripheral speed cuts more efficiently than low speed because the metal is heated rapidly and becomes soft for easy metal removal. For wet cutting, rubber-bonded wheels operating around 8000 ft/min (40 m/s) are used. The surface speed is limited to this value to retain sufficient coolant on the wheel to prevent overheating. Cutting action depends entirely on the abrasive grains in the wheel and is unaffected by any metal softening. The finish and accuracy is 63 to 500 μin. (1600 to 12,000 nm) and $\pm$ ⅛ in. ($\pm$ 1.6 mm), which is better than steel friction blades.

Band Saws

The sawing machines described thus far are designed for straight cuts and for cutting off. Figure 10.7 illustrates a cutting saw of the *band type*, which is used for straight and curving work. Able to cut irregular lines, the applications for the band saw include cutting of dies, jigs, cams, templates, and other parts previously cut on other machine tools or by hand at greater expense. Suitable and accurate continuous filing and polishing, both

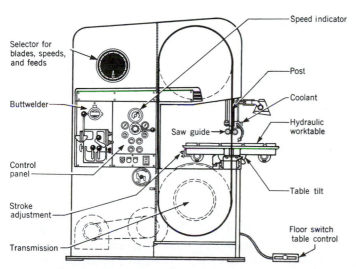

Figure 10.7
Production band saw.

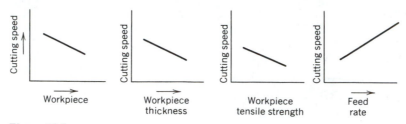

Figure 10.8
General relationships of cutting speed and feed for sawing and broaching.

necessary operations in contour finishing, can be accomplished by band saws. Band sawing machines have variable speeds of 50 to 1500 ft/min (0.3 to 7.6 m/s) to accommodate most materials. Blade specifications vary as in hacksaws, and specific recommendations are available.

Figure 10.8 shows the general relationships between material properties and cutting speed and feed. Computer control for band saws enables sawing intricate and complex parts, including drawing dies and mold sections.

Band friction cutting is another interesting process. High-speed band sawing machines designed for friction cutting have a surface speed range of 3000 to 15,000 ft/min (15 to 75 m/s). Saws for these machines are selected carefully, as pitch (number of teeth per inch) varies from 10 for thick materials to 18 for thin materials (0.4 to 0.7 teeth per mm). Band friction cutting is limited to relatively thin ferrous metals and some thermoplastic materials.

Diamond band cutting is another variation in band machines. Diamond-impregnated band sawing cuts glass, carbide, ceramic, dies, and hard semiconductor materials. The speeds and feeds must be precisely controlled, and the workpiece is flooded with a cutting fluid. These saws are usually small in size because band cost is high and contour cutting often involves a radical change in blade direction. It is not uncommon for the blade to be round with the teeth covering the surface; this is referred to as a wire blade.

Band filing and polishing is another variation for band machines. When the machine is to be used for filing work, a file band replaces a saw band. The file band is constructed of files mounted on a flexible steel band. A snap joint is provided for quick fastening and unfastening for internal filing. A light to medium pressure is used on contour filing, and the filing speeds range from 50 to 200 ft/min (0.3 to 1.0 m/s). An advantage of filing is the continuous downward stroke. The absence of a backstroke greatly lengthens the life of the file and helps in holding the work onto the table.

Files used on this machine have the same materials, forms, and styles found on standard commercial files. Single cut, *double cut*, and rasp cut are the terms used in describing the cut of the file. Rasp cut differs from the other two in that the teeth are disconnected from each other, each tooth being made by a single punch. The coarseness of the teeth is described by the terms rough, coarse, bastard, double cut, and smooth. File cross sections are indicated by such terms as flat, oval, half round, and mill.

10.2 BROACHING

Broaching is the operation of removing metal by an elongated tool. Since each successive tooth removes metal, either each tooth must be larger than the preceding one, or else each tooth is set higher than the previous one (see Figure 10.9). A part is broached in

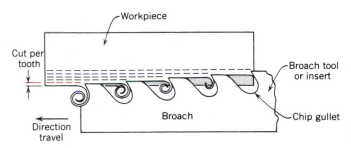

Figure 10.9
Broaching action.

one stroke, and the last teeth on the cutting tool conform to the final dimension. In most machines the broach is moved past the work, but equally effective results are obtained if the tool is stationary and the work is moved. Many cuts, both external and internal, can be made on machines at a high rate of production. An accuracy of $\pm$ 0.0005 in. ($\pm$ 0.013 mm) and a finish of 32 to 125 μin. (800 to 3000 nm) are common. Ferrous materials with a hardness up to Rockwell C40 can be broached. A broaching machine consists of a workholding fixture, broaching tool, drive mechanism, and a supporting frame.

Several variations in broach machine tool design are possible. Broaching can be pull broaching, where the tool is pulled through or across stationary work. Push broaching is where the tool is pushed through or across stationary work. In surface broaching, either the work or the broaching tool moves across the other. Finally, in continuous broaching the work is moved continuously against stationary broaches. The path of movement may be either straight or circular.

Most broaching machines are horizontal or vertical in design. See Figure 10.10A and B. The choice of the design depends on the size of part, size of broach, quantity, and type of broaching. Vertical machines having the tool supported on a suitable slide are adapted for surface broaching. Both pull- and push-type internal broaching machines are constructed with this design. Horizontal machines pull the broach and are often used for internal and surface broaching of small- and medium-size parts.

Most machines are hydraulically driven because of the large force requirements. Such a drive is smooth acting, economical, and adjustable for speed and length of stroke.

The surface broaching operation being performed on the vertical machine, shown in Figure 10.10A, is simple and quick. Most vertical machines are provided with a receding table, so that the fixture is loaded and unloaded while the broach is returning to its original position. The cycle is automatic and continuous except for the loading operation,

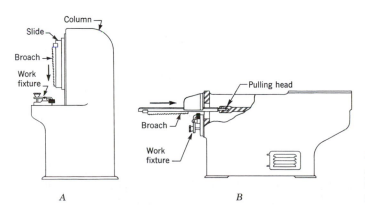

Figure 10.10
A, Vertical single-slide broaching machine. B, Horizontal broaching machine.

which, if cost warrants, can be made automatic. A vertical double-slide machine differs from the single-slide machine in that it has two slides that operate opposite each other. The work is held on shuttle tables that move out during the unloading and loading operations, while the ram returns to its starting position. While this is going on, the other ram is at work.

Although horizontal broaching machines have surface broaching applications, they are generally used for internal broaching of large- and medium-size parts. A diagrammatic sketch of a horizontal broaching machine adapted for surface broaching is shown in Figure 10.10B. Here the broach is pulled over the top surface of the workpiece held in the fixture.

Vertical pull-down broaching machines are adapted to internal broaching. The parts are placed in a fixture on the worktable, the pulling mechanism being in the base of the machine. Broaching tools are suspended above by an upper carriage. As the operation starts, the broaches are lowered through the holes to be broached and are automatically engaged by the mechanism, which pulls them through the part. Upon removal of the work the tools rise, are engaged by the upper holders, and return to their starting position. Machines of this type have the advantage over pull-up machines in that the positioning of the part is easier and large parts are handled with less difficulty.

Vertical pull-up broaching machines are also adapted to internal broaching and are frequently preferred for small parts. Although the general cycle of operation is similar to the pull-down machines, it is reversed.

Push broaching is shown in Figure 10.11A. The broach is finishing a round hole in a gear blank, this being faster than reaming or boring, and at the same time the hole is held to accurate limits. Finished holes are broached from holes previously drilled, punched, or cored. Push broaching requires comparatively short broaches of sufficient cross section to prevent column buckling resulting from the loads imposed during the operation. A crude form of push broaching is also possible on a simple utility press.

Figure 10.11B illustrates pull broaching on a horizontal machine. Large, heavy-duty horizontal broaching machines are used in the high-speed production of cylinder blocks, intake manifolds, bearing cap clusters, and aircraft turbine disks. The cutting speed approaches 200 ft/min (1.0 m/s) and the stock removal is up to ¼ in. (6.4 mm) per stroke.

Broaching internal keyways is one of the oldest uses of the broaching process. A *keyway* broach and its adaptor are shown in Figure 10.12.

Internal gear broaches are similar to spline broaches except for the involute contours

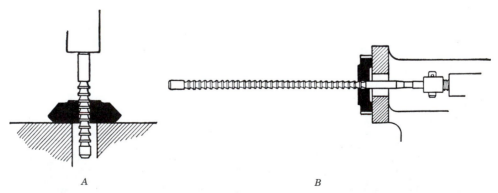

A   *B*

Figure 10.11
Round broaches for push- and pull-type machines. *A*, Push broaching. *B*, Pull broaching.

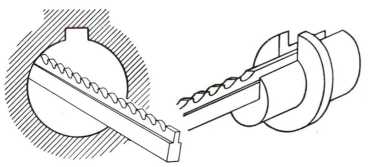

Figure 10.12
Keyway broaching.

on the sides of the teeth. They may be made to cut any number of teeth and are used for broaching as small as 48 diametral pitch. This method of gear cutting is known as the form tooth process, and the accuracy of the teeth depends on the accuracy of the form cutter.

An interesting development in horizontal broaching is cutting helical grooves or splines by pulling a broach through the part and at the same time either rotating the part or broaching tool according to the helix desired. This procedure has been adopted by many gun manufacturers in rifling small caliber and light cannon gun barrels.

Continuous or *tunnel broaching* machines are adapted only for surface broaching. This machine consists of a frame and driving unit with several workholding fixtures mounted on an endless chain that carries the work in a straight line past the stationary broaches. A view of a continuous machine equipped with a motor-driven conveyor for removing work from the machine is shown in Figure 10.13. Loading is done by an operator who drops the parts in the fixtures as they pass the loading station. The work is automatically clamped before it passes into the fixture tunnel in which the broaches are held. After the fixtures pass the broach, they are automatically released by a cam; at the unloading position the work falls out of the fixtures into the work chute. Six cylinder engine blocks can be surfaced in this manner at the rate of 120 per hour.

Broaching is a metal-cutting operation that has been adopted for mass production work because of the following features and advantages:

1. Both roughing and finishing cuts are completed with one pass of the tool.
2. Production rate is high because the actual cutting time is a matter of seconds. Rapid loading and unloading of fixtures minimizes total production time.
3. Internal or external surfaces can be broached.
4. Any form that can be reproduced on a broaching tool can be machined into the part.

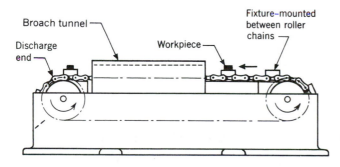

Figure 10.13
Continuous surface broaching machine.

5. Production tolerances are suitable to interchangeable manufacture.

6. Finished comparable to milling work can be obtained.

7. Burnishing shells incorporated as the final teeth on the broach improve the surface finish.

Broaching has high tool cost, however, particularly for large or irregular-shaped broaches, and short-run jobs are not advisable. Parts must be rigidly supported and capable of withstanding the broaching forces. The broached surface must be accessible. The method is not recommended for removing a large amount of stock. Broaching tools differ from most other production tools in that they are usually adapted to a single operation. The feed of the tool must be predetermined, and once a broach is made the feed remains constant.

Knowledge about the job, part material, and the machine are necessary before a broach can be made. In designing and constructing a broach the following information must be known. Most internal broaching is done with *pull broaches*, because they can take longer cuts and can remove more stock than push broaches. *Push broaches*, which are necessarily short to avoid buckling under load, are used principally in sizing holes in heat-treated parts and for short-run jobs. They are also used in broaching blind holes.

The simplest broaches designed for flat surfaces can be made with either straight or angular teeth. Angular teeth produce a smoother cutting action. Because the entire length of such broaches can be supported on a slide, it is possible to produce them in short sections. Heavy-duty broaches frequently have inserted teeth to reduce the initial cost and facilitate replacements. Specially designed fixtures are often necessary to hold the workpiece, because cutting forces may fracture or deform the part if it is not properly supported.

Figure 10.14A, showing a pull-type broach, illustrates some terms usually applied to broaches. Figure 10.14B shows an enlarged tooth form with terms and angles indicated.

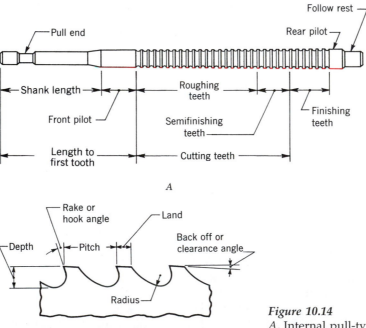

Figure 10.14
A, Internal pull-type broach. B, Tooth form on broach with principal terms and angles.

The top portion of a tooth is called the land and is ground to give a slight clearance. This angle, called backoff or clearance angle, is usually 1½° to 4° on the cutting teeth. Finish teeth have a smaller angle ranging from 0° to 1½°. Regrinding on the lands of most broaching tools should be avoided because it changes the size of the broach. Sharpening is done by grinding the face or front edge of the teeth. The angle to which this surface is ground corresponds to the rake angle on a lathe tool and is called the face angle, hook angle, undercut angle, or rake angle. The rake angle varies according to the material being cut and in general increases as the ductility increases. Values of this angle range from 0° to 20°, but for most steels a value of 12° to 15° is recommended. This angle has considerable effect on the force required to make the cut and the finish. A large angle might give excellent results, but from the standpoint of lengthening tool life a smaller angle is preferred. Frequently, the first cutting teeth are rugged in shape and have a small rake angle, whereas the finish teeth are given a large rake angle to improve the finish. Side rake angles of 10° to 30° are widely used on surface broaching to improve the finish of the cut.

For short broaches, the pitch is

$$P = K(L)^{1/2} \tag{10.1}$$

where

P = Pitch, in. (mm)
K = 0.35 for L, in.; 1.76 for L, mm
L = Length of cut, in. (mm)

For small broaches, the amount of material removed per tooth is approximately

$$D = \frac{0.10 \times (P - W)G}{L} \tag{10.2}$$

where

D = Depth of cut per tooth, in. (mm)
W = Length of land, in. (mm)
G = Gullet depth, in. (mm)

A longer broach will usually have at least two separate cutting areas, roughing and finishing; each is treated separately. To find the minimum force to operate a broach, a number of factors must be known. With a surface broach the force is calculated from

$$F = K_2 NDW \tag{10.3}$$

where

F = Force, lb (kg)
K_2 = Constant
N = Number of broach teeth cutting at one time
W = Width of cut, in. (mm)

Approximations are K_2 = 400 for mild steels and cast iron, and 200 for nonferrous material such as aluminum, zinc, and brass.

Sometimes holes are sized and then the finish is improved by *burnishing*. Burnishing is not a cutting operation, but consists of moving a very hard surface over the workpiece to remove irregularities and smear metal caused by prior machining. This is done with a burnishing broach or a regular broach that has several burnishing shells following the finishing teeth. The amount of stock left for burnishing should not exceed 0.001 in.

(0.03 mm); for ordinary steel 0.0005 in. (0.013 mm) is sufficient. This amount should be distributed over three or four burnishing shells. The operation is *cold working*, which produces a hard, smooth surface.

10.3 SHAPERS

The shaper is a machine having a reciprocating tool that takes a straight-line cut. By successive movement of the work across the path of the cutting tool, a plane surface is machined. Flatness of the machined surface does not depend on the accuracy of the tool as it does with a milling cutter, which can do similar work. By employing special tools, attachments, and devices for holding the work, a shaper is able to cut external and internal keyways, spiral grooves, gear racks, dovetails, and *T-slots*.

Power is applied to the machine by motor, either through gears or belt, or by hydraulic action. The reciprocating drive of the tool can be arranged in several ways. Some shapers are driven by gears or feed screws, but most shapers are driven by an oscillating arm and crank mechanism.

Figure 10.15*A* is a sketch of a plain horizontal shaper. It is also known as a horizontal push cut shaper because of the cutting action. Sometimes it is found in schoolroom shops. It is seldom used for production work. A horizontal shaper consists of a base and frame that support a horizontal *ram* and is simple in construction. The ram with its mounted tool is given a reciprocating motion equal to the length of the stroke. A quick-return mechanism driving the ram is designed to have the return stroke of the shaper faster than the cutting stroke. This reduces the nonmachining time. The toolhead at the end of the ram can be swiveled through an angle and is able to feed the tool into the work. A *clapper box* toolholder attached to the ram pivots at the upper end and flips up on the return stroke, so as not to dig into the work.

The worktable is supported on a crossrail in front of the shaper. A lead screw is mechanically linked with the crossrail and moves the work crosswise or vertically, either by power drive or manually.

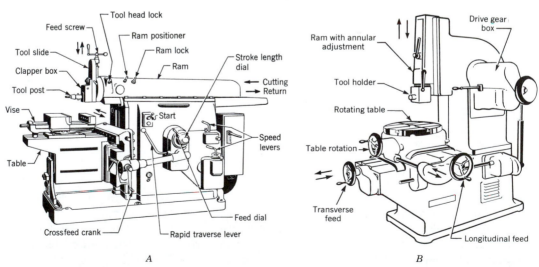

A   *B*

Figure 10.15
A, Plain horizontal shaper. *B*, Vertical shaper.

With horizontal draw cut shapers the tool is pulled across the work by the ram instead of being pushed. Horizontal draw cut shapers are useful for heavy cuts. Applications include cutting large die blocks and machining large parts in railroad shops. During the cut the work is drawn against the adjustable back bearing or face of the column, thereby reducing the strains on the crossrails and saddle bearings. There is little vibration or chatter as a tensile stress is applied to the ram during the cut.

Vertical shapers or *slotters* (Figure 10.15*B*) are used principally for internal cutting and planing at angles and for operations that require vertical cuts because of the position in which the work must be held. Applications are found in die work, metal molds, and metal patterns. A key seater is another type of vertical shaper and is designed for cutting keyways in gears, pulleys, and cams.

The shaper is driven by a mechanical quick-return mechanism (Figure 10.16*A*) or by a hydraulic system (Figure 10.16*B*). The mechanical system consists of a rotating crank driven at a uniform speed connected to an oscillating arm by a sliding block that works in the center of the oscillating arm. The crank is contained in the large gear and can be varied by a screw mechanism. To change the position of the stroke, the clamp holding the connecting link to the ram screw is loosened and the ram positioner is turned. By turning the positioner screw the ram can be moved backward or forward to the desired cutting position, as shown in Figure 10.16. Stroke length is varied by changing the length of the crank. The ratio of return to cutting speed is about 3:2.

Quick return in the hydraulic drive shaper is accomplished by increasing the flow of hydraulic oil during the return stroke. The hydraulic shaper is replacing mechanical shapers, because the cutting stroke has a more constant velocity and less vibration is induced in the hydraulic shaper. The cutting speed is generally shown on an indicator and does not require calculation. Both the cutting stroke length and its position relative to the work may be changed quickly without stopping the machine. Ram movement can

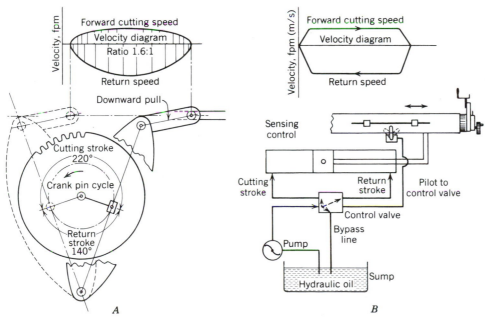

Figure 10.16
Shaper drives. *A*, Mechanical. *B*, Hydraulic.

be reversed instantly anywhere in either direction of travel. The hydraulic feed operates while the tool is clear of the work. The maximum ratio of return to cutting speed is about 2:1.

Cutting speed on horizontal shapers is defined as the average speed of the tool during the cutting stroke and depends on the number of ram strokes per minute and the length of the stroke. If the stroke length is changed and the number of strokes per minute remains constant, the average cutting speed is changed. The ratio of cutting speed to return speed enters into the calculation, as it is necessary to determine the proportion of time the cutting tool is working. With the ratio of cutting stroke to return stroke as 3:2, the cutter is working three-fifths of the time and the return stroke two-fifths of the time. The average cutting speed is determined by the following formula.

$$V_c = \frac{2LN}{12C} = \frac{LN}{6C} \tag{10.4}$$

where

$$V_c = \text{Cutting velocity, ft/min (m/min)}$$
$$N = \text{Stroke per minute (spm)}$$
$$L = \text{Stroke length, in. (mm)}$$
$$C = \text{Cutting time ratio,} \frac{\text{cutting time}}{\text{total time}}$$

The number of strokes per minute for a desired cutting speed is then

$$N = \frac{V_c \times 6C}{L} \tag{10.5}$$

To determine the number of strokes and the total time,

$$S = \frac{W}{f} \tag{10.6}$$

where

$$S = \text{Total number of strokes}$$

$$T = \frac{S}{N} \tag{10.7}$$

where

$$T = \text{Total time, min}$$
$$W = \text{Width of work, in. (mm)}$$
$$f = \text{Feed, in. (mm)}$$

An expression to determine the total time knowing the desired cutting speed and length of stroke is

$$T = \frac{SL}{V_c \times 6C} \tag{10.8}$$

10.4 PLANERS

A *planer* is a machine tool designed to remove metal by moving the work in a straight line against a single-edge tool. Similar to work done on a shaper, a planer is adapted for much larger work. The cuts, which are mainly plane surfaces, can be horizontal, vertical, or an angle. In addition to machining large parts, it is also possible to mount many small parts in line on the table. Planers are seldom used for production work, as most plane surfaces are machined by milling, broaching, or grinding. However, they are used for special purposes.

Hydraulic and mechanical power are used for planers. Uniform cutting speed is attained throughout the cutting stroke. Acceleration and deceleration of the table take place in a short distance of travel and does not influence the time to machine.

Double-housing planers consist of a long heavy base on which the table reciprocates. The upright housing near the center on the sides of the base supports the crossrail on which the tools are fed across the work. Figure 10.17A illustrates how the tools are supported both above and on the sides, and their adjustment for angle cuts. They are fed by power in either a vertical or a crosswise direction.

Open-side planers (Figure 10.17B) has the housing on one side only. The open side permits machining wider workpieces. Most planers have one flat and one double V-way, which allows for unequal bed and platen expansions. Adjustable dogs at the side of the bed control the stroke length of the platen.

Although the planer and shaper are able to machine flat surfaces, there is little overlapping in their application. They differ greatly in construction and in the method of operation. The planer is especially adapted to large work; the shaper can do only small work. On the planer the work is moved against a stationary tool; on the shaper the tool moves across the work, which is stationary. On the planer the tool is fed into the work; on the shaper the work is usually fed across the tool. The drive on the planer table is either by gears or by hydraulic means. The shaper ram also can be driven in this manner, but many times a quick-return link mechanism is used.

Most planers differ from shapers in that they approach more constant-velocity cutting speeds. Tools used in shaper and planer work are single point as used on a lathe, but are heavier in construction. The holder is designed to secure the tool bit near the centerline of the holder or the pivot point rather than at an angle as is customary with lathe toolholders. With the tip of the tool back, it tends to dig into the metal less and cause chatter.

Cutting tool shapes for planer operation are usually tipped with high-speed steel, cast

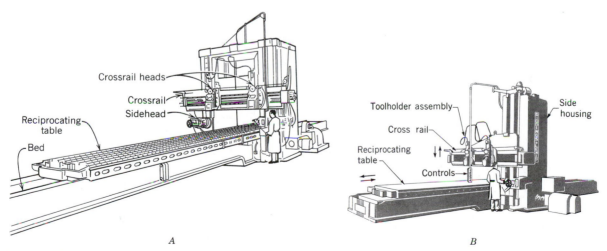

Figure 10.17
A, Double-housing planer. B, Open-side planer.

alloy, or carbide inserts. High-speed steel or cast alloys are commonly used in heavy roughing cuts and carbides for secondary roughing and finishing.

Cutting angles for tools depend on the tool used and the workpiece material. They are similar to angles used on other single-point tools, but the end clearance does not exceed 4°. Cutting speeds are affected by the rigidity of the machine, how the work is held, tool, material, and the number of tools in operation. Worktables on planers and shapers are constructed with T-slots to hold and clamp parts that are to be machined.

QUESTIONS

1. Give an explanation of the following terms:

Raker tooth saw   T-slots
Steel friction disks   Ram
Double cut   Clapper box
Keyway   Slotters
Tunnel broaching   Planer

2. How is the feed obtained on a shaper? On a planer?

3. How is the length of stroke changed on an oscillating arm shaper?

4. In a draw cut shaper the ram is in tension. In a push-cut unit the ram is in compression. Which one can take the heavier cut? Why?

5. How is return speed regulated on a hydraulic shaper?

6. How is the return speed regulated on a Whitworth drive-type shaper?

7. For what type of work are vertical shapers used?

8. Are there disadvantages in employing carbide tools on a shaper or planer?

9. What advantage does an open side planer have over a double housing planer?

10. In the conventional nonhydraulic shaper mechanism, describe why the speed of cutting is less than the return speed.

11. What type of tool is necessary on a horizontal push-cut shaper to cut a keyway in a gear?

12. How is a dial indicator used on a shaper or planer to ensure that a flat surface is perpendicular to the path of the ram?

13. Describe how kerf is obtained for a saw cut.

14. What factors are considered in the selection of the pitch for a hacksaw blade?

15. State the advantages and disadvantages of uniform feed on a hacksaw.

16. What kind of disks are used in abrasive cutoff machines? At what surface speeds are they operated?

17. Should a steel friction disk or an abrasive disk be used for

a. Cutting a steel pipe?

b. Cutting a steel I-beam?

c. Cutting aluminum pipe?

d. Cutting very hard steel rod?

e. Cutting glass?

18. List the work stock material for the skip tooth band saw.

19. Give an application for saw blades having sets such as

a. Raker.

b. Wave.

c. Straight.

20. What type of quick-return mechanism is employed on a hydraulic broaching machine? Why is a quick-return mechanism an economical addition to a broaching machine?

21. State the advantages and limitations of the broaching process.

22. How is rifling on a gun manufactured by using a broaching tool?

23. What advantage does a vertical pull-down machine have over a vertical pull-up machine?

24. What type of broaching machines do you recommend for broaching the following parts: keyway in gear, involute teeth on gear segment, top of engine cylinder, splines in gears, and cutting cap from connecting rod?

25. What is burnishing and how is it done on a broaching machine? What accuracy and surface finish is expected?

PROBLEMS

10.1. Sketch a method of holding a part to T-slots.

10.2. If the ratio of return to cutting speed is 3:2 on a shaper and the return speed is 200 ft/min, how much time is required for a 10-in. cut? What is the cutting speed? How does the feed rate affect the problem?

10.3. If the feed is 0.125 in. and the shaper is performing at 100 spm, how much time is required to surface a 7.5-in.-wide workpiece? How does the tool width affect your answer?

10.4. Suppose the feed is 0.5 in. on a planer and the tool width is 0.25 in. What is the surface appearance of the workpiece after machining?

10.5. If the feed on a shaper is ⅟₃₂ in. per stroke, how much time is required to machine a square part 1½ ft wide if the shaper is performing at 37 spm?

10.6. Sketch a pull-type surface broach that is able to machine a dovetail slot in a part that had a rectangular slot milled in it. Label the principal parts.

10.7. Sketch a quick-return mechanism that is used for a reciprocating saw.

10.8. For a shaft rotating at 900 rpm, find the diameter of an abrasive wheel for minimum peripheral speed.

10.9. A 14-in. diameter circular saw has 120 teeth, feed rate of 0.003 in. per tooth, and a cutting speed of 60 ft/min. Determine the cutting time for sawing a 3.375-in. diameter steel bar.

10.10. If an abrasive wheel cuts at 4000 ft/min, what are the revolutions per minute if the wheel is 260 mm in diameter?

10.11. A reciprocating hacksaw using a 16-in. (406 mm) high-speed steel blade with 12 teeth per inch (0.5 teeth per mm) is to cut a 3-in. (75 mm) steel bar. The machine operates at 80 spm with a uniform feed rate of 0.010 in. (0.25 mm) per stroke. If the tooth pressure will not exceed 8.5 lb (37.8 N) per tooth, what is the maximum pressure on the saw blade and the cutting time?

10.12. A rectangular piece of cold-rolled steel 2 in. (50.8 mm) thick is to be cut on a band saw having a 14 pitch (0.6 pitch per mm) blade and operating at 150 ft/min (0.76 m/s). If the average metal removed per tooth is 0.0001 in. (0.003 mm), how long is required to cut through a width of 3 in. (75 mm)?

10.13. A horizontal broaching machine has a cutting speed of 24 ft/min, return speed of 36 ft/min, and a stroke of 38 in. If the starting and stopping time is 4 s and the loading time 10 s, find the machine output per hour for a production efficiency of 90%.

10.14. If the tooth length of a 16-in. (406 mm) hacksaw blade is 13.2 in. (335 mm), how many teeth are on the blade?

10.15. A short 15-in. broach is to be made to broach mild steel. What pitch should be used? For this broach plot the material removed per tooth versus gullet depth for a land length of 0.125 in.

10.16. How much force in pounds are required to operate the broach described in problem 10.15 if the gullet depth is 0.375 in.?

10.17. What is the material removed (i.e., the total depth of cut) for the conditions for a 15-in. broach for mild steel if the gullet depth is 0.375 in. and the land length is 0.125 in.?

PRACTICAL APPLICATION

There are many possibilities for practical applications. Sawing, one of the minor manufacturing operations, remains important in several ways. Study various saw blades for cutting diverse materials. Visit a shop or a hardware store for examples of saw blades. Call a machine tool supplier for information. For your study consider a range of materials, such as metal (hard, soft), nonmetals (wall board, for example, wood), working application (manufacturing production, home construction), and the blade design. Discuss the similarities and differences of these saw blades with the single-point orthogonal-cutting tool. Note the importance of the working application in your study.

CASE STUDY: ACME BROACH COMPANY

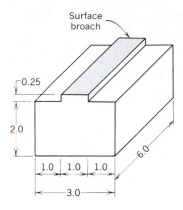

Figure C10.1
Case study.

The aluminum part shown in Figure C10.1 has a broached finish on the top surface. As the broach designer, you are to choose the type and size of broaching machine, minimum force, and specify the design of the broach to fabricate the part. The total amount of material to be removed is 0.031 in.

CHAPTER 11

MACHINING CUTTERS, OPERATIONS, AND PERFORMANCE

In the production of fabricated parts, machine tools are important, but they are not the entire story. Chapter 8 introduced metal-cutting theory, whereas Chapter 9 studied turning, drilling, boring, and milling machine tools.

In this chapter, attention is directed to the cutting tools and the performance of these *integrated production systems*. It is the production system as a whole, including the machine tool, cutter, attachments and tooling, and its operation, that are important when one evaluates performance. If any part of the system is subpar, the system suffers. Cutting tools must run optimally, operators need to be trained and skilled, fixtures and attachments require design and construction, and the selection of the best machine by the process planner is vital to economic success.

11.1 CUTTING TOOLS

In Chapter 8, cutting theory, machinability, tool life, materials for tools, and force and velocity patterns were discussed. Some general recommendations were given for single-point turning tools. This chapter expands on tools; that is, the point at which the material removal is ongoing, for turning, drilling, reaming, tapping, boring, milling, and other applications.

Some cutting tools are versatile and can be used on several machine tool types. Drills, for example, can be worked with lathes, drill presses, and milling machines. However, some tools are designed with the machine tool in mind, such as a shell end mill. Cutting tools are basically single-point or multipoint cutters. A turning or boring point is single edge, but a drill or milling cutter is multipoint.

Turning Tools

For turning operations, *single-point cutters* are used and the understanding of cutters begins there. Figure 11.1A and B are examples of *inserts* that are clamped on a tool holder. Figure 11.1A is a triphase TiC/TiC-N/TiN composition, which is a chemical vapor surface deposit on a cemented carbide substrate. This insert is used for interrupted cuts and heavy and moderate roughing of carbon and alloy steels, tool steels, stainless steels, and alloy cast iron. Figure 11.1A is triangular insert. Figure 11.1B, a diamond-shaped insert, is a cermet material, or a CERamic with METallic binders. It finds application with precision turning or boring of carbon, alloy, and stainless steels, and malleable and ductile cast irons.

A   *B*

Figure 11.1
A, Coated-carbide metal cutting insert machining steel workpiece. *B*, Cermet insert performing a finish turning operation on alloy steel workpiece.

Single-point tools are shown by the three types in Figure 11.2. Problems of cemented carbide tips that were brazed to the toolholder, as shown in Figure 11.2*A*, led to the disposable clamp-on insert shown in Figure 11.2*B*. When a cutting edge becomes worn or broken, the insert is unclamped and indexed to the next cutting edge. This saves time and avoids removing the cutter body and regrinding, inserting, and adjusting to tolerance and dimension. Where negative rake is incorporated into the toolholder, both sides of the tool insert are used. A square insert would have eight cutting edges. These inserts are often referred to as *throwaways*. It is cheaper to throw them away after their use than diamond regrinding.

Carbide tools do not have as much resistance to shock as high-speed steel. Figure 11.2*A* demonstrates the advantages of both materials. In this case, a pocket is milled out

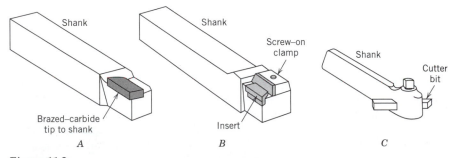

Figure 11.2
Toolholders. *A*, Brazed-carbide tip. *B*, Clamp-on type. *C*, Solid high-speed steel insert bit.

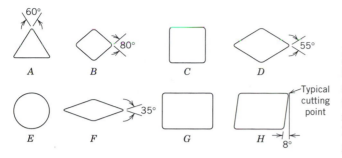

Figure 11.3
Standard shapes for indexable and disposable-insert carbide cutting tools. *A*, Triangular. *B*, Diamond. *C*, Square. *D*, Diamond. *E*, Round. *F*, Diamond. *G*, Rectangular. *H*, Parallelogram.

of a high-carbon steel shank and a tungsten carbide tip is brazed in place. *Brazed tip tools* are used for low-requirement turning, drilling, reaming, and boring. As single-point turning tools, they are adaptable when some job requires a special configuration. It may be necessary to grind the tip by diamond cutters. Tools with a brazed-carbide tip have brazing strains and have only one effective cutting corner.

The solid high-speed steel insert is often ground by hand and placed in the toolholder as shown in Figure 11.2C. Back rake is usually included in the holder to simplify grinding. This tool configuration is used for educational classes. It is not used in production.

Common insert shapes are shown in Figure 11.3. Inserts also can be provided with built-in chip breakers. They can be solid, screw, or clamp-on types. A great variety of designs, materials, and applications are available for selection.

Tool geometry and the nature of the angles were discussed in Chapter 8. *Tool designations and angles* for high-speed steel material tool blanks are shown in Figure 11.4. The total designation is called a *tool signature*. This right-hand tool travels from right to left. Positioning of the shank in the toolholder affects these angles.

Effects of these tool angles are shown in Figure 11.5. The end-cutting edge angle (ECEA) is formed to have only a small part of the end of the tool in contact with the workpiece. If a large part of the end of the tool is in contact with the work, there is vibration or chatter, which produces a poor surface finish and can initiate premature tool

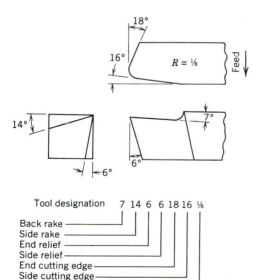

Figure 11.4
Cutting tool designations known as a tool signature.

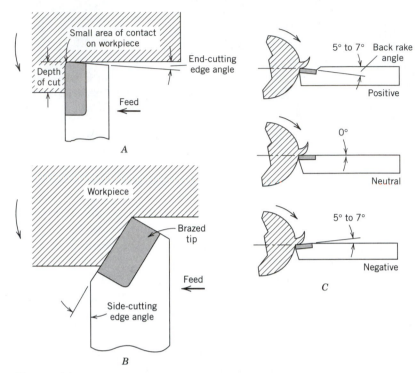

Figure 11.5
Effects of tool angles. *A*, ECEA should be small. *B*, SCEA can vary from
0° to 90°. As SCEA increases, the chip thickness is reduced and width is
increased. *C*, BRA controls direction of chip flow.

failure. For most operations the ECEA is limited to 5°. In Figure 11.5*B*, the angle can
vary from 0° to almost 90°. One effect of increasing the side-cutting edge angle (SCEA)
is to decrease the thickness of the chip and to increase its width. With a longer and
thinner chip, the downward pressure is spread over greater length, increasing tool life
because it is less likely that the cutting edge will be broken off by the force of the chip.
The back rake angle (BRA) of Figure 11.5*C* controls the direction of the chip flow. A
positive angle moves the chip away from the machined surface of the workpiece, and a
negative angle moves the chip toward the machined surface. However, negative angles
have the advantage of a strengthened tool, sometimes important with interrupted cuts.

Unlike cemented carbide inserts, high-speed steel cutter bits are often hand ground
to form. Refer to Figure 11.6 for various shapes. A lathe toolholder will have straight,
left- and right-hand shanks. Left-hand shanks permit machining operations close to the
lathe chuck or faceplate. Right-hand shanks are used for facing and machining operations
that are close to the tailstock.

Drills

A *drill* is a rotary end-cutting tool having one or more cutting edges and corresponding
flutes that continue the length of the drill body. The flutes, which can be either straight
or helical, provide passages for the chips and cutting fluid. Although most drills have
two flutes, three or four *fluted drills* are known as a core drill. These drills do not start
a hole, but enlarge or finish drilling a hole or a cored hole in an iron or steel casting.

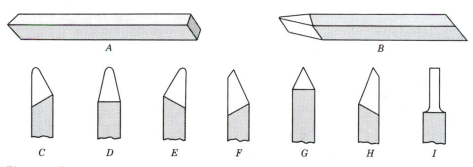

Figure 11.6
Lathe tool cutter bits. *A*, Cutter bit not ground. *B*, Cutter bit ground to form. *C*, Left-hand turning tool. *D*, Round-nose turning tool. *E*, Right-hand turning tool. *F*, Left-hand facing tool. *G*, Threading tool. *H*, Right-hand facing tool. *I*, Cutoff tool.

The most common drill is the *twist drill*, which has two flutes and two cutting edges. It is shown in Figure 11.7 with the drill terms indicated. The drill can be provided with either a straight or a tapered shank. Tapered shank drills are held and properly centered in the tapered socket of the drilling machine spindle. Drilling tools have a *Morse taper* of ⅝ in./ft, which also is standard for reamers and other similar tools. The tang at the end of the taper fits into a slot in the spindle socket to prevent slipping of the tapered surfaces. Straight shank drills are generally held and properly centered in a drill chuck, but many are tanged and used with taper split sleeves. These drills, cheaper than those having a tapered shank, are used only for sizes up to ½ in. (12.7 mm).

Several drills, which vary as to the number and angle of the flutes, are shown in Figure 11.8. Single-fluted drills originate holes and do deep-hole drilling. A two-fluted drill is the conventional type used for drilling holes. Either *interior oil channels* or external oil channels are used in production drilling. Three- and four-fluted drills principally enlarge previously drilled holes. Both have greater productivity and improved finish than two-fluted drills. Other drills with various flute angles are available to give improved drilling to special materials and alloys.

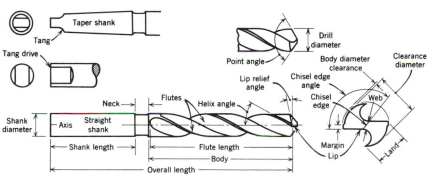

Figure 11.7
Standard twist drill and terms.

Taper shank oil drill hole

Drill for molded plastics

Four-fluted drill

Three-fluted drill

High helix drill for aluminum

Straight shank twist drill for mild steel

Figure 11.8
Types of drills.

Gun Drills

There are two kinds of *straight-fluted gun drills* used for deep hole drilling, as shown in Figure 11.9. One, known as a trepanning drill, has no dead center and leaves a solid core of metal. As the drill advances, the core acts as a continuous center guide at the point of cutting. This prevents the drill from running to one side, and hole accuracy is easier to maintain. The other type, known as a center-cut gun drill, has been the conventional drill for many years. It is still used for deep-hole drilling, such as the drilling of blind holes where a *core-type drill* cannot be used. Both types are generally carbide-tipped, as shown in Figure 11.9.

Gun drills operate at smaller feeds than conventional twist drills, but the cutting speeds are higher. Each drill has only a single cutting edge with a straight flute running throughout its length. Oil under considerable pressure is brought to the tip of the drill through the hole in the lip. The chips are carried out the hole along the flute of the drill as rapidly as they are formed. Greater accuracy and finish are obtained in deep holes by the subsequent use of special reamers or broaching tools.

Gun drills and gun bores have a single-lip cutting action, which is counteracted by bearing areas and lifting forces generated by coolant pressure. The supporting bearing arrangement forces the cutting edge to cut a true circular pattern and follow the direction of its own axis. Because of the single-point cutting action, the gun drill requires either a bushing or an accurate prebored start at the beginning of the hole. The tip of a gun drill has a V-shaped cutting edge and is usually a brazed carbide insert or a high-speed steel edge. Wear strips are located 180° and 90° from the cutting edge and consist of a wear-resistant material such as carbide.

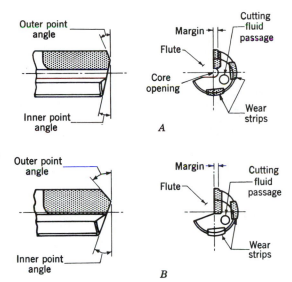

Figure 11.9
Straight-fluted gun drills. *A*, Trepanning drill. *B*, Center-cut gun drill.

Trepanning is used for large diameter holes where an annular groove is cut axially through the workpiece, leaving a core of material called a *slug*. It is efficient and cost-effective, particularly if the slug is reusable.

Both gun drills and trepanning tools require a high-pressure stream of cutting fluid to lubricate and cool the cutting surface and to carry away the chips. Chips are removed by two methods, external or internal passages. External passages are usually a straight flute; fluid passes through the shank and out the top and carries chips away through the flute. With an internal chip tool the chips pass through a chip mouth to the center of a tubular shank. Chip size must be regulated carefully to prevent clogging of the cutting tool. By using a heavy feed it is possible to install chip breakers to limit chip size. The internal chip tool is more torsionally rigid and cuts faster than the external design.

Special Drills

For drilling large holes in pipe or sheet metal, twist drills are not suitable because the drill tends to dig into the work or the hole is too large to be cut by a standard-size drill. Large holes are cut in thin metal by a hole cutter, as shown in Figure 11.10*A*. Saw-type cutters of this design are available in a wide range of sizes. For very large holes in thin metal a *fly cutter* is used. This cutter, as shown in Figure 11.10*B*, consists of tool bits held in a horizontal holder that can accommodate a range of diameters. Both cutters cut in the same path but one is set slightly below the other.

Spade drills, Figure 11.11, are another method to make large-diameter holes in the 1½- to 15-in. (38.1- to 380-mm) diameter range. For holes more than 3½ in. (90 mm) in diameter they are the only drills provided as stock items. Materials used for spade drills are high-speed steel or carbide tipped.

A drill designed for hardened steel operates at high speed and develops sufficient friction to anneal the steel and permit cutting without softening the drill point. Such drills, often carbide-tipped, are made with point angles of 118° or greater. Some drills are made in combination with other tools, such as the *combination drill and tap* or the *drill and countersink*. Double-margin drills and step drills are available to produce accurate holes, as shown in Figure 11.12.

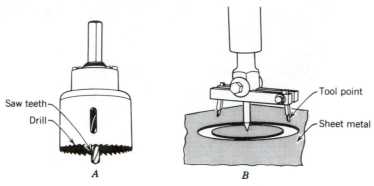

Figure 11.10
Cutters for holes in thin metal. *A*, Saw cutter. *B*, Fly cutter.

Drill Point Angle

To obtain good service from a drill it must be properly ground. The *drill point angle* should be correct for the material that is to be drilled. The usual point angle on most commercial drills is 118°, which is satisfactory for soft steel, brass, and most metals. For harder metals larger point angles, meaning the included angle, give better performance.

In Figure 11.13, two drills with point angles of 140° and 80° are shown. The thickness and width of the chips obtained from these drills are indicated by the letters T and W. Comparison of the two chips shows that the thickness T_1 for the 140° point angle is thicker than T_2 on the 80° point angle. Metal removed in the form of thick chips usually requires less energy per unit volume than when the same amount of metal is removed in the form of thin chips. In drilling hard and difficult-to-machine metals, the thicker chips allow some saving in power. It also may be noted that the width W_1 for the 140° point angle is less than W_2 for the smaller point angle. The larger width W_2, having a longer cutting edge, is useful in drilling materials creating some abrasive wear. The abrasive wear is distributed over a longer cutting edge and the cutting force per unit length is reduced. In addition, the corner angle for the 80° point drill (140°) is greater than that on the 140° point drill (110°), resulting in greater wear resistance to the drill at the corners. Materials such as soft cast iron and most plastics can be drilled best with point angles smaller than 118°.

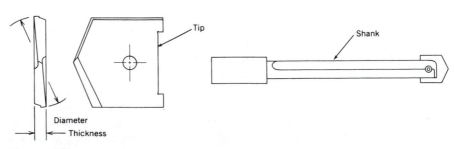

Figure 11.11
Spade drill and shank holder.

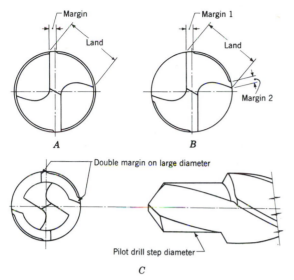

Figure 11.12
End views of conventional (*A*) and double-margin (*B*, *C*) drills showing construction differences. *C*, Double-margin step drill. Pilot diameter not relieved, which improves piloting action.

Drill Helix Angle

Drill performance is affected by the helix angle of the flutes. Although this angle may vary from 0° to 45°, the usual standard for steel and most materials is 24°. The smaller this angle is made, the greater the torque necessary to operate at a given feed. As the angle is increased appreciably, the life of the cutting edge is reduced for some materials. *Drill efficiency* is increased if the proper helix angle is used. For example, the angle for drilling copper, magnesium, and soft plastics should be approximately 35° to 45°, copper alloys 20° to 25°, hard plastics 17°, and soft to medium steel 24° to 32°. Tests show that there is a slight reduction in both torque and thrust as the helix angle is increased, but it is of minor importance as far as the overall performance of the drill is concerned.

Drill Point

On most conventional drills there is a chisel edge at the end of the *web* that connects the two cutting lips (Figure 11.7). This chisel edge does not cut efficiently because of the large negative rake that exists, not only at the center but also all along the chisel edge. Actually, there should be a slight crown to the chisel edge, which if sufficient will stabilize the drill when it is entering the work.

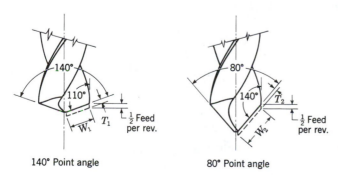

Figure 11.13
Point angle variation influences drill performance.

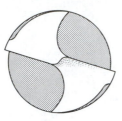
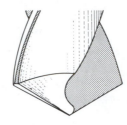

Figure 11.14
Spiral-point drill.

To improve drilling efficiency and reduce thrust, a self-centering drill point having a spiral edge has been developed, shown in Figure 11.14, which has much better cutting action close to the drill axis. Although this drill is self-centering and has less negative rake, it is difficult to grind and requires a special grinder. An easier way to reduce the end thrust is by web thinning and point splitting, as shown in Figure 11.15. The split point drill is used for drilling tough work-hardening steels and super alloys. Both of the drill point designs produce accurate holes with a minimum of oversize, and the thrust is much less than for chisel point drills.

Reamers

Reaming is the enlarging of a machined hole to proper size with a smooth finish. A *reamer* is an accurate tool and is not designed to remove much metal. Although reaming can be done on a drill press, other machine tools also are adapted to perform reaming. The material removed by a reamer depends on hole size and the material type. While 0.015 in. (0.38 mm) is probably a good average for stock removal, it can be as little as 0.005-in. (0.13 mm) for small reamer, or up to ⅟₃₂ in. (0.8 mm) for a large one. When reaming strain-hardened alloys, enough stock—never less than 0.005 in. (0.13 mm)—must be removed to avoid rubbing on the work-hardened surface. Because of the small amount of stock removed by this process, reamed holes are round and have a smooth surface. The terms applying to reamers are shown in Figure 11.16.

Illustrations of several reamers are shown in Figure 11.17. The hand reamer is a finishing tool designed for the final sizing of holes. They are ground with entry taper to provide easy access of the reamer into the hole.

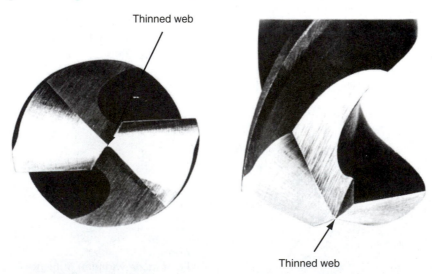

Figure 11.15
Split-point drill with thinned web to reduce end thrust.

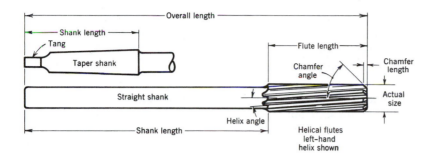

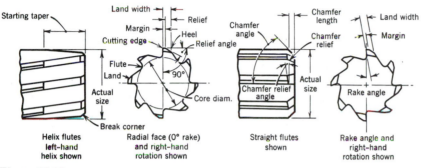

Figure 11.16
Standard reamer and terms.

Chucking reamers also are made with both straight and spiral flutes. *Spiral chucking reamers* that have a free reaming action are used for materials difficult to ream. Straight-fluted chucking reamers are commonly used in turret lathes, drill presses, and lathes. Both reamers have a slight chamfer of 45° on the end. *Rose reamers,* which do all cutting on the beveled end, find some use in reaming cored holes. The main difference between

Arbor for shell reamer

Fluted chucking reamer

Carbide-tipped shell expansion reamer

Taper pin reamer

Hand reamer

Figure 11.17
Types of reamers.

the rose reamer and the fluted chucking reamer is that the rose reamer is relieved only on the chamfered end. Shell reamers consist of a shell-type end mounted on a tapered arbor. Slots in the reamer engage lugs on the arbor to obtain a positive drive. *Shell reamers*, used principally on turret lathes, are not recommended for removing excessive amounts of stock. Taper pin reamers, usually of small diameter, are quite long. They are made in both straight and spiral flute types, the latter being better for machine reaming. *Taper pin reamers* must be constructed as sturdily as possible, since the cutting edges are engaged in the cut through most of their length. Expansion reamers can be adjusted to compensate for wear or purposely to ream oversize holes. *Adjustable reamers* differ in that they can be manipulated to take care of a considerable range in size.

Taps and Threading Tools

A *tap* is a cutter to thread internal holes. It has a shank and a round body with several radially placed chasers. It may be thought of as a screw with teeth and hardened to cut metals. Small taps are solid where large taps are solid or adjustable. A tap has two or more flutes that may be straight, spiral, or helical. They may be operated by hand or machine. Also, they are usually made of high-speed steel for rapid production and long wear.

When two parts are to be fastened together by a machine screw or cap screw, the hole in the untapped part is drilled larger than the major or outside diameter of the screw. The drill used for this operation is called a clearance drill, because it provides a slight clearance between the part and the screw.

Boring Tools

Boring enlarges holes previously drilled or bored. Drilled holes are frequently bored to eliminate any possible eccentricity. *Boring tools* finish holes to improved dimension and tolerance as is frequently done on large holes or on odd-size holes for which a reamer size is unavailable.

Counterboring is enlarging one end of a drilled hole. The enlarged hole is concentric with the original one and is flat on the bottom. The tool is provided with a pilot pin that fits into the drilled hole to center the cutting edges. Counterboring is used principally to set bolt heads and nuts below the surface. To finish off a small surface around a drilled hole is known as *spot facing*. This is a customary practice on rough surfaces, such as the pebbly surface of a casting, to provide smooth seats for bolt heads. If the top of a drilled hole is beveled to accommodate the conical seat of a flat-head screw, the operation is called *countersinking*.

Tools used in horizontal boring machines are mounted in either a heavy bar or a boring head that, in turn, is connected to the main spindle of the machine. Most boring operations use a cemented carbide single-point cutter, as shown in Figure 11.18, because they are simple to set up and maintain. The bar transmits power from the machine spindle to the cutter, as well as hold the boring tool rigidly during the cutting operation. The workpiece is normally stationary and the rotating cutter is fed through the hole. It is often necessary to provide additional support for the bar, as shown in the figure. The bar must be long enough to reach the end support and must provide the necessary longitudinal traverse for the machining operation.

For precision boring work on milling machines, jig borers, or drill presses, it is necessary to use a tool having *micrometer adjustment*. Such tools are held in a cutter head and

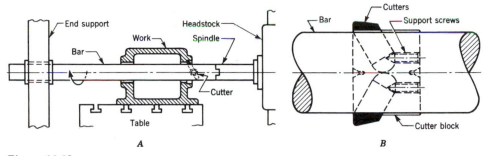

Figure 11.18
A, Straight boring on horizontal boring machine using line bar and end support.
B, Block-type boring cutter.

rotate. Hence, any increase in hole size is obtained by adjusting the tool radially from its center.

The most popular double-cutter arrangement is the *block-type boring cutter,* shown in Figure 11.18*B*, which consists of two opposing cutters resting in grooves on the block. Screws are provided to lock the cutters in position as well as to adjust them. The entire assembly fits into a rectangular slot in the bar and is keyed in place. Cutters are ground while assembled in the block and are held in alignment by the center holes provided.

The boring tool commonly used in small lathes is a single-pointed tool, supported in a manner that permits its entry into a hole. This tool, shown in Figure 11.19*A*, is forged at the end and then ground to shape. It is supported in a separate holder that fits into a lathe tool post. For turret lathes, slightly different holders and forged tools similar to the one shown in Figure 11.19*B* are used. A modification of this tool is the boring bar shown in Figure 11.19*C*, which is designed to hold a small high-speed steel tool bit at the end. The bar supporting the tool is rigid and may be adjusted according to the hole length. Although the clearance, rake, and cutting angles of these tools should be similar to those recommended for lathe work, these angles cannot be used if the holes are small. Greater end clearance is necessary because of the curvature of the hole surface, and back rake is difficult to obtain because of the position of the tool.

In production work boring cutters with multiple cutting edges are widely used. These

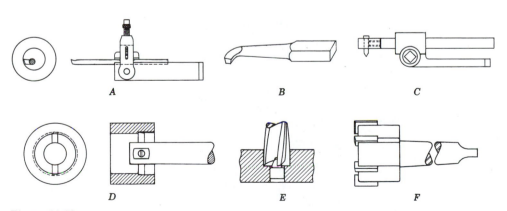

Figure 11.19
Types of boring tools. *A*, Light boring tool with bent shank. *B*, Forged boring tools.
C, Heavy boring tool. *D*, Double-ended cutter or boring tool. *E*, Counterboring tool with
pilot. *F*, Multiple-cutter boring tool.

cutters, shown in Figure 11.19*F*, resemble shell reamers in appearance but are usually provided with inserted cemented carbide tooth cutters, which may be adjusted radially to compensate for wear and variations of diameter. Boring tools of this type have a longer life than single-pointed tools and hence are more economical for production jobs. The counterboring tool in Figure 11.19*E*, provided with pilots to ensure concentric diameters, is designed to recess or enlarge one end of a hole.

Milling Cutters

The milling machine is versatile because of the large variety of cutters and the kinematic versatility of the machine tool components. The *milling cutter* is multitooth, unlike turning and boring tools.

Milling cutters consist of four basic groups of materials: tungsten carbide, cermet, ceramic, and polycrystalline. The tungsten carbide group can be uncoated, chemical vapor deposition (CVD) coated, and physical vapor deposition (PVD) coated materials. Each coated grade consists of various substrates of unalloyed (straight WC/Co) and alloyed (WC/TaC/TiC/NbC/Co) compositions.

The *cermets* (CERamic METals) are comprised mostly of TiN with a metallic binder. The ceramic-cutting tools are divided into alumina-based (aluminum oxide) ceramics and silicon nitride-based (sialon) ceramics. Polycrystalline materials are polycrystalline diamond (PCD) and polycrystalline cubic boron nitride (CBN).

Replaceable PVD-coated inserts mounted in a tool body, such as shown in Figure 11.20, are common. The PVD-coating process permits a sharp insert edge to retain full strength, allowing milling, cutoff, and grooving applications.

In addition, milling cutters are made of high-carbon steels and various high-speed steels (HSSs). High-carbon steel cutters have a limited use, because they dull quickly if high cutting speeds and feeds are used. Soft materials, or special-shaped cutters, or low-quantity requirements may use cutters made of HSSs, which maintain a keen cutting edge at temperatures lower than 1000° to 1100°F (500° to 600°C). Consequently, they may be used at cutting speeds 2 to 2½ times those recommended for carbon steel cutters. However, high carbon and HSS materials for milling cutters are declining in commercial application.

Figure 11.20
Milling cutter tooled with PVD-coated metal cutting insert.

Besides material used in the teeth, milling cutters are usually classified according to their general shape, method of mounting, or the method used in grinding the teeth. Methods of mounting milling cutters are

1. *Arbor cutters.* These cutters have a hole in the center for mounting on an arbor.
2. *Shank cutters.* Cutters have either a straight or tapered shank integral with the body of the cutter. These cutters are mounted in the spindle-end cavity.
3. *Face cutters.* These cutters are bolted or held on the end of the spindle or short arbors and are generally used for milling plane surfaces.

Teeth in milling cutters are made in two general styles according to the method in sharpening. Profile cutters are sharpened by grinding a small land back of the cutting edge of the tooth; this provides the necessary relief at the back of the cutting edge. Formed cutters are made with the relief (back of the cutting edge) of the same contour as the cutting edge. To sharpen these cutters the face is ground so as not to destroy the tooth contour.

Milling Cutter Design

The cutters most generally used, shown in Figure 11.21, are classified according to their general shape or the type of work they will do.

1. *Plain milling cutter.* A plain cutter is a disk-shaped cutter having teeth only on the circumference. The teeth may be either straight or helical if the width exceeds ⅝ in. (15.9 mm). Wide helical cutters used for heavy slabbing work may have notches in the teeth to break up the chips and facilitate their removal; see Figure 11.21C and F.

2. *Side-milling cutter.* This cutter is similar to a plain cutter except that it has teeth on the side. Where two cutters operate together, each cutter is plain on one side and has teeth on the other. Side-milling cutters may have straight, helical, or staggered teeth; see Figure 11.21H.

3. *Metal-slitting saw cutter.* This cutter resembles a plain or side cutter except that it is made very thin, usually ³⁄₁₆ in. (4.8 mm) or less. Plain cutters of this type are relieved by grinding the sides to afford clearance for the cutter; see Figure 11.21I.

4. *Angle-milling cutter.* All angle-shaped cutters come under this classification. They are made into both single- and double-angle cutters. The single-angle cutters have one conical surface, whereas the double-angle cutters have teeth on two conical surfaces. *Angle-milling cutters* are used for cutting ratchet wheels, dovetails, flutes on milling cutters, and reamers; see Figure 11.21D.

5. *Form milling cutters.* The teeth on these cutters are given a special shape. They include convex and concave cutters, gear cutters, fluting cutters, corner rounding cutters, and many others; see Figure 11.21J.

6. *End mill cutters.* These cutters have an integral shaft for driving and have teeth on both periphery and the end. The flutes may be either straight or helical. Notice Figure 11.21A, K, and L. Large cutters called shell mills have a separate cutting part held to a stub arbor, as shown in Figure 11.22. Owing to the lower cost of high-speed steel, as compared to carbide, this design results in a saving in material cost. End mills are used for surfacing projections, squaring ends, cutting slots, and in recess work such as die making.

Figure 11.21
Types of milling cutters. *A*, Spiral-end mill. *B*, T-slot milling cutter. *C*, Plain
cutter with helical teeth. *D*, Angle-milling cutter. *E*, Woodruff key-seat cutter.
F, Plain milling cutter. *G*, Inserted-tooth cutter. *H*, Side-milling cutter.
I, Metal-slitting saw cutter. *J*, Form-relieved cutter for gear teeth. *K*, Spiral
double-end mill. *L*, Extra-long spiral-end mill.

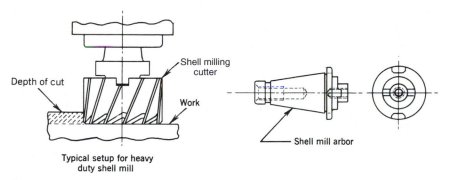

Figure 11.22
Shell mill and arbor.

7. *T-slot cutters.* Cutters of this type resemble small plain or side-milling cutters, which have an integral straight or tapered shaft for driving. They are used for milling T-slots. A special form is the Woodruff key-seat cutter, which is made in standard sizes for cutting the round seats for Woodruff keys. Notice Figure 11.21*B* and *E*.

8. *Inserted-tooth cutters.* As cutters increase in size, it becomes economical to insert the teeth made of expensive material into a less expensive body, perhaps made of HSS. Expensive milling teeth are replaced when worn or broken. Notice Figures 11.20 and 11.21*G*.

Milling Cutter Teeth

A typical milling cutter with various angles and cutter nomenclature is shown in Figure 11.23. For most high-speed cutters positive radial rake angles of 10° to 15° are used. These values are satisfactory for most materials and represent a compromise between good shearing or cutting ability and strength. Milling cutters made for softer materials such as aluminum can be given much greater rake with improved cutting ability.

Usually only saw-type and narrow, plain milling cutters have straight teeth with zero axial rake. As cutters increase in width, a positive axial rake angle is used to increase cutting efficiency.

For milling with carbide-tipped cutters, negative rake angles (both radial and axial) are generally used. Improved tool life is obtained by the resultant increase in the lip angle; also, the tooth is better able to resist shock loads. Plain milling-type cutters with

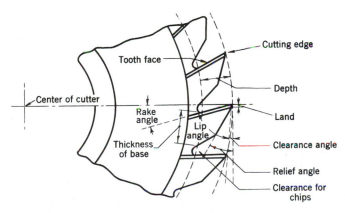

Figure 11.23
Milling cutter with nomen-clature.

teeth on the periphery usually are given a negative rake of 5° to 10° when steel is being cut. Alloys and medium-carbon steels require greater negative rake than soft steels. Exceptions to the use of negative rake angles for carbide cutters are made when soft nonferrous metals are milled.

The clearance angle is the included angle between the land and a tangent to the cutter from the tip of the tooth. It is always positive and should be small so as not to weaken the cutting edge of the tooth. For most commercial cutters more than 3 in. (75 mm) in diameter, the clearance angle is around 4° to 5°. Smaller diameter cutters have increased clearance angles to eliminate tendencies for the teeth to rub on the work. Clearance values also depend on the various work materials. Cast iron requires values of 4° to 7°, whereas soft materials such as magnesium, aluminum, and brass are cut efficiently with clearance angles of 10° to 12°. The width of the land should be kept small; usual values are ¹⁄₃₂ to ¹⁄₁₆ in. (0.8 to 1.6 mm). A secondary clearance is ground back of the land to keep the width of the land within proper limits.

Research on cutter form and size has proved that coarse teeth are more efficient for removing metal than fine teeth. A coarse-tooth cutter takes thicker chips and has free cutting action and more clearance space for the chips. As a consequence, these cutters provide increased production and decreased power consumption for a given amount of metal removed. Also, fine-tooth cutters have a greater tendency to chatter than those with coarse teeth, but they are recommended for saw cutters used in the milling of thin material.

11.2 OPERATIONS

Turning, drilling, boring, and milling operations are performed in many diversified ways. Even though these machines and operations started the industrial revolution, they continue to prove their great worth.

Lathe Operations

Operations on a lathe include turning, boring, facing, threading, and taper turning. For these operations a single-point cutter is moved slowly along the revolving workpiece. Drilling and reaming require other types of cutters, but are performed by the lathe group also. A brief description of some of the turning operations follows.

Cylindrical Turning

Figure 11.24 shows various applications for right- and left-hand tools. Recall that the tools are ground with specific applications in mind. Although the tool material of Figure 11.24 is HSS, there are similar turning tools that are suitable with cemented carbide grades.

A common way to support work on a lathe is to mount it between centers, as shown in Figure 11.25A. Holding work between centers allows the turning of the cylindrical surface along its length.

When a flat surface is to be cut, the operation is known as facing. The work is generally held in a faceplate or chuck, as illustrated in Figure 11.25B. In some cases, facing also is done with the workpiece between centers. Since the cut is at right angles to the axis of rotation, the carriage is locked to the lathe bed to prevent axial movement.

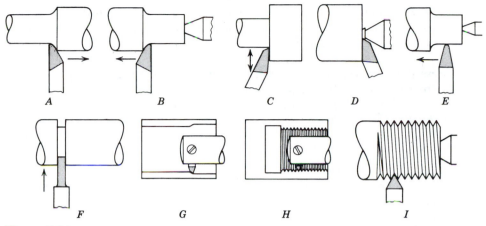

Figure 11.24
Lathe tools and applications. *A*, Left-hand turning tool. *B*, Right-hand turning tool. *C*, Right-hand turning tool. *D*, Left-hand facing tool. *E*, Round nose turning tool. *F*, Cutoff tool. *G*, Cutoff tool. *H*, Boring tool. *I*, Inside-threading tool.

Taper Turning

Many parts and tools made in lathes have tapered surfaces, varying from the short steep tapers found on bevel gears and lathe center ends to the long gradual tapers found on lathe mandrels. The shanks of twist drills, end mills, reamers, arbors, and other tools are examples of tapers. Such tools, supported by taper shanks, are held in true position and are easily removed.

Several taper standards found in commercial practice are as follows:

1. *Morse taper.* Largely used for drill shanks, collets, and lathe centers. The taper is ⅝ in./ft.

2. *Brown and Sharpe taper.* Used principally in milling machine spindles, ½ in./ft.

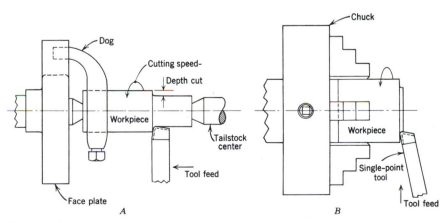

Figure 11.25
Lathe operations. *A*, Single-point tool in a turning operation. *B*, Facing cut.

3. *Jarno and Reed tapers.* Used by some manufacturers of lathe and small drilling equipment. Both systems have a taper of 0.6 in./ft, but the diameters are different.

4. *Taper pins.* Used as fasteners. The taper is ¼ in./ft.

Each of these standards is made in a variety of diameters and designated by a number. Accurate external tapers may be cut on a lathe in several ways:

1. Numerical control (NC) machines cut tapers as a matter of course. In addition, a straight chamfer, which is a type of taper, and radius operations are possible. See Chapter 28 for a discussion on computer numerical control.

2. A taper-turning attachment on the lathe, illustrated in Figure 11.26*A*, is bolted on the back of the lathe and has a guide bar, which is set at the desired angle or taper. As the carriage moves along the lathe bed, a slide over the bar causes the tool to move in or out according to the setting of the bar. Thus, the taper setting of the bar is duplicated on the work.

3. The compound rest on the lathe carriage, shown in Figure 11.26*B*, has a circular base and may be swiveled to any desired angle with the work. The tool is then fed into the work by hand. This method is especially adapted for short tapers.

4. Setting the tailstock center over, shown in Figure 11.26*C*, is another method. If the tailstock is moved horizontally out of alignment ¼ in. (6.4 mm) and a cylinder 12 in. (305 mm) long is placed between centers, the taper will be ½ in./ft. However, a cylinder 6 in. (152 mm) long will have a taper of 1 in./ft. Hence, the amount of taper obtained on a given piece depends on the length of the stock as well as on the amount the center is set over.

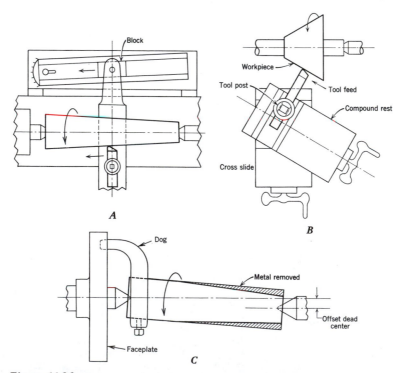

Figure 11.26
Taper turning. *A*, With attachment. *B*, With compound rest. *C*, Offsetting tailstock center.

If the tailstock is to be set over from the center, the set-off in inches is given by the length of the workpiece in feet multiplied by the taper in inches per foot divided by two.

Internal tapers are machined on a lathe by computer NC or by using the compound rest or the taper-turning attachment. Small holes for taper pins are first drilled and then reamed to size with a taper reamer.

Thread Cutting

Although it is possible to cut all forms of threads, the engine lathe is usually selected when only a few threads are cut or when special forms are desired, and only a few units are required. The engine lathe will have a threaded feed rod below the apron, which is essential for the cutting of threads. The form of the thread is obtained by grinding the tool to the proper shape using a suitable gage or template. Figure 11.27 shows a cutter bit ground for cutting 60° V-threads and the gage, which is used for checking the angle of the tool. Special form cutters are used for cutting these threads. These cutters are previously shaped to the correct form and are sharpened by grinding on the top face only.

In setting up the tool for V-threads, there are two methods of feeding the tool. It may be fed straight into the work, the threads being formed by taking a series of light cuts, as shown in Figure 11.27A. Cutting action occurs on both sides of the tool bit. Some back rake may be obtained, but it is impossible to provide any side rake on the cutting tool. This method is satisfactory on materials such as cast iron or brass, where little or no side rake is possible. The second method, for cutting steel threads, feeds the tool in

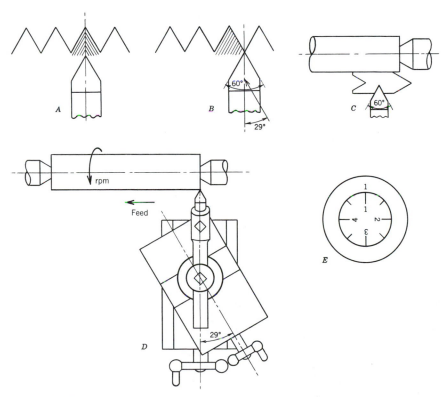

Figure 11.27
Method of setting a tool for thread cutting on a lathe. *A*, Straight feed. *B*, Feed at angle. *C*, Use of center gage for setting up threading tool. *D*, Method of setting up lathe for cutting V-thread. *E*, Threading dial.

at an angle as shown in Figure 11.27*B* and *D*. The compound rest is turned to an angle of 29° and, by using the cross feed on the compound rest, the tool is fed into the work so that all cutting is done on the left-hand side of the tool. The tool bit, being ground to an angle of 60°, allows 1° of the right-hand side of the tool to smooth off that side of the thread.

It is necessary that the tool be given a positive feed along the work at the proper rate to cut the desired number of threads per inch. This is accomplished by a train of gears located on the end of the lathe, which drives the lead screw at the required speed with relation to the headstock spindle. This gearing may be changed to cut any desired pitch of screw. The lead screw, in turn, engages the half-nuts on the apron of the lathe, providing positive drive for the tool.

After the engine lathe is set up, the crossfeed screw is set at some mark on the micrometer dial and a light cut is taken to check the pitch of the thread. At the end of each successive cut, the tool is removed from the thread by backing off the crossfeed screw. This is necessary since any back play in the lead screw would prevent the tool from returning in its previous cut. The tool is returned to original position. The crossfeed screw is set at the same reference mark, the tool is fed the desired amount for the next cut, and another cut is taken. These operations are repeated until the thread is cut to a proper depth.

Many engine lathes are equipped with a *thread dial indicator,* shown in Figure 11.27*E*. Close by the dial is a lever that is used to engage and disengage the lead screw with a matching set of half-nuts in the carriage. At the end of each cut the half-nuts are disengaged and then reengaged at the correct time so that the tool always follows in the same cut. The indicator is connected to the lead screw by a small worm gear, and the face of the dial, which revolves, is numbered to indicate positions at which the half-nuts may be engaged.

Turret Lathe Operation

Once a turret lathe is properly set up, an experienced machinist is not required for operation. However, skill is necessary in the selection, mounting, and adjustment of the tools. In small-lot production, it is important that this work be done quickly so as not to consume too much of the total production time, which consists of setup, workpiece handling, machine handling, and cutting time.

Setup time is reduced by having the tools in condition and readily available. For short-run jobs, a permanent set up of the usual bar tools on the turret is a means of reducing time. The tools selected are standard and when permanently mounted they may be quickly adjusted for various jobs. The loading and unloading time, which is the time consumed in mounting or removing the work, depends largely on the workholding devices used. For bar work this time is reduced to a minimum by using bar stock collets.

The time it takes to preposition the tools ready for cutting is part of the cycle time. This is reduced by having the tools in proper sequence for convenient use and by taking multiple or combined cuts whenever possible. The balance of the machine-handling time is made up of the time necessary to change the speeds and feeds.

Cutting time is controlled with proper cutting tools, feeds, and speeds. However, additional time is often saved by combining cuts, as shown in Figure 11.28. *Combined cuts* refer to the simultaneous use of both slide and turret tools. In *bar work* combined cuts are especially desirable, as additional support is given to the workpiece thereby reducing spring and chatter. In *chucking work,* internal operations such as drilling or boring may frequently be combined with turning or facing cuts from the square turret. Time is also saved by taking *multiple cuts*; that is, having two or more tools mounted on

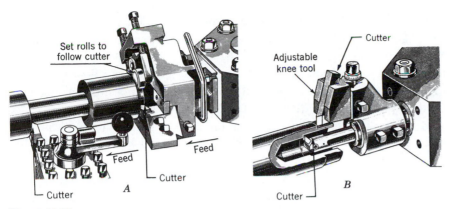

Figure 11.28
A, Combining cuts on bar work. *B*, Multiple cuts from hexagon turret.

one tool station. Figure 11.28*B* shows both boring and turning tools set up on one station of the turret.

For outside turning, a single-cutter turner or box tool (see Figure 11.28*A*) has been developed. As bar stock is supported only at the collet, additional support must be provided for heavy cuts to be taken. This is done by means of two rollers that contact the turned diameter of the stock to counter the thrust of the cutting tool.

To illustrate the method of tooling and sequence of operations for a given job, a hexagon turret set up for making necessary internal cuts on a threaded adaptor is shown in Figure 11.29. This shows the details of the internal cuts required to machine the adaptor. The sequence of the operations for the hexagon turret is

1. The bar stock is advanced against the *combination stock stop and start drill* and clamped in the collet. The start drill is then advanced in the combination tool, and the end of the work is centered.

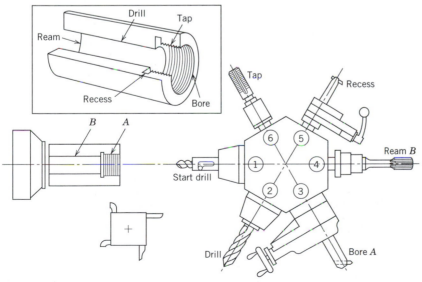

Figure 11.29
Hexagon turret setup illustrating the sequence of operations to handle required internal cuts on threaded adapter shown in insert.

2. The hole through the solid stock is drilled to the required length.

3. The thread diameter is bored to correct size for the threads specified. A stub boring bar in a slide tool is used.

4. The drilled hole is reamed to size with the reamer supported in a floating holder.

5. A groove for thread clearance is recessed. For this operation a quick-acting slide tool is used with a recessing cutter mounted in a boring bar.

6. The thread is cut with a tap held in a clutch tap and die holder. This operation is followed by a cutting off operation not shown in Figure 11.29. Cut off uses the square turret on the carriage.

A few of the tools used in turret lathe work are illustrated in Figure 11.30. These tools are so designed that they may be mounted quickly in the turret and adjusted for use. In addition to the usual operations of drilling, boring, reaming, and internal threading shown in the figure, various other threading, centering, and turning operations are available. Internal threading is frequently done with collapsible taps to facilitate quick removal. For the same reason automatic die head chasers, which open at the end of the thread, are used for external threads.

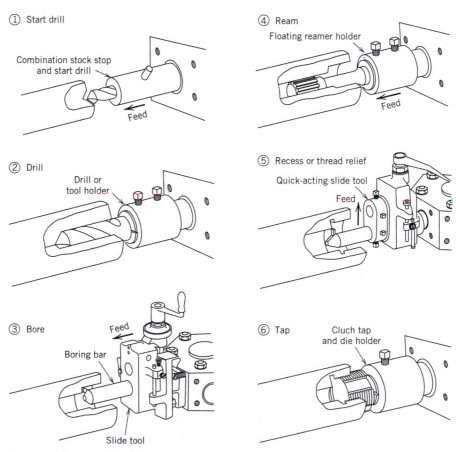

Figure 11.30
Setup for matching internal operations on threaded adapter.

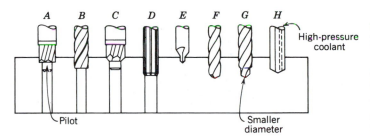

Figure 11.31
Operations for various drills. *A*, Counterboring. *B*, Core drilling. *C*, Countersinking. *D*, Reaming. *E*, Center drilling. *F*, Drilling. *G*, Step drilling. *H*, Gun drilling.

Drilling Operations

As hole-making is important in manufacturing, the choice of the right tool depends on the application. Production of holes with drills is seen in Figure 11.31. Notice that for some holes, a pilot is used to guide the hole cutting. In some cases there is a minor and smaller diameter that precedes the larger hole sizing.

In evaluating drill performance, drill material must not be overlooked. High-speed steel material tools will accept about twice the cutting speed of carbon tool steel. For hard and abrasive materials such as cast iron, drills tipped with tungsten carbide give improved results, but for some hard steels and other materials these drills are not satisfactory. High carbon, cobalt bearing, super HSSs capable of drilling steels having a hardness of Rockwell C68 are used for drilling tough stainless steels and aerospace alloys. Also, many drills are given a thin, hard case surface treatment or are chrome plated to provide a hard-wearing surface.

Drill Cutting Fluids

To obtain best performance and long life for cutting edges, some cutting fluid should be used. The cutting fluid improves the cutting action between the drill and the work, facilitates removal of chips, and cools the work and tool. In production drilling, cooling is most important, because the drill can overheat and soon loses its keen edge. To ensure long life, a cooling medium is selected to carry away the heat at the same rate that it is generated. Suggested coolants for some materials are

Aluminum: mineral, lard oil mix
Brass: dry, mineral, lard oil mix
Cast iron: dry, air jet
Malleable iron: soluble oil
Steel, soft: soluble oil, sulfurized oil
Steel, tool: lard, soluble oil

Milling

Two methods of feeding work to the cutter are shown in Figure 11.32. Feeding the work against the cutter, as indicated in Figure 11.32*A*, is usually recommended, because each tooth starts its cut in clean metal and does not break through possible surface scale. However, tests have proved that when the work is fed in the same direction as the cutter rotation, as shown at Figure 11.32*B*, the cutting is more efficient, larger chips are removed, and there is less tendency for chatter. This method is frequently used in production work, where large cuts are to be taken and the surface of the work is free from scale.

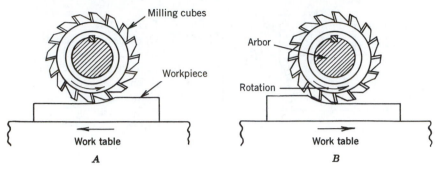

Figure 11.32
Methods of feeding work on milling machine. *A*, Conventional or up milling.
B, Climb or down milling.

Feed on milling machines is expressed in either of two ways. On some machines it is expressed in thousandths of an inch per revolution (mm/r) of cutter. Such machines have feed changes ranging from 0.006 to as high as 0.300 in. (0.15 to 7.62 mm). The other way expresses the feed of the table in inches per minute, ipm (mm/s), the usual range being from ½ to 20 in. per minute (0.20 to 8.5 mm/s).

Milling Cutter Operations

Figure 11.33 shows the common types of milling operations. These operations show the location of the cut and the position of the workpiece with its feed movement. In Figure 11.33*A*, peripheral or slab milling is shown. Face milling is shown in Figure 11.33*B*, and the cutter is mounted to the end of the spindle. In Figure 11.33*C* the end, or shell milling cutter, is mounted through the taper shank into the corresponding taper in the end of

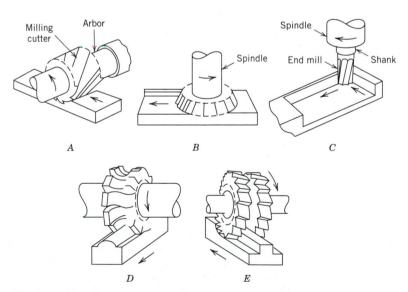

Figure 11.33
Milling operations. *A*, Slab or plain. *B*, Face. *C*, End. *D*, Contour or form. *E*, Straddle.

the spindle. Gang milling employs the combination of straight, side, step, angular, or form cutters all mounted together on a single arbor so as to cut the contour simultaneously. In Figure 11.33D a form or contour cutter is being used. Notice Figure 11.33E, which is called straddle milling.

Milling operations are joined with *box jigs,* fixtures, and table attachments. These attachments are used for special designs. Notice Figure 11.34A, which uses an index head for cutting gears. An index head or dividing head is a complex set of gears housed in a box. The geared shaft rotates work through a predetermined number of degrees, or a fraction of revolution. The index head and its tailstock employ an arbor with a spur gear mounted. A gear cutter is shaping the gear teeth. This method of gear production is limited to school shops for instructional purposes. Gear production is discussed in Chapter 21. Index heads are also controlled by computer numerical control, shown in Figure 11.34B, which is able to do the work faster than mechanical index heads.

A

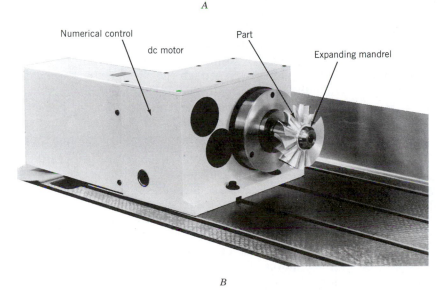

B

Figure 11.34
Index head. *A,* Mechanical index head for milling a spur gear. *B,* NC dividing head.

11.3 PERFORMANCE

Once the turning, drilling, boring, and milling machines are operating and effectively making parts, attention turns to the performance of the operation. *Performance* can imply the time that is required, or the power, tool wear, or surface integrity. Some of these issues have been settled in earlier chapters, and others, such as the estimated cost of the operation, are discussed in Chapter 31. Various measures of performance are discussed in this section.

At this stage several questions are raised. "How much time does the metal removal operation require?" In addition, "how much metal is being removed?" is vital. Other questions are stated. If the performance is poor, then other processes must be substituted to reduce the cost. Short- and long-term cost is the most important driving force in manufacturing.

Cutting Speed

The amount of metal removal is a function of both the cutting speed and feed. The cutting speed, expressed in ft/min (m/s), is a measure of the peripheral speed as indicated by the following expression:

$$V_c = \frac{\pi D N}{12} \tag{11.1}$$

where

V_c = Cutting velocity of surface, ft/min for workpiece, milling cutter, or drill OD
D = Diameter of work or cutter, in.
N = Revolutions per minute, rpm

This equation was introduced in Section 8.8. The equation is applicable for turning, drilling, boring, milling, and other processes. In turning operations, the diameter D is the outermost dimension of the work being turned. For example, if a bar is rough turned, the outside dimension is used. If the rough pass reduces the diameter by 1 in. (25.4 mm), the next calculation employs a smaller diameter, for example, a finish pass. In facing or cutoff operations the outer diameter is used, even though the diameter is continually reducing during the cut. For boring, the diameter is the internal dimension of the hole before being bored. In milling, the outermost diameter of the cutter serves as the diameter D, since it is rotating.

Usually, the *cutting speeds*, V_c, are known from information based on machinability testing and experience. Table 8.2 gives typical values for turning with HSS and carbide-type materials. With V_c given, the calculation provides N, the rotary speed, which is necessary for the operation of the equipment. N, as a machine operation performance number, is a posted or digital readout value given on the lathe, drill press, or milling machine, and knowledge of this value aids the efficiency of the equipment in the shop.

The tool or workpiece moves relative to another element of the machine tool. The reader may want to refer to Chapter 7. This movement is called feed. For example, a turning tool mounted on the carriage moves along the bed, and its rate of velocity, or carriage *feed*, is expressed in inch per revolution, or *ipr*. The revolutions referred to are the rotary revolutions of the bar stock. In milling if the cutter is fixed (meaning that it is not traversing, although it is rotating), the table feed is moving or feeding at a velocity expressed as inches per minute.

The time to machine is simply the distance that the tool travels divided by the feed velocity. For turning-type operations

$$t_m = \frac{L}{fN} = \frac{L\pi D}{12V_c f} \qquad (11.2)$$

where

t_m = Machining time, min
L = Length of cut for metal cutting, in.
D = Diameter, in.
f = Feed rate, ipr
N = Rotary cutting speed, $\dfrac{12V_c}{\pi D}$, rpm

Another feed rate for turning in in./min can be found as

$$f_m = fN \qquad (11.3)$$

where

f_m = Feed rate, in./min

The time of the operation as

$$t_m = L/f_m \qquad (11.4)$$

Turning and lathe operations are single-point operations. They differ from milling operations. Milling cutters are multitooth and the feed rate depends on the number of teeth in the cutter as well as the cutter diameter. Milling is unlike turning or boring operations as the diameter for V_c is the diameter of the milling cutter.

$$t_m = \frac{L\pi D_c}{12V_c n_t f_t} \qquad (11.5)$$

where

n_t = Number of teeth on milling cutter
f_t = Feed per tooth, in.
D_c = Diameter of milling cutter, in.

Note that for milling the feed is per tooth. Thus, milling cutter specification will indicate the diameter of the cutter, number of teeth, material of the insert, and the like.

A small list of values of V_c and f is given as Table 11.1 for operations of turning, drilling, reaming, face milling, slab or plain milling, and end milling. Manufacturers should be consulted for more thorough tables. Materials are HSS and tungsten carbide with the exception of drilling and reaming, which is HSS. Two depths of cut values are given for each material. The larger value is a rough depth. For deeper metal removal thickness, the speeds and feeds differ from those that have shallower or finish depths. Drilling and reaming are listed on the basis of tool diameter. The feed for milling entries is for each tooth. End milling uses a 1-in. end mill as the diameter for the table values.

Rate of Metal Removal

The rate of metal removal is found for turning by

$$Q = 12 \times t \times f_r \times V_c \qquad (11.6)$$

where

Q = Rate of metal removal, in.^{3}/min
t = Depth of cut, in.
f_r = Feed, ipr

Table 11.1 **Typical Machining Speeds and Feeds**

Material	Depth of Cut	High-Speed Steel		Carbide	
		V_c	f	V_c	f
Turning					
Carbon steels					
1010, 1025	0.150	125	0.015	400	0.020
	0.025	225	0.007	700	0.007
1040	0.150	70	0.015	350	0.020
	0.025	200	0.007	600	0.007
Free machining	0.150	100	0.015	400	0.020
1112	0.025	250	0.007	675	0.007
Stainless	0.150	70	0.015	325	0.015
	0.025	125	0.007	550	0.007
Gray cast iron	0.150	75	0.015	250	0.020
	0.025	120	0.007	450	0.010
Aluminum	0.200	150	0.015	400	0.020
	0.025	350	0.007	700	0.010
Drilling					
Carbon steels	1/16 OD	100	0.001		
	1/4	110	0.005		
	1/2	125	0.010		
	1	125	0.018		
Reaming					
Carbon steel	1/16 OD	130	0.005		
	1/4	130	0.008		
	1/2	150	0.012		
Face milling[a]					
Medium carbon steel	0.150	115	0.009	400	0.012
	0.025	155	0.006	450	0.008
Stainless steel	0.150	90	0.006	350	0.010
	0.025	120	0.005	465	0.008
Cast iron, gray	0.150	150	0.016	500	0.020
	0.025	195	0.010	665	0.010
Slab milling[a]					
Medium-carbon steel	0.150	120	0.008	400	0.012
	0.025	165	0.006	450	0.008
Stainless steel	0.150	85	0.006	350	0.010
	0.025	110	0.005	460	0.008
Cast iron, gray	0.150	105	0.011	500	0.020
	0.025	140	0.009	650	0.010
End milling[a]					
Medium-carbon steel	0.050	120	0.005	375	0.007
	0.015	180	0.005	485	0.006
Stainless steel	0.050	110	0.005	325	0.007
	0.015	145	0.004	420	0.006
Cast iron, gray	0.050	60	0.005	200	0.006
	0.015	75	0.004	260	0.005

[a] Milling feed is feed per tooth.

Q gives a measure of the amount of material that is being cut per unit of time. It is an important factor in machining, because it is related to the energy and forces involved in metal cutting.

As an example, a 2-in. diameter gray cast iron workpiece is rough turned with a disposable carbide insert tool point. From Table 11.1, $V_c = 250$ ft/min and $f = 0.020$ ipr, and the depth of cut $t = 0.150$ in. Then, rpm $= 478$ ($= 12(250)/(\pi \times 2)$), and $Q = 9$ in.^{3}/min ($= 12 \times 0.150 \times 0.020 \times 250$).

For drilling the rate of metal removal is given as

$$Q = \frac{D_d^2 \pi f_m}{4}$$ (11.7)

where

$$D_d = \text{Diameter of drill, in.}$$

The rate of metal removal for milling is

$$Q = w \times t \times f_m$$ (11.8)

where

$$w = \text{Width of cut, in.}$$
$$f_m = \text{Feed rate, in./min}$$

and

$$f_m = f_t \times n_t \times N$$ (11.9)

In milling the term "feed per tooth" or "feed per cutting edge" is used. It is a function of not only the feed velocity but also the rpm as well as the number of cutting edges on the cutter.

Consider the example of slab milling cast iron with coated and cemented carbide tool inserts with $V_c = 500$ ft/min and $f_t = 0.020$ in. per tooth (from Table 11.1) and a 4-in. OD cutter with the number of teeth $n_t = 8$ inserts. The cutter rpm $= 477$ ($= 12(500)/(\pi \times 4)$). Then, $f_m = 76.3$ in./min ($= 477 \times 0.020 \times 8$), which is the velocity of the table, as it passes the workpiece in either "up or down" milling. Now let the depth of cut $t = 0.25$ in., and width $w = 3$ in. The volumetric rate of machining $= Q = 57.2$ in.^{3}/min ($= 3 \times 0.25 \times 76.3$).

Horsepower

The horsepower required at the spindle can be found using the rate of metal removal for turning, drilling, and milling, or

$$HP_s = Q \times P$$ (11.10)

where

$$P = \text{Unit horsepower, horsepower per in.}^3/\text{min}$$

However, in determining the power required for metal cutting, there is also the efficiency of the machine tool when no cutting is taking place. Power is necessary for idling, or *tare horsepower*. Efficiency is a measurement of the portion of energy finally delivered to the cutter in the form of useful work. A machine that transforms 85% of the energy supplied to it has an efficiency of 85%. Efficiency is output work divided by input work, and power is the time rate of doing work, which is work divided by time.

Previously, in Chapter 8,

$$HP_s = \frac{F_c \times V_c}{33,000} \qquad (11.11)$$

where

$$F_c = \text{Cutting force, lb}$$

Let

$$HP_m = \text{Horsepower at motor, hp (watts)}$$
$$= \frac{HP_s}{E} \qquad (11.12)$$

where

$$E = \text{Efficiency of spindle drive, \%}$$

Sometimes it is important to understand that *unit power* equations can be calculated, given the information.

$$P = \frac{HP_s}{Q} \qquad (11.13)$$

where

$$P = \text{Unit power, hp/in.}^3\text{/min (W/mm}^3\text{)}$$

Some unit power requirements helpful for solving problems are listed in Table 11.2. These values are for turning, drilling, and milling. A similar, yet brief table is given as Table 8.1, which deals with turning only.

Torque

The torque at the spindle is found using

$$T_s = \frac{63,030 \, HP_s}{N} \qquad (11.14)$$

where

$$T_s = \text{Torque at spindle, in.-lb}$$

Table 11.2 **Average Unit Power Requirements for Turning, Drilling, and Milling**

Material	Hardness	Turning	Drilling	Milling
Steels	35–40 R_c	1.4	1.4	1.5
Cast irons	110–190	0.7	1.0	0.6
Stainless steels	135–275	1.3	1.1	1.4
Nickel alloys	80–360	2.0	1.8	1.9
Aluminum alloys	30–150	0.25	0.16	0.32
	(500 kg)			

Length of Cut

The *length of cut, L,* is the distance that the cutting tool or table is moving at feed f velocity. The general relationship for length L is given as

$$L = L_s + L_a + L_d + L_{ot} \tag{11.15}$$

where

L_s = Safety length, in. (mm)
L_a = Approach length due to cutter geometry, in. (mm)
L_d = Design length of workpiece requiring machining
L_{ot} = Overtravel length due to cutter geometry, in. (mm)

The *total length value L* is used in finding the time t_m or cost for metal cutting.

Feed, or ipr, is much less than the *rapid traverse velocity,* which may be 100 to 1000 in./min. During this "rapid traverse" movement, the machine elements, such as headstock, table, carriage, column, and base, are moving relative to one another, but there is no cutting. This rapid-traverse movement is done quickly, so to position the machine tool structural elements and present the cutting tool next to the workpiece ready to cut at the much slower rate, expressed as ipr. L_s, *safety length,* is necessary for any stock variations in length, for if the cutter is in rapid traverse velocity and the trips, which cancel the fast mode of table or cutter movement, are set too short, it is possible for the cutter to bang into the workpiece, causing damage and safety problems. It is called safety stock, because it implies that the metal-cutting mode is enabled and the machine is cutting air, so to speak, and not chips. Safety stock may vary from $\frac{1}{64} \leq L_s \leq \frac{1}{2}$ in.

Approach length, L_a, depends on the cutter workpiece geometry. Approach length is considered negligible for turning.

Milling approach distance depends on the diameter of the cutter D_c, depth of cut t, and the type of milling, whether it is vertical spindle with a shell end mill (see Figure 11.20) or a horizontal arbor-mounted slab cutter (see Figure 11.21*F* and *H*). Notice Figure 11.35*A* and *B* for milling operation approach and overtravel.

In Figure 11.35*A*, the view is from the top, and you are seeing the diameter D_c of the end or shell mill. The operation of the vertical milling operation is to end mill a width that requires two passes, as the machined slot width is greater than the diameter of the end mill. The circle, representing the D_c diameter of the cutter, is rotating, and the beginning station is 1. The part moves from station 1 to 2 to 3 to 4 and back to station 1, where it is in ready position for the next part. The feed f is shown with an arrow indicating direction. As the width of the machined dimension is greater than the diameter D_c, the end mill, it is necessary to move the cutter over a distance such that the full width is machined. Finally, a *rough milling pass* will always require an approach, but an overtravel may not be required, and the cutter center line would stop at the edge of the part line. If the milling cutter is performing a *finishing pass,* it is customary to move the cutter off the workpiece to avoid rough tool marks at the end of the workpiece. Remember that the cutter at station 1 needs to move $\frac{1}{2}D_c$ to have full width cutting of the shell end mill.

Figure 11.35*B* illustrates a side, saw, slab, or plain milling cutter (see Figure 11.21*C*) approach.

$$L_a = L_{ot} = \sqrt{\left(\frac{D_c}{2}\right)^2 - \left(\frac{D_2}{2} - t\right)^2} = \sqrt{t(D_c - t)} \tag{11.16}$$

where

D_c = Cutter diameter, in. (mm)
t = Depth of cut, in. (mm)

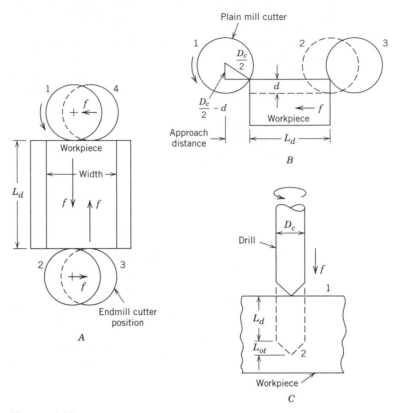

Figure 11.35
Approach and overtravel length. *A,* End milling (top view). *B,* Plain
milling (front view). *C,* Drilling (front view).

Notice if t, the depth of cut is increased, the length L_a is also increased. Or, if the diameter of the shell end mill is increased, the value of L_a increases.

For drilling, the *drill point* adds to the length of cut, and the trigonometric relationship gives the distance accounting for the 118° angle and drill point overtravel as

$$L_a = \frac{D_c}{2 \tan 59} = 0.3\, D_c \qquad (11.17)$$

where

$$D_c = \text{Drill diameter, in.}$$

Usually, however, the *drill-point approach* is about half the diameter. Other drill point angles will have different approaches. Note Figure 11.35C for the sketch of drill point overtravel.

L_d is the length of cut that is found from the *engineering drawing.* Typically, it is necessary to increase this for cutter geometry, depth of cut, and the nature of machining. Example: The top of a steel block, 40 in. long by 6 in. wide, is machined in a two-pass system, rough and finish milling. A shell end mill, $D_c = 6$ in. in diameter, is selected. The cutting insert material is PVD TiN-coated cemented carbide. The tool clamps $n_t = 8$ inserts. The depth of cut t for rough pass is 0.25 in. and for the finish pass is 0.025 in. The tool and workpiece material and the heavy vertical milling machine allow for milling rates of (V_c = ft/min, f_t = in./tooth) of (400, 0.012) and (450, 0.008) for *rough and finish milling.*

The length of cut for rough milling includes L_s = 0.5 in., L_a = 3 in., L_d = 40 in., and for rough milling L_{ot} = 0. The path length L for rough milling = 43.5 in. (= 0.5 + 3 + 40 + 0), whereas finish milling = 46.5 in. The time to rough mill = 1.78 ((= 43.5π6)/ (12 × 400 × 8 × 0.012)) and 2.53 min are required for finish milling. The rotary speed N = 254 rpm (= (12 × 400)/(π × 6)) for rough milling and 287 rpm for finish milling. A *cycle* will include time, in addition to machining time, for handling, inspection, and machine manipulation, which are often greater than cutting time. This is discussed in Chapter 31.

The feed rate f_m = 24.4 in./min (= 0.012 × 8 × 254) and 18.4 in./min for rough and finish milling. The volumetric rate of metal removal Q = 36.6 in.^{3}/min (= 6 × 0.25 × 24.4) for rough milling.

The horsepower at the spindle = 55 HP (= 36.6 × 1.5), where unit horsepower P = 1.5, from Table 11.2. Motor horsepower = 64.5 (= 0.55/0.85). As the machine is 100 HP rated, the operation is within capacity.

QUESTIONS

1. Give an explanation of the following terms:

Brazed tip tools
Tool signature
Fluted drills
Morse taper
Interior oil channels
Gun drills
Trepanning
Fly cutter
Spade drills
Drill point angle
Chucking reamers
Block-type boring cutter

Arbor cutters
Angle-milling cutters
Thread dial indicator
Combination stock stop and
    start drill
Box jigs
Cutting speeds
Tare horsepower
Unit power
Length of cut
Rapid traverse velocity
Approach length

2. What is wrong with a lathe that turns slightly tapered surfaces instead of cylindrical surfaces? Assume that the small end is on the tailstock end.

3. Briefly describe the steps in cutting a V-thread on an engine lathe.

4. Explain the terms "combined cuts" and "multiple cuts."

5. How does the box tool prevent bending of the bar stock when machining?

6. Describe the advantages and disadvantages of thin versus thick chips in drilling.

7. Sketch the coring, spot facing, counterboring, reaming, and countersinking operations.

8. Define a core drill and list its applications.

9. When the same amount of metal is removed, does it require more or less energy with thick chips than with thin chips?

10. What effect does the helix angle have on drill performance?

11. How is the coolant applied in deep-hole drilling?

12. What is the SI unit for the Morse taper?

13. List the types of operations that can be performed on a milling machine, lathe, and a drill press.

14. What are the differences between profile and formed tooth cutters?

15. Describe down milling. What are the advantages?

16. Compare the rake angles on milling and lathe cutters.

17. Sketch a plain milling cutter and indicate the rake angle, clearance angle, tooth face, land, and tooth depth.

18. Why is the effective length of cutting greater than the design requirements of the workpiece for machining length L?

PROBLEMS

11.1. If a milling cutter is 8-in. OD, $V_c = 400$, $f_t = 0.014$ and $n_t = 10$ cemented carbide inserts, what is the rotary speed N? Find the feed rate f_m.

11.2. A block is 20 in. long by 6 in. wide. A shell end mill $D_c = 6$ in. Find the length of cut for rough and finish milling.

11.3. Find the taper in in./ft if a 1-m bar were turned with the tailstock set over 3 mm.

11.4. A 14-in. (360-mm) length of stock is tapered so that the diameter on one end is 3 in. (75 mm) and on the other end 2⅞ in. (73 mm). How much should the tailstock be set over to cut this taper?

11.5. Using carbide cutters what are the speeds for rough-turning cast iron and stainless steel of the same diameter?

11.6. A bar of medium-carbon steel 2 in. (50 mm) in diameter is machined at 90 ft/min (0.5 m/s). What spindle speed is necessary? Repeat for 125 ft/min. Repeat for 4 in.

11.7. Find the rotary speed rpm for a 4 in. diameter bar stock being turned on a lathe for the following:

Stock Material	Tool Material	Type Cut
Stainless steel	HSS	Rough
Gray cast iron	HSS	Finish
Medium-carbon steel	HSS	Rough
Stainless steel	Carbide	Rough
Stainless steel	Carbide	Finish

11.8. Calculate the cutting time for a 3-in. initial outside diameter and turning length of 20 in. for the following:

Stock Material	Tool Material	Type Cut
Stainless steel	Carbide	Finish
Gray cast iron	HSS	Rough
Medium-carbon steel	Carbide	Finish
Stainless steel	Carbide	Rough
Stainless steel	HSS	Finish

11.9. Stainless steel material is rough and finish turned. Diameter and length are 4 in. by 30 in. Speed and feed for tungsten carbide tools are (350, 0.015) and (350, 0.007). Determine rough and finish time to machine. Find the rotary velocity of the bar stock. The rough pass removes ½ in. on the diameter.

11.10. Find the length of cut for the following conditions. Dimensions are given in inches. Assume safety stock = 0.25 in.

Cutter Type	D_c	Width of Cut	L_d	t	Type
End mill	2	2	20	0.1	Finish
End mill	1	2	20	0.25	Rough
Slabbing mill	6	6	30	0.50	Rough
Slabbing mill	8	16	30	0.75	Finish
Drill	1	—	3	—	—
Drill	0.5	—	2	—	—
Tapping	1	—	3	—	—

11.11. The length of a machining pass is 20 in. and the part diameter is 4 in. OD. Velocity and feed for this material are 275 ft/min and 0.020 ipr. Find the time to machine.

11.12. A workpiece is 10 in. long and is milled to a depth of ⅛ in. by a plain spiral cutter 6 in. in diameter. If the feed rate is 6 in./min, how much time is necessary to make the cut?

11.13. Stainless steel is rough and finish milled. Find the time to machine a flat of 10 in. by 20 in. for a cutter having a path width of 5 in. The number of teeth for this plain carbide cutter is 12 and the diameter is 4 in. What rpms are suggested for the rough and finish passes? The rough pass removes 1 in. and the finish pass removes 1/16 in.

11.14. A stainless steel part is tap drilled 5/16 in. for a depth of 1 in. It is followed by a tap 3/16–16 to a depth of ⅞ in. Find the drilling length and drilling and tapping time. Repeat for steel. (Hint: Assume that tapping time is 1.5 times drilling.)

11.15. Find the rate of metal removal for an 11.3 mm long slot, 15 mm deep, and 5 mm wide cut at a feed of 0.15 mm per second. If cutting occurs 40% of the time, how much metal is removed in 8 h?

11.16. Find the length of cut for the following dimensions, given in customary units:

L_s	L_a	L_d	L_{ot}
0.1	0.3	20	0.3
0.05	0.5	30	0.5
0.1	0.5	25	0

11.17. The top of a square 250-mm block has a 65-mm slot machined on it. The end mill is 25 mm in diameter and only one pass is traced over the slot. Safety stock is 5 mm. Find the length of cut.

11.18. Find the length of cut for a 2-in. plain milling cutter removing ½ in. thickness for a workpiece 10 in. in length. Safety stock is ¹⁄₁₆ in. The cutter is to leave the stock free of milling marks. Repeat for ¼ in. thickness.

11.19. Using a carbide tool, find the time to rough machine an aluminum shaft 3 in. in diameter and 24 in. long. The cutting speed is 600 ft/min and a tool has a feed of 0.004 ipr.

11.20. A rod 2½ in. in diameter and turning at 120 rpm is cut off by a tool having a feed of 0.004 ipr. What is the cutting time? Find the cutting time if the diameter is 4 in. and feed is 0.005 ipr.

11.21. A 3 in. in diameter AISI 1112 steel workpiece is finish turned with a disposable carbide insert tool point. The depth of cut $t = 0.005$ in. Find rpm and the volumetric rate of metal removal.

11.22. Stainless steel material is rough-milled with a tungsten carbide tool diameter of 8 in. The plain milling cutter has 18 teeth and a face width of 2 in. The depth and width of cut is 0.75 in. and 1.5 in. The cutter is mounted on a horizontal arbor, and a knee and column milling machine is selected. Find the rpm, velocity of the milling table, volumetric rate of machining, and if the length of the design is 25 in., what is the time for the metal removal portion of the operation? How much horsepower is required at the spindle and motor?

11.23. Stainless steel is finish turned on a lathe having a power efficiency of 85% and a tare horsepower of 0.6 HP. Use Table 11.2 for average unit power requirement, Table 11.1 for cemented carbide, and finish turning for cutting conditions. Depth of cut is 0.010 in. Determine the motor horsepower for this operation. Repeat for high-speed steel material.

MORE DIFFICULT PROBLEMS

11.24. A large quantity of a machinist vise jaw, as shown by Figure P11.24, is required. The material is medium-carbon cold rolled steel, and initial stock size is 0.625 in. square and 5.0 in. long. Milling and drilling operations are performed. Select the cutting tools, their material, and size, and choose the type of milling machine and drill press. Determine the method to hold the part on the table. Find the rpm, length of cut, and the machining time for each machining cut.

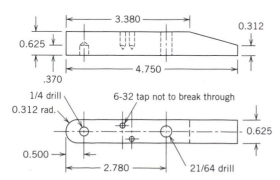

Figure P11.24

11.25 A surface, 6 in. wide by 20 in. long, is rough milled with a depth of cut of ¼ in. with a 4-in. OD cutter. A 16-tooth cemented carbide cutter face mill, 6 in. diameter, is used. Workpiece material is cast iron. What is the cutting length? Estimate the cutting time if $V = 120$ ft/min and

tooth chip load = 0.0012 in./tooth. Find the cutter rpm. What is the horsepower and torque requirement? Repeat for $V = 150$ ft/min.

11.26. Assuming a 0.015-in. feed and 0.062-in. depth of cut in machining a 2-in. bar of SAE 1020 steel, what is the metal removal rate in in.³/min as a function of the cutting speed? What is the approximate horsepower at the spindle? If the spindle efficiency = 92%, what motor horsepower is expected?

11.27. A 4-in. OD HSS plain milling cutter having a 6 in. face and 8 teeth is used for a ⅜-in. depth cut. The drawing length is 11 in.; safety stock ¼ in.; and work stock material is medium-carbon steel. If the cutter is required to have overtravel, what is the length of feed for the cutter? How much time is necessary for a rough pass? If the finish depth of cut is ⅛ in., how much time is necessary?

11.28. A machinist clamp screw is shown by Figure P11.28. Material is stainless steel and initial stock size is ⅝ in. in diameter by 16 ft long. Quantity is large and a numerical control turret lathe capable of handling bar stock is selected to make the part. After turning is completed, the cross hole is drilled. The lathe has a four- and eight-position turret for the cross and end turret locations. There is a single station upright drill press available for making the part after the lathe work is complete. List the sequence of machining cuts for the turret lathe. Specify the tools, their type, material, and diameter. Sketch the turrets and indicate the position of the tools. Find the length of cut, rpm,

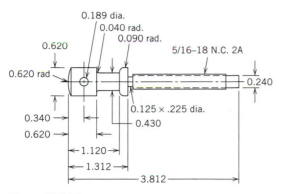

Figure P11.28

and time for each machining cut. (Hints: Use engineering computation paper for the sketch. Do not find the rpm for the threading operation.)

11.29. A 6-in. diameter cutter is milling steel with a Bhn of 225 and a unit horsepower of 0.75. The milling cutter has 10 teeth and the cutting velocity is 150 ft/min and 0.006 ipr per tooth. Depth of cut is 0.5 in., and the width of the cut is 3 in. The machine has a power efficiency of 85% and a tare horsepower of 0.65. Find the rpm of the cutter, feed rate of the workpiece and table, volumetric rate of metal removal, horsepower required at the cutter, and the total motor horsepower.

11.30. An equation models a slab milling operation for a 6-in. cutter OD and is given as $F_c = 30,000wt^{0.8}f^{0.7}$, where F_c = tangential force on cutter, lb; w = width of cut = 7 in.; t = depth of cut = 0.3 in.; f = feed = 0.012 in./tooth. If the cutting speed is 500 ft/min, find the tangential force of the cutter, torque transmitted by cutter, and total motor horsepower if $E = 80\%$.

PRACTICAL APPLICATION

Many possibilities exist for practical applications using the objectives of this chapter. For example, consider the following:

1. Fabricate Figures P11.24 or P11.28 in the school shop.

2. Conduct machining studies to determine the acceptability of Table 11.1 for V_c and f with either a lathe or a milling machine.

3. Visit and survey industrial machining centers and construct a table similar to Table 11.1.

CASE STUDY: NUMERICAL CONTROL LATHE MACHINED FORGING

Dick Crawford, manufacturing representative for Mestas Tooling, is retained as a consultant to improve the productivity of a 4140 steel housing forging being machined on an NC turret lathe (see Figures C11.1 and C11.2). The first few pieces of a 6000 quantity lot have been run and a cycle time of 24.3 min is time studied. A total of 61 lb of metal is removed.

Dick observes that the first three elements are rough face, rough turn taper, and rough bore at 0.015 ipr and 0.190 in. depth. Dick suggests a change in the cemented carbide tool inserts, an increase of the feed rate to 0.035 ipr 360 ft/min, and combining two roughing passes to one for a 0.380-in. dimension. His recommendations are given by the operational data in Table C11.1.

To obtain these data, Dick retools the job and provides a turret and part layout. Dick says, "If you do it this way, there will be free-flowing, open-type chips that will reduce the feed force. Also, we use some of the turret stations for double duty."

Reconstruct the original forging section and show the progressive stock removal by various "cross-hatching" styles. What is the saving resulting from Dick's recommendations if labor and the turret lathe cost $75/h? What is the saving for the first three elements if the original time is 6.6 min?

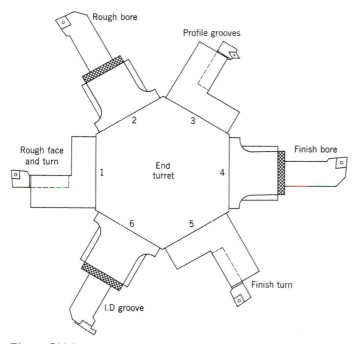

Figure C11.1
Turret tooling for machining 4140 steel forging.

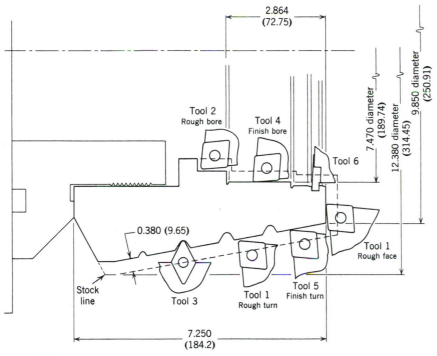

Figure C11.2
Position of tooling on workpiece.

Table C11.1 Data for Case Study

Elements	ft/min (m/s)	ipr (mm/r)	Depth, in. (mm)	Diameter, in. (mm)	hp (kW)	Removal Rate, in.³/min (mm³/s)	Cumulative Time (min)
Rough face	360 (1.8)	0.030 (0.76)	0.250 (6.35)	9.85 (250.2)	29 (22)	33 (0.009)	0.45
Rough-turn taper	360 (1.8)	0.035 (0.89)	0.380 (9.65)	12.0 (350)	42 (31)	57.4 (0.016)	1.50
Rough bore	360 (1.8)	0.030 (0.76)	0.250/0.350 (6.35/8.89)	7.47 (189.7)	32 (24)	45.4 (0.012)	2.60
Profile grooves	390 (2.0)	0.010 (0.25)	0.250 (6.35)	Variable	Machined in four places		7.2
Finish bore	600 (3.0)	0.015 (0.38)	0.030 (0.76)	Variable	—	—	8.4
Finish face and taper	600 (3.0)	0.020 (0.51)	0.030 (0.76)	Variable	—	—	11.3
Groove ID	400 (2.0)	0.010 (0.25)	0.125 (3.18)	Variable	—	—	11.6

CHAPTER 12

GRINDING AND ABRASIVE PROCESSES

Grinding means to *abrade*, to wear away by friction, or to sharpen. In manufacturing it refers to the removal of metal by a rotating abrasive wheel. Wheel action is vaguely similar to a milling cutter, which is considered a multiple cutting point tool. However, a grinding wheel is composed of many small grains bonded together, each one acting as a miniature cutting point. The grinding wheel is considered to have an infinite number of cutting points.

Precision grinding is the last, or close to last, process performed on a part. Grinding is usually a finishing operation, and the part has accumulated most of its cost value.

Because heat, small chips, and loose grinding wheel grit are released during grinding, *coolants* are almost always directed to the interface between the grinding medium and the workpiece. Coolants clean the abrasive wheel or belt to keep it *free cutting*. Coolants are usually water-based, with soluble oil additives to prevent rust, and add to the cooling effectiveness. Synthetic coolants also are available.

12.1 GRINDING AND ABRASIVE PRACTICES

Grinding has the following advantages:

1. Because of the many small cutting points inherent in the wheel, grinding produces fine finishes. Surface roughness of 8 to 90 μin. (0.4 to 2200 μm) are possible with 16 to 24 μin. commonplace.
2. Grinding can finish work to accurate dimensions. Because a small amount of stock is removed, the grinding machines require close wheel control. It is possible to grind work to $\pm$ 0.0001-in. ($\pm$ 0.005-mm) tolerance. Roundness and straightness tolerances under 0.00005 in. are possible.
3. It is technologically possible to grind materials that have a hardness of greater than Rockwell C50.
4. Varying amounts of material are removed in the process. Parts having high hardness are usually machined or formed in the annealed state, hardened, and then ground.
5. Grinding pressure is minimal. This feature allows the holding of parts with magnetic chucks.

Grinding is being challenged by "hard turning," and there are trends to keep work off grinding machines because of the processing cost.

Abrasive machining is a primary stock removal process. It is not a finishing operation, such as conventional grinding or lapping. Not limited to bonded-wheel operations, abrasive machining includes *coated-belt and free abrasive processes*. Coated abrasive belts consist of the abrasive grain, backing, and bond. Consider the grain as the tool, the backing as the toolholder, and the bond as the agent clamping the tool to the holder.

Each *grit* removes a chip. As each abrasive grit initially touches the workpiece, the depth of cut is zero. In addition, as the workpiece moves and the wheel revolves, the depth of cut increases to a maximum, as shown in Figure 12.1. A more complex situation occurs in grinding cylindrical pieces. Some grits remove a maximum-size chip and others practically none.

Figure 12.1 is a simple explanation of grinding. Grit height h is much smaller than its width w after the wheel is dressed with a diamond point, although initially the grit has a random slope. Thus,

$$r = \frac{w}{h} \tag{12.1}$$

where

r = Ratio of width to height of grit
w = Width of grit, in.
h = Height of grit, in.

A sharp grit will cut a chip, but a dull grit is pulled from the grinding wheel and washed away. An abrasive grain is selected based on its sharpness, hardness, and friability. *Friability* is the opposite of toughness. It is the tendency for an abrasive grain to fracture under pressure. As a grit cuts metal, it dulls, and before it dulls completely, grinding wheel manufacturers want it to break away and thus expose a new sharp grit. From Figure 12.1, it is seen

$$L = (D_c d)^{1/2} \tag{12.2}$$

where

L = Length of average chip, in.
D_c = Diameter of grinding wheel, in.
d = Depth of grind, in.

If the grinding wheel moves over a given length, the total volume of metal removed is expressed as

$$Q = 12v \times W \times d \tag{12.3}$$

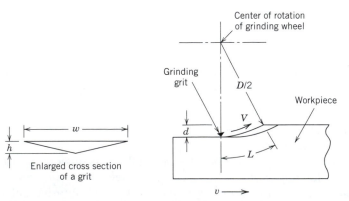

Figure 12.1
Grit and wheel geometry for a grinding wheel.

where

Q = Quantity of material removed, in.^{3}/min
v = Translational velocity of workpiece past the wheel, ft/min
W = Width of wheel, in.

The value of r is usually 10 to 20, and the average volume per chip is

$$q = \tfrac{1}{4}w \times h \times L \tag{12.4}$$

where

q = volume of an average individual chip, in.^{3}

The number n of chips per unit of time is

$$n = VWc \tag{12.5}$$

where

V = Peripheral wheel velocity, in./s
c = Number of grinding wheel active grits per in.^{2}

A simple way to find c is to roll a dressed wheel over a lightly smoked piece of plate glass and count the number of voids or impressions in the smoked surface layer.

Combining the previous equations, and where v is in./s, the mean *chip thickness* is approximately

$$t = \sqrt{\frac{4v}{rcV}\left(\frac{d}{D}\right)^{1/2}} \tag{12.6}$$

where

t = Chip thickness, in.

These equations lead to several conclusions about grinding. Because the horsepower or energy required is proportional to the volume of chips removed, the effect of almost several variables can be estimated. While these equations are based on rough assumptions, notice that smoother finishes are obtained as t decreases. The effect of the variables on surface finish is discernible.

12.2 PROCESSES

Grinding machines finish parts having cylindrical, flat, or internal surfaces. The surface of the part largely selects the grinding machine. A machine grinding cylindrical surfaces is called a *cylindrical grinder*. Machines designed for special functions, such as tool grinding or cutting off, are designated according to their operation.

Cylindrical Grinder

The cylindrical grinder is used primarily for grinding cylindrical surfaces, although tapered and simple-formed surfaces are also ground. They are classified according to the method of supporting the work. Figure 12.2 illustrates the essential difference in supporting the work between centers and centerless grinding.

These machines are computer numerical controlled, and perform multidiameters on the same setup. Angles, in addition to several diameters, are handled.

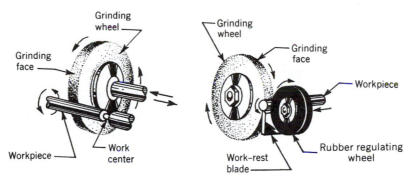

Figure 12.2
Methods of supporting work between centers and centerless type of cylindrical grinding.

A hydraulic, center-type cylindrical grinding machine is illustrated in Figure 12.3. Three movements are incorporated.

1. Rapid rotation of the grinding wheel at the proper grinding speed, usually 5500 to 6500 ft/min (28 to 33 m/s).
2. Slow rotation of the work against the grinding wheel at a velocity to give best performance. This varies from 60 to 100 ft/min (0.30 to 0.51 m/s) in grinding steel cylinders.
3. Horizontal traverse of the work back and forth along the grinding wheel wide enough to cover the desired surface. The work should be traversed nearly the entire width of the wheel during each revolution. In finishing, the traverse is reduced to one-half the width of the wheel and the depth of cut reduced.

The depth of cut is controlled by feeding the wheel into the work. Roughing cuts are about 0.002 in. (0.05 mm) per pass, but for finishing stock removal should be reduced to about 0.0002 in. (0.005 mm) per pass or less. In selecting the amount of infeed, consideration is given to the size and rigidity of the work, surface finish, and whether a coolant is used.

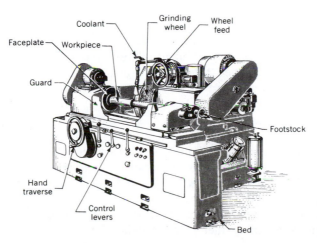

Figure 12.3
A 10 × 36-in. (250 × 915-mm) plain cylindrical grinder.

Where the width of the grinding wheel is wider than the part to be ground, it is not necessary to traverse the work. *Plunge cut grinding* is movement of the wheel into the work without traversing.

The grinding speed of the wheel is stated in terms of surface feet per minute.

$$V = \pi D_c \times N \tag{12.7}$$

where

V = Grinding wheel peripheral velocity, ft/min
D_c = Diameter of grinding wheel, ft
N = Revolutions of the wheel per minute, rpm

The cylindrical grinder traverses the work, to and fro, in repeated passes along the length of the diameter, and the time to traverse is found using:

$$\text{time/pass} = \frac{L_s \times T_s \times D}{(WP)2f_i\pi v} \tag{12.8}$$

where

L_s = Length of ground dimension on workpiece, in.
T_s = Total rough or finish stock depth removed from diameter, in.
D = Original workpiece diameter, in.
W = Wheel width, in.
P = Traverse for each work revolution in fraction of wheel width
f_i = Infeed of wheel per pass, in./pass
v = workpiece peripheral velocity, in./min

Representative values for aluminum-oxide material grinding wheel are given as

Material	Condition	v	Rough f_i	Finish f_i	Rough P	Finish P
Tool steel	Hardened	50	0.001	0.0002	¼	⅛
Steel	Hardened	70	0.001	0.0004	¼	⅛
Steel	Annealed	100	0.001	0.0004	½	⅙

Typically, in two-pass rough and finish grinding about 0.002 in. is left for *finish stock removal. Rough grinding* removes most of the stock material.

The *tool post grinder* is used for miscellaneous and small grinding work on a lathe. The grinder is held on the tool post and fed across the work, the regular longitudinal or compound rest feed being used. A common application of this grinder is the truing of lathe centers.

In the centerless-type grinder, the work is supported by the *work rest,* regulating wheel, and the grinding wheel. Both cylindrical and *centerless grinders* use plain grinding wheels with the grinding face touching the outside diameter of the part.

Centerless grinders are designed so that they support and feed the work by using two wheels and a work rest, as illustrated in Figures 12.2 and 12.4. The large wheel is the grinding wheel and the smaller one is the pressure or regulating wheel. The regulating wheel is a rubber-bonded abrasive having the frictional characteristics to rotate the work at its own rotational speed. The speed of this wheel, which may be controlled, varies

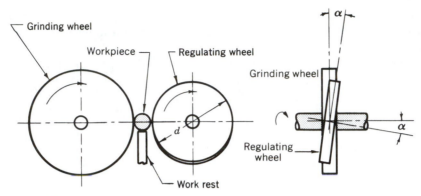

Figure 12.4
Principle of centerless grinding.

from 50 to 200 ft/min (0.25 to 1.02 m/s). Both wheels are rotated in the same direction. The rest assists in supporting the work while it is being ground, being extended on both sides to direct the work travel to and from the wheels. The axial movement of the work past the grinding wheel is obtained by tilting the wheel at a slight angle from horizontal. An angular adjustment of 0° to 10° is provided in the machine for this purpose. The actual feed is calculated by this formula:

$$f = \pi D_r N \sin \alpha \tag{12.9}$$

where

f = Feed, in./min
N = rpm
D_r = Diameter of regulating wheel, in.
α = Angle inclination of regulating wheel, degree

Centerless grinding is applied to cylindrical parts of one diameter. In production work on such parts as piston pins, a magazine feed is arranged and the parts sequence through several machines in series before the part is finished. Each grinder removes from 0.0005 to 0.002 in. (0.013 to 0.05 mm) stock.

Where parts are not uniformly the same diameter or where they require form grinding as does a ball bearing (Figure 12.5), an infeed centerless grinder is used. The method of operation corresponds to the plunge cut form of grinding. The length of the section to be ground is limited to the width of the grinding wheel. The part is placed on the work rest and is moved against the grinding wheel by the regulating wheel. Upon completion the gap between the wheels is increased and the work is ejected from between the wheels. There are several advantages to centerless grinding.

1. Holding the work in collets, chucks, and mandrels is avoided.
2. The work is rigidly supported. There is no chatter or deflection of the work.
3. The process is rapid and especially adapted for production work.
4. Accuracy is easily controlled.
5. A true floating and unstressed part condition exists during grinding, and less grinding stock is required.
6. Low labor costs are possible.

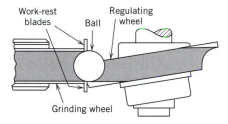

Figure 12.5
Centerless grinding of ball bearings.

Disadvantages include the following:

1. Workpieces having flats and keyways cannot be ground.
2. Concentricity between internal and external diameters is difficult for very close tolerances.
3. Work having several diameters is not easily handled.

Internal Grinders

The work done on an internal grinder is shown in Figure 12.6. Tapered holes or multidiameter holes are accurately finished in this manner. There are several variations of internal grinders.

1. The wheel is rotated in a fixed position while the work is slowly rotated and traversed back and forth.

2. The wheel is rotated and at the same time reciprocated back and forth through the length of the hole. The work is rotated slowly but otherwise has no movement.

3. The work remains stationary and the rotating wheel spindle is given an eccentric motion according to the diameter of hole to be ground. This is called a *planetary grinder* and used for work that is difficult to rotate.

4. Work is rotated on the outside diameter by driving rolls, thus grinding the bore concentric with the outside diameter. This arrangement lends itself to production work because loading is simplified and magazine feed is possible. A sketch of a *centerless internal grinder* is shown in Figure 12.7. Three rolls—regulating, supporting, and pressure—support and drive the work. Centerless grinders of this type are designed for automatic loading and unloading by swinging the pressure roller out of the way at

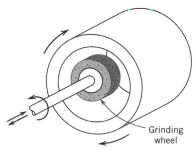

Figure 12.6
Sizing to dimension by internal grinding.

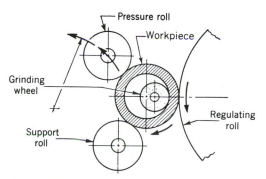

Figure 12.7
Centerless internal grinding.

the end of the cycle. Advantages of internal centerless grinding include elimination of workholding fixtures and ability to grind both straight and tapered holes.

Because internal grinding wheels are small in diameter, the spindle speed is much higher than for cylindrical grinding to attain surface speeds up to 6000 ft/min (30.5 m/s).

Toolroom grinding is done dry. Common practice on production work, however, is to grind steel wet with a coolant, and bronze, brass, and cast iron are ground dry. Metal allowances for internal grinding depend on the size of the shaft or hole to be ground. *Allowance* in this setting implies the amount of stock that is left following a turning or boring operation. Standard practice suggests an allowance around 0.010 in. (0.25 mm).

Surface Grinding

Grinding plane surfaces is known as *surface grinding.* Planer-type machines with a reciprocating table and those having a rotating worktable are the principal machine tool designs. Each machine has the possible variation of a horizontal or vertical positioned grinding wheel spindle. Figure 12.8 shows four possibilities of construction.

A surface grinder with the principal parts labeled is shown in Figure 12.9. This machine has hydraulic control of table movements and wheel crossfeed. Wheels are straight or recessed wheels (types 1, 5, and 7 in Figure 12.17). Grinding is on the outside face or circumference. Surface grinders recondition dies, grind machine tool ways, and other long surfaces.

Reciprocating table grinders have a vertical-spindle design, the grinding being done by a segment or ring-shaped wheel. These machines grind gear faces, thrust washers, cylinder head surfaces, and other flat-surfaced parts.

A high-powered vertical rotary surface grinder set up for grinding a large part is shown in Figure 12.10. Similar machines can remove 600 lb of metal per hour (270 kg/h). The spindle of the machine can be tilted while grinding to reduce the area of wheel in contact with the work, which produces deeper penetration, less heating, and better utilization of spindle horsepower. The spindle is returned to a perpendicular position, presenting

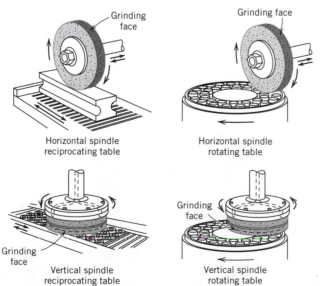

Horizontal spindle
reciprocating table

Horizontal spindle
rotating table

Vertical spindle
reciprocating table

Vertical spindle
rotating table

Figure 12.8
Types of surface grinding machines.

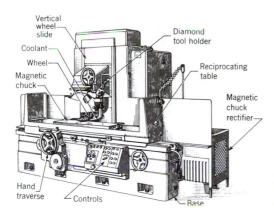

Figure 12.9
Horizontal spindle, reciprocating table surface grinder.

a flat wheel surface to the workpiece for finish grinding. Flatness accuracy of 0.005 in. (0.013 mm) over the full diameter is obtainable.

Tool and Cutter Grinders

In grinding tools by hand (known as *offhand grinding*), a bench or pedestal type of grinder is used. The tool is hand-held and moved across the face of the wheel continually to avoid excessive grinding in one spot. This type of grinding is used to a large extent for single-point tools. Quality depends on the skill of the operator.

For sharpening miscellaneous cutters, a universal-type grinder is used. It is equipped with a universal head, vise, headstock and tailstock, and attachments for holding tools and cutters. Although essentially designed for cutter sharpening, a universal-type grinder is also used for cylindrical, taper, internal, and surface grinding. Accuracy is of paramount importance in toolroom work, particularly when grinding form tool cutters and special

Figure 12.10
High-powered vertical spindle rotary surface grinder for rapid metal removal.

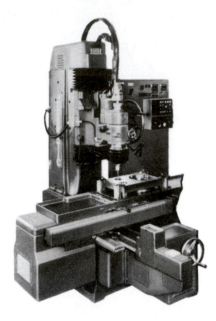

Figure 12.11
Precision jig grinder.

shapes. Some grinders have optical magnifications of 10×, 20×, and 50× for monitoring progress and accuracy.

A *jig grinder* (shown in Figure 12.11) is used for prototype and toolroom work. This grinder is considered the best for obtaining hole location and size accuracies. The jig grinder was originally developed for grinding drill jigs and tools where extreme accuracy is necessary. Accuracies of ± 0.0001 in. (± 0.003 mm) are achievable. The speed range is from 6700 to 175,000 rpm. The machine is available with computer numerical control. It has the appearance of a jig borer, but operates at speeds too high for drilling or boring.

Honing

Honing is a *low-velocity abrading* process. Because material removal occurs at lower cutting speeds than in grinding, heat and pressure are less, resulting in excellent size, surface finish, and metallurgical control of the surface. The cutting action is obtained from abrasive sticks (aluminum oxide or silicon carbide) mounted on a metal mandrel, as shown in Figure 12.12. Distortion is minimized because the workpiece is not clamped

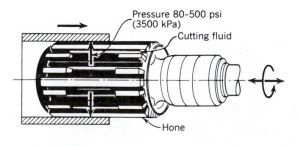

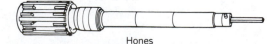

Figure 12.12
Honing using a mandrel with honing sticks on an expanding core.

or chucked and thus is said to "float." For small-diameter bores, a one-piece mandrel having a U-shaped cross section is used. There are two integral shoes and a narrow honing stone that provide a three-line (unevenly spaced) contact with the work circle. The abrasive stone is mounted on a wedge-actuated holder.

The work is given a slow reciprocating motion as the mandrel rotates, thus generating a straight and round hole. Parts honed for finish remove only 0.001 in. (0.03 mm) or less. However, certain inaccuracies, such as bell mouth, can be corrected in amounts up to 0.020 in. (0.51 mm). Coolants are essential to flush away small chips and to keep temperatures uniform. Sulfurized mineral base or lard oil mixed with kerosene is generally used.

Lapping

Lapping produces geometrically true surfaces, corrects minor surface imperfections, improves dimensional accuracy, or provides a very close fit between two contacting surfaces. Although lapping is a material-removing operation, it is not cost-effective for that purpose. The material removed is usually less than 0.001 in. (0.03 mm).

Lapping is used on flat, cylindrical, spherical, or specially formed surfaces in contact with a lap. The workpiece and the lap have relative motion between each other in such a way that fresh contacts between a grit and the part are constantly being renewed. Loose abrasive is carried in some vehicle, such as oil, grease, or water. Sometimes the abrasive is in the form of a bonded wheel, and the lapping operation is similar to centerless and vertical-spindle surface grinding. Metal *laps* must be softer than the work and are usually made of close-grain gray iron. Other materials, such as steel, copper, lead, and wood, are used where cast iron is not suitable. By having the lap softer than the work, the abrasive particles (usually boron carbide, silicon carbide, aluminum oxide in fine-screened sizes, or flour) embed in the lap and cause the greater wear to occur on the hard surface. In lapping carbide tools and jewels, diamond particles permanently embedded in copper laps are most successful.

Commercial accuracy can be held to 0.000024 in. (0.00061 mm) and to even closer limits if needed. Products commonly finished by this process include gages, piston pins, valves, gears, roller bearings, thrust washers, and optical parts.

Superfinishing

Machining operations and usual grinding processes leave a surface coated with fragmented, noncrystalline, or *smear metal*. Although easily removed by sliding contact, the surface may result in excessive wear, increased clearances, noisy operation, and lubrication difficulties. *Superfinishing* is a surface-improving process that removes this undesirable fragmentation metal, leaving a base of solid crystalline metal. It is somewhat similar to honing in that both processes use an abrasive stone, but it differs in the motions given to the stone. This process, which is a finishing process and not a dimensional one, can be superimposed on other finishing operations.

In *cylindrical superfinishing* (Figure 12.13A), a bonded-form abrasive stone, having a width about two-thirds of the diameter of the part to be finished and the same length, is operated at low speed and pressure. The motion given to the stone is oscillating with an amplitude of 1/16 to 1/4 in. (1.6 to 64 mm) at about 450 cycles per minute (cpm) (7.5 Hz). The stone pressure is 3 to 40 psi (21 to 275 kPa). If the part is of greater length than the stone, an additional longitudinal movement of either stone or work is necessary. The work is rotated at a speed of about 50 ft/min (0.25 m/s). During the operation, the work is flooded with a light oil that carries away the minute particles abraded from the

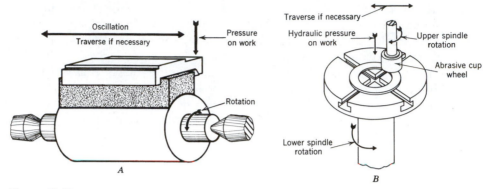

Figure 12.13
Motions between abrasive stone and work. *A*, Cylindrical superfinishing. *B*, Flat superfinishing.

surface by the short, oscillating stone strokes. The stone action is similar to a scrubbing movement and removes all excess and fragmented metal on the surface.

Superfinishing flat surfaces is illustrated in Figure 12.13*B*. A rotating cup-shaped abrasive stone is used with the work resting on a circular table carried by a rotating spindle. An additional oscillating movement is given to the stone, but because both it and the work are rotating, this action is not so important in developing a continually changing path of the abrasive particles. Superfinishing spherical surfaces is similar to that used for flat surfaces, except that a formed-cup spindle is at an angle to the work spindle and no oscillating motion can be used.

Abrasive-Belt Grinding

Abrasive machines, powered by 300-hp (200 kW) motors, remove metal from forgings, castings, and various stock shapes. The process provides fast metal removal, fair surface finish, rough size control, little need for holding fixtures, and adaptation to automation. Although up to ½ in. (12.7 mm) of metal can be removed in a matter of seconds, there is the problem with heat dissipation generated in abrasive machining.

This method is used for stock removal and surface preparation. Sometimes it is termed *high-energy grinding*. It is performed using a tensioned abrasive belt over precision pulleys at speeds between 250 and 6000 ft/min (1.27 to 30.48 m/s). Tension is maintained by springs or a hydraulic ram as the belt heats and stretches. The surface speed of a belt is approximately the same as the drive roller that pulls the belt. There may be one or more idler rollers of the same or a different size. The linear belt speed is

$$V_b = \pi k D_{br} N \qquad (12.10)$$

where

V_b = Belt velocity, ft/min
D_{br} = Diameter of belt roller, ft
N = rpm of drive motor
k = Constant for belt slippage, usually 0.85 to 0.96

Heat is caused by belt slippage on the drive roller and by the abrasive cutting. *Air and liquid coolants* may be employed. There are about 1 million abrasive grains on a

50-grit belt 8 in. (203 mm) wide and 96 in. (2438 mm) long. Each grain passes the workpiece about 500 times a minute. If only 1 grit in every 1000 removed a chip, an enormous number of chips and hence prodigious quantities of metal are capable of being removed. With heavy-duty machines, this process can be almost as economical as a milling process for removing metal.

Figure 12.14 shows a *belt-grinding machine*. Major areas of application include preparing flats, tubing and extrusion, and finishing partially fabricated stampings, forgings, and castings. Some belt-grinding machines use wet belts and impervious plastic-bonded cloths. Surface finish is comparable to light milling and turning, and there is minimal work hardening and warpage resulting from heat generation. In some machines employing up to 150 hp (112 kW), the table rotates and the individual parts may have oscillatory motion as well.

On these machines a 24-in. (600 mm) or wider *abrasive belt* can attain stock removal rates of 30 in.^{3}/min-in.^{2} (13 mm^3/s-mm^2) of contact on cast iron. Stock removal depth of 0.100 to 0.250 in. (2.54 to 635 mm) is possible.

Mass Media Finishing

Barrel finishing or *tumbling* is a controlled method of processing parts to remove burrs, scale, flash, and oxides as well as to improve surface finish. Widely used as a finishing operation for many parts, it attains a uniformity of surface finish not possible by hand finishing. It is generally the most economical method of cleaning and surface conditioning large quantities of small parts. Materials that may be barrel-finished include all metals, glass, plastics, and rubber. Parts are placed in a rotating barrel or vibrating unit (Figure 12.15), with an abrasive medium, such as water or oil, and usually some chemical compound or detergent to assist in the operation. There is a rough similarity to home laundry tubs and their action. As the barrel rotates slowly, the upper layer of the work is given a sliding movement toward the lower side of the barrel, causing the abrading or polishing action to occur. A speed of 15 rpm will produce light action, whereas a speed of 30 rpm

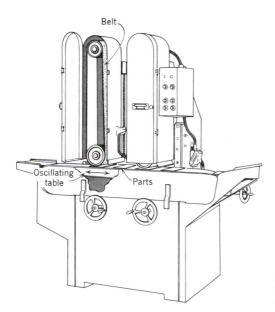

Figure 12.14
Dual-head, belt-grinding machine.

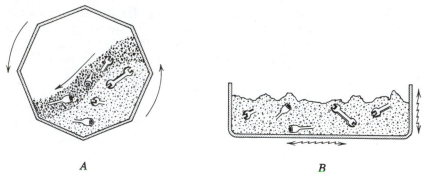

Figure 12.15
Two methods of barrel finishing. *A*, Slide action. *B*, Vibratory finishing machining.

produces rapid abrasion. The time employed varies from a few minutes to several hours. The same results also may be accomplished in a vibrating unit in which the entire contents of the container are in constant motion.

Machines for tumbling can be tub-shaped where batches of parts are processed. Also, *mass media finishing* can be continuous instead of batch type. Parts enter the conveyor tub of a vibratory finishing machine and eventually move along an open oscillating chute to a separating screen where the abrasive medium drops away. The tumbling abrasives are returned automatically to the starting point. These systems can be integrated with automatic handling equipment.

Abrasive-Media Flow Deburring Machine

These machines deburr internal edges or surfaces or cavities by a controlled-force flow of an *abrasive-laden semisolid grinding medium.* Compare this material to a ''silly putty'' with abrasive dispersed in the media. A range of viscosities are available, and velocity through the cavity to be deburred can be controlled. For example, this material is able to deburr or abrade the rough internally cored cast-iron gas engine block. Abrasive media is a semisolid plasticlike substance that conforms to the shape and contour of the volume being abraded. Notice Figure 12.16*A*, which shows the media oozing through openings in the workpiece. Figure 12.16*B* and *C* shows the before and after effect of this method on internal cavity roughness.

Fixtures hold the workpiece in a position to contain, direct, or restrict the flow of the abrasive flow media to areas of the workpiece where abrasion is desired. The flow of the medium is three-dimensional. In application, the number of holes or surface area does not necessarily influence the cycle time. Abrasion occurs in areas where medium flow is restricted.

Abrasive-media flow machines deburr hidden or secondary burrs and improve internal surface finish. These machines remove burrs that are inconvenient (or impossible) to reach by manual deburring methods. The deburring results in a smoother and brighter finish.

Different media may be selected for different surface finish and burr removal. Hydraulic force is also used to clamp the workpiece and to give the desired medium pressure. Passage sizes range from 1/64 in. to 2 in.

 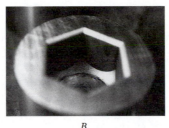 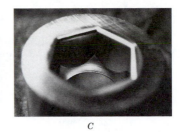

A   B   C

Figure 12.16
A, Abrasive media is a semisolid plasticlike substance that conforms to the shape and contour of the area being abraded. *B,* Before abrasive media polishing. *C,* After abrasive media polishing.

Miscellaneous Finishing Operations

Wire Brushing. Rotating brushes with wire *bristles* are used to clean castings and to remove scratches, scale, sharp edges, and other surface imperfections. Tampico brushes with an abrasive compound may be used if the material is not too hard. Little metal is removed by *brushing,* and a satinlike finish appears on the surface. Usually a buffing operation is necessary if high polish is required.

Polishing. Cloth wheels or belts coated with abrasive particles are used for polishing operations. Although not considered precision metal-removing processes, such operations can remove sufficient metal to blend scratches and other minor surface imperfections. Both wheels and belts are flexible and will conform to irregular and rounded areas. Wide belts are used to polish plates, sheets, and other large metal parts. The amount of metal removed and the surface finish are controlled by the characteristics of the material being polished, belt speed, pressure, and grit size.

Polishing wheels are produced in disks of cotton cloth, canvas, leather, felt, or similar materials glued or sewed together to provide the required face width. Metal side plates may be used for stiffening. These wheels are coated with glue or a cold cement and immediately rolled in a trough containing abrasive grains. After drying a second application of glue and abrasive may be given. When the wheel is dry, the hard, abrasive-coated surface is cracked by striking diagonally across the wheel face with an iron bar. This action breaks up the layer of glue and abrasive into small areas that provide flexibility and good polishing action to the wheel. Aluminum oxide and silicon carbide of various grit sizes are used as abrasives. Usually a part is passed over several wheels of decreasing grit size before the final polish is obtained.

Buffing. Buffing is a final operation to improve the polish of a metal and to bring out maximum luster. The wheels are similar to polishing wheels and are generally made of cotton, hemp cloth, flannel, linen, or sheepskin. They are charged with a fine abrasive such as rouge, tripoli, or amorphous silica. Buffing is frequently performed before plating.

12.3 ABRASIVES, GRINDING WHEELS, AND STONES

Abrasives for grinding, honing, and superfinishing are bonded to a tool suitable for specific processes. They are hard materials that have been processed to cut or wear away softer materials.

Materials

A classification of the common abrasive materials used for wheels and special shapes follows.

A. Natural

1. Sandstone or solid quartz
2. Emery, 50% to 60% crystalline, Al_2O_3, plus iron oxide
3. Corundum, 75% to 90% crystalline Al_2O_3, plus iron oxide
4. Diamond
5. Garnet

B. Manufactured

1. Silicon carbide, SiC
2. Aluminum oxide, Al_2O_3
3. Cubic boron nitride, CBN
4. Zirconium oxide, ZrO_2

For many years it was necessary to rely on natural abrasives in manufacturing grinding wheels. Sandstone wheels, though available, are rarely used. Garnet is used for wood and has application in wood sanding.

Corundum and emery have long been used for grinding purposes. Both are composed of crystalline aluminum oxide in combination with iron oxide and other impurities. Like sandstone, these materials lack a uniform bond and are seldom used in production work.

Diamond wheels made with a resinoid bond are useful in sharpening cemented carbide tools, ceramics, marble, glass, and granite. In spite of high initial cost, diamond wheels have proved economical because of their rapid cutting ability, slow wear, and free-cutting action. Very little heat is generated with their use, which is an added advantage.

Silicon carbide was discovered during an attempt to manufacture precious gems in an electric furnace. The hardness of this material, according to Mohs' scale is slightly more than 9.5, which approaches the hardness of a diamond. Raw materials consisting of silica sand, petroleum coke, sawdust, and salt are heated in a furnace to around 4200°F (2300°C) and held there for a considerable period of time. The product consists of a mass of crystals surrounded by partially unconverted raw material. After cooling, the material is broken up, graded, and crushed to grain size. Silicon carbide crystals are very sharp and extremely hard. Their use as an abrasive is limited because of brittleness.

Aluminum oxide is a synthetic material. It is manufactured by first refining the claylike mineral bauxite, which is the main source of aluminum, in an electric furnace. A charge of coke and iron is added. In the furnace fusion occurs at 4000°F that results in a rocklike mass that is crushed into grain sizes usable in a grinding wheel. Aluminum oxide is slightly softer than silicon carbide, but it is much tougher. Most manufactured wheels are made of aluminum oxide. Part of the versatility of aluminum oxide comes from its versatility to chemically bond with other materials in the fusion process.

To grind ferrous materials, another abrasive was developed, called cubic boron nitride (CBN). The compound is the second hardest material currently known; diamond is harder.

Grinding Wheels

The process of making grinding wheels is similar for all materials, sizes, and shapes. The procedure is as follows:

1. The material is reduced to small size by being run through roll and jaw crushers. Fines are removed by passing the material over screens between crushing operations.
2. The material is passed through magnetic separators to remove iron compounds.
3. Dust and foreign material is removed by a washing process.
4. The grains are graded by being passed over vibrating standard screens. A standard 30-mesh screen has 30 meshes per in. or 900 openings per in.[2] (0.595-mm nominal sieve opening). The No. 30 size material passes through a No. 30 screen and is retained on the next finer size, which in this case is No. 36.
5. Grains are mixed with bonding material, molded or cut to proper shape, and heated. The heating or burning procedure depends on the bonding agent.
6. The wheels are bushed, trued, tested, and given a final inspection.

Bonding Processes

The bonding process is performed to join the abrasives into a usable form. Because grinding operations are accomplished at high speeds and the wheels are dense, high bonding strength is necessary for structure and safety. The principal processes are described as follows.

1. *Vitrified process.* The abrasive grains are mixed with claylike ingredients that are changed to a glasslike material upon being "burned" at high temperature. The addition of water is required and the wheels are shaped in metal molds under a hydraulic press. Wheels produced by these methods are dense and accurately shaped. The time for burning varies with the wheel size, being anywhere from 1 to 14 days. The process is similar to firing tile or pottery.

Vitrified wheels are porous, strong, and unaffected by water, acids, oils, and climatic or temperature conditions. About 75% of all wheels are made by this process. Recommended wheel peripheral speeds vary from 1500 ft/min (6.72 m/s) when grinding titanium to a maximum safe velocity of 12,000 ft/min (60.96 m/s) when grinding titanium to a maximum safe velocity of 12,000 ft/min (60.96 m/s) for specially reinforced grinding wheels. The maximum standard speed generally used is 6500 ft/min (33.02 m/s).

2. *Silicate process.* In this process, sodium silicate is mixed with the abrasive grains and the mixture is tamped in metal molds. After drying several hours the wheels are baked at 500°F (260°C) from 1 to 3 days.

Silicate wheels are milder acting than those made by other processes and wear more rapidly. They are suitable for grinding edges of tools where the heat must be kept to a minimum. This process also is recommended for large wheels because they tend not to crack or warp in the baking process.

3. *Shellac process.* The abrasive grains are first coated with shellac by being mixed in a steam-heated mixer. The material is then placed in heated steel molds and rolled or

pressed. Finally, the wheels are baked a few hours at a temperature around 300°F (150°C). This bond is adapted to thin wheels as it is strong and elastic.

4. *Rubber process.* Pure rubber with sulfur as a vulcanizing agent is mixed with abrasive by feeding the material between heated mixing rolls. After it is rolled to thickness, the wheels are cut out with dies and vulcanized under pressure. Very thin wheels can be made by this process. Wheels having this bond are used for high-speed grinding at 9000 to 16,000 ft/min (45 to 81 m/s) because they afford rapid removal of the stock. They are used as snagging wheels in foundries and as cutting-off wheels.

5. *Bakelite or resinoid process.* The abrasive grains in this process are mixed with a thermosetting synthetic resin powder and a liquid solvent, then molded and baked. This bond is very hard and strong; wheels made by this process can be operated at speeds of 9500 to 16,000 ft/min (48 to 81 m/s). They are used for general-purpose grinding and in foundries and billet shops for snagging purposes because of their ability to remove metal rapidly.

Wheel Selection

Selection of a grinding wheel for a definite purpose is important. Many factors complicate the choice among a variety of wheels. The most important ones are listed here.

1. *Size and shape of wheel.* Standard shapes are shown in Figure 12.17. Different wheel faces are available.

2. *Type of abrasive.* The choice of silicon carbide or aluminum oxide largely depends on the physical properties of the material to be ground. Silicon carbide wheels are recommended for materials of low tensile strength, such as cast iron, brass, stone, rubber, leather, and cemented carbides. Aluminum oxide wheels are best used on materials of high tensile strength, such as hardened steel, high-speed steel, alloy steel, and malleable iron.

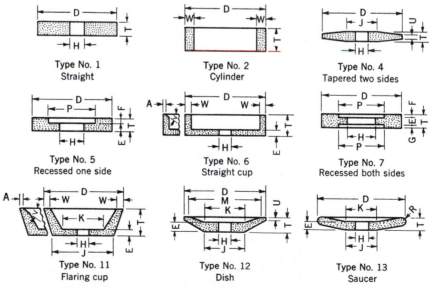

Figure 12.17
Standard grinding wheel shapes.

3. *Grain size of abrasive particles.* In general, coarse wheels are used for fast removal of materials. Fine-grained wheels are used where finish is important. Coarse wheels may be used for soft materials, but generally a fine grain should be used for hard and brittle materials.

4. *Grade or strength of bond.* The grade depends on the kind and hardness of the bonding material. If the bond is very strong and capable of holding the abrasive grains against the force tending to pry them loose, it is said to be hard. If only a small force is needed to release the grains, the wheel is said to be soft. Hard wheels are recommended for soft materials, and soft wheels are recommended for hard materials.

5. *Structure of grain spacing.* The structure refers to the number of cutting edges per unit area of wheel face as well as the number and size of void spaces between grains. Soft, ductile materials require a wide spacing. A fine finish requires a wheel with a close spacing of the abrasive particles.

6. *Type of bond material.* The vitrified bond is most commonly used. When thin wheels are required or high-operating speed or high finish is necessary, other bonds are more advantageous.

A standard system of marking grinding wheels, adopted by the American National Standards Institute, is shown in Table 12.1. The code indicates wheel performance and construction.

Table 12.1 **Standard Marking System Chart**

Sequence		1	2	3	4	5	6
	Prefix	Abrasive Type	Grain Size	Grade	Structure	Bond Type	Manufacturer's Record
	51 —	A —	36 —	L —	5	— V	— 23

MANUFACTURER'S SYMBOL INICATING EXACT TYPE OF ABRASIVE. (USE OPTIONAL)

MANUFACTURER'S PRIVATE MARKING TO IDENTIFY WHEEL

(USE OPTIONAL)

			Very	Dense to open	
Coarse	Medium	Fine	fine	1	9
10	30	70	220	2	10
12	36	80	240	3	11
14	46	90	280	4	12
16	54	100	320	5	13
20	60	120	400	6	14
24		150	500	7	15
		180	600	8	Etc.

V-VITRIFIED
S-SILICATE
R-RUBBER
B-RESINOID
E-SHELLAC
O-OXYCHLORIDE

ALUMINUM OXIDE-A

SILICON CARBIDE-C

(USE OPTIONAL)

Soft   Medium   Hard

A   B   C   D   E   F   G   H   I   J   K   L   M   N   O   P   Q   R   S   T   U   V   W   X   Y   Z

GRADE SCALE

Coated Abrasives

When abrasive particles are glued to paper or other flexible backings, they are known as *coated abrasives*. Common products include "sandpaper," abrasive disks, and belts such as those used on belt-grinding machines. Any of the abrasives used in wheel manufacture may be applied in this way. The most common type is sandpaper, which frequently identifies this entire group. The abrasive in sandpaper is a flint quartz that is mined in large lumps and then crushed to size and graded. Another natural abrasive is garnet, a red mineral. Of the several kinds of garnet known, the one called almandite is the best for abrasive coatings. It is much harder and sharper than flint and, when broken down, breaks into crystals with many cutting edges. Other natural abrasives are emery and corundum. The two manufactured abrasives that have wide application are silicon carbide and aluminum oxide.

Types of *backing* include paper, cloth, fiber, paper cloth, cloth fiber, and plastic. Each backing has characteristics that give advantages for an application.

An important phase in the manufacture of coated abrasives is the application of the abrasive particles to the backing material. The abrasive grains must be securely held to give the best cutting action. For severe service, *closed coating* where the grains completely cover the surface of the backing is recommended. When increased flexibility is required and there is no tendency for the particles to become loaded or clogged, an *open coating* is used. A method of coating, known as the *electrocoating* process, is illustrated in Figure 12.18. This method uses the principle that opposite-charged particles attract each other. The abrasive particles pass on a conveyor belt into an electrostatic field. As the particles enter the field they first stand on end, aligning themselves in the direction of the flow of the electric force, and are then attracted to the glue-covered backing, which is traveling in the same direction as the other belt. The abrasive grains are embedded on end and are equally spaced on the backing.

Most coated abrasive machines utilize either belts or disks. On *disk grinders* a metal disk forms the backing for the coated abrasive. The belt or abrasive belt machines pass over a contact wheel and an idler, and the grinding or polishing is done against the contact wheel. When the grinding is done between the wheel and the idler, the belt is supported by a platen. Contact supporting wheels are made of steel, rubber, phenolic

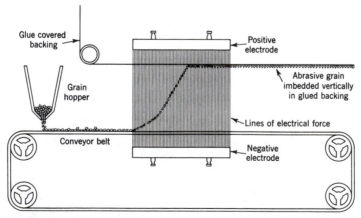

Figure 12.18
Electrocoating method for applying abrasive grains to glue-coated backing.

resin, or cloth depending on the type of work to be done. Hard contact wheels provide faster stock removal rate and soft wheels provide a better finish.

Not all coated abrasives are graded in the same manner. Flint paper and emery cloth have their own systems, whereas garnet and manufactured abrasives are graded by another system. Table 12.2 will be of assistance in the selection of the proper grades to use.

Mass Media Abrasives

An important factor in the successful operation of barrel finishing is selection of the abrasive. Manufactured aluminum oxide and silicon carbide media in various sizes and shapes have proved to be satisfactory because of their cutting ability. Selection of the abrasive depends on the size and shape of the part to be processed, workpiece material, and the surface finish desired.

Abrasives are obtained in the same range of grit sizes as is available for wheels. When plastic parts are to be processed, rubber-lined barrels are recommended. In finishing rubber articles, parts are quickly frozen by adding a CO_2 refrigerant so that the flash breaks off easily. The fatigue strength of the steel parts is raised with barrel finishing.

Table 12.2 **Table Comparative Grit Sized—Approximate Comparison of Grit Numbers[a]**

	Durite Metalite	Adalox Garnet	Flint	Emery Cloth	Emery Polishing Paper
Extra fine	600				4/0
					3/0
	500				2/0
	400	400(10/0)			0
	360				
	320	320(9/0)			½
Very fine	280	280(8/0)			1
	240	240(7/0)			1-G
	220	220(6/0)	Extra fine		2
Fine	180	180(5/0)		Fine	3
	150	150(4/0)			
	120	120(3/0)	Fine		
Medium	100	100(2/0)		Medium	
	80	80(0)	Medium		
	60	60(½)		Coarse	
Coarse	50	50(1)	Coarse		
	40	40(1½)			Very coarse
Very coarse	36	36(2)	Extra coarse		
	30	30(2½)			
	24	24(3)			
Extra coarse	20	20(3½)			
	16	16(4)			
	12				

[a] Table prepared by Behr-Manning Division of Norton Company. Metalite and Adalox are made from aluminum oxide manufactured abrasive. Durite is made from silicon carbide manufactured abrasive. Garnet, flint, and emery are natural abrasives.

Figure 12.19
Miscellaneous sizes and shapes of ceramic-based tumbling nuggets.

The *tumbling nugget* sizes vary from 3/32 to 2 in. (2.4 to 50 mm) and come in a variety of shapes, as Figure 12.19 shows. The shapes are classed as random or preformed. The shapes of the preforms are triangular, rodlike, or spherical. Random shapes are preferred where part geometry avoids lodging of the nugget and where the vigor of the tumbling action provides reasonable abrasive life.

QUESTIONS

1. Give an explanation of the following terms:

Friability   High-energy grinding
Chip thickness   Barrel finishing
Tool post grinder   Coated abrasive
Work rest   Electrocoating
Honing   Tumbling nugget

2. Discuss the approximate grinding wheel surface speeds.

3. Discuss the principal advantages of grinding as a machine operation.

4. If a part is to be hardened to Rockwell C54, is the grinding operation normally before or after heat treatment? Why?

5. Approximately what grinding wheel speeds are used for

a. Offhand grinding?

b. Centerless grinding?

c. Abrasive-belt grinding?

d. Grinding between centers?

e. Surface grinding?

6. What inaccuracies are eliminated by honing? Does honing remove material?

7. What is Mohs' hardness and how does it relate to manufactured abrasives?

8. Make a table showing the various bonding materials for grinding wheels and show their comparisons for speed, elasticity, approximate cost, and principal use.

9. What can be said about the surface speed of a grinding wheel as compared to the surface speed of the workpiece of a cylindrical grinder, a centerless grinder, and surface grinders?

10. Select a suitable grinding wheel for each of the following applications, indicating wheel type, wheel size, bonding material, kind of abrasive, and wheel speed: sharpening high-speed steel milling cutter; grinding flat hardened steel; cutting off thin-walled aluminum tubing; finishing flat plates of soft steel; snagging iron casting; cutting glass bar stock.

11. How is the "through speed" or feed regulated on a centerless grinder? What are the ways through speed can be doubled? Examine the equations.

12. What are the principal limitations when considering a centerless grinder for grinding shafts?

13. What are the characteristics of the following grinding wheels: 32-C-100-P-11-B and 47-A-24-E-14-R?

14. What is a soft grinding wheel? What kind of work is it used for?

15. In lapping, what would be the consequence of having a situation in which the workpiece was softer than the lap?

16. How does the process of mass media tumbling compare to grinding? What kinds of work do you perform with tumbling?

17. What type of process is most adaptable to deburring?

18. State the limitations of natural abrasives.

PROBLEMS

12.1. Plot a curve of surface speed for a 12-in. diameter grinding wheel versus grinding wheel revolutions per minute. Wheel speeds range between 1000 and 2500 rpm.

12.2. If the maximum safe speed of a grinding wheel is 8000 ft/min, develop a table giving maximum safe revolutions per minute for grinding wheels from 4 in. to 24 in. in 2-in. increments.

12.3. Determine a graph showing the feed on a centerless grinder versus regulating wheel revolutions per minute of a 5° angle of inclination. Use increments of 100 rpm to a value of 2000. Regulating wheel is 6-in. OD.

12.4. Construct a graph giving the feed of a centerless grinder versus angle of inclination for an 8-in. OD regulating wheel revolving at 700 rpm.

12.5. If on a surface grinder using a 10-in. (254-mm) wheel a depth of cut of 0.009 in. (0.229 mm) is being ground, what is the average length of chip generated? For a wheel width 1 in. and a width of a grit equal to 0.01 in., find the volume of an average chip. The number of active grits equals 150/in.^{2} Wheel speed is 75 ft/min. Translational velocity is 50 in./min. Find the average chip thickness.

12.6. Plot a curve showing the effect of workpiece velocity on chip thickness.

12.7. A grinder is equipped with a 20-in. OD by 2-in. wide wheel. Depth of cut is 0.002 in. and relative velocity is 500 ft/min. A grit is 0.005 in. wide and its height is 0.008 in. The number of grits on the surface of the grinding wheel is 750/in.^{2}. Find the average chip thickness.

12.8. A surface grinding operation is performed with the following conditions: wheel, 8 in. by ¾ in.; table speed 4 ft/min; number of grits 1900/in.^{2}; depth of cut 0.003 in.; grit is 0.005 in. wide and 0.008 in. high; and peripheral wheel velocity 500 ft/min. Find the average chip thickness.

12.9 A hardened steel shaft is turned to ϕ 1.512/1.515 in. and is 13 in. long. The grinding wheel width is 2 in. Final

diameter is 1.4982/1.4987 in. There are 8 and 10 rough and finish passes, respectively. Find the time for rough and finish grinding. (Hint: Use 0.002 in. for finish stock removal)

12.10. Tool steel is turned to ϕ 2.512/2.515 in. and is 18 in. long. There is sufficient material on one end to use a driving dog. There are 14 rough passes and 10 finish passes. The grinding wheel width is 1 in. Final stock is ϕ 2.4982/2.4987 in. Find the total time for rough and finish grinding.

12.11. An annealed plain carbon steel shaft is turned to ϕ 0.412/0.415 in. and it is 8 in. long. There is sufficient material for grinding wheel clearance. The aluminum oxide grinding wheel diameter is 12 in. and its width is 1 in. Final stock is ϕ 0.4092/0.4097 in. There are 4 and 5 rough and finish passes. Assume that maximum material is removed. Find the total time for grinding.

12.12. A 9-in. driving-wheel belt grinder requires a surface speed of 15 m/s. Specify the rpm if $k = 0.9$.

12.13. A belt grinder is chosen to surface finish aluminum sheet. Velocity of the belt = 600 ft/min, motor rpm = 120, and belt slippage is assumed to be 0.87. Find the diameter of the belt roller.

12.14. A grinding wheel and initial work stock diameter are 12 in. and 5 in. Radial infeed for the grinding operation is 0.002 in. Find the length of arc of contact and the chip area.

12.15. If the regulating wheel of a centerless grinder is 12 in. in diameter and is tilted to an angle of 5°, and is rotating at 30 rpm, what is the feed in in./min?

12.16. How much time is required to remove ¹⁄₁₆ in. from the top surface of a case-hardened iron plate ½ by 4 by 12 in. on a coated-abrasive machine?

12.17. To grind off 0.012 in. on the diameter of a cylinder 14 in. long, what depth of cut and approximate feed is required for a 1-in. wide wheel? (Hint: Consider a centerless type of grinder.)

PRACTICAL APPLICATION

What type of grinding machines are used for grinding or finishing the following parts: piston pins, carbide tools, flat washers, ball bearings, inside and outside of tapered roller bearing housings, form gear cutters, and large rolls? Conduct a telephone survey of practicing grinding-process professionals involved with companies that perform grinding operations. You may substitute other products for your survey.

CASE STUDY: KARLTON GRINDING COMPANY

The Karlton Grinding Company is a full-service grinding job shop. A shaft connector, shown in Figure C12.1, is hardened steel and previously machined and heat treated. It is slightly oversized. All the surfaces, except the ends and angle, are ground. There are 20,000 pieces. Karlton needs work for its centerless grinding machines, but can these parts be used for the centerless machine? Prepare an operations plan similar to that shown and choose the grinding equipment.

It is recognized that when a part reaches the shop floor, it is simply "sitting in a stock bin" 75% of the elapsed time. Less than 25% of the time the part is being worked on. At the conclusion of grinding, about 1% of the parts are lost to poor quality and are scrapped. What is your estimate of Karlton's cost for this job?

What is the advantage of a continuous process if a car manufacturing company adopts the part for mass production? (Hints: Although your estimate of times are, perhaps, inaccurate, they are useful for an approximation of cost. Figure labor, grinding, and inspection time at $50/h. A shop helper's time costs $20/h. The shop helper moves the parts from one machine to another.)

OPERATIONS PLAN

Part name: _____

Date raw   Number received: _____
material received: _____
Date sent to shop: _____   Number sent: _____

Operation Number	Machine	Estimated Time/pc	Date Started	Date Finished	% Passing Inspection

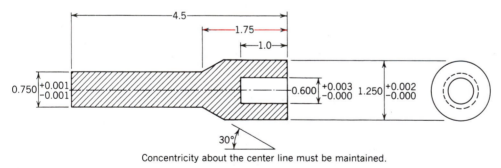

Concentricity about the center line must be maintained.

Figure C12.1
Case study.

CHAPTER 13

WELDING AND
JOINING PROCESSES

Different methods are used by manufacturing systems to join materials permanently. This chapter concentrates on the most popular techniques to join metals: welding and brazing. Attention is given to adhesive bonding that, although commonly used to join plastics, is applied to a variety of engineering materials. Soldering, another very important joining process, is discussed briefly here and, because of its critical importance for electronic manufacturing, it is discussed in more detail in Chapter 19.

Welding is a metal-joining process in which coalescence is obtained by heat and/or pressure. In welding, the metallurgical bond is accomplished by the attracting forces between atoms, but before these atoms can bond, absorbed vapors and oxides must be eliminated from contacting surfaces. The number one enemy to welding is oxidation, and, consequently, many welding processes are performed in a controlled environment or shielded by an inert atmosphere. If force is applied between two smooth metal surfaces to be joined, some crystals will crush through the surfaces and be in contact. As more and more pressure is applied, these areas spread out and other contacts are made. The brittle oxide layer is broken and fragmented as the metal is deformed plastically.

Many welding processes have been developed. They differ widely in the manner that heat is applied and in the equipment used. Some of these processes require hammering, rolling, or pressing to effect the weld; others bring the metal to a fluid state and require no pressure. Processes that use pressure generally require bringing the surfaces of the metal to a temperature sufficient for cohesion to take place. This is usually a subfusion temperature; however, if the fusion temperature is reached, the molten metal must be confined by surrounding solid metal. Most welds are made at fusion temperature and require the addition of weld metal in some form. Welds are also made by casting when metal is heated to a high temperature and poured into a cavity between the two pieces to be joined.

13.1 FUNDAMENTALS OF A WELDING SYSTEM

In any welding, coalescence is improved by cleanliness of the surfaces to be welded. Surface *oxides* should be removed, because they tend to become entrapped in the solidifying metal. *Fluxes* are often employed to remove oxides in fusible slags that float on the molten metal and protect it from atmospheric contamination. In electric-arc welding, flux is coated on the electrode, whereas in gas or forge welding a powder form is used. In other processes a nonoxidizing atmosphere is created at the point where the welding is done. Since oxidation takes place rapidly at high temperature, *speed* in welding is important.

Perhaps one of the most significant variables controlling the amount of porosity is *weld bead shape*. Welds that are narrow and have a high crown tend to trap hydrogen (main cause of porosity), because individual pores must rise a long distance before escaping to the surface. Welding *position* is another important factor: Overhead welds

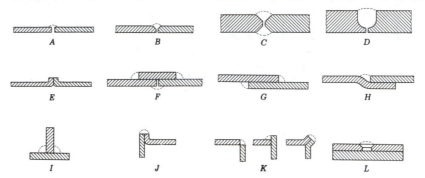

Figure 13.1
Types of welded joints. *A*, Butt weld. *B*, Single vee. *C*, Double vee (heavy plates). *D*, U-shaped (heavy casting). *E*, Flange weld (thin metal). *F*, Single-strap butt joint. *G*, Lap joint (single- or double-fillet weld). *H*, Joggled lap joint (single or double weld). *I*, Tee joint (fillet welds). *J*, Edge weld (used on thin plates). *K*, Corner welds (thin metal). *L*, Plug or rivet butt joint.

present five times more defects than flat welds. *Storage time* after cleaning and prior to welding affects the quality of the weld. Porosity susceptibility, for example, increases as the storage time of a cleaned workpiece is lengthened prior to welding. *Welding current* and *welding speed* are two variables of fundamental importance in welding processes. In arc welding of aluminum, for example, fast welding speeds suppress the formation and growth of pores in the welds. Other variables that substantially affect the degree of hydrogen absorption in welds deposited by arc welding are arc length, shielding gas flow rate, and electrode dressing.

Design Fundamentals of Welded Joints

Different welding processes require different joint designs. For example, lap and butt joints are ordinarily used in resistance welding. Resistance-welded joints must be prepared more accurately and be cleaner than those for other processes. Both gas and arc welding use the same kind of joints. Figure 13.1 illustrates a few common weld types. Joints for forge welding differ in their manner of preparation and do not resemble those shown in the figure. Material imposes limitations on joint designs as well. For example, the minimum welding thickness for aluminum is about 0.030 in. (0.76 mm) and 0.015 in. (0.38 mm) for steel.

Knowledge of welding symbols is important for proper understanding of weld design. A *weld symbol* indicates the required type of weld or braze. The *welding symbol*, however, is an arrow that includes the weld symbol and some of the following elements: reference

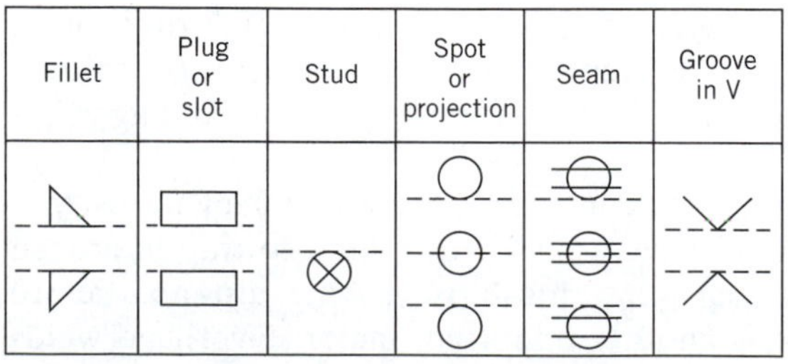

Fillet	Plug or slot	Stud	Spot or projection	Seam	Groove in V

Figure 13.2
Basic weld symbols for six common types of welds.

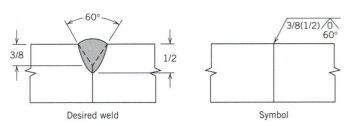

Figure 13.3
Example of the welding symbol for a single-V-groove weld.

line, dimensions and other data, supplementary symbols, finish symbols, specifications, process, and other references. Figure 13.2 shows the weld symbol for six common types of welds, whereas Figure 13.3 illustrates the welding symbol for a single-V-groove weld. The desired weld has a preparation depth of ⅜ in. (9.5 mm), zero root opening, a 60° groove angle, and a groove size of ½ in. (12.7 mm), which is always shown in parentheses.

To be efficient a welded joint must be designed properly for the intended service. If strength and rigidity are important, tension, compression, bending, shear, and torsion analyses are crucial to evaluate the strain.

For example, assume that you are faced with the problem of obtaining adequate stiffness in a cantilever beam welded at the fixed end. The magnitude of vertical deflection at the end of the beam, with a concentrated load at the end, is determined by using the following deflection formula

$$\Delta = \frac{FL^3}{3EI} \tag{13.1}$$

where

Δ = Deflection, in. (mm)
F = Applied force or load, lb (N)
L = Length of beam, in. (mm)
E = Modulus of elasticity, psi (Pa)
I = Area moment of inertia of the beam section, in.^{4} (mm^4)

Since the goal is to have the least amount of deflection, E and I should be as large as possible. This is why steel ($E = 30 \times 10^6$ psi) is such a common structural metal.

13.2 ARC WELDING PROCESSES

Arc welding is a process in which coalescence is obtained by heat produced from an electric arc between the work and an electrode. The electrode or filler metal is heated to a liquid state and deposited into the joint to make the weld. First, contact is made between the electrode and the work to create an electric circuit. Then an arc is formed by separating the conductors. The electric energy is converted into intense heat in the arc, which attains a temperature around 10,000°F (5500°C).

Either direct or alternating current can be used for arc welding, direct current (d-c) being preferred for most purposes. A d-c welder is simply a motor generator set of constant energy type (constant potential may be used also) having the necessary characteristics to produce a stable arc. There should not be too great a current surge when the short circuit is made, and the machine should compensate to some extent for varying lengths of the arc. Direct-current machines are built in capacities up to 1000 A, having an open-circuit voltage of 40 to 95 V. A 200-A machine has a rated current range of 40 to 250 A according to the standard of the National Electrical Manufacturers Association (NEMA). While welding is going on, the arc voltage is 18 to 40 V. In *straight polarity* the electrode is connected to the negative terminal, whereas in *reverse polarity* the electrode is positive.

Welding speed is a sensible variable in arc welding because it influences the volume of porosity. Travel speed, V (in mm/s), is related to the total arc energy input, H, by the empirical equation

$$V = \frac{P}{H} \qquad (13.2)$$

where

P = Welding power, W
H = Arc energy, BTU

Carbon Electrode Welding

The first methods of arc welding, which employed only carbon electrodes, are still in use to some extent for both manual and machine operation. The carbon arc is used only as a source of heat and the flame is handled in a fashion similar to that used in gas welding. Filler rods supply weld metal if additional metal is necessary. The twin-carbon arc method was one of the first used. The arc is between the two electrodes and not with the work. In operation, the arc is held ¼ to ⅜ in. (6.4 to 9.5 mm) above the work, and best results are obtained with the work in a flat position. The use of this method is limited to brazing and soldering.

A second process utilizing a single carbon electrode is considerably simpler. The arc is created between the carbon electrode and the work, and any weld metal needed is supplied by a separate rod. Such an arc is easy to start as the electrode does not stick to the metal. Straight polarity must always be used, because a carbon arc is unstable under reverse polarity.

Metal Electrode Welding

Shortly after the development of carbon electrode welding, it was discovered that a metal electrode with the proper current characteristics could be melted to supply the necessary weld metal (a patent for this process, still in general use, was issued to Charles Coffin in 1889). An arc is started by striking the work with an electrode and quickly raising it a short distance. New powdered-metal electrode designs involve a drag technique where the coated electrode rides lightly on the work. In both, the electrode end is melted by the intense heat. Most of the heat is transferred across the arc in the form of small globules to a molten pool. A small amount of metal is vaporized and lost. Some globules are deposited beside the weld as spatter. The arc is maintained by uniformly moving the electrode along the work at a rate that compensates for metal that has been melted and transferred to the solid weld. At the same time, the electrode is gradually moved along the joint.

For ordinary welding, there is little difference in the weld quality made by alternating current (a-c) and d-c equipment, but polarity causes great variation in weld quality. Most a-c machines consist principally of static transformers. Their efficiency is high, their loss at no loads is negligible, and their maintenance and initial costs are low. Since there is less magnetic flare of the arc or "*arc blow*" with a-c than d-c equipment, the former is preferred in welding heavy plates or fillet welding. Alternating-current welders are built in six sizes as specified by NEMA and are rated at 150, 200, 300, 500, 750, and 1000 A. For hand welding, requiring 200 A or higher, a-c equipment is recommended.

Welding speed and the ease of welding with a-c and d-c welders are similar. However, on heavy plate using large-diameter rods, a-c welding is faster. Direct-current machines are used with all types of carbon and metal electrodes, because the polarity can be changed to suit the electrode. Most of the nonferrous metals and many of the alloys cannot be welded with a-c equipment because electrodes have not been developed for this purpose. With a-c welders the alternating current is constantly reversing with every cycle, and electrodes have to be selected to operate on both polarities. Alternating-current welders operate at slightly higher voltages, hence the danger of shock to the operator is increased.

Electrodes. There are three types of metal electrodes (or "rods"): *bare*, *fluxed*, and *heavy coated*. Bare electrodes, normally used with straight polarity, are used in welding wrought iron and low- or medium-carbon steel. Welds are improved by applying a light coating of *flux* on the rods with a dusting or washing process. The flux assist in eliminating undesirable oxides and in preventing their formation.

Figure 13.4 is a diagrammatic sketch showing the action of an arc using a heavily coated electrode. In the ordinary arc with bare wire the deposited metal is affected by the oxygen and nitrogen in the air. This causes undesirable oxides and nitrides to be formed in the weld metal. The effect of heavy coatings on electrodes provides a gas shield around the arc to eliminate such conditions and covers the weld metal with a protective slag coating that prevents oxidation of the surface metal during cooling. Welds made from rods of this type have superior physical characteristics.

Electrode Coating. Electrodes coated with slagging or fluxing materials are necessary in welding alloys and nonferrous metals. Some of the elements in these alloys are unstable and are lost if there is no protection against oxidation. Heavy coatings permit the use of larger welding rods, stronger current, polarity variation, and higher welding speeds. In summary, *electrode coatings* perform the following functions: provide a protecting atmosphere, provide slag of suitable characteristics to protect the molten metal, facilitate overhead and position welding, stabilize the arc, add alloying elements to the weld metal, perform metallurgical refining operations, reduce spatter of weld metal, increase deposition efficiency, remove oxides and impurities, influence the depth of arc penetration, influence the shape of the bead, and slow down the cooling rate of the weld.

Coating compositions are classified as organic and inorganic, and sometimes a combination of the two is used. Inorganic coatings are further subdivided into flux compounds and slag-forming compounds. The term *"contact electrode"* is given to electrodes having a thick coating with a high metal powder content. These electrodes are suitable for welding with a drag or contact technique. Most electrodes have an automatic striking or

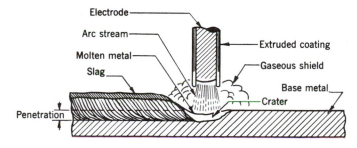

Figure 13.4
Diagrammatic sketch of arc flame.

self-igniting characteristic because of the iron in the coating. By adding metal powder to the coating and increasing its thickness, the deposition rate is increased. Some of the principal coating constituents are:

1. *Slag-forming constituents.* SiO_2, MnO_2, and FeO. Al_2O_3 is sometimes used but it makes the arc less stable.
2. *Constituents to improve arc characteristics.* Na_2O, CaO, MgO, and TiO_2.
3. *Deoxidizing constituents.* Graphite, aluminum, and wood flour.
4. *Binding material.* Sodium silicate, potassium silicate, and asbestos.
5. *Alloying constituents to improve strength of weld.* Vanadium, cesium, cobalt, molybdenum, aluminum, zirconium, chromium, nickel, manganese, and tungsten.

Atomic Hydrogen Arc Welding. In this process, a single-phase a-c arc is maintained between two tungsten electrodes and hydrogen is introduced into the arc. As the hydrogen enters the arc, the molecules are broken into atoms, which recombine into molecules of hydrogen outside the arc. This reaction is accompanied by intense heat of about 11,000°F (6100°C). Weld metal may be added to the joint by a welding rod. The operation is similar to the oxyacetylene process. The atomic hydrogen process differs from other arc-welding processes in that the arc is formed between two electrodes (Figure 13.5) rather than between one electrode and the work. This makes the electrode holder a mobile tool that travels without extinguishing the arc.

The advantage of this process is its ability to provide high heat concentrations. The hydrogen also acts as a shield and protects the electrodes and molten metal from oxidation. Filler metal of the same analysis can be used with manual and automatic equipment. It is used successfully for many alloys difficult to weld by other processes. Most applications of atomic hydrogen arc welding are accomplished by the inert-gas–shielded-arc process.

Inert-Gas–Shielded-Arc Welding. In this process, coalescence is produced by heat from an arc between a metal electrode and the work shielded by an atmosphere of argon, helium, CO_2, or a mixture of gases. Although CO_2 is not an inert gas, it ionizes at welding temperatures and acts like one. Two methods are employed: one uses a tungsten electrode with filler metal as in gas welding (*TIG welding*, tungsten inert gas), and the other uses a consumable metal wire as the electrode (*MIG welding*, metal inert gas). Both methods are adaptable to manual or automatic machine welding, with no flux or wire coating required for protection of the weld.

The first method, also known as *gas tungsten arc welding*, GTAW, uses a single nonconsumable tungsten electrode (see Figure 13.6). The weld zone is protected by inert gas fed through the water-cooled electrode holder. Argon is frequently used, although

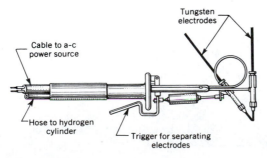

Figure 13.5
Torch for atomic hydrogen welding.

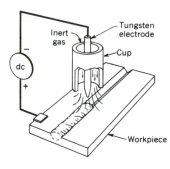

Figure 13.6
Schematic diagram of the gas tungsten arc welding (GTAW) process.

helium or a mixture of the two is employed. The kind of metal to be welded determines the kind of current to be used. Direct current with straight polarity is required for steel, cast iron, copper alloys, and stainless steel, whereas reverse polarity is rarely used. However, a-c is used for aluminum, magnesium, cast iron, and a number of other metals. Operating costs make TIG welding better suited for welding of light-gage work rather than for welding heavier materials.

The other method, also known as *gas metal arc welding*, GMAW, is accomplished by employing a shielded arc between the consumable bare wire electrode and the workpiece. Because filler material is transferred through the protected arc, greater efficiency is obtained, resulting in more rapid welding. The metal is deposited in an atmosphere that prevents contamination. In MIG welding (Figure 13.7), a wire is fed continuously through a gun to a contact surface that imparts a current to the wire. Direct current reverse polarity provides a stable arc and offers the greatest heat input at the workpieces; it is generally recommended for aluminum, magnesium, copper, and steel. Straight polarity with argon has a high burnoff rate but the arc is unstable with high splatter. Alternating current, being inherently unstable, is seldom used in MIG welding.

Carbon dioxide gas is used widely in welding plain-carbon and low-alloy steels. Since the gas decomposes into carbon monoxide and oxygen at high temperature, safety regulations require that some gas-evolving flux be provided in or on the wire. Since MIG welding has excellent penetration, it produces sound welds at high speeds. This makes the process especially adapted to automatic operation.

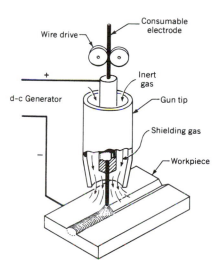

Figure 13.7
Schematic diagram of the gas metal arc welding (GMAW) process.

Arc Spot Welding. An application of inert-gas arc welding is to make spot, plug, or tack welds by an argon-shielded electric arc using consumable electrodes. To effect a weld a small welding gun with pistol grip is held tightly against the work. As the trigger is released, the argon valve is opened. The current is allowed to pass through the electrode for a preset interval (2–5 s), and then both are shut off. Gas tungsten arc welding guns using direct current are employed in making spot welds. An advantage of this equipment is that spot welds can be made on thin sheets from one side of the work. Being a low-cost process, it is particularly useful in welding large or irregular-shaped assemblies that are difficult to spot weld with resistance equipment.

Submerged-Arc Welding. This process is so named because the metal arc is shielded by a blanket of granular, fusible flux during the welding. Aside from this feature, its operation is similar to other automatic arc-welding methods. A bare electrode is fed through the welding head into the granular material (Figure 13.8). This material is laid along the seam to be welded and the entire welding action takes place beneath it. The arc is started by striking beneath the flux on the work or by initially placing some conductive medium, such as steel wool, beneath the electrode. The intense heat of the arc immediately produces a pool of molten metal in the joint and melts a portion of the granular flux. This material floats on top, forming a blanket that eliminates spatter losses and protects the welded joint from oxidation. Upon cooling the fused slag solidifies and is easily removed. Granular material not fused is recycled and used again.

This process is limited to flat welding, although welds can be made on a slight slope or on circumferential joints. It is advisable to use a backing strip of steel, copper, or some refractory material on the joint to avoid losing some of the molten metal. The process uses strong current (300–4000 A), which permits a high rate of metal transfer and welding speeds. Deep penetration is obtained and most commercial-thickness metal plates are welded with one pass. As a result, thin plates are welded without preparation, whereas a small vee is required on others. Most submerged-arc welding is done on low-carbon and alloy steels, but it may be used on many of the nonferrous metals.

Stud Arc Welding. Stud arc welding is a d-c arc-welding process developed to end-weld metal studs to flat surfaces. It is accomplished with a pistol-shaped welding gun that holds the stud or fastener to be welded. When the trigger of the gun is pressed, the stud is lifted to create an arc and then forced against the molten pool by a backing spring. The entire operation is controlled by a timer, preset according to the size of the stud being welded. The arc is shielded by surrounding it with a ceramic ferrule, which also confines the metal to the weld area and protects the operator from the arc.

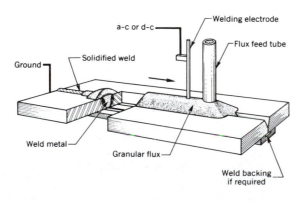

Figure 13.8
Schematic sketch of submerged-arc welding.

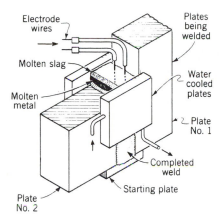

Figure 13.9
Diagram of the electroslag welding process.

A percussion-type gun has a small projection on the end of the stud. As the stud moves forward, the projection touches the work and is vaporized, causing an arc to form. The weld is completed as the stud is driven onto the work surface.

Electroslag Welding. This is a metal arc welding process preferred in welding heavy plates having the joint to be welded in a vertical position. Heat is obtained from the resistance of current in an electrically conductive molten flux. One or more electrodes are fed continuously into a pool of molten slag, which maintains a temperature in excess of 3200°F (1800°C).

A schematic diagram of the *electroslag welding* process is shown in Figure 13.9. In starting the weld, an arc is created between the electrode and the bottom plate and continues until a sufficiently thick layer of molten slag is formed. The current then flows through the slag, which maintains sufficient temperature to melt the wire electrodes and the surfaces of the workpiece. On either side of the joint are water-cooled copper slides that confine the molten metal and slag. As the metal solidifies, the copper plates automatically move upward. The rate is determined by the speed at which the electrodes and base metal are melted. The lower part of the metal bath is solidified by cooling from the plates; thus, the welding joint is formed. Slag may be added continuously from an overhead hopper or by using a flux-cored wire. The latter method is preferred, because the entering flux is always carried to the hottest part of the pool.

Advantages of the electroslag welding process include the ability to weld metals of great thickness in a single pass, minimum joint preparation, high welding speed, and good stress distribution across the weld with little distortion. The weld metal is protected at all times from contamination.

13.3 RESISTANCE WELDING PROCESSES

In a resistance welding (RW) process, a strong electric current is passed through the metals to cause local heating at the joint and then pressure is applied to complete the weld. A transformer in the welding machine reduces the a-c voltage from either 120 or 240 V to around 4 to 12 V and raises the amperage sufficiently to produce a good heating current. When the current passes through the metal, most of the heating takes place at the point of greatest resistance in the electrical path, which is at the interface of the two sheets. It is here that the weld forms. The amount of current necessary is 30 to 40 kVA/

in.^{2} (47 to 62 MVA/m^2) of the area to be united based on a time of about 10 s. The necessary pressure to effect the weld will vary from 4000 to 8000 psi (28 to 55 MPa).

Resistance welding is suitable to the joining of light gage metals. Usually, the equipment is adapted for only one type of weld and the work is moved to the machine. It is the only process that uses a pressure action at the weld during an accurately regulated heat application. Practically all metals can be welded by resistance welding, although a few, such as tin, zinc, and lead, present great difficulty.

In all RW processes, the parameters with the most impact are governed by the formula

$$H = I^2Rt \qquad (13.3)$$

where

$$
\begin{aligned}
H &= \text{Heat, BTUs} \\
I &= \text{Welding current, A} \\
R &= \text{Resistance, ohms} \\
t &= \text{Time, s}
\end{aligned}
$$

or equivalently

$$H = P \times t \qquad (13.4)$$

since power is expressed as $P = I^2R$. The heat distribution is illustrated in Figure 13.10. The greatest heat is generated at (b) in the figure, because the resistance is much greater at that point than at (a) and (c); thus, the far greater power is consumed in making the weld. The resistance of the parent metal between (a) and (c) is partially utilized in effecting the weld. Power is dissipated at (a) and (c) by the water-cooled electrodes.

The amperage of the *secondary* or *welding current* is determined by the transformer. To provide possible variation of the secondary current, the transformer is equipped with a regulator on the primary side to vary the number of turns on the primary coil (see Figure 13.11). For good welds current, resistance, and time must be carefully considered and determined by factors such as material thickness, kind of material, and type and size of electrode.

The timing of the welding current is important. An adjustable delay is needed from the moment that pressure is applied to the moment that the weld is started. The current is turned on by the timer and held a sufficient time for the weld. It is then stopped, but the pressure remains until the weld cools; thus, the electrodes do not arc and the weld is protected from discoloration. The pressure on the weld may be obtained manually,

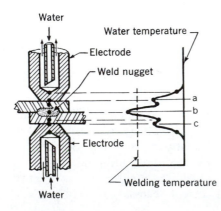

Figure 13.10
Temperature distribution in spot welding.

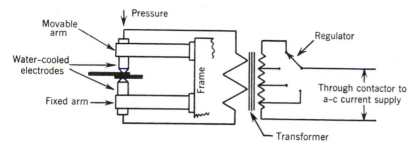

Figure 13.11
Diagram of spot welder.

by mechanical means, air pressure, springs, or hydraulic means. Its application is controlled and coordinated with the welding current.

Spot Welding

In this form of resistance welding, two or more sheets of metal are held under pressure between metal electrodes (Figure 13.11). In spot welding, there are five zones of heat generation: one at the interface between the two sheets, two at the contact surfaces of the sheets with the electrodes, and the other two in each piece of metal (Figure 13.10). The contact resistance at the interface of the two sheets is the point of highest resistance where the weld formation starts. Contact resistance at this point, as well as between the electrodes and the outsides of the sheets, depends on surface conditions, magnitude of electrode force, and the size of the electrodes. If the sheets are the same in both thickness and analysis, the heat balance will be such that the weld nugget will be at the center. Uneven sheet thicknesses or welding sheets of different thermal conductivities may necessitate using electrodes of different size or conductivity to obtain a proper heat balance.

The welding cycle is divided in four phases. During the first phase, known as *squeeze time*, the metal is under pressure and the electrodes are in contact with it until current is applied. In the second phase, a low-voltage current of sufficient amperage is passed between the electrodes causing the metal in contact to rise rapidly to a welding temperature. When the needed temperature is reached, the pressure between the electrodes squeezes the metal together and completes the weld. This period, usually 3 to 30 cycles of a 60-Hz current, is known as the *weld time*. Next, while the pressure is still on, the current is shut off for a period called the *hold time*, during which the metal regains some strength by cooling. The pressure is then released and the work is either removed from the machine or moved so that another portion can be welded; this is known as the *off time*.

Spot Welding Machines. Three general types of machines exist: stationary single spot, portable single spot, and multiple spot. *Stationary machines* may be classified further as rocker arm and direct-pressure types. The rocker-arm type generally is limited to machines of small capacity. It is so designated because the motion for applying pressure and raising the upper electrode is made by rocking the upper arm. The larger machines usually employ direct straight-line motion of the upper electrode. This arrangement permits them to be used for projection welding.

Portable spot welders are connected to the transformer by long cables so they are moved to any location and then used. Where all welds cannot be made by a single machine setup, welding jig assemblies are served best by portable welders. The principal

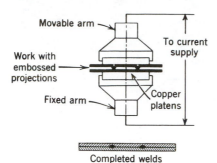

Figure 13.12
Projection welding.

differences among them are the manner of applying the pressure and the shape of the tips. Pressure is applied manually, pneumatically, or hydraulically, depending on the size and type of gun.

For production work, *multiple spot welding machines*, capable of producing two or more spots simultaneously, exist. A system known as indirect welding is used, where two electrodes are in series and the current passes through a heavy plate underneath the sheets and between the electrodes.

Projection Welding

Projection welding is illustrated in the line diagram shown in Figure 13.12. Projection welds are produced at localized points in workpieces held under pressure between suitable electrodes. Sheet metal is first put through a punch press that makes small projections or buttons in the metal. These projections are made with a diameter on the face equal to the thickness of the stock and extend from the stock about 60% of its thickness. Such projection spots or ridges are made at all points where a weld is desired. This process is also used for crosswire welding and for parts where the ridges are produced by machining. With this form of welding a number of welds can be made simultaneously. The only limit is the ability of the press to furnish and distribute equally the correct current and pressure. Results are generally uniform and weld appearance often is better than in spot welding. Electrode life is long, because only flat surfaces are used and little electrode maintenance is required.

Seam Welding

This method is in effect a continuous spot-welding process in which current is regulated by the timer of the machine. Seam welding consists of a continuous weld on two overlapping pieces of sheet metal that are held together under pressure between two circular electrodes. Coalescence is produced by heat obtained from the resistance to flow of current that passes through the overlapping sheets. In high-speed seam welding using continuous current, the frequency of the current acts as an interrupter.

The heat at the electrode contact surfaces is kept to a minimum by the use of copper alloy electrodes and is dissipated by flooding the electrodes and weld area with water. Heat generated at the interface by contact resistance is increased by decreasing the electrode force. Another variable that influences the magnitude of the heat is the weld time, which in seam welding is controlled by the speed of rotation of the electrodes. The amount of heat generated is decreased with an increase in welding speed.

Three types of seam welds used in the industry are illustrated in Figure 13.13. The

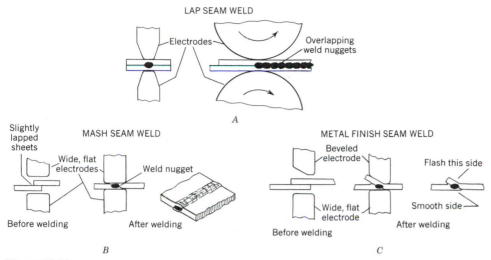

Figure 13.13
Types of seam welds.

most common is the *simple lap seam weld* (Figure 13.13*A*). This weld consists of a series of overlapping spot welds with sufficient overlap of the weld nuggets to provide a pressure-tight joint. If pressure-tight quality of the lap seam weld is not required, the individual nuggets are spaced to give a stitch effect. This process is known as *roll spot welding*. Another type of weld, the *mash seam weld* (Figure 13.13*B*), is produced by reducing the amount of sheet lapping to a small value. Broad-faced, flat electrodes forge the sheets together while welding. This forging action by the electrodes is known as "*mashdown*" and occurs simultaneously with the fusing on the sheets. Because the joint is covered above and below by the electrodes and on either side by the sheets, any extrusion or spitting from the weld is prevented. Micrographs showing longitudinal transverse sections of two pieces of 0.050-in. (1.27-mm) steel are shown in Figure 13.14. Normally the mashdown takes place on one side only if the electrode face contour is modified and if the amount of lap is increased. The mashdown is finished later, leaving no trace of the joint. This type of mash seam welding is often referred to as *finish seam welding* (Figure 13.13*C*) and finds application where the product is normally viewed from one side only.

Seam welding is used in manufacturing metal containers, automobile mufflers and fenders, refrigerator cabinets, and gasoline tanks. Advantages of this type of fabrication include improved design, material saving, tight joints, and low-cost construction.

Figure 13.14
Longitudinal and cross section of mash seam welds made with 0.050-in. (1.27-mm) steel at 80 in./min (33.8 mm/s). Current 19,000 *A*, electrode force 1500 lb (6600 N), and initial overlap 150%. Magnification ×12.5.

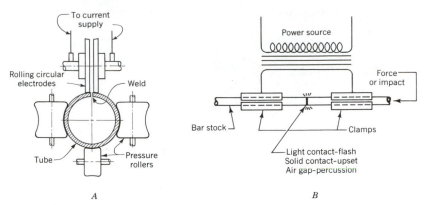

Figure 13.15
Butt-welding methods. *A*, Continuous-resistance butt welding of steel pipe.
B, Sketch illustrating forms of butt-welding bar stock.

Butt Welding

This form of welding, illustrated in Figure 13.15, is accomplished by gripping two pieces of metal that have the same cross section and pressing them together while heat is being generated in the contact surface by electrical resistance. Although pressure is maintained while heating takes place, at no time is the temperature sufficient to melt the metal. The joint is upset somewhat by the process, but this defect is eliminated by subsequent rolling or grinding. When parts to be welded have the same resistance, uniform heating at the joint is attained. If two dissimilar metals are to be welded, the metal projecting from the die holders must be in proportion to the specific resistance of the materials to be welded. The same treatment is used when materials of different cross sections are butt-welded.

In actual operation, the work is first clamped in the machine and pressure is applied on the joints. The welding current is then started and heating takes place; the rate depends on the pressure, the material, and the condition of surfaces. Because the contact resistance varies inversely with the pressure, the pressure is less at the start and is then increased to whatever is necessary to effect the weld. The pressure is usually about 2500 to 8000 psi (17 to 55 MPa) when the welding temperature is reached. This type of welding is especially adapted to rods, pipes, small structural shapes, and many other parts of uniform section. Areas up to 70 in.^{2} (0.05 m^2) are welded successfully, but generally the process is limited to small areas because of current limitations.

Figure 13.15 illustrates a special type of butt seam welding used in pipe manufacturing. Two rolling electrodes bring a high-amperage current across the joint that generates the heat in the contact surfaces by electrical resistance. For thin-walled tubes high-frequency current from an induction coil can be used to generate the heat in place of electrodes.

Flash Welding

Butt and flash butt welding, although similar in application, differ somewhat in the manner of heating the metal. For flash welding, the parts must be brought together in very light contact. A high voltage starts a flashing action between the two surfaces and continues as the parts advance slowly and the forging temperature is reached. The heat generated for welding results from the arcing between surfaces. The weld is completed by the application of a sufficient forging pressure of 5000 to 25,000 psi (35 to 170 MPa) to effect a weld.

Welding of small areas is usually done by the butt-welding method; large areas are done by the flash butt method. However, there is no clear demarcation between the two. The shape of the piece and the nature of the alloy are frequently the determining factors. Areas ranging from 0.002 to 50 in.^{2} (1 to 32,000 mm^2) are successfully welded by flash welding. Advantages of this process include: less current is required than in ordinary butt welding, there is less metal to remove around the joints, the metal that forms the weld is protected from atmospheric contamination, the operation consumes little time, and end-to-end welding of sheets is possible. These advantages make flash welding more widely used than the ordinary butt or upset processes. Although many nonferrous metals are flash-welded satisfactorily, alloys containing high percentages of lead, zinc, tin, and copper are not recommended for this process.

Percussion Welding

Some experts do not consider percussion welding a RW process because, like in flash welding, it relies on arc effects for heating rather than on the resistance in the metal. Pieces to be welded are held apart, one in a stationary holder and the other in a clamp mounted in a slide and backed up against heavy spring pressure (see Figure 13.15). When the movable clamp is released, it moves rapidly, carrying with it the piece to be welded. When the pieces are about $\frac{1}{16}$ in. (1.6 mm) apart, there is a sudden discharge of electric energy, causing intense arcing over the surfaces and bringing them to a high temperature. The arc is extinguished by the percussion blow of the two parts coming together with sufficient force to effect the weld.

The electric energy for the discharge is built up in one of two ways. In the electrostatic method, energy is stored in a capacitor and the parts to be welded are heated by the sudden discharge of a heavy current from the capacitor. The electromagnetic welder uses the energy discharge caused by the collapsing of the magnetic field linking the primary and secondary windings of a transformer or other inductive device.

The action of this process is so rapid (about 0.1 s) that there is little heating effect in the material adjacent to the weld. Heat-treated parts may be welded without being annealed. Parts differing in thermal conductivity and mass are joined successfully because the heat is concentrated only at the two surfaces. Examples are welding Stellite tips to tools, copper to aluminum or stainless steel, silver contact tips to copper, cast iron to steel, lead-in wires on electric lamps, and zinc to steel. These welds are made without any upset or flash at the joint. The principal limitation of the process is that only small areas up to $\frac{1}{2}$ in.^{2} (323 mm^2) of nearly regular sections can be welded.

High-Frequency Resistance Welding

This process, used primarily to manufacture structural steel shapes, utilizes an electric current at about 400,000 Hz. Such shapes as I-beams can be welded at high speeds with relatively low heat input, thus reducing grain distortion. Such a beam would consist of two flanges welded to the web section by means of applying high-frequency current and passing the flanges between pressure rolls and effecting the weld to the separating web section. No special cleaning or fluxing is required. Even dissimilar materials are joined at rates up to 170 ft/min (51 m/min). The carbon content of steels welded by this process must be under 0.40%, but structural shapes are typically low in carbon.

13.4 OXYFUEL GAS WELDING PROCESSES

Oxyfuel gas welding (OGW) includes all the processes in which a combination of gases is used to obtain a hot flame. Acetylene, natural gas, and hydrogen in combination with oxygen are the most common blends. Oxygen is produced by both electrolysis and liquification of air. *Electrolysis* separates water into hydrogen and oxygen by passing an electric current through it. Most commercial oxygen is made by liquefying air and separating the oxygen from the nitrogen. It is stored in steel cylinders at a pressure of 2000 psi (14 MPa), as noted in Figure 13.16.

Oxyacetylene Welding

This is the most common of all OGW processes. An oxyacetylene weld is produced by heating the parts with a flame that reaches temperatures of 6300°F (3500°C) and with or without the use of a filler metal. The flame is obtained from the combustion of oxygen and acetylene. Most often the joint is heated to a state of fusion and, as a rule, no pressure is used.

Acetylene gas (C_2H_2) is obtained by dropping lumps of calcium carbide in water. The gas bubbles through the water, and any precipitate is slaked lime. The reaction that takes place in an acetylene generator is

$$CaC_2 + 2H_2O \rightarrow Ca(OH)_2 + C_2H_2$$

This gas is not stored safely at pressures higher than 15 psi (0.1 MPa), so it is stored with acetone. Acetylene cylinders are filled with a porous filler saturated with acetone in which the acetylene gas can be compressed. These cylinders hold 300 ft³ (9 m³) of gas at pressures up to 250 psi (1.7 MPa).

A schematic sketch of a welding torch and its gas supply is shown in Figure 13.17. Gas pressures are controlled by regulating valves, and final adjustment is done manually at the torch. The perfect gas law describes the amount of gas available at a regulated pressure from a pressurized cylinder. That is,

$$\frac{P_1V_1}{T_1} = \frac{P_2V_2}{T_2} \tag{13.5}$$

where P_1, V_1, and T_1 refer to the pressure, volume, and absolute temperature in the cylinder; P_2, V_2, T_2 refer to the regulated pressure, volume, and temperature. The absolute temperature is $T = 460 + t°F$. If the cylinder and regulated temperatures are the same

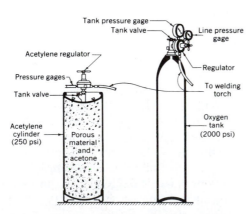

Figure 13.16
Cylinders and regulators for oxyacetylene welding.

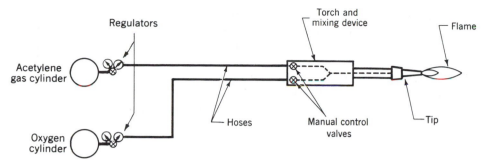

Figure 13.17
Schematic sketch of oxyacetylene welding torch and gas supply.

or nearly so, then

$$P_1V_1 = P_2V_2 \tag{13.6}$$

EXAMPLE 13.1

If an acetylene cylinder contains 300 ft³ of gas at 250 psi, it can deliver a much larger volume of gas at a regulated pressure of 7 psi above the atmospheric pressure (14.7 psi). From Equation (13.6) that volume is

$$V_2 = \frac{P_1V_1}{P_2} = \frac{250 \times 300}{7 + 14.7} = 3456 \text{ ft}^3 \tag{13.7}$$

Regulation of the proportion of the two gases affects the characteristics of the flame. The chemical reaction that occurs when acetylene and oxygen are burned in a neutral flame occurs in two stages, one at the inner cone and one at the outer envelope (Figure 13.18). The heat generated in the inner cone is approximately

$$H_{ic} = 150 \text{ BTU} \times V \tag{13.8}$$

where

$$H_{ic} = \text{Heat generated in the inner cone}$$
$$V = \text{Volume of regulated gas, ft}^3$$

and

$$H_{oc} = 960 \text{ BTU} \times V \tag{13.9}$$

where

$$H_{oc} = \text{Heat generated in the outer envelope}$$

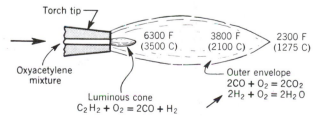

Figure 13.18
Sketch of neutral flame showing temperatures attained.

When there is an excess of acetylene used, there is a decided change in the appearance of the flame. Three zones are found instead of the two just described. Between the luminous cone and the outer envelope is an intermediate white-colored cone, whose length is determined by the amount of excess acetylene. This flame, known as a *reducing* or *carburizing flame*, is used in the welding of Monel metal, nickel, certain alloy steels, and many of the nonferrous, hard-surfacing materials.

If the torch is adjusted to give excess oxygen, a flame similar in appearance to the neutral flame is obtained, except that the inner luminous cone is much shorter and the outer envelope appears to have more color. This oxidizing flame may be used in fusion welding brass and bronze, but it is undesirable in other applications.

The advantages and uses of oxyacetylene welding are numerous. The equipment is comparatively inexpensive and requires little maintenance. It is portable and can be used with equal facility in the field and in the factory. With proper technique practically all metals can be welded and the equipment used for welding as well as for cutting, as explained later in this chapter.

Oxyhydrogen Welding

Oxyhydrogen welding was the first gas process to be developed commercially. Hydrogen is produced by either the electrolysis of water or passing steam over coke. Because oxyhydrogen burns at 3600°F (2000°C), a much lower temperature than oxygen and acetylene, it is used primarily for welding thin sheets and low-melting alloys and in some brazing work. While the same equipment can be used for both processes, flame adjustments are more difficult in hydrogen welding because there is no distinguishing color to judge the gas proportions. A reducing atmosphere is recommended, and the process is characterized by the absence of oxides formed on the surface of the weld. The quality of these welds is equal to that obtained by other processes.

Air Acetylene Welding

The torch used in the air acetylene welding process is similar in construction to a Bunsen burner, in which air is drawn into the torch as required for proper combustion. Use of this type of welding is limited because the temperature attained is the lowest of all OGW processes. Some applications include lead welding and low-temperature brazing or soldering operations.

Pressure Gas Welding

In pressure gas welding, the abutting areas of parts to be joined are heated with oxyacetylene flames to a welding temperature of about 2200°F (1200°C) and pressure is applied. Two methods are in common use. In the first, known as the *closed-joint method*, the surfaces to be joined are held together under pressure during the heating period. Multiflame, water-cooled torches designed to surround the joint are used. During the heating operation the torches are oscillated slightly to eliminate excessive local heating. As the heating progresses, the slightly beveled ends close. When the correct temperature is reached, an additional upsetting pressure is applied. For low-carbon steel, the initial pressure is below 1500 psi (10 MPa) and the upsetting pressure is around 4000 psi (28 MPa).

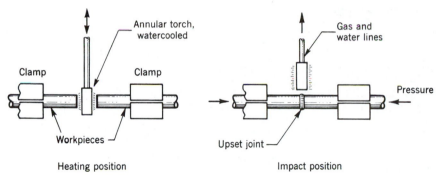

Figure 13.19
Schematic sketch illustrating pressure-gas butt welding.

The second method, or *open-joint method* (Figure 13.19) employs a flat multiflame torch placed between the two surfaces to be joined. This torch uniformly heats the surfaces until there is a film of molten metal over each of them. The torch is quickly withdrawn, and the surfaces are forced together and held at about 4000 psi (28 MPa) until solidification takes place. No filler metal is used.

13.5 SOLID-STATE WELDING PROCESSES

Solid-state derives its name from solid-state physics: the science of the crystalline solid. In solid-state welding (SSW) processes, such as co-extrusion welding, hot pressure welding, roll welding, and those explained in this section, metals do not melt. Adhesion is produced by a metallic bonding of energized crystals. Normally the size of fragments transferred between the parts grows until they become a continuous layer of plasticized metal, but no evidence of a molten state is ever present.

Cold Welding

Cold welding is a method of joining metals at room temperature by the application of pressure alone. The pressure applied causes the surface metals to flow, producing the weld. It is a solid-state bonding process in which no heat is supplied from an external source. The type of bond obtained in a lap weld is shown in Figure 13.20. Butt welds of wire and rods are made by clamping the ends in special dies and bringing them together under a load sufficient to produce plastic flow at the joint. Before a weld is made, the surfaces or parts to be joined must be wire-brushed thoroughly at a surface speed around 3000 ft/min (15 m/s) to remove oxide films on the surface. Other methods of cleaning seem to be unsatisfactory. In making a weld the pressure is applied over a narrow area so that the metal can flow away from the weld on both sides. It is applied either by impact or with a slow squeezing action; both methods are equally effective. Ring welds, continuous-seam welds, and spot welds can be made. Spot welds are rectangular in shape and are approximately t by $5t$ in area, where t represents the metal thickness. This method of welding is used primarily with aluminum and copper. Pressure required for aluminum is 25,000 to 35,000 psi (170 to 240 MPa). Lead, nickel, zinc, and Monel can also be joined by cold pressure.

Figure 13.20
A macrograph of an aluminum cold weld showing lines of flow.

Ultrasonic Welding

Ultrasonic welding is a solid-state bonding process for joining similar or dissimilar metals, generally with an overlap-type joint. High-frequency vibratory energy is introduced into the weld area in a plane parallel to the surface of the weldment. The forces set up oscillating shear stresses at the weld interface that break up and expel surface oxides. This results in metal-to-metal contact, permitting the intermingling of the metal and the forming of a sound weld nugget. No external heat is applied, although the weld metal does undergo a modest temperature rise.

A diagrammatic sketch of one spot-type welding system is shown in Figure 13.21. The machine is preset for clamping force, time, and power, and the overlapping pieces are placed on the anvil or support member. As the welding cycle starts, the *sonotrode*

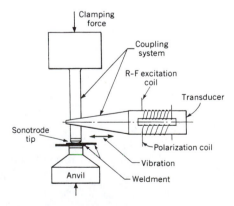

Figure 13.21
Ultrasonic welding.

(vibrating element equivalent to an electrode in resistance welding) is lowered to the weldment, and the clamping force builds to the desired amounts. Ultrasonic power of sufficient intensity is then introduced through the sonotrode for the preset time. The power is cut off automatically and the weldment is released.

Continuous-seam welding is accomplished with a rotating disk tip and with either a counter-rotating roller anvil or a table-type anvil. The entire transducer coupling tip assembly rotates, so that the peripheral speed of the tip matches the traversing speed of the workpiece. Continuous seams are produced with essentially no slippage between the tip and the workpiece.

The only machine settings for welding are power, clamping force, and weld time or welding rate for continuous-seam welding. These settings vary according to the type and thickness of the material to be joined. Thin foils or fine wires require only a few watts of power and a few ounces of clamping force, while heavier and harder materials require several thousand watts and several hundred pounds. The weld time for spot-type welds is very short, usually less than 1 s.

This process is used for materials up to about ⅛ in. (3 mm) in thickness, although equipment is being developed to extend this range. There is no minimum limit to the sheet thickness that can be welded. Ultrasonic welding is excellent for joining thin sheets to thicker sheets, because the thickness limitation applies only to the thinner piece.

The welds obtained are characterized by local plastic deformation at the interface and mechanical intermingling of the contacting surfaces. They are metallurgically sound and often exhibit strengths superior to those obtained by other joining processes. Ring-type and continuous-seam welds can be used for hermetic sealing. The process has a variety of applications in the electrical and electronic industries, in sealing and packaging, in foil splicing, in aircraft, in missiles, and in the fabrication of nuclear reactor components.

Explosive Welding

Explosive welding or *cladding*, as it is often called, brings together two metal surfaces with sufficient impact and pressure to bond. Pressure is developed by a high-explosive shot placed in contact with, or in close proximity to, the metals, as illustrated in Figure 13.22. Of the two arrangements, the left one is preferred. In some instances a protective material, such as rubber, is placed over the upper panel to prevent damage to the surface. The entire assembly is placed on a buffer plate or anvil to absorb energy generated during the joining operation.

To obtain a metallurgical bond, atoms from both surfaces must come into intimate contact. The oxides and films always present on the surface of metals are broken or dispersed by the high pressure or dissolved in the molten region. The explosive force brings the clean surfaces together and produces a sound bond. Figure 13.23 shows the development of a *high-velocity jet* emanating from the collision point, a phenomenon

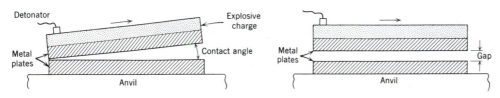

Figure 13.22
Common arrangements used to produce explosive clads. Buffer materials such as rubber are often used between the explosive and the metal.

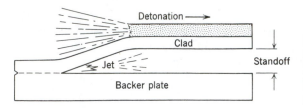

Figure 13.23
Explosion-bonding process showing high-velocity jet emanating from the collision point because of upstream pressure.

that occurs in most explosive welds. The jet is formed by the surfaces of both plates flowing ahead of the collision point into space between them.

The general appearance of most metal bonds is shown in Figure 13.24. The wavy profile developed as a result of the surface jetting effect is a combination of direct metal-to-metal bonding and periodic melted areas. The surface jetting is a result of a compressive stress wave progressing across the surface of the plates as they collide. This action flushes the impurities from the surfaces and enhances the weld. In general, the wavy-type bond is preferred because it localizes any solidification defects. A rather straight layer bond occurs with some metals as a result of a previously continuous molten surface zone that has solidified rapidly.

The principal use of explosive welding is the uniting of large-area sheets. Areas up to 7 ft by 20 ft (2 m by 6 m) have been bonded. In addition to area welds, seam, spot, lap, and edge welds are possible variations, as well as internal cladding of tubes and pressure vessels. Advantages for this process include simplicity, rapidity, close thickness tolerance, and the ability to unite dissimilar metals. Metals with low melting points and low impact resistance cannot be bonded effectively.

Diffusion Welding

In *diffusion welding*, clean flat parts with fine surface finish are brought together in a vacuum or in an inert gas atmosphere with pressure. Usually, the temperature is lower than the melting temperature of the base metal. The process temperature is about

$$T_p = 0.7T_m \tag{13.10}$$

where

T_p = Process temperature
T_m = Melting point of the lowest temperature of base metal being welded

Actually, the equation is accurate only if absolute temperatures are employed. For practical purposes, conventional temperatures are used. Pressure required varies from 6 to 50 psi (41 to 345 MPa), but pressures just under the yield stress at the operating temperature are preferred.

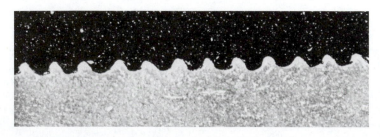

Figure 13.24
Direct metal-to-metal bond between explosively welded Monel 400 (top) to ASM-A 516-70 steel showing sawtooth bond. Magnification ×10.

In liquid diffusion welding, the temperature is slightly above the melting temperature. The process creates a metallic bond as a result of the pressure causing microdeformation of the surfaces into each other, and diffusion of atoms between the two materials being joined. Variations of this process use a thin film of a third metal to effect a weld between dissimilar materials. This is not always necessary, because almost any metal and many plastics can be joined by this method.

One common application is "coating" cutting tools by diffusion welding as opposed to brazing. However, the principal applications are in high technology, particularly in the atomic, aerospace, and electronic industries because of cost considerations. The process is slow and costly because (1) surfaces must be flat and prepared to a roughness of about 8 μm, which is costly; (2) extreme cleanliness must be maintained, which requires a "clean room"; (3) use of an inert atmosphere requires a containment vessel; and (4) time required varies from seconds to 20 h.

Forge Welding

Forge welding was the first form of welding and for many centuries was the only one in general use. Briefly, the process consists of heating the metal in a forge to a plastic condition and then uniting it by pressure. The heating is usually done in a coal- or coke-fired forge, although modern installations frequently employ oil or gas furnaces. The manual process is limited to light work, because all forming and welding is accomplished with a hand sledge. Before the weld is made, the metal is prepared in a process known as *scarfing* and the pieces are formed to correct shape. This permits parts to unite at the center first when they are welded. As they are hammered together from the center to the outside edges, any oxide or foreign particles are forced out.

Forge welding is slow, and there is danger of an oxide scale forming on the surface. Oxidation can be counteracted somewhat by using a thick fuel bed and by covering the surfaces with a fluxing material that dissolves the oxides. Borax in combination with sal ammoniac is commonly used. Heating must be slow when there are unequal section thicknesses. When heated to a desired temperature, the parts are removed to the anvil and pounded together.

For this type of welding, low-carbon steel and wrought iron are recommended, because they have a broad welding temperature range. Weldability decreases as the carbon content increases.

Friction Welding

In friction welding coalescence is produced from the heat generated by rotating one piece against another under controlled axial pressure. Although this process is sometimes considered a solid-state welding process, the two surfaces actually reach the melting temperature and the adjacent material becomes plastic. At this point the relative motion between the two is stopped and a forging pressure is applied, upsetting the joint slightly. This pressure may equal or exceed the pressure during heating, and its value depends on the surface material being welded. The flash developed during the process carries out surface oxides and impurities from the joint.

In the friction welding process, no special preparation of the weld surfaces is necessary except to have them clean and reasonably smooth. Equipment must be capable of holding one piece securely while the other rotates, then stopping rotation quickly when the welding temperature is reached. Figure 13.25 shows diagrammatically the setup for welding two round shafts. A similar arrangement is used in welding tubes. Other setups permit

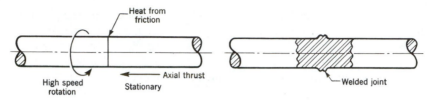

Figure 13.25
Illustrating a process that uses heat generated by friction to produce a weld.

welding of rods or tubes abutted to a flat surface. Rotational speeds and contact pressure depend on the work size and type of material. For example, recommendations for a 10-in. (25-mm) carbon steel bar call for a relative rotational speed of 1500 rpm and an axial pressure of 1500 psi (10 MPa), while the same size stainless-steel rod requires 3000 rpm and 12,000 psi (85 MPa).

The primary limitation of the process is that few configurations can be welded. Advantages claimed for the process are simple equipment, rapid production of sound welds, little preparation required for joints, and low total energy requirements. Also, many dissimilar metals can be welded and the proper welding cycle easily programmed into the machine. Friction welding finds considerable use in welding plastics.

13.6 SPECIAL WELDING PROCESSES

Induction Welding

Coalescence in induction welding is produced by the heat obtained from the resistance of the weldment to the flow of an induced electrical current. Pressure is frequently used to complete the weld. The inductor coil is not in contact with the weldment; the current is induced into the conductive material. Resistance of the material to this current flow results in the rapid generation of heat.

In operation a high current is induced into both edges of the work close to where the weld is to be made. Heating to welding temperature is extremely rapid and the joint is completed by pressure rolls or contacts.

A form of induction welding known as *high-frequency welding* is similar except that the current is supplied to the conductor being welded by direct contact. Frequencies ranging from 200,000 to 500,000 cps (Hz) are used in high-frequency work, whereas frequencies of 400 to 450 cps (Hz) are satisfactory for induction welding of most metals. High-frequency current flows near the surface of the metal and, because heating is almost instantaneous, there is no chance for harmful oxides to form. Vacuum tube oscillators are the source of power for most high-frequency welding.

Induction welding is used successfully for most metals and for some dissimilar metals. Applications include the butt and seam welding of pipe, sealing containers, welding expanded metal, and fabricating various structural shapes from flat stock.

Electron Beam Welding

In electron welding, coalescence is produced by bombarding the workpiece with a dense beam of high-velocity electrons. The metal is joined by melting the edges of the workpiece or by penetrating the material. Usually no filler metal is added. This process is used to

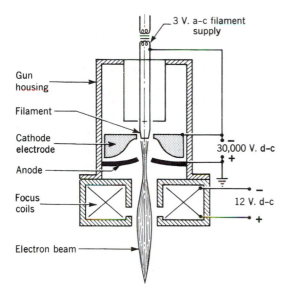

Gun housing

Filament

Cathode electrode

Anode

Focus coils

Electron beam

3 V. a–c filament supply

30,000 V. d–c

12 V. d–c

Figure 13.26
Diagrammatic sketch of electron gun used for electron beam welding.

join common metals, refractory metals, highly oxidizable metals, and various super alloys that previously have been impossible to weld.

Figure 13.26 is a diagrammatic sketch of an electron gun. This gun is placed within a vacuum chamber and arranged so that it may be raised, lowered, or moved in a horizontal plane. The gun can be positioned while the chamber is evacuated prior to the welding operation. After the chamber is evacuated to a pressure around 10^{-4} mm Hg, the beam circuit is energized and directed to the desired spot on the weldment. The beam generally remains stationary, and the work is moved at a desired speed past the electron beam. The temperature range of this electron gun is sufficient to vaporize tungsten or any known material.

In this process heat energy is imparted to the weldments at a much higher rate than it can be conducted away from the electron beam zone, resulting in a high depth-to-width ratio (up to 20:1). Figure 13.27 is a photomicrograph of a 0.5-in. (12-mm) joint in 2024-T4 aluminum that was welded at 32 in./min (2 m/s). Three-inch 5083 aluminum has been welded at 30 kV, 500 mA, and 15 in./min (0.9 m/s). Adjacent material is almost

Figure 13.27
Photomicrograph of a 0.5-in. (12.7-mm)-thick joint in 2024-T4 aluminum accomplished by electron beam welding. Magnification ×4.

unaffected because of the high welding speeds. This characteristic makes the process suitable for numerical control welding of many electronic components such as the "cans" for condensers. In this case, several units are placed in a special holding fixture and, after evacuation, the beam is directed around each unit at high speed to effect the weld.

Electron beam welding is performed in air, sometimes in near vacuum conditions or, with some limitations, under a blanket of inert gas. The electron beam is formed in a chamber arrangement similar to the vacuum machine, then passes through a special orifice, and finally through argon or helium to the workpiece. The maximum effectiveness of the beam is 1 in. (25 mm), with a workpiece limitation of ½ in. (12.7 mm). Although welding speed is increased, welds are not free from contamination and the weldment size is much smaller than those obtained by the vacuum method. The nonvacuum unit supplements the conventional vacuum-welding process and increases the range of welding that can be done with electron beam welding.

Laser Welding

The theory of the various types of lasers is beyond the scope of this book, but a brief description of an optical laser and its range of capabilities is given in Chapter 26. For a welding application, the laser is finely focused as a high-collimated beam of photons that is referred to as a *coherent beam*. This monochromatic beam is capable of delivering up to 30,000 W/in.[2]. It allows deep penetration welds without affecting the base metal because the energy bond is small and because broad heating does not take place. Thus, either thin or thick materials are welded successfully. Joint designs for laser welds are similar to those discussed earlier except that tolerances are tighter and the surface finish is better because of the accuracy of the laser beam. This type of welding makes excellent, precise welds, usually no more than 0.001 in. (0.025 mm) wide, on all metal (even dissimilar ones), and the base metal temperature rises almost imperceptibly. The most common laser in welding is the CO_2 type, which can weld $\frac{1}{32}$-in. (0.8-mm)-thick stainless steel. New gas dynamic lasers can weld up to ¾-in. (19-mm)-thick stainless. The principal disadvantages of this process are cost and size of equipment and the safety hazards associated with high-energy lasers.

An example of laser equipment, shown in Figure 13.28, is an 80-W unit with a wavelength of 10.6 λ. The lasing material is a mixture of carbon dioxide, helium, and nitrogen. This unit accomplishes soldering, welding, and cutting as well as other operations such as drilling, slitting, and perforating.

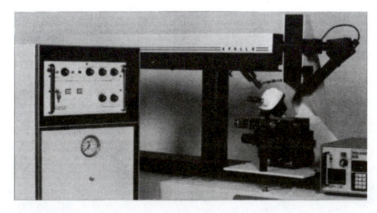

Figure 13.28
Apollo industrial laser.

Thermit Welding

Thermit welding, named after the company that invented it, is the only welding process employing an exothermal chemical reaction for the purpose of developing a high temperature. It is based on the fact that aluminum has a great affinity for oxygen and can be used as a reducing agent for many oxides. The usual Thermit mixture or compound consists of finely divided aluminum and iron oxide mixed at a ratio of about 1:3 by weight. The iron oxide is usually roll mill scale. This mixture is not explosive and can be ignited only at a temperature of about 2800°F (1500°C), so a special ignition powder is used to start the reaction. The chemical reaction requires about 30 s and attains a temperature of around 4500°F (2500°C). The mixture reacts according to the chemical equation

$$8Aln + 3Fe_3O_4 \rightarrow 9Fe + 4Al_2O_3$$

The resultant products are highly purified iron (actually steel) and an aluminum oxide slag that floats on top and is not used. Other reactions also take place, since most Thermit metal is alloyed with manganese, nickel, or other elements.

Figure 13.29 illustrates the method of preparing the material for such a weld. Around the break, where the weld is to be made, a wax pattern of the weld is built up. Refractory sand is packed around the joint and necessary provision is made for risers and gates. A preheating flame is used to melt and burn out the wax, to dry the mold, and to bring the joint to a red heat. The reaction is then started in the crucible and, when it is complete, the metal is tapped and allowed to flow into the mold. Because the weld metal temperature is approximately twice the melting temperature of steel, it readily fuses in the joint. Such welds are sound because the metal solidifies from the inside toward the outside, and all air is excluded from around the mold.

There is no limit to the size of welds that can be made by Thermit welding. It is used primarily for repairing large parts that would be difficult to weld by other processes.

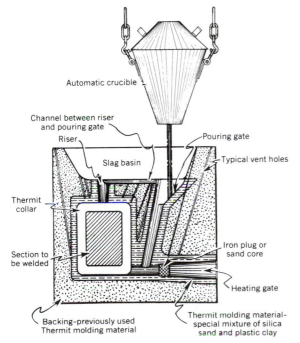

Automatic crucible

Channel between riser
and pouring gate
Riser
Pouring gate
Slag basin
Typical vent holes
Thermit
collar
Section to
be welded
Iron plug or
sand core
Heating gate
Backing-previously used
Thermit molding material
Thermit molding material-
special mixture of silica
sand and plastic clay

Figure 13.29
Line drawing of mold and crucible for a Thermit weld.

Flow Welding

Flow welding is defined as a welding process in which coalescence is produced by heating with molten filler metal poured over the surfaces until the welding temperature is attained and the required filler metal has been added. Flow welding is used in joining thick sections of nonferrous metals using a filler of the same composition as the base metal.

In operation the weld area is first properly prepared and preheated. Molten filler metal is then poured between the ends of the material until melting starts. At this point the flow is stopped and the joint, filled with metal, is allowed to cool slowly. To ensure the fusion of the top edges and provide a good weld, the level of the molten filler metal is kept higher than the surfaces being welded.

13.7 WELDING QUALITY AND SAFETY

The integrity of a weld is critical in the reliability of an assembly. Any fusion discontinuity alters stresses in direct proportion to the area change in the joint creating a potential for an accident of major proportions. For this reason it is important to be aware of the kinds of discontinuities and cracks affecting a welding system.

Cracks and fusion discontinuities can be related to the welding process, joint design, metallurgical behavior of the base metal, or the combination of all three. *Incomplete fusion*, for example, often results from poor joint preparation, wrong joint design, or incorrect weld parameters. *Porosity* results from an oily, wet, or dirty base metal; insufficient gas shielding; or dirt and heavy oxide on the filler rod. *Poor penetration*, however, results from too little heat input. Welding on cool base material requires low travel speed; as the workpiece absorbs heat and temperature rises, speed may increase. Table 13.1 presents a summary of discontinuities, and Figure 13.30 illustrates some discontinuity examples for a fillet and a butt profile weld.

Discontinuities are identified by several methods. *Visual inspection* reveals incorrect weld size, poor penetration, discolorization, surface porosity, and cracks. *Radiography* detects inclusions, internal cracks, lack of fusion, and internal porosity. *Liquid penetrant,*

Table 13.1 **Fusion Discontinuities in a Welded Joint**

Type	Discontinuity	Processes Most Affected
Process related	Inadequate penetration, undercuts, overlaps, slag inclusions	Arc welding processes and Thermit welding
	Incomplete fusion	All processes
	Warpage, shrinkage	All processes
	Burn-throughs	All processes

Metallurgical behavior	Cracks
	    Hot
	    Cold
	    Lamellar tearing
		All processes except for explosion, friction, and ultrasonic welding

	Porosity	Arc welding and gas welding processes
	Base plate laminations	All processes
	Heat alteration of the microstructure	All processes
Design related	Weld joint type	All welding processes
	Changes in section	All processes
	Stress concentration	

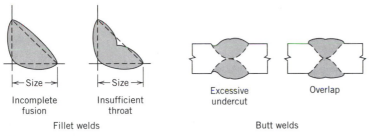

Figure 13.30
Unacceptable weld profiles.

drawn into surface defects by capillary action, reveals flaws that are open to the surface. Other nondestructive inspection methods, explained in Chapter 26, include magnetic particle, ultrasonic, and eddy current.

Safety is a major concern in welding. Gas tungsten arc welding, for example, performed with argon shielding, is the highest emitter of ultraviolet radiation and would burn the cornea and the skin if proper protection is not worn. A welding helmet with a number 12 safety lens for welding of ferrous material and number 11 for nonferrous alloys is mandatory for arc welding. Goggles with a number 2 lens and side shields underneath the helmet are recommended as well. Opaque shields or tinted curtains of polyvinyl-chloride plastic sheet should enclose the welder to protect bystanders from exposure to radiation.

Beyond head protection, welders should cover themselves with dark-colored leather or wool to ensure total skin protection. This must be done to ensure that natural ventilation is provided. In confined spaces, local-exhaust ventilation and respiratory protection is required. Otherwise, toxic fumes like ozone, nitrogen dioxide, and carbon monoxide pose risk of injury to welders' respiratory systems.

In handling gas cylinders and electrical equipment, welders should observe the following precautions: store gas cylinders upright in a safe area, shielding gas at the welding area should be limited to one day's consumption, 40 lb or more cylinders should be moved with a cart or truck, safety cap and valves must be secured tightly when cylinders are not in use, power supplies should be turned off when changing electrodes to prevent electrical shocks, cables and lines should be checked regularly to identify cracks, frayed cable covers, and worn or missing insulators.

13.8 OTHER JOINING PROCESSES

Adhesive bonding, soldering, and brazing are processes that have wide commercial use in uniting small assemblies and electrical parts. Adhesive bonding allows joining of dissimilar materials, whereas soldering and brazing unite metals with a third joining metal introduced into the joint in a liquid state and allowed to solidify.

Soldering

In soldering, two pieces of metal are joined with another metal that is applied between the two in a molten state. In this process, a little alloying with the base metal takes place and additional strength is obtained by mechanical bonding. Lead and tin alloys having a melting range of 350° to 700°F (180° to 370°C) are principally used. The strength of the joint is determined largely by the adhesive quality of the alloy, which never reaches

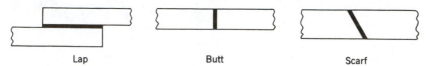

Figure 13.31
Common welding and brazing joints.

the strength of the materials being joined. Because solder has little strength, the solder should be as thin as possible. Due to the importance of soldering in electronic manufacturing, Chapter 19 discusses this process in more detail.

Brazing

In *brazing* a nonferrous alloy is introduced in a liquid state between the pieces of metal to be joined and allowed to solidify. The *filler metal*, having a melting temperature of more than 840°F (450°C) but lower than the melting temperature of the parent metal, is distributed between the surfaces by capillary action. *Braze welding* is similar to ordinary brazing except that the filler metal is not distributed by capillary attraction. In this process, filler metal is melted and deposited at the point where the weld is to be made. In both, special fluxes are required to remove surface oxide and to give the filler metal the fluidity to wet the joint surfaces completely. Alloys of copper, silver, and aluminum are the most common brazing filler metals.

The basic joint types found in brazing are the lap, butt, and scarf designs, as shown in Figure 13.31. Many modifications of these joints are used in production brazing, depending on the shape of the parts to be joined. Of the three joints shown, the lap joint is the strongest because it has greater contact area. Joint clearance is important, because sufficient space must be allowed for capillary attraction to provide filler metal distribution. When joining two pieces the lap or butt should be three times the thickness of the smaller piece. To determine the correct length of a lap the following formulas are used:

$$\text{For a flat joint:} \qquad L = \sigma t\, FS/\tau \qquad\qquad (13.11)$$

$$\text{For a tubular joint:} \quad L = \sigma t(D - t)\, FS/\tau \qquad (13.12)$$

where

L = Length of lap, in. (mm)
FS = Factor of safety
σ = Tensile strength of thinner member, psi (Pa)
t = Wall thickness of thinner member, in. (mm)
τ = Shear stress of braze filler metal, psi (Pa)
D = Diameter of lap, in. (mm)

EXAMPLE 13.2

It is required to braze a 0.025-in. plate to a 0.05-in. plate of the same type of metal. If their tensile strength is 50 ksi and the filler material has a shear strength of 5 ksi, what is the length of a lap joint with a safety factor of 1.2?

Equation (13.11), $L = \sigma t\, FS/\tau$, is used to obtain the lapping length $L = 50 \times 0.025 \times 1.2/5 = 0.3$ in.

Table 13.2 **Brazing and Braze Welding**

Dipping	Furnace	Torch	Electric	Welding
1. Metal	1. Gas	1. Oxyacetylene	1. Resistance	1. Torch
2. Chemical	2. Electric	2. Oxyhydrogen	2. Induction	2. Arc
			3. Infrared	

In any brazing operation the joint first must be cleaned of all dirt, oil, or oxides. Then the pieces are fitted together with appropriate clearance for the filler metal. Mechanical or chemical cleaning may be necessary in the joint preparation in addition to the flux used during the process. Borax either alone or in combination with other salts is commonly used as a flux.

To facilitate speed in brazing, the filler metal is prepared frequently in the form of rings, washers, rods, or other special shapes to fit the joint being brazed. This ensures having the proper amount of filler metal available for the joint as well as having it placed in the correct position.

The processes that have wide commercial use in uniting small assemblies and electrical parts are listed in Table 13.2. Advantages of the brazing process are the effecting of joints in materials difficult to weld, in dissimilar metals, and in exceedingly thin sections of metal. In addition, the process is rapid and results in a neat-appearing joint requiring a minimum of finishing. Brazing is used for the assembly of pipes to fittings, carbide tips to tools, radiators, heat exchangers, electrical parts, and the repair of castings.

Adhesive Bonding

Adhesive bonding is rapidly replacing other joining operations because (1) the operation can be more economical, (2) machining operations such as threading can be eliminated, (3) there is no change in material properties or surface roughness, (4) dissimilar materials can be joined, (5) lighter gage materials can be employed, and (6) the elastomeric nature of some adhesives give shock and vibration protection. The principal disadvantages are that (1) the surfaces to be joined usually must be chemically or mechanically scrubbed; (2) some adhesives are toxic and flammable; (3) the joints assembled are seldom effective at more than 400°F (240°C) and may start losing strength at much lower temperatures; (4) the use of adhesives has been difficult to incorporate in high-production lines, primarily because of curing time; and (5) some of the adhesives are unstable and have a short pot or shelf life.

The surface finish of the part to be bonded depends to a large extent on the ability of the plastic to wet the surface without asperities or air voids; cleanliness is usually more important than surface finish. Often the best surfaces are obtained by sandblasting or bead blasting followed by ion etching, which is accomplished by bombarding the surface with inert-gas ions.

Most of the plastics used for bonding are discussed in more detail in Chapter 18. Thermosetting plastics are used more often than thermoplastics because of the temperature softening characteristics of the latter. Design for adhesive bonds should be such that the joint is shear rather than tension. The ideal joint is shown in Figure 13.32 as a bevel lap joint. There is less stress concentration and less cleavage stress at the bond's edges. A joint with less adhesive area may be stronger than one with a larger area because of this phenomenon, but, in general, the more area involved the greater the strength. It is preferable to increase the width of the joint rather than the overlap for maximum strength

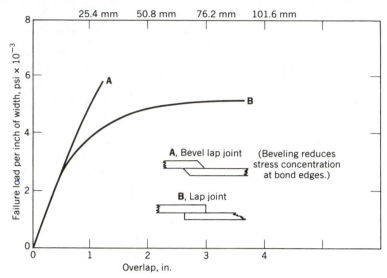

Figure 13.32
Strength and effect of adhesive joints.

per unit of bonded area. Most of the adhesives have a shear strength from 3500 to 7000 psi (25 to 28 MPa), although some heat-cured ones have a shear strength of more than 10,000 psi (68.9 MPa). The materials most used are: *epoxies* (one and two part), *anaerobics* (one part), *acrylic* (one part), *cyanoacrylates* (one part), and *urethanes* (one part thermoplastic and two part thermosetting). The one-part types usually require temperatures on the order of 300°F (149°C) for curing or an absence of oxygen. Many of the newer adhesives are made in films or pressure-sensitive tapes.

13.9 ALLIED PROCESSES

Welding technology and equipment are transferred successfully to other applications, particularly in metal cutting, as explained below.

Oxyacetylene Torch Cutting

Cutting steel with a torch is an important production process. A simple hand torch for flame cutting differs from a welding torch in that it has several small holes for preheating flames surrounding a central hole through which pure oxygen passes. The preheating flames are exactly like the welding flames and are intended to preheat the steel before cutting. The principle of flame cutting is that oxygen has an affinity for iron and steel. At ordinary temperatures this action is slow, but eventually an oxide in the form of rust materializes. As the temperature of the steel is increased, this action becomes more rapid. If the steel is heated to a red color, about 1600°F (870°C), and a jet of pure oxygen is blown on the surface, the action is almost instantaneous and the steel is actually burned into an iron oxide.

The amount of oxygen at regulated pressure necessary to burn or remove a cubic inch of iron can be expressed approximately as

$$V_o = 1.3 \times V_s \qquad (13.13)$$

where

$$V_o = \text{Oxygen required, ft}^3 \text{ (mm}^3)$$
$$V_s = \text{Amount of steel removed, ft}^3 \text{ (mm}^3)$$

or

$$V_o = 1.3 \times d \times l \tag{13.14}$$

where

$$d = \text{Thickness of steel, ft (mm)}$$
$$l = \text{Length of cut, ft (mm)}$$

Metal up to 30 in. (762 mm) in thickness is cut by this process.

Underwater cutting torches are provided with connection for three hoses: one for preheating gas, one for oxygen, and one for compressed air. The latter provides an air bubble around the tip of the torch to stabilize the flame and displace the water from the tip area. Hydrogen gas is generally used for the preheating flame, as acetylene gas is not safe to operate under the high pressures necessary to neutralize the pressure created by the depth of water.

Many *cutting machines* have been developed that automatically control the movement of the torch to cut any desired shape. Such a machine cutting several parts simultaneously is shown diagrammatically in Figure 13.33. In all such machines, some control or sensing device is provided to guide the torches along a predetermined path. This control in its simplest form may be a hand-guided pointer following a drawing or held against a template. Modern machines are driven electrically and are provided with magnetized drive spindles that follow a steel template and control the movement of the machine at the proper cutting speed. Also, there are electronically controlled sensing devices provided with an electric eye capable of following a line drawing, thus eliminating the need for template construction.

Numerical control can be adapted to cutting machines to provide greater accuracy and productive output in flame-cutting operations. All functions of the machine such as speeds, control of preheating, sequence of cutting, piercing, regulation of torch height, and travel from one piece to the next can be programmed. These operations are translated into control language from simple sketches as explained in Chapter 28.

Many parts that previously required shaping by forging or casting are now cut to shape by this process. Flame-cutting machines, which replace many machining operations where accuracy is not paramount, are used widely in the shipbuilding industry, structural fabrication, maintenance work, and the production of numerous items made from steel

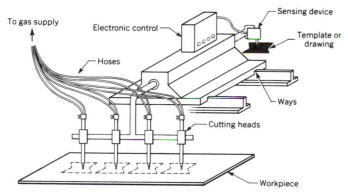

Figure 13.33
Schematic sketch of oxyacetylene torch cutting machine.

sheets and plates. Cast iron, nonferrous alloys, and high-manganese alloys are not readily cut by this process.

Transferred-Arc Cutting

In a plasma torch gas is heated by a tungsten arc to such a high temperature that it becomes ionized and acts as a conductor of electricity. In this state, the arc gas is known as *plasma*. The torch generally is designed so that the gas is confined closely to the arc column through a small orifice. This increases the temperature of the plasma and concentrates its energy on a small area of the workpiece, which melts the metal rapidly. As the gas jet stream leaves the nozzle, it expands rapidly, removing the molten metal continuously as the cut progresses. Because the heat obtained does not depend on a chemical reaction, this torch can be used to cut any metal. Temperatures approach 60,000°F (33,000°C), roughly 10 times that possible by the reaction of oxygen and acetylene.

Plasma-generating torches are of two general designs, one known as a *transferred plasma torch* and the other as a *nontransferred plasma torch*. In the nontransferred plasma torch, the arc circuit is completed within the torch and the plasma is projected from the nozzle. Such torches utilized for metal spraying are described in Chapter 22. Transferred-arc plasma torches used for cutting are diagrammed in Figure 13.34. The workpiece becomes the anode and the arc continues to the workpiece in the jet of gas. With the workpiece as one of the electrodes, the intensity of heat transfer and efficiency are increased, making it more suitable for metal cutting than the nontransferred plasma torch.

The Heliarc cutting torch, illustrated in Figure 13.34A, does not constrict the arc and is used for either welding or cutting most common metals. When used as a cutting torch, an argon–hydrogen gas mixture is used and the current density is increased over good welding conditions. Good kerf quality is obtained on one side of the cut, but the cutting thickness is limited to ½ in. (12.7 mm). For high-speed accurate cutting the constricted-arc torch, as illustrated in Figure 13.34B, is more satisfactory. The arc is constricted in a narrow opening at the end of the torch, creating a high-velocity arc that readily melts a narrow kerf through both ferrous and nonferrous metals. Higher temperatures are obtained than in the nonconstricted torch, and any metal up to 4 in. (100 mm) thick is cut with this method.

While gases used in plasma torch cutting include argon, hydrogen, and nitrogen, a combination of argon and nitrogen gives the best results. For manual operation, the arc is started in an atmosphere of argon followed by the right proportion of hydrogen blended

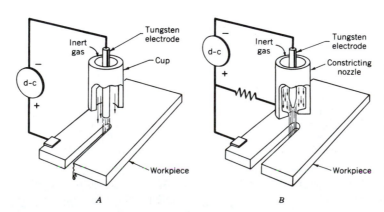

Figure 13.34
Schematic comparison of the two basic gas tungsten electrode arc cutting processes. *A*, Heliarc cutting—nonconstricted transferred-arc cutting. *B*, Plasma arc cutting—constricted transferred-arc cutting.

into the gas stream. A mixture of 80% argon and 20% hydrogen is used in cutting operations up to 400 A. For higher current proportion of 65% and 35% are recommended. Nitrogen is recommended only for the mechanized cutting of stainless steels, making sure that an exhaust system is provided to extract the toxic fumes.

Plasma arc, either in manual or mechanized operations, is particularly useful in cutting aluminum, stainless steel, copper, and magnesium. The process has little effect on the metallurgical characteristics or physical properties of adjacent metal because of the rapidity of its action.

QUESTIONS

1. Give an explanation of the following terms:

Straight polarity	Acetylene gas
Flux	Sonotrode
Electrode coatings	Diffusion welding
TIG welding	Scarfing
Electroslag welding	Thermit metal
Hold time	Filler metal
Roll spot welding	Plasma
Electrolysis	

2. Define the term "coalescence." How is it obtained in electric-arc welding and percussion welding?

3. Which welding joint is the strongest? Why?

4. What effect does carbon content of steel have on weldability?

5. Describe the process of furnace brazing two parts.

6. List the distinctive steps required for a spot weld.

7. How is acetylene made? Describe the differences between oxygen and acetylene cylinders.

8. Of the types of welded joints shown in Figure 13.1, which is suitable for spot welding?

9. What type of oxyacetylene flame is used for welding

a. Monel?

b. Nickel?

c. Steel?

d. Bronze?

10. Compare the processes of laser, electron beam, and ultrasonic welding.

11. What welding temperatures are attained in oxyhydrogen welding, oxyacetylene welding, plasma arc cutting, electric-arc welding, and Thermit welding?

12. Describe the chemical process that enables the cutting of ferrous materials with oxyacetylene.

13. How does the oxyacetylene cutting torch operate?

14. What are the advantages and disadvantages of an a-c welding machine?

15. What are the advantages and disadvantages of a d-c welding machine?

16. Describe the applicability of oxyacetylene cutting to gray cast iron.

17. What are the purposes of a coating on an arc-welding electrode?

18. How is underwater oxyacetylene cutting accomplished?

19. How does the exothermic chemical reaction apply to Thermit welding?

20. What are the hazards associated with oxyacetylene welding, electron beam welding, and laser welding?

21. List, sketch, and discuss the five zones of heat generation in a spot weld.

22. What process do you recommend for each of the following: butt weld band saw blades; lap weld foil material; welding two 12-in. (305-mm) diameter ferrous alloys; welding a crack in an aluminum automobile crankcase?

23. What are the materials for the spot-welding electrodes?

24. How and why are spot-welding electrodes cooled?

25. Describe how two cylindrical cylinders can be joined together, one inside the other, by explosive welding.

26. Explain the high-velocity jet phenomenon in explosive welding.

27. Why do welds made in steel tend to harden otherwise soft materials?

28. What safety precautions are necessary when cutting stainless steel within a nitrogen atmosphere?

29. What causes weldments to crack? Explain the reasons and suggest remedies.

30. What happens to weldability of steel as the carbon content is increased?

31. How can a lathe be equipped to weld short pieces of ¼-in. (6.3-mm) round bar stock end-to-end frictionally?

32. What advantages do coated electrodes give to electrode arc welding? How does manual skill prevent problems in arc welding?

33. What is contact electrode? List its advantages.

34. Give a range in seconds to spot weld.

35. What are the principal disadvantages of diffusion welding?

36. Describe five safety precautions needed during welding.

37. Name the sources of discontinuities in welded joints and give some examples. What detection methods do you know?

38. How does the strength of an adhesive bond compare with diffusion welding?

39. What is the principal reason that prevents adhesive joining of ceramic-to-titanium heat shields for spacecraft?

40. What are the commercial and economical applications for electron beam welding?

PROBLEMS

13.1. How many watts (volts × amperes) are necessary to weld ¼ in. (6.4 mm) diameter with a spot welder?

13.2. An electron beam welding machine costs about $300,000. What applications would justify its use?

13.3. Describe the steps in oxyacetylene cutting a ½-in. steel plate. If the kerf is ¼ in. wide, how many pounds of steel are wasted in cutting 1000 part of 24-in. circles?

13.4. There are three diffusion welding stations. One welds steel, the second one welds aluminum, and the last one welds titanium. You can choose one station to spend one week of training. Which one would you select? Why? (Hint: Melting point of metals is given in Table 2.1.)

13.5. Specify the projection dimensions on stock 0.125 in. (3.18 mm) thick that is to be projection welded.

13.6. If 3500 ft³ of regulated acetylene are used, determine the heat generated in the inner cone and in the outer envelope of the oxyacetylene flame.

13.7. A 1-in. solid ceramic shaft needs to be attached to a 1-in. solid aluminum shaft in end-to-end fashion. What adhesive joint design would you recommend? Which would probably be more expensive, diffusion welding or adhesives?

13.8. If on each pass of a ⅛-in. (3.18-mm)-diameter, 12-in. (305-mm)-long steel welding rod, 1 in. (25.4 mm) of rod is used for every ½ in. (12.7 mm) of weld, how many pounds of rod are used in making four passes on 600 lineal ft of welding? The last 2 in. of each rod are unusable.

13.9. A 4-ft-long steel bar is welded to a structure at one end. The free end supports a 250-lb load. What is the deflection if the cross section of the bar is 1 in. wide by 2 in. tall? (Area moment of inertia = $bh^3/12$.)

13.10. A manufacturer wants to lap braze a white cast-iron tube to a 0.04-in.-thick lug. If the chosen filler material has a shear strength of 2.5 ksi and the diameter of the tube is 1 in., estimate the lap length using a 1.1 factor of safety. (Hint: Tensile strength values are given in Table 2.1.)

13.11. Calculate the amount of oxygen to cut 1000 pieces of 30-in. (76-mm) by 1¼-in. (318-mm) steel plate 23 in. (538 mm) long. Assume that the cut is ⅜ in. (9.5 mm) wide.

MORE DIFFICULT PROBLEMS

13.12. Calculate the heat generated in a spot weld if the current is 300 A, the resistance is 2400 ohms, and the time is 20 cycles of 60-cycle current. Plot a curve for the heat generated as a function of current if the current is 100, 150, 200, 250, 300, and 350 A.

13.13. A V-groove weld requires a depth of preparation equal to 75% of the weld size. The groove angle is 45°, the groove weld size is half the thickness of the base metal, and the root opening is of 3 mm. Sketch the weld and the welding symbol.

13.14. A T-joint is formed by two steel plates each ¼ in. thick and 8 in. wide.

a. If the plates are welded with fillet welds of equal legs and the size of the weld is ⁵⁄₁₆ in., sketch the weld and the welding symbol.

b. If GMAW is used, with a consumable electrode's allowable tensile strength of 20 ksi, and the maximum load expected at the joint is 60,000 lb, what should the size of the weld be at each edge? What leg size alternatives would you recommend?

PRACTICAL APPLICATION

A 4- by 10- by ½-in. mild steel plate is butt-welded to a 4- by 12- by 20-in. high-carbon steel casting where the 10 by ½-in. dimension is the connecting edge. Two thousand assemblies are welded each month in this industrial operation. Recommend a welding method with safety procedures. Think through your own ideas, but follow up by calling a welding fabricator for suggestions.

CASE STUDY: WELDING SPACECRAFT HEAT SHIELDS

In your capacity as a designer in the Godeep Aerospace Company, a new heat shield for a spacecraft is being designed. It requires the welding of a thin section of titanium to a thick aluminum plate.

The thickness of titanium is 0.031 in. and the aluminum is 0.75 in. The unit makes a 12- by 12-in. plate about 1.06 in. thick. Six thousand units are required. Units are curved, and buckling and blisters must be avoided.

Describe a method for producing these heat shields. Sketch the production line and indicate the equipment.

CHAPTER 14

HOT WORKING
OF METAL

An ingot of steel has little commercial value until it is formed into other products; nor is continuous strand casting of molten steel a concluding operation. Similarly, a cold ingot cannot be converted economically into other shapes. Hot working is the important process that continues the conversion of the molded ingot or the continuous cast strand into shapes that are widely used.

Hot working is the shaping of an *ingot* or *strand* into structural products, bar stock, sheet or plate. Using intermediate shapes, hot working is a reforming process. With the ingot, strand, or the intermediate shape hot, they are hammered, pressed, rolled, or extruded into other shapes.

Scaling and oxidation exist at the high temperatures of hot working, and most ferrous metals are finish processed later by cold working to obtain improved surface finish, higher dimensional accuracy, and better mechanical properties. This chapter involves forming materials, mostly steel, at temperatures above the *recrystalline temperature*. These significant processes are called hot working and are different from shaping materials by casting, molding, machining, or welding.

The two principal types of mechanical work in which material undergoes *plastic deformation* and is changed in shape are hot working and cold working. Cold working is discussed in Chapter 15.

14.1 PLASTIC DEFORMATION

As with other metallurgical practices, differences between hot and cold working are not easy to define. When metal is hot worked, the forces for deformation are less and the mechanical properties are relatively unchanged. When a metal is cold worked, greater forces are required and the strength of the metal is increased. In hot working the thickness of the material is changed substantially, but in some cold working operations, such as the finish rolling of sheet metal, the thickness remains approximately the same.

In the manufacture of metal components, the basic alternatives available for the production of a designed shape include casting, machining, welding, and deformation processes. Hot working is a deformation process. *Metal deformation* exploits an interesting fact of metals: their ability to *flow plastically* in the solid state without accompanying deterioration of properties. Moreover, in forcing the metal into a desired shape there is little or no waste of material.

Hot working is the plastic deformation of metals above their recrystallization temperature, which varies with different materials. Hot working does not necessarily imply high absolute temperature. For example, lead and tin are hot worked at room temperature.

Recrystallization temperatures of common metals are given as

Recrystallization Temperatures of Metals

Metal	°F (°C)
Aluminum	300 (150)
Copper	390 (200)
Gold	390 (200)
Iron	840 (450)
Lead	Below room temperature
Magnesium	300 (150)
Nickel	1100 (590)
Silver	390 (200)
Tin	Below room temperature
Zinc	At room temperature

Although hot working causes plastic deformation above the recrystallization temperature, it does not produce *strain hardening*. Also, the hot worked metal does not possess a greater elastic limit or become stronger, and the metal usually experiences a decrease in yield strength; that is, a point where additional strain occurs without any increase in stress load on the material. *Ductility*, which is the ability of a material to be deformed plastically without fracture, is impaired. Thus, it is possible to alter the shape of the metals drastically with moderated forces by hot working and without causing fracture.

The recrystallization temperature of a metal determines whether or not hot or cold working is being accomplished. For steel, recrystallization starts around 950° to 1300°F (500° to 700°C), although most hot working of steel is at temperatures considerably above this range. Variations are caused by various alloying effects. There is no tendency for hardening by mechanical work until the lower limit of the recrystalline temperature range is reached. Some metals, such as lead and tin, have a low recrystalline range and can be hot worked at room temperature. Most commercial metals, however, require heating. Alloy composition of these metals influences the proper working temperature range. The typical result raises the recrystalline range temperature; this range also is increased by prior cold working.

During hot working operations the metal is in a plastic state and is formed readily by pressure. In addition, hot working has the following advantages:

1. Porosity in the metal is largely eliminated. Most ingots contain many small blow holes. These are pressed together and eliminated.

2. Impurities in the form of inclusions are broken up and distributed throughout the metal.

3. Coarse or columnar grains are refined. Since this hot work is in the recrystalline temperature range, it should be continued until the low limit is reached to provide a fine grain structure.

4. Physical properties are generally improved owing principally to grain refinement. Ductility and resistance to impact are improved, strength is increased, and greater homogeneity is developed in the metal. The greatest strength of rolled steel exists in the direction of metal flow.

5. The amount of energy necessary to change the shape of steel in the plastic state is far less than that required when the steel is cold.

Hot working processes present a few disadvantages that cannot be ignored. Because of the high temperature of the metal, there is rapid oxidation or *scaling* of the surface with accompanying poor surface finish. As a result of scaling, close tolerances are not practical. Hot working equipment and maintenance costs are high, but the process is economical if compared to working metals at low temperatures and the objective of the operation is similar.

The term *"hot finished"* refers to steel bars, plates, or structural shapes that are purchased in the "as rolled" condition from the hot working operation. Some descaling is done but, otherwise, the steel is ready for bridges, ships, railroad cars, and other applications where close dimensional tolerances are not required. The material has good weldability and machinability because the carbon content is less than 0.25% for these products.

14.2 ROLLING

Wrought steel is converted to useful products in two steps:

1. The molded ingot or the continuous strand is rolled into intermediate shapes— *blooms, billets,* and *slabs*.
2. These intermediate slabs are rolled further into plates, sheets, bar stock, structural shapes, strips, expanded metal grates, and the like.

The ingot remains in the mold until the solidification is about complete and the mold is stripped from the ingot. While still hot, the ingots are placed in gas-fired furnaces called *soaking pits*, where they remain until they have attained a uniform working temperature of about 2200°F (1200°C) throughout. The ingots are moved to the rolling mill where, because of the variety of finished shapes to be made eventually, they are rolled first into intermediate shapes such as blooms, billets, or slabs.

Remember, however, that continuous casting eliminates some of the requirements for the ingot process. The *continuous cast strand* also follows the same rolling processes. The two, ingot casting and strand casting, have similar rolling operations to the final end product.

A bloom has a square cross section with a minimum size of 6 by 6 in. (150 by 150 mm). A billet is smaller than a bloom and may have any square section from 1½ in. (38.1 mm) up to the size of a bloom. Slabs may be rolled from either an ingot or a bloom. They have a rectangular cross-sectional area with a minimum width of 10 in. (250 mm) and a minimum thickness of 1½ in. (38.1 mm). The width is always three or more times the thickness, which may be as much as 15 in. (380 mm). Plates, skelp, and thin strips are rolled from slabs.

One effect of a hot working rolling operation is the *grain refinement* brought about by recrystallization, which is shown in Figure 14.1. Coarse grain structure is broken up and elongated by the rolling action. Because of the high temperature, recrystallization starts immediately and small grains begin to form. These grains grow rapidly until recrystallization is complete. Growth continues at high temperatures, if further work is not carried on, until the low temperature of the recrystalline range is reached.

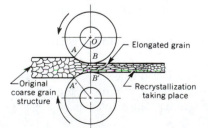

Figure 14.1
Effect of hot rolling on grain structure.

In Figure 14.1, AB and A'B' are the contact arcs on the rolls. The wedging action on the work is overcome by the frictional forces that act on these arcs and draw the metal through the rolls. In the process of rolling, stock enters the rolls with a speed less than the peripheral roll speed. The metal emerges from the rolls traveling at a higher speed than it enters. At a point midway between A and B, metal speed is the same as the roll peripheral speed. Most deformation takes place in thickness, although there is some increase in width. Temperature uniformity is important in all rolling operations since it controls metal flow and plasticity.

In rolling, the quantity of metal going into a roll and out of it is the same, but the area and velocity are changed. Thus,

$$Q_1 = Q_2 = A_1V_1 = A_2V_2 \tag{14.1}$$

where

Q_1 = Quantity of metal going into roll
Q_2 = Quantity of metal leaving the roll
A_1 = Area, ft², of an element in front of roll
A_2 = Area, ft², of an element after roll
V_1 = Velocity, ft/s, in element before the roll
V_2 = Velocity, ft/s, in element after the roll

$$\frac{A_1}{A_2} = \frac{V_2}{V_1} \tag{14.2}$$

In the process of becoming thinner, the rolled steel becomes longer and may become wider, but it is constrained by vertical rolls set to restrict this sideways growth. As the cross-sectional area is decreased, the velocity increases as does the length of the material. For example, a heated slab 7 in. thick weighing more than 12 tons is reduced to a coil of thin sheet in a matter of minutes. The delivery velocity of the hot rolled product may be more than 3500 ft/min at the conclusion of rolling.

Most primary rolling is done in either a two-high reversing mill or a three-high continuous rolling mill. In the two-high reversing mill, Figure 14.2A, the piece passes through the rolls, which are then stopped and reversed in direction, and the operation is repeated. At frequent intervals, the metal is turned 90° on its side to keep the section uniform and to refine the metal throughout. About 30 passes are required to reduce a large ingot into a bloom. Grooves are provided on both the upper and the lower rolls to accommodate the various reductions in cross-sectional area. The two-high rolling mill is versatile, since it has a range of adjustment as to size of pieces and rate of reduction. It is limited by the length that can be rolled and by the inertia forces that must be overcome each time

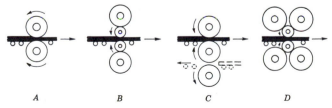

A   B   C   D

Figure 14.2
Various roll arrangements used in rolling mills. *A,* Two-high mill, continuous reversing. *B,* Four-high mill with backing-up rolls for wide sheets. *C,* Three-high mill for back-and-forth rolling. *D,* Cluster mill using four backing-up rolls.

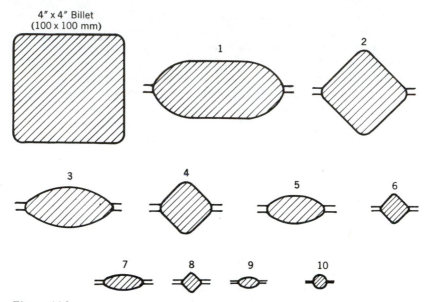

Figure 14.3
Diagram illustrates number of passes and sequences in reducing the cross section of a 4- by 4-in. (100- by 100-mm) billet to round bar stock.

a reversal is made. These are eliminated in the *three-high mill*, Figure 14.2C, but an elevating mechanism is required. The three-high mill is less expensive to manufacture and has a higher output than the reversing mill.

Other arrangements of rolls used in rolling mills are shown in Figure 14.2. Those that have four or more rolls use the extra ones for backing up the two that are doing the rolling. In addition, many special rolling mills take previously rolled products and fabricate them into such finished articles as rails, structural shapes, plates, and bars. Such mills usually bear the name of the product being rolled and, in appearance, are similar to mills used for rolling blooms and billets. A mill specializing in rolling rails is able to roll a rail continuously almost ¼ mile long.

Billets could be rolled to size in a large mill for blooms, but usually this is not done for economic reasons. Frequently, they are rolled from blooms in a continuous billet mill consisting of six or more rolling stands in a straight line. The steel makes but one pass through the mill and emerges with a final billet size, approximately 2 by 2 in. (50 by 50 mm), which is the raw material for many final shapes such as bars, tubes, and forgings. Figure 14.3 illustrates the number of passes and the sequence in reducing the cross section of a 4- by 4-in. (100- by 100-mm) billet to round bar stock.

14.3 FORGING

Forging is defined as the controlled, plastic deformation or working of metal into predetermined shapes by pressure or impact blows, or a combination of both. Forging aligns the grain to the contour of the die, and thus to the part, increasing its strength.

An estimate of the total deformation energy required for standard forgings is given by

$$E_d = 2000\pi \times V \times F_i \times S_f \qquad (14.3)$$

where

E_d = Deformation energy required for forging, ft-lb
V = Volume of billet, in.^{3}
F_i = Forgeability index for material
S_f = Shape factor found from square root of ratio of average width and average thickness

Find the energy in forging a 2-in. diameter inconel bar stock that is 2 in. long. Let F_i = 2.5 and S_f = 1.4. Then E_d = 44,000 ft-lb (= 2000π × 1 × 2 × 2.5 × 1.4).

Hammer or Smith Forging

This type of forging consists of hammering the heated metal either with hand tools or between flat dies in a steam hammer. *Hand forging* as done by the blacksmith is the oldest form of forging. The nature of the process is such that accuracy is not obtained nor can complicated shapes be made. Forgings ranging from a few pounds to more than 200,000 lb (90 Mg) are made by *smith forging*.

Forging hammers are made in the single- or open-frame type for light work, whereas the double-housing type is made for heavier service. A typical steam hammer is shown in Figure 14.4*A*. The force of the blow is controlled closely by the hammer operator and considerable skill is required.

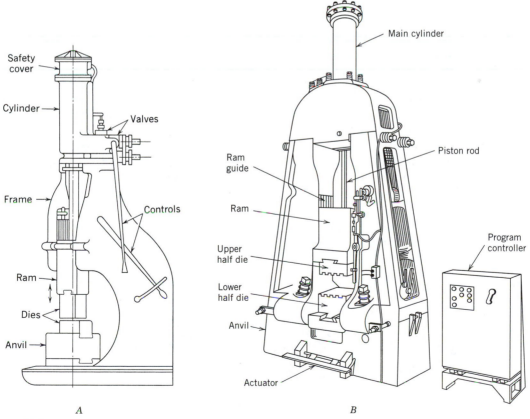

A   B

Figure 14.4
A, Open-frame steam hammer. B, Piston lift gravity drop hammer.

Drop Forging

Drop forging differs from hammer forging in that *closed-impression* rather than open-face or flat dies are used. The dies are matched and attached separately to the movable ram and the fixed anvil. The forging is produced by impact or pressure, which compels the hot and pliable metal to conform to the shape of the dies, as shown in Figure 14.5. In this operation there is drastic flow of the metal in the dies caused by the repeated blows on the metal. To ensure proper flow of the metal during the intermittent blows, the operation is divided into a number of steps. Each step changes the form gradually, controlling the flow of the metal until the final shape is obtained. The number of blows required varies according to the size and shape of the part, the forging qualities of the metal, and the tolerances required. For products of large or complicated shapes a preliminary shaping operation, using more than one set of dies may be required.

Approximate forging temperatures are steel, 2000° to 2300°F (1100° to 1250°C); copper and its alloys, 1400° to 1700°F (750° to 925°C); magnesium, 600°F (315°C); and aluminum, 700° to 850°F (370° to 450°C). Closed-die steel forgings vary in size from a few ounces to 22,000 lb (10 Mg).

The two principal types of drop-forging hammers are the *steam hammer* and the *gravity drop* or *board hammer*. In the former the ram and hammer are lifted by steam, and the force of the blow is controlled by throttling the steam. These hammers operate at more than 300 blows per minute. The capacities of steam hammers range from 500 to 50,000 lb (2 to 200 kN). They usually are of double-housing design, with an overhead steam cylinder assembly providing the power for actuating the ram. For a given weight ram a steam hammer will develop twice the energy at the die as can be obtained from a board or gravity-drop hammer.

In the gravity-type hammer, the impact pressure is developed by the force of the falling ram as it strikes on the lower fixed die. A piston lift gravity drop hammer is shown in Figure 14.4B; it utilizes air or steam to lift the ram. This type of hammer permits the preselection of a series of short- and long-stroke blows. The operator is relieved of the responsibility of regulating stroke heights, and greater uniformity in finished forgings results. Hammers of this type are available for ram weights of 500 lb (225 kg) up to, and including, ram weights of 10,000 lb (4500 kg). The board drop hammer has several hardwood boards attached to the hammer for lifting purposes. After the hammer has fallen, rollers engage the boards and lift the hammer up to 5 ft (1.5 m). When the stroke is reached, the rollers spread and the boards are held by dogs until they are released by the operator. The force of the blow is entirely dependent on the weight of the hammer, which seldom exceeds 8000 lb (35 kN). The board hammer is not so quick lifting as the air or steam unit. Gravity hammers find extensive use in industry for articles such as

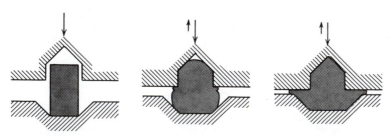

Figure 14.5
Drop forging with closed dies.

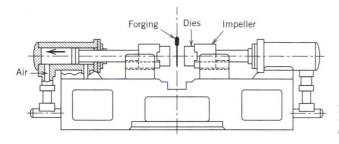

Figure 14.6
Horizontal-impact forging machine.

hand tools, scissors, cutlery, and implement parts. Operator protection for hands, eyes, and ears is required for both types of hammers.

The *impact forging hammer*, Figure 14.6, has two opposing cylinders in a horizontal plane, which actuate the dies toward each other. Stock is positioned in the impact plane in which the dies collide. Its deformation absorbs the energy, and there is less shock or vibration in the machine. With this process the stock is worked equally on both sides, there is less time of contact between stock and die, less energy is required than with other forging processes, and the work is held mechanically.

A forging will have a thin projection of excess metal, or *flash*, extending around it at the parting line, which is removed in a separate trimming press. Small forgings may be trimmed cold, although care must be taken in the trimming operation not to distort the part. The forging usually is held uniformly by the die in the ram and pushed through the trimming edges. Punching operations also may be done while trimming is taking place.

Figure 14.7 shows the dies for forging the main landing gear outer cylinders for a large aircraft. The dies weigh more than 31 tons (28 Mg). Some forging operations require reheating of the part between die stations.

Figure 14.7
Forging die for main landing gear outer cylinder.

Upon completion, all forgings are covered with scale and must be cleaned. This can be done by *pickling in acid, shot peening*, or *tumbling*, depending on the size and composition of the forgings. If some distortion has occurred in forging, a sizing or straightening operation may be required. Controlled cooling is usually provided for large forgings and if certain physical properties are necessary, provision is made for heat treatment.

Advantages of the forging operation include a fine crystalline structure of the metal, closing of any voids, reduced machining time, and improved physical properties. Forging is adaptable to carbon and alloy steels, wrought iron, copper, and aluminum and magnesium alloys. Disadvantages include scale inclusions and the high cost of dies that prohibit short-run jobs. Die alignment is sometimes difficult to maintain, and care is required in die design to prevent cracks from occurring in the forging because of the metal's folding over during the operation. *Closed-impression die* forgings have better utilization of material than *open-flat dies*, better physical properties, closer tolerances, higher production rates, and less operator skill.

Press Forging

Press forging employs a slow squeezing action in deforming the plastic metal as contrasted to the rapid impact blows of a hammer. The squeezing action is carried completely to the center of the part being pressed, thoroughly working the entire section. These presses are the vertical type and may be either mechanically or hydraulically operated. The mechanical presses that are faster operating can exert 500 to 10,000 tons (4 to 90 MN) of force.

The pressure necessary to form steel at forging temperature varies from about 3000 psi to 27,000 psi (20 to 190 MPa). These pressures are based on the cross-sectional area of the forging when measured across the surface of the die at the parting line.

The press capacity is expressed as

$$P = \frac{F}{A \times 2000} \tag{14.4}$$

where

P = Pressure required, psi (usually about 15,000 psi for mild steel)
F = Press capacity, tons
A = Area of the forging at the parting line, in.^{2}

For small press forgings closed-impression dies are used, and only one stroke of the ram is normally required to perform the forging operation. The maximum pressure is built up at the end of the stroke, which forces the metal into shape. Dies may be mounted as separate units, and one or two or more cavities may be cut into a single block. There is some difference in the design of dies for different metals. Copper-alloy forgings can be made with less draft than steel; consequently, more complicated shapes can be produced.

In the forging press, a greater proportion of the total energy input is transmitted to the metal than in a drop hammer press. Much of the impact of the drop hammer is absorbed by the machine and foundation. Press reduction of the metal is faster, and the cost of operation is consequently lower. Most press forgings are symmetrical in shape with surfaces that are smooth, and they provide a closer tolerance than is obtained by a drop hammer. However, many parts of irregular and complicated shapes can be forged more economically by drop forging. Forging presses often are used for sizing operations on parts made by other processes.

Upset Forging

Upset forging entails gripping a bar of uniform section in dies and applying pressure on the heated end, causing it to be upset or formed to shape, as shown in Figure 14.8.

The maximum length of stock to be upset is expressed as

$$L = \frac{kP}{\pi} \tag{14.5}$$

where

> L = Maximum length of stock to be upset, in.
> P = Perimeter of cross section, in.
> k = Constant with values of 2 or 3, usually about 2.6 for steel

The length of the stock to be upset cannot be more than two or three times the diameter or the material will bend rather than bulge out to fill the die cavity.

For some products the heading operation is completed in one position, although in most cases the work is progressively placed in different positions in the die. The impressions may be in the punch, in the gripping die, or in both. In most instances the forgings do not require a trimming operation. Machines of this type are an outgrowth of smaller machines designed for cold heading nails and small bolts.

Progressive piercing, or internal displacement, is the method frequently employed on upset forging machines for producing parts such as artillery shells and radial engine cylinder forgings. The sequence of operations for a cylinder forging is shown in Figure 14.9. Round blanks of a predetermined length for a single cylinder are first heated to forging temperature. To facilitate handling the blank, a porter bar is pressed into one end. The blank is upset and is progressively pierced to a heavy bottom cup. In the last operation, a taper-nosed punch expands and stretches the metal into the end of the die, frees the porter bar, and punches out the end slug. Large cylinder barrels weighing more than 100 lb (45 kg) can be forged in this manner. Parts produced by this process range from small to large products weighing several hundred pounds. The dies, not limited to upsetting, also may be used for *piercing*, punching, trimming, or extrusion.

To produce more massive shapes by this method, a continuous upsetting machine has been developed. This machine can feed induction heated bar stock to the die cavity where rapid blows of the upsetting die build up the part. Some of these machines have a hollow upsetting die so that long lengths of constant cross-sectional shape can be produced.

Another variation to upset forging is *metal gathering*. Rather than form an opening in heated bar stock, an operation of forging a conical shape, similar to Figure 14.8, is followed.

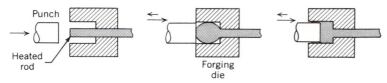

Figure 14.8
Upset forging.

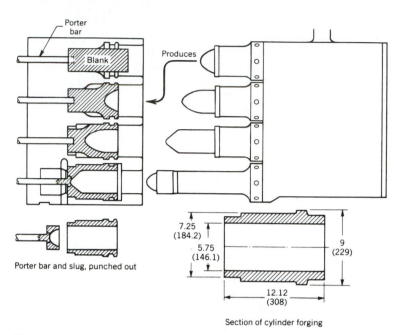

Section of cylinder forging

Figure 14.9
Sequence of operations for a cylinder forging on an upset forging machine.

Roll Forging

Roll forging machines are adapted primarily to reducing and tapering operations on short lengths of bar stock. The rolls on the machine, shown in Figure 14.10, are not completely circular but are 25% to 75% cut away to permit the stock to enter between the rolls. The circular portion of the *rolls* is grooved according to the shaping to be done. When the rolls are in open position, the operator places the heated bar between them, restraining it with tongs. As the rolls rotate, the bar is gripped by the roll grooves and moved forward. When the rolls open, the bar is pushed back and rolled again or is placed in the next groove for subsequent forming work. By rotating the bar 90° after each roll pass, there is no opportunity for flash to form.

In rolling wheels, metal tires, and similar items, a roll mill of somewhat different construction is used. Figure 14.11 shows how a rough forged blank is converted into a finished wheel by the action of the various rolls circumventing the wheel. As the wheel rotates, the diameter is gradually increased while the plate and rim are reduced in section.

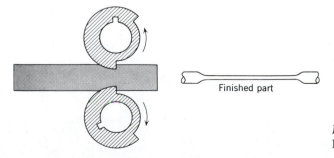

Finished part

Figure 14.10
Principle of roll forging.

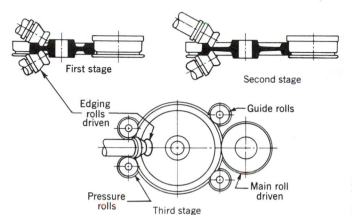

Figure 14.11
Wheels formed by hot-roll forging.

When the wheel is rolled to its final diameter, it is transferred to a press and given a dishing and sizing operation.

Roll forging is sometimes used for axles, blanks for airplane propellers, crowbars, knife blades, chisels, tapered tubing, and ends of leaf springs. Parts made in this fashion have a smooth finished surface and tolerances equal to other forging processes. The metal is hot worked thoroughly and has good physical properties.

14.4 EXTRUSION

Metals that can be hot worked are extruded to uniform cross-sectional shape by the aid of pressure. The principle of *extrusion*, similar to the act of squirting toothpaste from a tube, has long been utilized in processes ranging from the production of brick, hollow tile, and soil pipe to the manufacture of macaroni. Some metals, notably lead, tin, and aluminum, may be extruded cold, whereas others require the application of heat to render them plastic or semisolid before extrusion. In the actual operation of extrusion, the processes differ slightly, depending on the metal and application, but in brief they consist of forcing metal (confined to a pressure chamber) through specially formed dies or orifices. Rods, tubes, molding trim, structural shapes, brass cartridges, and lead-covered cables are typical products of metal extrusion.

Most presses used in conventional extruding of metals are a horizontal type and hydraulically operated. Operating speeds, depending on temperature and material, vary from a few feet a minute up to 900 ft/min (4.6 m/s).

The advantages of extrusion include the ability to produce a variety of shapes of high strength, good accuracy, and surface finish at high production speeds with a relatively low die cost. More deformation or shape change can be achieved by this process than by any other process, except casting. Almost unlimited lengths of a continuous cross section can be produced, and because of low die costs production runs of 500 ft (150 m) may justify its use. The process is about three times as slow as roll forming and the cross section must remain constant. There are several variations of this process.

Direct Extrusion

Direct extrusion is illustrated in Figure 14.12. A heated round billet is placed into the die chamber, and the dummy block and ram are placed into position. The metal is

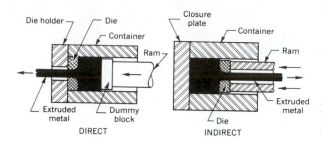

Figure 14.12
Diagram illustrating direct and indirect extrusion.

extruded through the die opening until only a small amount remains. It is then sawed off next to the die and the butt end removed.

Indirect Extrusion

Indirect extrusion, Figure 14.12, is similar to direct extrusion except that the extruded part is forced through the ram stem. Less force is required by this method since there is no frictional force between the billet and the container wall. The weakening of the ram when it is made hollow and the difficulty of providing good support for the extruded part constitute limitations of this process.

Impact Extrusion

In impact extrusion, a punch is directed to a slug with a force that the metal from the slug is pushed up and around it. Most *impact extrusion* operations, such as the manufacture of collapsible tubes, are cold working ones. However, there are some metals and products, particularly those in which thick walls are required, that have the slug heated to elevated temperatures. Impact extrusion is covered in Chapter 15.

14.5 PIPE AND TUBE MANUFACTURE

Pipe and tubular products may be made by butt or electric welding, formed skelp, piercing, and extrusion. Piercing and extrusion methods are used for *seamless tubing,* which is found in high temperature and pressure applications as well as for transporting gas and chemical liquids. Seamless steel pipe up to 16 in. (400 mm) in diameter has been manufactured. Extruded tubes also are used for gun barrels, since the process can be adapted to internal configurations such as rifling and grooves. Butt-welded pipe is the most common and is used for structural purposes, posts, and for conveying gas, water, and wastes. The *electric-welded pipe* is used primarily for pipelines carrying petroleum products or water.

Butt Welding

Intermittent and continuous *butt-welding methods* are used. Heated strips of steel known as *skelp,* which have slightly beveled edges, meet accurately when formed to a cylindrical shape. In the intermittent process, one end of the skelp is trimmed to a V shape to permit the entry into the welding bell, as shown in Figure 14.13*A*. When the skelp is brought up to welding heat, the end is gripped by tongs that engage a draw chain. As the tube

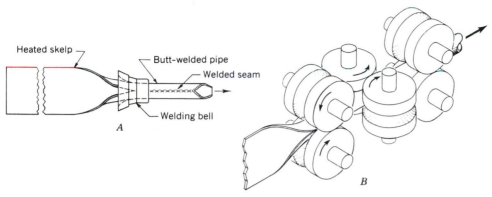

Figure 14.13
Producing butt-welded pipe. *A*, Drawing skelp through a welding bell. *B*, Skelp being formed into a continuous butt-welded pipe.

is pulled through the welding bell, skelp is formed to a cylindrical shape and the edges are welded together. A final operation passes the pipe between sizing and finishing rolls for correct sizing and scale removal. Continuous butt welding of pipe is accomplished by supplying the skelp in coils and providing a means for flash welding the coil ends to form a continuous strip. As the skelp enters the furnace, flames impinge on the edges of the strip to bring them to welding temperatures. Leaving the furnace the skelp enters a series of horizontal and vertical rollers, which form it into pipe. A schematic view of the rollers showing how the pipe is formed and sized is shown in Figure 14.13*B*. As the pipe leaves the rollers, it is sawed into lengths that finally are processed by descaling and finishing operations. Butt-welded pipe is made by this method in sizes up to 3 in. (75 mm) in diameter.

Electric Butt Welding

The electric butt welding of pipe necessitates cold forming of the steel plate to shape prior to the welding operation. The form is developed by passing the plate through a continuous set of rolls that progressively change its shape. This method is known as *roll forming*. The welding unit, placed at the end of the roll forming machine, consists of three centering and pressure rolls to hold the formed shape in position and two electrode rolls that supply current to generate the heat. Immediately after the pipe passes the welding unit shown in Figure 13.15*A*, the extruded flash metal is removed from both inside and outside the pipe. Sizing and finishing rolls then complete the operation by giving the pipe accurate size and concentricity. This process is adapted to the manufacture of pipe up to 36 in. (915 mm) in diameter with wall thickness varying from $\frac{1}{8}$ to $\frac{1}{2}$ in. (3.2 to 12.7 mm). Pipes of larger diameter are usually fabricated by submerged-arc welding after being formed to shape in large, specially constructed presses.

Lap Welding

In the lap welding of pipe, the edges of the skelp are beveled as it emerges from the furnace. The skelp is then drawn through a forming die or between rolls, to give it cylindrical shape with the edges overlapping. After being reheated, the bent skelp is

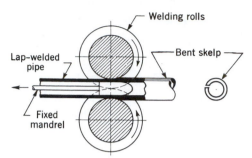

Figure 14.14
Method of producing lap-welded pipe from skelp.

passed between two grooved rolls, as shown in Figure 14.14. Between the rolls is a fixed mandrel to fit the inside diameter of the pipe. The edges are lap welded by pressure between the rolls and the mandrel. Lap-welded pipe is made in sizes 2 to 16 in. (50 to 400 mm) in diameter.

Piercing

To produce seamless tubing, cylindrical billets of steel are passed between two conical-shaped rolls operating in the same direction. Between these rolls is a fixed point or mandrel that assists in the piercing and controls the size of the hole as the billet is forced over it.

The entire operation of making seamless tubing by this process is shown in Figure 14.15. The solid billet is first center punched and then brought to forging heat in a furnace

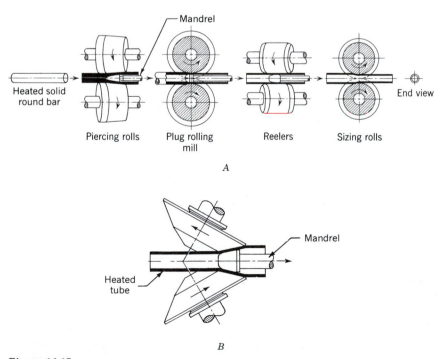

Figure 14.15
A, Principal steps in the manufacture of seamless tubing. B, Rotary seamless process for large tubing.

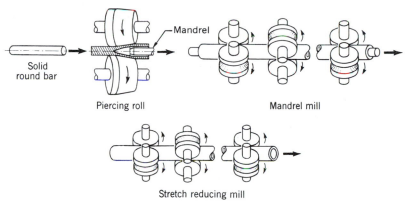

Figure 14.16
Principal steps in the manufacture of continuous tubing.

before being pierced. It is then pushed into the two piercing rolls, which impart both rotation and axial advance. The alternate squeezing and bulging of the billet open a seam in its center, the size and shape of which are controlled by the piercing mandrel. As the thick walled tube emerges from the piercing mill, it passes between grooved rolls over a plug held by a mandrel and is converted into a longer tube with specified wall thickness. While still at working temperature, the tube passes through the reeling machine, which further straightens and sizes it and gives the walls a smooth surface. Final sizing and finishing are accomplished in the same manner as with welded pipe.

This procedure applies to seamless tubes up to 6 in. (150 mm) in diameter. Larger tubes up to 14 in. (355 mm) in diameter are given a second operation on piercing rolls. To produce sizes up to 24 in. (610 mm) in diameter, reheated, double-pierced tubes are processed on a rotary rolling mill, as shown in Figure 14.15*B*, and are finally completed by reelers and sizing rolls, as described in the single-piercing process.

In the continuous method shown in Figure 14.16, a 5½-in. (139.7-mm) round bar is pierced and conveyed to a mandrel mill, where a cylindrical bar or mandrel is inserted. These rolls reduce the tube diameter and wall thickness. The mandrel is then removed and the tube reheated before it enters a stretch-reducing mill. This mill reduces not only the wall thickness of the hot tube but also the tube diameter. Each successive roll speeds up to produce a tension sufficient to stretch the tube between stands. The maximum delivery of this mill is 1300 ft/min (6.6 m/s) for pipe around 2 in. (50 mm) in diameter.

Tube Extrusion

The usual method for *tube extrusion* is seen in Figure 14.17. It is a form of direct extrusion but uses a mandrel to shape the inside of the tube. After the billet is placed inside, the

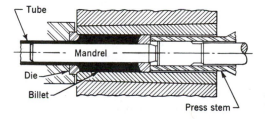

Figure 14.17
Extruding a large tube from a heated billet.

die containing the mandrel is pushed through the ingot. The press stem then advances and extrudes the metal through the die and around the mandrel. The entire operation must be rapid, and speeds of up to 10 ft/s (3 m/s) are used in making steel tubes. Low-carbon steel tubes can be extruded cold, but for most alloys the billet is heated to around 2400°F (1300°C).

14.6 DRAWING

For products that cannot be made with conventional seamless rolling mill equipment, the process illustrated in Figure 14.18 is used. A bloom is heated to forging temperature and, with a piercing punch operated in a vertical press, the bloom is formed into a closed-end hollow forging. The forging is reheated and placed in the hot draw bench consisting of several dies of successively decreasing diameter mounted in one frame. The hydraulically operated punch forces the heated cylinder through the full length of the draw bench. For long, thin-walled cylinders or tubes, repeated heating and *drawing* may be necessary. If the final product is a tube, the closed end is cut off and the tube is processed through finishing and sizing rolls similar to those used in the piercing process. To produce closed-end cylinders similar to those used for storing oxygen, the open end is swaged to form a neck or reduced by hot spinning.

14.7 SPECIAL METHODS

Hot Spinning

Hot spinning of metal is used commercially to dish or form heavy circular plates over a rotating form and to neck down or close the ends of tubes. In both cases a form of lathe

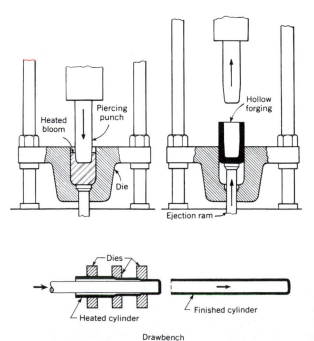

Figure 14.18
Drawing thick-walled cylinders from heated bloom.

Figure 14.19
Thermo-forged socket-head cap screw.

is used to rotate the part rapidly. Shaping is done with a blunt pressure tool or roller that contacts the surface of the rotating part and causes the metal to flow and conform to a mandrel of the desired shape. Once the operation is started, considerable frictional heat is generated, which aids in maintaining the metal at a plastic state. Tube ends may be reduced in diameter, formed to some desired contour, or may be closed completely by the spinning action.

Warm Forging

A process known as *thermo-forging* or *warm forging* utilizes a temperature between that normally used for cold and hot working. There are no metallurgical changes in the metal and no surface imperfections often associated with metal worked at elevated temperatures. Figure 14.19 is a photograph of a cross section of an acid-etched socket-head cap screw. The continuous fiber structure indicative of high strength is visible. Since the flow lines follow the contour of the part, stress concentrations are reduced. The

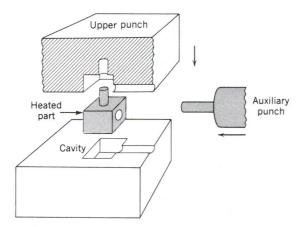

Figure 14.20
Use of an auxiliary punch in die forging.

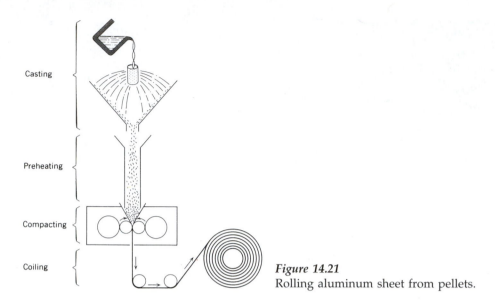

Casting

Preheating

Compacting

Coiling

Figure 14.21
Rolling aluminum sheet from pellets.

temperature of the metal and the forging pressures and speeds must be accurately controlled, since the metal is below the recrystallization temperature.

Additional Methods

High-energy rate forming is usually associated with cold working operations but some high-velocity presses are driven by various mechanisms, explosive charges, or capacitor discharges. Most parts formed in this manner are completed with one blow. Since the operation is fast, thin sections can be forged before the heat is lost. Because of the impact load and the rapid temperature increase of the die associated with this operation, die life is relatively short. The process is useful in forging high temperature, difficult-to-form alloys.

Because of the specialized problems encountered in mass-produced parts, some forging presses are fitted with auxiliary rams or punches that move within or through conventional ones. Figure 14.20 shows the use of an auxiliary punch to produce a hole in the forging. Usually, punches of this kind are delayed in their operation until the dies have either almost or fully completed their work. Because of the complexity of such operations, only mass-production runs are considered. Metals that are difficult to forge (e.g., titanium) can be cast in a press surrounded by inert gas. This process, known as *environmental hot forming*, eliminates most oxidation and scaling and tends to prolong die life. For very large forgings, the inert gas is introduced into the forming area alone, but in case of small presses they are totally enclosed by a cabinet into which argon is admitted.

Small aluminum pellets, smaller than grains of rice, can be rolled into sheets. Figure 14.21 shows how molten aluminum is poured into a revolving perforated cylinder. It is transported by air to a preheating chamber, hot rolled into sheet, and coiled. This process is adaptable to high-volume production with a minimum outlay of equipment. Theoretically, sheets of unlimited length can be formed by this process.

QUESTIONS

1. Give an explanation of the following terms:

Recrystalline   Progressive piercing
Blooms   Piercing
Soaking pits   Metal gathering
Three-high mill   Roll forging
Smith forging   Skelp
Board hammer   Tube extrusion
Impact forging   Drawing
Press forging   Hot spinning
Upset forging   Warm forging

2. Contrast hot and cold working methods in a table.

3. Why is porosity largely eliminated when metal is hot worked?

4. Define recrystalline temperature.

5. Describe the difference between hot working and hot finishing.

6. Describe the difference between bars of steel finished by cold working and hot working processes.

7. Give general statements about the physical properties of steel as a result of hot working.

8. Why are steel ingots not cooled and then reheated before rolling?

9. A board hammer receives its name because of what feature?

10. Describe the following shapes used in connection with the rolling of steel: ingot, bloom, slab, and billet.

11. How would the following size rolled shapes be classified: 15 by 15 in., 20 by 2 in., 3 by 1 in., and 3 by 3 in.?

12. List the advantages of impact forging.

13. Compare a forging press to a drop hammer press and discuss the energy transmitted to the work.

14. Why is it necessary to cool the rolls of a rolling mill with water when this causes scale to form on the metal and reduces its temperature?

15. Why can open-die forgings be made larger than closed-die forgings?

16. Describe progressive piercing and why it is done.

17. What advantages does press forging have over drop forging?

18. Why does the press forging process give greater efficiencies as far as input-output work effort is concerned?

19. Why can metals such as aluminum and tin be extruded cold?

20. Describe why continuous upsetting is a competitive process with extrusion.

21. List products that are shaped by roll forging.

22. Describe the process for hot rolling a metal wheel.

23. Why is it necessary in forming tubing that each successive set of rollers run at a higher speed?

24. Describe the rollers in a stretch-reducing mill.

25. Describe the continuous method of making seamless tubing.

26. List the advantages and disadvantages of heating dies for making forgings.

27. Give the purpose of forging in an inert gas atmosphere.

28. Why is it necessary to have the extruded shape's cross section remain constant?

29. What is the reason for double piercing large-size tubes?

30. Compare the processes of forming aluminum sheet by pellet rolling, by the continuous casting process, and by the conventional method.

31. At what percentage of the melting temperature are the following usually forged: aluminum, steel, and copper?

32. Given a material with the same recrystallization temperature as nickel, and a very high yield strength, what recrystallization temperature and method of hot working would you use?

PROBLEMS

14.1. Sketch a general curve indicating how temperature affects the energy required for rolling, plasticity of metal, and the rate of oxidation.

14.2. Sketch the shape of the rolls in rolling an I-beam from an ingot and describe the operations. (Hints: A sketch is a careful engineering description. It is not CAD or mechanical drawing using drafting instruments.)

14.3. If about 30 passes are required to reduce a large ingot to a bloom, about what percentage reduction per pass is attained? (Hint: Use Figure 14.3 to estimate the percent reduction for each pass. Plot on graph paper.)

14.4. Sketch a method to extrude a lead sheath on wire cable.

14.5. Design a process for producing square tubing. Sketch the roll system. (Hint: Confine your work to one sheet of engineering paper.)

14.6. What size press is needed to forge an open-end wrench that has a "flat size" cross section of 9.3 in.^{2}?

14.7. A 2-in.^{2} steel cross section is upset. What is the maximum length that can be upset?

14.8. Sketch five methods for the manufacture of pipe.

14.9. Find the deformation energy in forging a 4-in. OD slug of steel that is 2 in. long. It has a forgeability index = 1.5 and a shape factor = 1. Repeat for 5-in. slug that is 2 in. long.

PRACTICAL APPLICATION

Check the yellow pages of the telephone book to determine "hot working operations" within your region. Make inquiries about a plant visit. Plan for the tour and assess the importance of product opportunities for hot working within your area.

CASE STUDY: THE YUNGK COMPANY

Steel forgings are produced by the Yungk Company. Material cost is an important part of the price and is estimated carefully. Opportunities exist for the improvement of material yield, and George Yungk recognizes that die design and construction, choice of stock size, and minimizing tong hold, flash, and scrap make for an attractive price and desirable profit. If George does not have a good yield, he loses money.

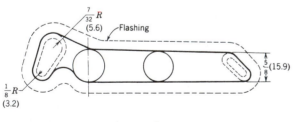

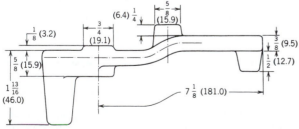

Figure C14.1
Case study.

The Yungk Company, in business for more than 60 years, has developed its own approach in estimating material cost. The shape-volume contained within the forging drawing, including openings, is found. A cut-weight is the weight of the material at the press to produce one forging. It is equal to the shape weight, plus allowances for flashing, tong hold, and scale losses. In addition, crop end losses, because the required forging raw material bar stock may be an uneven multiple of purchased bar stock length, are added to the scrap estimate.

George knows that using the proper bar size to make a forging is important. If the bar OD is too large, excessive *fullering* must be done to reduce the size. This results in low production and excessive die wear. When the bar size is too small, it becomes impractical to fill the heavier sections of the part.

An arm forging print, Figure C14.1, has been received. Shape-volume has been determined to be 3.55 in.^{3}, density = 0.283 lb/in.^{3} Studying the print George believes that the forging will require a $\frac{15}{16}$-in. OD and will have flashing 0.075 in. thick by 1 in. wide for a periphery of 20 in. "A tong hold of 1 inch is sufficient," George says. "End losses for cropping should be about 2% and scrap will be 3%."

Determine the total weight for one forging. What is the forging material cost if bar stock costs $0.426/lb? What is the percent yield of finished forging? What percent scrap can George tolerate if he makes 25% profit on the product?

CHAPTER 15

COLD WORKING
OF METAL

When a metal is rolled, extruded, or drawn at a temperature below the recrystallization temperature, the metal is cold worked. Most metals are cold worked at room temperature. Cold work distorts the grain and does little toward reducing the dimensions of the material. Cold work improves strength, machinability, dimensional accuracy, and surface finish of metal. Because oxidation is less for cold working than for hot working, thinner sheets and foils are rolled.

Even though the forming causes a local temperature rise at the point of processing, the temperature is not as high as hot working. Hot working, however, is performed on metal in the plastic state while refining the grain structure. Similar processes and equipment are used for both hot and cold work, but the required forces and the results are different.

15.1 COLD WORKING

To understand the action of *cold working*, one must have knowledge of the structure of metals. Metals are crystalline in nature and are composed of irregularly shaped grains of various sizes. They may be seen using a microscope if the metal has been properly polished and etched. Notice earlier chapters where pictures of *metallograph specimens* are shown.

Each grain is constructed of atoms in an orderly arrangement known as a *lattice*. The orientation of the atoms in a grain is uniform but differs in adjacent grains. When material is cold worked, the change in material shape brings about marked alterations in the grain structure. Structural changes that occur are grain fragmentation, movement of atoms, and lattice distortion. Slip planes, shown in Figure 15.1, develop through the lattice structure at points where the atom bonds of attraction are the weakest and whole blocks of atoms are displaced. When slip occurs, the orientation of the atoms is not changed. In cases where there is reorientation a phenomenon called *twinning* occurs. In twinning, the lattice on one side of the plane is oriented in a different fashion from the other, but the atoms have shapes identical to adjacent atoms. Slip is the more common result of deformation.

Greater pressures are needed for cold working than for hot working. The metal, being in a more rigid state, is not permanently deformed until stresses exceed the elastic limit. Since there can be no recrystallization of grains in the cold working range, recovery from grain distortion or fragmentation does not occur. As grain deformation proceeds, greater resistance to this action results in increased strength and hardness. The metal is strain hardened and, for some metals that will not respond to heat treating, it is the only known method of changing physical properties such as hardness and strength. Several theories have been advanced to explain this occurrence. In general, they refer to resistance built up in the grains by atomic dislocation, fragmentation, or lattice distortion, or a combination of the three phenomena.

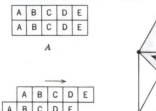
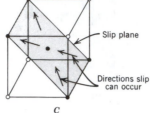

Figure 15.1
Representation of slip in a body-centered lattice system.

The amount of cold work that a metal will withstand depends on its ductility. The higher the ductility of a metal, the more it is able to be cold worked. Pure metals withstand a greater amount of deformation than metals having alloying elements, since alloying increases the tendency and rapidity of *strain hardening*. Large-grain metals are more ductile than smaller-grained metals.

When metal is deformed by cold work, severe stresses known as *residual stresses* are set up. To remove these undesirable stresses, the metal is reheated to slightly below the recrystalline range temperature. In this range the stresses are rendered ineffective without appreciable change in physical properties or grain structure. Heating into the recrystalline range eliminates the effect of cold working. Sometimes it is desirable to have residual stresses in the metal. The fatigue life of small parts may be improved by *shot peening*, which causes the surface metal to be in compression and the material below the surface to be in tension.

Advantages and Limitations

Many products are cold finished after hot rolling to make them commercially acceptable. *Hot rolled* strips and sheets are soft and have surface imperfections. They lack dimensional accuracy and desirable physical properties. The cold-rolling operation reduces size only slightly, permitting accurate dimensional control. Surface oxidation does not result from the process, and a smooth surface is obtained. Strength and hardness are increased. For metals that do not respond to heat treatment, cold work is a possible method to increase hardness.

Ductile materials can be extruded at temperatures below the recrystallization range. However, higher pressure and heavier equipment are needed for cold-working operations than for hot-working operations. Brittleness results if the metal is overworked, and an annealing operation then becomes necessary. In general, cold working produces the following effects:

1. Stresses are set up in the metal, which remain unless they are removed by subsequent heat treatment.
2. Distortion or fragmentation of the grain structure is caused.
3. Strength and hardness of the metal are increased with a corresponding loss in ductility.
4. Recrystalline temperature for steel is increased.
5. Surface finish is improved.
6. Close dimensional tolerance can be maintained.
7. The process is economical and produces parts in high-volume applications.

15.2 PROCESSES

The effects previously discussed are not achieved by all cold-working processes. Operations involving bending, drawing, and squeezing metal result in grain distortion and changes in physical properties, whereas shearing or cutting operations change only form and size. Cold-working processes pertain primarily to rolling, drawing, or extrusion.

Tube Finishing

Tubing, which requires dimensional accuracy, smooth surface, and improved physical properties, is finished by either cold drawing or a *tube reducer*. Tubing that has first been hot rolled is treated by pickling and washing to remove scale. Before the cold tube-finishing operation, a lubricant is applied to prevent galling, reduce friction, and increase surface smoothness. *Tube drawing* is done in a *drawbench*, as shown in Figure 15.2. One end of the tube is reduced in diameter by a *swaging* operation to permit it to enter the die, then it is gripped by tongs fastened to the chain of the drawbench. In this operation, the tube is drawn through a die smaller than the outside diameter of the tube. The inside surface and diameter are controlled by a fixed mandrel over which the tube is drawn. This mandrel may be omitted for small sizes or for larger sizes if the accuracy of the inside diameter is not important. Drawbenches require a pulling power ranging from 50,000 to 300,000 lb (0.2 to 1.3 MN) and may have a total length of 100 ft (30 m).

The operation of drawing a tube is severe. The metal is stressed above its elastic limit to permit plastic flow through the die. The maximum reduction for one pass is around 40%. This operation increases the hardness of the tube so much that if several reductions are desired, the material must be annealed after each pass. This method also produces tubes having smaller diameters or thinner walls than can be obtained by hot rolling. Hypodermic tubing is produced in this manner with an outside diameter of less than 0.005 in. (0.13 mm).

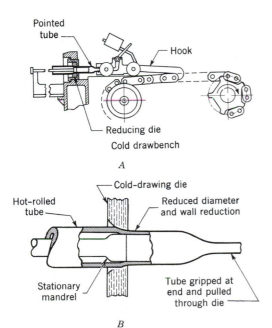

Pointed tube

Hook

Reducing die

Cold drawbench

A

Cold-drawing die

Hot-rolled tube

Reduced diameter and wall reduction

Stationary mandrel

Tube gripped at end and pulled through die

B

Figure 15.2
Process of cold-drawing tubing.

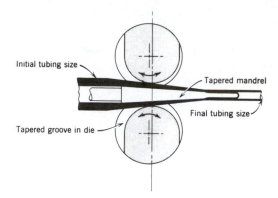

Figure 15.3
Schematic of a tube reducer.

The tube reducer has semicircular dies with tapered grooves through which the previously hot-rolled tubing is alternately advanced and rotated. The dies, Figure 15.3, rock back and forth as the tubing moves through them. A tapered inside mandrel regulates the size to which the tube will be reduced. The tube reducer can make the same reduction in one pass that might take four or five passes in a drawbench, but its chief advantage is the much longer lengths of tubing that can be produced.

Tubes that are finished by either method have all the advantages found in cold-worked metals and can be made in longer lengths and thinner walls than is possible by hot working.

Wire Drawing

Wire is made by cold-drawing hot-rolled wire rod through one or more dies, as shown in Figure 15.4, to decrease its size and increase the physical properties. The wire rod, about $\frac{7}{32}$ in. (6 mm) in diameter, is rolled from a single billet and cleaned in an acid bath to remove scale, rust, and coating. A coating is applied to prevent oxidation, neutralize any remaining acid, and to act as a lubricant and a coating to which a lubricant applied later may cling.

Both single-draft or continuous-drawing processes may be used. In the first method a coil is placed on a reel or frame, and the end of the rod is pointed so that it will enter the die. The end is grasped by tongs on a drawbench and pulled through to such length as may be wound around a drawing block or reel. From there on the rotation of the draw block pulls the wire through the die and forms it into a coil. These operations are repeated with smaller dies and blocks until the wire is drawn to its final size.

In the continuous-drawing process, Figure 15.5, the wire is fed through several dies and draw blocks arranged in series, which permits drawing the maximum amount in one

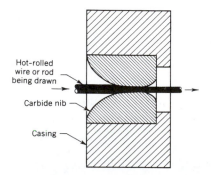

Figure 15.4
Section through a die used for drawing wires.

Figure 15.5
A continuous wire-drawing system of rolls.

operation before annealing is necessary. The number of dies in the series will depend on the kind of metal or alloy being processed and may vary from 4 to 12 successive drafts. The dies are usually made from tungsten carbide, although diamond dies can be used for drawing small diameters.

Wire drawing reduction is given as

$$\% \text{ Reduction in area} = \frac{(A_o - A_f)}{A_o} \times 100 \tag{15.1}$$

$$\% \text{ Elongation} = \frac{L_f - L_o}{L_o} \times 100 \tag{15.2}$$

where

A_o, A_f, L_o, and L_f are original and final area and lengths.

The continuity equation also holds for wire drawing:

$$Q_o = Q_f = A_o V_o = A_f V_f \tag{15.3}$$

where

Q_o, Q_f = Quantity of metal entering and leaving a die
V_o, V_f = Velocity of wire entering and leaving a die

Foil Manufacture

Foils are made from a broad variety of pure metals and alloys by cold rolling to thickness as thin as 0.00008 in. (0.0020 mm). Most foil manufacturers find it necessary to control the raw material so closely that they employ their own vacuum-melting equipment. Only the purest metals are charged into the melting furnaces and alloys are added when required. In most instances the foil is continuously cast, cooled, and rolled as it comes from the furnace.

One producer of aluminum foil casts the material by the Hunter process, in which the molten metal is forced through a nozzle or tip onto water-cooled rolls where it solidifies 0.250 in. (6.35 mm) thick in less than 3 s. In one continuous series of operations the aluminum is rolled to about 0.006 in. (0.15 mm) in thickness at velocities approaching 2000 ft/min (10 m/s). Thinner foil is produced by additional rolling.

Foil thickness is produced by a combination of roll pressure and controlled tension on the material. Most foils can be produced with two sides bright or one side bright and the other a satin finish. The latter is produced by pack or double thickness rolling. When two sheets are passed through the rollers at the same time, the faces contacting the rollers are bright and the mating faces have the satin finish when stripped apart.

Metal Spinning

Metal spinning is the operation of shaping thin metal by pressing it against a form while it is rotating. The nature of the process limits it to symmetrical articles. This work is done on a speed lathe, similar to the wood lathe, except that in place of the usual tailstock, it is provided with some means of holding the work against the form, as shown in Figure 15.6. The forms are usually turned from hardwood and attached to the face plate of the lathe, although smooth steel chucks are recommended for production jobs.

Parts are formed with the aid of blunt hand tools that press the metal against the form. The cross slide has a hand or compound tool rest in the front for supporting the hand tools and some means for supporting a trimming cutter or forming roll in the rear. Parts may be formed either from flat disks of metal or from blanks that have been drawn previously in a press. The latter method is used as a finishing operation for many deep-drawn articles. Most spinning work is done on the outside diameter as shown in Figure 15.6, although inside work is also possible. Figure 15.7 shows spinning a ⅝ in. thick, 140-in. diameter plate, which, when finished, is a 120-in. diameter, elliptical-shaped head.

Bulging work on metal pitchers, vases, and similar parts is done by having a small roller, supported from the compound rest, operate on the inside and press the metal out against a form roller. The part must be drawn first and often is given a *bulging operation* beforehand, as spinning cannot be done near the bottom.

Lubricants such as soap, beeswax, white lead, and linseed oil reduce the tool friction. Since metal spinning is a cold-working operation, there is a limit to the amount of drawing or working the metal will stand, and one or more annealing operations may be necessary.

Simple shapes can be formed from soft nonferrous metals up to ¼ in. in thickness and from low-carbon steel up to 3/16 in. in thickness. Tolerances up to ±1/32 in. can be maintained easily for diameters under 18 in.

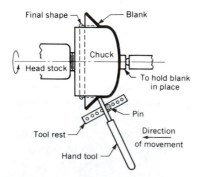

Figure 15.6
Metal-spinning operation.

Figure 15.7
Metal spinning a 120-in. (3-m)-diameter head for missile.

Spinning lends itself to short-run production jobs of about 5000 pieces or less, although it has many applications in quantity production work. *Spinning* has several advantages over press work in that tooling costs are lower, a new product can be brought to the production stage sooner, and for very large parts the high cost of a press capable of doing the job may be prohibitive. Labor costs are higher for spinning than for press work and the production rate may be much less. This process is used frequently in making bells on musical instruments and for light fixtures, kitchenware, reflectors, funnels, and large processing kettles.

Shear Spinning

In spinning thick metal plates, power-driven rollers must be used in place of conventional hand-spinning tools. This operation is called *shear spinning*. Figure 15.8 shows the progressive steps of a shear-spinning operation, where a conical shape is formed from a flat plate. The plate initially is held against the mandrel by a holder. The roll formers force the plate to conform to the mandrel, maintaining a uniform wall thickness from the

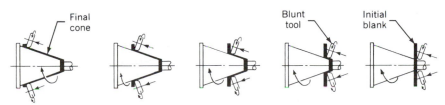

Figure 15.8
Progressive forming in a shear-spinning operation in which a conical shape is formed from a flat plate.

starting point until completion. This relationship is given as

$$t_f = t_s(\sin \alpha/2) \qquad (15.4)$$

where

t_f = Final thickness, in.
t_s = Starting thickness, in.
α = Included angle of cone, °

The wall thickness obtained is equal to the plate thickness times sin $\alpha/2$, α being the cone included angle. Parts having a cone angle α less than 60° require a conical preform. Reduction in wall thickness up to 80% is possible although, in some cases, the reductions are much smaller.

In conventional spinning, the wall thickness remains about the same throughout the operation. Spinning tools merely bend or flare the metal into a new contour and do not cause a plastic flow or reduction in wall thickness. In shear spinning, the metal is reduced uniformly in thickness over the mandrel by a combination of rolling and extrusion. Advantages claimed for shear spinning include increased strength of part, material savings, reduction in cost, and good surface finish.

Most metal can be formed by this process. Although heat is sometimes applied throughout the cycle, it is not required for steel alloys and most nonferrous metals.

Stretch Forming

In forming large sheets of thin metal involving symmetrical shapes or double-curve bends, a *metal stretch press* can be used effectively. Figure 15.9A shows one of the simpler hydraulically operated presses. A single die mounted on a ram is placed between two slides that grip the metal sheet. The die moves in a vertical direction and the slides move horizontally. Large forces of 50 to 150 tons (0.5 to 1.3 MN) are provided for the die and slides. The process is a stretching one and causes the sheet to be stressed above its elastic limit while conforming to the die shape. This is accompanied by a slight thinning of the sheet, and the action is such that there is little *springback* to the metal once it is formed.

Adapted to both production and short-run jobs, inexpensive dies of wood, kirksite, plastic, or steel are used. Large double curvature parts, difficult by other methods, are easily made with this process. The process is used with many hard-to-form alloys, there is little severe localized cold working, and the problem of unequal metal thinout is minimized. Scrap loss is high since material must be left at the ends and sides for trimming, and there is a limitation to the shapes that can be formed.

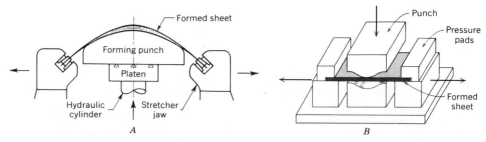

Figure 15.9
Stretch processes. *A*, Stretch forming. *B*, Stretch draw forming.

Stretch forming requires the metal to be stretched to a point greater than its yield strength and less than its tensile strength. For estimating the pressure required for stretch forming, the following formula can be used:

$$P = 1.25 Y_s A \tag{15.5}$$

where

P = Stretch forming pressure, lb
Y_s = Yield strength of metal, psi
A = Cross-sectional area, in.^{2}
1.25 = Empirical constant

A combination of *stretch and draw forming*, Figure 15.9*B*, can be employed. Manufacturers of stretch-draw forming presses maintain that all metals and alloys become unusually ductile when stretched from 2% to 4% and can be formed with about one-third of the force normally required. Despite waste losses caused by the necessity of gripping the material in jaws during stretching, the process is used not only for short-run aircraft parts of aluminum but also is employed in the automotive industry to make steel roof panels, hood covers, rear deck lids, and door posts. Titanium and stainless steel sheet can also be formed in this manner.

Swaging and Cold Forming

These terms refer to methods of cold working by a compressive force or impact, which causes the metal to flow into a predetermined shape such as the design of a die. The metal conforms to the shape of the die, but it is not restrained completely and may flow at some angle in the direction to which the force is applied.

Sizing, the simplest form of cold forging, is the process of slightly compressing a forging, casting, or steel assembly to obtain close tolerance and a flat surface and/or a flash removal operation. The metal is confined only in a vertical direction.

Small pinions, less than 1 in. in diameter, are cold-extruded. *Rotary swaging*, Figure 15.10, reduces the ends of bars and tubes by rotating dies, which open and close rapidly on the work so that the end of the rod is tapered or reduced in size by a combination of pressure and impact. Mechanical pencils, metal furniture legs, and umbrella poles are examples of parts made by this process. Since swaging action is rather severe, the material hardens and an annealing operation is necessary if much reduction is desired.

Cold forming, or *cold heading*, or *upsetting* of bolts, rivets, and other similar parts is done on a cold-forming machine and is another form of swaging. Since the product of the cold header is made from unheated material, the equipment must withstand the high pressures that develop. Also, alignment of the upsetting tool with the dies must be accurate so that the work turned out will be free from defects. A solid die machine of

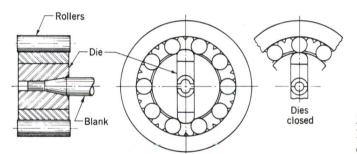

Figure 15.10
Illustrating the operation of dies in a swaging machine.

Figure 15.11
Single-stroke, cold-header machine producing 36,000 parts per hour from 3/16-in. (4.8-mm) coiled steel wire.

this type is illustrated in Figure 15.11. The rod is fed by straightening rolls up to a stop, and then it is cut off and moved into one of the four types of cold-header dies shown in Figure 15.12. The heading operation may be either single or double, and on completion the part is ejected from the dies. Production rates of 600 parts per minute are not uncommon.

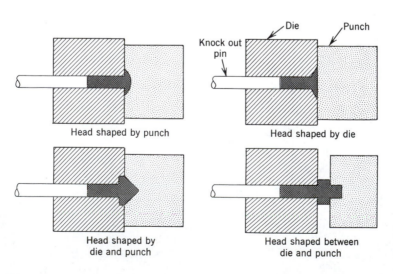

Figure 15.12
Types of cold-header dies.

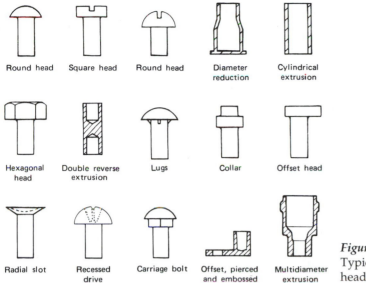

Round head   Square head   Round head   Diameter reduction   Cylindrical extrusion

Hexagonal head   Double reverse extrusion   Lugs   Collar   Offset head

Radial slot   Recessed drive   Carriage bolt   Offset, pierced and embossed   Multidiameter extrusion

Figure 15.13
Typical parts made by cold heading from wire stock.

Nails, rivets, and small bolts are forged cold from coiled wire, whereas large bolts require heating of the end of the rod before the heading operation. In nail making, the head is formed before shearing the wire. The wire is fed forward, clamped, headed, and pinched or sheared off to form the point and is finally expelled. The nails may be tumbled together in sawdust to remove the lubricants and "whiskers" of metal before packaging. Figure 15.13 shows examples of cold-headed parts.

Bolt-making machines are available, which completely finish the bolt before it leaves the machine. The operation consists of cutting off an oversize blank, extruding the shank, heading, trimming, pointing, and roll threading. All operations are carried on simultaneously, and outputs range from 50 to 300 pieces per minute.

Intraforming squeezes metal at a pressure of about 300 tons (4000 MPa) or less, onto a die or mandrel to produce an internal configuration. A mandrel, workpiece, and finished part are shown in Figure 15.14. Ferrous and nonferrous forgings, powdered metal parts, and tubing are produced. Splines, internal gears, special-shaped holes, and bearing retainers are produced by this method. Tooling is inexpensive, die life is from 1000 to 100,000

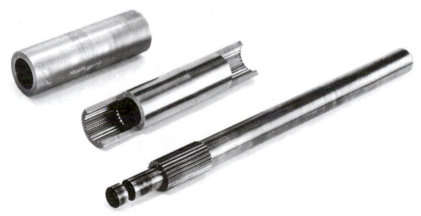

Figure 15.14
Intraforming workpiece, finished part, and mandrel.

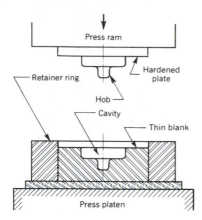

Figure 15.15
Die hob producing a mold cavity by pressing into soft steel.

parts, and good surface finishes and high accuracy may be obtained. The maximum ID of the workpiece is about 4 in. (100 mm).

Hobbing

Mold cavities, Figure 15.15, are produced by forcing a hardened steel form or hob into soft steel. The *hob*, machined to the exact form of the piece to be molded, is heat treated to obtain the necessary hardness and strength to withstand the pressures involved. Pressing the hob into the blank requires care, and frequently several alternate pressings and annealings are necessary before the job is complete. During the hobbing operation, the flow of metal in the blank is restrained from appreciable lateral movement by a heavy retainer ring placed around it. The actual pressing is done in hydraulic presses having capacities ranging from 250 to 8000 tons (2 to 70 MN).

The advantage of hobbing is that multiple identical cavities can be produced economically. The surface of the cavity has a highly polished finish, and machine work is unnecessary other than to remove surplus metal from the top and sides of the blank. This process is used in producing molds for the plastic and die-casting industries.

Coining and Embossing

The operation of *coining*, Figure 15.16, is performed in dies that confine the metal and restrict its flow in a lateral direction. Shallow configurations on the surfaces of flat objects, such as coins, are produced in this manner. *Knuckle joint presses* that develop high pressures are required in this operation, and its use is limited to relatively soft alloys.

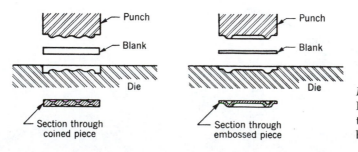

Figure 15.16
Illustrating the difference between coining and embossing.

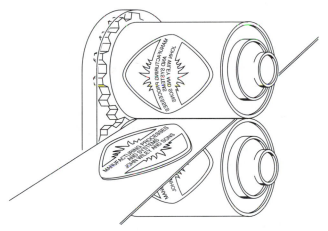

Figure 15.17
Rotary embossing.

Obviously, coining is an operation that produces billions of coins annually throughout the world.

Embossing is more of a drawing or stretching operation and does not require the high pressures necessary for coining. The punch is usually relieved so that it touches only the part of the blank that is being embossed. The major use for embossing is making name-plates, medallions, identification tags, and aesthetic designs on thin sheet metal or foil. The embossed design is raised from the parent metal. The mating die, Figure 15.16, conforms to the same configuration as the punch so that there is very little metal squeezing in the operation and practically no change in the thickness of the metal.

Rotary embossing, using cylindrically shaped dies, is used extensively on thin sheet metal and foils. The metal is fed through the rolls, as shown in Figure 15.17.

Riveting, Staking, and Stapling

Riveting, staking, and stapling fasten parts together, as illustrated in Figure 15.18. In the usual *riveting operation*, a solid rivet is placed through holes made in the parts to be fastened together, and the end is pressed to shape by a punch. Hollow rivets may have the ends secured by curling them over the edges of the plate. Explosive rivets have a powder charge that expands the nonhead end.

Staking is a similar operation in that the metal of one part is upset to cause it to fit tightly against the other part. A staking punch may have one or more projections, as shown in Figure 15.18, or it may be in the form of a ring with sharp chisellike edges. Both operations can be performed on small presses because not much pressure is required. *Stapling* is used to join the two or more sheets of metal and to join sheet to wood.

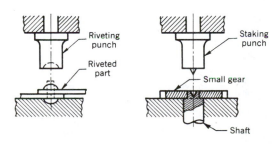

Figure 15.18
Illustrating the difference between riveting and staking processes.

Roll Forming

Cold roll forming machines are constructed with a series of mating rolls that progressively form strip metal as it is fed continuously through the machine at speeds ranging from 50 to 300 ft/min (0.3 to 1.5 m/s). A machine is shown in Figure 15.19 in which tubular sections are being produced by five pairs of rolls.

The tubular section enters a resistance welder after being formed and is welded continuously. The number of roll stations depends on the intricacy of the part being formed. For a simple channel four pairs may be used, whereas for complicated forms several times that number may be required. In addition to the mating horizontal rolls, these machines are frequently equipped with guide rolls mounted vertically to assist in the forming operation and straightening rolls to "true up" the product as it emerges from the last forming pass.

Figure 15.20 shows typical parts that are roll formed. Note the sequence of forming operations for a window screen section in Figure 15.20B. In forming the sequence, the vertical center or pass line is established first so that the number of bends on either side is about the same. Forming usually starts at the center and progresses out to the two edges as the sheet moves through the successive roll passes. The amount of bending at any one roll station is limited. If the bending is too severe, it carries back through the sheet and affects the section at the preceding roll station. Corner bends are limited to a radius equal to the sheet thickness.

In terms of capacity for working mild steel, standard machines form strips up to 0.156 in. (3.96 mm) thick by 16 in. (400 mm) wide. Special units have been made for much heavier and wider strip steel. The process is rapid and is applicable to forming products having sections requiring a uniform thickness of material throughout their entire length. Unless production requirements are high, the cost of the machine and tooling cannot be justified.

Figure 15.19
Cold roll tube-forming machine. Strip enters machine from coil (not shown) and is bent to tubular shape by five pairs of rolls before being welded.

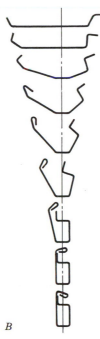

Figure 15.20
Cold-rolled formed parts. *A*, Miscellaneous parts formed from coiled strip. *B*,
Sequence of forming operation for window screen section.

Tolerances of roll forming are affected by the size of the section, material, product, and the gage and gage tolerances of the material. In addition, the length of the piece must be toleranced, and it is influenced by the speed of operation, length of the piece to be cut, and accuracy of the cutting device. Attainable tolerances are: piece length, ± 1/64 to 1/8 in.; length, ± 1/64 to 1/8 in.; straightness and twist, ± 1/64 to 1/4 in. in 10 ft; cross-sectional dimension, ± 1/64 to 1/16 in.; and angles, ± 1° to 2°.

Several guidelines are followed when designing a product to be cold roll formed. A slight angle in a section is more desirable rather than long vertical side walls and blind corners. Sharp radii should be avoided to prevent inaccuracy resulting from rolls without control features. Smaller bend radii are easier and less expensive to make than those that are larger.

Plate Bending

Another method of bending metal plates and strips into cylindrical shapes is by a *roll-bending* machine, as illustrated in Figure 15.21. This machine consists of three rolls of the same diameter. Two of them are held in a fixed position and one is adjustable. As

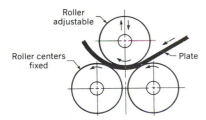

Figure 15.21
Plate-bending rolls.

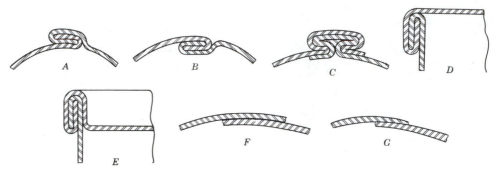

Figure 15.22
Seams used in container manufacture: *A*, Longitudinal outside lock seam. *B*, Longitudinal inside lock seam. *C*, Compound longitudinal seam. *D*, Double seam for flat containers. *E*, Double seam for recessed-bottom containers. *F*, Adhesive-bonded top seam. *G*, Resistance-welded seam.

a metal plate enters and goes through the rolls, its final diameter is determined by the position of the adjustable roll. The closer it is moved to the other rolls, the smaller the diameter. Machines of this type are made in capacities ranging from those that form small gage thicknesses to others that form heavy plates up to 1¼ in. (30 mm).

Seaming

In the manufacture of metal drums, pails, cans, and numerous other products made of light-gage metal, several types of *seams* are used. The most common seams are shown in Figure 15.22.

The lock seam used on longitudinal seams is adapted for joints that do not require absolute tightness. After the container is formed, the edges are folded and pressed together. The compound seam, sometimes called the Gordon or box seam, is stronger and tighter than the lock seam and is suitable for holding fine materials. Both these joints are formed and closed on either hand or power-seaming presses.

Bottom seams, which are similar to the longitudinal seams, are made in either flat or recessed styles. Flat-bottom seaming is limited to one end of a container, as the container must be open to make the joint. Double seaming with recessed bottoms can be done on both ends of a container. Edge flanging, curling, and flattening, the operations necessary to make a recessed double seam, are shown in Figure 15.22.

Double-seaming machines may be hand operated, semiautomatic, or automatic. Semiautomatic machines must be loaded and unloaded by the operator, but the operation of the machine is automatic. In automatic machines, the cans are brought to the machine by conveyor and the ends are supplied by magazine feed. The cans feed from the conveyor to a star wheel, which transfers them to an automatic delivery turret. The delivery turret feeds them into position with the seaming heads, and the closing seam is made.

15.3 HIGH-ENERGY RATE FORMING

High-energy rate forming (HERF) includes a number of processes in which parts are formed at a rapid rate by extremely high pressures. While the term inaccurately describes the process, it is accepted by industry for what would be better called *high-velocity forming*. The deformation velocities of several processes are noted in Table 15.1.

By imparting a high velocity to the workpiece, the size of equipment to form large

Table 15.1   Approximate Deformation Velocities

	Velocity	
Process	ft/s	m/s
Hydraulic press	0.10	0.03
Brake press	0.10	0.03
Mechanical press	0.1–2.4	0.03–0.73
Drop hammer	0.8–14	0.24–4.3
Gas-actuated ram	8–270	2.0–82
Explosive	30–750	9–230
Magnetic	90–750	27–230
Electrohydraulic	90–750	27–230

parts is reduced, and certain materials, which may not lend themselves to conventional forming methods, can be processed.

Die costs are low, good tolerances can be maintained, and the production costs are minimized. While development of these processes has centered on forming relatively thin metal, applications of high-energy rate forming include compacting metal powders, forging, cold welding, bonding, extruding, and cutting.

Explosive Forming

Methods for applying energy at a high rate are diagrammed in Figure 15.23. *Explosive forming* has proved to be an excellent method of utilizing energy at a high rate, since the gas pressure and rate of detonation can be controlled.

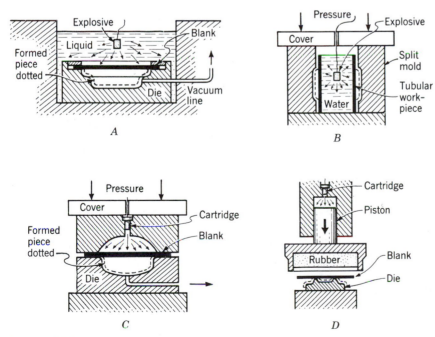

Figure 15.23
Methods in high-energy rate forming. *A,* Direct forming with fluid pressure. *B,* Bulging operation. *C,* Direct forming by gas pressure. *D,* Gas-actuated drop hammer.

Figure 15.24
Bellows explosively formed using a 12-gage shotgun shell.

Both low and high explosives are used in the various processes. With low explosives, known as cartridge systems, the expanding gas is confined, and pressures may build up to 100,000 psi (700 MPa). High explosives, which need not be confined and which detonate with a high velocity, may attain pressures of up to 20 times that of low explosives. Explosive charges, whether exploded in air or in liquid, set up intense shock waves that pass through the medium between the charge and the workpiece, but decrease in intensity as the waves spread over more area.

Springback is minimized in explosive forming but does exist. Less springback occurs with the use of sheet explosives close to the workpiece and high clamping forces on the hold-down areas, and, in the absence of lubricants, thick materials exhibit less springback than do parts made from thin blanks.

Aside from the generation of gas pressure by powder, high gas pressures also may be attained by the expansion of liquefied gases, explosion of hydrogen–oxygen mixtures, spark discharges, and the sudden release of compressed gases. A compressed nitrogen, which on release, accelerates a heavy ram to a speed of 2000 in./s (50 m/s) or less. Attaining very high pressures, it is used for open end extrusion, impact extrusion, forging, and for compacting powders.

Figure 15.23C and D show examples of expanding gas methods. In Figure 15.23C, the gas presses against the workpiece and forces it to conform to the die. In Figure 15.23D, the gases act against a piston, which forces the confined rubber punch over the blank and die. This method is similar to that of a drop hammer but is much more rapid. Thin-wall tubing may be formed by "slow" explosive forming using a powder that deflagrates rather than detonates. The expanding gases are trapped inside a boot within the tubing, and the expanding boot forces the tubing into the configuration of the die. Figure 15.24 is an example of such forming using a shotgun shell.

Electrohydraulic Forming

Electrohydraulic forming, also known as *electrospark forming*, is a process whereby electrical energy is converted directly into work. The forming equipment for this process is similar to Figure 15.23A or B, but pressure is obtained from a spark gap instead of

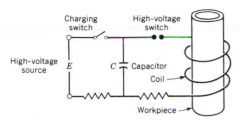

Figure 15.25
Schematic of an electromagnet-forming circuit.

an explosive charge. A bank of capacitors is first charged to a high voltage and then discharged across a gap between two electrodes in a suitable nonconducting liquid medium. This generates a shock wave that travels radially from the arc at high velocity supplying the necessary force to form the workpiece to shape.

This process is safe to operate and has low die and equipment cost. The energy rates also can be controlled closely.

Magnetic Forming

Magnetic forming is another example of the direct conversion of electrical energy into useful work. At first it served primarily for swaging-type operations, such as fastening fittings on the ends of tubes and crimping terminal ends of cables. Other applications are embossing, blanking, forming, and drawing, all using the same power source but differently designed work coils.

Figure 15.25 is a sketch illustrating how *electromagnetic forming* works. The charging voltage E is supplied by a high voltage source into a bank of capacitors connected in parallel. The amount of energy stored can be varied either by adding capacitors to the bank or by increasing the voltage. The latter is limited by the insulating ability of the dielectric material on the coils. The charging operation is rapid and, when complete, a high voltage switch triggers the stored electrical energy through the coils, establishing a high-intensity magnetic field. This field induces a current into the conductive workpiece placed in or near the coil, resulting in a force that acts on the workpiece. This force, when it exceeds the elastic limit of the material being formed, causes permanent deformation.

Three different forming possibilities are shown in Figure 15.26. In Figure 15.26A, the coil surrounds a tube that when energized forces the material tightly around the fitting. The same principle applies if a conducting ring were placed around a number of wire ends. If the coil is placed inside an assembly as indicated in Figure 15.26B, the force expands the tube into the collar. By changing the design of the coil as at Figure 15.26C, flat plates may be embossed or blanked. The process is used to assemble fragile parts such as the swaging of an aluminum dial on a plastic knob.

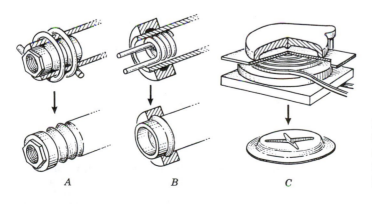

A   B   C

Figure 15.26
Different applications of magnetic forming. *A*, Swaging. *B*, Expanding. *C*, Embossing or blanking.

Both permanent and expandable coils are used in this process. Since the forces act on the coils as well as the work, the coils and insulation must be capable of withstanding the forces or else they will be destroyed. Dielectric insulators fail at around 10 kV and are expensive, whereas the cost of expendable coils is less. As a result, expendable coils generally are used when a high-energy level is required. In magnetic forming, the conducting metal being formed is rapidly accelerated so that much of the forming takes place after the magnetic impulse. Since the magnetic field will not be restricted by nonconducting materials, it is possible to form a part within a container.

The pressure on the work is uniform, production rates are rapid, reproducibility is excellent, lubricants are unnecessary, there are no moving mechanical parts to the machine, and relatively unskilled labor is required. The process is limited in that complex shapes may be impossible to form, pressures cannot be varied over the workpiece, and present units are limited to 60,000 psi (400 MPa) pressure.

15.4 OTHER METHODS

Impact Extrusion

An interesting example of *impact extrusion* is in the manufacture of collapsible tubes for shaving cream, toothpaste, and paint pigments. These extremely thin tubes are pressed from slugs, as illustrated in Figure 15.27. The punch strikes a single blow, causing the metal to squirt around the punch.

The outside diameter of the tube is the same as the diameter of the die, and the thickness is controlled by the clearance between the punch and die. The tube shown in Figure 15.27 has a flat end, but any desired shape can be made by properly forming the die cavity and the end of the punch. For toothpaste tubes, a small hole is punched in the center of the blank and the die cavity is shaped to form the neck of the tube. On the upstroke the tube is blown from the ram with compressed air. The entire operation is automatic with a production rate of 35 to 40 tubes per minute. The tubes are then threaded, inspected, trimmed, enameled, and printed.

Zinc, lead, tin, and aluminum alloys are worked in this fashion. Some tubes are lined

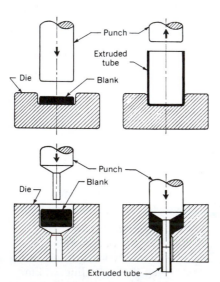

Figure 15.27
Methods of cold-impact extrusion for soft metals.

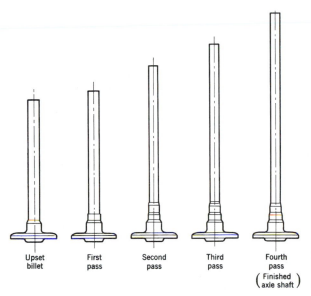

Upset billet	First pass	Second pass	Third pass	Fourth pass (Finished axle shaft)

Figure 15.28
High-speed cold extrusion operations used to manufacture axle shafts.

with foil of a different material that has been clad previously to the blank. Both the foil coating and base metal are formed contiguously.

Impact extrusions are low in cost, have excellent surface quality, and are adapted to production rates of from 100,000 to 20 million parts per year. The process is used to make shell cases, soft drink cans (but not the very popular 12-oz beverage containers), and hollow mechanical parts that have one end totally or partially closed.

The lower half of Figure 15.27 illustrates a variation of what is known as the Hooker process for extruding small tubes or cartridge cases. Small slugs or blanks are used as in the impact-extrusion process, but in this case the metal is extruded downward through the die opening. The size and shape of the extruded tube are controlled by the space between the punch end and die cavity wall.

Copper tubes having wall thicknesses of 0.004 to 0.010 in. (0.10 to 0.25 mm) can be produced in lengths of about 12 in. (300 mm). The process may work-harden the material to an extent that intermediate annealing must be done.

High-speed cold extrusion is used to manufacture axle shafts, as shown by Figure 15.28. Upset billets are loaded onto a three-station cascade-type loading magazine. From this station on, the feeding through the four workstations is automatic. The cold-forming extrusion process elongates the shaft. It is accomplished by forcing a ring die over the axle shaft; the diameter of the die is less than the shaft. The total overall diameter reduction is accomplished in the four stages.

Advantages of this method include: (1) improved surface finish that extends the fatigue life of the material, (2) less stock removal for final finishing is required, and (3) work-hardening of the surface improves the physical characteristics of the material.

Shot Peening

This method of cold working improves the *fatigue* resistance of the metal by setting up compressive stresses in its surface. This is done by blasting or hurling a rain of small shot at high velocity against the surface to be peened. As the shot strikes, small indentations are produced, causing a slight plastic flow of the surface metal to a depth of a few thousandths

Figure 15.29
Surface character of 45 Rockwell C steel that has been shot peened with steel shot. A 10½-in. (490-mm)-diameter Wheelabrator unit was used at a speed of 2250 rpm.

of an inch. This stretching of the outer fibers is resisted by those underneath, which tend to return them to their original length, thus producing an outer layer having a compressive stress while those below are in tension. In addition, the surface is slightly hardened and strengthened by the cold-working operation. Since fatigue failures result from tension stresses, having the surface in compression greatly offsets any tendency toward such a failure.

Shot peening uses an air blast or some mechanical means such as centrifugal force for hurling steel shot on the work at a high velocity. Figure 15.29 is an example of the surface obtained by shot peening. *Surface roughness* or finish can be varied according to the size of shot. Stress concentrations caused by the roughened surface are offset because indentations are close together and sharp notches do not exist at the bottom of the pits. Intense peening is undesirable as it may cause weakening of the steel.

This process adds increased resistance to fatigue failures of working parts and is used on parts of irregular shape and on local areas that may be subject to stress concentrations. Surface hardness and strength are also increased, and in some cases the process is used to produce a commercial surface finish. It is not effective for parts subjected to reversing stresses nor is its effect appreciable on heavy metal sections.

QUESTIONS

1. Give an explanation of the following terms:

Cold working	Coining
Residual stresses	Embossing
Shot peening	Staking
Tube drawing	Seams
Drawbench	Explosive forming
Swaging	Electrospark
Metal spinning	Magnetic forming
Sizing	Impact extrusion
Cold heading	

2. Refer to the body-centered space lattice structure described in Chapter 2. Show how a material made of this structure might develop a slip plane.

3. Give reasons that prevent hot rolling of foils.

4. What effect does the elastic limit of a material have on its ability to be cold worked?

5. Explain the process of strain hardening.

6. How are residual stresses removed from cold-worked metals?

7. Why are pure metals more easily cold worked than alloys?

8. Give the history of cold-drawn tubing from ore to finished product.

9. List the principal advantages for the cold drawing and tube reducer methods of tube production.

10. Explain the production of hypodermic tubing.

11. Discuss the speed of travel of wire feeding through a group of successive dies.

12. Compare spinning over press work for sheet metal products.

13. Explain how bright and satin finishes are produced on foil.

14. What are the characteristics that give advantages in stretch draw forming?

15. Describe the physical properties of a metal that has been cold swaged.

16. What is coining and why is it sometimes used in preference to embossing?

17. What other processes are competitive with die hobbing?

18. How does coining differ from embossing? Which requires more energy at the point of impact?

19. What is the difference between die casting and intraforming?

20. Describe how the fatigue resistance of metal can be improved by cold working.

21. List the high-energy rate forming operations and state the type of work for which each is adapted.

22. Discuss the term "springback" by using concepts from Chapter 2 on stress versus strain.

23. Why do thick materials exhibit less springback than thin ones when explosively formed?

24. For what purposes is shot peening used? List the advantages and disadvantages.

PROBLEMS

15.1. Convert the following pressures to SI units: 40,000 psi, 10 psi, 750 psi.

15.2. Sketch a mandrel to intraform a small internal gear in a forged tube. (Hint: Use engineering paper. A sketch is not an instrument or CAD drawing. Be neat and provide rough dimensions.)

15.3. Prepare a series of sketches for form-rolling house gutters. (Hint: Call up a "seamless gutter" contractor and seek practical advice on the rolling action.)

15.4. Sketch the seam for an oil drum and bucket.

15.5. Describe the process of making a nail in a cold header machine. Sketch a set of dies and cutters. (Hints: Here is an opportunity for "reverse engineering." Examine a nail, the bigger the better, and notice the die gripping marks on the shaft and the die marks on the head of the nail.)

15.6. Design a small tube for holding and dispensing mashed potatoes for space flights. Describe the steps for its manufacture and filling.

15.7. If a ⅛-in.-thick stock is shear spun and the cone angle is 70°, what is the final thickness?

15.8. Design a die and explosive forming system that makes a small metal drinking cup.

15.9. What stretch-forming pressure is required for a material having a yield strength of 105,000 psi for a part that has an area of 194 in.²?

15.10. An aircraft panel that has a compound form has an area of 63 in.² This material has a yield strength of 140,000 psi. How many pounds force is required?

15.11. A 4-in. round pipe with a 1-in. inside diameter is drawn into a 2-in. diameter pipe. About how many passes are required?

15.12. Approximately how many feet of 0.005 in. foil 3 ft. wide can be made from 3 lb of lead? (Hint: The density of lead is given in Chapter 4.)

15.13. Calculate the wall thickness of a large nose cone made from 25.4-mm plate. The included angle of the cone is 60°.

15.14. If it is desired that steel shot leave the wheel at 500 ft/s, what size wheel is needed if it operates at 3600 rpm? What is 500 ft/s in SI units?

15.15. A metal is shear-spun to a thickness of 0.098 in., and a cone angle of 65° is necessary. What original stock thickness is required?

15.16. Select processes to manufacture Figure P15.16 for

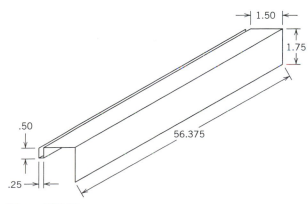

Figure P15.16

the following conditions:

	Material	Thickness	Quantity
a.	CRS	18 ga.	5000
b.	Aluminum	14 ga.	150

PRACTICAL APPLICATION

A class team, as directed by the instructor, is to prepare a list of cold-worked products and processes, as described in this chapter, that are available by local manufacturers. Individual visits and plant reports are encouraged.

CASE STUDY: CENTRIFUGAL FAN FATIGUE PROBLEM

The Wayne Company fabricates high-velocity, high-pressure industrial blowers. Until recently, assembly of the steel fan impellers was completed by welding the blades to the impeller sides, as shown in Figure C15.1. Welding causes the blades to deform slightly and dynamic balancing of the impeller is difficult because of weld grinding. To eliminate the balancing operation, manufacturing superintendent Dave Hock proposes to the company owner that he buy a press to bend a 15-mm flange on both edges of the blades to allow riveting to the sides, thus eliminating welding. Dave claims that the bending and riveting operations will avoid the dynamic problems experienced with welding.

The process to bend the flange on the impeller blades is installed, and, after the usual amount of machine adjustments, the process seems to operate as well as Dave had claimed. The bending and riveting process yields consistent parts that require no dynamic balancing.

Two fans are tested on the test stand for performance. After 1500 hours of operation both fans develop severe vibrations and are shut down for troubleshooting.

Dave, being interested in the success of the project, takes the lead in vibration analysis. He finds that the vibrations are caused by the absence of large pieces that have broken off from the ends of a few of the fan blades. After studying the crack pattern, Dave determines that the failure is caused by fatigue cracks that originated at the flange bends on the blades and that the failure must be related to the new bending process.

Help Dave with this problem. You must determine, given a fatigue crack originating at the bend, the possible alternatives for altering the manufacturing process or material preparation to eliminate the cause of fatigue failure. What additional provision should you consider? How about reinstituting the balancing operation? Consider the solution of a different material thickness. Examine library sources for stress patterns that result from different radius used for bends.

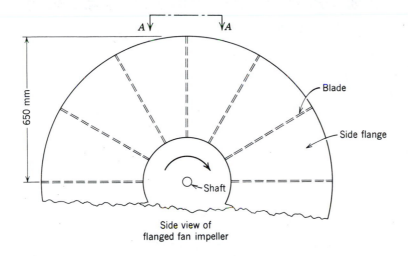

Side view of
flanged fan impeller

650 mm

A ↓   ↓A

Blade

Side flange

Shaft

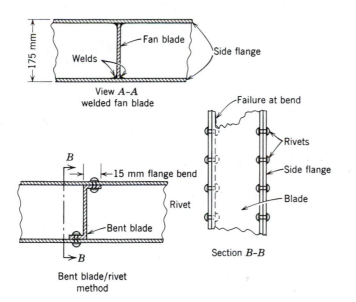

175 mm

Fan blade

Welds

Side flange

View A–A
welded fan blade

Failure at bend

Rivets

Side flange

Blade

Section B–B

B

←15 mm flange bend

Rivet

Bent blade

B

Bent blade/rivet
method

Figure C15.1
Case study.

CHAPTER 16

PRESSWORKING AND OPERATIONS

Pressworking of metal is an essential part of modern manufacturing systems. It allows fast and simple production of millions of products throughout the world. Simply, *pressworking* is cutting, forming, and drawing of sheet metal materials using the punch and die.

In the cutting of sheet metal, "shear forces" describe the forces applied to the metal. The pressure applied by the punch and die is called shear stress. The resistance of the metal is called the shear strength. In order for the metal to be cut, the shear stress must be greater than the shear strength.

The machine used for most cold-working operations and some hot-working operations is known as a *press*. It consists of a machine frame supporting a bed and a ram source of power, and a mechanism to move the ram at right angles to the bed. A press is equipped with dies and punches designed for producing parts in pressworking operations. These tools are necessary for forming, ironing, punching, blanking, slotting, and the many operations that use pressworking equipment.

16.1 PRESSES

Some presses are better adapted for certain classes of work than others. However, most forming, punching, and shearing operations can be performed on any standard press if the proper dies and punches are employed. Presses are versatile, and the same press can be used for different jobs and operations because of the *interchangeable tools*.

Presses are capable of rapid work since the operation time is only one stroke of the ram, plus the time necessary to load the stock and unload the part. Production costs are low. The process is suitable for mass-production methods. For example, pressworking is found in the manufacture of automotive and aircraft parts, hardware specialties, and kitchen appliances. Sometimes, pressworking plants are called *stamping plants*.

An unambiguous classification of press machines is difficult to make. It is not entirely correct to call one press a bending press, another an embossing press, and still another a blanking press as these three operations can be done on one machine. However, some presses are especially designed for one type of operation and are known by the operation name. Examples are a punch press and a coining press. A simple classification is according to the source of power, either manual or power operation. Manual-operated machines are used on thin sheet metal, particularly in jobbing work, but most production machines are power operated, normally by mechanical or hydraulic means. Presses can be grouped according to the number of rams or method of operating the rams. Most manufacturers name them according to the general design of the frame, although they may also be designated according to the means of transmitting power.

Several factors are considered in selecting a press. The operation, size of part, power required, and the speed of operation are important. For most punching, blanking, and trimming operations, *crank- or eccentric-type presses* are generally used. In these presses the energy of the flywheel is transmitted to the main shaft either directly or through a gear train. For coining, squeezing, or embossing operations, the *knuckle joint* is ideal. It has a short ram and is capable of exerting a tremendous force, especially at the most extended position of the ram. Hydraulic presses for drawing operations have slower speeds than those employed for punching and blanking. The standard practice is not to exceed 50 ft/min (0.25 m/s) when drawing mild steel; aluminum and other nonferrous metals may be worked up to 150 ft/min (0.75 m/s).

Presses are given a *tonnage rating*, which is expressed as "the energy expended on the work at each stroke of the press must equal the tonnage required times the distance through which this tonnage must act." This energy is available from that stored in the flywheel and is given to the work as the flywheel slows down. This energy is calculated by

$$E = \frac{N^2 D^2 W}{5.25 \times 10^9} \tag{16.1}$$

where

E = Energy (ton-in.) available at 10% slow down from normal rpm of the flywheel
N = Rotary speed of flywheel, rpm
D = Flywheel diameter, in.
W = Weight of flywheel, lb.

Tonnage is also limited by motor restrictions.

Inclined Press

An *inclinable open-back press* with a gap frame is shown in Figure 16.1. This press, although shown in the vertical position, can be tilted backward to permit the parts and scrap to slide off the back side. Parts slide by gravity into a tote box. Incoming material may feed by chute into the dies. Most presses of this type are adjustable and vary from vertical to a steep angle position. Inclinable presses often are used in the production of small parts involving bending, punching, blanking, and similar operations.

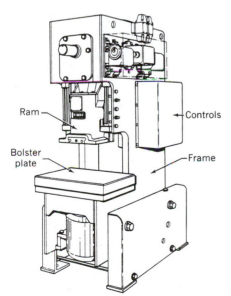

Figure 16.1
Open-back inclinable press.

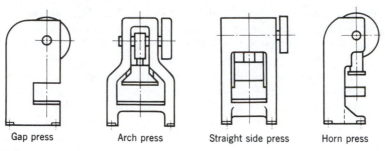

Figure 16.2
Typical frame designs used in presses.

Gap Press

Gap or C-frame presses are named because of the open arrangement of the press frame, as shown in Figures 16.1 and 16.2. Gap presses provide excellent clearance around the dies and permit the press to be used for long or wide parts. Stamping operations are performed on a gap press.

Arch Press

The *arch press*, shown in Figure 16.2, is named for the particular shape of its frame. The lower part of the frame near the bed is widened to permit the working of large area sheet metal. The crankshaft is small in relation to the area of the slide and press bed, as these presses are not designed for heavy work. Arch presses do blanking, bending, and trimming.

Straight-Side Press

As the capacity of the press is increased, it is necessary to increase the strength and rigidity of the frame. *Straight-side presses* are stronger since the heavy loads are taken in a vertical direction by the massive side frame, and there is little tendency for the punch and die alignment to be affected by the strain. These presses are available for capacities in excess of 1250 tons (11 MN).

Straight-side presses are manufactured with various means of supplying power and different methods of operation. For the smaller presses a single crank or eccentric is usually employed, but as the size of the press increases, additional cranks are needed to distribute the load on the slide uniformly. The slide is guided by either one, two, or four guides or points of suspension. *Double-acting presses*, used extensively in drawing operations, have an outer ram that precedes the punch, which clamps the blank slowing its entry into the die before the drawing operation. The outer ram is usually driven by a special link motion or cams, whereas the inner ram carrying the punch is crank driven. The inner ram controls the punch, which then forces itself into the metal giving the shape of the component.

A large straight-sided, enclosed, double-action *toggle press* is shown in Figure 16.3. Pressure is applied on the slide in four places. There is an advantage in large-area presses because such construction prevents tilting of the slide with unbalanced loads. The *toggle mechanism* in this machine controls the motion of the blank holder. A toggle mechanism, shown in Figure 16.12, is a grouping of two or more bars or linkages. When joined

Figure 16.3
Turret top completely formed with one stroke
of enclosed toggle press.

together end to end, they are not in line except when the "knee" is straightened; then great force is achieved on the ends. The time that the force is applied and when there is no motion on the blank holder is known as the dwell period. Dwell is necessary for blank holding on drawing operations, and it frequently is advisable to have a slight dwell on the punch to allow the metal to adjust itself properly under pressure. Straight-side frames also are used on hydraulic presses where heavy loads are encountered, such as forming heavy-gage material, press forging, coining, and deep drawing.

Horn Press

Horn presses, shown in Figure 16.2, ordinarily have a heavy shaft projecting from the machine frame instead of the usual bed. Where a bed is furnished, provision is made to swing it to one side when the *horn* is used. This press is used principally on cylindrical objects involving seaming, flanging edges, punching, riveting, and embossing. Part accessibility is possible with the horn.

Knuckle Joint Press

Presses designed for coining, sizing, and heavy embossing must be massive to withstand the large concentrated loads. The knuckle joint press, shown in Figure 16.4, is designed for this purpose and is equipped with a *knuckle joint mechanism* for actuating the slide. The upper link or knuckle is hinged at the upper part of the frame at one end and fastened to a wrist pin at the other. The lower link is attached to the same wrist pin and the other end to the slide. A third link is fastened to the ends of the wrist pin and acts in a horizontal direction to move the joint, as illustrated in Figure 16.12. As the two

Figure 16.4
Knuckle joint press, 600-ton
(5.3 MN) capacity, where
frame is cast iron.

knuckle links are brought into a straight-line position, tremendous force is exerted by the slide.

This press is used widely for striking coins. A copper penny (which, by the way, only looks as if it is copper, since the core is 99.2% zinc and 0.8% copper, whereas the surface is plated with pure copper) requires 40 tons/in.[2]. Sizing, cold heading, straightening, heavy stamping, and similar operations can also be performed. As the stroke of this press is short and fast, it is not adapted to drawing or bending operations.

Press Brake

Press brakes are used to brake, form, seam, emboss, trim, and punch sheet metal. These presses have a maximum width of 30 ft (9 m) and will form thickness to ⅝ in. (16 mm). Capacities range from ten to several thousand tons of bending capacity. The bed is stationary and is the mounting surface for the lower die. Figure 16.5 is a mechanical brake that will form, punch, and cut to length steel parts.

The pressure capacity required of a press brake is determined by the length of work it will take, thickness of the metal, and the radius of the bend. For bending operations, the pressure required varies in proportion to the tensile strength of the material. Press brakes have short strokes and are generally equipped with an eccentric-type drive mechanism.

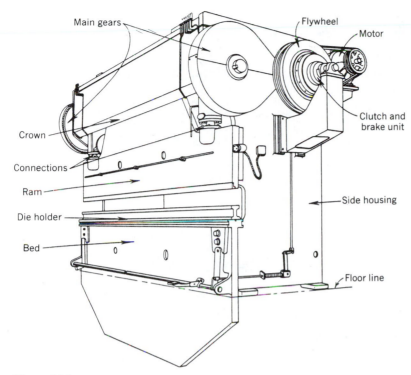

Figure 16.5
Power-press brake.

Figure 16.6 illustrates two unusual presses used with heavier gage material for the production of 30- and 36-in. (760- and 915-mm) pipe for gas transmission systems. The press-type brake bends the huge plate into a U shape as a first step in the operation. From this shape it is squeezed in an "O" press at pressures up to 18,000 psi (125 MPa) into a tubular form. Following this series of forming operations, the pipe is resistance welded, cleaned, and inspected.

Squaring Shears

This machine shears sheets of metal and is made in both power- and manual-operated types. Sheets to a width of 18 ft (5.4 m) can be accommodated. Hydraulic hold-down plungers are provided every 12 in. (300 mm) to prevent any movement of the sheet during the cutting. In operation, the sheet is advanced on the bed so that the line of cut is under the shear. When the foot treadle is depressed, or the computer numerical control (CNC) activated, the hold down plungers descend and the shearing blade cuts progressively across the sheet. The shear angle of the blade can be flat or have a slight angle, which reduces the force of cutting. *Shearing* makes a straight-line cut and does not waste material.

Turret Press

Turret presses are especially adapted to the production of sheet metal parts having varied hole patterns of many sizes. In conventional presses of this kind a template guides the punch, and the hole size and shape are selected by rotating a turret containing the

Figure 16.6
Steps in press-forming large-diameter pipe.

punches. Figure 16.7 illustrates a 30-ton (0.3-MN) CNC turret punch press that will handle plate or sheet metal sizes to 48 in. by 72 in. (1200 mm by 1830 mm). The sheet metal is positioned under a punch with a table speed of 300 in./min (0.1 m/s). Numerical control directs the location of the sheet stock at the punch and die position. Holes up to 4¾ in. (120 mm) can be punched in ⅜-in. (9.4-mm)-thick steel at a rate of more than 30 hits per minute to an accuracy of 0.005 in. (0.13 mm). The press is called a ''turret'' because it will garrison a large number of punch and die sets.

Hydraulic Press

Hydraulic presses have longer strokes than mechanical presses and develop full tonnage throughout the entire stroke. The tonnage and length of stroke of these presses is adjustable. Maybe only a fraction of the capacity may be necessary. The presses are adapted to deep drawing operations because of their slow uniform motion. This characteristic makes for more consistent parts.

They briquette powdered metals, extrude, laminate, plastic mold, and press forge. They are not recommended for heavy blanking and punching operations as the break-through shock is detrimental to the press. Maintenance is higher than for mechanical presses even though the operation of the press is much slower. Small hydraulic presses resemble straight-side presses. For large area work, the post or four-column type of construction is used.

The hydraulic press, Figure 16.8, is designed for making deep draws in sheet metal.

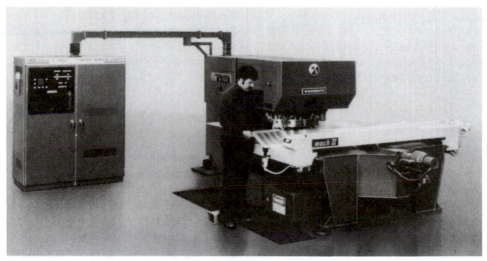

Figure 16.7
Tape-controlled turret punch press.

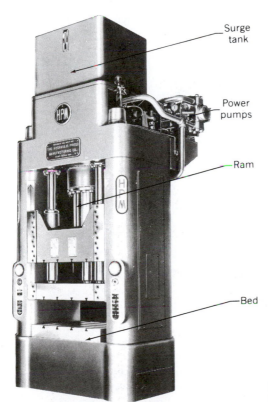

Figure 16.8
Double-action metal-drawing press.

The main draw punch mounted on the upper slide moves in tandem with the blankholder slide and ring below it until the blank is contacted. The die rests on the bolster plate; below it is a die cushion that can assist in maintaining pressure on the blank or ejecting the formed part. By locking the blankholder to the main slide and the die cushion idle, the press acts as a single-action hydraulic press.

Transfer Press

Transfer presses are fully automatic and are capable of performing consecutive operations simultaneously. Material is fed to the press by rolls or as blanks from a stack feeder. In operation the stock is moved from one station to the next by a mechanism synchronized with the motion of the slide. Each die is a separate unit and is provided with a punch that may be independently adjusted from the main slide. Figure 16.9 is a 250-ton (2.2-MN) unit that produces 1600 starter end plates per hour.

Cost-effective use of transfer presses depends on quantity production, as their usual production rate is 500 to 1500 parts per hour. Sheet metal products made on this equipment have a sequence of shearing, blanking, and forming operations. Some presses have 1200 strokes per minute.

Fourslide

The *fourslide machine* has many advantages over the punch press for complex forming operations on small parts made of sheet metal or wire. The basic machine has four power-driven slides set 90° apart, which are separately controlled by cams to move progressively through a cycle. Figure 16.10 shows the sequence of operations for forming a spring clip.

The fourslide machine can be equipped with punches, cutting off and blanking tools, lifting and shifting devices, drills, and, in some cases a vertical punch. The tools can be

Figure 16.9
Transfer press with capacity of 250 tons (2.2 MN) produces 1600 starter end plates per hour.

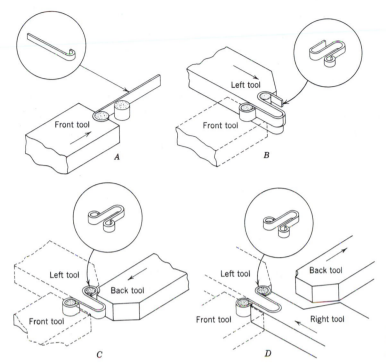

Figure 16.10
Sequence of operations
on a fourslide machine.

made to pivot or open and close. If wire, strip, or coil stock is fed into the machine, it is first straightened by passing it through rollers attached to the frame. Hoppers and precision locators can be used if the parts are preblanked. Spot and butt welders also can be employed. The process is almost automatic and lends itself to mass-produced parts.

Laser Cutting

Laser-cutting machines utilize a high-powered CO_2 or $ND:YAG$ laser to melt or vaporize certain materials. The laser produces an invisible infrared beam of radiation, which is

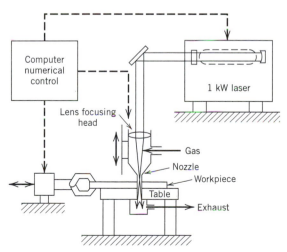

Figure 16.11
Laser cutting.

focused on the workpiece by a ZnSe lens. The power ranges from 200 watts to 1.5 kW. The lens focuses the beam on a very small area (0.3-mm spot diameter). A laser power of 1 kW gives an average density of 1.4 MW/cm^2 that will boil instantly any material if exposed for a length of time. This pin pointing allows for a narrow kerf.

With the numerically controlled table, arbitrary contours can be cut with tight tolerances. Lasers cut material from soft rubber to titanium alloys. Materials having a high reflectivity, such as copper, silver, and gold, are limitations. From the same nozzle of the laser, there are assisting gases, such as CO_2, applied to the cutting area. These high-pressure gases assist the burning reaction, remove the molten metal, and protect the lens from splatter. Notice Figure 16.11 for the diagram of the process.

Metals up to ½ in. and nonmetals up to 1 in. can be cut. These machines are useful for very small quantities and for irregular geometries.

16.2 DRIVE MECHANISMS FOR PRESSES

Drive mechanisms for transmitting power to the slide are shown in Figure 16.12. The most common drive is the single crank, which gives a slide movement approaching simple harmonic motion. On a downstroke, the slide is accelerating. Reaching its maximum velocity at midstroke, it then decelerates. Most press operations occur near the middle of the stroke at maximum slide velocity.

The *eccentric drive* gives a motion like that of a crank and is often used where a shorter stroke is required. Its proponents claim it has greater rigidity and less tendency for deflection than a crank drive. Cams are used when some special movement is desired, such as a dwell at the bottom of the stroke. This drive has some similarity to the eccentric drive except that roll followers are used to transmit the motion to the slide.

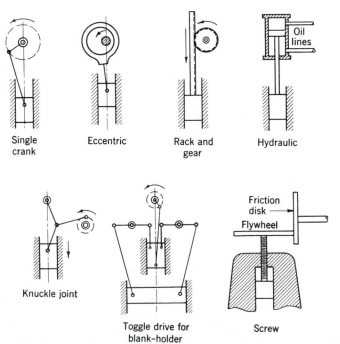

Figure 16.12
Drive mechanisms used on presses.

Rack and gear presses are used for applications requiring a very long stroke. The movement of the slide is slower than in crank presses and uniform motion is attained. These presses have stops to control the stroke length and are equipped with a quick return feature to raise the slide to starting position. The manual-controlled arbor press is a familiar rack and gear example.

Hydraulic drive is used in presses for a variety of work. It is especially adapted to large pressures requiring slow speed for forming, pressing, and drawing operations.

In the screw drive, the slide is accelerated by a friction disk that engages the flywheel. As the flywheel moves down, greater speed is applied. From the beginning to the end of the stroke the slide motion is accelerated. At the bottom of the stroke the amount of stored energy is absorbed by the work. The action resembles that of a drop hammer but is slower and there is less impact. Presses of this type are known as percussion presses.

Several link mechanisms are used in press drives because of the motion or because of their mechanical advantage. The knuckle joint mechanism is common. It has a high mechanical advantage near the bottom of the stroke when the two links approach a straight line. Because of the high load capacity of this mechanism, it is used for coining and sizing operations.

Eccentric or hydraulic drives may be substituted for the crank. Toggle mechanisms used primarily to hold the blank on a drawing operation are made in a variety of designs. The auxiliary slide in the figure is actuated by a crank, but eccentrics or cams may be used. This mechanism obtains a motion having a dwell so that a blank can be held effectively.

16.3 FEED MECHANISMS

Safety is a paramount consideration in press operation, and every precaution must be taken to protect the operator. Wherever possible, material is fed to the press that eliminates any chance of the operator having his or her hands near the dies. In long-run production, these features are designed into the operation. Feeding devices applied to medium-size and small presses have the advantage of rapid, uniform machine feeding in addition to the safety features.

One common feeding mechanism is the double-roll feed utilizing coiled stock and scrap reels. The operation of the feed rolls is controlled by an eccentric on the crankshaft through a linkage to a ratchet wheel, which pulls the material across the die. Each time the ram moves up, the rolls turn and feed the correct amount of material for the next stroke. By providing the machine with a variable eccentric, the amount of stock fed through the rolls is varied. An automatic high-speed press equipped with a slide feed is shown in Figure 16.13. The scrap from this press, instead of being rerolled on a scrap reel, is sheared to small lengths for easy handling. For heavy materials, straightening rolls also act as feeding rolls.

Another feeding device is the dial station feed. This method is designed for single parts previously blanked or formed in some other press. The indexing is controlled by an eccentric on the crankshaft through a link mechanism to the dial. For each stroke the dial indexes one station. Feeding by the operator takes place at the front of the machine away from the dies.

Light parts can be stacked in a magazine and placed in the die position by a suction device. A blank is lifted off the top of the stack by suction fingers and placed against a stop gage on the die. Magazine feeds are also used with a reciprocating mechanism that feeds blanks from the bottom of the stack. Gravity feed is sometimes used on inclined presses, the blank sliding into a recess at the top of the die.

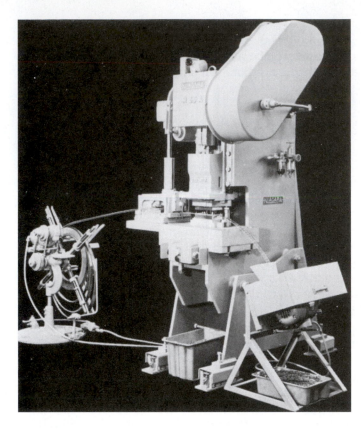

Figure 16.13
Automatic high-speed press
with propeller shaft-driven
slide feed. Capacity 35 tons
(0.3 MN).

16.4 OPERATIONS

Operations that make up press work are varied, but can be broadly classified as shearing, bending, and drawing. Pressworking tools are called *punches* and *dies*. The punch of the assembly is attached to the ram of the press and is forced into the die cavity. The die is usually stationary and rests on the press bed or a bolster plate. The die has an opening to receive the punch, and the two must be in alignment for a quality operation. Punches and dies are not interchangeable; they are mated for a particular operation. A single press may do a large variety of operations depending on the dies used.

Shearing

Shearing is a general description for most sheet metal cutting, but in a specific case, it is the cut along a straight line completely across a strip, sheet, or bar.

Cutting metal involves stressing it in shear above its ultimate strength between adjacent sharp edges, as shown in Figure 16.14. As the punch descends on the metal, the pressure first causes a plastic deformation to take place, as shown in Figure 16.14*B*. The metal is highly stressed adjacent to punch and die edges, and fractures start on both sides of the sheet as the deformation continues. When the ultimate strength of the material is reached, the fracture progresses and if the clearance is correct and both edges are of equal sharpness, the fractures meet at the center of the sheet, as shown in Figure 16.14C.

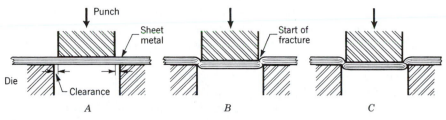

Figure 16.14
Process of shearing metal with punch and die. *A*, Punch contacting metal.
B, Plastic deformation. *C*, Fracture complete.

Notice Figure 16.15 for common pressworking operations. Cutting off implies the shearing of a piece from a strip with a cut along a single line. Parting implies that waste is removed between two pieces to part them.

Blanking cuts an entire piece from sheet metal, but there is stock entirely around the contour of the part. The good part is called a blank. The remaining material is often termed a *skeleton* or waste. If the operation is to cut a hole, and the material in the hole is waste, then the operation is *piercing* or *punching*.

Nibbling is an operation that cuts stock using a machine called a nibbler. The machine has a small circular or triangular punch that oscillates vertically and rapidly in and out of a mating die. The material is guided such that the nibbles are overlapping holes that punched. Eventually a path is nibbled such that a part is defined. Nibbling is a low-quantity method and does not require tools. Templates or scribe lines are used as guides. Some nibbling machines can fold, bead, louvre, flange, and slot. Sizes of nibbling machines are rated by throat depth from the punch to the C-frame and maximum thickness of mild steel that can be cut.

Shaving is a finishing or sizing process, and very little material is removed from the edges. Usually, the edge has been pressworked previously. Shaving is intended to produce tight tolerance holes or blanks.

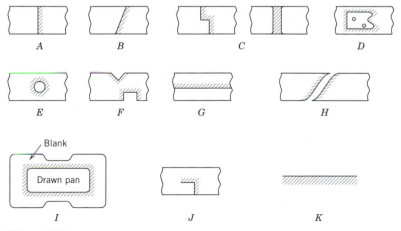

Figure 16.15
A, Shearing. *B*, Cutting off. *C*, Parting. *D*, Blanking. *E*, Piercing. *F*,
Notching. *G*, Slitting. *H*, Nibbling. *I*, Trimming. *J*, Lancing. *K*, Symbol
indicating the shear cut.

Slotting is the cutting of elongated holes, or slots. *Perforating* is the cutting of a small and evenly spaced grouping of holes. *Notching* is the cutting of material from the side of a strip or sheet or blank. Trimming is the cutting away of excess material in a flange or flash from a piece.

Blanking, as shown in Figure 16.16*B*, is the operation of cutting flat areas to a desired shape. It is usually the first step in a series of operations. In this case, the punch should be flat and the die given some shear angle so that the finished part will be flat. Punching or piercing holes in metal, notching metal from edges, or perforating are all similar operations. For these operations the shear angle is on the punch and the metal removed is waste.

Slitting is the lengthwise cutting of coil or sheet stock into narrower widths. *Lancing* makes a cut part way through a blank; notice Figure 16.16*D*.

Clearance between the punch and die diameters plays an important part in die design. If improper clearance is designed, the shear fractures between the top and bottom of the stock do not meet, and instead the punch travels the entire sheet thickness, using more power. For insufficient clearance, a secondary shear is caused by fractures. Too much clearance causes excessive deformation.

The designed distance of this clearance depends on the thickness, hardness, and strength of the material. For thin material the punch should be a close sliding fit. For heavier stock the clearance is larger to create the proper shearing action on the stock and to prolong the life of the punch. It is important that the clearance between punch and die be properly dimensioned for the desired clean cut.

Clearance, then, is the distance per side between the punch and die. The advantage of designating clearance as the space on each side is necessary with dies of irregular form or shape. Whether clearance is added or deducted from the diameter of the punch or to the diameter of the die depends on the operation, either blanking or piercing. If a blank has a required size, the punch is made smaller and the die dimension is to size in the operation of blanking. When a hole dimension is required, as in piercing, the punch is ground to the required diameter and the die is made larger. Piercing is the simple punching of holes.

Notice Figure 16.17. In Figure 16.17*A*, the blanking dimension is the size of the die opening, and the punch is reduced by 2*C*, where *C* = the clearance factor times the thickness. In Figure 16.17*B*, which is for piercing, the hole diameter is the punch diameter.

For 1100 and 5052 aluminum alloys, the clearance is 4½%. With 2024 and 6061 aluminum alloys, brass, cold-rolled steel (CRS), the average clearance is 6% of the thickness, whereas for hard steels the clearance is 7½%. For example, a 0.562-in. hole is pierced from 18 ga (0.048 in.) CRS. The *slug* from the hole is waste. One-side clearance is 0.0029

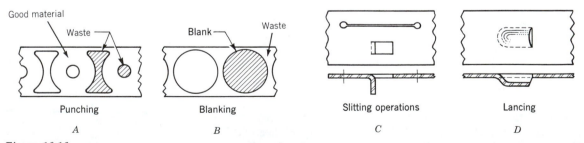

Figure 16.16

A, B, Illustrating the difference between punching and blanking operations. *C, D,* Examples of slitting and lancing operations.

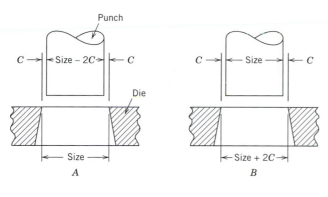

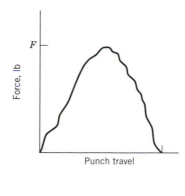

Punch

C → |← Size – 2C →| ← C

Die

|← Size →|

A

C → |← Size →| ← C

|← Size + 2C →|

B

Figure 16.17
Punch and die. *A,* Blanking clearance where slug is desired part. *B,* Piercing clearance where slug is waste.

($= 0.06 \times 0.048$), the punch diameter is 0.5620 in., and the die opening is 0.5678 ($= 0.5620 + 2 \times 0.0029$). Punch and die diameters are ground to closer ten-thousandth tolerances than the part dimension.

Notice Figure 16.18 for diagrams of the force exerted by a punch-penetrating sheet stock. As the punch enters the material, the force quickly builds up, as indicated in Figure 16.18*A.* If the clearance is correct, the material breaks suddenly. If the break clearance is not correct, the force does not dissipate quickly, and the force curve is given by Figure 16.18*B.* In either case the energy is represented by the area under the curve. In practice, however, the situation requires that friction be overcome. Simple relationships can be constructed to find horsepower.

The distance of *penetration,* given as a percent p times the thickness of the stock t in in., is the x-axis of Figure 16.18. The maximum force F in pounds is equal to the product of the length of cut in inches times the thickness of the stock times the shearing strength S, given in psi. If the piece is round, then

$$F = \pi D t S \qquad (16.2)$$

where

$$F = \text{Blanking or punching force, lb}$$
$$D = \text{Diameter of punch, in.}$$
$$t = \text{Stock thickness, in.}$$
$$S = \text{Shear strength of material, psi}$$

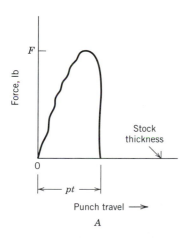

F

Force, lb

Stock thickness

0

|← pt →|

Punch travel →

A

F

Force, lb

Punch travel

B

Figure 16.18
Force exerted by a punch shearing sheet metal. *A,* Correct punch and die clearance. *B,* Improper clearance.

Any odd-shaped perimeter can be used by substituting the length L for πD. The energy E is estimated from an empirical formula

$$E = 1.16Fpt/12 \qquad (16.3)$$

where

E = Energy, ft-lb
p = Penetration of punch into stock, %

Shearing strength and percent penetration are properties of the material. Typical values are 8000 psi and 60% for aluminum, 48,000 psi and 38% for 0.15% carbon steel annealed, and 71,000 psi and 24% for 0.5% carbon steel annealed.

If the punch makes N strokes per minute, the power in horsepower is

$$P = EN/33,000 \qquad (16.4)$$

where

N = Strokes per minute

Flat punches and dies, as shown in Figure 16.17, require a maximum of power. To reduce the shear force the punch or die face should be ground at an angle so that the cutting action is progressive. Think of using scissors that cut the stock, such as paper. Punches and dies that have a die face that angles slightly behave in the same way. This distributes the shearing action over a greater length of the stroke and can reduce the power required by up to 50%.

Bending and Forming

Bending and forming may be performed on the same equipment as that used for shearing; namely, crank, eccentric, and cam-operated presses. Where *bending* is involved, the metal is stressed in both tension and compression at values below the ultimate strength of the material without appreciable changes in its thickness. Notice Figure 16.19.

In a press brake, simple bending implies a straight bend across the sheet of metal. Other bending operations, such as curling, seaming, and folding are similar, although the process is slightly more involved. Bending pressures are determined using the following

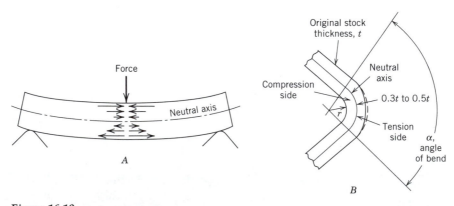

Figure 16.19
A, Simplest form of bending with deflection within elastic limits. B, Effects of bending on length of bend.

empirical relationship:

$$F = \frac{1.33LSt^2}{W}$$
(16.5)

where

F = Bending force, tons
L = Length of bend, in.
S = Ultimate tensile strength, tons/in.^{2}
t = Metal thickness, in.
W = Width of V-channel or U lower die, in.

A V-channel or a U-die is a punch-and-die arrangement. These punch and dies often form a 90° internal angle between the faces of the metal. The empirical constant of 1.33 is a die-opening factor proportional to metal thickness.

In designing a rectangular section for bending, one must determine how much metal should be allowed for the bend, since the outer fibers are elongated and the inner ones shortened. During the operation the neutral axis of the section is moved toward the compression side, which results in more of the fibers being in tension. The thickness is slightly decreased. Although correct lengths for bends can be determined by empirical formulas, they are influenced considerably by the physical properties of the metal.

The minimum inside radius of a *bend* is usually limited to the thickness of the material. The stretching of a bend causes the neutral axis along which the stock is not strained to move to a distance of $0.3t$ to $0.5t$ from the inside of the bend. An average $0.4t$ is used for calculations. Notice Figure 16.19B. The arc length of the bend is found as

$$L = 2\pi(r + 0.4t)\alpha/360$$
(16.6)

where

L = Length of bend, in.
r = Inside radius of bend, in.
α = Angle of bend, degree

Metal that has been bent retains some of its original elasticity, and there is some elastic recovery after the punch is removed, as shown in Figure 16.20A. This is known

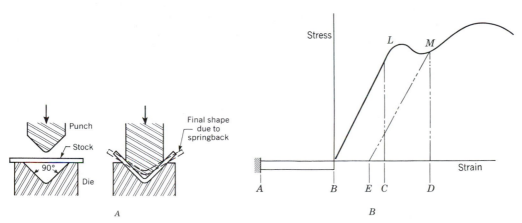

Figure 16.20
A, Springback in bending operations. B, Springback and its relationship to stress–strain.

as *springback*. Figure 16.20*B* shows a generalized stress–strain curve that explains the importance of the yield point in bending operations. If a part *A–B* is stretched to a length *A–C* and released, it will return to its original length *A–B*. If *A–B* is stretched past the proportional limit, *L*, on the stress–strain curve to a length *A–D* and the load removed, then its final length is *A–E*. The springback in this case is *E–D*. The line *M–E* is parallel to *L–B*. Thus, depending on the stress–strain characteristics of the metal and the load employed, an indication of the extent of springback is possible.

Springback may be corrected by overbending an amount such that when the pressure is released, the part will return to its design dimensions. Springback is more pronounced in large-bend radii. The minimum-bend radius varies according to the ductility and thickness of the metal.

Forming is an operation that reproduces flat stock into the shape of the punch and die with little or no plastic flow of the metal. An example is a punch and die designed to form a flat strip of steel into a U-shape.

Drawing

Drawing is seen in sheet metal pots, automobile panels, cartridge and shell cases, and pans. Drawing employs plastic flow of sheet metal where a deeply recessed hole exceeds the final part diameter. Think of a bullet shell. Another example is the beverage can container industry, where the 12-oz. cans are drawn and ironed. *Ironing* results in the thickness of the shell wall reduced and its surface smoothed. In ironing, the wall thickness is less than the original stock thickness, but the bottom of the formed shell remains the same as the original stock thickness. Drawing is mostly done cold.

Three bent flanges are shown in Figure 16.21. Sketch *A* is the simple straight bend. The straight flange has no longitudinal stresses imposed on the material except at the bend radius.

The shrink and stretch flange, shown in Figure 16.21*B* and *C*, involve metal plastic flow that does not take place in a straight bend flange. This plastic flow or adjustment of metal is characteristic of all drawing operations. Stresses are involved that exceed the elastic limit of the metal to permit the metal to conform to the punch and die. However, these stresses cannot exceed the ultimate strength without developing cracks and tears. If the stretch flange in Figure 16.21*B* is considered to be a section of a circular depression that has been drawn, the metal in arc *aa* must be stretched to *a'a'*. The action is a thinning one and must be uniform to avoid tears or cracks. In the *shrink flange* in Figure 16.21*C* the action is just the opposite, and the metal in the flange is thickened. For the shrink flange, it is important to avoid wrinkles. Most drawn parts start with a flat plate of metal. As the punch is forced into the metal, severe tensile stresses are induced into the sheet

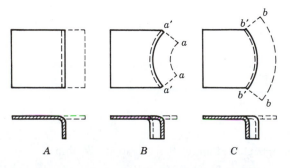

Figure 16.21
Types of flanges. *A*, Straight. *B*, Stretch. *C*, Shrink.

being formed. At the same time, the outer edges of the sheet that have not engaged the punch are in compression and undesirable wrinkles tend to form. This must be counteracted by a blank holder or pressure plate, which holds the flat plate firmly in place.

Most drawing, involving the shaping of thin metal sheets, requires double-acting presses to hold the sheet in place as the drawing progresses. Notice Figure 16.8. Presses of this type usually have two slides, one within the other. One slide controlling the blank holding rings moves to the material ahead of the other to provide resistance for the blank. The intention is to provide resistance to the forming of the blank into the cavity. The motion of the blank holding slide is controlled by a toggle or cam mechanism in connection with the crank. Hydraulic presses are well adapted for drawing because of their relatively slow action, close speed control, and uniform pressure.

However, these presses need tools, both standard and designed. Components of die drawing are shown in Figure 16.22. In Figure 16.22A, there is the punch, which moves downward in the press cycle, forcing the circular blank and a circular draw ring, which has a radius, to encourage smooth transition of the metal into the die. In simple drawing operations of relatively thick plate, the plate thickness may be sufficient to counteract wrinkling, and a smooth cup is drawn.

In Figure 16.22B and C, the double-acting hydraulic press slide of Figure 16.8 has the first slide engage the blank holding ring, and the second slide controls the action of the punch. The arrangement in Figure 16.22B is more suitable for thin sheet stock. The punch then forces the sheet metal blank into the shape of the punch and die, and the cup is ejected through the bottom of the die.

Differences between drawing and ironing are shown in Figure 16.23. In Figure 16.23A, the flat blank has an initial thickness of t_1. A partially drawn cup is given in Figure 16.23B, and the thickness relationship is $t_2 < t_1$. Ironing is the operation in which the thickness of the shell wall is reduced and its surface smoothed. Remember the example of the beverage can. In Figure 16.23C, $t_3 < t_1$ and the draw is complete to a cup. The diameter relationships are $D_1 > D_2 > D_3$ and height is $h_2 < h_3$.

The necessary force applied to the punch to draw a shell or a cup is equal to the product of the cross-sectional area and the yield strength, S, of the metal. A constant that covers the friction and bending is necessary in this relationship. The force for a cylindrical shell is expressed by the empirical equation

$$F = \pi dt S(D/d - 0.6) \tag{16.7}$$

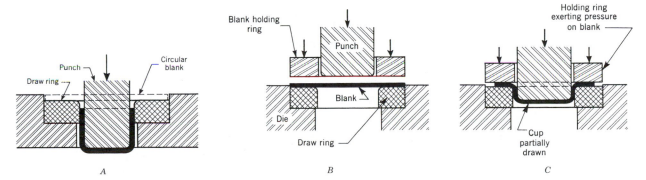

Figure 16.22
A, Arrangement of punch and die for simple drawing operations. B, C, Action of blankholder and punch in a drawing operation.

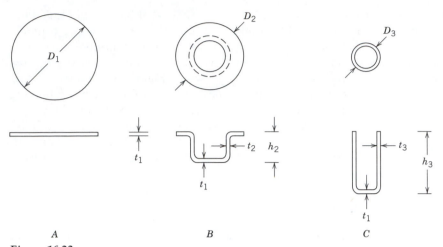

Figure 16.23
A, Circular blank. B, First draw. C, Continued drawing and ironing.

where

$$D = \text{Blank diameter, in.}$$
$$d = \text{Shell diameter, in.}$$
$$t = \text{Metal thickness, in.}$$
$$S = \text{Tensile strength, psi}$$

Rubber-Pad Processing

In *rubber-die processing*, a rubber or urethane pad confined in a container replaces the die or punch. Under pressure, the rubber pad flows around a form block or punch, and the blank is punched or formed. Advantages of these methods are lower cost tooling for short-run production. In the *Guerin process*, the rubber pad is in a boxlike container mounted on the ram side. As the *platen* moves down, the force of the ram is exerted evenly in all directions, resulting in the sheet metal being pressed against the die block, as shown in Figure 16.24. Cutting die blocks are merely steel templates of the part to be made and need not be more than ⅜ in. (10 mm) thick. Forming dies may be made of hardwood, aluminum, or steel. Aluminum sheet can be cut in thicknesses up to 0.051 in. (1.30 mm); for bending and forming the usual limit is approximately ³⁄₁₆ in. (5 mm) thick.

A Marform process permits deeper drawing and the forming of irregularly shaped

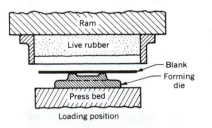

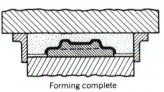

Figure 16.24
Method for forming sheet metal using single die and rubber pad.

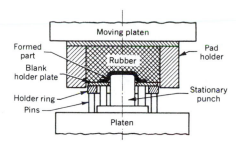

Figure 16.25
Arrangement of the components in a forming operation with the Marform process.

parts. In the operations shown in Figure 16.25, a flat piece of metal is placed on the blank holder plate. As the platen descends, the rubber pad contacts the blank and clamps it against the top of the punch and surrounding plate. As the downward movement continues, the blank is formed over the end of the punch and sufficient pressure is exerted over the unformed portion to prevent metal wrinkle. During the drawing operation the downward movement of the blank is opposed by controllable pressure pins. In forming aluminum, sheet thicknesses up to 0.675 in. (17.15 mm) have been processed.

Deep drawing can be achieved by another process known as Hydroform, which employs a rubber cavity in the ram containing hydraulic fluid. Lowering the ram clamps the blank between the flexible rubber die member and a blank holder on the bolster. A punch attached to a hydraulic cylinder assembly moves upward and increases the diaphragm pressure on the blank drawing it around the punch. Figure 16.26 shows the elements of the process.

Steel Rule Dies

The *steel rule die* process employs *steel ribbons* mounted in hardwood as the die. The steel rule die is similar to a cookie cutter. The process eliminates the solid metal die section used in blanking sheet metal stock. The steel rules are shaped to the outer configuration of the part to be blanked and are inserted in a hard or specially laminated wood backing. The shearing action and a steel rule are shown in Figure 16.27. The punch may be of 1045 flame-hardened steel or ground-gage stock. The strip of stock used in steel rule dies varies from 0.056 to 0.166 in. (1.42 to 4.22 mm) thickness and the shearing edge side is ground at a 45° bevel. Steel rule dies are useful for low quantity blanking of up to ⅜ in. (10 mm) of steel or 0.55 in. (15 mm) of aluminum.

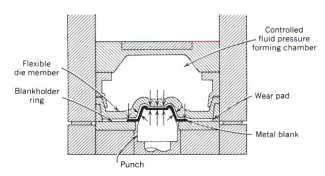

Figure 16.26
Principle of fluid forming.

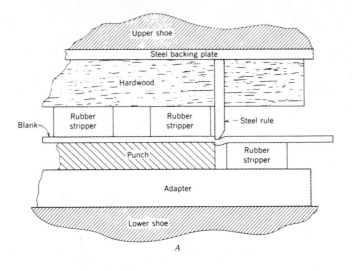

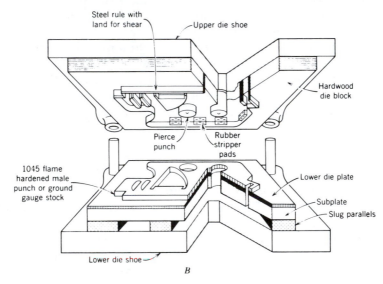

Figure 16.27
Steel rule die process.
A, Cross section of
shearing action of steel
rule. *B*, Assembled steel
rule die.

16.5 EFFICIENT USE OF MATERIALS

Sheet metal products are usually material intensive, meaning that the cost of material is the major part of product value. Stamping firms look for every way to reduce the cost of the product, and the material cost is given close scrutiny. The design dictates the size, gage, grade, finishes, appearance, and selection of the product, and the raw material is firmly guided by those specifications.

Raw material for first pressworking operations starts as coil stock, sheets, strips, and even bars. Efficient material utilization involves the principle of minimum waste, which means that the skeleton, or the leftover after the part has been blanked out, is minimum. Stock layout is one of the first steps in die engineering.

Virtually every stamping design incurs losses. These losses are called waste, scrap, and shrinkage. *Waste* is the losses of the manufacturing operation, such as the skeleton strip stock left over after the blanking operation. *Scrap* is a loss of material because of mistakes in stamping, either design or manufacturing operations. An example is worn punches that are unable to hold tolerances. *Shrinkage* is the loss of material that is caused by changes in material quality, perhaps resulting from aging. Layout of the part in the wrong way on the strip because of the grain alignment caused by directional rolling in the steel mill, and then the stress release of the part after blanking, resulting in loss is an example.

Yield is the effective material compared to original material of the stamping operation, and is found as

$$E_s = S_t/S_a \times 100 \qquad (16.8)$$

where

E_s = Shape yield of the operation, %
S_t = Theoretical material in part design, lb or in.^{2}
S_a = Actual material used, lb or in.^{2}

Notice Figure 16.28, which shows two potential layouts of a ϕ 1.500 in. blank. The die designer geometrically evaluates the yield, which is the material in blank to material requirements of the strip. Advance is the term that indicates the distance that the stock is pulled though the die for one blanking cycle or ram stroke to shear the stock completely, allowing it to separate freely from the strip. The advance for Figure 16.28*A* and *B* is 1.625 and 0.813 in. In Figure 16.28*B*, the blank circles are laid with an angle of 30°. Excluding any waste at the beginning and end of the strip, the *yield efficiency* of Figure 16.28*A* is 58% and the yield of *B* is 65%. It is evident that more pieces on the width can reduce the cost of the part. This is not always true, however, as a wider strip stock can increase tooling cost, require more equipment capacity, and require greater maintenance. Therefore, more factors are required before it can be stated with certainty that wider strip sizes will always reduce the cost of the part.

Minimizing losses for sheet metal work are guided by experience. One rule of thumb in stamping fabrication is to restrict the distance between blanks as 0.75 by gage thickness. Margins between the edge of the strip and the blank are restricted to 0.90 by thickness for each side. Usually, laying out the blank design at an angle, and having multiple blanks per advance increases the yield.

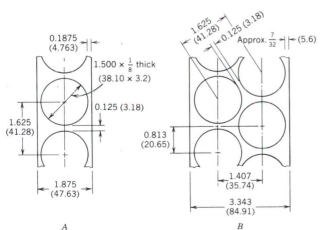

Figure 16.28
Strip stock layouts.

QUESTIONS

1. Give an explanation of the following terms:

Knuckle joint	Slitting
Double-acting presses	Lancing
Toggle mechanism	Springback
Horn	Shrink flange
Shearing	Guerin process
Turret presses	Platen
Transfer presses	Steel rule die
Fourslide machine	Scrap
Perforating	Yield efficiency

2. Explain the differences between a die and a punch.

3. Give the principal advantage of an inclined press over a vertical one. Also, over one with a rigid frame.

4. Define blanking and how it differs from other processes.

5. What type of press is used for making automobile license plates? Why is it used?

6. When do you recommend a turret press?

7. List the differences between hydraulic and mechanically driven presses.

8. Describe a transfer press. List 10 products that can be produced on a transfer press.

9. Select a part to be made economically on a transfer press and sketch the sequence of operations.

10. When is the hydraulic press adopted for making parts?

11. Consider the ultimate strengths of aluminum and mild steel, and then discuss the action in shearing.

12. Discuss the clearance between a punch and die for shearing operations. Why is it important?

13. List the advantages of the five types of press drives.

14. Explain the purpose of double acting presses.

15. Describe the following operations: blanking, punching, shaving, slitting, and lancing.

16. What role does the rubber pad play in processes using only a single die?

17. Describe a steel rule die. Why are they limited to short run work?

18. What press do you recommend for each of the following jobs: forming steel tops for automobiles, stamping coins, screw tops for glass jars, nose cones for small missiles, and file-case drawer fronts?

19. List the differences in finding clearance between a punch and die for blanking and piercing.

PROBLEMS

16.1. Convert the following customary press sizes to SI units: 400 ton, 1200 ton, and 3 ton.

16.2. Sketch the operation of a knuckle joint mechanism. What products use this press? (Hint: A sketch is done with pencil, not mechanical drawing instruments or CAD.)

16.3. Sketch a process to form a small metal cup.

16.4. Sketch an arch press and describe the work for which it is designed. (Hints: Use a 0.5-mm pencil and engineering computation paper, and label the major structural parts of the press. Do only an elevation view. Remember that a sketch is not a finished engineering drawing.)

16.5. Construct a flow diagram for manufacturing license plates. (Hints: Start with the raw material. Sketch a stamping plant using a block layout. Assume that the license plate is the only product. A *flow diagram* is a sequence of the operations and handling leading to shipment of the product to distributors. Confine your work to one page of engineering computation paper.)

16.6. Sketch the stages of converting a ribbon of thin metal 0.020 in. thick and ½ in. wide into a paper clip on a fourslide machine.

16.7. What size press in tons is needed if a 6-in. diameter is formed at a pressure of 20 tons per square inch?

16.8. A press flywheel has the diameter of 28 in., weight of 690 lb, and it operates at 360 rpm. Find the available energy of the flywheel.

16.9. Find the total energy of the flywheels in ton-in. if each of two wheels is 36 in. in diameter, has a thickness of 4.25 in., operates at 420 rpm, and the density is 0.29 lb/in.^{3}.

16.10. Large, flat, round washers, ϕ 6.000 in., have their ϕ 2.000-in. ID centers pierced. Raw material is low-carbon annealed steel, 0.125 in. thick.

a. Sketch and dimension a cross section of a punch and die only. (Hints: Find the clearance for this material. Use engineering computational paper.)

b. Repeat for the blanking of the outside diameter.

16.11. Large, flat, round washers, 5.000 $^{+0.006}_{-0.000}$-in. OD, are to have their 1.000 $^{+0.008}_{-0.000}$-in. ID centers pierced using a piercing die and in secondary operation, the OD is blanked. The washer raw material is CRS, 0.0625 in. thickness.

a. Dimension a sketch of the cross section of a punch and die for piercing and blanking.

b. Repeat for aluminum material.

c. Repeat for stainless steel. (Hints: Work the design from the nominal dimension of the hole, and then apply tolerances to the punch and die. The tolerances for the punch and die are 10% of the hole tolerance.

16.12. Find the force to punch the strip stocks shown by Figure 16.28. Cold-rolled steel shear strength is 42,000 psi.

16.13. Assume that the material of Figure 16.28 is stainless steel and has a shear strength of 122,000 psi. Find the force.

16.14. Find the punching force for 2024-T grade of aluminum having a shear strength of 40,000 psi. The blank perimeter is 17 in. and thickness is 0.070 in.

16.15. Steel grade AISI 1050, 0.93 in. thick, is pierced for a 2.8-in. ID hole using a punch press with 119 strokes per minute. Find the force, energy in ft-lb, and horsepower. Repeat for 0.75-in. stock.

16.16. Aluminum grade 1100-0 stock, 0.25 in. thick, has an irregular periphery of 17.3 in. for a blank. A punch press is capable of 179 ram strokes per minute. Find the force, energy, and horsepower. (Hint: Assume a shear strength of 13,000 psi.)

16.17. Annealed carbon steel, AISI 1018, 0.093 in. thick, is pierced for a 1.950-in. hole. The punch press is scheduled for 210 strokes per minute. Find the force, energy, and horsepower.

16.18. A press brake bends a cold-rolled steel sheet along a 48-in. dimension. The ultimate tensile strength of the material is 95,000 psi and the thickness is ¼ in. What is the bending force if the width of the lower die is 1 in.? What is the importance of this number?

16.19. A ½-in. plate is 96 in. by 36 in. and will have a bend in both directions. Each corner is notched to 6 by 6 in. to permit bending of the plate into a boxlike container. The dimension of the U-die is 1 in. The ultimate tensile strength of the material is 135,000 psi. Find the maximum bending force.

16.20. Find the drawing pressure for a beverage can shown in Figure C16.1. The tensile strength of this grade of aluminum is 27,500 psi.

16.21. Find the minimum inside radius that usually will apply if a 1-in.-thick plate is being bent into a 90° corner. Also calculate for a ¼-in.-thick plate. What is the length of the bend?

16.22. A bending operation works on a ¼-in. aluminum plate. The radius of the forming punch is minimum. The corner is 90°. What is the length of the bend? What type of material stresses are in the inside and outside fibers of the material?

16.23. Find the drawing force for the following material and design. Assume a free draw with clearance sufficient to prevent ironing, and a maximum reduction of 50%, a deep draw of steel stock ⅛ in. thick with a tensile strength of 50,000 psi into a shell 10 in. in diameter. The blank diameter is 12 in.

16.24. Consider the following: Metal having a tensile stress of 40,000 psi, mean diameter of shell 4 in., blank diameter of 9.5 in., and thickness of ¹⁄₁₆ in. Find the force applied to the punch.

16.25. What is the maximum area of the blank that can be formed in a 700-ton (6.2-MN) press if the pressure necessary to form the material is 14,000 psi (96.5 MPa)?

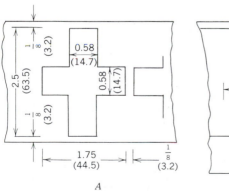

A

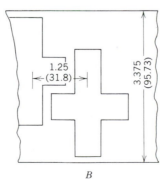

B

Figure P16.28
Layouts of blanks.

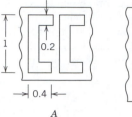

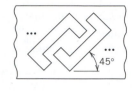

A   *B*   *C*   *Figure P16.34*

16.26. Suppose 1-in. (25.4-mm) disks are punched from a 1.25-in. (31.8-mm) flat coiled stock. What is the yield? (Hint: It is necessary to assume the advance of the disks.)

16.27. A force of 98 tons (0.9 MN) is required to coin clear impressions in half dollars. Find the pressure that this force causes.

16.28. Find the yield for Figure P16.28*A* and *B*.

16.29. Reconsider Figure P16.28 and attempt various angle layouts and determine the yield.

16.30. Consider Figure P16.28. Material cost = $0.57/lb ($1.27/kg). In addition to blanking losses, there is an additional loss of 4% for waste. Density = 0.278 lb/in.^{3} (7692 kg/m^3). Salvage of the material skeleton and waste is sold back at 5% of the original price of the material.

a. Find the cost of one blank for both designs in English units.

b. Find the cost of one blank for both designs in SI.

16.31. Two potential blank layouts are given in Figure P16.28. Material is cold-rolled steel having a tensile strength of 40,000 psi. Thickness is ⅛ in.

a. Determine punching pressure. Density = 0.278 lb/in.^{3} (7692 kg/m^3). Cost of the material is $0.75/lb ($1.65/kg). Waste loss in addition to the skeleton is 4%.

b. Find the economic and physical efficiency of strip stock layout in Figure P16.28*A* and *B*. Which design is preferred?

c. Repeat in metric units.

16.32. A large quantity of flat rectangular blanks, 3.5 by 6.5 in., are sheared from CRS 4 by 8-ft sheet stock. The stock thickness is 0.048 in.

a. Find the preferred raw material layout. What is the total number of pieces from one sheet? Find the yield.

b. Repeat for a 36 by 96-in. sheet.

16.33. A large quantity of rectangular blanks of 4.5 by 7.6 in. are sheared from CRS 4 by 10-ft sheet stock. The stock thickness is 0.048 in. Find maximum yield per sheet.

16.34. Find the most effective layout of the designs given by Figure P16.34. Material is low-carbon steel, Browne and Sharpe gage no. 10, 0.102 in. (Hints: Suggest an advance for the die, separation distance, and distance to the edge of the strip stock, and then find the yield. You may find it helpful to provide more dimensions in a simple sketch.)

MORE DIFFICULT PROBLEMS

16.35. Plot a velocity diagram of the slide for a single crank drive as a function of the crank position.

16.36. Consider Figure P16.28 again. The material cost is $0.75/lb ($1.65/kg). In addition to blanking losses of the skeleton, add a 5% loss for overall waste.

a. Find the losses for both designs. Density = 0.278 lb/in.^{3} (7692 kg/m^3).

b. Salvage of the waste and skeleton is recovered at 10% of the original value. Find the economic yield of both designs. Which one is cheaper?

c. Repeat in SI units.

16.37. Refer to Figure P16.28. Cold-rolled steel is available in a strip width of 24.75 in. A slitting operation prepares suitable widths as required by the design without loss in material, except that which is not an even multiple of the design width. In addition to the blanking losses, add 5% loss for scrap and ends. Find the yield of the material. Determine the unit cost for designs in Figure P16.28*A* and *B*. (Hints: The strip weighs 5 lb/ft^2 and costs $0.57 per lb. Density of this material is 0.278 lb/in.^{3}.)

16.38. Small, flat, round washers, 1.000 $^{+0.006}_{-0.000}$-in. OD, are to have their 0.750 $^{+0.008}_{-0.000}$-in. ID centers pierced using a piercing die in the first operation and the OD blanked in a second operation from strip stock 1.100 in. wide, where the advance of the strip stock is 1.125 in. The washer raw material is CRS, 0.125 in. thickness. Shearing clearance is 5%, and the maximum yield psi of the steel is 68,000. The lot quantity is 125,000.

a. Sketch and dimension a cross section of the punch and die for the piercing and blanking operations. Show the stock, clearances, and shape of the punch and die.

b. Find the yield.

16.39. A pattern is coined on the surface of a 2-in. diameter bronze medallion. Approximate the tonnage rating of the press. Determine the rpm of a 38-in. diameter 3-in. thick flywheel. (Hints: Yield strength = 18 kspi and density of the flywheel = 0.29 lb/in.^{3}.)

PRACTICAL APPLICATION

This chapter concentrates on pressworking operations and stamping plants, where many operations are seen. These are significant manufacturing businesses. Under the direction of your instructor, plan a plant visit to a stamping plant. Before the visit, ask about the variety of their operations and review the principles of these processes. Give attention to the total character of a stamping plant and see if it is effective. Report your findings.

CASE STUDY: TWELVE-OUNCE BEVERAGE CONTAINER COMPANY

This company manufactures beverage containers in the popular 12-fluid-oz. (0.36-liter) size. With production exceeding 10,000,000 cans daily, material efficiency is absolutely necessary. Refer to Figure C16.1 for dimensions.

The can is composed of three pieces: body, top, and pull ring. The rivet connecting the pull ring to the top is formed from the top during the mechanical joining process. The container body is blanked from 3004-H19 aluminum coils, and the layout is shown in Figure C16.1. An intermediate cup is formed without any significant change in thickness. The cup is drawn in a horizontal drawing machine, and metal is "ironed" to a sidewall thickness of 0.0055 in. (0.140 mm). Bottom thickness remains unchanged. The can is trimmed to a final height of 5.25 in. (133.4 mm) to give an even edge for rolling to the lid.

Determine the "metal efficiency"; that is, the final metal in the can to the original metal in the coil. What is the metal efficiency of the can to the 5.700-in. (144.78-mm) OD blank? (Hints: Disregard the 0.05-in. (1.3-mm) radius in any calculations. Also notice that these questions are concerned with the can body, not the top or the tab.) If the 3004-H19 aluminum coil stock costs $1.00/lb ($2.222/kg) and the density = 0.0981 lb/in.^{3} (2715 kg/m^3), find the cost of the can if waste is recovered and recycled at $0.25/lb ($0.556/kg). What is the material cost lost in waste per can? For a production of 1,000,000 cans daily, what is the material cost lost in waste? Describe some alternatives for improving the yield. If assigned by the instructor, repeat in SI units.

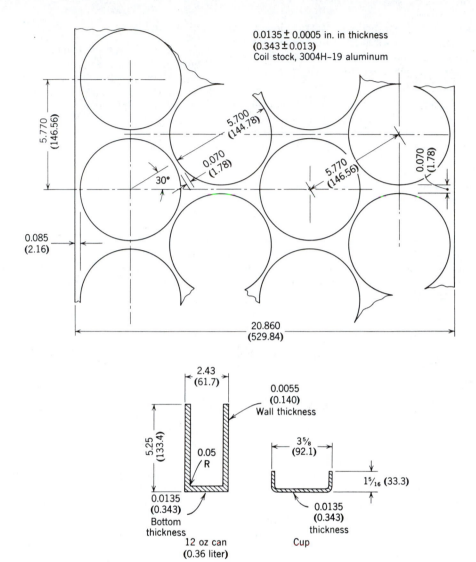

0.0135 ± 0.0005 in. in thickness
(0.343 ± 0.013)
Coil stock, 3004H–19 aluminum

5.770
(146.56)

5.700
(144.78)

0.070
(1.78)

30°

5.770
(146.56)

0.070
(1.78)

0.085
(2.16)

20.860
(529.84)

2.43
(61.7)

0.0055
(0.140)
Wall thickness

5.25
(133.4)

0.05
R

0.0135
(0.343)
Bottom
thickness

12 oz can
(0.36 liter)

3⅝
(92.1)

1⁵⁄₁₆ (33.3)

0.0135
(0.343)
thickness

Cup

Figure C16.1

CHAPTER 17

HEAT TREATING

The understanding of heat treatment is embraced by the broader field of metallurgy. *Metallurgy* is the physics, chemistry, and engineering of metals from ore extraction to final product. *Heat treating* is the operation of heating and cooling a metal in its solid state to change its physical properties. For example, steel can be hardened to resist cutting action and abrasion, or it can be softened to permit machining. With heat treatment internal stresses may be removed, grain size reduced, toughness increased, or a hard surface produced on a ductile interior.

This chapter deals mostly with steel, both plain carbon and alloy. The chemical analysis of the steel must be known because small percentages of certain elements, notably carbon, greatly affect the physical properties during the heat-treating operation. Alloy steels owe their properties to the presence of one or more elements other than carbon, namely nickel, chromium, manganese, molybdenum, tungsten, silicon, vanadium, and copper. Because of alloy steels' improved physical properties, they are used commercially in many ways not possible with carbon steels.

This study is aided by photographs of the grain structure, where magnification of 100 to 500 diameters is possible with a *metallograph*, a microscope capable of seeing the details of highly polished metal specimens. The study of microscopic structures of metals is called *metallography*.

17.1 IRON–IRON CARBIDE DIAGRAM

Under conditions of equilibrium the knowledge of steel and its structure is best summarized in the partial iron–iron carbide diagram shown in Figure 17.1.

This diagram is found in the following way. If a piece of 0.20% carbon steel is heated slowly and uniformly and its temperature recorded at definite intervals of time, a curve as shown in Figure 17.2 is obtained. This curve is called an *inverse rate curve*. The abscissa is the *heating rate* or the time required to heat or cool the steel 10°F. The curve is a vertical line, except at those points where the heating or cooling rates show marked change. It is evident that at three temperatures there is a definite change in the heating rate. In a similar fashion these same three points occur upon cooling, but at slightly lower temperatures. Where structural changes occur, these points are known as critical points and are designated by the symbols Ac_1, Ac_2, and Ac_3. The letter "c" is the initial letter of the French word *chauffage*, meaning "to heat." The points on the cooling curve are designated by Ar_1, Ar_2, and Ar_3. The "r" is taken from the word *refroidissement*, meaning "to cool." The Ac_2 and Ar_2 values are believed to have little effect on the understanding of the metallurgy of iron and steel.

The changes that take place at these critical points are called *allotropic*. Although the chemical content of the steel remains the same, its properties are altered, as the atomic arrangement in the space lattice system changes. Principal among these are changes in electrical resistance, atomic structure, and loss of magnetism. Allotropic change is a reversible change in the atomic structure of the metal with a corresponding change in the properties of the steel. These critical points should be known, because most heat-

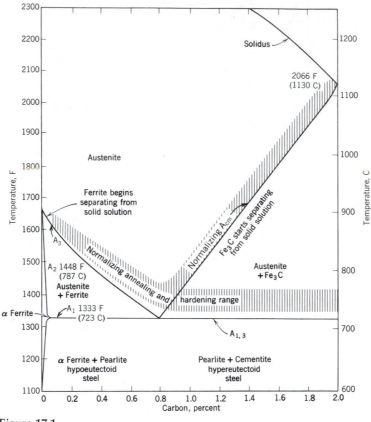

Figure 17.1
Partial iron–iron carbide phase diagram.

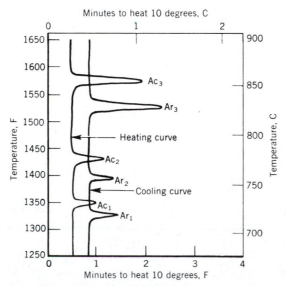

Figure 17.2
Inverse rate curve for SAE 1020 steel.

treating processes require heating the steel to a temperature above this range. Steel cannot be hardened unless it is heated to a temperature above the lower critical range and in certain instances above the upper critical range.

If a series of time–temperature heating curves are made for steels of different carbon contents and the corresponding critical points plotted on a temperature-percentage carbon curve, a diagram similar to Figure 17.1 is obtained. This diagram, which applies only under slow cooling conditions, is known as a *partial iron–iron carbide diagram*. The diagram is called "partial" because the full range of carbon is not given.

The diagram, Figure 17.1, provides much information. For example, the proper quenching temperatures for any carbon steel may be observed in this diagram; or the relationship between the amount of carbon and the grain structure in steel can be understood more clearly by studying this diagram.

Consider again the piece of 0.20% carbon steel that has been heated to a temperature around 1600°F (870°C). Above the Ar_3 point this steel is a solid solution of carbon in gamma iron and is called *austenite*. The iron atoms lie in a face-centered cubic lattice and are nonmagnetic. Upon cooling this steel, the iron atoms start to form a body-centered cubic lattice below the Ar_3 point. This new structure being formed is called *ferrite* or *alpha iron* (also α) and is a solid solution of carbon in alpha iron. The solubility of carbon in alpha iron is much less than in *gamma iron*. At the Ar_2 point the steel becomes magnetic and, as the steel is cooled to the Ar_1 line, additional ferrite is formed. At the Ar_1 line the austenite that remains is transformed to a new structure called *pearlite*. This constituent is lamella in appearance under high magnification. The lamellae are alternately ferrite and iron carbide. Called pearlite because of its "mother-of-pearl" appearance, pearlite is shown under high magnification in Figure 17.3.

As the carbon content of the steel increases above 0.20%, the temperature at which the ferrite is first rejected from the austenite drops until at about 0.80% carbon, where no free ferrite is rejected from the austenite. This 0.80% carbon steel is called *eutectoid*

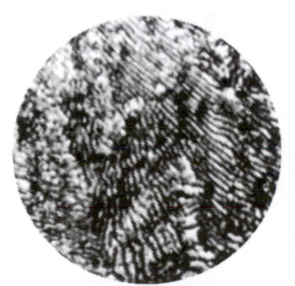

Figure 17.3
Structure of SAE 1095 steel, furnace cooled from 1550°F. Etched in 5% picral. Showing lamellae of cementite and ferrite in pearlite. Magnification ×1200.

steel and is 100% pearlite in structure composition. The eutectoid point in any metal is the lowest temperature at which changes occur in a solid solution. It marks the lowest temperature for equilibrium decomposition of austenite to ferrite and cementite. If the carbon content of the steel is greater than the eutectoid, a new line is observed in the iron–iron carbide diagram labeled *Acm*. The line denotes the temperature at which iron carbide is first rejected from the austenite instead of the ferrite. The iron carbide (Fe_3C) is known as cementite and is extremely hard and brittle. Steels containing less carbon than the eutectoid are *hypoeutectoid steels*, and those with more carbon content are called *hypereutectoid steels*.

The structures of these steels are shown in a series of photomicrographs in Figure 17.4. Figure 17.4*A* shows pure iron or ferrite. As the carbon content increases up to 0.80% carbon (Figure 17.4*E*), the dark areas of the pearlite form and increase in quantity while the white background area of ferrite decreases and the sample is nearly all pearlite. In the sample containing 1.41% carbon (Figure 17.4*F*), the pearlitic area is smaller and the white background area is now cementite and ferrite. The maximum amount of cementite at 1.41% carbon would be about 11%. All these iron–carbon alloys have been cooled slowly to produce the constituents just described.

Regarding steel heat treating, the partial diagram shown in Figure 17.1 is sufficient because 2.0% is the limit of carbon content in steel. If the diagram is extended to include the cast irons with carbon contents up to 6.67%, it will appear as shown in Figure 17.5.

(A)   (B)   (C)

(D)   (E)   (F)

Figure 17.4
Photomicrographs of iron–carbon alloys showing the effect of increasing amounts of carbon on structure of the metal. *A*, High-purity iron; *B*, 0.12% carbon; *C*, 0.40% carbon; *D*, 0.62% carbon; *E*, 0.79% carbon; *F*, 1.41% carbon.

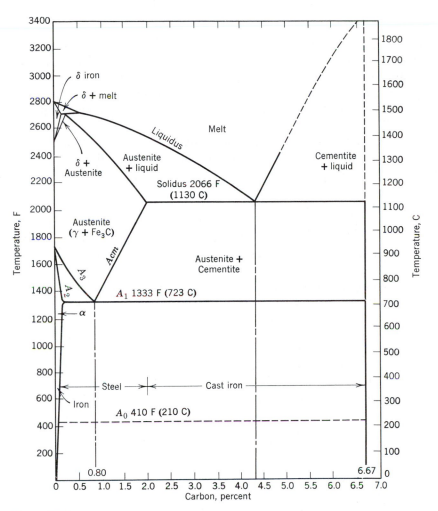

Figure 17.5
Iron–iron carbide diagram.

It is not shown beyond this point because 6.67% carbon is the carbon content of cementite. Actually, most commercial cast irons have a carbon content from about 2.25 to 4.50%.

The amount of ferrite, pearlite, and cementite can be calculated as follows using the *lever rule*. For a hypoeutectoid steel, the percentage pearlite is

$$\%Pe = \frac{\%C}{0.8} \tag{17.1}$$

where %C is the percentage carbon in the specimen. The remainder is ferrite.

For a hypereutectoid steel the percentage cementite can be expressed as

$$\%Ce = 100 \left[\frac{1 - (6.67 - \%C)}{5.87} \right] \tag{17.2}$$

The remainder is pearlite.

17.2 GRAIN SIZE

Upon cooling, molten steel starts solidifying at many small centers of nuclei. The atoms in each group tend to be positioned similarly. The irregular grain boundaries, seen under the microscope after polishing and etching of the specimen, are the outlines of each group of atomic cells that have the same general orientation. The size of these grains depends on a number of factors, the principal one being the furnace treatment it has received.

Coarse-grained steels are less tough and have a greater tendency for distortion than those having a fine grain. However, they have better machinability and greater depth-hardening power. The fine-grained steels, in addition to being tougher, are more ductile and tend to distort less or crack during heat treatment. Control of grain size is possible through regulation of composition in the initial manufacturing procedure, but after the steel is made, the control is by proper heat treatment. One scheme, the addition of aluminum, when used as a *deoxidizer*, is the most important controlling factor during the manufacturing period, because it raises the temperature at which rapid grain growth occurs.

When a piece of low-carbon steel is heated, there is no change in the grain size up to the Ac_1 point. As the temperature increases through the critical range, the ferrite and pearlite are gradually transformed to austenite, and at the upper critical point Ac_3, the average grain size is at a minimum. Further heating of the steel causes an increase in the size of the austenitic grains, which, in turn, governs the final size of the grains when cooled. Quenching from the Ac_3 point results in a fine structure, whereas slow cooling or quenching from a higher temperature yields a coarser structure. The final grain size depends to a large extent on the prior austenitic grain size in the steel at the time of quenching.

Not all steels start growing large crystals immediately upon being heated above the upper critical range. Some steels can be heated to a higher temperature with little changes in their structure. A temperature known as a coarsening temperature is eventually reached, and grain size increases rapidly. This is characteristic of medium-carbon steels, many alloy steels, and steels that have been deoxidized with aluminum. The coarsening temperature is not a fixed temperature, and it may be changed by prior hot or cold working and heat treatment. Hot working of steel is started at temperatures well above the critical range with the steel in a plastic state. Hot working refines the grain structure

Figure 17.6
Crystalline separation and excessive grain size.
Magnification ×300.

and eliminates any coarsening effect caused by the high temperature. Moreover, hot forging or rolling should not continue below the critical temperature.

The principal method of determining grain size is by microscopic examination, although it may be roughly estimated by examination of a fracture. Low-carbon steels have ferrite precipitated from the austenite upon a slow cooling, and the outlines of these grains can be highlighted clearly by polishing and etching. Because a very slow cooling rate may produce too much primary ferrite to permit evaluation of prior austenitic grain size, a cooling rate must be employed such that the proeutectoid constituent is restricted merely to outlining the pearlitic regions. Likewise, for medium-carbon steels the former austenitic grain size would be represented roughly by the pearlitic area plus one-half the surrounding ferrite. Hypereutectoid steels will have the grain boundaries outlined by the cementite that is precipitated.

An example of a large-grained steel is shown in the photomicrograph in Figure 17.6. This specimen has been heated to an excessively high temperature, resulting in large-grain growth and some crystalline separation. Steel that is "burned" shows this separation as a result of oxidation at the grain boundaries, and this characteristic cannot be remedied by heat treatment. The steel is rendered fit for commercial use by remelting.

17.3 ISOTHERMAL TRANSFORMATION DIAGRAMS

The iron–iron carbide phase diagram in Figure 17.1 is useful in selecting temperatures for parts to be heated for various heat-treating operations. It also shows the type of structure to expect in slowly cooled steels. Although very useful in heat-treating operations, the diagram does not give much information concerning effects of cooling rate, time, grain structure, or structures obtainable when the quench is interrupted at certain elevated temperatures. *Isothermal transformation* diagrams, also known as *time–temperature transformation* (TTT) diagrams or *S curves*, provide this information; see Figure 17.7 for an example. The axes are log time, typically in units of seconds, and temperature. Two cooling curves are shown, one for the surface and the other for the center of the steel piece. This diagram shows the way austenitized steel changes if heating is held at some constant temperature. Knowing this temperature, the times at which the transformation starts and ends may be determined. The resulting structure is indicated

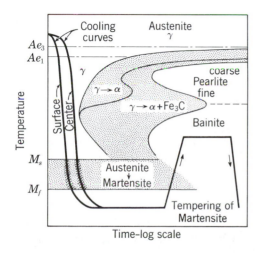

Figure 17.7
Transformation diagram illustrating the formation of tempered martensite.

on the diagram. To obtain a martensitic structure the steel must be quenched with sufficient rapidity so that the cooling curve does not intersect the left-pointing nose of the transformation curve. This is indicated in Figure 17.7, which shows the cooling curve passing through the M_s and M_f lines (start and finish of austenite transforming to *martensite*).

The general shape of the TTT curve differs for each steel, depending on the carbon content, alloys present, and austenitic grain size. Most alloying elements in steel shift the curves to the right, thus allowing more time to harden the steel fully without hitting the bend in the curve. This increase in *hardenability* of the steel permits the hardening of thicker sections. In carbon steels, lowering the carbon content moves the curve to the left and raises the M_s and M_f temperature lines. This makes it very difficult to produce martensite by quenching a hypoeutectoid steel. Carbon steel having an eutectoid composition responds well to hardening treatments. Steels having a fine austenitic grain size displace the curve to the left, thus making it more difficult to harden a fine-grained steel than one having coarse grains. However, coarse-grained steels are more apt to crack or distort during quenching, so their increased hardenability is of little advantage.

17.4 HARDENING

Hardening is the process of heating a piece of steel to a temperature within or above its critical range and then cooling it rapidly. If the carbon content of the steel is known, the proper temperature to which the steel should be heated may be obtained by reference to Figure 17.1, the iron–iron carbide phase diagram. However, if the composition of the steel is unknown, preliminary experimentation may be necessary to determine the range. A good procedure to follow is to heat-quench a number of small specimens of the steel at various temperatures and observe the results, either by hardness testing or by microscopic examination. When the correct temperature is obtained, there is a marked change in hardness and other properties.

In any heat-treating operation the rate of heating is important. Heat flows from the exterior to the interior of steel at a definite rate. If the steel is heated too fast, the outside becomes hotter than the interior and uniform structure cannot be obtained. If a piece is irregular in shape, a slow rate is more essential to eliminate warping and cracking. The heavier the section, the longer the heating time must be to achieve uniform results. Even after the correct temperature is reached, the piece should be held at that temperature for a sufficient period of time to permit its thickest section to attain a uniform temperature.

The hardness obtained from a given treatment depends on the quenching rate, carbon content, and work size. In alloy steels the kind and amount of alloying element influences only the hardenability (the ability of the workpiece to be hardened to depths) of the steel and does not affect the hardness except in unhardened or partially hardened steels.

Reference to the isothermal transformation diagram in Figure 17.7 indicates that a very rapid quench is necessary to avoid intersecting the nose of the curve and to obtain martensitic structure. For low and medium plain-carbon steels quenching in a water bath is a method of rapid cooling that is common practice. For high-carbon and alloy steel, oil generally is used as the quenching medium because its action is not as severe as water. Various commercial oils, such as mineral oil, have different cooling speeds and, consequently, impart different hardness to steel on quenching. For extreme cooling, brine or water spray is most effective. Certain alloys can be hardened by air-cooling, but for ordinary steels such a cooling rate is too slow to give an appreciable hardening effect. Large parts are usually quenched in an oil bath, which has the advantage of cooling the part down to room temperatures rapidly and yet not being too severe. The temperature of the quenching medium must be kept uniform to achieve uniform results. Any quenching bath used in production work should be provided with means for cooling.

Steel with low carbon content will not respond appreciably to hardening treatments. As the carbon content in steel increases up to around 0.60%, the possible hardness obtainable also increases. Above this point, the hardness is increased only slightly because steels above the eutectoid point are made up entirely of pearlite and cementite in the annealed state. Pearlite responds best to heat-treating operations; any steel composed mostly of pearlite can be transformed into a hard steel.

As the size of parts to be hardened increases, the surface hardness decreases somewhat even though all other conditions remain the same. There is a limit to the rate of heat flow through steel. No matter how cool the quenching medium may be, if the heat inside a large piece cannot escape faster than a certain *critical rate*, there is a definite limit to the inside hardness. However, brine or water quenching is capable of rapidly bringing the surface of the quenched part to its own temperature and maintaining it at or close to this temperature. Under these circumstances, there would always be some finite depth of surface hardening regardless of size. This is not true in oil quenching, when the surface temperature may be high during the critical stages of quenching.

Hardenability of Steel

Hardenability refers to the response of a metal to quenching and may be measured by the *Jominy end-quench* test, as illustrated in Figure 17.8. A normalized steel specimen is machined to a diameter of 1 in. (25 mm) and a length of 4 in. (100 mm) and then heated to its austenitizing temperature. It is quickly placed in the quenching fixture, where it is held above a flowing orifice of water until the specimen is cool.

Upon removal from the fixture, two flats, each 0.015 in. (0.38 mm) deep, are ground on opposite sides of the specimen. Rockwell hardness readings are then taken at ¹⁄₁₆-in. (1.6-mm) intervals from the bottom of the specimen and plotted as shown in Figure 17.8. The reading next to the bottom shows the greatest hardness because that portion has been most severely quenched. Readings away from the quenched end show progressively lower hardnesses as the heat must pass through the specimen by conduction during cooling. A steel with high hardenability shows high hardness readings for some distance

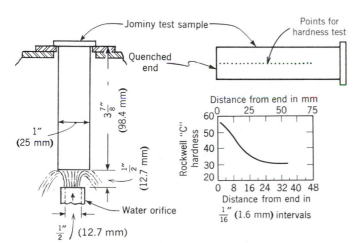

Figure 17.8
The Jominy end-quench test for measuring hardenability.

from the quenched end, while the hardness readings for low-hardenability steel show a sharp drop only a short distance from the end.

This test compares the *depth-hardening ability* of different steels. Alloys increase the hardenability of steel and make it possible to harden small pieces uniformly from the outside to the inside. Because the effect of an alloy on the isothermal transformation diagram is to move the curve to the right, it is easier to quench such a steel without intersecting the curve. Because of this characteristic, it is possible to harden alloy steels at a slower rate than plain carbon steels. This permits alloy steels to be hardened effectively by quenching in oil instead of water.

Constituents of Hard Steel

As has been previously stated, austenite is a solid solution of carbon in *gamma iron*. All carbon steels are composed entirely of this constituent above the upper critical point. The appearance of austenite under the microscope is shown in Figure 17.9A, 18-8 stainless steel, at a magnification of ×125. Extreme quenching of a steel from a high temperature will preserve some of the austenite at ordinary temperatures. This constituent is about half as hard as martensite and is nonmagnetic.

If a hypoeutectoid steel is cooled slowly, the austenite is transformed into ferrite and pearlite. Steel having these constituents is soft and ductile. Faster cooling will result in a different constituent and the steel will be harder and less ductile. A rapid cooling such as a water quench will result in a martensitic structure, which is the hardest structure that can be obtained. Cementite, although somewhat harder, is not present in its free state, except in hypereutectoid steels, and then only in such small quantities that its influence on the hardness of the steel can be ignored.

The essential ingredient of any hardened steel is martensite. Martensite is obtained by the rapid quenching of carbon steels and is the transitional substance formed by the rapid decomposition of austenite. It is a supersaturated solution of carbon in *alpha iron*. Under the microscope it appears as a needlelike constituent, as seen in Figure 17.9*B*. The hardness of martensite depends on the amount of carbon present and varies from Rockwell C45 to C67. It cannot be machined, is brittle, and is strongly magnetic.

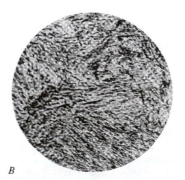

A   B

Figure 17.9
A, Structure of 18-8 stainless steel, water-quenched to show austenite. Lines caused by hot rolling. Magnification ×125. *B*, Structure of SAE 1095 steel water-quenched. Etched with Villella's reagent to show martensite. Magnification ×562.

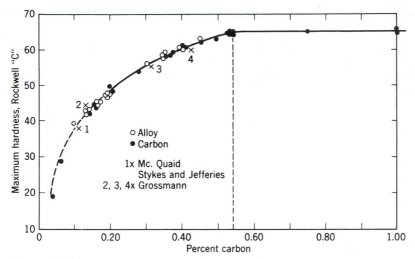

Figure 17.10
Maximum hardness versus carbon content.

If steel is quenched at slightly less than the critical rate, a dark constituent with somewhat rounded outlines will be obtained. This constituent is called *fine pearlite*. Under the microscope at usual magnifications it appears as a dark unresolved mass, but at very high magnification a fine lamella structure is seen. Fine pearlite is less hard than martensite. It has a Rockwell C hardness varying from 34 to 45, but is tough and capable of resisting considerable impact. As the quenching rate is further reduced, the pearlite becomes coarser and is lamellar under high magnification at slow rates of cooling.

Maximum Hardness of Steel

Maximum hardness in steel depends on the carbon content. Although various alloys such as chromium and vanadium increase the rate and depth-hardening ability of alloy steels, their maximum hardness will not exceed that of carbon steel having the same carbon content. This is illustrated in Figure 17.10, where Rockwell C hardness is plotted against percentage of carbon. This curve shows the maximum hardness that is possible for a given carbon percentage. To obtain maximum hardness the carbon must be completely in solution in the austenite state when quenched. The critical quenching rate, which is the slowest rate of cooling that will result in 100% martensite, should be used. Finally, austenite must not be retained in any appreciable percentages because it will soften the structure.

Figure 17.10 is made up of test points from both alloy and carbon steels, and little variation in the results is evident. However, the same quenching rate cannot be used for both alloy and carbon steels of the same carbon content. The maximum hardness obtained in any steel represents the hardness of martensite and is approximately 66 to 67 Rockwell C. Carbon content equal to or in excess of 0.60% is necessary to achieve this level.

17.5 TEMPERING

Steel that has been hardened by rapid quenching is brittle and unsuitable for most uses. By *tempering* or *drawing*, the hardness and brittleness is reduced to the desired point for service conditions. As these properties are reduced, there is also a decrease in tensile strength and an increase in the ductility and toughness of the steel. The operation, as depicted in Figure 17.11, consists of reheating quench-hardened steel to some temperature below the critical range followed by any rate of cooling. Although this process softens steel, it differs considerably from annealing in that the process lends itself to close control of the physical properties and in most cases does not soften the steel to the extent that annealing would. The final structure obtained from tempering a fully hardened steel is called tempered martensite.

Tempering is possible because of the instability of the martensite, the principal constituent of hardened steel. Low-temperature draws, from 300° to 400°F (150° to 205°C), do

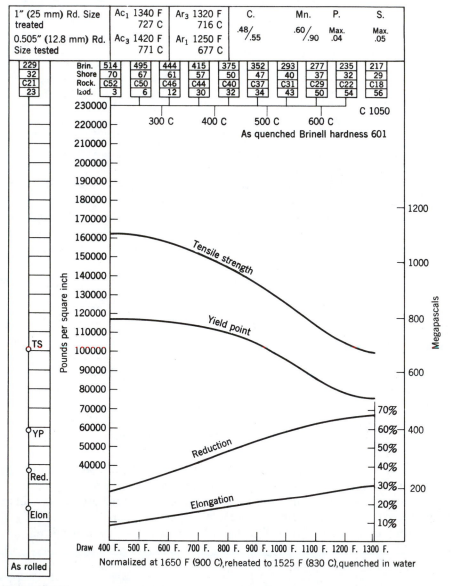

Figure 17.11
Physical properties chart (average values) of AISI 1050 steel, fine grain, water-quenched.

not cause much decrease in hardness and are used principally to relieve internal strains. As the tempering temperatures are increased, the breakdown of the martensite takes place at a faster rate, and at about 600°F (315°C) the change to tempered martensite is very rapid. The tempering operation is described as one of precipitation and agglomeration or coalescence of cementite. A substantial precipitation of cementite begins at 600°F (315°C), which produces a decrease in hardness. Increasing the temperature causes coalescence of the carbides with continued decrease in hardness. Figure 17.11 shows a typical set of property curves for AISI 1050 steel, giving the tensile strength, hardness, percentage elongation, and percentage reduction in area for various tempering or draw temperatures. The influence of tempering on the physical properties of the steel is shown by these curves. Alloying elements have a profound influence on tempering, the general effect being to retard the softening rate so that alloy steels will require a higher tempering temperature to produce a given hardness.

In the process of tempering, some consideration is given to time as well as to temperature. Although most of the softening action occurs in the first few minutes after the temperature is reached, there is some additional reduction in hardness if the temperature is maintained for a prolonged time. Usual practice is to heat the steel to the desired temperature and hold it there only long enough to have it uniformly heated.

Two special processes using interrupted quenching are a form of tempering. In both, the hardened steel is quenched in a salt bath held at a selected lower temperature before being allowed to cool. These processes, known as *austempering* and *martempering*, result in products having certain desirable physical properties.

Austempering

The interrupted quenching process (Figure 17.12A) is known as austempering. It is an isothermal transformation that converts austenite to a hard structure called *bainite*. Parts must be quenched rapidly to the correct holding temperature so that the cooling curve is not permitted to intersect the nose in the transformation diagram. The steel is held at a temperature above the M_s line but below 800°F (430°C). When held at the constant

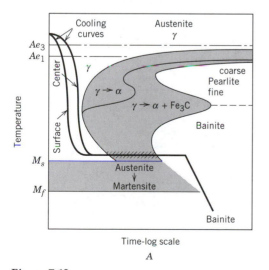
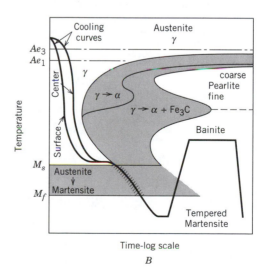

Figure 7.12

Transformation diagrams illustrating the interrupted quenching processes for austempering and martempering. *A*, Austemper. *B*, Martemper.

temperature for sufficient time to complete the transformation, a structure called bainite is obtained. This structure resembles martensite when viewed in a microscope. Although this microstructure exhibits approximately the same hardness, it is relatively tougher than in the quenched and tempered condition. The process is limited to small parts with good hardenability such that pearlite is not formed during the initial quench.

Martempering

In the process known as martempering the steel is quenched rapidly from the austenite region to a temperature just above the M_s line (see Figure 17.12B). Here, the steel is held long enough to enable the surface and the center of the piece to arrive at the same temperature. When this occurs, the piece is usually cooled in air to room temperature, thus forming martensite. The steel is reheated to a temperature varying with the carbon and alloy content, although for plain-carbon steels containing around 0.40% carbon, the temperature is 700°F (370°C). The main purpose of martempering is to minimize distortion, cracking, and internal stresses that result from quenching in oil or water. Although the resulting product is similar to tempered martensite, a subsequent tempering operation is generally performed.

17.6 ANNEALING

The primary purpose of *annealing* is to soften hard steel so that it may be machined or cold worked. This is usually accomplished by heating the steel to slightly above the Ac_3 critical temperature, holding it there until the temperature of the piece is uniform throughout, and then cooling at a slowly controlled rate so that the temperature of the surface and center are approximately the same. This process, illustrated in Figure 17.13A, is known as *full annealing,* because it wipes out all trace of previous structure, refines the crystalline structure, and softens the metal. Annealing also relieves internal stresses set up previously in the metal.

When hardened steel is reheated to above the critical range, the constituents are

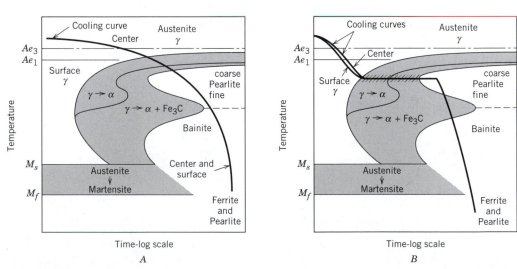

Figure 17.13
Transformation diagram. *A*, Full annealing. *B*, Isothermal annealing.

changed back into austenite, and slow cooling then provides ample time for complete transformation of the austenite into the softer constituents. For the hypoeutectoid steels these constituents are pearlite and ferrite. It may be noted by referring to the equilibrium diagram that the annealing temperature for hypereutectoid steels is lower, being slightly above the A_1 line. There is no reason to heat above the Acm line, as it is at this point that the precipitation of the hard constituent cementite is started. All martensite is changed into pearlite by heating above the lower critical range and slowly cooling. Any free cementite in the steel is unaffected by the treatment.

The temperature to which a given steel should be heated in annealing depends on its composition. For carbon steels it is obtained from the partial iron–iron carbide equilibrium diagram in Figure 17.1. The heating rate should be consistent with the size and uniformity of sections, so that the entire part is brought up to temperature as uniformly as possible. When the annealing temperature is reached, the steel should be held there until there is a uniform temperature throughout. This usually takes about 45 min for each inch (25 mm) of thickness of the largest section. For maximum softness and ductility the cooling rate should be very slow, such as allowing the parts to cool down with the furnace. The higher the carbon content, the slower this rate must be.

Isothermal Annealing

Isothermal annealing, as illustrated in Figure 17.13*B*, provides a short annealing cycle. Steel is quenched rapidly to the temperature at which austenite transforms to a relatively soft ferrite carbide aggregate in the shortest possible time. It is then held for the time necessary to transform the austenite completely to pearlite. After the transformation is complete, the part is cooled in any manner. Isothermal annealing results in giving pearlite a more uniform structure than that obtained by other annealing processes. Fineness depends on the transformation temperature used.

Process Annealing

Process annealing, practiced in the sheet and wire industry between cold-working operations, consists of heating the steel to a temperature a little below the critical range and then cooling it slowly. This process is more rapid than *spheroidizing* and results in the usual pearlitic structure. It is similar to the tempering process but will not give as much softness and ductility as a full anneal. Also, at the lower heating temperature there is less tendency for the steel to scale or decarburize.

17.7 NORMALIZING AND SPHEROIDIZING

The process of *normalizing* consists of heating the steel about 50° to 100°F (10° to 40°C) above the upper critical range and cooling in still air to room temperature. This process is used principally with low- and medium-carbon steels as well as the alloy steels to make the grain structure more uniform, to relieve internal stresses, or to achieve desired results in physical properties. Most commercial steels are normalized after being rolled or cast.

Spheroidizing is the process of producing a structure in which the cementite is in a spheroidal distribution, as shown in Figure 17.14. If a steel is heated slowly to a temperature just below the critical range and held there for a prolonged period of time, this structure is obtained. It also may be accomplished by alternately heating and cooling between temperatures that are just above and below the Ac_1 range. The globular structure

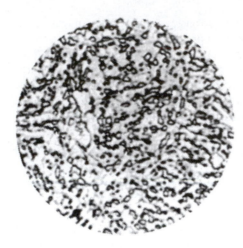

Figure 17.14
SAE 1095 steel quenched from 1550°F and tempered at 1250°F for 8 h. Structure is spheroidized cementite in a ferritic matrix. Magnification ×900.

obtained gives improved machinability to the steel. This treatment is particularly useful for hypereutectoid steels that must be machined.

17.8 SURFACE HARDENING

Several methods of surface hardening are used to alter the chemistry of steel. Sometimes these processes involve dangerous gases or salts. Specifically, carbon monoxide is utilized in *pack carburizing*, ammonia is applied for gas nitriding, and sodium cyanide is used in *liquid carburizing*. Their safety is discussed in Chapter 30.

Carburizing

The oldest known method of producing a hard surface on steel is case hardening or carburizing. Steel is heated above Ac_1 while in contact with some carbonaceous material, which may be solid, liquid, or gas. Iron at temperatures close to and above its critical temperature has an affinity for carbon. The carbon is absorbed into the metal to form a solid solution with iron and converts the outer surface into a high-carbon steel. The carbon is gradually diffused to the interior of the part. The depth of the case depends on the time and temperature of the treatment. *Pack carburizing* consists of placing the parts to be treated in a closed container with some carbonaceous material, such as charcoal or coke. The actual carburization is done by the carbon monoxide, not the charcoal. It is a time-consuming process and produces fairly thick *cases* from 0.030 to 0.160 in. (0.76 to 4.06 mm) in depth. A schematic of a pack carburizing system is given in Figure 17.15A.

For shallower case depths of 0.005 to 0.030 in. (0.13 to 0.76 mm), gas carburizing is frequently used, employing such hydrocarbon fuels as natural gas or propane. Gas carburizing is adapted to the case hardening of small parts that may be surface hardened by direct quenching from the furnace at the end of the heating cycle.

In liquid carburizing the steel is heated above the Ac_1 in a cyanide salt bath, causing the carbon and some nitrogen to diffuse into the surface, which is called a "case." The parts are immersed into the molten baths. Notice Figure 17.15C. Liquid carburizing involves cyanide, which is chemically composed of carbon and nitrogen. Most carburizing

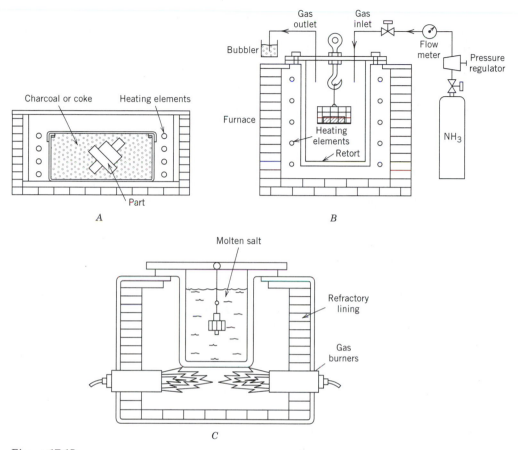

Figure 17.15
A, Pack carburizing. *B*, Gas carburizing. *C*, Liquid carburizing.

salts have sodium cyanide, NaCN, as the active ingredient, and occurs with the following reaction: $2NaCN \rightarrow Na_2CN_2 + C$.

The cyanide supplies both carbon and nitrogen to the steel surface. Alloying the steel with nitrogen lowers the transformation temperature and reduces the transformation rate. Modification of the carburizing process by diffusing of both carbon and nitrogen into the surface layer enables the process to be carried out at lower temperatures than with carbon alone.

It is similar to cyaniding, except that the case is higher in carbon and lower in nitrogen. Liquid carburizing can be used for case thickness up to 0.250 in. (6.35 mm), although the thickness seldom exceeds 0.025 in. (0.64 mm). This method is best suited for case hardening small- and medium-sized parts.

Steel for carburizing is usually a low-carbon steel of about 0.15% carbon that would not respond appreciably to heat treatment. The process converts the outer layer into a high-carbon steel, with a content ranging from 0.9% to 1.2% carbon.

A steel with varying carbon content and, consequently, different critical temperatures requires a special heat treatment. Because there is some grain growth in the steel during the prolonged carburizing treatment, the work should be heated to the critical temperature of the core and then cooled, thus refining the core structure. The steel should then be reheated to a point above the transformation range of the case (Ac$_1$) and quenched to produce a hard, fine structure. The lower heat-treating temperature of the case results because hypereutectoid steels are normally austenitized for hardening just above the lower critical point. A third tempering treatment may be used to reduce strains.

Carbonitriding

Carbonitriding, sometimes known as *dry cyaniding* or *nicarbing*, is a case-hardening process in which the steel is held at a temperature above the critical range in a gaseous atmosphere from which it absorbs carbon and nitrogen. Any carbon-rich gas with ammonia can be used. The wear-resistant case thickness ranges from 0.003 to 0.030 in. (0.08 to 0.76 mm).

An advantage of carbonitriding is that the hardenability of the case thickness is significantly increased when nitrogen is added, permitting the use of low-cost steels.

Cyaniding

Cyaniding, or *liquid carbonitriding* as it is sometimes called, is also a process that combines the absorption of carbon and nitrogen to obtain surface hardness in low-carbon steels that do not respond to ordinary heat treatment. The part to be case hardened is immersed in a bath of fused sodium cyanide salts at a temperature slightly above the Ac_1 range, the duration of soaking depending on the depth of the case. See Figure 17.15C. The part is then quenched in water or oil to obtain a hard surface. Case depths of 0.005 to 0.015 in. (0.13 to 0.38 mm) are obtained by this process. Cyaniding is used principally for the treatment of small parts.

Nitriding

Nitriding is somewhat similar to ordinary case hardening, but it uses a different material and treatment to create the hard surface constituents. In this process the metal is heated to a temperature of around 950°F (510°C) and held there for a period of time in contact with ammonia gas. Nitrogen from the gas is introduced into the steel, forming very hard nitrides that are finely dispersed through the surface metal.

Nitrogen has greater hardening ability with certain elements than with others, and special nitriding alloy steels have been developed. Aluminum in the range of 1% to 1.5% has proved to be suitable in steel, in that it combines with the gas to form a stable and hard constituent. The temperature of heating ranges from 925° to 1050°F (495° to 565°C).

Liquid nitriding utilizes molten cyanide salts and, as in gas nitriding, the temperature is held below the transformation range. Liquid nitriding adds more nitrogen and less carbon than either cyaniding or carburizing in cyanide baths. Case thicknesses of 0.001 to 0.012 in. (0.03 to 0.30 mm) are obtained, whereas for gas nitriding the case may be as thick as 0.025 in. (0.64 mm). In general, the uses of the two nitriding processes are similar.

Nitriding develops extreme hardness in the surface of steel. This hardness ranges from 900 to 1100 Brinell, which is considerably higher than that obtained by ordinary case hardening. Nitriding steels, by virtue of their alloying content, are stronger than ordinary steels and respond readily to heat treatment. It is recommended that these steels be machined and heat treated before nitriding, because there is no scale or further work necessary after this process. Fortunately, the interior structure and properties are not affected appreciably by the nitriding treatment and, because no quenching is necessary, there is little tendency to warp, develop cracks, or change condition. The surface effectively resists corrosive action of water, saltwater spray, alkalies, crude oil, and natural gas.

Induction Hardening

The induced electric current is a method of heating. Called *induction hardening*, it is a method of heating with applications in the melting of metals, hardening and other heat-treatment operations, preheating metals for hot work, and heating for sintering, brazing, and similar operations. High frequency alternating current obtained from motor generator sets, mercury arc converters, or spark gap oscillators are used for this type of heating. Although there are different powers and frequency limitations for each type of equipment, most do not exceed 500,000 cps (Hz). For shallow surface thickness of hardness, high frequencies are used, whereas for intermediate and deep thickness of hardness, low frequencies give better results.

Induction heating has proved satisfactory for surface-hardening operations on crankshafts and similar wearing surfaces. It differs from ordinary case hardening practice in that the analysis of the surface steel is unchanged, because the hardening is accomplished by an extremely rapid heating and quenching of the wearing surface, which has no effect on the interior core. The hardness obtained as a result of induction hardening is the same as that obtained in conventional treatment and depends on carbon content.

An inductor block acting as a primary coil of a transformer is placed around but not touching the journal to be hardened. A high-frequency current is passed through this block inducing a current in the surface of the bearing. The heating effect is caused by induced eddy current and hysteresis losses in the surface material. As the steel is heated to the upper critical range, the heating effect of these losses is gradually decreased, thereby eliminating any possibility of overheating the steel. The inductor block surrounding the heated surface has water connections and numerous small holes on its inside surface. As soon as the steel is to the proper temperature, it is automatically spray quenched under pressure.

An important feature of this hardening is its quickness because it requires only a few seconds to heat steel to a depth of ⅛ in. (3.2 mm). The actual time depends primarily on the frequency, power input, and depth of hardening required. Although the equipment cost is high, it is offset by the advantages of the process which include fast operation, freedom from scaling, clean, little distortion, no manual handling of hot parts, and low treating costs. Medium-carbon steel has proved satisfactory for parts, and the nature of the process eliminates the necessity of using costly alloy steels. Figure 17.16 illustrates the local heating obtained in a hardened crank pin bearing that is induction hardened.

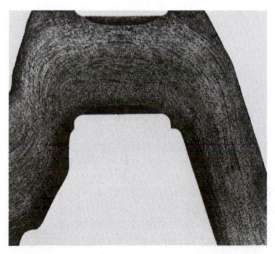

Figure 17.16
Section of an induction-hardened crankpin bearing.

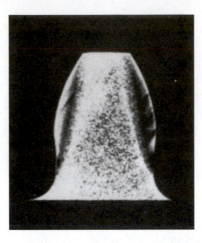

Figure 17.17
Section through a gear tooth showing the structure obtained by flame hardening.

Flame Hardening

Flame hardening, like the induction-hardening process, is based on rapid heating and quenching of the wearing surface. The depth of hardness of the case depends entirely on the hardenability of the material, as no other elements are added or absorbed during the process. Heating is with an oxyacetylene flame, which is applied for enough time to heat the surface above the critical temperature range of the steel. Integral with the flame head are water connections that cool the surface by spraying as soon as the desired temperature is reached. With proper control the interior surface is unaffected. The depth of the case is a function of the heating time and flame temperature. Figure 17.17 shows an etched cross section of a gear tooth and the hardened area. With this process hard surfaces with a ductile backing are obtained, large pieces are treated without heating the entire part, the base depth is easily controlled, the surface is free of scale, and the equipment is portable.

There are various torch combinations that allow coating powders to be applied after the flame, and surfaces can be laid on that vary in hardness, strength, and permeability. Usually when materials are added, the part must be machined or ground after application. The advantage of this technique is that parts such as shafts can be built up at the same time they are treated.

17.9 HARDENING NONFERROUS MATERIALS

Precipitation hardening, often called *age hardening*, can be achieved only with those alloys, in which there is a decreasing solubility of one material in another as the temperature is reduced. Such a situation is illustrated in Figure 17.18, where an enlarged portion of an aluminum–copper constitutional diagram is shown. If the alloy is cooled slowly from a temperature at *A*, the Al_2Cu material is precipitated out of the solid solution because the solubility of copper is decreased greatly at low temperature. An alloy in this condition will not respond to a hardening treatment.

To harden the alloy shown in Figure 17.18, two treatments are necessary. First, the alloy is heated to *A* above the solubility line *CD* to allow the constituents to enter into a solid solution. The alloy is held there for sufficient time until a homogeneous alloy is obtained. It is then cooled rapidly to room temperature, leaving the alloy in a supersatu-

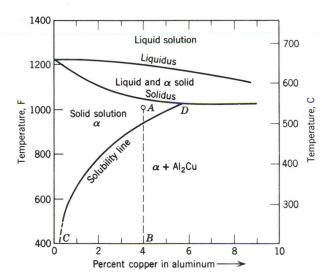

Figure 17.18
Partial constitutional diagram for aluminum–copper alloy system.

rated, unstable state. In this state the alloy is soft but, being unstable, the excess constituents will precipitate out of solution with the lapse of time. This process, if accomplished at room temperature, is known as natural aging. Particles that are precipitated from the solid solution form at grain boundaries and slip planes, producing a "keying" action that reduces slippage between the crystals. The hardness resulting from the phenomenon is influenced by the size of the precipitated particles. It is increased as the particle size is increased, but after a critical size is reached, further growth results in brittleness and loss of strength. The process of precipitation hardening can be delayed by refrigeration. An example of this is holding aluminum rivets in a refrigerator until they are ready for use. In this way they remain soft for the riveting operation and then gradually harden with time.

Artificial aging differs slightly in that an additional heat treatment is used. In this case the alloy is heated to some elevated temperature that accelerates the precipitation of some constituent from the supersaturated solution. As the temperature is increased, the precipitation and increase in hardness is more rapid. An equilibrium condition is finally reached that decreases the strength as a result of the large size of the precipitate crystals. This condition, which should be avoided, is known as overaging.

Numerous alloys are of a composition that can be age hardened, the only requirement being that solubility must decrease with decreasing temperature, permitting the formation of a supersaturated solid solution. In this group the nonferrous metals, which are most important, include many of the aluminum, copper, nickel, and magnesium alloys. Some stainless steels also respond to this treatment.

17.10 FURNACES FOR HEAT TREATING

The *car bottom furnace* (Figure 17.19A) is used for large parts that are to be heat treated and particularly for annealing and normalizing. Such furnaces get their name from their design, which employs a flatbed car on railroadlike wheels that is pushed into either a gas-fired or electrically heated, well-insulated, roomlike furnace. They vary in size from about 320 to 8000 ft² (30 to 740 m²).

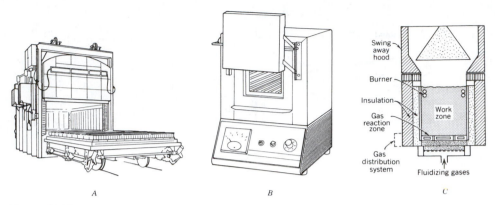

Figure 17.19
Typical heat-treating furnaces. *A*, Car bottom. *B*, Small box. *C*, Fluidized bed.

The bell top furnace is made so that once parts are loaded on a platform the furnace can be lifted up and placed over the parts. The box furnace, in which steel pellets are loaded into a furnace by a forklift truck, is a variation of the car bottom furnace. Small versions of the box furnace are called a muffle furnace (Figure 17.19*B*) and can sit on a bench.

Continuous furnaces employ a set of rollers, skids, or a walking beam on which the castings are moved systematically through the furnace. One type utilizes a carousel arrangement. Continuous furnaces are seldom used for normalizing. Because the size of the part regulates the time spent in the furnace, such furnaces are often single-purpose units.

Salt bath furnaces are *resistance heated* using salt as the resistance, and the magnetic field around the electrodes causes the bath to circulate. Because of the environmental problems associated with the vapors and the disposal of the chlorides of barium, cadmium, and sodium salts, these furnaces are declining in favor. Notice Figure 17.15*C*.

For small parts the *fluidized bed* (Figure 17.19*C*) is increasingly used. It consists of a container filled with pellets of alumina or sand, which are heated by a high-velocity, air-gas flame that not only heats the bed but also "fluidizes" the particles. The particles become almost airborne, and parts immersed in the pellets are surrounded by the pellets much as if they were in a liquid. If a stoichiometric mixture of air and gas is used, an almost inert atmosphere is present in the container. Some furnaces are electrically heated and an inert gas is forced upward to fluidize the pellets.

When parts are vacuum heat treated, a furnace is heated by electrical means or hot gases are circulated around it. The parts are placed inside and a vacuum is pulled. These furnaces offer operator safety, economy, uniform heating, and low capital cost, and they present no environmental problems. This type of operation is expensive but is often used for heat-treating die-casting dies and similar highly machined parts.

QUESTIONS

1. Give an explanation of the following terms:

Chauffage	Drawing
Austenite	Austempering
Pearlite	Annealing
Hypoeutectoid steels	Spheroidizing
Martensite	Normalizing
Hardenability	Nitriding
Fine pearlite	Car bottom furnace

2. What is martensite? How does it appear under the microscope?

3. Describe the lattice structure of a steel with 0.4% carbon as it goes from a molten to a cool condition. At what temperature does the transformation take place?

4. What determines the maximum hardness that is obtained in a piece of steel?

5. What happens if more than 6.67% carbon is present in a batch of molten steel?

6. What is the difference between the eutectic and eutectoid points?

7. Why is oil instead of water used as a quenching medium for many steels? What other materials or processes can be used as quenching media?

8. What is the relationship between the amount of pearlite in a piece of steel and its ability to be hardened?

9. Explain what happens to surface hardness for a given quench if the part is greatly increased in size. Why?

10. If 1050 steel is heat treated to yield a tensile strength of 160,000 psi, what other physical characteristics can you predict?

11. What is the value of an isothermal transformation diagram?

12. Describe the process of carburizing.

13. How is a time–temperature transformation diagram changed by varying the carbon content or by adding alloys?

14. Why does tempering cause a decrease in tensile strength?

15. What effect does carbon content have on hardenability?

16. What can be said about the physical characteristics of martensite?

17. Discuss the instability of martensite with temperature.

18. After steels are rolled or cast, which heat treatment is usually applied?

19. How does liquid nitriding differ from cyaniding?

20. What factors affect hardness as obtained by induction hardening?

21. Explain why Duralumin (4% copper) can or cannot be age hardened.

22. A box of aluminum rivets located in the storeroom was found to be age hardened. Explain whether or not the rivets can be salvaged.

23. Describe the advantages of a fluidized-bed furnace.

24. What information is available in the isothermal transformation diagram that is lacking in the iron carbon equilibrium diagram?

25. Briefly describe the mechanism by which austenite transforms to pearlite.

26. Explain the reason for the pearlite nose on the S curve.

27. Describe the operations for hardening, annealing, or normalizing steel.

28. If a steel possesses high hardenability, does this necessarily mean that the steel is extremely hard in the quenched condition? Explain.

29. Explain martempering.

30. On what size specimen is Figure 17.8 based? What is the result for a larger specimen? A smaller specimen?

PROBLEMS

17.1. How much pearlite is in pure iron? In a 1080 steel?

17.2. Sketch the microstructure of an etched specimen of 1040 steel and a 1080 steel.

17.3. Find the percentage of pearlite in a 1030 steel.

17.4. To what temperature must a 1050 steel be elevated if it is to be hardened? Sketch the constituents of an annealed 1050 steel.

17.5. To what temperature must the following steels be elevated for hardening: 1040, 1060, 1080?

17.6. To what temperature must the following steels be elevated for normalizing: 1040, 1060, 1080?

17.7. What is the maximum attainable hardness to which the following steels can be heat treated: 1020, 1040, 1060, 1080?

17.8. Suppose in Problem 17.7 the steels were alloy steels but with the same carbon content. What is the maximum attainable hardness?

17.9. What are the critical temperatures of a 1050 steel?

17.10. Calculate the percentages of pearlite and cementite in the following annealed steels: 1020, 1040, 1080.

17.11. Plot a curve showing percentage pearlite as a function of carbon content.

17.12. Sketch a design for an electrically heated, fluidized-bed furnace. (Hints: Assume a simple shape, think of the quantity involved, roughly estimate a transit time for the process, and then size the internal dimensions.)

17.13. How can a 1050 steel be heat treated to yield a tensile strength of 160,000 psi? 120,000 psi? Give the approximate Rockwell hardness of the two strengths of steel?

17.14. Calculate the percentages of pearlite and cementite in the following annealed steels: 1015, 1035, 1070.

17.15. Sketch and label a simulated photomicrograph of AISI 1020 steel.

17.16. Find the percentage of pearlite and the cementite in 10110 steel at room temperature.

17.17. A 1.5% carbon steel is cooled from 3200°F to room temperature. Find the composition at 3200°F, 2400°F, 2000°F, 1400°F and room temperature?

17.18. For C 1050 steel find the drawing temperature if the yield point is 100,000 psi. Also, what are the values of tensile strength, reduction, and elongation? If the drawing temperature decreases 200°F, what is the difference in values of yield point, tensile strength, reduction, and elongation?

PRACTICAL APPLICATION

Professional and technical societies are important to manufacturing. There are many excellent ones, such as American Foundrymen's Association, American Society of Mechanical Engineers, American Society of Metals, Institute of Electrical and Electronic Engineers, Institute of Industrial Engineers, National Association of Industrial Technology, Society of Automotive Engineers, Society of Manufacturing Engineers, and Society of Plastic Engineers, to name a few. Your instructor will know other groups and he or she is a member of one or more organizations that is devoted to manufacturing education and professional activities. These societies welcome student inquiry. Visit the Internet homepage of the American Society of Metals, http://www.asm-intl.org, and check their technical information. Conclude with a page of their activities.

CASE STUDY: HEAT TREATING

As chief engineer of Real Temperature Company, you are asked to bid on heat treating 500,000 parts per year for an automotive parts company. The part weighs about 1.9 lb and is shaped like a 4-in. OD saucer. The material is AISI 1050 steel and must be heat treated to have a tensile strength of 145,000 psi. It is your job to reply to the bid request by stating the method you will use to heat treat parts and to report on the physical properties of the finished product. Discuss in your report the type of furnace.

CHAPTER 18
PLASTIC MATERIALS
AND PROCESSES

Plastics, often called *resins*, are a common material in modern manufacturing systems. The term "plastics" usually refers to a *synthetic organic material* made from chemical raw materials called monomers. A monomer (such as ethylene) is reacted with other monomer molecules into long chains of repeating groups to form a polymer (such as polyethylene). Sometimes the term *polymer* (from the Greek *poly*: many, *meres*: parts) is used to refer to any substance in which several or many thousands of molecules are joined into larger more complex molecules. Such long chains of molecules combine atoms of oxygen, hydrogen, nitrogen, carbon, silicon, fluorine, and sulfur.

Plastics, then, begin as a gas (monomer), approach the liquid state for forming, and are shaped by heat and/or pressure to become solid in their final form. Adding reinforcements to plastics creates a different class of materials known as composites (see Chapter 2). This chapter considers the types of plastics, some methods of processing them, and fundamental design characteristics of plastic products.

18.1 RAW MATERIALS AND PROPERTIES

The basic plastic molecule is carbon. Raw materials for plastics are petroleum and carbon-based gases. Resins, necessary for the plastic materials, are produced by chemically reacting *monomers* to form long-chain molecules called polymers. This process is called *polymerization*. Two methods are used to achieve polymerization: *addition polymerization*, where two or more similar monomers directly react to form these long-chain molecules; and *condensation polymerization*, where two or more dissimilar monomers react to form long-chain molecules plus the by-product water.

The raw materials for plastic compounds are various agricultural products and numerous minerals and organic materials, including petroleum, coal, gas, limestone, silica, and sulfur. During the process of manufacture, additional ingredients, such as color pigments, solvents, lubricants, plasticizers, and filler material are added. Wood powder, flour, cotton, rag fibers, asbestos, powdered metals, graphite, glass, clays, and diatomaceous earth are the major materials used as fillers. Products such as outdoor chair seats, plastic cloth, garbage cans, machine housings, luggage, safety helmets, and equipment parts are examples of products that utilize fillers. Their use reduces manufacturing costs, minimizes shrinkage, improves heat resistance, provides impact strength, or imparts other desired properties to the products. *Plasticizers* or solvents are used with some compounds to soften them or to improve their flowability in the mold. Lubricants also improve the molding characteristics of the compound. All these materials are mixed with the granulated resins before molding.

Products can be made from plastic resins rapidly and with close dimensional tolerance. Surface finish is excellent. They are often a substitute for metals when lightness, moisture or corrosion resistance, and dielectric strength are factors. These materials can be either transparent or in colors. They are able to absorb vibration and sound. Usually, they are easier to fabricate than metals, and the final cost is usually cheaper than metal parts.

The use of plastics is limited because of comparatively low strength, low heat resistance, low dimensional stability, and often material costs may be higher than other raw materials. Compared to metals, plastics are softer, less ductile, and more susceptible to deformation under load. Plastics under load are *viscoelastic*, meaning that the material has a viscous and elastic response to applied loads. Unlike metals, which fail by plastic deformation, or slip off the molecules under load, plastics fail because of viscoelastic deformation. When a load is applied on a plastic, there is a combination of rapid elastic change or elastic response, and a slow change that is the viscous response. The viscoelastic deformation is primarily caused by the long-chain molecular structure of plastics. Under the applied load, the long chains slide past each other and the amount of movement is determined by the type of bond. Plastics with weak bonds deform more easily than plastics with strong bonds.

Plastics are low-density materials and have extensive application, such as heat and electrical insulators, because of their low thermal and electrical conductivity. Even the simplest plastic parts are subjected to stresses caused by assembly, handling, temperature variations, and other environmental effects. Simple analysis using information in Chapter 2 can be used to ensure that parts withstand these stresses.

Stresses

When parts made out of materials with different coefficients of thermal expansion are joined by any method that prevents relative movement, there is potential for thermal stress. This happens when nonreinforced thermoplastic parts are joined with materials such as metals, glass, or ceramics that have lower coefficients of thermal expansion.

The change in a linear dimension, such as length or diameter, is proportional to the change in temperature of the object, ΔT, its length L, and the coefficient of expansion α; that is,

$$\Delta L = \alpha L \Delta T \tag{18.1}$$

If the plastic component cannot expand or contract, the strain ε_T, induced by a temperature change, is

$$\varepsilon_T = \frac{\Delta L}{L} = \alpha \, \Delta T \tag{18.2}$$

and, as explained in Chapter 2, the stress is calculated by multiplying the strain ε_T by the tensile modulus of the material.

EXAMPLE 18.1

An 8-ft long copper wire is insulated with ASA unreinforced resin at 15°C. What is the required length of the insulator to guarantee that at 50°C an excess of 1 in. of insulator is left on each end?

For Cu: $\alpha = 17 \times 10^{-6}$ cm/cm°C, then

$$\Delta L_{Cu} = \alpha_{Cu} L_{Cu} \Delta T_{Cu} = 17 \times 10^{-6}/°C \times 8 \text{ ft } (50-15)°C = 0.00476 \text{ ft}$$

The required final length of the insulator must be

$$L_{Cu} + \Delta L_{Cu} + 2 \text{ in.} = 8.17 \text{ ft}$$

Table 18.1   **Average Properties of Some Common Thermoplastics***

Base Resin	Specific Gravity		Impact Strength[a] (ft-lb/in.)		Water Absorption 24 hr (%)		Tensile Strength (10^3 psi)		Thermal Expansion (10^{-5}) in./in. °F	
ASA[b]	1.03	**1.06**	6	**10**	0.3	**0.25**	5.0	**5.85**	5.5	**5.9**
Synthetic rubber[c]	—	**1.80**	—	**5**	—	**0.01**	—	**8**	—	**—**
Polymide	1.90	**1.43**	15	**0.7**	0.2	**0.39**	28	**10**	0.8	**2.6**
Rigid PVC[d]	1.55	**1.45**	0.8	**5**	—	**0.2**	15	**6.8**	1.2	**3.0**

* Lightface values for glass-reinforced resins. Boldface values for unreinforced resins for comparison.

[a] Izod, notched.

[b] Acrylic-Styrene-Acrylonitrile.

[c] Polybutadiene.

[d] Polyvinyl chloride.

Therefore, the initial length of the insulator is

$$L_{ASA} = 8.17 \text{ ft} - \Delta L_{ASA}$$

Since

$$\Delta L_{ASA} = \alpha_{ASA} L_{ASA} \Delta T_{ASA}$$

then

$$L_{ASA} = 8.17 \text{ ft} - \alpha_{ASA} L_{ASA} \Delta T_{ASA}$$

Solving for L_{ASA}

$$L_{ASA} = 8.17/(1 + \alpha_{ASA} \Delta T_{ASA})$$

From Table 18.1,

$$\alpha_{ASA} = 5.9 \times 10^{-5} \text{ in./in.°F or } (1.06 \times 10^{-4}/°C)$$

Therefore,

$$L_{ASA} = \frac{8.17 \text{ ft}}{(1 + 1.06 \times 10^{-4}/°C \times 35°C)} = 8.14 \text{ ft}$$

EXAMPLE 18.2

A plastic part is mounted to a metal part. Both components expand with changes in temperature. The plastic imposes insignificant load to the metal but considerable stress is generated in the plastic. The reader can show that the approximate thermal stress σ_T in the plastic is given by

$$\sigma_T = (\alpha_m - \alpha_{pl}) E_{pl} \Delta T \tag{18.3}$$

where

α_m = Coefficient of thermal expansion of the metal
α_{pl} = Coefficient of thermal expansion of the plastic
E_{pl} = Tensile modulus of the plastic at the temperature involved

When temperature increases, most plastics expand more than metals and their tensile modulus drops. This results in buckling because of a compressive load in the plastics. Conversely, when the temperature drops, the plastic shrinks more than the metal and the tensile modulus increases. This can cause tensile rupture of the plastic part. Such thermal stress is minimized by providing clearances around fasteners. Allowances must be made for temperature changes, especially with large parts. This is accomplished by providing room for relative motion ΔL_{rel} between the two materials,

$$\Delta L_{rel} = (\alpha_{pl} - \alpha_m)\, L\Delta T \tag{18.4}$$

Types of Plastics

The term *elastomer* is used to designate all flexible materials that can be stretched up to about double their length at room temperature and can return to their original length when released. Plastics are broadly classified as *thermosetting* and *thermoplastic*. *Thermosetting plastics* are formed to shape with heat, with or without pressure, resulting in a product that is permanently hard. The heat first softens the granules, or resins, but as additional heat or special chemicals are added, the plastic is hardened by a chemical change known as polymerization and then cannot be resoftened. Processes used for the thermosetting plastics include compression or transfer molding, casting, laminating, and impregnating. See Section 18.5.

Thermoplastic materials do not undergo chemical changes in molding and do not become permanently hard with the application of pressure and heat. They remain soft at elevated temperature until they are hardened by cooling. They may be remelted repeatedly by successive application of heat, as in the melting of paraffin. *Thermoplastic elastomers* are often used in place of rubber, and also may be used as additives to improve the impact strength of rigid thermoplastics. Thermoplastic materials are processed by injection, blow molding, extrusion, thermoforming, calendering, and others. See Section 18.6.

18.2 THERMOSETTING COMPOUNDS

Thermosets are not heat reversible. In other words, after they are molded, all the molecules are interconnected with a strong, permanent physical bond. In a sense, curing a thermoset is like cooking an egg. Once it is cooked, reheating does not cause remelting, so it cannot be remolded. The most common thermosets are described next and some of their properties are listed in Table 18.1.

Phenolics

These resins are popular for thermosetting applications. The synthetic resin, made by the reactions of phenol with formaldehyde, forms a hard, high-strength, durable material that is capable of being molded under a variety of conditions. The material has high heat and water resistance, and can be colored in a variety of ways. It is used in manufacturing coating materials, laminated products, grinding wheels, and metal and glass bonding agents, and can be cast into molded cases, bottle caps, knobs, dials, knife handles, electronic appliance cabinets, and numerous electrical parts. Phenolic resins are used with wood-particle chip boards, and in the foundry as a sand bond for cores and molds. Phenolic compounds may be molded by compression or transfer methods.

Amino Resins

The most important resins are urea-formaldehyde and melamine-formaldehyde. These thermosetting compounds can be obtained in the form of molding powders or as a liquid solution for bonding and adhesive applications. Both are compounded with a variety of fillers to improve mechanical or electrical properties. The good flow characteristics of *melamine* make transfer molding useful for items such as tableware, ignition parts, knobs, and electric-shaver housings. Urea resins, suitable for compression and transfer molding, have a hard surface and high dielectric strength and can be produced in all colors. Products include electric appliance housings, circuit breaker parts, and buttons. Both resins are used in the application of adhesives and for laminating wood or paper.

Furane Resins

Furane (C_4H_4O) resins are obtained by processing waste farm products, such as corn cobs, hulls from rice, and cotton seeds, with certain acids. The products from the thermosetting resins are dark in color, water resistant, and have good electrical properties. Furane resins are used for core sand binders, hardening additives for gypsum plaster, as well as bonding agents for floor compositions and graphite products.

Epoxides

Epoxy resins are used for casting, laminating, molding, *potting* (the encasing of electrical parts often by pouring the solution into a container), and as paint ingredients and adhesives. Cured resins are low in shrinkage, have good chemical resistance, have excellent electrical characteristics, have strong physical properties, and adhere well to both glass and metal. As adhesives, they replace other forms of fastening. Epoxies are used in manufacturing laminates and with glass fibers to make panels for printed circuits, tanks, jigs, and dies. Their resistance to wear and impact suggest uses in manufacturing press dies for metal-forming dies.

Silicones

These polymers differ materially from most other plastics that are based on the carbon atom. Silicones possess a desirable combination of properties for a large group of industrial products, such as oils, greases, resins, adhesives, and rubber compounds. Their outstanding properties include stability, resistance to high temperatures over long periods of time, good low temperature and high electric characteristics, and water repellence.

Some oils and greases operate well over a temperature range of $-40°$ to $500°F$ ($-40°$ to $260°C$). The silicone resins may be molded, used in laminates or as coatings, or may be processed into foam sheets or blocks. Silcone rubbers are used in molding, extrusion, gaskets, electrical encapsulation of electronic components, glass cloth, electrical connectors, or as shock-absorption material. Silicones are available as a liquid for casting and laminating resins and as a powdered molding compound for foam products. Because of their high cost, the use of silicone products is often limited to designs where their unusual properties are the most useful. Silicone-based polymers are processed by compression or transfer molding, extrusion, and casting.

18.3 THERMOPLASTIC COMPOUNDS

Thermoplastics, unlike thermosets, are long polymer chains with no physical connections. They resemble long intertwined bundles of spaghetti. Thermoplastics can be classified by their structures into three categories: (1) *crystalline*, if the polymer chains are packed together in an organized way; (2) *amorphous*, if the chains have no organized pattern; and (3) *liquid crystalline polymers* (LCPs), if the chains are organized in rodlike structures. Crystalline thermoplastics (such as acetal, nylon, polyethylene, polypropylene, and polyester) are stronger and stiffer, have higher melting temperatures, higher shrinkage, and higher warpage factors than amorphous plastics. Amorphous thermoplastics (such as polycarbonate; polystyrene; ABS—acrylonitrile, butadiene, and styrene; and PVC—polyvinyl chlorides) are more resistant to impact. However, LCPs (like all the aromatic copolyesters) have good dimensional stability and maintain significant order during the melt phase, which gives them the lowest shrinkage and warpage factors.

Cellulosics

These thermoplastics are prepared from various treatments of cotton and wood fibers. They are very tough and can be produced in a variety of colors.

Cellulose acetate is a durable compound having considerable mechanical strength and can be fabricated into sheets or molded by injection, compression, and extrusion. Display packaging, toys, knobs, flashlight cases, bristle material for paint brushes, radio panels, film for recording tape, and extruded strips are made successfully from this compound.

Cellulose acetate-butyrate molding compound is similar to cellulose acetate, and both are produced in all colors and by the same processes. In general, cellulose acetate-butyrate is recognized for its low moisture absorption, toughness, dimensional stability under various atmospheric conditions, and ability to be continuously extruded. Typical butyrate products include steering wheels, football helmets, goggle frames, trays, belts, furniture trim, insulation foil, sound tapes, buttons, and extruded tubing for gas and water.

Ethyl cellulose has the lowest density of the cellulose derivatives. In addition to its use as a base for coating materials, it is employed extensively in the various molding processes because of its stability and resistance to alkalies.

Polystyrene

This thermoplastic material has the outstanding characteristic of low specific gravity (1.07), availability in colors from clear to opaque, resistance to water and most chemicals, dimensional stability, and insulating ability. *Polystyrene* is an excellent rubber substitute for electrical insulation. Styrene resin is molded into battery boxes, dishes, radio parts, lenses, flotation gears, foundry patterns, ice chests, packaging waste, insulated and disposable cups, and wall tile. Polystyrene is especially adapted to injection molding extrusion and is formable in dies.

Polyethylene

These materials are flexible at room and low temperatures, waterproof, unaffected by most chemicals, capable of being heat-sealed, and can be produced in a variety of colors. Polyethylene, which floats on water, has a density range from 0.91 to 0.96. It is one of the inexpensive plastics, and its moisture-resistant characteristics ensure its use for packaging and squeeze bottles. Other products are ice-cube trays, developing trays, fabrics, film for packaging, collapsible nursing bottles, garden hose, coaxial cable, and insulating parts for high-frequency electrical fields. Polyethylene products can be made by injection molding, blow molding, or extruding into sheets, film, and monofilaments.

Polypropylene

Polypropylene is produced by all thermoplastic techniques. It has excellent electrical properties, high impact and tensile strength, and is resistant to heat and chemicals. Monofilaments of polypropylene are used in making rope, nets, and textiles. Other products are hospital and laboratory ware, toys, luggage, furniture, film for food packaging, television cabinets, and electrical insulation.

ABS

ABS stands for acrylonitrile, butadiene, and styrene, which are combined to make the ABS plastic. This plastic can be compounded to have a degree of hardness or great flexibility and toughness. The ABS plastics are used in applications that require abuse resistance, colorability, hardness, electrical and moisture properties, and limited heat resistance (220°F, 105°C). These plastics are processed by thermoforming injection, flow, rotational, and extrusion molding. Applications include household piping, cameras, electrical hand tool housings, telephone handsets, and canoes.

Polyimide

These thermoplastics are produced in the form of solids, films, or solutions. They have unusual heat-resisting properties up to 750°F (400°C), low coefficient of friction, high degree of radiation resistance, and good electrical properties. Products include sleeve bearings, valve seats, tubing, and various electrical components. The films, tough and strong, are used for wire insulation, motor insulation, and printed circuit backing. The solutions are used in varnishes, wire enamels, and coated glass fabrics.

Another grade is the *nylons*, which are used in molding and extruding as well as in the textile fiber and filament field. Molded and extruded products of nylon include bearings, gears, valves, tubing, kitchen accessories, and luggage. Nylon monofilaments are used for hosiery, glider tow ropes, and brush bristles.

Acrylic Resin

This resin has the property of light-transmitting power, ease of fabrication, and resistance to moisture. The acrylic resin commonly used is methyl methacrylate, better known by the commercial name as Lucite or Plexiglas. It can be formed by casting, extruding, molding, stretch-forming for airplane windows, shower doors, toilet articles, and covers where visibility is desirable.

Vinyl Resins

The vinyl resins commercially available include polyvinyl chlorides (PVCs), polyvinyl butyrates, and polyvinylidene chloride. These thermoplastic materials can be processed by compression or injection molding, extrusion, or blow molding. Vinyl resins are suitable especially for surface coating and for flexible and rigid sheeting. Polyvinyl butyrate is a clear tough resin, which is used for interlayers in safety glass, raincoats, sealing fuel tanks, and flexible molded products. It has moisture resistance, great adhesiveness, and stability toward light and heat. Polyvinyl chloride has a high degree of resistance to many solvents and will not support combustion. Industrially, it is used for rubberlike products, including raincoats, packaging, and blow-molded bottles. Polyvinylidene chloride is used for saran films and pipe. Cellular vinyl-foamed products include floats, upholstery, and protective pads for sport uniforms.

Synthetic Rubber

Many of the industrialized nations have no source of natural rubber, which led to the development of synthetics, such as GR-S, nitrile, thiokol, neoprene, butyl, and silicone rubbers. The *synthetic rubber* GR-S is produced in the largest quantity, is very similar to natural rubber, and is combined with it for tire use. It is a copolymer of butadience and styrene and is cured to any degree of rubber hardness. Its strength is improved by adding carbon black. The butadience-acrylonitrile copolymers (known as Buna N or *nitrile* rubbers) are employed principally because of their resistance to oils, and are used in oil hose, gaskets, and diaphragms. They also serve to some extent as a blending material with phenolics and vinyl plastics. The organic polysulfides, known as *Thiokols*, are very resistant to gasoline, oils, and paints, as well to sunlight, and are used to manufacture hose, shoe heels and soles, coated fabrics, and insulation coatings. Resilient solid objects can be molded in conventional machines used for other plastics.

The chloroprene polymer, known as *neoprene*, is produced from coal, limestone, water, and salt. Calcium carbide, a product of coal and limestone, when added to water forms acetylene gas (C_2H_2). This gas, when combined with hydrogen chloride, forms chloroprene, which is changed to neoprene by polymerization. Neoprene has good resistance to oils, heat, and sunlight, and is used for conveyor belts, shoe soles, protective clothing, insulation, hose, printing rolls, tires, and tubes and as a bonding material for abrasive wheels. Neoprene has a wider application than other synthetic rubbers and can replace natural rubber. *Butyl*, an isobutylene copolymer, has many of the properties and characteristics of natural rubber. Because of its strength, resistance to abrasion, and low permeability to gases, it is used for inner tubes. Other applications include steam hose, conveyor

Table 18.2   Average Properties of Some Common Thermosets*

Base Resin	Specific Gravity		Impact Strength[a] (ft-lb/in.)		Water Absorption 24 hr (%)		Tensile Strength (10^3 psi)		Thermal Expansion (10^{-5}) in./in. °F	
Alkyd amino resin	2.05	**2.2**	27	**28**	0.08	**0.07**	8.0	**7.5**	2	**2**
Standard epoxy[b]	1.9	**1.16**	0.45	**1.1**	0.7	**0.55**	9.5	**4.5**	1.5	**3.5**
Phenolics	1.87	**1.4**	5.0	**0.37**	0.6	**0.55**	9	**7**	0.88	**2.1**
Silicone[c]	1.52	**1.20**	2.5	**14**	0.12	**0.15**	23	**8.7**	0.93	**3.75**
Polyester thermoset[d]	6.9	**1.27**	5.5	**0.29**	0.63	**0.4**	7.5	**7**	1.5	**4.75**

* Lightface values for glass-reinforced resins. Boldface values for unreinforced resins for comparison.

[a] Izod, notched.

[b] Cast, flexible.

[c] General purpose.

[d] Cast, rigid. Compare with polyester (PBT) thermoplastic in Chapter 2.

belting for heated materials, and tank linings. Silicone (polysiloxane) rubbers are extremely resistant to both high and low temperatures as well as to lubricating oils, dilute acids, and sunlight; they are used when other synthetics are not capable of performing satisfactorily. O-rings and seals for oil and gas lines, sealing doors on airplanes, and wire and cable insulation are some of the applications. Other commercial synthetic rubbers are polybutadiene (used for tires), polyacrylate (used for oil hose and automotive gaskets), and urethane elastomers (used as shock-absorbing pads, forming pads in press work, conveyor rolls, and solid tires).

Material properties of thermoplastics are listed in Chapter 2 and in Table 18.2.

18.4 PROCESSING PLASTICS

The plastic industry consists of the manufacturers who produce the resins, chemicals, and fabricators. The manufacturers make raw material, which includes powders, granules, liquids, and standard forms such as sheets, bars, tubes, flats, rolls, and laminates. The fabricators get the product ready for distribution to industries or the consumer.

In most cases, it is necessary to mix and prepare the raw materials for a final product. This step is referred to as *compounding* or *preforming*. The compounding process is normally carried out in a muller into which any number of additives (see Table 18.3) are mixed. Materials that have been mixed and sometimes melted are placed into the feed hoppers of injection, extrusion, or calendering machines. Some thermoplastic materials are preformed into small pellets of the proper size and shape for a given mold cavity. All preforms are of the same density and weight, and the operation avoids waste of material in loading molds and speeds up production with no possibility of overloading

Table 18.3   **Common Additives for Resins**

Additive	Reason to Be Added
Fillers	Added to reduce cost and/or increase strength. They include wood flour, quartz, limestone, cotton, rag fibers, powdered metals, graphite, clays, and diatomaceous earth.
Reinforcements	Added for physical strength. They include sisal, jute, glass, graphite whiskers, ceramics, and nylon, cotton, and orlon fabrics.
Flame retardants	Added to impair burning of the finished product. The principal flame retardant is a phosphate ester.
Stabilizers	Added to keep a plastic from oxidizing or degrading during use. Zinc soap is added to vinyls and phenols to the styrenes.
Antistatics	Added to prevent a buildup of an electric charge on powders during processing when required amines are added.
Colorants	Added to improve the acceptance of the product by the consumer. Colors are added by mulling with the resin or liquid coloring is metered into the plastic in some injection-molding and extrusion machines. Colors are available as organic and inorganic pigments and dyes as a solid or liquid. Many resins are precolored.
Lubricants	Added to improve processing by promoting better flow in molding. They include waxes, zinc, and calcium stearates.
Plasticizers	Added to improve flexibility of the final product. Vinyl is very brittle unless phthalate is added.
Ultraviolet protector	Added to vinyls, styrenes, polyesters, and fiberglass to improve their life span when subjected to sunshine. Carbon black is one such additive.

Figure 18.1
Rotary preforming press used for making disk pellets of various molding compounds.

the molds. In the preforming operation the thermosetting powder is cold-molded, and no curing takes place. Preforms are used in compression and transfer-molding processes.

A *rotary preforming* press used in making disk pellets of various molding compounds is shown in Figure 18.1. The powder is fed by gravity into the mold cells and excess powder is scraped off. The amount of material fed into each cavity is controlled by regulating the lower punch. As the table revolves, pressure is applied uniformly on both sides compressing the powder charge, and at the end of the cycle the tablet is ejected. Reciprocating machines, which differ from rotary machines, may have more than one cavity and more than one preform is ejected with each cycle. A formula for the production rate of a reciprocating machine is

$$\text{min/preform} = (\text{spm} \times \text{no. of cavities in die})^{-1} \tag{18.5}$$

where

$$\text{spm} = \text{Strokes per minute}$$

18.5 PROCESSING THERMOSETS

Compression Molding

For *compression molding* thermosetting material, usually in powder or granular form, is placed in a heated die. The upper half of the die compresses the material, which melts and fills the die cavity. After compression, the part solidifies (polymerizes or *cures*) and the upper half of the die retracts or opens, and the part is removed. The molding process is illustrated by Figure 18.2. The pressure causes the liquid material to fill and conform to the die shape. Pressures used in compression molding vary from 100 to 8000 psi (0.7 to 55 MPa), depending on the nature of the material and the size of product. The

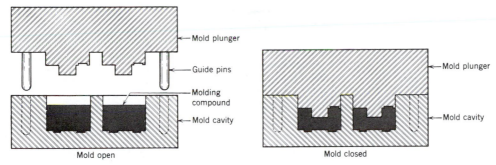

Figure 18.2
The compression-molding process.

temperature range is 250° to 400°F (120° to 205°C). Heat is very important for thermosetting resins, since it is used first to plasticize and then to polymerize or make the product permanently hard.

Some thermoplastic materials are processed by compression, but the cycle of rapid heating and cooling of the mold adds to the difficulty in using such material. Unless the mold is sufficiently cooled before ejection, distortion of the piece is likely.

Transfer Molding

Transfer molding is similar to compression molding, as heat and pressure are used. In this method, when the powder or granules are in a semiliquid state, the plastic is forced or transferred to the die cavity through a sprue. After curing the plastic in the mold, the mold is opened and the part is removed. The curing time for transfer molding is usually less than for compression molding. The loading time is also shortened. The method is adaptable for parts requiring small metal inserts. Intricate parts and those having large variations in section thickness are produced by this method. Figure 18.3 is a sectional view of the transfer mold before and after the transfer of the plastic. Losses caused by the runner, sprue, and well are limitations of the method. Die costs may be greater than for compression molding.

Injection Molding

Thermosetting materials are injection molded by a process known as *jet molding*. With a few changes many of the standard thermoplastic machines can be converted for jet molding. The nozzle, which is the important part of the machine, must be both heated and cooled during the molding cycle. The resin is first heated in the cylinder surrounding the plunger, making it plastic, although not polymerized appreciably. As the plunger forces the resin through the nozzle to the mold, additional heat is applied. When the mold is full, the nozzle is cooled rapidly by water to prevent further polymerization.

An alternate to jet molding is the *reciprocating screw injection machine* (also used for thermoplastic materials), shown in Figure 18.4. Material is fed by gravity to a rotating screw, where it is heated by contact with the heated barrel and the frictional heat developed by the rotating screw. As the screw revolves, plasticized material is built up ahead of the screw, being blocked from entering the transfer chamber by the *transfer ram* in the upper position until a sufficient amount is accumulated. The ram then returns to the lowered position and the screw, not revolving on the forward stroke, forces the

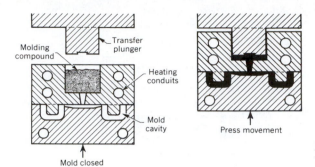

Figure 18.3
The transfer-molding process.

material into the transfer chamber, where the plunger pushes it upward into the mold cavities. Precuring of the material is prevented by a water-cooled band around the end of the cylinder. This process is similar to transfer molding, except that it is automatic in operation.

Reinforced Plastics Processes

Reinforced plastics include a wide range of products made from thermosetting resins with random or woven fibers. Although glass fibers predominate, asbestos, cotton, graphite, and synthetic fibers also are used. Polyester resins are low in cost and their properties are good. Epoxies provide extra strength and chemical resistance, whereas silicones find

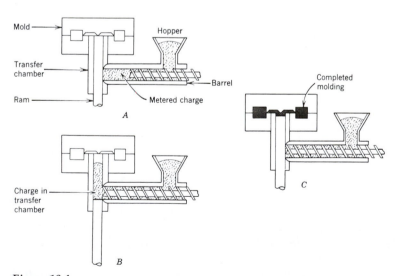

Figure 18.4
Screw injection molding cycle. *A*, Screw retracts while revolving as molding material feeds into barrel by gravity. *B*, The screw while not revolving forces material into the vertical plunger chamber. *C*, Hydraulic plunger forces plasticized material into mold.

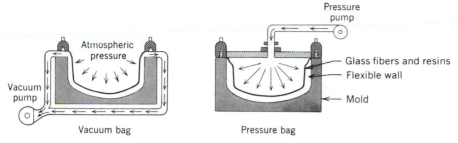

Figure 18.5
Examples of producing reinforced plastic product by the open-mold process.

use where heat resistance and electrical properties are important. Other resins are available for special properties and applications.

Fiberglass and other reinforced plastics are made by various processes, but in general all are classified as open and closed molding. The *open-mold process* with a single cavity mold, either male or female, makes a product with little or no pressure. Fiberglass boat bodies are a good example, as the process adapts well to fabricating large objects where only one side is finished. First, the finish coat of paint is sprayed in the mold, and then glass fibers and resin are placed into the mold and rolled to compress and remove air. Such molds normally cure in air, but either a vacuum or pressure bag is used against the layup to provide additional pressure (Figure 18.5). For still more pressure, the assembly is placed in a steam autoclave at pressures up to 100 psi (0.7 MPa). Other products of the open-mold process include aircraft parts, luggage, truck and bus components, and large containers.

The *closed-mold* or *matched-die process* uses two-part molds usually made of metal. Both sides are finished and good detail is obtained. The labor cost is low and because the molds are heated, a high production rate is possible. Products obtained from this process include luggage, helmets, trays, and machinery housing. In general, small-size products are made by this process because of the high cost of closed molds.

Several other techniques find commercial use in the manufacture of reinforced plastics, one of which is shown in Figure 18.6. In the *sprayup process*, fiberglass and resin are deposited simultaneously in a mold by spray guns. Boats and other large objects are fabricated in this manner. In *filament winding,* single strands of fiber are fed through a bath of resin and wound on a mandrel, as illustrated in Figure 18.7. This process is used in making pressure cases, tubing, and missile bodies where high strength is a requirement.

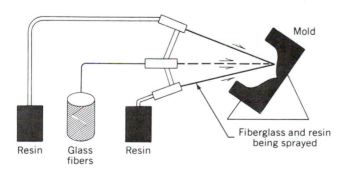

Figure 18.6
Fiberglass and resin being simultaneously deposited in mold by spray guns.

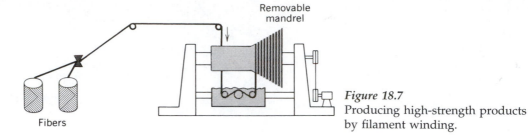

Figure 18.7
Producing high-strength products by filament winding.

Processes for Laminated Plastics

Laminated plastics consist of sheets of paper, fabric, asbestos, wood, or similar materials that are first impregnated or coated with resin and then combined under heat and pressure to form commercial materials. These materials, which are hard, strong, impact resisting, and unaffected by heat or water have desirable properties for numerous electrical applications. The final product may consist of either a few sheets or more than a hundred, depending on the required thickness and properties. Although most laminated stock is made in sheet form, rods and tubes as well as special shapes are available. The material has good machining characteristics, which permit its fabrication into gears, handles, bushings, and furniture.

In manufacturing laminated products, the resinoid material is dissolved by a solvent to convert it into a liquid varnish. Rolls of paper or fabric are then passed through a bath for impregnation (Figure 18.8). To facilitate lamination the sheets are then cut into convenient sizes and stacked together in numbers sufficient to make up the thickness of the final sheet. Tubes made by machine winding strips of the prepared stock around a steel mandrel are cured by being placed in a circulating hot air oven or are subjected to both heat and pressure in a tube mold. Paper base laminates are used in electrical products. Fabric base materials, which are stronger and tougher, are better for stressed parts. Gears made of a canvas base are quiet in operation. Fiberglass cloth is recommended for heat-resisting and low water absorption uses. Thin sheets of wood are laminated to produce a light material equal in strength to some metals and resistant to moisture. Safety glass is, in effect, a laminated plastic product, since thermoplastic layers are used between the glass sheets to make it nonshattering. The four resins used most are phenolics, silicones, epoxies, and melamines.

Casting

Thermosetting materials used for casting include the phenolics, the polyesters (notice that there are thermoplastic polyesters as well), epoxies, and the alkyl resins. The last

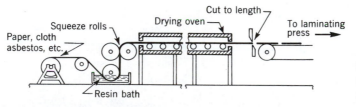

Figure 18.8
Sketch of process for preparing sheet material for lamination.

is especially useful for optical lenses and other applications requiring excellent clarity. These resins have a wider use in casting than the thermoplastics, as they have greater fluidity in pouring. However, some thermoplastics, such as ethyl cellulose and cellulose acetate butyrate, are used where impact strength and rigidity are needed (e.g., drop hammer and stretch dies). Acrylics and thermoplastics are used in casting transparent articles and flat sheets.

Plastics are cast when the number of parts desired is not sufficient to justify making expensive dies. Frequently, open molds of lead are formed by dipping a steel mandrel of special shape into molten lead and stripping the shell from the sides of the mandrel after it solidifies. Cores of lead, plaster, or rubber may be introduced if desired. Hollow castings are produced by the slush-casting method. Solid objects may be made from molds of plaster, glass, wood, or metal. When parts have numerous undercuts, the molds are made of synthetic rubber.

18.6 PROCESSING THERMOPLASTICS

Injection Molding

Injection-molding machines are somewhat similar to those used for die casting. Thermoplastic material is converted from a granular material to a liquid and then injected into a mold, where it solidifies. This material can be changed repeatedly from solid to liquid without chemical change, making it ideal for rapid processing.

Injection-molding machines are specified by the tonnage with which the dies may be clamped and the amount of material injected per cycle. Most machines of this type have a 50- to 2500-ton (0.4- to 22-MN) clamping force, and the shot capacity varies from less than 1 oz to about 300 oz (9 kg). The machine shown in Figure 18.9 is a 2500-ton (22-MN) hydraulic clamp machine capable of molding 300 oz (9 kg) per cycle. The plastic is plasticized up to 400 lb/h (0.05 kg/s) in the machine before being injected at rates up

Figure 18.9
Injection-molding machine for plastics.

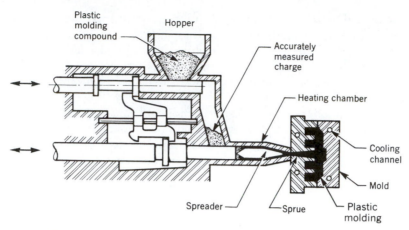

Figure 18.10
Sketch of an injection-molding machine.

to 5000 in³/min (0.081 m³/s). Figure 18.10 is a schematic sketch showing the operation of injection-molding machines.

There are nine basic steps of *injection molding*. First, the granular plastic material is loaded into a heating cylinder. The material is then compressed and air is forced out by a ram. Second, the material is transferred to the heated section by the ram where it melts. The moving mold part is now matched to the fixed mold part by a locking device on the machine. The ram pressure injects the softened material through the nozzle of the machine, displacing the air content of the mold cavity. The plastic in contact with the cold sidewalls of the mold stiffens first in the gate sealing the cavity. As cooling takes place, the mass is hardened, permitting the removal of the finished part. After cooling, the mold is opened and the part is removed by the ejector system.

The most important factors of injection molding are the outer and inner pressure and temperature of the material and mold. Figure 18.11 shows a pressure-time diagram of injection molding. The outer pressure, created by the ram pushing the plastic into the mold, ensures that the entire cavity is filled. The inner pressure, originating in the heating cylinder, keeps the product from decreasing in size as it cools. The temperature of the

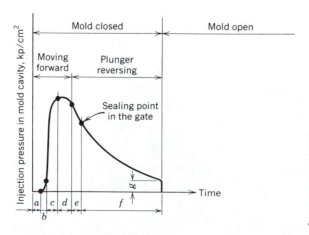

Figure 18.11
Pressure–time diagram of injection molding. *a*, Cavity filling period; *b*, pressure rises; *c*, post-pressure period; *d*, unloading; *e*, cooling till sealing; *f*, cooling till mold is opened; *g*, residual pressure.

material determines the viscosity of the liquid plastic. A high temperature, low viscosity material fills a highly detailed mold quite easily. The temperature of the mold itself affects the time needed for cooling; the cooler the mold, the quicker the material hardens. Most molds are maintained between 16° to 200°F (74° to 93°C) by circulating water; otherwise, after a short time the temperature would be so high that the mold would never harden. The different combinations of these four basic factors determine the type of part produced.

The heating chamber construction of most injection machines is cylindrical in shape with a *torpedo-like spreader* in the center so that the incoming material is kept in a layer thin enough to be heated both uniformly and rapidly. The heating chamber temperature ranges from 250° to 500°F (120° to 260 °C), depending on the material being charged and the mold size. Heat is furnished by a series of electrical resistance coils. These chambers must be of substantial construction, as injection pressures may reach 30,000 psi (200 MPa).

Injection molding is faster than compression molding. A production cycle of two to six shots per minute is possible. Mold costs are lower since fewer cavities are necessary to maintain equivalent production. Articles of difficult shape and of thin walls are produced successfully. Metal inserts such as bearings, contacts, or screws can be placed in the mold and cast integrally with the product. Material loss in the process is low because sprues and gates can be reused. Although the capacities of injection machines vary from 2 oz to 8 lb (0.06 to 3.6 kg), small parts machines of 8 to 16 oz (0.23 to 0.45 kg) capacity are most popular.

Also employed in injection molding of many thermoplastics is the in-line reciprocating screw injection machine described in Section 18.5.

Extruding

Thermoplastic materials, such as the cellulose derivatives, vinyl resins, polystyrene, polyethylene, polypropylene, and nylon may be extruded through dies into simple shapes of any length. A schematic diagram of an *extruding* press is shown in Figure 18.12. Granular or powdered material is fed into a hopper and forced through a heated chamber by a spiral screw. In the chamber, the material becomes a thick viscous mass that is forced through the die. As it leaves the die, it is cooled by air or water, or by contact with a chilled surface and fully hardens as it rests on the conveyor. Long tubes, rods, and many special sections are readily produced in this manner. Such products as conduits for electric conductors and chemicals are made by this process, because thermoplastic extrusions are bent or curved to various shapes after extrusion by immersion in hot water. Thermosetting compounds are not well adapted to this type of extrusion because they harden too rapidly; however, they are used to a limited extent in the production of thick-walled tubes.

A machine for the extrusion of thermosets utilizes a ram instead of a screw to force the material through the die. Material is fed from a hopper at the rear of the cylinder,

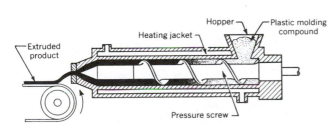

Figure 18.12
Plastic extrusion press.

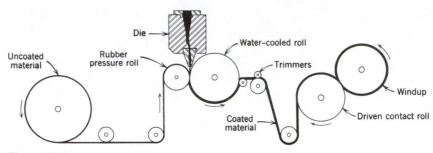

Figure 18.13
Extrusion coating process.

and by repeated strokes of the ram is forced into a long, tapered die that has heated zones. Additional heat results from frictional resistance as the material is forced through the cylinder and die. Curing is complete as it reaches the forward end. Products include tubes, rods, moldings, bearings, brake linings, and gears. Cross-sectional tolerances of 0.005 in./in. (0.4%) can be maintained.

A process known as extrusion coating is used extensively for coating paper, fabrics, and metal foil. A thermoplastic material is extruded through a flat die (Figure 18.13) onto a sheet passing beneath it. The *extrudate*, while soft, blends onto the *substrate* and is contacted by a rubber roll that holds it against the steel roll at a desired pressure. The edges of the sheet are trimmed prior to the windup. Although any thermoplastic material can be extruded as a coating, the ones used most are the vinyls, polyethylene, and polypropylene. Another important extrusion coating application is that of insulation on wire and cable.

Rotational Molding

Rotational molding employs the simultaneous rotation of thin-walled molds about two axes, primary and secondary, which are perpendicular to each other. After charging with appropriate plastic material, the molds are heated while in rotation, causing the particles to melt on the inner surface of the mold, depositing in layers until all the material is fused. The molds are cooled while still rotating and opened so that the finished article can be removed and the molds recharged. The process is intended primarily for hollow objects from thermoplastic materials. The toy industry uses *plastisols* and polyethylene with rotational molding to make squeeze toys.

The rotational powder method differs from other molding processes in that whereas the others all require both heat and pressure to plasticize the resin, rotational powder molding requires only that the mold be heated.

Thin cast-aluminum molds normally are used in rotational molding but electroformed copper or sheet metal is also satisfactory. The mold sections must fit tightly together so that no moisture enters the mold to cause warping. The rotational speeds of the two mold axes are generally controlled by separate motors; normally, a ratio of 3:1 exists between the major and minor axes. The rotational speed of the major axis is generally under 18 rpm while mold temperature ranges from 500° to 700°F (260° to 370°C).

The principle of rotational molding is shown schematically in Figure 18.14. In one case a single mold is shown, whereas in the other four molds are assembled on a single arm unit. In both the arm is pivoted so that it can be swung into a heating oven, after which it can be directed to a cooling chamber, as shown in Figure 18.15. Some designs

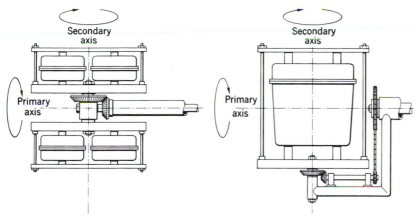

Figure 18.14
Schematic of rotational molding showing two mold mounting systems.

have the motors and drive spindles mounted on a track that permits them to move from the oven to the cooling chamber and the unloading station.

Advantages for rotational molding include low initial investment, a flexibility that allows a variety of parts to be made on the same equipment, low tooling costs, totally enclosed and open-end pieces, fine detail, excellent surface finish, and low cost per unit produced. Products made by powder rotational molding are often of considerable size such as children's chairs, 55-gal (0.2 m³) drums for food storage, phonograph cases, machinery guards, garbage containers, and gasoline tanks. The same equipment is used for either thermoplastic powder or plastisol molding.

Blow Molding

Blow molding is used primarily to reproduce thin-walled hollow containers from thermoplastic resins. A cylinder of plastic material, known as a *parison*, is extruded as rapidly as possible and is positioned between the jaws of a split mold, as shown in Figure 18.16.

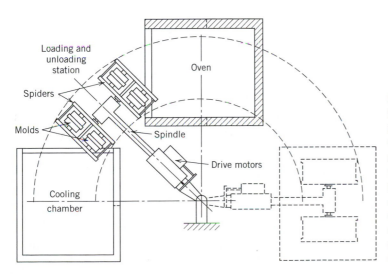

Figure 18.15
Schematic of pivoting-arm mold moving system. Spindle and molds swing in a 90° arc between heating and cooling chambers. Addition of a second cooling chamber, pivot, and spindle (dotted lines) decreases molding cycle time.

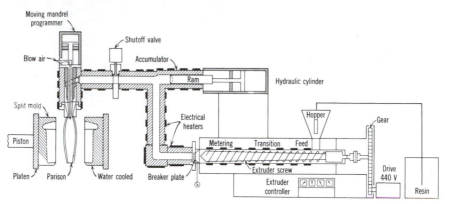

Figure 18.16
Typical blow-molding machine.

As the mold is closed, it pinches off the parison and the product is completed by air pressure, forcing the material against the mold surfaces. Molds should be vented adequately to eliminate poor surface finish. As soon as the product is cooled sufficiently to prevent distortion, the mold opens and the product is removed. The entire operation is similar to that used for forming bottles in the glass industry.

Figure 18.17 illustrates an eight-station machine for continuously blowing bottles by the *pinch-tube* process. A tube of thermoplastic material is extruded from a plasticizer into an open mold. Each end of the plastic tube is pinched shut by the closing of the mold, and air pressure is fed into the hollow tube by a core tube in the crosshead of the mold. The air pressure expands the plastic to conform with the walls of the mold. After a short cooling cycle, during which air pressure is maintained, the pressure is released, the mold opens, the bottle is ejected, and the mold is made ready to begin the cycle again. For some plastics, the bottle must be cooled to room temperature by a water spray. The top and bottom of the bottle must be trimmed to remove the scrap, but no

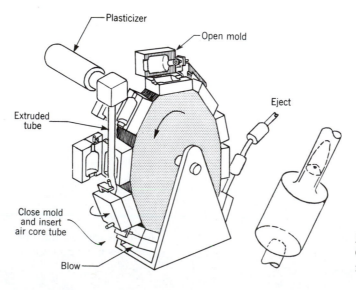

Figure 18.17
Continuous-tube process for making plastic containers.

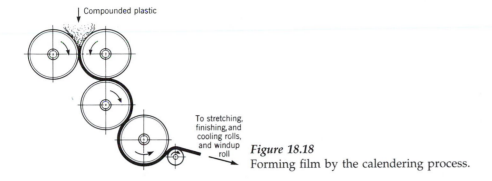

Figure 18.18
Forming film by the calendering process.

further processing is necessary. This process is repeated continuously for each of the eight mold stations.

Blow-molded products include cosmetic packaging, bottles, floats, automobile heater ducts, liquid detergent containers, and hot water bottles. Polyethylene, polypropylene, and cellulose acetate are some of the plastics that can be formed by blowing.

Film and Sheet Forming

The basic methods for producing film or thin sheets are calendering, extruding, blowing, and casting. The one chosen depends on the type of thermoplastic resin selected that, in turn, governs the required properties of the product. *Calendering* is the formation of a thin sheet by squeezing a thermoplastic material between rolls, as shown in Figure 18.18. The material, composed of resin, plasticizer, filler, and color pigments, is compounded and heated before being fed into the calender. The thickness of the sheet produced depends on the spacing between the rollers that stretch the plastic. Before the film is wound, it passes through water-cooled rolls. Vinyl, polyethylene, cellulose acetate films and sheeting, and vinyl floor tile are products of calendering. The same process is used for rolling out uncured rubber stock tire manufacture.

In making sheets of polypropylene, polyethylene, polystyrene or ABS, an *extrusion* process is used; Figure 18.19 is a schematic diagram of this process. After the material has been compounded, it is placed in the feed hopper. The material is heated to not more than 600°F (315°C) and forced into the die area at pressures of 2000 to 4000 psi (14 to 28 MPa) by the screw conveyor. By the combination of the choker bar and die opening, the thickness of the sheet is controlled. After extrusion, the sheet passes through

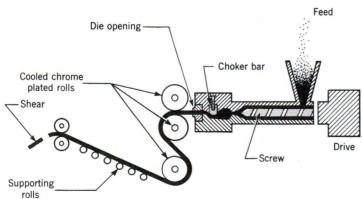

Figure 18.19
Extrusion of thin sheets and film.

oil or water cooled chromium-plated rolls before being cut to size. Oil cooling is recommended since the temperatures should be maintained at approximately 250°F (120°C) for proper curing. Sheet material made in this manner can vary in thickness from 0.001 to 0.125 in. (0.03 to 3.18 mm).

Blown tubular extrusion produces film by first extruding a tube vertically through a ring die and then blowing it with air into a large diameter cylinder. The blown cylinder is air cooled as it rises vertically and is flattened ultimately by driven rolls before it reaches the winder. This process permits the extrusion of thin film used for items such as trash bags and packaging materials.

In *film casting* the plastic resins are dissolved in a solvent, spread on a polished continuous belt or a large drum, and conveyed through an oven, where they are cured and the solvent is removed. In *cell casting* a cell is made up of two sheets of polished glass, separated according to the sheet thickness desired, and gasketed about the edges to contain the liquid catalyzed monomer. The cell is then raised to the proper temperature in an oven where it remains until curing takes place. Cell casting is used in the production of most acrylic transparent sheet.

Thermoforming

Thermoforming, or *sheet molding*, consists of heating a thermoplastic sheet until it softens and then forcing it to conform to some mold either by differential air pressure or mechanical means. Many techniques have been developed for applying pressure to the sheet, such as free forming by differential pressure (either pressure or *vacuum forming*), *vacuum snapback* forming, *vacuum drawing* or blowing into a mold, *drape forming*, *plug assist* vacuum or pressure forming, and matched-mold forming. Several techniques are shown schematically in Figure 18.20.

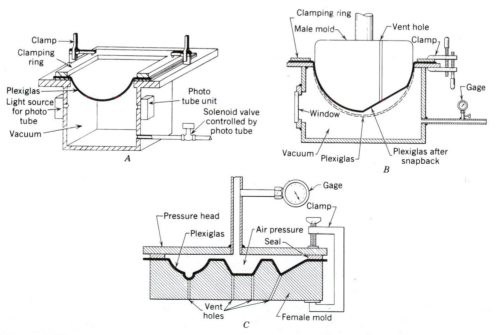

Figure 18.20
Forming methods used for heated thermoplastic sheets. *A*, Free forming. *B*, Vacuum snapback forming. *C*, Positive-pressure molding.

Figure 18.20*A* illustrates what is generally known as *free forming*, a technique that uses air pressure differential and where no male or female form is required. The drawn or blown section retains its shape on cooling. A somewhat similar process known as *vacuum snapback* forming is illustrated in Figure 18.20*B*. After the heated sheet is clamped, a vacuum is created in the chamber, which causes the sheet to be drawn down, as shown by the dotted lines. The male mold is then introduced into the formed sheet, and the vacuum is gradually reduced, causing the sheet to snap back against the mold form. A setup in which sheets are formed to shape by air pressure and are actually blown into the mold is shown in Figure 18.20*C*. This process is used for more complicated shapes where possible surface defects are not objectionable. By using special synthetic greases in the mold, the tendency for marks to show on the formed part is decreased.

In *drape forming* the plastic sheet is clamped and then drawn over the mold, or the mold is forced into the sheet. Plug assist forming first heats and seals the sheet over the mold cavity. A plug, somewhat smaller than the mold form, pushes the plastic sheet into a near bottom position. Vacuum or air pressure is applied to complete formation of the sheet. *Matched mold* forming is the same as the forming of sheet metal in dies. Molds are made of wood, plaster, metal, or plastics. This technique requires care to avoid marring the sheet surface.

18.7 OTHER PROCESSES

As the plastics industry evolves, more processes and techniques are developed. A few of these processes are described.

Tape Placement

Thermoset of thermoplastic, fiber-reinforced unidirectional tape is laid automatically by a programmed dispensing machine to form a desired shape. A gantry system provides the necessary lay-down motions of the tapehead, producing either flat or contoured surfaces. Lay-down rates are substantially higher than hand lay up, with improved placement accuracy and reduced human error. Equipment is relatively expensive and stiff tapes limit the complexity of finished parts.

Fiber Placement

Fiber placement is similar to automated tape placement using resin-impregnated fiber rovings. Rovings are automatically placed on a complex mold surface, which can include both positive and negative surfaces. It is also applied to surfaces of rotation, like filament winding. Fiber roving flexibility allows more complex shapes than tape placement. Unlike filament winding, rovings are placed and not wound under tension, diminishing consolidation, and structural properties, such as tensile strength.

Autoclave Molding

After hand lay up, winding, or other fabrication techniques, mold and composite parts are placed in an autoclave. Heat and pressure are applied via steam, consolidating and curing part. Two versions of the autoclave molding process, hydroclave and thermoclave, are usually combined with a vacuum-bag bleeder and release cloth. *Hydroclave* uses water as the pressure media, whereas *thermoclave* uses powdered silicone rubber, which acts as a fluid under heat/pressure.

Rubber Processes

Several techniques, variations of processes described for thermoplastic compounds, are used in fabrication of rubber parts. Table 18.4 presents a summary of the four most common methods.

Lamellar Injection Molding

Lamellar injection molding (LIM) is a microlayer injection-molding technology that produces molded parts from multiple resins in distinct microlayers. The resultant lamellar or layered morphology contrasts with the conventional alloy/blend dispersion, or the skin-and-core structure common of coinjected parts. The process combines a variety of dissimilar polymers as long as their viscosities are within a factor of three and their melt temperatures are close. This allows one to create a layered structure with enhanced physical properties. For example, polycarbonate/polybutylene terephthalate microlayer parts produced by LIM have 40° to 50°F higher heat-distortion resistance because of the inherent support offered by the stacked planar interfaces. Likewise, dimensional stability, optical clarity, and environmental stress crack resistance are improved. Figure 18.21 illustrates a LIM technique using a feedblock and layer multipliers to combine melt streams from dual injection cylinders into a microlayer structure. (The flow direction in the feed stream close-ups is toward the reader.) In general, the molding process employs a feed block that combines the original flows from a dual-barrel injection machine into

Table 18.4   **Processes for Rubber Compounds**

Process	Advantages	Limitations
Compression molding. An excess amount of uncured compound is placed in mold cavity; mold is closed and heat and pressure is applied, forcing compound to fill mold cavity; heat cures (*vulcanizes*) compound and mold is opened to remove hardened parts.	Good surface finish; parts are made in almost any hardness, shape, and size; relatively low cost; little waste; most compounds suitable.	Close tolerances difficult to achieve; flash has to be removed; extreme intricacy difficult; slow production rate.
Transfer, injection molding. Similar to compression molding, except that mold is closed empty and rubber compound is forced into it through sprues, runners, and gates.	Good dimensional accuracy; no flash removal; good for intricate parts; good finish and uniformity; rapid production rate.	High mold costs; not all compounds suitable; high scrap loss because of sprues, runners, and gates.
Extrusion. Similar to plastic extrusion in that heated material is forced through a die having desired cross section. However, vulcanization does not take place in mold cavity; extruded lengths are cured in a steam vulcanizer and either used as-is or cut into sections.	Low operation cost; high variety of complex shapes possible; rapid production rate.	Close tolerances difficult; only uniform cross section (along length) possible; openings must be in direction of extrusion.
Die cut. Parts are stamped or cut from vulcanized sheet or slab with inexpensive steel dies.	Any rubber material suitable; low cost; economical for small lots.	Thickness of parts very limited; only flat parts possible.

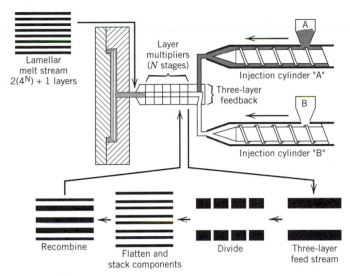

Figure 18.21
Illustration of the lamellar injection-molding process.

a three-layer feed stream. A five-layer feed stream would result from the addition of a third material, such as an adhesive tie layer for incompatible resins. From the feed block, the precisely metered melt runs through tailor-made *layer multipliers*. Within the multiplier, a set of channels repeatedly divides prior feed streams and then combines them into new laminar streams consisting of more numerous but thinner layers.

18.8 DESIGN FUNDAMENTALS

Molds for Plastics

Molds for both the compression and injection processes are made of heat-treated steel. The production of these molds demands the same type of machine work and the usual precision required for die casting. There are some differences in construction because of varying characteristics in the materials being processed. *Injection molds* are explained in detail in Chapter 25.

Compression molds are made in hand and semiautomatic types with *positive molds*, semipositive molds, and flash mold designs. The hand molds are charged and unloaded on a bench. Heating and cooling are accomplished by plates on the presses, which are provided with the necessary circulating facilities. The semiautomatic molds are fastened rigidly to the presses and are heated or cooled by adjacent plates. As they open, work is ejected automatically from the molds, which are single or multiple cavity.

In mold design and in determining wall thickness, design and manufacturing engineers must evaluate the ability of plastics to flow into the narrow mold channels. This flowability depends on temperature and pressure to some extent, but varies for different materials. Table 18.5 shows some typical nominal wall thicknesses for various types of thermoplastics.

In addition, if the plastic part is loaded, stress and deflection analysis is crucial. When stress or deflection is too high, it is recommended that ribs or contours be used to increase section modulus. Other alternatives are to use a higher-strength, higher-modulus (fiber-

Table 18.5   **Typical Nominal Wall Thicknesses for Various Classes of Thermoplastics**

| Thermoplastics Group | Working Range | | | |
| | (in.) | | (mm) | |
	Min.	Max.	Min.	Max.
ABS	0.045	0.14	1.143	3.556
Acetal	0.03	0.12	0.762	3.048
Acrylic	0.025	0.15	0.635	3.81
Liquid-crystal polymer	0.008	0.12	0.2032	3.048
Nylon	0.01	0.115	0.254	2.921
Polycarbonate	0.04	0.15	1.016	3.81
Polyester	0.025	0.125	0.635	3.175
Polyethylene	0.03	0.2	0.762	5.08
Polypropylene	0.025	0.15	0.635	3.81
Polystyrene	0.035	0.15	0.889	3.81
Polyurethane	0.08	0.75	2.032	9.05
PVC	0.04	0.15	1.016	3.81
Styrene-acrylonitrile (SAN)	0.035	0.15	0.889	3.81

reinforced) material or to increase the wall thickness, if possible. Where space allows, adding or increasing rib density improves structural integrity without thickening walls.

Design for Assembly

Plastic parts are assembled by methods such as molded-in snap-fits, press-fits, pop-ons, and thread fasteners.

Snap-fit designs are widely used and all have a molded part that must flex like a spring, usually past a designed-in interference, then return to its unflexed position to hold the parts together. There must be sufficient holding power without exceeding the elastic or fatigue limits of the material. Beam equations are used to calculate the maximum strain during assembly. If the stress is kept below the yield point of the material, the flexing finger returns to its original position.

In some cases, the calculated bending stress exceeds the yield point stress considerably if the movement is done rapidly. It is common, then, to evaluate snap-ins by calculating strain instead of stress. Dynamic strain ε, for the straight beam, is calculated as

$$\varepsilon = \frac{3Yh_o}{2L^2} \tag{18.6}$$

where

Y = Maximum deflection
L = Length of the snap finger
h_o = Width

The calculated values should be compared with the permissible dynamic strain limits for the material in question, if known. A tapered finger provides more uniform stress distribution and is recommended where possible. Sharp corners or structural discontinuities that will cause stress concentrations on fingers must be avoided.

Other assembly methods include

1. *Chemical bonding*, which involves fixtures, substances, and safety equipment. It does not create stresses and is suited to leak-tight applications. Its main limitation is that adhesives and solvents are flammable and preparation and cure times are long.

2. *Thermal welding* methods include ultrasonic, hot-plate, spin, induction, and radio-frequency energy. Special equipment is required and materials must be compatible and have similar melting temperatures.

3. *Mechanical fasteners.* There are many fasteners designed specifically for plastics but those designed for metals are generally usable. Typical are bolts, self-tapping and thread-forming screws, rivets, threaded inserts, and spring clips. Creep can result in loss of preload if care is not taken to avoid overstressing the parts.

QUESTIONS

1. Give an explanation of the following terms:

Thermosetting	Torpedo-like spreader
Thermoplastic	Extruding
Potting	Rotational molding
Polystyrene	Calendering
Synthetic rubber	Film casting
Rotary preforming	Sheet molding
Compression molding	Drape forming
Filament winding	Positive molds

2. Describe the process of rotational molding and give its advantages.

3. How are the gates and sprues of thermoplastic materials salvaged? Thermosetting materials?

4. Are furane resins organic or inorganic materials? Why?

5. What is meant by polymerization?

6. List the processes used in forming plastics. Give the type of plastic that may be formed in each.

7. How are plastics compounded?

8. What is the advantage of a preform? List its applications.

9. Why is it difficult to process thermoplastic materials by compression molding?

10. Describe the process of compression molding.

11. What advantages does transfer molding have over compression molding?

12. Which type of plastic molding is similar to die casting?

13. List the plastic materials generally used in injection molding. What properties make them desirable?

14. Give the purpose in cooling the nozzle in jet molding.

15. Name five products that can be made by extrusion.

16. What processes do you recommend for producing the following products: boat hulls, squeeze toys, garbage containers, film, packaging food, and radio cabinets?

17. Give the purposes for laminated plastics.

18. Describe the vacuum snapback method of forming.

19. Can thermosetting compounds be easily extruded into tubes and rods? Why?

20. How are plastic bottles made?

21. List the reasons that make the pinch-tube method in bottle production attractive.

22. Select a product and construct a flow chart using the calendering process.

23. Give the difference between positive- and flash-type molds.

24. How are plastic molds vented?

25. List the advantages of plastic parts over metal parts and vice versa.

26. How does film casting differ from the calendering process?

27. What allowances for plastic molds are similar to those for sand castings?

28. Why are gears sometimes made from plastic having a canvas base or filler?

29. If a plastic part is loaded in such a way that stress and deflection are expected to be high, what would you recommend to increase the reliability of the part?

30. List the properties of silicone-based polymers that make them desirable as engineering materials.

31. List some plastics that are used as manufacturing materials.

32. Which plastic materials are used in electrical wire insulation, printed circuit boards, knobs, and chassis? What are the principal properties that enter into your selection?

PROBLEMS

18.1. How much plastic needs to be plasticized each 24-h day for a 30-oz. (0.85-kg) machine making 2 cycles per min?

18.2. If a 1-kg machine has 40 cycles per minute, how many kilograms of plastic are plasticized each 8-h shift?

18.3. A thermoplastic plastic material having low density and high insulation qualities is required for a design. Suggest several candidate materials.

18.4. What temperature pressure combination is required to manufacture a thin-walled, highly intricate molding?

18.5. What temperature–pressure combinations would be best to manufacture thick undetailed objects rapidly, such as dog food dishes?

18.6. Find the time in minutes per unit for the production of preforms. Each die has four cavities and the machine is rated at 40 spm.

18.7. A copper bar has a hole of diameter 10 cm at 30°C. An unreinforced phenolic plug has a diameter of 9.98 cm. At what temperature must the assembly be maintained to guarantee a snug fit? (Hint: See Example 18.1 and Table 18.1 for α_{Cu} and α_{pl}.)

18.8. If the maximum deflection allowed in a straight Plexiglas snap-in finger is 5°, calculate the dynamic strain. The finger's dimensions are: length, 4 in.; width, 0.25 in.

MORE DIFFICULT PROBLEMS

18.9. A reciprocating preform production machine is rated at 50 spm. The die unloads four preforms with each stroke. One operator tends three machines that are similar. A part uses 60 g and each preform is about 10 g. The number of parts required is 25,000. Find the production rate for one machine, one operator, and the amount of premixed material for the lot. How many elapsed hours are required to run the quantity for the lot?

18.10. A part requires 18 g and each preform will be about 8 g. The cavities per stroke are three, and the excess is required for the sprue and runners of the injection machine. One operator tends one machine, which has 80 strokes per minute. The order requires 12,000 parts, and the machine, which does the prepacking of the material, costs $60/h. The blended material will cost $2/kg. Find the production rate for the operator, machine, and the cost for the part and operator. If the billing rate is two times this cost, what is the price the owner will charge for the lot?

18.11. A computer part is molded of clear polycarbonate plastic two at a time. A partially dimensioned sketch, Figure

P18.11, gives part dimensions, sprue, runners, tabs, and the two parts. Density = 0.0404 lb/in.³ (1119 kg/m³).

a. In customary units find the weight of one part and the shot requirements for the sprue, runners, tabs, and two parts.

b. Determine the yield of the part to total material.

c. The cost of the material is $3.10/lb ($6.89 per kg) and waste is recovered at 10% of original value. Find the cost of the part, including a fair share of the waste products.

d. The cycle time for one operator and one injection molding machine is 45 s. For a labor rate of $16.50/h find the labor cost per unit.

e. What is the total cost per unit?

f. Repeat parts a. to d. in metric units.

18.12. An aluminum sheet has a 2-in. diameter hole at 70°F. The hole is insulated with ABS. The assembly is heated to 300°F. What is happening to the assembly? Redesign the assembly if necessary. (Hint: $\alpha_{Al} = 12.2 \times 10^{-6}$ in./in.°F. α_{ABS} is given in Table 18.1 and typical ABS wall thickness is given in Table 18.5.)

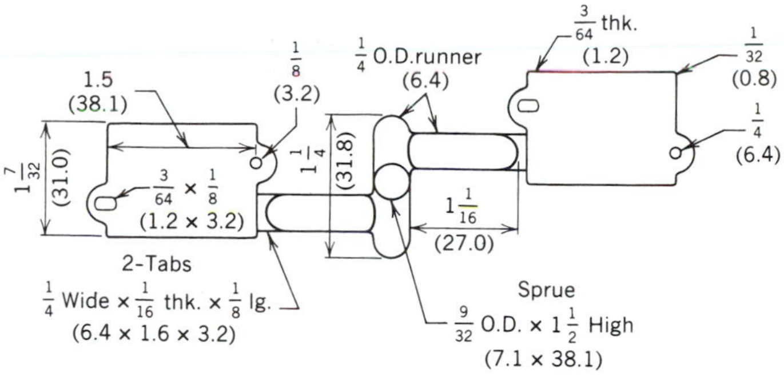

Figure P18.11
Typical molded part showing sprue, runners, and tabs.

PRACTICAL APPLICATION

Polymers continue to advance into engineering applications that range from automotive and aerospace to medical and electrical. Choose three plastic materials (e.g., thermosets, thermoplastics, acetal resins, etc.) and investigate their manufacturing, physical, mechanical, thermal, electrical, and frictional properties. Then select one of them to produce an article of your choice (e.g., a football helmet). Identify a local distributor or a local manufacturer that you can interview to gather more information.

CASE STUDY: THE GENERAL PLASTICS COMPANY

Rich Hall, die designer for General Plastics, mutters to himself, "What counts in this problem is minimum plastic volume in the runner system." Rich knows that for this plastic mold design it will be impractical to reuse the scrap, because the plastic part will be colored and the value of the scrap runners represents a small fraction of the virgin material cost.

The part to be molded is roughly 25 mm in diameter and 10 mm thick, similar to a preform, except for the novelty impressions on the surface.

Rich, recently hired, has learned that full-round runners are preferred. They have a minimum surface-to-volume ratio, thus reducing heat loss and pressure drop. Balance runner systems are preferred because they permit unifor-

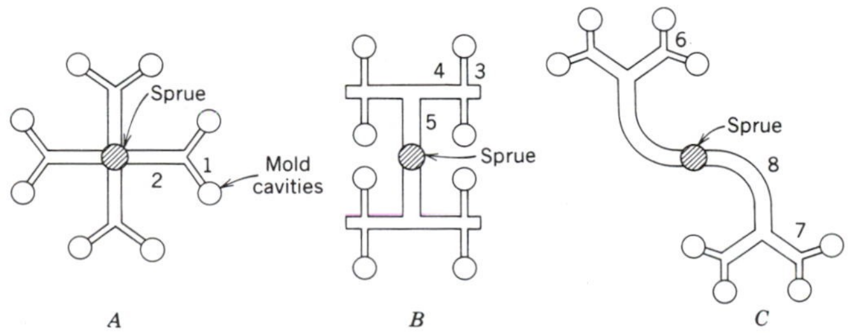

Figure C18.1
Case study. Three design configurations: A, Star pattern. B, "H" pattern. C, Sweep pattern.

mity of mass flow from the sprue to the cavities, since the cavities are at an equal distance from the sprue. Main runners adjacent to the sprue are larger than secondary runners.

Rich has designed three configurations, as shown in Figure C18.1, and he will select the one that uses a minimum of runner material. The time factor is not critical as the three arrangements provide identical number of parts per shot. The sprue volume for the three arrangements is equal. Die data are shown below. Determine which arrangement has a minimum of material for the runner system. If the plastic cost $0.60/kg and density is 1.05, what is the prorated loss per unit? Determine the overall material efficiency if

$$\text{Efficiency} = \frac{\text{material in parts}}{\text{material in shot}}$$

Arrangement	Runner Section	Diameter, mm	Section Length, mm
A	1	5	5
	2	6	25
B	3	5	12
	4	6	75
	5	8	25
C	6	5	8
	7	6	175
	8	10	100

CHAPTER 19
ELECTRONIC FABRICATION

As electronic products evolve, they become smaller, lighter, faster, more capable, and less power hungry. Electronics packaging techniques have evolved from discrete circuits, through dual-in-line integrated circuits, to surface mount integrated circuits. New emerging steps are multichip modules and chip-on-board techniques.

Electronic fabrication, however, does not include only electronic packaging concerns. Soldering materials and soldering processes, fluxes, adhesives, thermal management, electromagnetic shielding, electrostatic discharge protection, and cleaning alternatives are crucial for competitive production. This chapter gives an overview of these elements and provides some design fundamentals of electronics manufacturing.

19.1 COMPONENTS AND DEFINITIONS

By definition an *integrated circuit* (IC) is a group of inseparably connected circuit elements fabricated in place within a substrate. A *substrate* is a waferlike piece of insulation material that may serve as a physical support or base and thermal sink for a printed pattern of circuitry. An IC is basically a single functional block containing many individual devices (transistors, resistors, capacitors, etc.), which is known as a chip. Most often the substrate or base of the chip is made of *silicon*.

The structure of the IC is complex in the topography of its surface and in its internal composition. Each element of this device has an intricate *three-dimensional* architecture that must be reproduced identically in every circuit. The structure is composed of layers, each of which is a detailed pattern. Some of the layers lie within the silicon wafer and others are stacked on top. The manufacturing process consists in forming this sequence of layers precisely in accord with the plan of the circuit designer.

A large-scale integrated circuit contains tens of thousands of elements, yet each element is so small that the complete circuit is typically less than a quarter of an inch on a side. Production of these circuits is to fabricate them many at a time on a larger *silicon wafer* 3 or 4 in. in diameter. A *wafer* is passed through many stages, where a complete microelectronic circuit is composed on this substrate; these circuits are separated into individual dice or chips. These circuits or chips are "packaged" or fastened to a metal stamping. Fine wire leads are then connected from the bonding pads to the electrodes of the package, and a plastic cover is molded around each die. The units are separated from the metal strip and later inserted into the printed circuit board individually. This is discussed later. Figure 19.1 illustrates the wafer and IC manufacturing flow.

Fabrication of the Wafer

Raw silicon is first reduced from its oxide, which is the main component of raw sand. A series of chemical steps are taken to purify it until the purity level reaches 99.999999%. A charge of pure silicon is placed in a crucible and brought to the melting point of silicon, 2588°F (1420°C). An inert gas prevents the addition of unwanted impurities at this point. However, desired impurities known as *dopants* are added to the silicon to

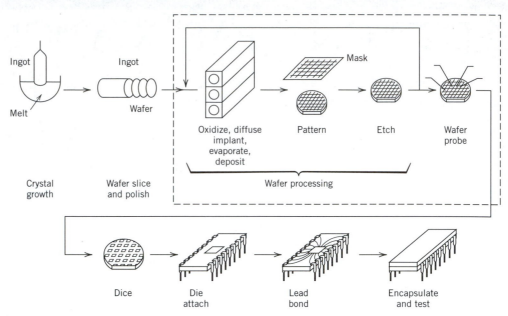

Figure 19.1
Integrated circuit manufacturing process flow.

produce a specific type of conductivity characterized by either *positive* (*p type*) charged *carriers* or *negative* (*n type*) ones.

A large single crystal is grown from the melt by inserting a perfect single crystal "seed" and slowly withdrawing it. Single crystals 3 to 4 in. in diameter and several feet long can be pulled from the melt. The uneven surface of a crystal as it is grown is ground to produce a cylinder of standard diameter, typically 3.94 in. (100 mm). The cylinder is cut into wafers with a thin high-speed diamond saw. The wafers are ground on both sides and then polished on one side. The final wafer is about 0.002 in. (0.5 mm) thick. These final steps of the *wafer-shaping* process are carried out in an absolutely clean environment. There can be no defects, polishing scratches, or chemical changes on the final surface.

Silicon has a dominant role as the material for microelectronic devices as its *oxide, silicon dioxide*, forms on the surface when heated in the presence of oxygen, or water vapor. The film serves as an insulator. It can be used as a mask for the selective introduction of dopants. This layer of silicon dioxide (SiO_2), about 0.75 micron thick, is formed on the chip to prevent such materials as arsenic, antimony, and boron from diffusing into the silicon chip. The thickness of silicon dioxide is gased onto the wafers as the wafers are loaded in boats or fixtures, which are inserted into ovens at about 2000°F (1100°C). The silicon dioxide thickness of 0.004 in. (0.1 mm) will grow in 1 h at a temperature of 1920°F (1050°C) in an atmosphere of pure oxygen. Several hundred wafers can be oxidized simultaneously in the fixture.

Component Categories

There are three categories of components: (1) active, (2) passive, and (3) conductive. *Active* components are semiconductor devices that amplify, switch, or rectify electronic signals. Examples of this category are transistors, diodes, and integrated circuits. *Passive* components are conductive devices that alter signals without amplifying, switching, or rectifying. Included in this category are resistors, capacitors, and inductors. The last category includes *conductive devices* that simply pass signals, unaltered by the device itself, along prescribed pathways from one functional circuit to another.

The Printed Circuit Board

The *microcircuit* begins with conceptualization of the circuit design. When the design is completed, sets of plates called *photomasks* are produced from the circuit layout. Each photomask contains a pattern for a layer on the circuit. These patterns are transferred to the silicon wafer by the photolithographic process, thus building layer upon layer on the microelectronic circuit.

An early step in circuit fabrication is the *layout*. This phase is often called the *artwork*. To print the circuit, first a photographic mask must be made. The mask is a photographic reduction of a set of cut and strip masters typically 250 to 500 times larger than the actual size of the circuit. At each stage of this process, including the final stage when the entire circuit is completed, the layout is checked by means of detailed computer drawn plots. Since the individual circuit elements can be as small as a few micrometers across, the checking plots must be magnified. Thus, these plots are 500 times larger than the final size of the circuit. Eventually, the photomask becomes a glass plate about 5 in./side (125 mm) and has a single circuit pattern repeated many times on its surface. These plates are transferred to the wafer fabrication facility, where they will be used to produce the desired sequence in a physical structure.

Another step in the construction of the mask is *photography*. High-precision camera equipment must be used to ensure a sharp, clear image on the reduction. Also, the camera and master must be clean and rigidly supported. A dust spot could render the final circuit inoperational. A *clean room* where conditions can be controlled is necessary. Long exposures are often taken. Therefore, the camera must not be affected by any vibrations.

Photolithography in Electronic Fabrication

Photolithography is a process by which a microscopic pattern can be transferred from a photomask to a material layer in an actual circuit on a silicon wafer.

The most basic photolithographic process involves etching the pattern into the silicon dioxide using contact photolithography. Once a layer of silicon dioxide is produced on the silicon wafer, the wafer is covered by a material called *photoresist*. This is done by applying a drop of the photoresist to the wafer and then spinning the wafer rapidly until a thin film of material is affected by exposure to ultraviolet radiation. Note these steps in Figure 19.2.

The resists are applied to a thickness of about 2 microns, usually by spinning, and then prebaking. In the next operation, where the chip and photographic plate are used together, cleanliness is extremely important. A vacuum printer is used to prevent contamination. The photographic plate then serves as a stencil. The wafer is exposed to a carbon arc light source for several minutes. After exposure, the resist is developed and then washed in deionized water to remove the unpolymerized material. The remaining resist is then baked at 300°F (150°C) for 10 min. To remove the unprotected oxide layer, the chip is etched with hydrofluoric acid.

The mask contains the *opaque printed pattern*. The mask either may be brought into contact with the wafer or it may be held slightly above the wafer. The selection of the

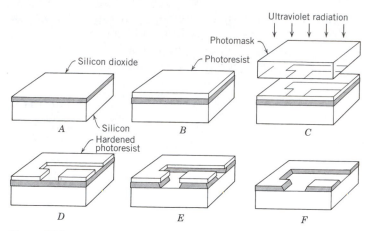

Figure 19.2
A, Oxidized layer on wafer of pure silicon. B, Coated with layer
of light-sensitive material called photoresist. C, Exposed to ultraviolet light through photomask. D, Exposure renders the photoresist insoluble in a developer solution and a pattern of photoresist is left wherever mask is opaque. E, Wafer immersed in
hydrofluoric acid solution, which attacks silicon dioxide only.
F, Photoresist pattern removed by another chemical treatment.

technique is based on conditions of mask life and sufficient resolution. The structure is
now flooded with ultraviolet radiation.

Under the opaque areas of the mask the photoresist is unaffected. However, in the
transparent areas the photoresist becomes insoluble in the developer solution. Thus,
when the structure is washed in the developer solution, those areas under the opaque
pattern of the mask are removed, whereas the rest of the structure is unaffected.

Photoresists are of two types: *negative resists,* where polymerization is caused by
exposure to ultraviolet light, and *positive resists,* where polymerization is degraded by
exposure to light. A photosensitive resist is a lacquerlike material that, when exposed
to light, is converted to a film that adds to the support and resists chemical action. Where
it is not exposed to the light, it washes away in the developer. When the wafer is exposed
to an etching solution of hydrofluoric acid, the areas of silicon dioxide exposed by the
missing photoresist are removed.

The remaining photoresist is removed by another chemical solution, the desired pattern
in the silicon dioxide is obtained, and the wafer is prepared again for the next masking
process. By this method successive layers can be built up on the wafer.

The use of silicon as the base for these circuits is important to the photolithographic
process. Silicon dioxide can be etched away in certain areas while leaving the silicon
base unaffected, thus allowing the formation of the specified patterns on the wafer.
Once all the patterns have been photolithographed onto the wafer and the circuits are
completed, the elements are tested and then packaged. Throughout the entire fabrication
of the circuit, extreme cleanliness must be maintained in the fabrication facility, since
even minute dust particles can ruin a microelectronic circuit.

The process is repeated as many times as necessary with the final step of the depositing
of a thin layer of aluminum over the entire integrated circuit. The aluminum is etched

selectively to leave the desired conductor pattern interconnecting the proper devices on the integrated circuit as well as to provide connecting pads, which will be wirebound later to the external leads prior to packaging.

19.2 FROM COMPONENTS TO PRODUCTS

Printed Circuit Board Stuffing

The term *stuffing* refers to the insertion of electronic components in the board. Two major methods dominate the industry: the *insertion mounting technology* (*IMT*) and the *surface mounting technology* (*SMT*).

Insertion Mounting Technology

Insertion mounting technology, also known as through-hole insertion (*THI*), requires lead preparation, insertion of the component, clipping of the extra length of the lead, and folding over of the lead to hold the chip or electronic component against the underside of the PCB. The work is the act of inserting the leads through the holes in the printed circuit board (PCB). This work may be done manually for low volume or occasional rework of PCB assemblies. Machines, such as the radial sequencer-inserter of Figure 19.3, are used for high-volume production.

The worker will start by adding each element to the PCB one at a time. Every element is checked carefully for its identification number. After an element is in place on the board, it is then soldered to the board. One or more parts may be soldered to the board at the same time.

Figure 19.3
Radial lead component sequencer-inserter machine.

Once all the parts are attached to the board, the worker will clean it by brushing it with alcohol. The board is then dried with an air hose. Next, the PCB is sent to the inspector, where it must pass an inspection. If any errors are found, they are fixed and it is reinspected. When the board has passed the inspection, it is chemically sealed.

Surface Mounting Technology

Surface mounting technology derives its name from the way its components are attached to printed wiring boards (PWBs). In SMT active and passive circuit elements are packaged in a variety of shapes, sizes, and configurations with one common feature among them: each is mechanically attached and electrically connected to the PWB through their multiple leads soldered to matching, coplanar pads on one or both surfaces of the PWB. Single-plane attachment makes it possible and convenient to subminiaturize components, and to automate the fabrication of the SMT assemblies.

The design of these printed circuits, however, presents numerous problems. Material compatibility with the devices and the process depends on dimensional stability (to maintain tolerances), electrical parameters (impedance, capacitance), assembly techniques (mounting, soldering, cleaning, testing, inspection), and cost justification.

Fine-line boards require a fresh approach in terms of material selection, production processes, and equipment. For this and other practical reasons, SMT devices (SMD) are mounted on conventional boards with lower densities. An increase in total component density can be gained by adding surface mounted components to the unused bottom of the board. This trend is driven by the cost pressures and production capabilities.

19.3 THE SOLDERING SYSTEM

The electromechanical reliability of interconnections depends on the soundness of their solder joints. Overall integrity of PWBs is almost solely dependent on the overall reliability of the soldering system, which includes consumable and nonconsumable materials, processes, and joints design.

The range of methods available for soldering electronic components is wide and varied. Components can be soldered by hand or with programmed machines. Solder, in the form of paste or preforms, can be applied to PWBs prior to positioning components or applied after, as in the case of wave soldering. Basic solder activation heat-transfer methods (convection, conduction, radiation, or a combination of the three) can be accomplished by using a choice of diverse heat-transfer media that include molten metal, energized light, hot gas, hot saturated vapor, radiant energy, or hot soldering irons.

Whatever the method used, the basic solder connection is formed as shown in Figure 19.4. First, the flux solution settles above the oxidized metal surface (point A). Then the boiling flux solution removes the oxide film (point B) and the bare metal enters in contact with the flux (point C). Liquid solder now replaces the fused flux (point D) to allow tin to react with the basis metal and form a new alloy (point E) that creates a solder bridge (point F). Note that in Figure 19.4 the soldering iron tip is not touching the base metal for reasons of clarity only.

Vapor-phase, infrared, laser, conduction belts and plates, convection, soldering irons, and hot bars are soldering methods most likely to be encountered in SMT.

Wave soldering is the method of choice for boards having a mixture of SMT and IMT components. A wave soldering machine (see Figure 19.5) applies a layer of molten solder to the underside of a printed circuit board. After solidification, the solder electrically

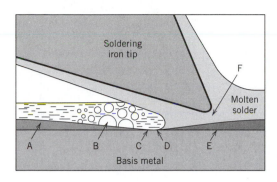

Figure 19.4
Diagram of solder connection being formed.

secures all the inserted components into place. The machine consists of a conveyor that mechanically transports the circuit boards through the system. Machines vary, but a typical process for the board is: (1) flux that applies a liquid layer to remove surface oxides and to give the filler metal the fluidity to wet the joint surfaces completely, (2) heating elements raise the temperature for proper solder adhesion, (3) air knives used to remove excess flux, (4) a pass over the crest of the liquid solder wave at a set height, and (5) the washing cycle.

Other soldering methods for high-volume SMT soldering operations are vapor-phase (VP) and infrared reflow (IR) soldering. *Infrared soldering* is suited ideally for assemblies with uniform distribution of components having gull-wing leads and for assemblies with uniformly distributed heat mass and components no closer than 5 mm to the edge of the board. Infrared reflow is a noncontact, medium-free, heat-transfer system that uses radiant energy as the primary heating source. *Vapor-phase* reflow soldering, however, is the method of choice for boards having a variety of leadless and leaded components, with uneven component placement, and uneven thermal mass and where the component—board-edge distance—can be closer than 5 mm. Vapor-phase is a condensation-soldering technique that uses an oxygen-free chemically inert vapor to engulf and heat electronic assemblies. It produces an even and high temperature that rapidly reflows the solder.

Combination soldering, wave-IR or wave-VP, is used when both SMT and IMT components are mounted on top, and the bottom is either blank or populated with SMT components. Figure 19.6 illustrates an in-line vapor-phase machine with IR preheaters that permit better control of ramp-up assembly temperatures. *Hand soldering* and *laser*

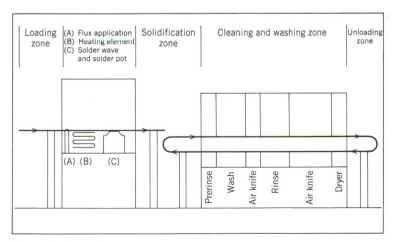

Figure 19.5
Automatic wave soldering machine, variable board sizes and speed rates.

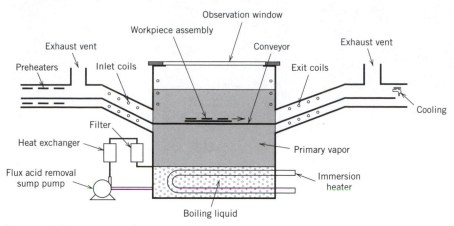

Figure 19.6
Diagram of an in-line vapor-phase machine.

soldering are used often in combination with the other soldering methods when special components and/or component terminations are not suitable for mass soldering. Some of these special components are connectors, transformers, power devices, and heat-sensitive devices.

System Variables

Soldering methods are influenced by a set of variables that must be understood and controlled. Such variables can be classified as related to the components, the board, the process, or the materials. See Table 19.1 for a summary of the most relevant variables of soldering systems.

To promote reliable soldering the temperature of metal surfaces involved in the solder joint must reach equilibrium by the time reflow begins. Metals, however, have two thermodynamic characteristics (thermal conductivity and thermal capacitance) that quantitatively differ from one metal to the other and, consequently, complicate the way

Table 19.1   **Classification of Relevant Soldering System Variables**

Board and/or Component-Related Variables	
Wettability of component and PWB terminations	Presence of constraining layers in PWBs
Lead pitch, configuration, and orientation	PWB size, thickness, and material type
Component placement density	SMT and IMT components mix

Material-Related Variables	Process-Related Variables
Thermal conductivity of materials	Presence of adhesives
Thermal density of components and board materials	Ability to clean all surfaces
Presence of solder masks	Need for off-line soldering
Composition of solder alloys	Rework and reparability
Composition of flux	Production quantity throughput
	Product life-cycle reliability
	Cost goals

temperature equilibrium is reached. Clear understanding of the time–temperature curve of the soldering cycle for each assembly is an important method to control these variables.

Cycle Characteristics

The soldering cycle can be divided in three periods: preheating, soldering, and cooling. The cycle is usually characterized by determining three elements: rate of temperature change, dwell times, and temperature levels. A cycle applicable to boards with high glass transition level is described next and is displayed in Figure 19.7.

1. *Preheating period*. Two steps are involved in this period: ramp-up and equalization.

   a. *Ramp-up*. As a general rule, the soldering cycle begins by raising the assembly temperature to 230°F (110°C) ± 10% at a rate of about 6°F/s (3°C/s). The rapid heating causes thermal gradients within the assembly. At around 230°F (110°C) heating is stopped to allow the accumulated thermal energy to even out. This reduces the possibility of thermal shocking sensitive components at the next higher temperature level. During this delay, flux activators within the *solder paste* start to reduce oxides from metal surfaces.

   b. *Equalization*. Assembly temperatures are raised at a rate of 1°F/s (0.5°C/s) until a temperature of 320°F (160°C). During this "soaking" period of about 1 min, materials with slower heat transfer characteristics catch up with materials that have faster transfer coefficients so that delamination can be minimized. Most of the remaining solder paste solvents are driven off. Fluxes are fully activated, cleaning pads and leads and promoting surface wetting.

2. *Soldering period* (*wave and reflow*). The temperature of PWBs and the components normally can be raised from 320°F (160°C) to 428°F (220°C)—the reflow level—without compromising reliability. Therefore, to minimize components' exposure to elevated temperature, the rate of thermal change is set at the more rapid pace of 6°F/s (3°C/s). Although the melting point of eutectic solder (63% tin/37% lead) occurs at 360°F (183°C), most processes operate with a solder reflow temperature of 428°F (220°C) to ensure quick and total reflow and the formation of acceptable solder joints. During this period, the flux removes oxides and other surface contaminants, such as sulfides and carbonates. A flux layer is formed over the newly cleaned surfaces, promoting good wetting and preventing the recurrence of oxide formation while the solder wets and holds. At the peak dwell, individual solder particles suspended within the paste begin to melt and form a mass that flows toward the hotter spots. This solder mass pushes the flux away from the cleaned surfaces, covering them until all the competing forces (i.e., surface tension,

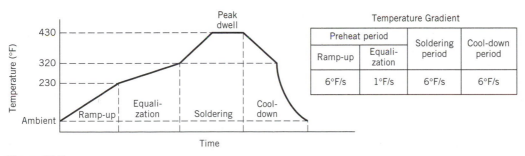

Figure 19.7
Soldering process cycle characterization.

metallurgical-chemical attractions, capillary action, component buoyancy and gravity) reach a state of equilibrium. At this time the final joint configuration has been formed and the last step, the cool-down period, begins.

3. *Cool-down period.* During this solidification period, the cool-down rate should be the same as the ramp-up rate. This cool-down rate should be maintained until the assembly reaches 320°F (160°C). Then the cool-down should be at a natural descent within the room environment. This cool-down rate allows for a tighter, smaller, better fatigue-inhibiting grain structure.

Flux

Tarnish is the reaction of metal products in air-forming oxides and sulfides. *Flux* is usually a liquid applied to the mating faces of conductors and the surface to be soldered to remove tarnish from metallic surfaces.

Specifically, flux performs four functions when heated: (1) chemically reacts with oxides, lifting them from the surface and forming soluble compounds that can then be removed, (2) protects surfaces from reoxidation, (3) aids in transfer and distribution of heat, and (4) lowers the cohesive force of solder to enable wetting. In addition, when paste solder is used, flux serves as the carrier and vehicle for solder particles, organic and inorganic solvents, flow modifiers such as waxes and oils, activators, and other additives.

The gum from pine trees, *Water White* (WW) *Rosin*, has served almost exclusively as the soldering flux for the electronics industry. It serves as the base for a solder-flux grading system that describes the activity levels of fluxes.

1. *Rosin fluxes.* Three types exist: *nonactivated*, *mildly activated*, and *fully activated* rosins (type R, RMA, and RA, respectively), which decompose at about 525°F (274°C). Rosin fluxes have lost popularity because their residues are removed by using solvent cleaners containing ozone depleting chlorofluorocarbons (CFCs).

2. *Organic water soluble fluxes.* Applicable to most metal soldering, these fluxes decompose at about 180°F (100°C) higher than resin-based fluxes. Three of the most common types are: *synthetic resin* (SR), *organic acid* (OA), and *synthetic activated* (SA) fluxes.

Synthetic resin fluxes have low solids content resulting in an inert residue that does not need to be removed. Organic acid, or intermediate fluxes (i.e., lactic, citric, and benzoic acids), are water soluble and equivalent to RMA fluxes in activity. They leave a corrosive residue that must be removed thoroughly by aqueous cleaning without detergents. Synthetic activated fluxes are soluble in fluorinated liquids (used in vapor-phase) but not in water. Therefore, they are not recommended for VP soldering because the deposits that result from the synthetic activator breakdown, build up on the surface of the heaters and subsequently release toxic products.

3. *Inorganic water soluble fluxes.* By far the most active and corrosive fluxes are used for structural soldering. Zinc, chromium, steel, and tinning of lead wires often require these types of fluxes.

Solder

The operating temperature of the equipment is the major consideration in selecting the solder because the composition determines the melting point. In selecting a solder, a number of considerations must be taken into account: liquidus and solidus temperature, strength, corrosion resistance, electrical and thermal conductivity, and thermal expansion.

The characteristics of most solders are shown in a phase diagram like the lead–tin alloys diagram shown in Figure 19.8. A Sn 50%, Pb 50% composition, common in most soldering, flows at 430°F (220°C) with complete solidification at the eutectic. The eutectic alloy, 63 Sn/37 Pb, is most widely used because of its unique feature of going directly from solid to liquid (when heated to reflow temperatures), bypassing a plastic range. To obtain higher physical properties or to reduce price, some solders contain other elements such as cadmium, silver, copper, or zinc. For SMT, for example, it is common to add silver to create the 62 Sn/36 Pb/2 Ag (tin/lead/silver) alloy.

Solder pastes have become an indispensable part of today's electronic manufacturing. Solder paste is made of small, spherically shaped particles of solder that are uniformly suspended within a mucilage composed of flux and other agents. Solder paste is commonly considered as a two-part mixture of metal (the collection of solder particles) and organics (the conglomerate of flux and other constituents). The weight-to-volume ratio between these two parts has a significant impact on paste shelf life, viscosity, and application methods and the volume of the final reflowed solder joint. The volatile component of solder pastes may be alcohol or water, either of which can be evaporated by a carefully designed preheat cycle leading up to the solder fusing cycle.

Epoxy Conductive Adhesives

Solder is structurally poor, fatigues quickly, is a nightmare when encountered by precious metals, adheres to very few materials, contains lead (U.S. Environmental Protection Agency's condemned material), and has a relatively low melting temperature. Soldering operations require very careful control of time and temperature, an inordinate amount of surface preparation, and demand certified operators. The process requires ventilation, expensive machine maintenance, and regulated cleanup. In response, conductive adhesives can be used.

Epoxy adhesives can be cured at either low or high temperatures. Low-temperature curing would resolve the problem with heat-sensitive components but would require longer cure time. High-temperature curing could use the same IR soldering machines and profiles as the normal soldering production throughput. This approach would, however, continue the high-temperature exposure for the components and materials. Epoxies cured at solder reflow temperatures require less than 10 min, and those cured at 257°F (125°C) require four to five times that amount. Assemblies with epoxy adhesive conductive joints can be repaired simply by applying 302°F (150°C) heat and lifting the faulty component. Replacement joints can be formed without removing adhesive residue.

Unlike solder, epoxy adheres to most materials with far less surface preparation. Like solder, however, silver, the conventional conductive filler, migrates from the adhesive to

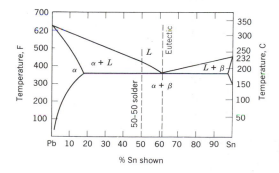

Figure 19.8
Lead–tin phase diagram.

adjacent conductors in the presence of moisture and a sufficient voltage differential. However, with the proper development, performance tests, and process tests, conductive adhesives can become the mass attachment technique that replaces solder.

19.4 SOLDER JOINT DESIGN

In general, each solder joint must meet six major design criteria: three *material-oriented* factors (electrical conductivity, mechanical stability, heat dissipation), which can be modified by selecting a different material system, and three *design-oriented* parameters (ease of manufacturing, simplicity of repair, fully inspectable), which affect reliability and cost. The designer must pay close attention to all these details to achieve fast, efficient soldering.

Analysis of Through-Hole Joint Configuration

Although THI work is losing popularity, it is still used widely by job shop manufacturers, so its joint structure must be understood. First of all, the solder fillet is a function of joint structure and varies from single-sided boards to plated-through holes. It also depends on lead wire arrangement and differs from straight through to bent-over leads. The selection of a single-sided versus a double-sided plated-through board is seldom based on the quality of solder joints. The decision to use a bent-over versus a straight lead must be made with great care. The analysis that follows, coupled with tensile force and electrical conductivity evaluation, will help the student determine whether a design meets an assembly's physical needs.

To simplify the analysis, five separate zones are identified on solder fillets (see Figure 19.9). Zones I to IV are shown on a bent lead placed in a plated-through hole. Zone V represents the base for a single-sided printed circuit board with a straight-through lead. A combination of these five zones can thus represent any known configuration of solder joints as follows: (1) plated-through hole with bent-over lead: Zones I, II, III, and IV; (2) plated-through hole with straight-through lead: Zones V, III, and IV; (3) single-sided board with bent-over lead: Zones I and II; (4) single-sided board with straight-through lead: Zone V.

Each zone is analyzed separately. Zone I is shown here as an example; students are encouraged to do the others as end-of-chapter problems. The analysis can be used to calculate if a plate-through hole fillet has to exceed, say, 75% of board thickness. In this case, one calculates the properties required (strength, conductivity, etc.) for the appropriate fillet sections. In our example, just omit Zone IV completely, take 75% of Zone III, and then Zones I and II (for bent-over leads). In the total fillet evaluation, all zones are arithmetically added.

Zone I represents the important component of a *clinched lead*. This part of the fillet is in reality a lap joint between the bent wire and the printed circuit board. By itself, a

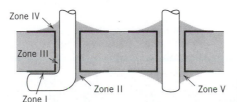

Figure 19.9
A schematic of the average fillet used.

lap solder joint is a preferred configuration for mechanical strength only, because the solder in a lap joint is stronger than in any other stress mode. However, in the total stress analysis of a plated-through hole, this does not hold true. The bent-over lead wire does not allow any degree of freedom in the upper direction, causing problems under slow cycle fatigue conditions. Note that Zone I is the only joint that is entirely independent of the hole-to-wire ratio.

Zone I, as defined in this analysis, does not include the filled portion of the hole, which is labeled as Zone II. If Zone I alone is used, without Zone II, only half of a partially open hole would exist. This is sufficient for conductivity and strength purposes. One must ensure that there is good wetting of the pad and the wire. Furthermore, the solder must extend at least half of the circumference of the hole, to give the copper foil added peel strength.

This solder joint is several orders of magnitude larger than the peel strength of the joint for 1 oz of copper. The superiority of the clinched lead joint has been confirmed by numerous tests throughout the electronic industry. Tests of solder joints in Zones I and II show that they can support between 2 and 6 lb of upward pull.

In addition to structural and conductivity requirements of a solder joint, an extensive set of board preparation and workmanship issues must be observed. A few examples follow.

EXAMPLE 19.1

Components Vertically Mounted in Series (See Figure 19.10).

When axial lead components are mounted in series, each component's body or weld bead (beginning of the lead) should be mounted 0.030 to 0.090 in. from the PWB, and a mechanical wrap should begin 0.090 in. from the component's body.

However, if the connection is in series but components are parallel to the board's surface, each component lead should be bent one complete wrap, which starts approximately 0.090 in. from the component body. There should be 0.060 in. from the completion of the wrap to the component body. The components may have lead bends starting up to 0.125 in. from its body if the complete wrap does not allow the solder fillet to wick closer than 1 lead diameter to the body. (See Problem 19.1.)

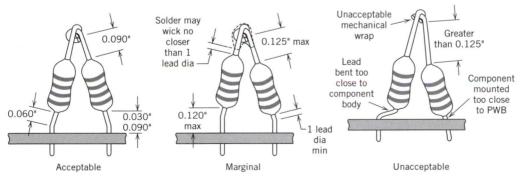

Figure 19.10
Components vertically mounted in series.

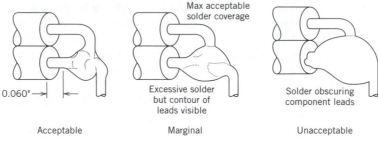

Figure 19.11
Wrapped leads soldering.

EXAMPLE 19.2

Wrapped Leads Soldering (See Figure 19.11).

The solder must be held approximately 1 lead diameter from the component body. The minimum solder is evidenced by a 75% fillet between the leads, provided no voids exist. The solder can be excessive to the extent that it is not severely convex or obscure the basic contour of the leads being soldered.

EXAMPLE 19.3

Mechanical Wraps of Lead Terminations (See Figure 19.12).

Before soldering, wires should be smoothly, tightly, and evenly wrapped around the connection between ¾ and 1.5 turns of wire, in a manner that avoids any possibility of electrical shorts. The wrap must be made in the same direction as the service loop.

EXAMPLE 19.4

Fiducial Marks (See Figure 19.13).

Fiducial marks are reference points in the surface copper pattern of a PWB that are used for the accurate alignment of assembly tooling and components. To accommodate vision-system requirements, fiducial targets are etched into the board to minimize dimen-

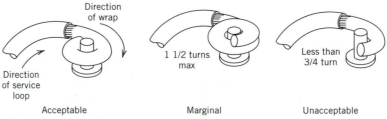

Figure 19.12
Mechanical wraps of lead terminations.

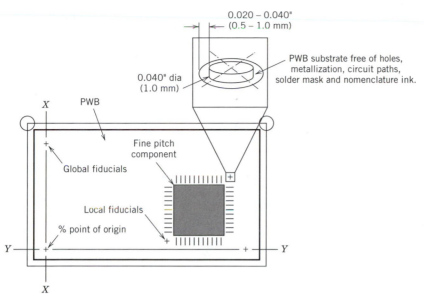

Figure 19.13
Recommended footprint for fine-pitch component fiducials.

sional tolerance accumulation with respect to the devices' land pattern geometries. Three global fiducial targets are required for each board side to place SMDs, and two additional targets for each fine-pitch device. The minimum size of the fiducial is 0.040 in. in diameter with a clearance of 0.020 in.

19.5 THERMAL CHARACTERISTICS

Thermal characteristics of IC packages are a major consideration in electronic manufacturing, because an increase in junction temperature (T_j) can have an adverse effect on the long-term operation life of an electronic device.

With the increased use of SMD technology, management of thermal characteristics remains a concern because not only are the SMD packages much smaller, but also the thermal energy is concentrated more densely on the printed wiring board. For these reasons, the designer and manufacturer of surface mount assemblies (SMAs) must be more aware of all the variables affecting T_j.

It is known that power dissipation (P_D) varies from one device to another and can be obtained by multiplying V_{CC}-Max by typical I_{CC}. Since I_{CC} decreases with an increase in temperature, maximum I_{CC} values are not used.

The ability of the package to conduct this heat from the chip to the environment is expressed in terms of *thermal resistance*. The term normally used is Theta JA (Θ_{JA}); Θ_{JA} is often separated into two components: thermal resistance from the junction to case and the thermal resistance case to ambient. Θ_{JA} represents the total resistance to heat flow from the chip to ambient and is expressed as

$$\Theta_{JC} + \Theta_{CA} = \Theta_{JA} \tag{19.1}$$

In the case of *conductive* heat transfer (Q), thermal resistance, for a plain slab whose two faces are maintained at temperatures T_1 and T_2, is calculated as

$$\Theta = \frac{(T_1 - T_2)}{Q} = \frac{t}{kA} \tag{19.2}$$

where

t = Thickness of the material
A = Area
k = Thermal conductivity of the material

Junction temperature (T_J) is the temperature of a powered IC measured at the substrate diode. When the chip is powered, the heat generated causes the T_J to rise above the ambient temperature (T_a). Junction temperature is calculated by multiplying the power dissipation of the device (P_D) by the thermal resistance of the package and adding the ambient temperature to the result.

$$T_J = (P_D \times \Theta_{JA}) + T_a \tag{19.3}$$

EXAMPLE 19.5

The square chip shown in Figure 19.14 is bonded to the die support paddle of a typical dual-in-line package (DIP). The constriction resistance from the chip to the paddle is known to be 2.4°C/W. The resistance from the junction at the top surface of the chip to the top side of the paddle needs to be estimated. Assume the chip is 20 mils thick and 0.150 in. (0.381 cm) wide. The layer of epoxy attaching the silicon to the metal paddle is 5 mils thick. The thermal conductivity of silicon is 153.5 W/m°C and for epoxy is 0.26 W/m°C.

The problem is done converting all linear units to meters and then using Equation (19.2). The resistance due to the chip is

$$\frac{5.08 \times 10^{-4}}{(3.81 \times 10^{-3})^2 \times 153.5} = 0.23°C/W$$

Similarly, the resistance due to the epoxy is calculated as 3.37°C/W.

According to Equation (19.1), the total resistance is the sum of the contributions of chip, epoxy, and the constriction resistance:

$$(0.23 + 3.37 + 2.4)°C/W = 6.0°C/W.$$

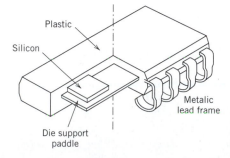

Plastic

Silicon

Metalic
lead frame

Die support
paddle

Figure 19.14
Chip bonded to the die support paddle of a typical DIP.

19.6 ELECTROMAGNETIC INTERFERENCE AND ELECTROSTATIC DISCHARGES

There are two other important issues to consider during electronic manufacturing. One is the electromagnetic interference (*EMI*) that, in many applications, can cause loss of communications, errors in data-processing equipment, and inaccurate presentation of images. The other issue is electrostatic discharges (*ESD*), which can destroy boards and components if care is not taken.

Electromagnetic Interference

Any circuit utilizing a time-varying voltage or current is capable of generating electromagnetic energy and, therefore, producing EMI. High levels of attenuation of low-frequency magnetic fields can be achieved using high-permeability materials as gaskets. The primary function of an EMI gasket is to reduce the interface resistance across a joint. The lower the resistance, the better the EMI seal. The higher the resistance, the poorer the seal. The relationship between interface resistance and EMI leakage for a particular housing, tested at 700 MHz, is shown in Figure 19.15.

Joint interface resistance is a function of several factors: the surface resistance of the flanges, the conformability of the gasket to the flange, the amount of pressure on the gasket, and the gasket's conductivity. If gasket conductivity degrades over time or during certain environmental conditions, EMI performance also will degrade. This is a primary source of EMI gasket failure; materials can start out meeting all electrical specifications and within a relatively short time become so resistive that they cannot function as an EMI insulator.

Electrostatic Discharges

Static charge accumulates on the human body and if it is not discharged prior to handling components, an ESD physically damages paths and chips.

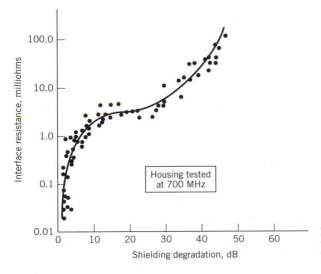

Figure 19.15
Relationship between interface resistance and EMI leakage.

The most common way of addressing the discharging of static is by using "ground or foot straps" that conduct the static safely to a ground and prevent any subsequent buildup while working. Basic requirements include wearing approved wrist straps connected to a solid ground, using antistatic waxes on the floors and antistatic floor mats, and receiving and transferring all components in antistatic bags and trays. These antistatic or grounded surfaces are required where inspection, touch-up, repair, cleaning, and rework are performed.

19.7 CLEANING PROCESS

With a very active flux omitting or delaying cleaning is detrimental to performance and reliability of electronic devices. Testing has shown that these fluxes must be removed within 30 min after soldering to prevent their corrosive action. CFC-113 and 1,1,1-trichloroethane have been used as universal cleaners and, unfortunately, there is not a known cleaning chemistry that can replace their cleaning and drying properties. There are, however, some alternatives.

Aqueous Cleaning

Water-based cleaning is the most mature technology that lends itself to batch and in-line cleaning. This alternative is generally the first choice because of the perceived low cost; however, disposal and water treatment costs can be considerable. Normally, a saponified bath (a 1% to 10% alkaline concentration diluted in water) is used. After the product is processed through the cleaner, the residual chemicals and particles are rinsed in high-quality water. The remaining water is dried with forced hot air.

Semiaqueous Cleaning

Good for batch and in-line cleaning, semiaqueous processes use a hydrocarbon product at a 100% concentration. Cleaning is followed by multiple rinse water stages to remove the hydrocarbon cleaner and any nonorganic product on the surface of the board and components. Since closed-loop control of the solvent and the rinse water is normally part of the system, waste treatment costs are low.

Water Miscible Cleaning

Like the previous two methods, water miscible cleaning (WM) lends itself to batch and in-line cleaning. This method uses an organic cleaning product fully miscible with water. Drying is simple, by forced hot air, but removing the cleaning chemistry from the rinse water is an expensive operation.

Solvent–Solvent Cleaning

Mostly applicable to batch cleaning, this technology uses a hydrocarbon solvent to clean parts and the same or another nonwater product to rinse them. The clean solvent residual is removed from the PCB with compressed air, and a final drying cycle with forced warm air is needed. This extra cycle and safety precautions raise the costs of operation.

Perfluorinated Rinse

This older technology operates like a vapor degreaser; however, the chemistry used seems to contribute to global warming. Parts are cleaned with a natural organic product and then perfluorinated hydrocarbon (PFC) is used to remove the cleaning chemistry. Final rinsing is done in a PFV vapor blanket, which later evaporates, leaving a clean, dry surface.

New Developments

Small volumes of high-speed fluid can be used to transport and support ceramic substrates by the force of their spray through the rinsing/drying process. Test jets are applied independently from the top and bottom, controlling the pace of movement as the panels travel on a hydraulic "pallet," avoiding contact with the tooling as they move.

In the system shown in Figure 19.16, the rinse fluids serve as both a transport and processing mechanism, reducing the amounts of water requirements by up to 75%. The advantages of such a system are important. Clips, guides, and fixtures can be eliminated for small boards since the force of the flow keeps them stationary and in line. By being totally enclosed, rinse water is used more efficiently compared to the flooding techniques used in aqueous and semiaqueous processes. In operation, high-velocity flow is directed very close to the product surfaces via special delivery heads (see Figure 19.17), thus reducing misting, or atomization, of the process fluid. In addition, fluids tend not to redeposit on substrate surfaces, favorably affecting throughput. Last, drying speed increases since most water is removed mechanically rather than evaporated.

The type of contaminants (fluxes, dirt, wet or dry films) present determine the type of cleaning and type of cleanliness monitoring required. To date there is no known universal dirt meter or method of testing, but MIL-P-28809 regulates a number of methods.

One known category of contaminants are the ionic materials, which promote electrical leakage, EMIs, and corrosion. These contaminants are amenable to electrical measurements, generally changes in resistivity. Surface insulation resistance (SIR), for example,

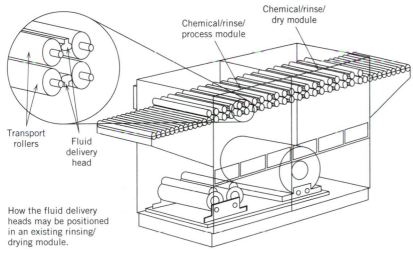

Figure 19.16
High-speed fluid transport cleaner.

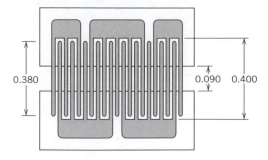

Figure 19.17
Shaded area is solder-masked. Conductors are 0.400 in. by 0.012 in. wide; spacing: 0.025 in. Unconnected ends of conductors are radiused (length of conductor overlap: 0.380 in.).

uses a special pattern to monitor the changes in board resistivity (measured in ohms, Ω) before and after cleaning. If the SIR measurement is more than a given minimum limit, cleanliness is complete.

To conduct the test, the specimen is conditioned under high humidity at an elevated temperature. A direct current potential is applied, and the leakage over the pattern is used to assess the relative value of the surface of the board.

EXAMPLE 19.6

A test to determine the effect of omitted cleaning of water-soluble paste residues is based on SIR measurements. The test is conducted in a humidity chamber at 90% relative humidity and 95°F (35°C). A pattern, such as the one in Figure 19.17, is in the oven for 24 h with a 48 VDC input and tested at 100 V. If the SIR measurement shows a minimum of 3×10^9 ohms, cleanliness is acceptable and the SIR of the PWB has not been affected. Results of the test using four different solder pastes are shown in Figure 19.18. Before cleaning, all four pastes are below the lower limit. After cleaning, Pastes D, A, and C (in descending order of removal) passed the test, while Paste B was not removed effectively.

19.8 EMERGING PACKAGING TECHNOLOGIES

As noted earlier, ICs or "chips" are essentially silicon wafers containing hundreds or thousands of transistor junctions interconnected to perform a variety of circuit functions. They are so small and delicate that they must be enclosed in one package before they can be handled and assembled into another package. This second package is the actual circuit board, which is then assembled into the electronic device or product.

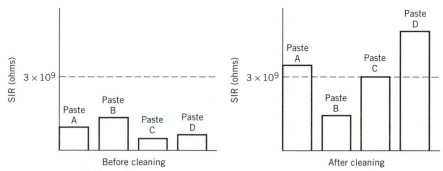

Figure 19.18
Surface insulation resistance test of four solder pastes.

Table 19.2   **Evolution of Integrated Circuit Packaging Technology**

Name	Description
Dual-in-line packaging (DIP)	Single chip in a rectangular plastic or ceramic package with electrical leads projecting from two opposite sides. May be soldered.
Surface mounting technology (SMT)	Single, very large chip in a square ceramic package with many leads projecting from all four sides or from the bottom around the perimeter. Usually requires machine placement and soldering.
Multichip module (MCM)	Multiple chips wired together and installed in a relatively large sealed module with leads projecting from two or four sides. Usually requires machine placement and soldering.
Chip-on-board (COB)	Unpackaged chip mounted directly to circuit boards using a chemical bond such as epoxy. There are no leads; wire bonds are attached between the chip and the circuit board. The chips are then chemically sealed. Requires machine placement and wire bonding.

The first two stages of the evolution of packaging technology concentrated on getting a more complex chip into the first package. The third stage put multiple chips into the first package, and the fourth stage concentrated on eliminating the package and assembling the chip directly onto the circuit board. These stages are described in Table 19.2 and are illustrated in Figure 19.19.

Packaging techniques, as currently practiced, will be altered significantly by new materials, processes, and technologies. Figure 19.20 depicts the circuit densities achieved and projected, assuming an average chip size of 0.25 in. per side. Some of the factors that can affect future packaging include:

1. *Three-dimensional packaging.* Flip-chip technology and total-plane interconnects will permit 3D stacking of chips and interplane wiring.

2. *Solder substitute.* Molten solder is being replaced by low-temperature organic conductive adhesives. Precision lasers will be commonplace on the assembly line for very-fine-pitched joints, rework and repair, assembly-line measurements, and automatic lamination alignment tools.

3. *Optoelectronics.* Fiber optics is replacing many wire conductors, even within localized systems and assemblies. Optical receivers and transmitters will be prominent components

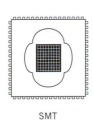
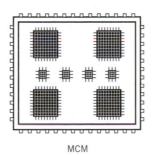
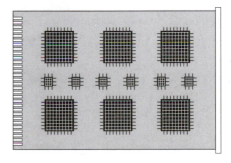

DIP   SMT   MCM   COB

Figure 19.19
Emerging electronics manufacturing technologies.

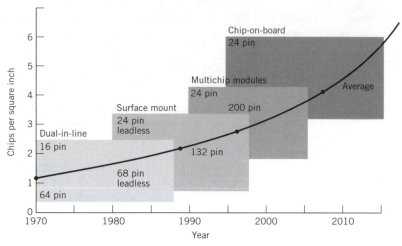

Figure 19.20
Advances in electronic circuit density.

on PWBs. Speed, elimination of crosstalk and propagation delays, security, and signal capacity will all force the use of optics. Relatively small, specialized metal packages, with superaccurate cavities and tailored thermal expansion coefficients, will be grown to precision dimensions for fiber-optic terminations. *Diamond film* is becoming a very important material used in optics, sensors, and thermal management. *Superconductors* are also playing a key role, but in a few specialized areas involving magnetics and super high speeds.

QUESTIONS

1. Give an explanation of the following terms:

Integrated circuit	THI
Silicon wafer	Vapor-phase
Dopants	Solder paste
Wafer shaping	Tarnish
Artwork	Water White Rosin
Photoresist	EMI
SMT	ESD

2. Describe the categories of electrical components and give examples.

3. How is the wafer covered by the photoresist?

4. Describe the two types of photoresists.

5. Describe the steps for producing microelectronic chips.

6. List applications for photoresists.

7. How are printed circuit boards made? Consider only the etching and drilling of the boards, not their loading.

8. Describe the function of opaque printed patterns in the photomask.

9. Describe the process for making the wafer.

10. What is the difference between SMT and IMT?

11. What is the function of a dopant?

12. Prepare the list of processes or steps for making a microelectronic chip.

13. Contrast the functions of a component sequencing machine and automatic stuffing.

14. Why is wave soldering superior to hand soldering printed circuit boards?

15. List the steps for wave soldering printed circuit boards.

16. Name five difficulties of surface mounting technology.

17. Explain how a solder connection is formed.

18. What is the best application for infrared soldering?

19. Study Table 19.1. Which do you think are the five most critical variables of a soldering system?

20. Explain the preheating period of a soldering cycle.

21. Why are rosin fluxes losing popularity?

22. What are four specific functions that flux performs?

23. What type of flux is not recommended for vapor-phase soldering? Why?

24. What solder alloy is the most widely used? Why?

25. What are solder pastes?

26. Name six disadvantages of using solder.

27. If you do not want to use solder, what alternatives do you have? Highlight some advantages of your alternate adhesive material.

28. What are the major design criteria that must be evaluated for solder joint design?

29. What are fiducial marks and what are their minimum dimensions?

30. Name four packaging technologies and explain their main differences.

31. Why is thermal management so important in packaging technology?

32. What are some factors that can affect future packaging techniques?

33. Name two cleaning techniques used in electronic manufacturing that are detrimental to the environment.

34. Explain two cleaning techniques suitable for in-line processes.

35. Why is electromagnetic interference a concern in electronic manufacturing?

36. What factors affect joint interface resistance?

PROBLEMS

19.1. Study Figure P19.1 and determine which connection is acceptable, which one is not, and explain why.

19.2. Describe the characterization of the IR reflow soldering profile shown in Figure P19.2.

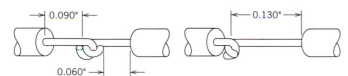

Figure P19.1
Component-to-component connections in series.

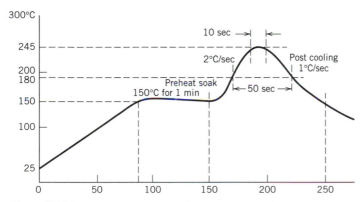

Figure P19.2
Infra-red reflow soldering temp/time profile.

19.3. Calculate the thermal resistance of a circular chip that is 20 mils thick and has a 0.5 in. diameter. The thermal conductivity of the chip is 160 W/m°C.

19.4. Calculate the maximum power dissipation allowable for a ceramic small outline chip that operates at an ambient temperature of 25°C. The typical thermal resistance junction to ambient is 100°C/W and the maximum operating junction temperature is 150°C.

19.5. The minimum bend radius, R, is defined as the radius of the arc formed when a PC board of length d bends. Derive an expression for the deflection, Y, as a function of d and R only. What is the maximum deflection for a 4-in. long board if the minimum bend radius is 5 ft?

MORE DIFFICULT PROBLEMS

19.6. Adhesive epoxy is used to mount 100 surface mount monolithic ceramic ICs on a 7 by 6 by ¼-cm cold plate that has a thermal conductivity of 300 W/m°C. Each IC is 5 mm long, 4 mm wide, and 2 mm thick, with a thermal conductivity of 155 W/m°C. Each chip's power is dissipated with a heat flux (Q) of 1.1 W. If the epoxy is 0.1 mm thick and there is no thermal interference among the chips, estimate the maximum junction temperature if the plate is cooled by forced air at 25°C. Assume that constriction resistance and convection resistance (due to air cooling)

amount to 6°C/W. The thermal conductivity of epoxy is 0.26 W/m°C. (Hints: review Example 19.5. $T_j - T_{air} = Q \cdot \Theta_{JA}$).

19.7. If the cold plate of Problem 19.6 can be cooled by water or other fluids to keep the wall temperature at 25°C under whatever power dissipation levels, what is the maximum allowable chip power if the junction should work at less than 65°C? Ignore the effect of constriction resistance.

19.8. Describe the characterization of the double-wave soldering profile shown in Figure P19.8.

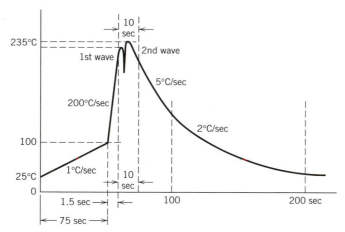

Figure P19.8
Double-wave soldering temp/time profile.

PRACTICAL APPLICATION

The electronic industry is constantly changing. New epoxies, environmentally safe cleaning systems, advanced packaging techniques, and better soldering materials are some areas of rapid expansion. Select any element of electronic manufacturing of your interest. Then identify a manufacturer, a distributor, or a consultant that you can visit for an interview/plant tour. During your conversation, ask about trends and emerging technologies. Prepare a report that describes the existing conditions and the advantages/disadvantages of implementing the new technologies. (Hint: SME, IIE, ASME, NAIT, and similar organizations can assist in identifying consultants in this field.)

CASE STUDY: THE HOT CHIP PROBLEM

Talk's Chip, Inc., is a manufacturer of voice-synthesized devices using SMT components. One rainy Monday afternoon in April, the CEO receives a phone call from his sister-in-law.

"Gabe!"

"Erica! Howzit goin'?"

"Listen, not so good. You remember that voice synthesizer you gave me for my birthday? Well I was trying to make a new funny message for my voice mail and somehow the darn thing quit on me. It just went 'whoomf!' and died."

"Okay, Erica. Why don't you bring it in this afternoon and I'll have one of the engineers take a look at it. Maybe it's something we should know about before we start mass production of that model."

Later that day Gabe brings the dead machine to you. After careful evaluation, you determine that the problem is with a chip that overheats. You decide to find the effect of four different chips and recommend the best two alternatives to Gabe. Data for your analysis are given as follows. Assume that the same type of epoxy will be used in all cases and that constriction resistance and all other conditions will be equivalent in all cases.

Chip Geometry	Cost per Chip ($)	Assembly Time (min)	Assembly Cost ($/hr)
Square: width = 0.050 in. $k = 154$ W/m°C	3.50	2.5	9.00
Rectangle: 0.075×0.050 $k = 110$ W/m°C	4.0	2.8	9.00
Circular: Diam. = 0.07 $k = 112$ W/m°C	3.60	2.7	8.85
Triangular: 0.075×0.1 $k = 160$ W/m°C	3.55	2.6	8.85

(Hint: Both the effect of thermal resistance and the cost of manufacturing are important to the company.)

CHAPTER 20

NONTRADITIONAL PROCESSES AND POWDER METALLURGY

The needs of current manufacturing systems have led to adaptations of very old techniques, such as the use of powder metals, and to the development of many special processes. Several manufacturing processes that do not fall under categories of traditional methods are considered first. The last portion of the chapter is dedicated to powder metallurgy, a technology that allows production of complicated shapes more economically than conventional metal processes.

20.1 SPECIAL MACHINING PROCESSES

The shaping of parts made from carbides and other metals difficult to machine has for many years been limited to diamond wheel grinding. Because of the expense of diamond wheels and the time required for grinding, effort and research have been directed toward the development of more economical methods. There are no less than four fundamental types of machining energy being used: mechanical, chemical, electrochemical, and thermoelectric. These can be subdivided into a number of special processes, each having some special use or advantage.

Ultrasonic Machining

Ultrasonic machining is a mechanical process designed to machine hard, brittle materials. As shown in Figure 20.1*A*, it removes material by abrasive grains that are carried in a liquid between tool and the work, and bombards the work surface at high velocity. This action gradually chips away minute particles of material in a pattern controlled by the tool shape and contour. A *transducer* causes an attached tool to oscillate linearly at a frequency of 20,000 to 30,000 Hz and at an amplitude of 0.0005 to 0.004 in. (0.013 to 0.10 mm). The tool motion is produced by being part of a sound wave energy transmission line that causes the tool material to change its normal length by contraction and expansion. The toolholder is threaded to the transducer and oscillates linearly at ultrasonic frequencies, thus driving the grit particles into the workpiece. The cutting particles boron carbide, silicon carbide, and aluminum oxide are of a 280-mesh size or finer depending on the accuracy and the finish desired.

The metal removal rate is slow or about 0.022 in.^{3}/min in tungsten carbide. Major factors influencing material removal rates, surface roughness, and accuracy are amplitude and frequency of the tool oscillation, impact forces, tool material, abrasive, and content of the slurry.

The tools are made of brass or soft steel and must match the surface to be machined. Tolerances of 0.002 in. (0.05 mm) can be maintained with 280 grit, or by using finer grit a tolerance of 0.0005 in. (0.013 mm) can be held. Operations include drilling, tapping, coining, and the making of openings in dies. Ultrasonic machining is used principally for

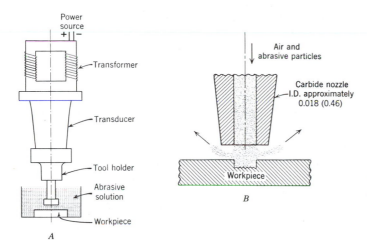

Figure 20.1
Ultrasonic and abrasive machining. *A*, Schematic diagram of the ultrasonic process. *B*, Abrasive jet cutting with aluminum oxide particle, 15 to 40 μm.

machining materials, such as carbide, tool steel, ceramic, glass, gem stones, and synthetic crystals. Advantages include the absence of thermal stresses, low tooling cost, and the use of semiskilled workers for precision work. Holes with a curved axis, nonround holes, or holes of any shape for which a tool can be made are candidates.

Abrasive Jet Machining

Abrasive jet machining is a mechanical process for cutting, deburring, and cleaning hard brittle materials. Cutting rates are generally slow, and focusing the stream of particles is a problem in tooling and motion control. The cutting action is cool because the fluid tends to cool the surface while carrying the abrasive. It is similar to sand blasting, but uses much finer abrasives with particle size and velocity under close control. The cutting action is shown schematically in Figure 20.1*B*. Air or CO_2 is the carrying medium for the abrasive particles that impinge on the workpiece at velocities about 500 to 1000 ft/s (150 to 300 m/s). Aluminum oxide or silicon carbide powders are used for cutting, while softer powders, such as dolomite or sodium bicarbonate, are used for cleaning, etching, and polishing. Powders are not recycled because of possible contamination, which is apt to clog the system.

Abrasive jet machining cuts fragile materials without damage. Other uses include frosting glass, removing oxides from metal surfaces, deburring, etching patterns, drilling and cutting thin sections of metal, and cutting and shaping crystalline materials. It is not suitable for cutting soft materials because abrasive particles tend to become embedded. Compared with conventional processes, its material removal rate is slow. For glass it is 0.001 in.^{3}/min (0.273 mm^3/s).

Water Jet Machining

Water jet machining or fluid jet machining is a process that utilizes a high-velocity stream of water as the cutting agent. Jet nozzles are approximately 0.0015 to 0.0067 in. (0.038 to 0.17 mm) in diameter and operate at velocities of 2000 to 3000 ft/s (600 to 900 m/s). Pressures range from 30,000 to 55,000 psi (207 to 380 MPa). Higher pressures reduce seal life from about 800 to 150 h. At these velocities the jet cuts through wood, plastics, textiles, and, in some cases, ceramics, steel, and titanium. A limitation of this process is the lack of suitable pumping equipment. The abrasive may be garnet or even silica, which reduces costs. Applications include aerospace beveling of edges for subsequent welding or cutting off sprues and gates by foundries.

Electrical Discharge Machining

Electrical discharge machining (EDM) is a process that removes metal with good dimensional control from any soft or hard metal. It cannot be used for machining glass, ceramics, or other nonconducting materials. The machining action is caused by the formation of an electrical spark between an *electrode* shaped to the required contour and the workpiece. The conventional EDM machine cuts metals by electrical discharge or "spark erosion" between the metal to be cut (negative (−) charge) and an electrode (positive (+) charge). This cutting takes place in a nonconductive fluid known as a *dielectric*.

Since the cutting tool does not touch the workpiece, it is made of a soft, easily worked material such as brass. The tool works in a fluid such as mineral oil or kerosene, which is fed to the work under pressure. The coolant serves as a dielectric, to wash away particles of eroded metal from the workpiece or tool, and to maintain a uniform resistance to current flow. The tank is filled with the dielectric fluid and the workpiece, and the electrode end is submerged. An electrode, chosen depending on the shape of the cut, is positioned on the top of the workpiece leaving a small gap.

After connecting the electrode to a positive charge, spark erosion takes place, creating a "miniature thunderstorm" between the two metals. Flashes of "lightning" in rapid succession occur. Each one produces a tiny crater in the surface of the two metals. *Metal evaporation* occurs where the flash strikes.

Equal amounts of material are *not* removed from both plates. By an appropriate choice of materials, electrode of copper, workpiece of steel, and a skillful selection of the opening and closing times of the automatic switch, more material is removed from the steel than from the copper.

During the process, the dielectric is flowing through the tank requiring filtration. Also, erosion creates heat so the dielectric has to be cooled. The capacity of work of the conventional EDM machine is measured by the rate of material removal, in.³/min.

Figure 20.2*A* is a diagram of a simple arrangement for electrical discharge machining. A condenser parallel with the electrode and workpiece receives a charge of direct current through a resistor. As the condenser is energized, its potential rises rapidly to a value sufficient to overcome the dielectric fluid between the electrode and work. The gap distance is *servo controlled* so as to maintain a fairly constant potential to bring about an electrical breakdown of the dielectric fluid between the electrode and work. This distance is only a few thousandths of an inch (about 0.05 mm), so hand control is difficult

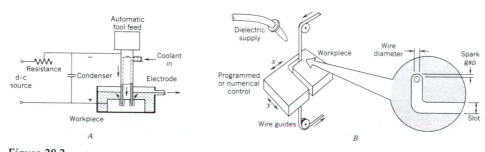

Figure 20.2
A, Diagram for electrical discharge machining. *B*, Traveling-wire electrical discharge machining.

to maintain. Regardless of the electrode tool area, sparking occurs at the point where the gap is the smallest. The current density at this point is high and of sufficient force to erode small particles from the workpiece. These small particles of metal are vaporized or melted by the spark, cooled by the electrolyte, and flushed from the gap between the electrode and the workpiece. The rate of metal removal is not fast as compared with commercial machining and may vary from a small fraction of 1 in.^{3} to around 15 in.^{3}/h (70 mm^3/s). The best surface finishes are obtained with slow rates of metal removal. Most machines have a frequency selector that controls the number of sparks per second between the electrode and the workpiece. As the spark frequency increases, the surface finish is improved owing to the reduction of energy per spark. Frequency of the sparks may range from 500 to 500,000 sparks per second.

The electrode is the cutting tool and, although it is not subject to much heat, it should have a high melting temperature and be a good electrical conductor. The wear rate of electrode materials varies with the material to be machined, but its selection is determined by the material cost and how it is made. Acceptable electrode materials include tungsten carbide, copper tungsten, graphite, copper, brass, and zinc alloys. The only requirement of the workpiece material is that it must be a good conductor of electricity.

In Figure 20.3, an electrical discharge machine is shown equipped with numerical control for the positioning with greater accuracy of the electrode from one hole to the next.

Electric discharge machining is used for salvaging hardened parts, machining carbide stock, producing dies and metal molds for stamping, forging, and jewelry manufacture, as well as for making numerous parts from hard metals. A broken tap inside a hole is

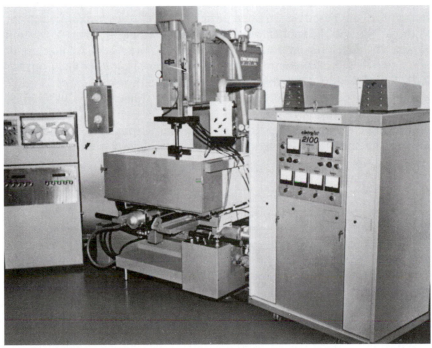

Figure 20.3
Electrical discharge machine (EDM) with power unit and numerical control system, especially designed for the production of multiple-hole or complex cavity patterns.

"salvaged" by EDM as it is possible to erode away the tap without damaging the tapped hole. Close tolerances and finishes of 8 to 10 μin. (200 to 250 nm) are possible.

Traveling-wire electrical discharge machining, a metal-cutting process that removes metal with an electrical discharge, is suited for production of parts having extraordinary workpiece configurations, close tolerances, the need of high repeatability, and hard-to-work metals. Wire electrical discharge machining produces a variety of parts such as gears, tools, dies, rotors, and turbine blades. It is appropriate for small to medium batch quantities. Actual machining times vary from half an hour to 20 h. It uses the heat of an electrical conductor to vaporize material; therefore, essentially no cutting forces are involved and parts can be machined with fragile, complex geometries. The sparks are generated one at a time in rapid succession (pulses) between the electrode (wire) and the workpiece. The sparks must have a medium in which to travel; thus, a flushing fluid (water) is used to separate the wire and workpiece. Hence, the one requirement is that the workpiece must be electrically conductive. A vertically oriented wire is fed into the workpiece continuously traveling from a supply spool to a take-up spool so that it is constantly renewed. Note Figure 20.2*B*.

A power supply provides a voltage between wire and workpiece and, by means of an adjustable setting pulse amplitude, pulse duration and the on and off times (μsec) are determined. On-time refers to metal removal; off-time is the period during which the gap is swept clear of removed metal via flushing. Therefore, both the intensity of the spark and the time it flows determine the energy expended and, consequently, the amount of material removed per unit time.

Electrochemical Machining

The electrochemical process is based on the same principles used in *electroplating*, except the workpiece is the anode and the tool is the cathode, as indicated in Figure 20.4. It is a *deplating operation*. The development of this process is due to the successful machining of hard and tough materials as well as complicated configurations. Low microinch surface finishes are standard, tool wear is negligible, and the stock removal rate exceeds other nontraditional machining processes. An average rate of stock removal is 1 in.³/min of metal (270 mm³/s) for each 10,000 A of current.

In this process, electrode accuracy is important since the surface finish of the electrode tool is reproduced in the surface of the workpiece. Copper is used frequently as the

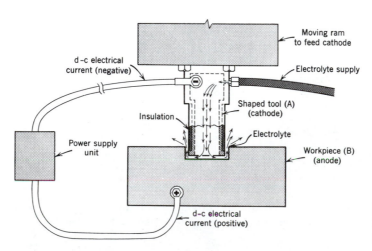

Figure 20.4
Diagram of electrochemical machining system.

electrode, but brass, graphite, and copper tungsten also find use. The tool must conduct the quantity of current needed, be easy to machine, and be corrosion resistant. Although there is no standard electrolyte, sodium chloride is used more generally than others.

The accuracy of the product is greatly influenced by the accuracy of the electrode form and its surface finish. It is also affected by irregularities in electrolyte flow or current flow. The electrolyte enters the gap between the electrolyte and the work at pressures ranging from 200 to 350 psi (1.4 to 2.4 MPa), whereas the flow rate reaches 150 gpm (0.01 m³/s) for high pressures. Current flows at a constant density if a uniform gap is needed. A temperature increase in the electrolyte improves the surface finish, but when temperature increases, the metal removal rate also is accelerated, increasing the size of the gap. This changes gap resistance so that less current flows and the metal removal rate is lowered to normal. When operating parameters are properly chosen, small variances are self-adjusting.

Electrochemical machining performs stress-free cutting of all metals, has high current efficiency, and can produce complex configurations difficult to obtain by conventional machining processes. Aside from cavity sinking as in die work, holes are drilled, external surfaces shaped, circular pieces turned, and contour machining performed. In the steel industry these machines extract test specimens from ingots, castings, and rolled shapes. In such processes, all metal is removed by electrochemical decomposition.

In these several electrical machining processes, there are advantages usually not found in conventional machining: Hard or soft conductible materials can be cut; the cutting tool can be of soft material since it does not touch the workpiece; there is no appreciable heating, so it avoids metallurgical changes or stresses caused by elevated temperatures; cutting forces are not involved; multiple operations can be carried out simultaneously; surface finishes can be maintained from 5 to 10 μin. (125 to 250 nm); and manual deburring is unnecessary.

Electrochemical Grinding

Electrochemical grinding, also known as *electrolytic grinding*, is similar to electrochemical machining. It differs slightly in that 10% of metal removal is done by some abrasive action, whereas 90% is done by electrochemical decomposition. Figure 20.5 illustrates the process.

In electrochemical grinding, a metal disk with embedded abrasive particles serves as the cathode. The workpiece is the *anode*, and the electrolyte, used in much the same fashion as a coolant, completes the electrochemical circuit. Figure 20.6A shows an electro-

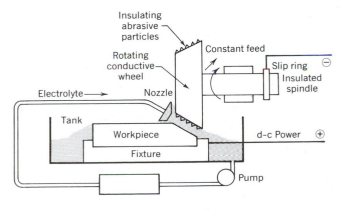

Figure 20.5
Schematic of electrochemical grinding machine.

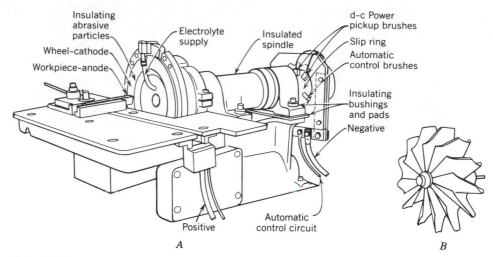

Figure 20.6

A, Setup for tool sharpening on electrochemical grinding machine. Short-circuiting between wheel and workpiece is prevented by abrasive particles projecting from wheel. A conductive electrolyte is used in the process. *B*, Curved form on Stellite turbine impeller ground by electrochemical grinder.

chemical grinder with the various parts labeled. In operation, the abrasive particles maintain the proper spacing between the disk and the workpiece. Diamond abrasive is ordinarily used, but since the electrochemical process accounts for 90% of the cutting action, wheel wear is not great. Although electrochemical grinding is particularly adaptable to the sharpening of carbide tools and the grinding of chip breaker grooves, many other grinding applications are possible, such as shown in Figure 20.6*B*. Electrochemical grinding is cool, burrs are not generated, and it produces a surface finish ranging from 8 to 12 μin. (200 to 300 nm). An average metal removal rate is 0.010 in.^{3}/min (2.7 mm^3/s) per 100 A.

Laser Beam Machining

The term *laser* is an abbreviation of "light amplification by stimulated emission of radiation." Simply stated, it is a very strong monochromatic beam of light that is highly collimated and has a very small beam divergence. Laser beam machining, a *thermoelectric process*, is accomplished largely by material evaporation, although some material is removed in a liquid state. Figure 20.7 is a pictorial view of a laser head. A relatively weak light flash is amplified in the ruby because certain of the chromium ions in the ruby emit photons as the light beam bounces back and forth in it. This released energy from the ruby accelerates the intensity of the beam of light, which leaves the rod and is focused on the workpiece. The ruby laser is most efficient when kept very cold, and liquid nitrogen at −320°F (−196°C) serves this purpose. The light flash operates best when warm; hence, hot air is circulated over it. The vacuum chamber between the ruby and the flash acts as an insulator and enables the two temperature climates to be maintained. The lamp operates from 1 flash every 3 min to 12 flashes per minute. The laser energy is applied to the workpiece in less than 0.002 s.

In addition to ruby lasers, other types are gaseous state lasers that utilize CO_2 and other gases, liquid-state lasers, and semiconductor lasers. All four involve the use of a

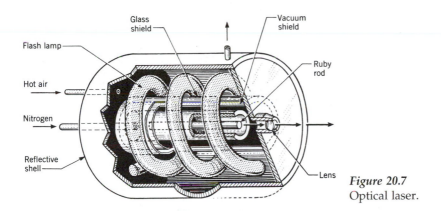

Figure 20.7
Optical laser.

well-defined beam of light. Since the metal removal rate is very small, they are used for such jobs as drilling microscopic holes in carbides or diamond wire drawing dies and for removing metal in balancing high-speed rotating machinery.

Lasers machine through transparent materials and can vaporize any known material. They have small heat-affected zones and work easily with nonmetallic hard materials. Major limitations to the process are the high cost of the equipment, low operating efficiency, difficulty in controlling accuracy, and its use primarily for small parts.

One important application of lasers is in the area of welding. Carbon dioxide (gas) lasers are used in welding, but they are limited to metal thickness of less than 0.02 in. (0.5 mm). Pulsed ruby lasers also have been used successfully for welding thicker materials. (Welding is discussed in Chapter 13.) Another application is in the area of metal cutting, where the carbon dioxide laser is most widely used. This laser, operating continuously, can cut any material if the beam is focused and a jet of gas is used to concentrate the beam. Figure 20.8 shows a *gas-assisted laser* cutting head cutting a steel plate.

Electron Beam Machining

Like laser beam machining, *electron beam machining* is a thermoelectric process. It is similar because of the high temperatures and high thermal energy densities that can be

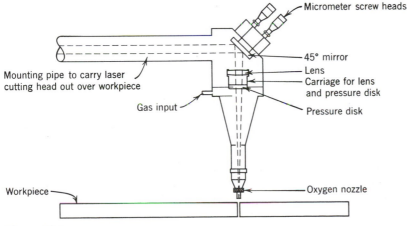

Figure 20.8
Gas-assisted laser cutting head.

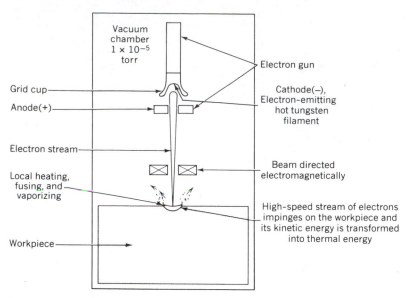

Figure 20.9
Electron beam machining.

achieved. In electron beam machining, heat is generated by high-speed electrons imping-ing on the workpiece, the beam being converted into thermal energy. At the point where the energy of the electrons is focused, it is transformed into sufficient thermal energy to vaporize the material locally. The process is generally carried out in a vacuum. Figure 20.9 illustrates the arrangement of the gun and beam directors, which focus the elec-trons magnetically.

Although the metal removal rate is approximately 0.0001 in.^{3}/min (0.002 cm^3/mm), the tool is accurate and is especially adapted for micromachining. Beams can be concen-trated on spots as small as 0.0005 in. (0.00127 mm) in diameter. There is no significant heat-affected zone or pressure on the workpiece, and extremely close tolerances can be maintained. Limitations of the process include high equipment cost, the need for skilled operators, and a vacuum chamber that restricts the workpiece size. The process results in X-ray emission, which requires that the work area be shielded to absorb radiation. The process is used for drilling holes as small as 0.002 in. (0.05 mm) in any known material, cutting slots and shaping small parts for the semiconductor industry, and machin-ing sapphire jewel bearings.

20.2 TEMPERATURE MACHINING

Elevated Temperature Machining

The shear strength of the metal is reduced when the workpiece is heated and plastic deformation ahead of the cutting tool is accomplished with less power. The chips tend to be continuous and the temperature at the *tool chip interface* does not increase in proportion to the temperature of the workpiece. Tool life is increased and cutting speeds may be doubled. Unfortunately, the expense of heating the workpiece is usually prohibi-tive. Heating may be accomplished by passing a current through the work, electric arc, acetylene flame, induction heating, or by a radio-frequency heating apparatus. Since the greatest efficiency comes from heating the shear zone just ahead of the cutting tool, radio-frequency resistance heating is the most economical for magnetic materials, whereas the tungsten inert gas torch can be used for heating nonmagnetic materials.

Figure 20.10A shows a radio-frequency heating apparatus that has been designed to concentrate heat in the area of the shear zone. Radio-frequency current tends to flow in the path of least impedance; hence, the return conductor is designed and placed in such a manner that the lowest impedance, and consequently the highest temperature, is focused ahead of the tool and in the shear zone.

The process is limited because the effects of elevated temperatures on the workpiece may cause a metallurgical change or distortion of the part, and heating the workpiece is expensive.

Plasma arc cutting torches employ a restricted tungsten arc through which gas is made to flow. A *plasma* is a gas that has been heated to a sufficiently high temperature to become partially ionized. Plasma cutting refers to metal cutting of the type accomplished by oxy-fuel torches. The operation normally is used for cutoff or rough shaping of plates or bars. These plasmas develop temperatures that approach 60,000°F (33,000°C). Such a torch can be used to replace rough machining operations, such as turning and planing. Observe Figure 20.10B, which shows the plasma working on a turned diameter. Although it is effective in cutting all metals regardless of metal hardness, there is a resulting rough surface and possible surface damage because of oxidation and overheating. Metal removal rates of 7 in.³/min (115 cm³/min) have been reported.

Cold Temperature Machining

At the opposite extreme of elevated temperature machining is the method of keeping metals at very low temperatures or *cold temperature machining*, which reduces tool tip temperatures. Ways to accomplish this include surrounding the area with a cold mist about −117°F (−83°C), using a dry-ice coolant or deep freezing the part itself.

Temperatures at the tool and work interface may sometimes exceed 2160°F (1180°C) when machining tough high alloy material. Problems arising from this heat include galling, seizing, work hardening, low-temperature oxidation, and short tool life. Conventional

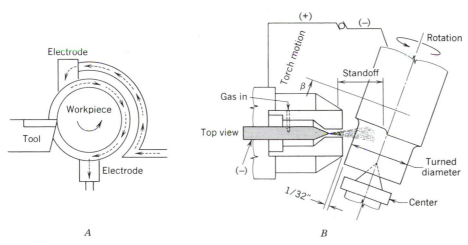

Figure 20.10
A, Elevated temperature machining using radio frequency heating. B, Plasma arc machining using torch.

practice reduces the surface speed to prolong tool life, but the resulting reduced output has spurred the search for a better way. With parts mist-cooled, tool life is improved for some alloys from 100% to 300% over conventional machining.

20.3 CHEMICAL ENERGY

Chemical Milling

Chemical milling or *chem milling* is a controlled etching process in which metal is removed to produce complex patterns, lightweight parts, tapered thickness sheets, and integrally stiffened structures. It is the adaptation of an old process to one that can remove metal successfully as in a machining process. The development of this process started in the aircraft industry, where many complicated machining problems exist in the removal of excess metal to reduce weight. Although most chem milling has been done with aluminum alloys, any metal for which an etching solution is available can be processed in this manner; observe Figure 20.11.

The process that is relatively simple consists first of thoroughly cleaning the sheet or part to be etched. It is then prepared for the etching process by masking those areas not to be affected with a chemically resistant coating. If the entire area is to be reduced, this is unnecessary. The part is then submerged in a *hot alkaline solution,* where metal in the unprotected area is eroded. The amount of metal removed depends mainly on the time the part is in the hot solution. Finally, the part is neutralized and rinsed, and the masking material is removed.

The success of chemical milling can be controlled by suitable masking employing organic coatings that are capable of resisting the action of hot alkaline solutions. Masking is applied by either a dip or an airless spray technique, and two or four coats are required depending on the materials. Coatings are then air-cured or baked to increase the etchant resistance of the mask. Patterns or templates outlining the area where metal is to be removed are placed on the mask and marked along the edges with a knife. After the mask is scribed, it is removed from the area where metal removal is to take place. Adhesive tapes applied to surfaces are excellent for chem milling panels to several thicknesses. Tapes are removed progressively after the etching begins if several panel thicknesses are required. Electroplating copper on the area to be masked also may be used, but it is somewhat expensive. Photosensitive coating masking is frequently used for complex designs such as that shown in Figure 20.12.

Where a uniform weight reduction is necessary, no masking is required. Similarly, sheet metal formed parts, which require a heavy sheet thickness for the operations involved, may be reduced uniformly in weight without masking. Another instance in which no masking is required is in the tapering of sheets (skins) for airplane use. Here

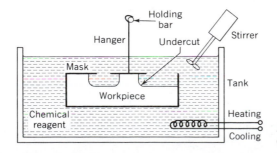

Figure 20.11
Chemical milling.

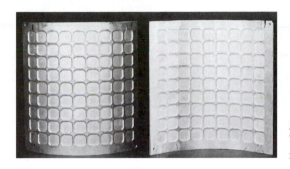

Figure 20.12
Wafflelike panels, an example of chemical milling.

the sheet to be tapered is immersed gradually or is withdrawn from the solution. The operation of chem milling is uniform on all exposed areas; neither are internal stresses developed nor is there any change in the metal structure. Tests indicate that the physical properties of the metal are not impaired by this process if proper control and etchants are used. However, bend and fatigue strength can be reduced somewhat by improper etching, the result of which is rough surface and consequent notch effects. Normally, a surface roughness of 50 to 60 μin. (1300 to 1500 nm) is obtained, which is comparable to surfaces on many die castings and ground parts.

One unusual problem involved in chemical milling is that of undercutting. As the chemical dissolves the bottom of a hole, it also attacks the side of the hole underneath the resist. The rate at which this takes place is called the *etch factor*, expressed as the ratio of the side penetration over the total depth of the cut, as shown in Figure 20.13. The etch factor may vary from less than ⅓ to more than 2, depending on the material and depth of penetration. It must be included in the design.

In comparison with machine milling, the following advantages are claimed for chemical milling: Material can be removed uniformly from all surfaces exposed to the etching solution; material can be removed after parts are formed to shape; sheets and structural members can be uniformly tapered; highly skilled operators are not required; close tolerance can be maintained and surface finish is good; and operating costs are generally less than in machine milling and equipment cost is less.

This process is limited in its applications for the following reasons: Aluminum is the only metal being chem milled on a commercial scale; on parts that can be machined while laying flat, conventional machining is more economical and accuracies greater than with chem milling; surface roughness is 50 μin. (1300 nm) higher; the depth of cut is limited when masking is used; gases collect under the mask and cause uneven etching; masking techniques for certain conditions are expensive; and gas generated in the process must be carried away.

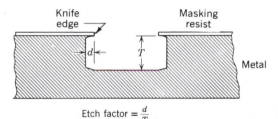

Etch factor $= \dfrac{d}{T}$

Figure 20.13
Undercutting in chemical milling.

Figure 20.14
Examples of thin parts produced by chemical blanking.

Chemical Blanking

This form of chemical material removal is used to produce thin metallic parts by chemical action, such as shown in Figure 20.14. A chemically resistant image of the part is first placed on the sheet, which is then exposed to chemical action by immersion or spraying.

Chemical metal removal does not require an electric current to carry the metal away in an electrolyte. It is known as *electroless etching*. In this type of etching the metal is converted chemically to a metallic salt, which is carried away as the etchant is replaced. In general, there are two procedures for chemical removal. In one there is no attempt to control the location of the metal removed, and the entire workpiece is affected. In the other, certain areas are covered with a protective coating to resist metal removal. The latter is generally spoken of as a selective metal removal process.

Chemical milling is a *selective metal-reducing* process for the purpose of weight reduction. Other selective processes include decorative etching, printed circuit etching, and chemical piercing and blanking. The sequence of operations in the selective processes is essentially the same except for chemical blanking, where a photographic-resistant image is placed on both sides of the metal blank.

The first step in chemical blanking (or *chem blanking*) is to prepare an accurate image of the part to be made. The metal should be chemically cleaned to eliminate all dirt, grease, and oxides. After cleaning, the metal is immersed in a tank containing the photographic resist and then hung up to drain and dry. The metal coated with this photographic resist material, when exposed to ultraviolet light, will polymerize and remain on the panel. When developed, this *polymerized layer* acts as a barrier to the etching solution. Both sides of the metal panel must be exposed simultaneously so that the metal is removed on both surfaces. After printing, the panel is developed in a spray to remove the coating, except in the areas of the workpiece that have been converted into etch-resistant images. Figure 20.15 shows the steps for blank preparation. In the etching

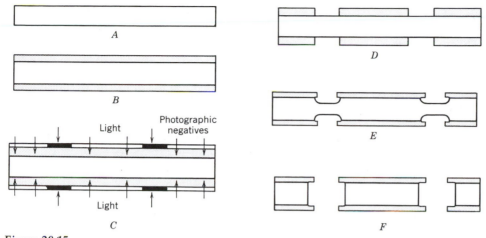

Figure 20.15
Steps in the photographic resist process. *A*, Clean metal. *B*, Metal coated with photo-resist on both sides. *C*, Photographic negative and exposure. *D*, Resist developed and hardened and partially removed. *E*, Partially etched. *F*, Fully etched and separated.

machine shown in Figure 20.16, both sides of the metal are sprayed and then washed and dried. The photoresist may or may not be removed from the parts.

The advantages of chemical blanking are extremely thin metal can be worked without distortion (most blanking is under 1/16 in. thick), no burrs are left on the edges, hard and brittle materials can be protected, setup and tooling costs are low, and design change costs are low.

Limitations of this process are skilled operators and good photographic facilities are required, etchant vapors are quite corrosive, and maximum metal thickness is small.

Chemical Engraving

Chemical engraving is used to produce such parts as nameplates and other parts that customarily are produced on a pantograph engraving machine. It is similar to chemical blanking, except that the lettering or design is on one side only. Figures or letters may

Figure 20.16
Horizontal conveyorized spray etcher used in chemical blanking.

be either depressed or raised. If the letters are depressed, they can be filled with paint. This process can be used on most metals, including hard-to-work metals such as stainless steel. Fine detail is possible and the process is less expensive than other methods generally used on flat work.

20.4 ELECTROFORMING

Electroforming is one of the special processes for forming metals. Parts are produced by *electrolytic deposition* of metal on a conductive removable mold or matrix. The mold establishes the size and surface smoothness of the finished product. Metal is supplied to the conductive mold from an *electrolytic solution*, in which a bar of pure metal acts as an anode for the plating current. The process differs from plating in that a solid shell is produced, which is later separated from the form on which it was deposited.

Electroforming is particularly valuable for fabricating thin-walled parts requiring a high order of accuracy, internal surface finish, and complicated internal forms that are difficult to core or machine. It also may be used to advantage in producing a small number of parts that would otherwise require expensive tooling.

The first step in production is to fabricate a *negative image* of the part. This is known as a matrix, mold, or pattern, which may be either permanent or expendable. Permanent molds can be used if there is sufficient draft to withdraw them without damage to the formed part, for example, in producing metal fountain pen caps, trumpet bells, and circular to rectangular transition-area waveguides. Such molds are generally machined from metal and are economical when many parts are to be made.

When it is impossible to use permanent patterns, *expendable* ones that are either chemically soluble or have a low melting temperature can be used. Soluble metals have the advantage of good internal finish and close tolerance. Also, they often may be made cheaply by die casting or plastic molding. Principal materials include aluminum, zinc alloys, and plastics. Low-melting materials, such as wax and lead tin bismuth alloys, can be molded at low cost but are easily scratched. Both the fusible and the soluble molds have their principal use in complex internal forms that would be difficult or impossible to make by other processes.

Since some of the materials used for forms are nonconducting, first they must be coated with a metallic film. This can be done in a variety of ways, including brushing, spraying, and chemical reduction. Wax molds can be coated with graphite. The conductivity of the film must not be too low, and good electrical contacts are important.

After the forms are prepared, they can be placed in the electrolytic solution and processed. In Figure 20.17, the interior of a tank for nickel electroplating solutions is shown. The tank is equipped with an automatic device to control the solution level and temperature. For rapid deposition the solution should be agitated, which is usually done by air. When sufficient time has elapsed to build up the required thickness, the part is removed from the bath, rinsed, and stripped from the mandrel.

Materials Used

All metal that can be used for plating also can be electroformed. Copper, nickel, iron, silver, zinc, lead, tin, cadmium, gold, aluminum, and a few others fall into this category. Copper, nickel, iron, and silver, used extensively in electroforming, contain properties such as good reproducibility, resistance to corrosion, electrical conductivity, good bearing surface, and adequate strength, which are required in most products. A *dendritic structure*,

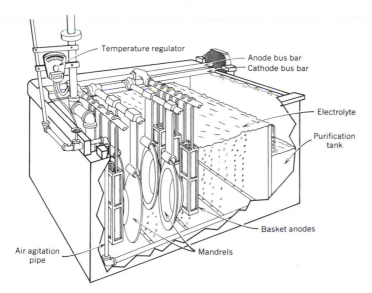

Figure 20.17
Tank interior for nickel electroplating solutions.

aligned normally to the conductive form surface (Figure 20.18), is common to all electroformed metals. Although the physical properties depend on the characteristics of the metal used, they also are influenced by the rate of deposition, plating temperature, and other bath variables. Dense structures can be obtained and, for some metals, the properties changed materially by heat treatment.

Figure 20.18
Cross section of copper deposit from acid sulfate bath. Sharp right angle shows weak plane formed by juncture of columnar crystals growing from sides of the angle. Magnification ×50.

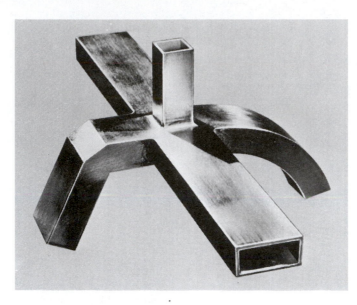

Figure 20.19
Intricate waveguide shape
electroformed with nickel.

Advantages and Limitations

Metallurgically, parts made by electroforming do not differ materially from parts made by other processes. The importance of electroforming as a process is its ability to produce complex parts requiring intricate detail that are almost impossible to make by other processes.

Some advantages of electroforming are: extreme dimensional accuracy can be held on surfaces next to the conducting form; identical parts can be made with practically no dimensional variation; surface finishes of 8 μin. (200 nm) or less can be maintained; parts of extreme thinness can be made; laminated metals can be produced; extreme metal purity can be obtained; intricate internal or external surfaces difficult to form by other processes can be produced (see, for example, Figure 20.19); and surfacing of parts to provide special physical or metallurgical properties is possible.

When compared with most of the traditional processes, electroforming is not an economical method of fabrication, except for precision parts that are costly to produce by other processes. Limitations of electroforming are: rate of production is slow, cost is high, accuracy of exterior surfaces cannot be controlled, process is confined to relatively thin products seldom exceeding ⅜ in. (9.5 mm), selection of materials is limited, and sharp internal angles should be avoided if design permits.

20.5 METAL SPRAYING

Several kinds of guns have been developed for spraying *molten metal* onto prepared base metals. The designs depend on the form of the metal when introduced to the gun, material being sprayed, and the temperature that is required. With the melting temperature range available almost any metal, alloy, or ceramic material can be sprayed. Materials are supplied as rods, wire, or in powder form.

Metallizing

Many metals in wire form are sprayed from a gun of the type shown in Figure 20.20. The wire is drawn through the gun and nozzle by a pair of rollers, melted by an oxyacetylene flame, and then blown by compressed air to the prepared surface. Although any

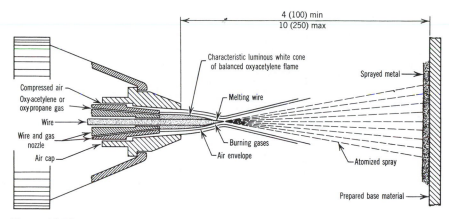

Figure 20.20
Nozzle cross section of metal spray gun.

metal that can be drawn into wire can be used in this gun, those most used are alloys such as steel, bronze, aluminum, and nickel.

Because the bond between the spray metal and the parent metal is entirely mechanical, it is important that the surface to be *metallized* be prepared properly before spraying. The surface is cleaned by blasting with sharp silica sand or angular steel grit. Cylindrical objects may be prepared by rough machining the surface. Either of these methods roughens the surface and provides the necessary *interlocking surfaces* or keys to allow plastic metal adherence. The molten metal is blown with force against the surface, causing it to flatten and interlock with surface irregularities and adjacent metal particles. The spray metal provides a suitable surface for successive coatings and permits building up a layer of considerable thickness.

The change in the physical properties of metal applied in this manner is an increase in porosity and a corresponding decrease in the tensile strength of the material. The reason is that the bond is mechanical and not fusile as it is for welding. Compressive strength is high and there is some increase in hardness.

Metal Powder Spraying

A process, sometimes referred to as *thermospray*, utilizes metal and other materials in a powder form. These are placed in a small container on top of the gun and fed by gravity to the gas mixture. Here they are carried to the nozzle and melted almost instantly by an oxyacetylene or hydrogen gas flame. No compressed air is required, as there is sufficient force from the gas flame to carry the atomized metal at high speed to the surface being sprayed.

Powder materials developed for this process provide unusual surface properties, such as a thermal barrier, resistance to oxidation or corrosion, high hardness, and extreme wearing ability. Materials that spray successfully include stainless steel, bronze, tungsten carbide, and various alloys.

Plasma Flame Spraying

A cross section of a plasma flame spray gun is shown in Figure 20.21. A gas (nitrogen, argon, or hydrogen) is passed through an electric arc, causing it to become ionized and

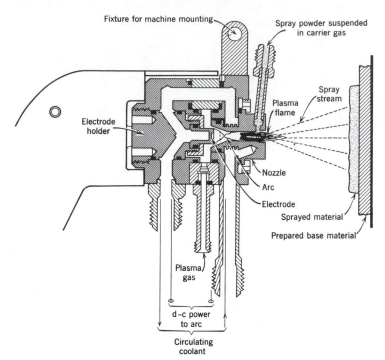

Figure 20.21
Plasma flame spray gun.

raised to temperatures in excess of 30,000°F (17,000°C). Material to be sprayed is introduced in this gas stream, melted, and transported to the object being coated. The high-velocity gas stream as it leaves the nozzle is known as a *plasma jet*, a stream of ionized conducting gas.

The applications of *plasma flame spraying* are similar to those of powder metal spraying. Because of the extremely high temperatures possible, the *plasma torch* is particularly useful in spraying high temperature metals and refractory ceramics. A few of the materials that may be applied include tungsten, zirconium oxide, cobalt, chromium, and aluminum oxide.

The success of metal spraying is largely a result of the surface properties obtainable and of how rapidly the metal is applied. There is no distortion in the parts being surfaced, nor do internal stresses develop. Practically any metal can be applied to any other metal surface and even to other base surfaces, such as wood and glass.

20.6 POWDER METALLURGY

Powder metallurgy is a technology to produce parts from metallic powders by applying pressure and heat. The application of heat, known as *sintering*, is done at temperatures below the melting point of the powder. This bonds the fine particles together, improving the strength and other properties of the finished product.

Important Characteristics of Metal Powders

The two important characteristics of a metal powder part are strength and machinability. Even though powder metallurgy is a process that has its chief economical advantage in minimizing machining, parts generally need some machining operation, such as broaching or tapping.

The particle size, shape, and size distribution of metal powders affect the characteristics and physical properties of the compacted product. Powders are produced according to specifications such as shape, fineness, particle size distribution, flowability, chemical properties, compressibility, apparent specific gravity, and sintering properties.

The *shape* of a powder particle depends largely on how it is produced. It may be spherical, ragged, dendritic, flat, or angular. A powder with good green strength can be produced only from irregularly shaped particles that will interlock on compacting.

Fineness refers to the particle size and is determined by passing the powder through standard sieves or by microscopic measurement. Standard sieves ranging from 100 to 325 mesh (45 to 150 μm) are used for checking size and for determining particle size distribution within a certain range. Because most powders are of irregular shape and not spheres, the size of individual units cannot be stated specifically. Perhaps the most used size is about 100 μm.

Particle *size distribution* has reference to the amount of each standard particle size in the powder. It influences flowability, apparent density, compressibility, final porosity, and mechanical properties, such as strength and elasticity. It cannot be varied appreciably without affecting the size of the compact.

Flowability is that characteristic of a powder that permits it to flow readily and conform to the mold cavity. It can be described as the rate of flow through a fixed orifice. Small and nearly spherical powders flow best. Very fine particles on the order of 1 to 10 μm have characteristics similar to a liquid when they are pressed and tend to "flow" into complex molds. The addition of stearate to the mix also increases flowability.

A specification of *chemical properties* has to do with the purity of the powder, amount of oxides permitted, and the percentages of other elements allowed. Alloys can be made in the crucible prior to manufacture of the powder, and various additions can be made to change the characteristic of the final product.

Compressibility is the ratio of the volume of initial powder to the volume of the compressed piece. It varies considerably and is affected by particle size distribution and shape. The green strength of a compact depends on compressibility. There is no formula for compressibility, but a general "rule of thumb" is that most fine powders compress to about two-thirds of their original filling depth. Every mold, depending on the powder, powder shape and size, lubrication, and compacting pressure, must be specifically designed.

The *apparent density* of a powder is expressed in lb/in.^{3} (kg/m^3). It should be kept constant so that the same amount of powder can be fed into the die each time.

Flowchart

The flowchart in Figure 20.22 details the steps to manufacture metal powder parts. The chart shows the sequence from raw material to finished parts. The same sequence is followed in this section to explain powder metallurgy. Be aware that tolerances shown in the flowchart are only approximate and apply to small uniform parts.

A *green compact* is unsintered and has little strength or resistance to abrasion. Products made by powder metallurgy are frequently mixed with different metal powders or contain nonmetallic constituents to improve the bonding qualities of the particles and properties of the final product. For example, cobalt or some other metal is necessary in bonding tungsten carbide particles, whereas graphite is added with bearing metal powders to improve the lubricating qualities of the finished bearing.

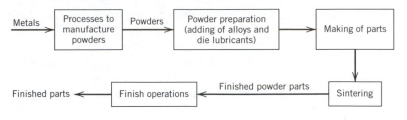

Dia. ±0.5 mil (±0.013 mm)

Length: ±1 mil (±0.03 mm)

Figure 20.22
Flowchart for fabricating metal powder parts.

Processes to Manufacture Powders

Although all metals can be produced in the powder form, only a few are widely applied in manufacturing pressed-metal parts. Some lack the desired characteristics or properties described previously, which are necessary for economical production. The two principal kinds are iron and copper base powders. Both lend themselves to powder metallurgy. Whereas bronze is used in porous bearings, brass and iron are found more often with small machine parts. Other powders of nickel, silver tungsten, and aluminum have a limited but important application in the field of powder metallurgy.

All metal powders, because of their individual physical and chemical characteristics, cannot be manufactured in the same way. The procedures vary widely, as do the sizes and structures of the particle obtained from the various processes. Machining results in coarse particles and is used principally for producing magnesium powders. Milling and grinding processes, utilizing various types of crushers, rotary mills, stamping mills, and grinders, break down the metals by crushing and impact. Brittle materials may be reduced to irregular shapes of almost any fineness by this method. The process is also used in pigment manufacturing of ductile materials where flake particles are obtained. An oil is used in the process to keep them from sticking together. Atomization or the operation of metal spraying is an excellent means of producing powders from many of the low-temperature metals, such as lead, aluminum, zinc, and tin. Iron powders with high-carbon content have been produced by this process, but the green strength is lower than for other methods. Iron powders also are produced by the following steps: (1) melt in a crucible; (2) atomize using air or water to break up the particles; and (3) grind to about 100 mesh, anneal, grind to 100 mesh a second time. The particles are irregular in shape and are produced in many sizes. A few metals can be converted into small particles by rapidly stirring the metal while it is cooling. This process, known as *granulation*, depends on the formation of oxides on the individual particles during the stirring operation.

Electrolytic deposition is a common means for processing silver, tantalum, tungsten, molybdenum, nickel, and cobalt. The reduction method reduces metal oxides to powder form by contact with a gas at temperatures below the melting point. For making iron powder, millscale (a form of iron oxide) is fed into a rotating kiln along with crushed coke. Near the discharge end the mixture is heated to around 1900°F (1050°C), causing the carbon to unite with the oxygen in the iron oxide. This forms a gas that is removed through a stack. With the oxygen removed the product remaining is a relatively pure iron having a spongelike structure. The chemical reaction is

$$2Fe_2O_3 + 3C \xrightarrow{heat} 3CO_2 \uparrow^{gas} + 4Fe \qquad (20.1)$$
$$\text{(millscale)} \quad \text{(coke)} \qquad\qquad\qquad \text{(pure iron)}$$

Various other methods involving precipitation, condensation, and other chemical processes exist for producing powdered metals.

Special Powder Preparation

Prealloyed Powders. Alloyed powder products obtained by blending pure metal powders do not provide some of the properties that are possible with prealloyed powders. Blended-powder products are cheaper to produce and require lower pressures to compact. Prealloyed powders, alloyed in the melting process, provide product properties similar to those possible from the melt composition when maximum density is achieved. This permits the production of alloys such as the stainless steels and other highly alloyed compositions that heretofore have not been successfully obtained by blending. Prealloyed metal powder products may give properties such as corrosion resistance, high strength, or resistance to elevated temperature.

Precoated Powders. Metal powders may be coated with an element by passing the powder through a carrier gas. Each particle is uniformly coated, thus producing a powder product that when sintered has certain characteristics of the coating. This permits a cheap bulk powder to be used as a carrier for the outside active materials. Products made from precoated powders that are sintered are more homogeneous than those produced by blending.

Making of Parts

Powder for a given product must be carefully selected to ensure economical production and to obtain the desired properties in the final compact. If only one powder is to be used and the particle size distribution meets specification, additional processing or blending is unnecessary before pressing. Sometimes various sizes of powder particles are mixed together to change the characteristics of flowability or density, but most powder is produced with sufficient particle size variation. Mixing or blending becomes necessary in production when the powders are alloyed or when nonmetallic particles are added. Any mixing or processing of the powder must be done under favorable conditions to prevent oxidation or defects.

Almost all powders have lubricants added in the blending operation to reduce die wall friction and to aid part ejection. Although these lubricants add to the porosity, they increase the production rate and are necessary in presses using automatic powder feed. Lubricants may include stearic acid, lithium stearate, and powdered graphite.

Pressing. Powders are pressed to shape in steel dies under pressures ranging from a few thousand to 200,000 psi (1400 MPa). Because the soft particles can be pressed or keyed together quite readily, powders that are plastic do not require so high a pressure as the harder powders to obtain adequate density. The density and hardness increase with pressure, but in every instance there is an optimum maximum pressure above which little advantage in improved properties can be obtained. Owing to the necessity for strong dies and large-capacity presses, production costs increase with high pressures. The density of the final product varies with compacting pressure, as shown in Figure 20.23.

Many commercial presses developed for other materials are adaptable to powder metallurgy. Although mechanically operated presses generally are used because of their high rate of production, hydraulic presses may be employed if the part is large and high pressures are required. The single-punch press and the high-speed rotary multiple-punch press are designed so that their operation, from filling the cavity with powder to ejection of the finished compact, can be either a continuous or single cycle. Rotary table presses have a high rate of production because they are equipped with a series of die cavities, each provided with top and bottom *punches*. In the course of production, the table indexes and the operations of filling, pressing, and ejecting the product are accomplished at the various stations. A simple punch and die arrangement for compacting metal powder is shown in Figure 20.24. Two punches are involved—an upper punch that conforms to the top shape of the part and a lower punch that conforms to the lower end of the die cavity. The lower punch also acts as an ejector to remove the *briquetted* part from the die. The die cavity must be smooth to reduce friction and must have a slight draft to facilitate removal of the part. Wall friction prevents much of the pressure from being transmitted to the powder, and if pressure is exerted only from one side, there will be a considerable variation in density from top to bottom. This accounts for using both top and bottom punches in most dies. The travel of the punches depends on the compression ratio of the powder, which for iron and copper is roughly 2.5 : 1. The die cavity is filled to a level about three times the height of the finished compact. The ejected part, known as a *green compact*, resembles the finished part but has only the structural strength

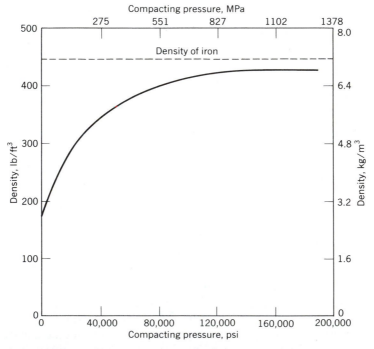

Figure 20.23
Approximate compressibility of iron powder.

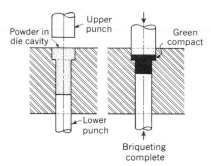

Figure 20.24
Punch and die arrangement for compacting metal powder.

derived from interlocking of the powder particles obtained by compression. Final strength is obtained by sintering. A tooling arrangement for compacting loose bronze powder is shown in Figure 20.25.

Figure 20.26 shows a press setup for compacting small pinions from metal powders. Many similar products are completed by the pressing operation and require no further processing other than sintering. Cold-bond pressures of 10 to 35 tons psi (150 to 500 MPa) are necessary on the part. The sintering operation increases the strength and improves the crystalline structure.

The size of metal powder parts is limited by the capacity of presses used in the compacting operation. Press capacity to 750 tons (7 MN) and more is available. The maximum area of a compact may be calculated readily by the following simple relationship:

$$\text{area } A = \frac{F}{P} \tag{20.2}$$

where

F = Press capacity, lb or N
P = Required compacting pressure, psi or Pa

The density of a powdered-metal product is one of its most distinguishing characteristics. An increased pressure on a compact causes a higher density part and, consequently, an increased tensile strength. The density may be increased by using a powder of smaller particle size.

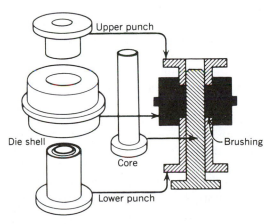

Figure 20.25
Tooling arrangement for briquetting or green compacting bronze powder into a brushing.

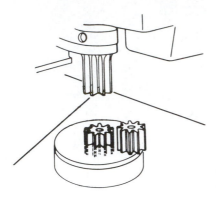

Figure 20.26
Pressing small pinions from metal powder.

Processes used in the making of parts include centrifugal compacting, slip casting, extruding, gravity sintering, rolling, isostatic molding, explosive compacting, and fiber metal processing. (The student is referred to the bibliography to expand on these subjects.)

Sintering

The operation of heating a green compact to an elevated temperature is known as sintering. It is the process by which solid bodies are bonded by atomic forces. With application of heat the particles are pressed into more intimate contact, and the effectiveness of surface tension reactions is increased. In other words, in sintering, the particles are fused together to increase density. During the process grain boundaries are formed, which is the beginning of recrystallization. Plasticity is increased and better mechanical interlocking is produced by building a fluid network. Also, any interfering gas phase is removed by the heat. The temperatures used in sintering are usually well below the melting point of the principal powder constituent, but may vary over a wide range up to a temperature just below the melting point. Tests have proven that there is usually an optimum maximum sintering temperature for a given set of conditions with nothing to be gained by going above this temperature.

For most metals the sintering temperatures can be obtained in commercial furnaces, but for some metals, requiring high temperatures, special furnaces must be constructed. There is considerable range in the sintering temperature, but the following temperatures have proved satisfactory: 2000°F (1095°C) for iron, 2150°F (1180°C) for stainless steel, 1600°F (870°C) for copper, and 2700°F (1480°C) for tungsten carbide. Sintering time ranges from 20 to 40 min for these metals. The time element varies with different metals, but in most the effect of heating is complete in a very short time, and there is no economy in prolonging the operation. Atmosphere is important because the product, made up of small particles, has a large exposed surface area. The problem is to provide a suitable atmosphere of some reducing gas or nitrogen to prevent the formation of undesirable oxide films during the process.

Furnaces for sintering may be either the batch or continuous type. The continuous type, which has a wire mesh belt to carry the compacts through the furnace, is shown in Figure 20.27. Pusher and roller hearth furnaces also are used and are similar in appearance. The dimensional change of growth or shrinkage may occur in sintering. What happens depends on the shape and particle size variation of the powder, the powder composition, sintering procedure, and briquetting pressure. Accurate size is maintained by compensat-

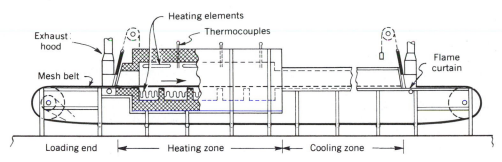

Figure 20.27
Continuous-type furnace for sintering powder metal compacts.

ing for the change in making the green compact and then maintaining uniform conditions. Finer powders sinter better and are particularly important for slip casting.

Most attempts to combine sintering with pressing have not been successful. An exception to this is a process known as *spark sintering*, which is a method of processing that combines pressing and sintering metal powders to a dense metal part in 12 to 15 s. In operation, a high-energy electrical spark removes surface contaminants from the powder particles. This causes the particles to combine as in conventional sintering, forming a solid, cohesive mass. Immediately following the spark, the current continues for about 10 s with the temperature well below the melting point of the material, which furthers crystal bonding between the particles. Finally, by hydraulic pressure the mass is compressed between the electrodes to increase its density.

A schematic sketch illustrating a spark-sintering setup for fabricating a bimetallic (carbide and steel) punching die blank is shown in Figure 20.28. The area labeled "sintered carbide" represents the carbide material sinter-bonded to the steel substrate. The powder is originally filled to the top of the graphite mold before being compressed. The vertical graphite centerpiece merely defines the doughnut shape of the punching die.

Final densification pressure is applied through the graphite electrodes. This process is applied to sintering aluminum, copper, bronze, iron, and stainless steel.

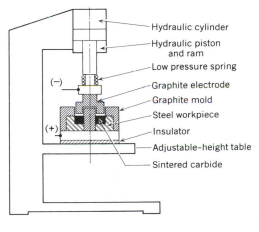

Figure 20.28
Schematic of spark-sintering machine using high-energy spark discharge to stop surface contaminants from powder particles. This is followed by 10- to 12-s spark discharge resistive-heating cycle.

Finishing Operations

Finished parts undergo additional operations, such as tumbling, welding, brazing, drawing, steam treatment, oil impregnation, or sometimes additional sintering.

Bearings made from porous metal represent one of the important products of powder metallurgy. Porosities ranging from 25% to 35% are generally used because higher values result in lower bearing strength. Impregnation is accomplished either by immersing the sintered bearing in heated oil for a period of time or by a vacuum treatment, which is much quicker. Such bearings retain oil by capillary force, the oil being released as the bearing is used. Most bearings are of porous bronze or iron composition.

Infiltration is the process of filling the pores of a sintered product with molten metal to decrease porosity or to improve physical properties. In this operation the melting point of the liquid metal must be considerably lower than the solid metal. Prior to the operation a chemical treatment is advisable to increase the extent of infiltration. Liquid metal is infiltrated into the part either by allowing it to enter from above or by absorbing it from below. As an example, copper placed on a piece of presintered iron and heated to 2100°F (1150°C) is drawn into the iron by capillarity.

Sizing and coining operations to close tolerances are regularly performed and may necessitate a final operation, such as repressing the part in a die similar to the one used for compacting it. Such sizing is a cold-working operation that improves surface hardness and smoothness as well as dimensional accuracy. The density of the part is also increased.

All pressed metal parts may be *heat-treated* by conventional methods, although the results do not always conform to those obtained in solid metals. Best results are obtained with dense structures. Porosity influences the rate of heat flow through the part and permits internal contamination if salt bath heat-treating pots are used in the process. For this reason, liquid carburizing is not recommended for surface treatment of metal powder parts.

High-density parts are *plated* by standard procedures, but medium- or low-density parts require some prior treatment to close the pores. Preparations such as peening, burnishing, or plastic resin impregnation will close surface pores and eliminate salt entrapment that causes the plate to blister. Standard plating procedures are then followed.

One of the characteristics of powder metal products is that they may be die-pressed into dimensionally accurate, finished shapes. However, products requiring such features as threads, grooves, undercuts, or side holes cannot be produced by powder metallurgy methods and must be finish machined. Tungsten carbide tools are recommended, although high-speed steel tools are used only when a few parts are to be machined. The use of coolants with water is not advisable for iron base parts, as corrosion may occur.

Metal Powder Products

Metal powder is increasingly being used to make machine parts (Figure 20.29). Most of these products are completed without machining. Following are some of the prominent powder metal products.

Permanent metal powder filters have greater strength and shock resistance than ceramic filters. Fiber metal filters, with a porosity up to 97%, are used for filtering hot or cold fluids and air. As illustrated in Figure 20.30, one use is in dehydrators for diffusing moisture-laden air around a drying agent such as silica gel. Another common use in gasoline tanks is for separating moisture and dirt from the fuel system. Metallic filters also are used for flame arresting and sound deadening. Filters used for removing foreign matter are cleaned by reversing the flow of the liquid.

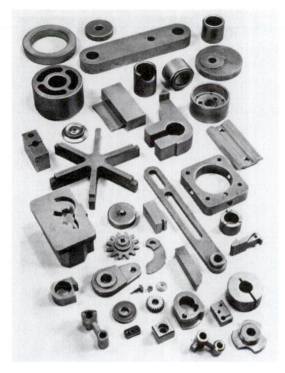

Figure 20.29
Variety of machine parts made from metal powders.

Tungsten carbide particles are mixed with a cobalt binder, pressed to shape, and then sintered at a temperature above the melting point of the matrix metal to produce cemented carbides that are used for cutting tools and dies.

Gears and pump rotors are made from powdered iron mixed with sufficient graphite to give the product the desired carbon content. A porosity of around 20% is obtained in the process. After the sintering operation, the pores are impregnated with oil to promote quiet operation.

Brushes for motors are made by mixing copper with graphite to give the compact adequate mechanical strength. Tin or lead may be added also in small quantities to improve wear resistance.

Most bearings are made from copper, tin, and graphite powders, although other metal combinations are used. After sintering the bearings are sized and then impregnated with oil by a vacuum treatment. Porosity in the bearings is controlled readily and may run as high as 40% of the volume.

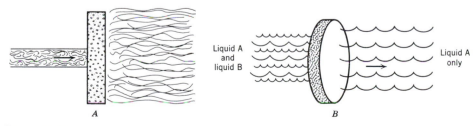

Figure 20.30
Special uses of permanent metal filters. *A*, Diffusing. *B*, Separating.

Excellent small magnets are produced from several compositions of iron, aluminum, nickel, and cobalt when combined in powder form. Alnico magnets, made principally from iron and aluminum powders, are superior to those cast.

Electric contact parts are adapted to powder metallurgy fabrication, because it is possible to combine several metal powders and maintain some of the principal characteristics of each. Contact parts must be wear resistant, somewhat refractory, and have good electrical conductivity. Many combinations such as tungsten–copper, tungsten–cobalt, tungsten–silver, silver–molybdenum, and copper–nickel–tungsten have been developed for electrical applications.

Numerous other parts, including clutch faces, brake drums, ball retainers, and welding rods, are produced by powder metallurgy. Other uses for powder metals include paint pigments, missile fuels, and Thermit welding. The addition of powdered metals to plastics increases strength and contributes metallic properties.

Advantages and Disadvantages

Metal in powder form is higher in cost than in solid form, and expensive dies and machines are required to adapt the process to mass production. This higher cost is often justified by the unusual properties obtained. Some products cannot be made by any other process; others may compete favorably with their counterparts made by other methods, because the close tolerances obtained eliminate the need of any further processing.

Some of the advantages of powder metallurgy include: no material losses reduce fabrication cost; skilled machinists are not required, so labor cost is low; parts with controlled porosity can be produced; powders that are available in a pure state produce items of extreme purity; and large-scale production is possible of many small parts that compete favorably with machined parts because of close tolerance and surface finish. In addition, sintered carbides and porous bearings can be produced only by this process, and bimetallic products can be formed from mold layers of different metal powders.

However, powder metallurgy has the following limitations: uniformly high-density products are difficult to fabricate; equipment costs are high; metal powders are expensive and sometimes difficult to store without some deterioration; intricate designs in products are difficult to attain because there is little flow of metal particles during compacting; and some powders (i.e., aluminum, magnesium, zirconium, titanium) in a finely divided state present explosion and fire hazards. In addition, some products can be made more economically by other methods, because the size of powder-fabricated parts is controlled by the capacity of the presses available and by the compression ratio of the various powders. Furthermore, some thermal difficulties appear in sintering operations, particularly with low-melting powders such as tin, zinc, and cadmium. Most oxides of these metals cannot be reduced at temperatures below the melting point; hence, if the oxides exist they will have detrimental effects on the sintering process and result in an inferior product.

QUESTIONS

1. Give an explanation of the following terms:

Ultrasonic machining	Electroforming
Abrasive jet machining	Electrolytic deposition
Water jet machining	Electrolytic solution
Electroplating	Metallized
Electrochemical	Thermospray
machining	Plasma flame spraying
Anode	Sintering
Thermoelectric process	Compressibility
Gas-assisted laser	Granulation
Plasma arc	Green compact
Cold temperature	Punches
machining	Briquetted
Chem milling	Infiltration
Etch factor	Sizing
Chem blanking	Heat-treated

2. What are the disadvantages of abrasive jet machining?

3. Prepare a brief description of the electroforming process.

4. Name at least six products that are made by electroforming.

5. What is a laser? Name three applications.

6. What are the advantages of the electroforming process?

7. What is plasma? If used in a cutting application, what metal removal rates are possible?

8. Describe the spark and where it occurs in electrical discharge machining.

9. In electrical discharge machining, why must the workpiece be an electrical conductor? Would the process work for machining ceramics?

10. How are hard surfacing materials such as carbides applied to oil well drilling tools?

11. What makes molten metal from a spray gun adhere to a revolving shaft?

12. What is the safety problem connected with electron beam machining?

13. Describe the ultrasonic machining process and state its advantages.

14. Describe in some detail how electroforming differs from the plating process.

15. What is the difference between electrochemical machining and electrochemical grinding? Is metal removed in the same way in both cases?

16. Explain how material is removed in the electrochemical machining process.

17. What is the purpose of a chemical resistant?

18. Describe the chem milling process.

19. How does chemical blanking differ from chemical milling?

20. Can electroformed parts be plated? Why?

21. How does metal powder spraying differ from the method of applying sintered carbides?

22. Describe the procedure for chemically blanking a part.

23. Describe the process in which hollow steel tubes are filled with hard surfacing material and then used as "welding rods."

24. How is the higher cost of powder metal products justified?

25. How would the following metal powder parts be made: porous sheet metal; long, uniformly shaped pieces; beryllium part; cemented carbides; and alnico magnet?

26. How is sintering accomplished? Why not hot-press all products and eliminate the sintering operation?

27. How and by what method are ferrous alloy powders made?

28. What kind of press is used in slip casting?

29. What is the principal disadvantage of slip casting?

30. What are the purposes of lubricants in powder metallurgy? What are their disadvantages?

31. In the manufacture of metal powder bearings, how may the porosity of the metal be controlled to a desired percentage? What advantage do these bearings have over cast bearings?

32. What is the advantage of irregular-shaped particles in powders?

33. What are the principal advantages of fine powders as compared to coarse powders?

34. What are the two most important characteristics of a powdered-metal part?

35. What accuracy can be expected on a part that has been coined or sized in a press?

36. Why is sintering usually accomplished in a controlled atmosphere?

37. Name three metal powder products that cannot be made by other processes.

38. How are bearings impregnated with oil?

39. Aluminum paint is often manufactured by mixing alu-minum powder in a lacquer base. How would you recommend making such powder?

40. What is infiltration and how is it done?

PROBLEMS

20.1. In an electrochemical grinding process, 0.040 in. (1.02 mm) is to be removed. Calculate the thickness removed by abrasive action. How long would it take to remove the 0.040 in. (1.02 mm) over a 1 in.^{2} (625 mm^2) area if the current is 200 A?

20.2. What is the approximate density in lb/in.^{3} of a part that is molded from a powder having a density of 100 lb/ft^3? Also calculate in SI metric units.

20.3. How would you define the compressibility of a powder that occupied two-thirds of the space after compaction, as compared to the space it occupies as a free-standing powder?

20.4. If the compacting pressure for stainless steel is 200,000 psi (1333 MPa), what is the maximum-diameter part that can be made in a 300-ton (3-MN) press?

MORE DIFFICULT PROBLEMS

20.5. A powdered-metal iron cylinder 1 in. (25.4 mm) in diameter and 1 in. (25.4 mm) long is to be compacted at 50,000 lb/in.^{2}. What is the approximate density of the part? What size press in tons is needed for the part? Calculate the approximate volume of raw powder needed for the part. Calculate the approximate length of mold that is necessary for the part. (Hint: Most fine powders compress to about two-thirds of their original filling depth. Use Figure 20.23.)

20.6. It is found experimentally that to make a 2-in. (50-mm) cube from powdered iron a press pressure of at least 80,000 psi (551 MPa) is needed on each surface. The vertical press force P is related to the formula $P = 0.6A$, where A

is the wall force. What is the force in lb (N) that will be exerted by the press on the powder? Where will the pressed powder be more dense, on the top surface or on the sides? (Hint: Most fine powders compress to about two-thirds of their original filling depth.)

20.7. Additional equipment is needed to make 10,000 sintered bearings of a 90% brass and 10% graphite mixture every month. The bearings are 1 in. long with a 0.50-in. OD and a 0.250 ± 0.002-in. ID. Densities are 535 lb/ft^3 and 79 lb/ft^3 for brass and graphite. The brass powders cost $3.13/lb and selling price is four times the material cost. If it is necessary to pay for the additional equipment in three years, how much can be invested in new equipment?

PRACTICAL APPLICATION

A gear for a small transmission system currently is produced by forging and then machined. Since the stresses during actual service are relatively low, the manufacturer is considering producing gears by powder metallurgy to facilitate higher production rates. Evaluate the idea, outline the operations needed to produce the gears, and write a report with your recommendations. Call or e-mail professionals that work with powder metals to gather supporting information. Create an association with the professional, and see if he or she will "mentor" your study.

CASE STUDY: ELECTROCHEMICAL MACHINING

"Well, I assume all of you have read the sales order to increase the titanium compressor parts from two per day to eight per day."

Charles Kaghn, president of Acme Tool and Die, a medium-size machine shop, was addressing Bill Wainwright, his production manager, and Lynne Lylle, vice president for manufacturing.

Bill Wainwright spoke up first. "I just don't see any way we can fill the orders. This part is a really tough machining job. It's about all we can do to produce the two per day we are making now."

Charles replied, "I know we'll need additional production capacity, so I asked Ms. Lylle to look into the possibility of using electrochemical machining for this larger order. Lynne, what have you found out?"

"A 10,000-A, 12-V electrochemical machine can produce this part in about 45 min. Allowing another 15 min for replacing the finished part with a new blank, we can easily make one per hour or eight per day. The basic machine costs $35,000. We would also need an electrolyte system for about $65,000, and a power supply unit for about $40,000. Installation costs are about $10,000. We need to add 500 ft (150 m) of water line at $5/ft ($16.40/m); 400 ft (120 m) of 440-V, 550-A three-phase power line at $25/ft ($82/m); and 150 ft (45 m) of drainage line at $75/ft ($245/m). Of course, before we can produce anything, we would have to make an electrochemical machining form of the part, which will cost an additional $10,000."

The president continued, "These are high initial costs, but once we get into production, we should only have to spend about $25/hr for labor, repair, and overhead and about another $5 per part for electric power. Bill, how does this compare with the current system?"

"Well, we spend about $98 per part for labor, repair, and overhead right now. I don't think any other material or administrative costs would change if we went to the electrochemical machining process. What do you think, Lynne?"

"I think that's a reasonable assumption. Is there any other information you need, Charles?"

"No, I think that's everything. Thank you both for your help. I'll get back to you if I need more input."

What is the total initial investment for the ECM machine? What is the cost savings per part between the current process and ECM? Given this cost savings, how many parts will it take to recover the initial investment? Assuming 300 work days per year, how many years will it take to recover the investment?

CHAPTER 21

THREAD AND
GEAR WORKING

Screw and gear technology are common applications in manufacturing systems. Their popularity is seen in many consumer products. The screw thread is an ages old invention whose origin is unknown. The threads of a screw are essentially an inclined plane, which is wrapped in a regular helical manner about the length of the screw.

Toothed gearing is one of the most widely used mechanisms of the modern world. Many of our daily activities depend on the rotation of shafts connected by toothed gears.

This chapter deals with the background for threads and gears. Very important, it discusses their manufacture.

21.1 SCREW THREADS

A *screw thread* is a ridge of uniform section in the form of a helix on the surface of a cylinder. The *terminology* relating to screw threads is shown in Figure 21.1. The sequence designating a screw thread is the *nominal size* (fractional diameter or screw number), *number of threads per inch*, *thread series symbol*, and *thread class*. The nominal size is the basic major diameter.

The Unified (UN) system of screw threads has a designation system, as does the *ISO or metric standard*. The tolerance class designation in the ISO system refers to the quality of thread. The first digit and letter determine the pitch diameter, a lowercase "g" being for external threads and an uppercase "G" for internal threads. Grade 6 is medium quality and Grade 5 is slightly less. The second series, 6g, refers to the quality of the thread crest; Grade 6 is again medium and Grade 9 is the highest. The letters may vary from "e" to "h" or "E" to "H," where "e" or "E" is for a large allowance, and "h" or "H" is for a thread with little or no allowance.

If left-hand threads are designated, the term "LF" appears at the end of the thread specification.

Pitch is expressed by a fraction, with 1 as the numerator and the number of threads per inch as the denominator. A screw having 16 single threads per inch has a pitch of ¹⁄₁₆. It should be remembered that only on single-threaded screws does pitch equal the lead. In SI, the pitch is the distance between corresponding points on adjacent profiles.

Lead is the length a screw advances axially in one revolution. On a double-threaded screw the lead is twice the pitch and on a triple-threaded screw the lead is three times the pitch.

Screw threads are used as fasteners and to transmit power, as illustrated in the screw jack. Threads also transmit motion when used in a lead screw on a lathe. Screw threads are employed in measuring devices such as micrometers. These definitions are shown by Figure 21.2.

Screw threads are standardized according to their cross-sectional form. The principal threads used worldwide are classified into two systems.

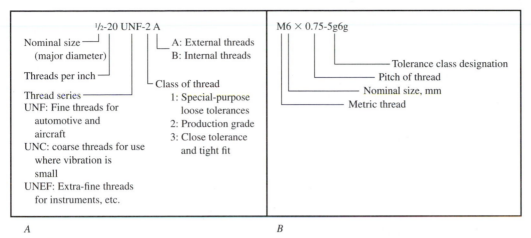

Figure 21.1
Symbols for threads. *A*, Unified system. *B*, ISO system.

1. *Unified screw standards.* With slight modification, this threading system was formerly known as the American National Standard series. These threads are sometimes called inch threads as distinct from metric threads.

2. *ISO standards.* The ISO system was promulgated by the International Organization for Standards and covers metric threads. See Figure 21.3 for a comparison between the Unified Series and the ISO standard. They are not interchangeable.

In the UN system there are a number of classifications, the principal ones being the fine, coarse, and extra-fine series. There are also eight other series of UN threads with constant pitches of 4, 6, 8, 12, 16, 20, 28, and 32 threads per inch. Whenever possible, selection should be made from the standard series (UN screw threads), with preference being given to the coarse- and fine-thread series. The coarse-thread series is generally used for the bulk production of screws, bolts, and nuts. The coarse-thread series (UNC) provide more resistance to internal thread stripping than the fine or extra-fine series. The fine-thread series (UNF) is used because of its high strength and in applications where vibrations occur. This series has less thread depth and a larger root diameter than the coarse-thread series. To prevent internal thread stripping a longer length of

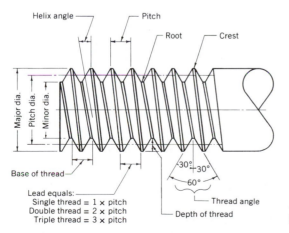

Figure 21.2
Screw thread terminology.

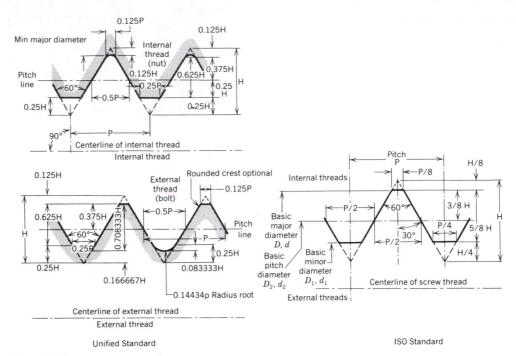

Figure 21.3
Profile comparison of Unified and ISO metric threads.

engagement is required for the fine series. The extra-fine series (UNEF) is used for equipment and threaded parts that require fine adjustment.

The ISO metric screw thread shown in Figure 21.3 has essentially the same basic profile as the UN screw thread basic form. Note the external and internal form is the same for ISO threads except when rounded roots are called for on external threads. Tables of inch-to-metric conversions, such as given in Table 21.1, should be used for comparative reference only. The biggest difference between the UN and ISO series is the number of threads per unit length. In Table 21.1, several major diameters of the UN

Table 21.1   **A Comparative Chart of UN Thread Series to ISO Metric Series**[a]

UNC Threads	UNF Threads	ISO Metric Threads
2–56 (2.18)[b]	2–64 (2.18)	M2 × 0.4 (2.00)
3–48 (2.51)	3–56 (2.51)	M2.5 × 0.45 (2.50)
4–40 (2.84)	4–48 (2.84)	M3 × 0.5 (3.00)
5–40 (3.18)	5–44 (3.18)	
6–32 (3.51)	6–40 (3.51)	
8–32 (4.17)	8–36 (4.17)	M4 × 0.7 (4.00)
10–24 (4.83)	10–32 (4.83)	M5 × 0.8 (5.00)
¼–20 (6.35)	¼–28 (6.35)	M6 × 1 (6.00)

[a] Threads are not mechanically interchangeable.

[b] Numbers in parentheses indicate major diameter in millimeters.

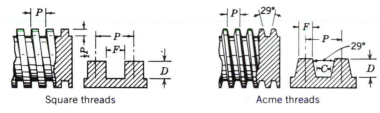

Figure 21.4
Screw forms used for transmitting power.

thread size are compared to the major diameter of the closest equivalent ISO metric thread size for sizes up to ¼ in. (6.35 mm) in diameter.

The basic profile for ISO and UN, as shown by Figure 21.3 is essentially the same. The principal differences are related to basic size, the magnitude and application of allowances and tolerances, and thread designations. The crest width of the nut is ¼ pitch, whereas that of the screw is ⅛ pitch. The increased flat on the nut makes production easier and at the same time the nut is as serviceable. The shape of the thread crest and root is not mandatory; it can be either flat or rounded.

Square threads, shown in Figure 21.4, are suitable for transmitting power when there is a large thrust on one side of the thread. These threads cannot be cut with taps and dies and must be machined on a lathe. Another type, similar to the square thread, is known as a buttress thread. It has one side that slopes 45°; the other is perpendicular. Although this thread does transmit power, the thrust is only in one direction. Acme screw threads (Figure 21.4) have the advantage that wear may be compensated for by adjusting half-nuts in contact with the screw. They can be cut with taps and dies. Worm threads, similar to the Acme standard except that they have a greater depth, are used exclusively for worm gear drives. The Acme thread is often seen on lathes, right below the feed rod.

21.2 PROCESSES FOR MAKING THREADS

External threads are produced by the following manufacturing processes:

1. Lathe single-point turning
2. Die and stock
3. Automatic die head
4. Automatic collapsible tap
5. Threading machine
6. Die casting
7. Grinding

Taps and Dies

Taps are used for the production of internal threads. Figure 21.5 illustrates a tap with the parts of the tool labeled. The tool itself is a hardened piece of carbon or alloy steel resembling a bolt with flutes cut along the side to provide the cutting edges. For *hand tapping* these are furnished in sets of three for each size. In starting the thread, the taper tap should be used because it ensures straighter starting and more gradual cutting action

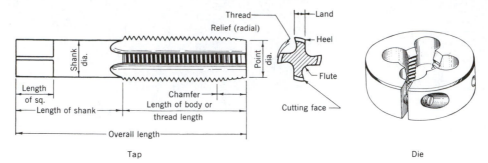

Figure 21.5
Tap and die.

on the threads. If it is a through hole, no other tap is needed. For closed or blind holes with threads to the very bottom, the taper, plug, and bottoming taps should all be used in the order named. Other taps are available and are named according to the kind of thread they cut.

Where a hole is to be tapped, the hole that is drilled before the tapping operation must be of such a size as to provide the necessary metal for the threads. Such a hole is a *tap size hole*. For many years a 75% thread has been recommended, and most published tables show *tap drill sizes* that will yield a 75% thread. Tests have shown that a greatly reduced percentage of thread (a larger tap size hole) will give adequate strength to the fastening and greatly reduce the high torque required for tapping perhaps causing fewer taps to break in the hole.

$$t = 0.47d \tag{21.1}$$

where

$$t = \text{Nut thickness, in. (mm)}$$
$$d = \text{Nominal bolt diameter, in. (mm)}$$

Because nuts are usually made of slightly softer metal, $t = \frac{7}{8}d$ is usually employed.

The most common method of cutting external threads is by the *adjustable die* shown in Figure 21.5. It can be made to cut either slightly undersize or oversize. When used for hand threading, the die is held in a die stock.

For successful operation of either taps or dies, consideration is given to the material to be threaded. No tool can work successfully for all materials. The shape and angle of the cutting face also influence the performance. Another important factor is proper lubrication of the tool during the cutting operation; this ensures longer life of the cutting edges and results in smoother threads.

Taps and dies also can be used in machine cutting of threads. Because of the nature of the cutting operation, they are held in a special holder, so designed that the tap or die can be withdrawn from the work without injury to the threads. This is frequently accomplished by reversing the *rotation* of the tool or work after the cut has been made. In some equipment, this reversing action is faster than the cutting.

In small-production work on a computer numerically controlled turret lathe, the tap is held by a special holder that prevents the tap from turning as the threads are cut. Near the end of the cut the turret holding the tool is stopped, and the tap holder continues to advance until it pulls away from a stop pin a sufficient distance to allow the tap to rotate with the work. The rotation of the work is then reversed and, when the tap holder is withdrawn, it is again engaged with the stop and held until the work is rotated from the tap. External threads can be cut with a die utilizing the same procedure, although most of the time such threads are cut with self-operating dies.

Thread Chasing

In production work, *self-opening dies* and *collapsible taps* are used to eliminate back-tracking of the tool and to save time. The tools have individual cutter dies known as *chasers* that are mounted in an appropriate holder and are adaptable to adjustment or replacement. With chasers more accurate work results, the cutters are kept in proper adjustment, and there is no danger of damaging the cut thread as the tool is withdrawn. In some instances, the tool is held stationary and the work revolves; in others, the reverse procedure is used. All precision screws require a *lead screw* feed to obtain accuracy.

Two types of automatic die heads are used. In one, the cutters or chasers are mounted tangential, as shown in Figure 21.6. In the other, they are in a radial position. Radial cutters can be changed quickly; consequently, they are used for threading materials that are hard to cut. The die head commonly used on most turret lathes is of the stationary type. The work rotates and the chasers open automatically at the end of the cut so that they can be withdrawn from the work without damage. In threading machines, the dies rotate and the work is fed to them but otherwise the operation is the same.

Tapping Machines

Although tapping is done on drill presses equipped with some form of tapping attachment, most production tapping is done on specially constructed automatic machines. Nuts to be threaded are fed from an oscillating hopper to the working position, spindles are reversed at double the tapping speed, and nuts are discharged to containers.

A common type of tapping machine has a multispindle arrangement provided with taps having extra-long shanks. The tap is advanced through the nut by the lead screw and, on completion of the threading, continues downward until the nut is released. The spindle then returns to its upper position with the tapped nut on its shank. When the shank has been filled with nuts, the tap is removed and the nuts are emptied.

Figure 21.6
Revolving tangent die head.

Thread Milling

Accurate threads of large size, both external and internal, can be cut with standard *hob-type cutters*. For long external threads, a threading machine similar in appearance to a lathe is used. Work is mounted either in a chuck or between centers, the milling attachment being at the rear of the machine. In cutting a long screw a single cutter is mounted in the plane of the thread angle and fed parallel to the axis of the threaded part, as shown in Figure 21.7. The feed (*f*) in thread milling is expressed as the cutter advance per tooth, or inches per cutter tooth, by the following formula:

$$f = \frac{ds}{nN} \tag{21.2}$$

where

d = Nominal diameter of thread, in.
s = rpm of work
N = rpm of cutter
n = Number of teeth in cutter

From this expression it is evident that the cutter load per tooth, which varies directly with the feed, can be changed by varying the cutter speed, work speed, or number of teeth in the cutter. This permits reducing the load on the cutter teeth so that deep threads can be cut in one pass.

For short external threads, a series of single-thread cutters is placed side by side and made up as one cutter, having a width slightly more than that of the thread to be cut. The cutter is fed radially into the work to the proper depth and, while rotating a little over one revolution, completes the milling of the thread. Proper lead is obtained by a feed mechanism that moves the cutter axially while it is cutting.

Milling machines of the planetary type are also used for mass production of short internal or external threads. The milling head carrying the *hob* is revolved eccentrically about the rigidly held work, which is rotated simultaneously on its own axis. It is advanced by means of a lead screw for a sufficient distance to produce the thread.

Thread Rolling

More than 90% of bolts up to 2 in. (51 mm) in diameter have *rolled threads*. Threads can be rolled using any material that has sufficient ductility to withstand the forces of cold working without disintegration. An elongation of 12% or more in a 2-in. (51 mm) gage length is considered to be a good index of rollability. *Rollability* is the behavior of a metal during the rolling process. Because of the behavior of various types of metal during the formation of threads, rollability cannot be determined accurately by any single physical property of the metal. Certain factors that influence a material's resistance to

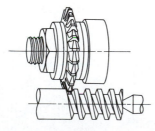

Figure 21.7
Single-thread milling cutter.

plastic deformation are material hardness or lack of ductility, internal friction developed during plastic deformation, yield point of the material that must be exceeded in the process, and tendency of the material to work harden. Steels that work harden rapidly require greater pressures and the die life is reduced. However, work-hardened threads may eliminate subsequent hardening and grinding operations. Materials that have good rollability include the basic open-hearth steels, sulfurized steels up to 0.13% sulfur, 1300 series manganese steels, nickel steel, and many of the nonferrous alloys.

In *thread rolling*, the metal on the cylindrical blank is cold forged under considerable pressure by the rolling action of the blank between either rotating cylindrical dies or reciprocating flat dies. The surface of the dies has the reverse form of the thread that is rolled. Rolling under pressure results in a plastic flow of the metal. The die penetrates to form the root of the thread and the displaced metal flows to form the crest. Less material is required for rolled threads over cut threads, as illustrated in Figure 21.8. This saving ranges from 16% to 25%. For example, the savings for a ½-in., no. 13 thread screw size is 19%. Beginning stock diameter is approximately equal to the pitch diameter indicated for several UNC threads.

Two methods are employed in rolling threads. In one the bolt is rolled between two flat dies, each being provided with parallel grooves cut to the size and shape of the thread. One die is held stationary while the other reciprocates and rolls the blank between the two dies. Figure 21.9 illustrates the method of rolling a screw between two soft boards under pressure. Each board has impressed into its surface a series of angular, parallel lines. By reversing this illustration and starting with similar grooves in hardened steel, threads are rolled into a rod placed between them.

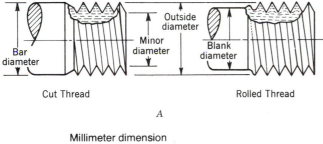

A

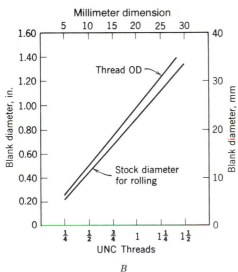

B

Figure 21.8

A, Illustrating stock material saving of rolled threads over cut threads. *B*, Blank diameter for rolled and cut threads.

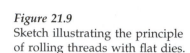

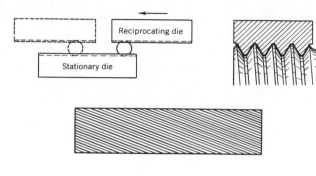

Figure 21.9
Sketch illustrating the principle
of rolling threads with flat dies.

The other method, shown in Figure 21.10, employs either two- or three-grooved roller dies. In the two-die method the blank is placed on the work rest between two parallel, cylindrical rotating dies, and the right-hand die is fed into the blank until the correct size is reached, returning then to its starting position. The three-die machine utilizes cylindrical rotating dies mounted on parallel shafts driven synchronously at the desired speed. They advance radially into the blank by cam action, dwell for a short interval, and then withdraw.

Equipment also is available for producing large forms and work threads, as shown in Figure 21.11. The 1½-in. pitch worm gears are produced at the rate of 20 per h. This machine, designed for forming operations requiring rolling forces of 20,000 to 50,000 lb (0.2 to 0.9 MN), is equipped with an air-operated workholding fixture. The fixture automatically retracts from the work area to unload the finished workpiece and insert a new blank. Because of the extreme depth of the teeth, the workpiece is passed back and forth between the rolls until the required penetration is obtained. Other recent developments in the field of cold rolling include the production of small gears and splines.

Following are some of the advantages of the thread rolling process.

1. Improved tensile, shear, and fatigue strength properties.
2. Finer surface finish of 4 to 32 μin. (100 to 800 nm).
3. Close accuracy.
4. Less material.
5. Cheaper materials usable because of improvement of physical properties during rolling.
6. High production rate.
7. Wide variety of thread forms.

Limitations of the thread rolling process include the following:

1. Necessary to hold close blank tolerance.
2. Uneconomical for low quantities.

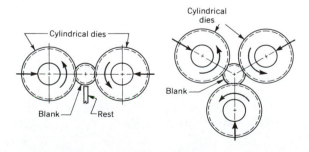

Figure 21.10
Thread rolling using either two or
three cylindrical dies.

Figure 21.11
Thread rolling and forming machine.

3. Can roll only external threads.
4. Cannot roll material having a hardness exceeding Rockwell C37.

Thread Grinding

Grinding is used as either a finishing or a forming operation on many screw threads where accuracy and smooth finish are required. This process is particularly applicable for hardened threads.

Two types of wheels used in thread grinding are illustrated in Figure 21.12. Shown in Figure 21.12*A* is a single wheel shaped to correct form that traverses the length of the screw. The wheel is rotated against the work, usually at speeds ranging from 750 to 10,000 ft/min (4 to 50 m/s), and at the same time traverses the length of the screw at a velocity determined by the pitch of the thread. With the feed ranging from 1½ to 10 ft/min (0.008 to 0.05 m/s), the surface speed of the work is determined by the depth of grind and material.

Short threads may be grounded by the *plunge cut* method, as shown in Figure 21.12*B*. The wheel is fed to the full thread depth before the workpiece is rotated. It then makes one revolution while traversing a distance equal to one pitch, thus completing the thread.

Most precision external threads are ground after heat treatment to eliminate possible distortion. Threads of 10 pitch or finer can be ground directly from solid hardened blanks. Coarse threads are previously roughed.

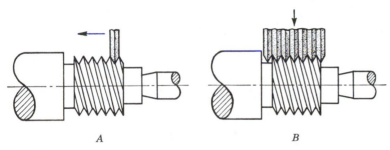

A   *B*

Figure 21.12
Methods used in thread grinding. *A*, Single-wheel traverse grinding. *B*, Plunge cut grinding.

21.3 GEARS

Gears transmit power and motion between moving parts. Positive transmission of power is accomplished by projections or teeth on the circumference of the gear. There is no slippage as with friction and belt drives, a feature most machinery requires, because exact speed ratios are essential. Friction drives are used in industry, where high speeds and light loads are required and where loads subjected to impact are transmitted.

When the teeth are built up on the circumference of two rolling disks in contact, recesses must be provided between the teeth to eliminate interference. The circumference on which the teeth are developed is known as the *base circle*. The pitch circle, however, is an imaginary circle with the same diameter as a disk that would cause the same relative motion as the gear. One of the most important gear design calculations is based on the diameter of the pitch circle. A portion of a gear is shown in Figure 21.13.

Gear Nomenclature

The system of gearing used today is known as the involute system, because the profile of a gear tooth is principally an involute curve. An *involute* is a curve generated on a circle, the normals of which are all tangent to this circle. The method of generating an involute is shown in Figure 21.14. Assume that a string having a pencil on its end is wrapped around a cylinder. The curve described by the pencil as the string is unwound is an involute, and the cylinder on which it is wound is known as the base circle. The portion of the gear tooth from the base circle at point *a* in the figure to the outside diameter at point *c* is an involute curve and is the portion of the tooth that contacts other teeth. From point *b* to point *a*, the profile of the tooth is a radial line down to the small fillet at the root diameter. The location of the base circle on which the involute is described is inside the pitch circle and is dependent on the angle of thrust of the gear teeth. The relationship existing between the diameter of the pitch circle D and base circle, D_b, is

$$D_b = D \cos \phi \tag{21.3}$$

where

$$\phi = \text{Pressure angle}$$

The two common standard systems have their pressure angles or lines of action at $14\frac{1}{2}°$ and $20°$. Other angles are possible, but with larger angles the radial force component

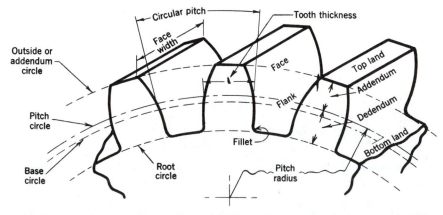

Figure 21.13
Nomenclature for involute spur gear.

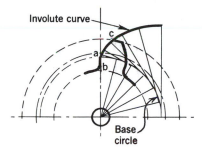

Figure 21.14
Method of generating an involute tooth surface.

tending to force the gears apart becomes greater. If a common tangent is drawn to the pitch circles of two meshing gears, the line of action or angle of thrust is drawn at the proper angles (14½°) to this line. The base circles on which the involutes are drawn are tangent to the line of action.

Standard gears transmitting power use the 20°, full-depth, involute tooth form. These gears have the same tooth proportion as the 14½° full-depth involute but are stronger at their base because of greater thickness. The 20°, fine-pitch involute gears are similar to the regular 20° involute and are made in sizes ranging from 20 to 200 diametral pitch. These gears are used primarily for transmitting motion rather than power. The 20° stub tooth gear has a smaller tooth depth than the 20° full-depth gear and is consequently stronger. Involute gears fulfill all the laws of gearing and have the advantage over some other curves in that the contact action is unaffected by slight variation of gear center distances.

The nomenclature of a gear tooth is illustrated in Figure 21.13. The principal definition and tooth parts for standard 14½° and 20° involute gears are discussed here.

The *addendum* of a tooth is the radial distance from the pitch circle to the outside diameter or addendum circle. Numerically, it is equal to 1 divided by the diametral pitch P.

The *dedendum* is the radial distance from the pitch circle to the root or dedendum circle. It is equal to the addendum plus the tooth clearance.

Tooth thickness is the thickness of the tooth measured on the pitch circle. For some standard cut gears, the tooth thickness and tooth space are equal. Cast gears are provided with some backlash, the difference between the tooth thickness and tooth space measured on the pitch circle.

The *face* of a gear tooth is that surface lying between the pitch circle and the addendum circle.

The *flank* of a gear tooth is that surface lying between the pitch circle and the root circle.

Clearance is a small distance provided so that the top of a meshing tooth will not touch the bottom land of the other gear as it passes the line of centers.

Table 21.2 gives the proportions of standard 14½° and 20° involute gears expressed in terms of diametral pitch P and number of teeth N.

Pitch of Gears

The *circular pitch p* is the distance from a point on one tooth to the corresponding point on an adjacent tooth and is measured on the pitch circle. Expressed as an equation

$$p = \frac{\pi D}{N} \tag{21.4}$$

where

D = Diameter of pitch circle
N = Number of teeth

Table 21.2   **American Gear Manufacturers Association Standard for Involute Gearing**

	20° Full Depth	14½° Full Depth	20° Fine Pitch	20° Stub Tooth
Addendum	$\dfrac{1}{P}$	$\dfrac{1}{P}$	$\dfrac{1}{P}$	$\dfrac{1}{P}$
Clearance	$\dfrac{0.250}{P}$	$\dfrac{0.157}{P}$	$\dfrac{0.2}{P} + 0.002$	$\dfrac{0.2}{P}$
Dedendum	$\dfrac{1.250}{P}$	$\dfrac{1.157}{P}$	$\dfrac{1.2}{P} + 0.002$	$\dfrac{1}{P}$
Outside diameter	$\dfrac{N+2}{P}$	$\dfrac{N+2}{P}$	$\dfrac{N+2}{P}$	$\dfrac{N+1.6}{P}$
Pitch diamter	$\dfrac{N}{P}$	$\dfrac{N}{P}$	$\dfrac{N}{P}$	$\dfrac{N}{P}$

Metric gearing is based on the module (m) instead of the diametral pitch P, as in the English system. The basic metric module formula is $m = D/N$ = amount of pitch diameter per tooth = millimeters per tooth measured on the pitch diameter. Also, $m = 1/P$ is expressed in millimeters, and $mP = 25.4$.

The *diametral pitch P*, often referred to as the pitch of a gear, is the ratio of the number of teeth to the pitch diameter. It may be expressed by the following equation:

$$P = \frac{N}{D} \tag{21.5}$$

Upon multiplying these two equations, the following relationship between circular and diametral pitch results.

$$p \times P = \frac{\pi D}{N} \times \frac{N}{D} = \pi \tag{21.6}$$

Hence, knowing the value of either pitch we may obtain the other by dividing into π.

Gears and gear cutters are standardized according to diametral pitch. This pitch can be expressed in even figures or fractions. Circular pitch, being an actual distance, is expressed in inches and fractions of an inch. A 6-pitch gear (6-diametral pitch) is one that has 6 teeth per in. of pitch diameter. If the pitch diameter is 3 in., the number of teeth is 3×6 or 18. The outside diameter (OD) of the gear is equal to the pitch diameter, plus twice the addendum distance, or OD = $D + 2(1/P)$ or 3.333 (= 3 + 2(0.166)).

Any involute gear of the same diametral pitch and a same cutting pressure angle will mesh properly with a gear of any other size of the same diametral pitch. For practical reasons the number of teeth in an involute gear should not be less than 12 because of undercutting.

Gear Speeds

The speeds in rpms of a pinion and a gear, s and S, of two meshing gears vary inversely with both the pitch diameters and the number of teeth. This may be expressed as

$$\frac{s}{S} = \frac{D}{d} = \frac{T}{t} \tag{21.7}$$

where D and d represent *pitch diameters*, as indicated in Figure 21.15. T and t represent number of teeth on the gear and pinion.

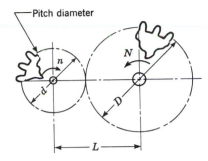

Figure 21.15
Nomenclature for meshing gear and pinion.

The speed ratio for a worm gear set depends on the number of teeth on the gear and the lead of the worm. For a single-threaded worm the ratio is

$$\frac{\text{rpm of worm}}{\text{rpm of gear}} = \frac{T}{1} \tag{21.8}$$

Kinds of Gears

The gears most commonly used are those that transmit power between two parallel shafts. Such gears having their tooth elements parallel to the rotating shafts are known as *spur* gears, the smaller of the two being known as a *pinion* (Figure 21.15).

If the elements of the teeth are twisted or helical, as shown in Figure 21.16*B*, they are known as *helical gears*. These gears may be for connecting shafts that are at an angle in the same or different planes. Helical gears are smooth acting because there is always more than one tooth in contact. Some power is lost because of end thrust, and provisions must be made to compensate for this thrust in the bearings. The herringbone gear is equivalent to two helical gears, one having a right-hand helix and the other a left-hand helix.

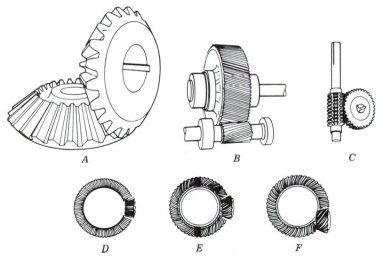

Figure 21.16
Special types of gears. *A*, Miter. *B*, Helical. *C*, Worm. *D*, Zerol. *E*, Helical zerol. *F*, Hypoid.

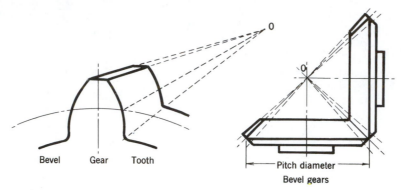

Figure 21.17
All elements of straight bevel gears converge at the cone apex of the gears.

Usually, when two shafts are in the same plane but at an angle with each other, a *bevel* gear is used. Such a gear is similar in appearance to the frustum of a cone having all the elements of the teeth intersecting at a point, as shown in Figure 21.17. Bevel gears are made with either straight or spiral teeth. When the shafts are at right angles and the two bevel gears are the same size, they are known as miter gears (Figure 21.16A). *Hypoid gears*, an interesting modification of bevel gears shown in Figure 21.16F, have their shafts at right angles but they do not intersect as do the shafts for bevel gears. Correct teeth for these gears are difficult to construct, although a generating process has been developed that produces satisfactory teeth. *Zerol* gears (Figure 21.16D) have curved teeth but have a zero helical angle; they are produced on machines that cut spiral bevels and hypoids. Worm gearing is used when a large speed reduction is desired. The small driving gear is called a worm and the driven gear is called a wheel. The worm resembles a large screw and is set in close to the wheel circumference, the teeth of the wheel being curved to conform to the diameter of the worm. The shafts for such gears are at right angles but not in the same plane. These gears are similar to helical gears in their application, but differ considerably in appearance and method of manufacture. A worm gear set is shown in Figure 21.16C.

Rack gears, which are straight and have no curvature, represent a gear of infinite radius and are used in feeding mechanisms and for reciprocating drives. They may have either straight or helical teeth. If the rack is bent in the form of a circle, it becomes a bevel gear having a cone apex angle of 180° known as crown gear. The teeth all converge at the center of the disk and mesh properly with a bevel gear of the same pitch. A gear with internal teeth, known as an *annular gear*, can be cut to mesh with either a spur or a bevel gear, depending on whether the shafts are parallel or intersecting.

21.4 PROCESSES FOR MAKING GEARS

Most gears are produced by some machining process. Accurate machine work is essential for high-speed, long-wearing, quiet-operating gears. Die and investment casting of gears has proved satisfactory, but the materials are limited to low-temperature-melting metals and alloys. Consequently, these gears do not have the wearing qualities of heat-treated steel gears. Stamping, although reasonably accurate, can be used only in making thin gears from sheet metal.

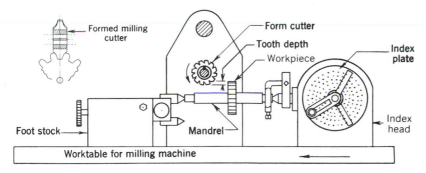

Figure 21.18
Setup for cutting a spur gear on a milling machine.

Formed Tooth Process

A formed milling cutter, as shown in Figure 21.18, is commonly used for cutting a spur gear. Such a cutter used on a milling machine is formed according to the shape of the tooth space to be removed. Theoretically, there should be a different-shaped cutter for each size gear of a given pitch as there is a slight change in the curvature of the involute. However, one cutter can be used for several gears having different numbers of teeth without much sacrifice in their operation. Each pitch cutter is made in eight slightly varying shapes to compensate for this change. They vary from no. 1, which is used to cut gears from 135 teeth to a rack, to no. 8, which cuts gears having 12 or 13 teeth. The eight standard involute cutters are listed in Table 21.3.

Setup of a milling machine to cut spur gears is illustrated in Figure 21.18. A discussion of this process is given in Chapter 9. Formed milling is an accurate process for cutting spur, helical, and worm gears. Although sometimes used for bevel gears, the process is not accurate because of the gradual change in tooth thickness. When used for bevel gears, at least two cuts are necessary for each tooth space. The usual practice is to take one center cut of proper depth and about equal to the space at the small end of the tooth. Two *shaving* cuts are then taken on each side of the tooth space to give the tooth its proper shape.

The formed-tooth principle also may be utilized in a broaching machine by making the broaching tool conform to the tooth space. Small internal gears can be cut completely in one pass by having a round broaching tool made with the same number of cutters as the gear has teeth. Broaching is limited to large-scale production because of the cost of cutters.

Table 21.3 **Standard Involute Cutters**

No. 1	135 teeth to a rack
No. 2	55 to 134 teeth
No. 3	35 to 54 teeth
No. 4	26 to 34 teeth
No. 5	21 to 25 teeth
No. 6	17 to 20 teeth
No. 7	14 to 16 teeth
No. 8	12 and 13 teeth

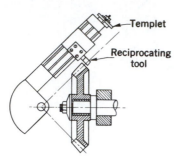

Figure 21.19
Schematic view of a bevel gear planer.

Template Gear Cutting Process

In the template process, the form of the tooth is controlled by a template instead of by a formed tool. The tool itself is similar to a *side-cutting shaper tool* and is given a reciprocating motion in the process of cutting. The process is especially adapted to cutting large teeth, which would be difficult with a formed cutter, and to cutting bevel gear teeth. Bevel gear teeth present a problem in cutting because the tooth thickness varies along its length, as may be seen in Figure 21.17. All elements of each tooth converge to point 0, the cone apex of the gear. In a bevel gear planer, diagrammatically shown in Figure 21.19, the frame carrying the reciprocating tool is guided at one end by a roller acting against a template, while the other end is pivoted at a fixed point corresponding to the cone apex of the gear being cut. Three sets of templates are necessary, one for the roughing cut and one for finishing each side of the tooth space. The gear blank is held stationary during the process and is moved only when indexed. This method of cutting produces an accurately formed tooth having the proper taper. Machines of this type are used only for planing teeth of very straight bevel gears. Most bevel gears are cut by various generating processes.

Cutter Gear Generating Process

The cutter gear generating process for cutting involute gears is based on the fact that any two involute gears of the same pitch will mesh together. Hence, if one gear is made to act as a cutter and is given a reciprocating motion as in a shaper, it will cut into a gear blank and generate conjugate tooth forms. A gear shaper cutter of this description is shown in Figure 21.20, and Figure 21.21 shows how it is mounted in a Fellows gear

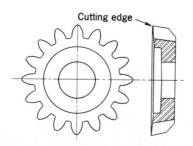

Figure 21.20
Gear shaper cutter.

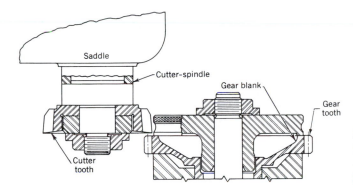

Figure 21.21
Sketch showing mounting of cutter and blank on a gear shaper.

shaper. In operation, both the cutter and the blank rotate at the same pitch line velocity and a reciprocating motion is given to the cutter. The rotary feed mechanism is arranged so that the cutter can be fed automatically to the desired depth while both cutter and work are rotating. The cutter feed takes place at the end of the stroke, at which time the work is withdrawn from the cutter by *cam action*. Figure 21.22 illustrates this process in cutting an internal gear. The operation of the machine is similar to that shown in Figure 21.21. In both, the generating action of the cutter and the blank is shown in Figure 21.23. The fine lines indicate metal removed by each cut in a given tooth space. Cutting action may occur on either the downstroke or the upstroke, whichever proves to be the best procedure. A *shaper* is shown in Figure 21.24.

The cutter gear method of generating gears is not limited to involute *spur gears*. By giving a spiral cutter a twisting motion on the cutting stroke, *helical gears* may be generated. Worm threads may be cut in a similar fashion. In addition, this process may be used to cut racks, sprocket wheels, gear-type clutches, cams, ratchet wheels, and other straight and curved forms.

An application of the cutter gear generator principle is the Sykes gear generating machine, known for its ability to cut continuous herringbone teeth. This machine employs two cutter gears mounted in a horizontal position, as shown in Figure 21.25. In cutting herringbone gears, the cutters are given a reciprocating motion, one cutting in one direction to the center of the gear blank and the other cutting to the same point when

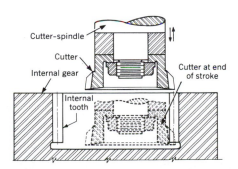

Figure 21.22
Sketch showing cutter setup for cutting an internal gear.

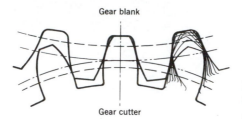

Figure 21.23
Generating action of Fellows gear shaper cutter.

the motion is reversed. The cutters not only reciprocate but also are given a twisting motion according to the helix angle. Both the gear blank and cutters slowly revolve, generating the teeth in the same fashion as the Fellows shaper. Machines of this type are built in various sizes up to those capable of cutting gears 22 ft (7 m) in diameter.

Bevel Gear Generators

Straight *bevel gears* can be produced by two different types of generators: the two-tool generator with two reciprocating tools, and the completing generator with two multiblade rotating cutters. The principle involved is based on the fact that any bevel gear will mesh with a crown gear of the same pitch whose center coincides with the pitch cone apex of

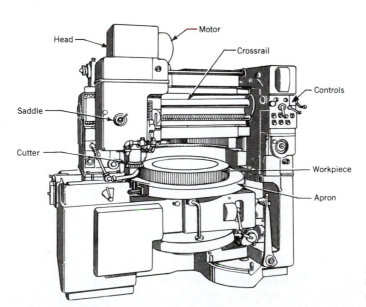

Figure 21.24
Gear shaper setup for cutting spur gears.

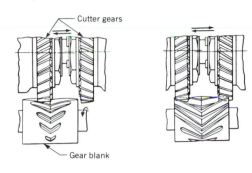

Figure 21.25
The cutters of Farrel–Sykes machines reciprocate, one cutting when the movement is in one direction and the other when the movement is reversed. Each ends its stroke at the center of the blank. These machines produce a herringbone gear.

the gear. In Figure 21.26, the two cutting tools represent the sides of adjacent teeth of a crown gear. These tools are mounted on a cradle that rotates about the axis of the crown gear. At the same time, the tools are given a reciprocating motion. The gear blank also is rotated about its axis at the rate it would have were it meshing with the crown gear. As the tools are simulating the respective positions taken by the crown gear, the correct form of tooth is cut. Both sides of a single tooth are cut on a single generating roll of the cradle; at the end of the generating roll the blank is withdrawn and indexed while the cradle returns to the starting position for the next cut. This cycle is repeated until all the teeth in the gear are cut. In general, the tooth spaces are roughed out in a separate operation so that only a small amount of metal is removed by the reciprocating tools when finishing.

An advantage of this process is that a prior roughing cut is unnecessary, thus saving one handling of the blank. Cutter life is longer, gear quality is improved, and setup requires less time. Both processes will produce a localized tooth bearing in straight bevel gears. A slight crowning on the tooth surface localizes the tooth bearing in the center three-quarters and eliminates load concentrations on the ends of the teeth.

The method of cutting spiral bevel gears also uses the generating principle, but the cutter in this case is circular and rotates as a face milling cutter. The cutter is similar to Figure 21.27, which is shown cutting a *hypoid pinion*. The spiral teeth on gears cut by this process are curved on the arc of a circle, the radius being equal to the radius of the cutter. The blades of the cutter have straight cutting profiles to correspond with the tooth profile of a crown gear tooth. As in the previous method, the teeth are first rough cut before the true shape is generated. The rotating cutters may be designed to cut only one or both sides of the tooth space, the latter type of cutter having the advantage of more rapid production. Spiral bevel gears have an advantage over straight bevel gears in that the teeth engage with each other gradually, eliminating any shock, noise, or vibration in their operation. The hypoid gear can be cut in the machine just described.

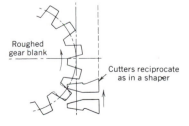

Figure 21.26
Cutting straight bevel gears with two reciprocating tools.

Figure 21.27
Close-up of a hypoid gear generating machine cutting a pinion.

Generating Gears with a Hob Cutter

Any involute gear of a given pitch will mesh with a rack of the same pitch. One form
of cutting gears utilizes a rack as a cutter. If it is given a reciprocating motion similar to
a cutting on a Fellows shaper, involute teeth will be generated on the gear as it rotates
intermittently in mesh with the rack cutter. This method is shown in Figure 21.28. Such
machines require a long rack cutter to cut all the teeth on the circumference of a large
gear, and for this reason they are used very little.

The hobbing system of generating gears is somewhat similar to the principle just
described. A rack is developed into a cylinder, the teeth forming threads and having a
lead as in a large screw. Flutes are cut across the threads, forming rack-shaped cutting
teeth. These cutting teeth are given relief and, if the job is viewed from one end, it looks
the same as the ordinary form of gear cutter. This cutting tool, known as a *hob*, may be
briefly described as a *fluted steel worm*. In Figure 21.29, a hob is shown in section and
end view as it appears when cutting a gear.

Hobbing, then, may be defined as a generating process consisting of rotating and
advancing a fluted steel worm cutter past a revolving blank. This cutting action is illustrated in Figure 21.30, where the teeth on a worm gear are being cut to full depth by a

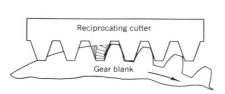

Figure 21.28
Rack-type cutter generating teeth for
a spur gear.

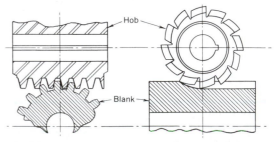

Figure 21.29
Cutting gear with a hob.

Figure 21.30
Gear hobbing machine.

rotating hob. In this process all motions are rotary, there being no reciprocating or indexing movements. In the actual process of cutting, the gear and hob rotate together as in mesh. The speed ratio of the two depends on the number of teeth on the gear and on whether the hob is single threaded or multithreaded.

An elementary gear train for a hobbing machine is illustrated in Figure 21.31. The rotating hob in this figure is shown cutting a spur gear. Because the figure is oversimplified and the machine as shown can only cut gears having a specified number of teeth, change gears must be introduced in the drive to the worm shaft so that gears of any number of teeth can be cut. The mechanisms for feeding the hob or adjusting it at various angles are not shown. The hob cutting speed is controlled by change gears that vary the speed of the main drive shaft.

At the start of operation the gear blank is moved in toward the rotating hob until the proper depth is reached, the pitch line velocity of the gear being the same as the lead velocity of the hob. The action is the same as if the gear were meshing with a rack. As soon as the depth is reached, the hob cutter is fed across the face of the gear until the teeth are complete, both gear and cutter rotating during the entire process.

Inasmuch as the hob teeth have a certain lead, the axis of the hob cannot be at right angles to the axis of the gear when cutting spur gears but must be moved an amount equal to the lead angle. For helical gears, the hob must be moved around an additional

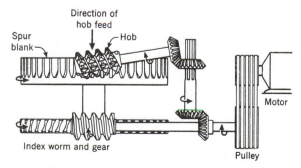

Figure 21.31
Hobbing machine gear train.

angle equal to the helix angle of the gears. Worm gears may be cut with the axis of the hob at right angles to the gear and the hob fed tangentially as the gear rotates.

Gear hobs based on the rack principle will cut gears of any diameter and eliminate the need for a variety of hobs for gears having the same pitch but varying in diameter.

21.5 FINISHING OPERATIONS FOR GEARS

The object of any finishing operation on a gear is to eliminate slight inaccuracies in the tooth profile, spacing, and concentricity so that the gears will have conjugate tooth forms and give quiet operation at high speeds. These inaccuracies are very small dimensionally, frequently not exceeding 0.0005 in. (0.013 mm), but even this amount is sufficient to increase wear and set up undesirable noises at high speeds.

To remedy these errors in gears that are not heat treated, such operations as shaving or burnishing are used. *Burnishing* is a cold-working operation accomplished by rolling the gear in contact and under pressure with three hardened burnishing gears. Although the gears may be made accurate in tooth form, the disadvantage of this process is that the surface of the tooth is covered with amorphous or "smear" metal rather than metal having true crystalline structure, which is desirable from a long-life standpoint. More accurate results may be obtained by a shaving process, which removes only a few thousandths of an inch of metal. This process is strictly a cutting and not a cold-working process.

Rolling the gear in contact with a rack cutter and using a rotary cutter are two methods of shaving. Either will produce accurately formed teeth. Both external and internal spur and helical gears can be finished by this process.

Heat-treated gears can be finished either by *grinding* or by *lapping*. Grinding may be done by either the forming or the generating process. The disadvantage of gear grinding is that considerable time is consumed in the process. Also, the surfaces of the teeth have small scratches or ridges that increase both wear and noise. To eliminate the latter defect, ground gears are frequently lapped.

Gear *lapping* is accomplished by having the gear in contact with one or more cast-iron lap gears or true shape. The work is mounted between centers and is slowly driven by the rear lap. It in turn drives the front lap, and at the same time both laps are rapidly reciprocated across the gear face. Each lap has individual adjustment and pressure control. A fine abrasive is used with kerosene or light oil to assist in the cutting action. The time consumed for average-size gears is ½ to 2 min per side of gear teeth. The results of lapping are demonstrated by longer wearing and quieter operating gears.

QUESTIONS

1. Give an explanation of the following terms:

ISO   Base circle
Lead   Involute
Chasers   Hypoid gears
Rollability   Annular gear
Plunge cut   Shaving

2. On a triple-threaded screw as often found on a fountain pen, what is the lead?

3. What is the principal advantage of multithreaded screws? What are the disadvantages?

4. Discuss the difficulty of comparing threads with the Unified and ISO standards.

5. What is the difference in shaving and burnishing?

6. Give the differences among a square thread, an Acme thread, and the buttress thread.

7. How is collapsible tap made?

8. What threading tool is used for internal threads in a blind hole?

9. How can you thread a piece of steel with a ¾-in., 10-thread die to have a ½-in. lead?

10. What is the purpose of a bottoming tap? On what type of work is it used?

11. How is the tap drill size for an internal thread determined?

12. Discuss the principal metallurgical and physical characteristics of a metal employed in thread rolling.

13. What advantage do cut threads have over rolled threads?

14. Describe the threads denoted by: ¼-20 UNC-3 BLF and M2.5 × 0.45-6G8F.

15. How are the following threads made?
a. ½ by 2½-in. machine bolts.
b. Lead screw on lathe.
c. Square threads on jack.
d. Internal threads in nuts.
e. Acme thread on end of rod.

16. What is the function of the fillet between the base and root circle of a gear?

17. How are most bevel gears cut?

PROBLEMS

21.1. Determine the pitch and depth of the following threads:
a. ½ in.-13 UNC.
b. ¾ in.-10 UNC.

21.2. Determine the feed per tooth of a 24-tooth thread milling cutter turning at 50 rpm and the 2-in. workpiece at 10 rpm.

21.3. What percentage saving in material cost is obtained in producing ⅝ in.-11 UNC threads on 2-in. stud bolts if thread rolling is used in place of cutting dies?

21.4. Approximately how much stock is wasted in chips if 10,000, ¾-16 UNC screws threaded 4 in. long were to be threaded by die as compared to rolling?

21.5. Determine the root diameter, pitch diameter, and lead angle for a ⅝-11 UNC thread.

21.6. What is the minimum nut thickness for the following diameter bolts: ¼ in., ½ in., ¾ in., 1 in., 2.5 mm, 6 mm, 12 mm?

21.7. A gear has a pitch diameter of 5 in. What is the difference in the diameter of the base circles for a 14½° and a 20° involute gear?

21.8. For a gear with a diametral pitch of 10, what is the difference in the addendum and total tooth depth for each of the types of gears listed in Table 21.2?

21.9. Sketch and show all diameters and dimensions of a 14½° involute gear that has a pitch of 10 and 32 teeth. It is ½ in. wide.

21.10. Sketch and show all diameters and dimensions of a 8-pitch, 20° stub gear that has 34 teeth.

21.11. Calculate the pitch of a gear that has a pitch diameter of 3.5 in. and 35 teeth.

21.12. A 14½° involute gear has a thickness of 1½ in., an outside diameter of 11 in., and 60 teeth. Calculate diameters and dimensions.

21.13. The diameter of the pitch circle of a 14½° involute gear is 4.0 in. What is the diameter of the pitch circle in SI units? What is the diameter of the base circle? (cos 14½° = 0.97.)

21.14. The circular pitch of a gear is 0.314 in. and the pitch diameter is 5 in. What standard gear cutter should be used?

21.15.

a. In the gear train shown in Figure P21.15 what is the pitch diameter of each gear if the circular pitch is 0.5283?

b. If shaft A turns at 1800 rpm, what is the rpm of B?

c. What is the centerline distance between shafts A and B?

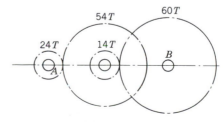

Figure P21.15

21.16. A spur pinion having 40 teeth is in mesh with a gear having 32 teeth. If a 10-pitch cutter is used in cutting the teeth, what is the correct center distance between the two gears?

21.17. A 10-pitch 14½° involute spur gear has 64 teeth. Determine the base circle diameter and outside diameter of this gear.

21.18. A 20° full-depth involute gear having 96 teeth and an outside diameter of 8.25 in. is cut with a single-tooth form cutter. Determine the pitch diameter, diametral pitch, tooth clearance, and addendum.

21.19. A standard involute 14½° gear has a pitch diameter of 5 in. and is cut with a 10-pitch cutter. How many teeth will the gear have and what is the total tooth depth?

PRACTICAL APPLICATION

Identify a nearby manufacturing plant involved with gear or thread making. Team up with one or two other students and solicit permission to visit the plant. Document the kinds of equipment used to manufacture either gears or threads (in various ways) and write a one-page report of your observations. (One individual report per student.)

CASE STUDY: LOTUS GEAR WORKS

The job shop Lotus Gear Works does specialty cutting of spur gears. Customers come to Lotus with "down" problems on equipment and expect quick turnaround. Most customers know very little about gears and do not give specifications. For example, "not all gears are cut on milling machines equipped with dividing heads" makes no sense to them.

A customer seeks Lotus' advice because she has a gear in the gear box of a large construction crane that has worn teeth. The company manufacturing the crane is out of business. The customer says, "I can only see the gear teeth as they go by; I don't know the gear's size, but I need another one just like it. The gear it meshes with has an OD of 15 in., 133 teeth, a width of 3 in., and has a 14½° gear pressure angle stamped on it. The centerlines of the two shafts holding the two gears are 22.07 in. apart. That's all I know. Can you make me a gear?"

Because Lotus has had jobs like this before, you are asked to specify the gear. What are the dimensions you decide on? Depth of cut? Number of teeth? What cutter should be used? What can be said about the speed of the two meshing gears? Your design should show a rough sketch of the gear and give calculations.

CHAPTER 22
FINISH PROCESSES

Several variables need to be considered when surface finish is needed. Some of these variables are type of surface preparation, kind of electroplating solution, plating procedure, process control, and inspection methods. When evaluating the type of surface preparation, two approaches exist: mechanical and chemical surface preparation.

Because of the scope of this textbook and their importance in today's environment, only blast finishing, impact blasting, and water rinsing preparation methods are covered. In addition, two plating procedures are illustrated, brush plating and metalizing of nonconductors. Finally, design considerations for finish methods are discussed.

22.1 MECHANICAL SURFACE PREPARATION

The following are common mechanical surface preparation methods: polishing, buffing wheels, mush buffing, surface conditioning abrasives, belt polishing, mass finishing, blast finishing, and impact blasting with glass beads. The last two methods are discussed next.

Blast Finishing

Blast finishing provides enough power to remove heavy surface materials or can be gentle enough to take print off paper without penetration. Table 22.1 presents a good summary of impact media.

As a finishing process, it is used routinely to remove surface contamination, roughen surfaces for the application of paint, remove surface irregularities, or create a specific finish. As a surface modification method, it is used to increase compressive stresses on the surface to provide increased fatigue life, decrease susceptibility to stress corrosion, correct distortion, and form structural steel and plate.

Blast cabinets are customarily self-contained units and the operator is completely isolated from the process. This provides increased operator safety. Dust removal and grit reclamation are integral parts of the blasting machine. Because it is an easy-to-learn process, little time is lost to training.

Many kinds of manufactured and natural abrasives, ranging from 20 to 6000 mesh, can be used in this process. Depending on the amount of direct pressure exerted through the blast-finishing nozzle and the surface being processed, each type of media can achieve different results.

The correct blasting pressure and impact angle must be determined to achieve the best possible blasting results. Correct pressure selection also will make any blasting operation more economical and cost efficient.

When using direct pressure units, 40 to 60 psi is usually the most economical pressure. Induction units operate well at 60 to 80 psi. The use of excessive pressure will only accelerate the breakdown of the abrasive, with minimal decrease in blasting time. For example, blasting at 100 psi may reduce a time cycle by 5%, as compared to blasting at 60 psi, but the abrasive may break down at a 50% faster pace.

Pressure selection also must take abrasive type into account. For example, if intricately designed jewelry is to be blasted, a fine abrasive with a soft texture would be used at a

Table 22.1   **Impact Media Comparison**

Media	Glass Beads	Coarse Abrasives	Metallic Abrasives	Fine Angular Abrasives	Organic Abrasives
Example		Sand	Steel and iron shot/grit	Aluminum oxide	Walnut shells
Application	Cleaning, finishing, deburring, peening (light and medium)	General cleaning where metal removal and surface contamination are not considered	Rough general cleaning and high-intensity peening	Cleaning where smooth finish and surface contamination are not important	Light deburring and cleaning of fragile items
Shape	Spherical	Granular	Spherical/irregular	Angular	Irregular
Color	Clear	Tan	Gray	Brown/white	Brown/tan
Specific gravity	2.45–2.50	2.4–2.7	7.6–7.8	2.4–4.0	1.3–1.4
Silica content	None	100%	None	<1%	None
Hardness (MOH)	5.5	7.5	7.5	9.0	1.0
Metal removal	Low/none	High	High/medium	High	None
Cleaning rate	High	Fast	Medium/high	Fast	Slow
Peening ability	High	None	High	None	None
Sizes	20–325 mesh	8–200 mesh	6–200 mesh	80–325 mesh	60–325 mesh
Consumption rate	Low	High	Low	High	High

pressure of 10 to 15 psi. However, the removal of scale from steel castings could require a coarse, hard abrasive and an air pressure of 60 to 80 psi. Amounts of air needed for different equipment are given in Table 22.2.

The *blast angle* is another important variable. For example, aluminum oxide at a 45° angle, results in maximum scuff, cut, and roughness, whereas a 30° angle produces the finest surface finish and a smoother scuff pattern.

The distance from the nozzle to the part being blasted should remain constant throughout the process, but this distance may vary from project to project. When artificial abrasives are used, the recommended distance is 6 to 12 in. More distance is required for heavy ferrous metals. Softer, natural abrasives should be blasted from a distance of 3 to 6 in., depending on the action needed. In air suction blast equipment, the discharge nozzle diameter should be about twice the diameter of the air jet.

Table 22.2   **Air Requirements (CFM)**

		40 psi	60 psi	30 psi	100 psi
Pressure blast	⅛″ nozzle	10	14	17	20
	³⁄₁₆″ nozzle	22	30	38	45
	¼″ nozzle	41	54	68	81
Induction/suction blast	¼″ nozzle, ⅛″ air jet	12	17	21	26
	⁵⁄₁₆″ nozzle, ⁵⁄₃₂″ air jet	19	27	34	42
	⁷⁄₁₆″ nozzle, ⁷⁄₃₂″ air jet	38	52	66	80

CFM, cubic feet per minute.

Impact Blasting with Glass Beads

Impact blasting with glass beads is an energy-efficient and environmentally acceptable method of metal finishing. When properly controlled, the system is safe for workers and spent media presents no disposal problems.

Glass beads are virtually chemically inert. This factor, combined with their spherical shape, minimizes media consumption. Table 22.3 compares consumption data of impacting media on different metal surfaces of varying hardnesses. On both metals tested, glass beads offer the lowest consumption per cycle. In addition, close tolerances are maintained and, as indicated in Figure 22.1, glass beads remove a minimal (if any) amount of surface metal. Glass beads also are used extensively as a peening medium, achieving a wide range of arc height peening intensities in a variety of applications and industries (Figure 22.2).

Impacted surfaces are free of smears, contaminants, and media embedments; high points are blended and press sealed. A wide range of finishes from mat to bright sating are achievable. The peening action of the media further imparts a layer of compressive stresses on the surface of the part. This increases fatigue life, decreases susceptibility of the part to stress corrosion, and enhances surface strength.

Process Control

System control via arc height peening intensity is applicable to all cleaning, finishing, peening, and deburring operations. In cleaning, the arc height technique can be used to maintain process speed. In finishing, profilometer measurements of RMS (root mean

Table 22.3 **Consumption Studies of Various Impacting Media Direct Pressure Dry Blasting System**[a]

Blasting Material	Target Material[b]	Hardness, Rockwell	Average Consumption per Cycle (%)
Glass beads	Al 2024	75B	2.1
(106–53 microns)	Al 2024	75B	
	CRS	30C	3.5
	CRS	30C	
Aluminum	Al 2024	75B	6.0
oxide (180 grit)	Al 2024	75B	
	CRS	30C	6.7
	CRS	30C	
Soda-lime silicate	Al 2024	75B	10.6
ground glass (125 microns and finer)	Al 2024	75B	
	CRS	30C	12.9
	CRS	30C	
Borosilicate	Al 2024	75B	11.4
ground glass (125 microns and finer)	Al 2024	75B	
	CRS	30C	13.9
	CRS	30C	

[a] Blasting parameters: ³⁄₁₆-in. ID nozzle; ⅛-in. ID grit stem; 90° target angle; 6-in. target distance; 0.006 N peening intensity; 25 PSI blasting pressure.

[b] Materials used: cold-rolled steel (CRS) and aluminum 2024 (Al 2024).

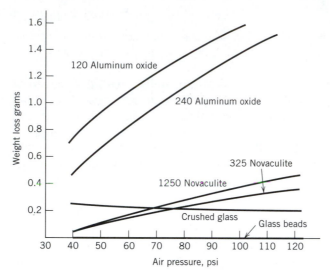

Figure 22.1
Weight loss from blasting with various media.

square) microinch finish can be correlated to peening intensity, thereby eliminating any subjective evaluation of performance. In peening, the degree of compressive stress induced is directly related to the arc height peening intensity. By such control, significant benefits are achieved in terms of labor productivity, reduced supervision requirements, and decline in the number of rejected parts.

As indicated in Table 22.1, both steel shot and glass beads are available for peening applications. Steel shot with its heavier density offers a deeper depth of compression, but requires more energy to propel while leaving dissimilar metallic smears (i.e., various forms of contamination) on the part's surface. Glass beads are often used as a secondary peening medium, removing contamination while improving surface texture and finish (lower RMS) of the part.

Typical glass bead peening applications take place before plating and after grinding and welding on aerospace, automotive, and machine tool components.

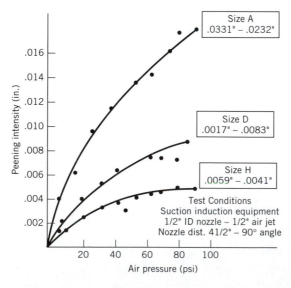

Figure 22.2
Experimental peening intensities achieved with solid glass beads.

22.2 CHEMICAL SURFACE PREPARATION

When chemical preparation is chosen, the following methods are available: vapor degreasing, metal cleaning, water rinsing, surface active agents, ultrasonic cleaning, and pickling and acid dipping. Because of the environmental importance that water rinsing has acquired, this method is discussed next.

Water Rinsing

The rinse must not cause a loss of quality to the product. It must not cause the work to lose activity prior to plating. It must not cause the precipitation of chemical products on the work, because that will interfere with proper conditioning in the next step. The rinse must not cause staining or etching of a final finish when the work is dried. The *"simple rinsing equation"* is the most useful expression for the solution of rinsing problems:

$$D \times C_t = F \times C_r \qquad (22.1)$$

where

D = Dragin to the rinse, in liters
C_t = Concentration of the dragin (the concentration of the solution in the preceding tank), g/liter
F = Flow through the rinsing tank, liters
C_r = Concentration in the rinse, g/liter

Dragin. *Dragin* (D) is determined after the water has been calibrated as follows: stop the flow in the rinse, agitate the water to be sure that it is completely mixed, then measure the conductivity. Process one or more racks through the preceding solution and then into the still rinse. Remove the work, agitate the rinse once more, and measure the conductivity again. The dragin volume can now be determined from the increase in conductivity.

Dragout. The *dragout* volume from the preceding tank depends on the viscosity of the solution, the surface tension, the withdrawal time, and the draining time. Allow some of the dragout to drain back into the tank and onto a drainboard sloping back to the tank, so that most of the dripping dragout is returned directly to the tank. A large portion of the dragout can also be returned to the tank by directing a quick spray rinse onto the work when it is withdrawn and held just above the tank.

Concentration in the Tank. Obviously, the concentration in the preceding tank (C_t) will be known when this solution is analyzed for routine control; therefore, it is useful to save a portion of the analytical sample to use for calibration of the tap water. This relates the chemical analysis to the conductivity measurement, and the information is used to relate that solution to the rinse as well as to relate it to the dragin.

Concentration in the Rinse. The concentration of a rinsing tank (C_r) can be read easily with a conductivity meter. Such readings will reveal the changing concentrations in the flowing rinse when they are related to the calibrated water.

Flow. The flow (F) can be calculated when D, C_t and C_r are known and when the flow, F, is in the same units as D to satisfy the volume rinsing ratio R_v. These units will usually be per period or per minute. At equilibrium, C_r is the contamination limit that is selected to ensure the quality of the work. The terms C_t and C_r are combined as the constant R_c

in the form of the rinsing equation that is used to calculate the flow; therefore, the flow is set in proportion to the dragin and R_c is the proportionality constant,

$$F = R_c D \tag{22.2}$$

Equilibrium and Effectivity. Equation (22.1) applies to a rinsing tank at equilibrium when the quantity of dissolved substances that are dragged into the rinse is equal to the quantity overflowing from the rinse, that is, salt in = salt out. Equilibrium lasts for a short time in a particular period and the tank seeks another equilibrium as the dragin or the flow changes. If the equilibrium exists, one element of Equation (22.1) can be calculated when the other three are known.

Sometimes, however, the simple rinsing equation does not apply, so it is useful to measure all four terms and calculate the *effectivity*, E, which then tells how the rinsing tank deviates from the equation. This can indicate a number of possibilities. If E is reasonably close to 1, the equation probably applies, and it often does apply within practical limits. If E is not close to 1, something may have changed. The rinsing tank may not have been at equilibrium. There may have been a change in the dragin, or the equation may not apply; however, a corrected equation may be used by making E the correction factor.

$$E = F \times C_r / D \times C_t \tag{22.3}$$

The maximum allowable concentration of chemicals in the rinse is known as the "*contamination limit.*" Then a dilution factor, known as the "*rinsing ratio*" (R), is defined as

$$R = C_t / C_r \tag{22.4}$$

The contamination factor is determined by experimentation, experience, or first guess. The limit that will avoid staining is established by rinsing the final work in waters with increasing concentrations of the plating solution in the rinse water. Such a series of tests showed that bright chromium, for example, will not stain at 0.040 g/L.

EXAMPLE 22.1

Consider the contamination limit 0.040 g/L applied to the rinsing of chromic acid containing 300 g/L of CrO_3. This means that the chromic acid on the surface of the work must be diluted from 300 to 0.040 g/L. The rinsing ratio given by Equation (22.4) is

$$R = \frac{C_t}{C_r} = \frac{300}{0.040} = 7500$$

One more bit of information is needed to calculate the flow of water through a rinsing tank at equilibrium: The amount of solution carried over by the work is known as D, the dragin. From Equation (22.1), calculate the flow as

$$F = D \times \frac{C_t}{C_r} \tag{22.5}$$

Now assume that the dragin was found to be 100 ml or, 0.1 L. Then the flow is

$$F = 0.1 \times \frac{300}{0.040} = 750 \text{ L}$$

At equilibrium, the flow is proportional to the dragin and R, the rinsing ratio, becomes the proportionality constant:

$$F = R \times D \tag{22.6}$$

Assume now that racks of work are entering and leaving the rinse at a rate of one every minute, that the rinsing tank is at equilibrium, and that the amount of chromic acid carried in on each rack is equal to the amount of chromic acid overflowing from the rinsing tank. Then the simple rinsing equation is

$$D \times C_t = F \times C_r$$
$$0.1 \times 300 = 750 \times 0.04$$
$$30g\ CrO_3\ in = 30g\ CrO_3\ out$$

The rinsing equation also can be written as two equal ratios, R_c and R_v,

where

$$R_c = \text{Concentration rinsing ratio}$$
$$R_v = \text{Volume rinsing ratio}$$

$$R_c = R_v \text{ or } \frac{C_t}{C_r} = \frac{F}{D} \tag{22.7}$$

Under ideal conditions the rinsing tank responds according to the equation, and the accuracy of the equation can be evaluated when the four terms in the equation are measured.

$$\frac{R_v}{R_c} = E \frac{F \times C_r}{D \times C_t} = \frac{\text{chemicals out}}{\text{chemicals in}} \tag{22.8}$$

When the rinse is operating according to the simple rinsing equation, $E = 1.00$.

The rinsing equation was derived with the assumption that mixing of the dragin and the water in the rinsing tank takes place instantly. Obviously this is not true, but simulated rinsing behavior with small tanks has shown that a rinsing tank usually performs approximately as predicted by the simple equation. This has also been confirmed many times by evaluation of operating rinse tanks on production lines.

Multiple Rinsing

Water is substantially saved with a double rinse and more is saved by using a triple rinse, but the savings diminish as more tanks are used.

The savings of water with more than three counterflow tanks is often questionable. However, when recovery of chemicals is important, the use of more than three tanks for recovery of chemicals in the first rinse becomes attractive.

Performance of a rinsing tank is important. Figure 22.3 shows a saw-tooth line for a theoretical rinsing tank approaching equilibrium (assuming instantaneous mixing). The ragged line shows the actual performance and the drop in concentration when the flow continues without the introduction of more work. Many experiments like this show that the simple rinsing equation is useful to calculate the performance of rinsing tanks.

Performance estimation aids in the design of a rinse tank. Figure 22.4 shows a suggested design for a triple counterflow rinsing tank with three double-wall overflows. When bulky

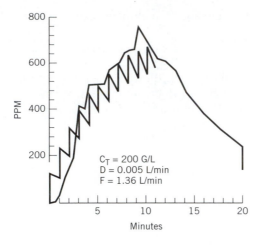

Figure 22.3
Profile of a theoretical rinsing tank with regular additions of dragin compared to an experimental profile with the same terms.

work is introduced into the tank, it causes water to flow backward through the double-wall section. The backflow can mix the more concentrated water in rinse 1 with the dilute water in rinse 2. This can be avoided if the work is entered slowly, but it is inconvenient. When bulky work does enter, the concentrated water may be confined temporarily by the double wall and then flow back to tank 1 when forward flow is resumed. A one-way flapper at the underflow will minimize backflow.

The double wall provides another important function. It guides the water down and up through the underflows and overflows to provide a long path that promotes mixing of the water. Notice that the inlet water in tank 3 is placed high to provide a diagonal flow across the tank.

A weir should be cut in the overflow of tank 1 to serve two purposes—to allow more overflow capacity for bulky work and to serve as a visual flow meter. The flow over the weir gives an indication of the flow and the weir can be calibrated.

Automatic Control

A conductivity controller, shown in Figure 22.5, known as a *rinse tank controller*, restricts the flow of water to the quantity needed to keep the concentration at the contamination level. C_r becomes a constant, K, when the rinse is controlled automatically and the flow is directly proportional to the dragin:

$$F = K \times D \tag{22.9}$$

Water will flow only in response to the need for dilution of the dragin, and the flow will be reduced automatically by any means that reduces the dragin.

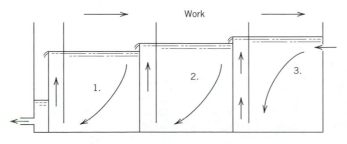

Work

Figure 22.4
A triple counterflow rinsing tank with double-wall overflows.

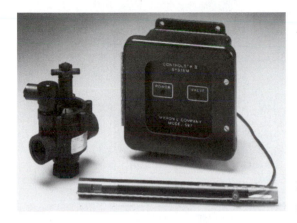

Figure 22.5
Transformer and solenoid valve of automatic rinse tank control.

22.3 PLATING PROCEDURES

Coatings are applied to base material to provide properties not inherent in the base. These include, but are not limited to, corrosion protection, wear resistance, conductivity, color, reflectivity, and solderability. The most common plating procedures are barrel plating, brush plating, metalizing nonconductors, hot dip galvanized coatings, electroforming, mechanical plating, vacuum metalizing, thermal sprayed coatings, electroless (autocatalytic, chemical) plating, and immersion plating.

Brush Plating

In the unetched condition, most brush-plated deposits are metallurgically dense and free of defects. Some of the harder deposits, such as chromium, cobalt-tungsten, and the hardest nickels, are microcracked much like hard tank chromium.

A few are deliberately microporous, such as some of the cadmium and zinc deposits. Microporosity does not affect the corrosion protection afforded, since these are sacrificial-type coatings. The microporous structure, however, permits trace amounts of hydrogen, which are imparted while plating, to be baked out naturally at ambient temperature or in a baking operation.

Etched brush-plated deposits show grain structures that vary, but parallel those of tank deposits. Table 22.4 shows the hardness of bath-plated deposits versus the hardness of brush-plated deposits.

Electroplates are often used to ensure good electrical contact between mating components on printed circuit boards, bus bars, and circuit breakers. A low contact resistance is desired in these applications.

EXAMPLE 22.2

To illustrate how brush plating is performed, consider building up a 6-in. diameter by 2-in. long bore in a housing that is oversized 0.002 in. on the radius.

The part is too large to be mechanically turned, so the operation will be done manually. The part is set up in a position that will be convenient for the operator and permit solutions and rinses to be controlled during processing.

Table 22.4   **Hardness of Deposits from Tank and Brush Plating**

Metal	Microhardness Bath	(DPH)[a] Brush
Cadmium	30–50	20–27
Chromium	280–1200	580–780
Cobalt	180–440	510
Copper	53–350	140–210
Gold	40–100	140–150
Nickel	150–760	280–580
Silver	42–190	70–140
Tin	4–15	7
Zinc	35–125	40–54

[a] DPH, diamond pyramid hardness number vickers.

The area to be plated is precleaned to remove visible films of oil, grease, dirt, and rust. Adjacent areas are solvent cleaned to ensure that masking will stick. Areas next to the bore are masked for about 2 in.

The base material is identified and the supplier's manual is examined to determine the preparatory procedure.

Several nickel and copper plating solutions are suitable for this application. If an alkaline copper solution is selected, a nickel preplate or "strike" is required to ensure adequate adhesion; therefore, a five-step preparatory and plating procedure is required.

Tools of suitable size and shape are selected for each step. Tools that cover about 10% of the total area are appropriate for preparing the base material. The plating tools should contact more surface area so plating proceeds faster.

Once the tools are selected, suitable amounts of solutions are poured into containers. There is enough plating solution to soak the anode covers and treat the area to be plated.

Prior to starting the job, the operator makes some calculations that will help ensure the job is carried out properly. The formulas used and sample calculations for this job are shown as follows.

Calculate the area to be plated (A) in in.^{2}. For a cylinder

$$A = 3.14 \times \text{diameter} \times \text{length}$$

then,

$$A = 3.14 \times 6 \times 2 = 37.7 \text{ in.}^2$$

Calculate the ampere-hours required as

$$\text{A-h} = F \times A \times T \tag{22.10}$$

where

F = A-h required to deposit the volume of metal equivalent to a 0.0001 in. thickness on 1 in.^{2} of area; assume 0.015 for this example
A = Area to be plated, in.^{2}
T = Thickness of deposit in terms of 0.00001 of an in.; that is, for a deposit thickness of 0.0020 in., T is 20

Then,

$$\text{A-h} = 0.015 \times 37.7 \times 20 = 11.31$$

Calculate the estimated plating amperage (*EPA*) as

$$EPA = CA \times ACD \tag{22.11}$$

where

    CA = Area of contact made by the plating tool on the workpiece, in in.^{2}
  ACD = Average current density, which varies depending on the solution; use 5 for this example

Then,

$$EPA = 6 \times 5 = 30 \text{ A}$$

Calculate the plating time required (*EPT*) as

$$EPT = (\text{A-h} \times 60)/EPA \tag{22.12}$$

where

EPA = estimated plating amperage. This plating current should result in a good combination of safety against burning and a high plating rate; 30 is a safe amperage for this solution

Then,

$$EPT = (11.31 \times 60)/30 = 22.6 \text{ min}$$

Calculate the rotation speed (*N*) in rpm, as

$$N = (\text{ft/min} \times 3.82)/\text{diameter} \tag{22.13}$$

where

ft/min = recommended anode-to-cathode rate of movement of the plating tool relative to the surface being plated; use 50 ft/min for the solution in this example

Then,

$$N = (50 \times 3.82)/6 = 31.8 \text{ rpm}$$

Finally, calculate volume of plating solution required (*V*) as

$$V = \text{A-h}/MRU \tag{22.14}$$

where

MRU = maximum recommended use of the solution so that no significant difference in deposit quality or plating characteristics are noted; use 44.5 A/L for this solution

Then,

$$V = 11.31/44.5 = 0.255 \text{ L}$$

Metalizing Nonconductors

The key to success of any process for *electroplating* on a nonconductor still involves very careful cleaning, removal of all cleaning compounds by proper neutralization, surface conditioning, sensitizing, and always using as much rinse water as possible between each step.

There are many parts whose functions are more adequately served when the properties of both a metal and nonmetal (plastic or ceramic) are combined. In these instances, the part is generally manufactured of the nonconductor and the metal is added to its significant surfaces to impart specific metallic properties such as strength, electrical conductivity, or appearance. Typical examples of this are

1. For strength, as in the plating of a thick copper envelope around a woman's high heel for added flexural strength and prevention of splitting from nails when being repaired with new ''lifts.''

2. For electrical conductivity, as in the printed circuit, where a patterned copper film on a nonconductor, such as plastic or ceramic, serves as the wiring in an electronic circuit.

3. For metallic appearance, as in metalizing of buttons, drawer pulls, door knobs, automotive and appliance hardware, toys, and the like, made of plastic.

Many nonconductive materials have been metalized successfully. Much attention has been given to acrylonitrile–butadiene–styrene (ABS) and polypropylene as plastic bases for metalizing.

In plating plastic parts consider the design before metalizing. Provide for the following:

1. The wall thickness, where the plating rack spring is to make contact, should be thick enough to withstand contact pressure. Generally, ⅛ in. is considered good; however, the temperature of the bath and physical properties of the plastic must be considered.

2. Use generous radii at corners and angles. An internal radius should be greater than an external one. Try to keep 1/16 in. R as a minimum.

3. Holes should be through the part wherever possible to minimize entrapment of treatment fluids. Avoid blind holes.

4. Avoid surface defects that would be identified as sink marks, ejection marks, blisters, parting lines, gate marks, and flash and weld lines.

Metalizing Processes

Conductive Paints. The part to be metalized is coated with a *conductive paint*. This is a lacquer or varnish in which a conductive pigment is suspended, such as graphite, copper, or silver. After the part has been dried thoroughly, it is generally plated in a standard acid–copper bath.

Particular care must be taken to preseal wood, plaster, and other porous parts, to prevent absorption of plating solution, which would gradually bleed out over the surface part.

Silver conductive paints are the forerunners of conductive plastics used in radio frequency (RF) shielding. Typically, they are silicone resins loaded with silver flake to provide conductivity and used in self-supporting sheet or molded form. *Copper (or bronzing)* paints, however, are often used for application as a conducting film. The mixture consists of nitrocellulose lacquer (1 fl oz), lacquer thinner (7 fl oz), and copper lining powder (2 oz).

Only enough mixture for immediate use should be prepared, since the metal powder often causes the lacquer to jell. If the copper powder is greasy, it should be washed with thinner before using. If sprayed, the copper paint should be applied with the gun held at a distance so that the film dries almost as soon as it reaches the surface. A glossy appearance indicates that the copper is coated with a layer of lacquer, which will prevent passage of current.

A good method of ensuring that the surface is conductive, is to dip the coated article in a solution of about 1 oz/gal silver cyanide and 4 oz/gal sodium cyanide. The absence of a silver deposit indicates that the conductive lacquer film has not been applied correctly. The silver deposit also provides a better conducting medium than the copper for subsequent electroplating.

Catalytic Deposition. Another method of applying a metal film on nonconductors is by catalytic deposition. *Electroless nickel* for example, is possible with thermoset plastics. Thermoplastics can be coated also, but only with caution because of the temperature–instability characteristic of this group of plastics.

The following preparatory procedure is suggested: First, clean and roughen the surface. Then, sensitize the part (with a 70 g/L stannous chloride and 40 g/L hydrochloric acid solution at 80°F, for example). Rinse thoroughly and immerse in cold palladium chloride solution (1 g/L), containing 1 ml of concentrated hydrochloric acid. Rinse again and immerse at 200°F in the following solution:

Nickel chloride	30 g
Sodium hypophosphite	10 g
Sodium citrate	10 g
Water	1000 g

The pH must be maintained at 4 to 6 and the plating rate at 0.2 mil/h. All these steps should be done in a moisture-free, nitrogen atmosphere. The appearance of the metal film will be a reflection of the substrate; lustrous on a smooth shiny surface, dull on a mat surface.

The literature now shows many adaptations of this procedure, which may be applied to cobalt, nickel–cobalt alloys, and iron–nickel alloys.

22.4 METAL DEPOSITION DESIGN CONSIDERATIONS

All electrolytic processes fundamentally follow Faraday's second law of electrolysis, which states that the material deposited is equal to the electrochemical equivalent of that material multiplied by the current flowing and the time for which it flows. In practice, the process is modified by an efficiency factor, especially when nondissolvable anodes are used.

$$\text{Material deposited } (m) = E \times i \times t \times \tau \qquad (22.15)$$

where

$$E = \text{Electrochemical equivalent, g per A-s}$$
$$i = \text{Current flowing, A}$$
$$t = \text{Time of current flow, s}$$
$$\tau = \text{Deposition efficiency, decimal}$$
$$m = \text{Mass, g}$$

For a given material and process, E and τ will remain substantially constant and, therefore, m is proportional to i by t. For this reason, most electrolytic processing plants are equipped with ammeters and timers that are read and controlled by the operator who presets the current and then operates the process for a given time. However, the current can change or fluctuate for a variety of reasons and control can require constant supervision, tying the operator to the plant or risking incorrect processing. Ampere-hour, Ampere-minute meters (or *current–time integrators*) are used, therefore, to (1) accurately

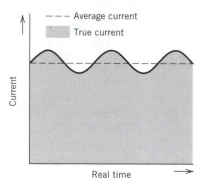

Figure 22.6
Digital integrators can account for ripple and measure the true current rather than the average.

measure the actual current–time product, and (2) control the current–time product for a given process cycle.

To control the process, modern current–time integrators have the advantage of measuring the actual current, taking into account the effect on electroplating of any ripple present, as shown in Figure 22.6 (the analog ammeter cannot do this accurately). They also detect and measure automatically any change in current that occurs. Since an electroplating bath load requires a fixed or predetermined amount of Ampere-minutes or Ampere-hours to obtain a particular average coating thickness, it also is possible to preset the current-time integrator to provide just this amount precisely and repetitively. The current supply can, thus, be terminated or reduced, an alarm triggered, or the whole load may be lifted out of the electroplating bath. Any overrun is measured and priced so that the true cost of the operation is known for auditing purposes.

Time and Tank Capacity Determination

The two graphs shown in Figures 22.7 and 22.8 offer visual guidelines for orientation and quick approximation of the quantity of solution in circular and rectangular containers. Figure 22.7 shows, for round containers, the given diameter in feet at the lower scale. For example, if $D = 4$ ft (118 cm), following the direction of the heavy arrows, one reads, at the height of the intersection of the corresponding vertical and the diagonal line, the solution volume per unit height (v/h) at the left ordinate. In this case, 95 gal/ft (11.2 L/cm).

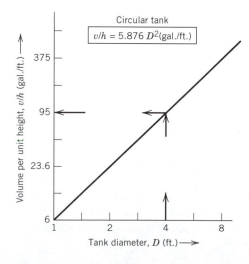

Figure 22.7
Determining circular tank capacity.

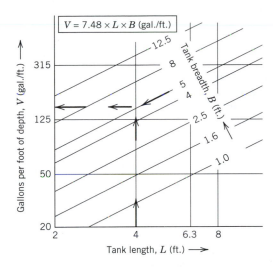

Figure 22.8
Determining rectangular tank capacity.

Figure 22.8 shows the given length and breadth of the tank at the lower scale and at the diagonals. For example, $B = 5$ ft and $L = 4$ ft. At the intersection of the corresponding vertical and diagonal line, one reads 150 gal/ft (19 L/cm) at the left ordinate.

The solution level (height) is measured either by reading fixed height markings at the inside tank wall or from a ruler dipped to the bottom of the tank. The total solution volume, V (gal), is then equal to the solution height, h (ft), multiplied by V/h (gal/ft).

It should be noted that the values shown on the graphs represent simplified evenly divided log scales involving inaccuracies up to about 3%. In most practices, this may be considered inconsequential. For numerically exact results, the use of the equation at the upper left of the corresponding graph is recommended.

Another process design consideration is *plating time*. Nomographs and tables exist in the literature. A convenient formula to estimate plating times is

$$t = \frac{F \times T \times 60}{CD} \tag{22.16}$$

where

t = Plating time, min
T = Plate thickness, mils
CD = Current density, A/ft^2
F = Conversion factor to deposit 1 mil of electroplate thickness, A-h/ft^2

The formula is based on 100% cathode efficiency. If the efficiency is less, for example 95% = 0.95, the theoretical plating time must be divided by this value.

22.5 THICKNESS TESTING

The amount of coating applied is critical to the final product's utility and cost. The determination of the amount of coating is therefore important in appraising its utility and assessing its cost.

Thickness is the most commonly used word to describe the amount of coating. A few of the methods used, such as the coating thickness gage shown in Figure 22.9, measure the linear depth of the coating directly. These include the micrometer, with variations using styluses attached to sensitive mechanical and electronic amplifiers, and the microscope, with various methods to expose the coating layers for measurement.

Figure 22.9
Coating thickness gage.

More commonly, gages estimating the weight per measured area are employed. The thickness is then calculated using the following equation:

$$T = \frac{m \times 10}{A \times d} \qquad (22.17)$$

where

T = Thickness, μm
m = Mass of coating, Mg
A = Area tested, cm^2
d = Density, g/cm^3

The instruments using the weight per unit area as the basis for their measurements are beta backscatter, coulometric, and X ray. Other methods, such as the magnetic and eddy–current methods, compare the magnetic and electrical properties of the base and coating materials to calibrated standards with similar properties. Another method, the drop test, is based on the rate of attack of certain chemical solutions. More generic methods, such as the gravimetric, microscopal, and X ray, are applied to almost all coating combinations.

With such a diversity of methods it is useful to consult a finishing handbook to help choose a measuring system for a particular requirement.

EXAMPLE 22.3

The coulometric method is based on Faraday's law. The law states that 1 g equivalent weight of metal will be stripped or deposited for every 96,500 coulombs (A-s) of electricity passed through the electrolyte. This law is so basic that it is used to define the international ampere.

The international ampere is defined as the unvarying electric current that, when passed through a solution of silver nitrate will remove 0.000118 g of silver per s from the anode. This figure is called the electrochemical equivalent of silver.

Equation (22.17) defines the weight of metal deposited according to Faraday's law [Equation (22.15)]. Therefore, to apply the coulometric method to thickness testing, four parameters must be controlled, namely, area, amperage, time, and anode efficiency. At 100% anode efficiency, by substituting the mass obtained from Faraday's law into the thickness formula, the thickness becomes:

$$T = \frac{E \times i \times t \times 10}{Ad}$$

The area to be measured is determined by a flexible rubber gasket. This area can range from 0.13 to 0.32 cm in diameter. The gasket is an integral part of the deplating cell that holds the solution during the test. The gasket must be flexible so that it will prevent leakage of the solution and yet be sufficiently rigid for precise maintenance of the area. A constant pressure device is included to aid in controlling the gasket pressure. Since this measurement yields weight per area, accurate control of this diameter is essential.

22.6 OTHER METALLIC COATINGS

As explained earlier, the coating process is the application of a finite thickness of some material over the metal or is the transformation of the surface by chemical or electrical means to an oxide of the original metal. A discussion of some other important processes follows.

Electroplating

Electroplating has long served as a means of applying decorative and protective coatings to metals. Most metals can be electroplated, but the common metals deposited in this way are nickel, chromium, cadmium, copper, silver, zinc, gold, and tin. In commercial plating, the object to be plated is placed in a tank containing a suitable electrolyte. The *anode* consists of a plate of pure metal. The object to be plated is the *cathode*. The tank contains a solution of salts of the metal to be applied. A direct current (d-c) having a density of 6 to 24 V is required for the plating operation. When the current is flowing, metal from the anode replenishes the electrolyte solution while ions of the dissolved metal are deposited on the workpiece in a solid state. The properties of the plated material and rate of deposition depend on current density, temperature of electrolyte, condition of surface, and properties of workpiece material.

Chrome Plating

For wear and abrasive resistance the outstanding metal for plating metallic surfaces is chromium. Coatings are seldom less than 0.002 in. (0.05 mm) thick and may be considerably more. Any measure of hardness or abrasive resistance is to some extent a function of the metal on which it is plated as well as of the *chromium deposit* itself.

The electrolytic process consists in passing an electric current from an anode to a cathode (the cathode being the object on which the metal is deposited) through a suitable chromium carrying electrolytic solution in the presence of a catalyst. The catalyst does not enter into the electrochemical decomposition. A solution of chromic acid with a high degree of saturation is used as the *electrolyte*. The surfaces must be polished thoroughly and cleaned before operations start. Since the rate of deposition is fairly slow, the work must remain in the tanks several hours for heavy plating.

Chromium has proved satisfactory for wear-resisting parts because its extreme hardness exceeds most other commercial metals. According to the Brinell scale, the hardness of plated chromium ranges from 500 to 900. This wide variation results not from the metal but from methods and equipment used.

Galvanizing

Galvanizing is a zinc coating used extensively for protecting low-carbon steel from atmospheric deterioration. It offers a low-cost coating that has reasonable appearance and good wearing properties. An improved appearance known as the spangle effect can be produced by small additions of tin and aluminum. Zinc baths are usually maintained at about 850°F (450°C).

Rolls, agitators, and metal brooms are used to remove the excess zinc from the product. Continuous and automatic processes are used for sheet and wire coating. Zinc coatings may be applied also by spraying molten zinc on steel by sheradizing, which is the tumbling of the product in zinc dust at elevated temperatures, and by electroplating. The zinc dust must be rigidly controlled through the use of Hydro Sonic cleaners or scrubbers because of the environmental problems it creates.

Galvanized steel is familiar as seen in highway guard rails, light poles, transmission towers, and the 2-g pail. A coating of 0.004 to 0.008 in. (0.10 to 0.20 mm) protects the steel from corrosive attack in atmospheres such as salt air. This protection continues even after small areas of base metal have been exposed as the *galvanic action* of the coat sacrifices itself as zinc is more electrochemically active than steel. The comparison of tin and zinc coatings is shown in Figure 22.10.

Tin Coating

Tin coatings often are applied to sheet steel to be used for food containers, with tin can manufacturers using approximately 90% of the tin produced. Although many tin coatings are now applied by *electrotinning*, a process where parts are immersed in an electrolyte and a current passed from the electrode to the work, the hot dip method is still used considerably. Tin can be applied easily without affecting the base metal by dipping at temperatures of approximately 600°F (300°C).

In most cases the tin coating is about 0.0001 in. (0.003 mm) thick as compared to only about 0.00003 in. (0.0008 mm) in electrotinned sheets. Porosity is greater in plated tin coatings, and when the containers are used for food, a lacquer seal is necessary.

Other Plating Metals

Copper, often used as an undercoat for subsequent nickel plating, provides good adhesion of the metal and improved appearance. It is seldom used alone as a plating except for

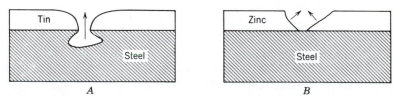

Figure 22.10
Sacrificial action of zinc continues to protect base metal after the area is exposed.

stop offs in selective carburizing and for electroforming heavy deposits. Nickel plating is most popular for protecting steel or brass from corrosion and for presenting a bright appearance. Lead has only limited commercial use, primarily as a protective coating against certain acids. For nonferrous articles used in food handling, silver plating is used widely.

Parkerizing

Parkerizing is a process for making a thin phosphate coating on steel to act as a base or primer for enamels and paints. In this process, the steel is dipped in a 190°F (90°C) solution of manganese dihydrogen phosphate for about 45 min. Bluing is a process of dipping steel or iron in a 600°F (300°C) molten bath of nitrate of potash (saltpeter) for about 1 to 15 min. There are many salts that can be used to color brass and steel by dipping at elevated temperatures, but most of these have limited application and differing degrees of permanence.

Anodizing

Anodizing is an oxidation process developed for aluminum. An electrolyte of sulfuric, oxalic, or chromic acid is employed with the part to be anodized as the anode. Since the coating is produced entirely by oxidation and not by plating, the oxide coating is a permanent and integral part of the original base material. Although the coating is hard, it is porous, which is an advantage from a decorative point of view. The oxide coating enables organic coatings and dyes to be applied successfully to the surface of aluminum. Colored aluminum tumblers and pitchers are examples of this process. Magnesium is anodized in a somewhat similar manner.

Calorizing

Calorizing is a process designed to protect steel from oxidation at high temperatures. Aluminum is diffused into the metal surface at elevated temperatures, forming a protecting film of aluminum oxide that prevents the underlying metal from oxidation. The process is used for treating parts for furnaces, oil refineries, dryers, and kilns.

Hard Surfacing

Hard surfacing is the application to a wearing surface of some metal or treatment that renders the surface highly resistant to abrasion. Such processes vary in technique. Some apply a hard surface coating by fusion welding; in others, no material is added and the surface metal is changed by heat treatment or by contact with other materials.

The several properties required of surfaces subjected to severe wearing conditions are hardness, abrasion resistance, and impact resistance. Hardness is determined easily by several known methods, and an accurate comparison of metals for this property can be obtained readily. Tests for wear or abrasion resistance have not been standardized and it is difficult to obtain meaningful results. In general, wear testing must simulate the service conditions for each type of hard facing material. The statement that "the wear resistance of a material is a function of the method by which it is measured" has been confirmed by both practical experience and research. All factors considered, hardness is probably the best criterion of wear resistance. Ability to withstand wear and abrasion usually increases as the hardness of the metal increases.

The classification of the various processes used for obtaining a hard surface does not point out that there is a great difference in the hardness that can be obtained, nor does it include heat-treating methods that are used to produce a hard surface or interior. The student is referred to the bibliography to expand in methods of producing hard surfaces.

QUESTIONS

1. Give an explanation of the following terms:

Dragin   Plating time
Dragout   Chrome plating
Contamination limit Electrolyte
Rinsing ratio   Parkerizing
Electroplating   Anodizing
Conductive paint   Calorizing
Silver conductive paints   Hard surfacing

2. Name five surface preparation methods.

3. What are the main variables of blast finishing?

4. What are some benefits of glass bead blasting?

5. Explain the components of the simple rinsing equation.

6. In water rinsing, how can dragin be determined after the water has been calibrated?

7. What are some variables that affect the dragout volume in water rinsing?

8. Explain the benefits and difficulties of multiple rinsing.

9. What considerations are needed to design a rinse tank?

10. What is the effect of using a rinse tank controller?

11. What is the purpose of applying coatings to base materials? Name five common plating procedures.

12. What are some keys to success of any process for electroplating on a nonconductor?

13. Give some examples when metalizing a nonconductor is recommended.

14. What design considerations should be made when a plastic is to be plated?

15. What is the difference between silver conductive paints and copper paints?

16. If an article has been coated with a conductive paint, how can you ensure that the surface is conductive?

17. Explain how catalytic deposition works.

18. Why are most electrolytic processing plants equipped with ammeters and timers?

19. Explain how metal deposition can be controlled with modern current–time integrators.

20. What metal has proved satisfactory for wear-resisting parts. Why?

PROBLEMS

22.1. The amount of chromic acid entering a rinse tank is 35 g. If the flow through the rinsing tank is 700 L and the concentration of chromic acid in the rinse is 0.045 g/L, estimate the tank's effectivity.

22.2. How long will it take to deposit 2.1 grams of silver if 10 A flow through the anode at a 90% efficiency? The electrochemical equivalent of silver is 1.18×10^{-4} g/A-s.

22.3. How many gallons of solution are there in a 7.8 ft-diameter circular tank if it can be filled 4.5 ft high?

22.4. Finishing Inc.'s labor and overhead rate is $6/h. It takes them one and a half minutes to polish and buff a new workpiece. Better results could be obtained if parts

were tumbled for 4.5 h. If it takes 18 min to load and unload the barrels and the cost of operating the equipment is $1.5/h, for what number of pieces is the cost the same by either method?

22.5. Size D solid glass beads at a blast pressure of 60 psi and a blast angle of 90° are used to peen aluminum panels. If the nozzle is at 4.5 in. from the target, its ID is 0.5 in. and the air pit is 0.25 in.

a. Determine the peening intensity achieved.

b. What air pressure is required to achieve the same intensity if 0.032-in. beads are used? What if 0.0048-in. beads were used? (Hint: Consult Figure 22.2.)

22.6. Parts are to be coated with a layer of nickel 1.6 mil thick. If the applied current density is 16 A/ft^2 and the A-h/ft^2 to deposit 0.001 in. of electroplate thickness is 19 for nickel, estimate the required plating time. What is the time if the cathode is 90% efficient?

MORE DIFFICULT PROBLEMS

22.7. Because of plant layout constraints, the maximum tank height for plating solutions is 4 ft. To control costs, you want to maintain less than 7000 L of solution in the tank. Calculate the diameter of a suitable circular tank. What are some possible dimensions of an equivalent rectangular tank?

22.8. Design a triple-flow cylindrical rinsing tank and determine its capacity per ft of depth if the dragin is 120 ml, the coating concentration on the surface work is 325 g/L, and the rinsing is 90% effective in Section 1, 95% effective in Section 2, and 98% effective in Section 3. Assume that work racks have an area of 4 ft^2, that loading and unloading take a total of 3 min, that each rack stays in the bath 75 s, and that coating concentration on the surface of the work must be diluted to 0.030 g/L. Production schedules require that at least 20 racks per hour be processed.

PRACTICAL APPLICATION

To meet environmental regulations, coatings with low volatile–organic compound (VOC) emissions are required more often. Powder coatings are a good alternative because they can be applied with minimal waste and give good-looking finishes with improved etch and chip resistance. Powder coatings, for example, are applied to oil filters, auto trim parts, wheels, and brackets. Evaluate this technology as a metal-coating method and substantiate your position about its use. (Hint: Visit a metal finishing company if possible or consult a metal finishing handbook. SME and its Association for Finishing Processes can give you a list of technical resources as well.)

CASE STUDY: STRIKE ONE

Your metal company, Steely Dan Inc., got an order to machine and brush plate a 6.5-in. diameter by 2.3-in. long bore in a housing that is oversized 0.003 in. on the radius. Your supervisor, Christy Copper, calls you and Long John Silver into her office and says, "O-kay guys, the base material for this part is cast iron, which requires electrocleaning, etch, and desmut as preplating steps."

"Good!," says John. "With cast iron we can use either nickel or copper plating solutions."

"Excellent!," replies Christy. "We always get a better deal from Big Ben Bronze when we buy copper, so let's go copper."

"But copper plating requires *striking* before plating or you won't get adequate adhesion!," you point out.

"No problem," your supervisor says looking at you seriously and adding, "Determine the volume of preplating solution required to 'strike' the iron surface and design a rectangular container to hold enough solution for a batch of 1000 parts.

"Prepare a process chart that includes preplating and plating steps, tools needed, solutions and sequence of use, voltages, estimated plating amperage, ampere-hours, and the like, necessary to process operations properly and without hesitation." (Hint: Review Example 22.2 to identify additional steps and operations. Use the data below as well.)

Additional Data

F (nickel) = 0.0149 A-h/in.^{2}
Average current density = 4.6 A/in.^{2}
Estimated plated amperage = 31 A
Anode-to-cathode rate of movement = 40 ft/min
Maximum recommended use of solution = 40 A/L

CHAPTER 23
OPERATIONS PLANNING

Production processes convert raw material into finished parts or products. Seldom will a manufacturer have only one manufacturing process. Indeed, there will be several to very many processes available for a coordinated system, which depends on the size of the firm. Even a simple part requires several operations. This chapter deals with the selection and order of these operations starting with raw material and ending with the product according to the instructions and the engineering drawings. Operations planning is an important part of manufacturing systems, as it integrates the shop floor into a cohesive working unit.

Henry Ford, the great inventor and genius of production, had many achievements with operations planning. He developed the assembly line, which is where you put "this" next to "that" in rapid and uniform sequence, and something better will emerge. For example, 7882 distinct tasks were required to assemble Ford's 1923 Model T touring car, and yet the whole is more than the sum of the parts. "In mass production," Ford said, "there are no fitters." By this he meant that the parts were so precisely made that anyone with a little training could put them together at the speed of an assembly line. This particular method of assembly is another example of operations planning.

Years ago the process of planning the sequence of the production for the part was handled by a *journeyman*, who would know firsthand the capabilities of the equipment. The equipment was general purpose, as specialized equipment was not in existence. This method of process planning and production built individuality into the product, but it was slow and cumbersome. As the worker was getting the material, setting up the workplace, making the tooling, and doing the planning, there would be idle equipment. It was quickly realized that by preplanning the work, securing the material, supplying information to the workers, greater production was realized. Now there are specialists in production planning, who are called *process engineers* or *process planners*.

This chapter describes the planning of the production process. The plan is also known as an "op sheet." It designates the sequence of the operations starting with raw material, or an intermediate product, and concludes with the part as specified by the design. Workstations, equipment, perishable and designed tools, and descriptions of the operations are part of this planning. For assembly, the operation involves fastening, joining, and bringing together the parts, subassemblies, and major assemblies into the final product. In this planning, the objective is to have the costs of the production competitive, and meet the quality and specifications of the engineering drawing.

23.1 BUSINESS OBJECTIVES

The systematic determination of the engineering processes and systems to manufacture a product competitively and economically is called *operations planning*. It is the stage between design and production. The plan of manufacture considers functional requirements of the product, quantity, tools and equipment, and eventually the costs for manufacture. In a sense, operations planning is a detailed specification and lists the operations, tools, and facilities.

Operations planning is a responsibility of the manufacturing organization. A number of functional staff arrangements are possible. This process leads to the same output despite organizational differences. The following are business objectives for operations planning.

1. *New product manufacture.* A new design may have not been produced before or, alternatively, new manufacturing operations may be introduced for the product. Unless there is planning, the product introduction will be helter-skelter.

2. *Sales.* Opportunity for greater salability of an existing or new product can develop from different colors, materials, finish, or functional and nonfunctional features. Sales and marketing departments provide advice to help manufacturing planning.

3. *Quantity.* Changes in quantity require different sequences, tools, and equipment. The *op planner* differentiates for these fluctuations. If volume increases, the chance is for lower cost. In contrast, if volume decreases, the cost should not increase out of reason. There may be a fortuitous opportunity for reduced cost, if economics and technology will allow substitution of new processes, training, and resources even if quantities are reduced. If quantity reduces too much, however, it is appropriate for the op planner to recommend that production may no longer be economical. Perhaps, a supplier may be the appropriate lower cost alternative.

4. *Effective use of facilities.* Operations planning often can find alternate opportunities for the plant's production facilities to take up any slack that may develop. Seasonal products, which might be popular in the summer, need an alternative product for the winter season. For example, companies that produce sporting equipment may use the same facilities to produce tennis rackets and skis.

5. *Cost reduction.* Various opportunities become available if the company has an ongoing cost reduction effort. Suggestion plans, value analysis, design for manufacturing (*DFM*), and directed and systematic effort involve operations planning.

23.2 SYSTEMS ANALYSIS

Before operations, tools and equipment, and labor types can be identified, several policy questions need to be resolved. The questions and answers are interrelated, so it may be necessary to backtrack several times in an iterative style.

Design considerations are important. For example, a part may be cast or welded, or an eyelet may be rolled or stamped, or they may be purchased or made internally. These are common choices. These contrasting questions occur by the hundreds in a moderate-size facility. However, the eventual selections have an important bearing on production operations.

Material specifications have a significant effect on production. High- or low-grade carbon steels or alloy steels affect the production operation. There may be common characteristics among the various parts. Production might be organized on parts that are made from sheet or bar stock. By contrast, material might be purchased because the company has machines that effectively make sheet metal products but are inefficient with bar stock. Tooling may be a deciding factor in the analysis.

Tooling consists of two types, perishable or capital. *Perishable tooling* is drill bits, small cutters, and the like. *Capital tooling* is jigs, dies, fixtures, and ancillary support tooling that requires tool design and construction. Perishable tooling is low cost as compared to capital tooling. A company may make a process decision on the availability of jigs and fixtures and the equipment to use the tooling. Avoiding new tooling cost is a popular policy because of its magnitude and the likelihood of tools becoming obsolete.

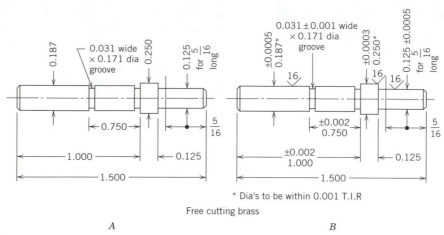

Figure 23.1
Shaft. *A*, Loose tolerances. *B*, Tight tolerances.

Tolerances are significant factors that influence the operation processing and cost. Notice Figure 23.1, which shows a pin with both loose and tight tolerances. There are extra and more expensive operations for the pin with the tight tolerances. However, it is the *functional engineering requirements* that set tolerances. Tolerances should be the largest values consistent with function, manufacturing, reliability, safety, appearance, and cost. Appropriate selection of tolerances is an important engineering activity.

23.3 OPERATIONS SHEET PREPARATION

The *operations sheet* is fundamental to manufacturing planning. It is also called "route sheet," "instruction sheet," "traveler," or "planner." Next to the product description, or the engineering print, the operations sheet is probably the most important document in manufacturing. There are many styles and each plant has its own form.

The purpose of the operations sheet is to select the machine, process, or bench that is necessary for converting raw material into product, provide a description of the operations and tools, and indicate the time for the operation. The order of the operations is special, too, as this sequence indicates the various steps in the manufacturing conversion. The shop will use these instructions in making the part.

The process description is a step-by-step set of instructions, which can be brief, or it can be very detailed. An *operation* is a step in the process sequence, and is defined as all work done at a workstation. A job to drill and tap a hole at one drill press would be one operation. If for some reason, however, it is necessary to drill the hole on one drill press, and then transfer the work to another drill press for tapping the hole, that would be two operations.

A *process* is one or more operations that are performed on the product from the time it leaves an inventory location until it returns to inventory. Thus, raw material is released from inventory, and eventually the processed part reaches the shipping dock, or some intermediate point where it is ready for sale and shipment or for additional work and integration.

The *sequence* of operations in the process depends on many factors. Indeed, some of these factors will be significant for one plant, but are minor for another plant. It depends on the plant that is producing the product. Plant layout is a factor because the operations should be in a sequence that minimizes the travel distance and time.

Engineering working drawings are essential, as these are used for manufacturing. Figure 23.2 is a simple example of a working drawing. There are many technical features that are removed from *working drawings*. Is the measurement unit metric or inch, for example? Are the *orthographic views* third or first angle? A panoply of information is removed from the title block: part number, part name, date, revision number, scale, sheet number, tolerance standards, material, and specifications. If the drawing is a detail or an assembly, it leads to different practices for developing the routing. Are geometric positioning tolerances applied, or is the tolerance system traditional? The answers to these questions guide the processes.

Dimensional integrity of the design with the equipment and tools is vital. Datum surfaces, geometric tolerances, ordinary and special designed tools that are compatible with the equipment, and part design are essential to producing a product that matches the engineering drawing. Difficulty of operations may be a consideration. For example, some plants want to do those operations that are critical and difficult early in the sequence, if scrap is factor. To perform difficult operations last might involve substantial value added and cost, only to find that the product is worthless. Obviously, the important principle is to avoid scrap and losses. Engineering and process planning need to avoid high-risk operations, if lower risk and lower cost operations are available. *Cost* is an important driver in the optimization of the process sequence.

Correct sequences will insist on drilling the hole before tapping, or the milling of the pocket before an assembly is welded to the pocket. Avoiding the tumbling of the part in a vibratory tumbler for burr removal, if there are external threads, is a foolish and costly blunder. Protection of the part finish after grinding will prevent the collection of

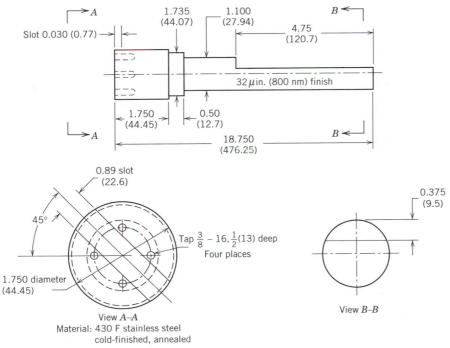

Figure 23.2
Example for process planning.

the parts in a barrel or a loose gathering of the shafts, such that they could bang on each other and mar the finish, are considerations. Operations planning requires this *practical thinking*.

With this general discussion as background, these steps can be identified in some order, remembering that a precise scheme is never achieved. For example, *process planning* is

1. Interpret and understand the engineering drawing and production specifications.
2. Select the equipment.
3. Position the part on the equipment relative to the tool.
4. Identify the workholders.
5. Specify the tools, perishable or design.
6. Simultaneously pick the operation sequence.
7. Write or enter the operations plan into the company form or computer.

23.4 INFORMATION

Before operations planning begins, one must have *engineering drawings*, marketing quantity, material specifications, and knowledge of the company's machines and processes. Job descriptions of the workers and their skills are necessary. Preparation of the operations sheet occurs at the same time as cost estimating, but that is deferred to Chapter 31.

Each operations sheet has a title block, indicating the material kind, part number, date, quantity, and operations planner. The part number and description are removed from the design and repeated on the title block. As each company's op plan is different, the amount and type of information in the title block varies. However, it takes more information than is found on the engineering drawing for the operations plan. It is necessary to know the anticipated quantities and if those quantities are for the lot, annual, or for the model life. This quantity information is supplied by marketing. With total information, however, the operations planner is able to proceed.

Notice Figure 23.3, which is an example of the process sheet developed for a pinion, given by Figure 23.2, which is a stainless steel shaft requiring several operations. The raw material for the part is a 12-ft bar of stainless steel, grade 430 Ferritic. Other lengths can be specified, and the selection of the optimum length is a choice for the process planner. As the plan progresses, it becomes apparent what the raw material bar stock length will be for maximum material yield. If the quantity is large, bar stock in longer length is usually cheaper than shorter stock length. The selection of the material is a choice for engineering design, but length of bar stock is an operations planner's selection.

23.5 SEQUENCE OF OPERATIONS

The operations sheet column titled "Description" in Figure 23.3 is an abbreviated instruction to the shop that they follow in making the part or subassembly. For "Op. No. 10" the instructions to the shop are listed on the form and correspond to Figure 23.3. Notice operations 20, 30, and 40. The vertical mill, horizontal mill, and a numerical control turret drill press complete the selections for the equipment or workstations.

The operations are customarily numbered as 10, 20, 30, and so on, the first time the operations are planned. Afterward, if an additional operation is necessary between 20 and 30, it might be numbered 25.

Part No. _4943806_   Ordering quantity _1000_   Material _430°F_

Part name _____   Lot requirement _200_   _Stainless 1.750 ± 0.003_

Operation planner _Ed_   _Cold finished 12-ft long_

Workstation	Op. No.	Description	Special or Standard tools
CNC turret lathe	10	Position for length, c'drill Face 0.015 off end Turn rough to 1.45 OD Turn rough to 1.15 OD Turn finish to 1.110 OD Turn 1.235 in OD Cut off to length 18.750	Carbide tools standard Use collet and support
Vertical mill	20	End mill 0.89 slot with 3/4 carbide end mill	Collet for vertical Hold
Horizontal mill	30	Slab mill 4.75 × 3/8	Nesting vise carbide milling cutter tool
NC turret lathe	40	Drill 3/8 holes for 3/8–16 tap	Collet fixture Drill Tap File
Deburring bench	50	Deburr Sharp edge	

Figure 23.3
Process sheet.

The description of operations is in the *jargon* or language of the shop. With the drawing and knowledge of the machine, the operator is able to execute the requirements. Tools or gages also can be listed with the description.

23.6 PINION OPERATIONS SHEET

Selection of the machine or process is made to manufacture the part. This selection is posted under the Workstation column, next to the operation number 10 in Figure 23.3.

A CNC turret lathe machine might be selected to face, turn, and cut off the part, as operation number 10. There will be other choices for the first operation equipment, but that depends on equipment availability within the plant. The initial raw stock often establishes first operation. Twelve-ft round stock is handled by a CNC turret lathe having capacity to handle this material, which is progressively machined and cut off.

A lathe will do cylindrical type of work. It is very unlikely that an engine lathe would be used for this operation, as these machines are not suitable for production quantities. The series of metal removal steps are given under Description. Probably each step requires a single tool, although it is also possible to gang a few of the dimensions and cuts together on one tool. A notation is given that the tool material is tungsten carbide. Special grades and characteristics of the tool cutting point will amplify the tool material. High-speed steel material is an unlikely possibility, since it wears quickly when machining stainless steel stock material. These cutting tools are mounted on the ram or the cross slide turret.

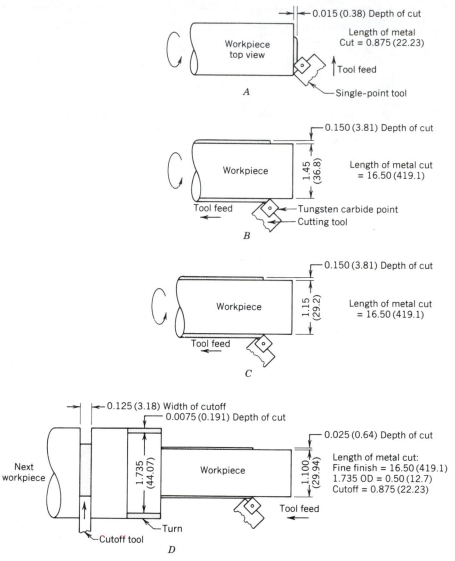

Figure 23.4
Machining cuts for operation 10 using turret lathe. *A*, Facing element. *B*, First pass, rough-turning element. *C*, Second pass, rough-turning element. *D*, Final turret lathe elements.

Notice Figure 23.4, where the operation number 10 machining cuts are illustrated. The facing cut, shown in sketch *A*, is an end cut that cleans the bar stock end, allowing for accurate datum dimensions from the end to other locations on the part. Face 0.015 in. means to remove 0.015 in. from the end face of the bar stock. The other passes shown in Figure 23.4 are indicated for operation number 10, which corresponds to Figure 23.3. Operation number 10 concludes the cylindrical work for the part.

Operation number 20 is an end mill machining pass that slots View *AA*, as shown in Figure 23.2. The workstation calls for a vertical mill having a vertical spindle. The part is held in a collet, which is mounted to the T-slots of the table. Inasmuch as the width

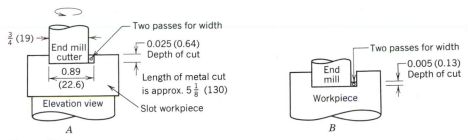

Figure 23.5
Vertical end mill operation on stainless-steel part. *A*, Rough pass. *B*, Finish pass.

of the slot (= 0.89 in.) is wider than the cutter diameter (= 0.75 in.), two passes are necessary for the width dimension. Then, too, a rough pass and a finish pass are necessary for depth. Positioning of the slot with respect to centerline of the shaft is important for this operation. Notice Figure 23.5 for vertical end milling.

Operation number 30 in Figure 23.3 uses a slab milling cutter to establish the flat on the right end, View *BB* in Figure 23.2. The workstation equipment is a horizontal milling machine with an arbor. The cutter is mounted on the arbor. The workholder is a fixture or a vise, where there is positioning of the part relative to the flat, because the keyway was machined previously in operation number 20. Notice Figure 23.6, which shows the relationship between the 6-in. wide, 4-in. OD, 8-tooth tungsten carbide cutter.

Operation number 40 is a drill and tap operation for the holes, as shown by View *AA* in Figure 23.2. There is a positioning relationship among the holes, the slot, and the flat. A collet workholder is necessary to hold the part vertically. There must be sufficient clearance for the part mounted on the table and the NC turret drill press. Remember that the part is 18.750 in. in length.

Typically, if there is machining operations, a *burr*, rolled edge, or a sharp edge is left from the milling cutter, for example, as it concludes after machining the surface. It is necessary to remove this sharp edge, and the process of burr removal is called deburring. This operation is processed by a variety of methods. The one selected here is "manual," hardly a high-tech approach. Other methods are certainly available, but a manual method with a hand file is chosen to focus attention that the operation is selected with the equipment and skills that are available within the plant. Because this is a *labor-intensive* operation, it will be a natural for cost reduction. However, the process planner may have no other option than to use this method.

A collet fixture and a nesting vise are required for operation numbers 20, 30, and 40. A *tool designer* will design tools to meet the needs of the op sheet.

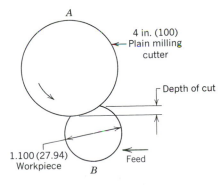

Figure 23.6
Horizontal mill operation on stainless-steel part. *A*, Cutter is 6-in. wide, 8-tooth, and one pass required. *B*, Workpiece has ⅜-in. depth of cut.

This operations sheet can be altered to consider single parts, simple assemblies, or complicated products, but the approach remains the same. It is a simple extension to have rpms and travel velocities of the cutters or the tables listed by an operations sheet.

23.7 WELDED STEEL ASSEMBLY OPERATIONS SHEET

Assemblies are processed using the standard op plan, but there are other practical considerations when compared to individual components. Notice Figure 23.7, which is a welded steel *assembly*. The title block tolerances are for 2-place dimension ± 0.010; 3-place dimension ± 0.001; and 4-place dimension ± 0.0005. The quantity is 35 total units. Raw material is hot-rolled steel (HRS) material, and is provided in strips that are 10-ft long and to widths and thicknesses that are 2 by 0.5 in. for part a; 1 by 0.25 in. for part b; and 2 by 0.625 in. for part c.

The assembly is welded with a flux cored wire that is matched to the chemistry of the HRS material. Anywhere there is a T joint, welding is performed for the length. Beside welding, there is milling of the slot and bottom, drilling, line reaming, and tapping.

Since there are three separate parts and a welded assembly, it is necessary to have four separate operations plans. For simplicity, these operations plans are gathered in Figure 23.8, and are indicated as "a," "b," "c," and "assembly." For 35 finished parts, 70 pieces of "a" and "b" and 35 pieces of "c" are required.

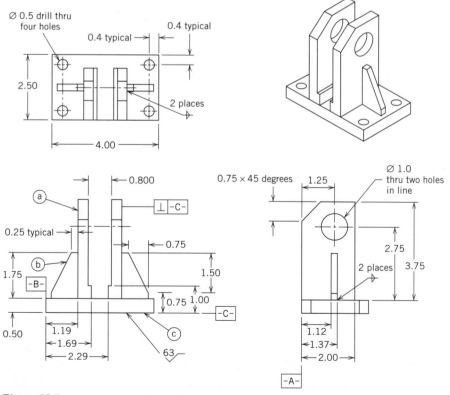

Figure 23.7
Welded steel assembly.

| Part No. _4943807_   Ordering quantity _35_   Material _Hot rolled steel_ |
| Part name _Base_   Lot requirement _____   _Strip 2×0.5; 1×1/4;_ |
| Operation planner _ed_   _2×5/8; 10 ft long_ |

Workstation	Op. No.	Description	Special or Standard tools
		Part a, 70 pieces	
Shear	10	Shear 2×½ ×3.75	Standard
Punch press	20	Notch 0.75×45° corner	
Punch press	30	Notch second corner	
Cleaning vat	40	Clean for weld	
		Part b, 70 pieces	
Shear	10	Shear 1.75 long	Standard
Punch press	20	Notch corner	
Cleaning vat	30	Clean for weld	
		Part c, 35 pieces	
Shear	10	Shear 2× 5/8 ×4 in.	Standard
Cleaning vat	20	Clean for weld	
		Assembly 35 finished pieces	
Welding bench & robot	10	Weld complete	Welding fixture
Heat treat furnace	20	Stress relieve	Standard sled
Hood	30	Shotblast off slag, dirt	
Horizontal mill	40	Mill 5/8 in. plate to 5 in.	HSS milling cutter
NC drill press	50	Drill & tap all holes Ream 0.999 hole	Hinged drift plate
Horizontal mill	60	Mill slot to 0.872	Special side milling cutter, 8 in. OD carbs
Tumbler	70	Tumble to remove burrs	

Figure 23.8
Process plan for Figure 23.7.

Raw stock is supplied in hot-rolled strips, which are purchased from a steel supplier to various widths and thicknesses, as needed by the design. The strips could be band sawed, hack sawed, or sheared to length. The tonnage required to shear soft steel can be calculated as suggested in Chapter 16. It is assumed that a guillotine squaring shear is available. Shearing will leave a smoother edge, and there is no loss of metal length for the kerf of the saw blade. Shearing leaves a more tidy bottom edge compared to sawing, and a deburring operation can be avoided. It is likely that shearing is cheaper than sawing. Notice Figure 23.8 where shearing to length is selected for each part. Notching of the corners on a punch press is next. The individual parts are cleaned for welding.

In view of the small quantity of 35, a welded design is chosen. Welding of the five-part assembly is a difficult operation, because of the vertical positioning of parts on the base. Weld tacking is necessary and a welding fixture is preferred. Certainly, simple blocks and clamps to hold the parts for assembly may avoid the cost of a fixture, but a simple fixture will avoid this fitting and should be cheaper.

Welding will cause the assembly to deform even though there is a holding fixture. A stress relieve operation is required. The assembly needs removal of dirt and any weld rod slag on the part. Discoloration by welding is a possibility, but the drawing did not indicate appearance requirements, for example.

Shot blasting the assembly is one way to prepare the part for milling of the ⅝-in. bottom plate to a 0.500-in. thickness. Shot blasting can give an appearance change, as well as remove slag. Drilling, reaming, tapping, and slot milling complete the part.

Notice that a 63 μin. surface is specified on the bottom surface. This is an ordinary mill surface roughness with finish milling conditions, and surface grinding the plane area is an over requirement, as well as more costly. A special 8-in. OD side milling cutter, with teeth that will cut on both sides of the cutter and that has a width dimension of 0.872 in. needs to be designed and built in preparation of the release of the lot to the shop. Remember that the diameter of the cutter, 8 in., has an arbor hole for attaching to the horizontal mill's arbor, and the depth of the milling slot of 3.75 in. stipulates that the diameter of the milling cutter needs to have clearance of the arbor above the top of the slot.

23.8 TRENDS

Traditional operations planning is done manually; that is, pad, pencil, and tables, or computer spreadsheets are adopted. Special software tools are available that automate the operations planning of parts and assemblies.

Obviously planning starts with the study and examination of the engineering drawing, and results in the completed operations plan. The previous experience of the process planner is critical to the plan. Planning is often described as an art form. This traditional approach results in differences between various planners even for near identical parts that have been processed over a period of time. Individuals have their own opinion as to an optimal routing. These manual methods often have similar operations; for example, shear strips, shear blanks, blank, and form are operations for sheet metal starting with sheets of steel. Methods are available that automate these processes.

Various programs have attempted to capture the logic, experience, and opinion that is so necessary for a manual preparation system. A *computer-aided process planning system* (CAPP) offers potential in reducing the clerical work, and can offer consistent routing that might be optimal. Computer systems are referred to as retrieval or generative. Many of these computer systems use expert systems or artificial intelligence in their scheme of preparation.

Retrieval systems use part classification systems, or group classification, discussed in Chapter 29, as a major consideration. In this way parts are classified into families of a similar part, and they are identified according to their manufacturing characteristics. For each part family, a standard routing is determined. This standard operations sheet is then stored in the computer for this parts family. Efficient retrieval is necessary and the parts classification is used as the call number. Variation from the standard plan is then thought through by the planner. If there is an exact match between the new part code number and an existing inventoried code, the retrieved op sheet is used for the new part.

Generative process plans use algorithms to create process plans from scratch. It is desirable that these plans do so without human assistance. Input to the system could include a complete set of specifications, perhaps even drawings. The analysis of the geometry, material, and reductions in the physical shape of the part leads to the selection of the sequence, machine, and time estimates. The generative plan is considered more difficult to develop, but it is also considered more advanced.

QUESTIONS

1. Give an explanation of the following terms:

Process planner

Perishable tooling

Capital tooling

Tolerances

Operations sheet

Sequence

Assembly planning

Process planning

Engineering design

Work stations

Retrieval system

Generative process plans

2. Discuss what types of equipment are typically required for small quantity production. How does this contrast to equipment that is used for large quantity?

3. Create a job description for the process planner.

4. List the steps in process planning as if the design is given to the process planner following the design. How does this list differ for concurrent engineering activities?

5. Write a paragraph about variant and generative process planning.

6. What did Henry Ford imply by saying, "In mass production there are no fitters."

7. Discuss the differences between process planning for assembly versus machining types of products.

8. Correct sequences of planning suggests that drilling precede tapping. Describe other sequences that have a precedence relationship.

9. How does scrap influence the sequencing of operations?

10. Discuss why the side milling cutter for the welded assembly needs to be 8 in. in diameter. See operation number 60 for Figure 23.8. (Hint: The machine is a horizontal milling machine and uses an arbor to hold the milling cutter.)

11. Discuss the differences in process planning for Figure 23.1.

PROBLEMS

23.1. Prepare an operations plan for Figure 23.1A. Let the raw material be 0.187-in. OD by 12-ft free cutting brass bar stock. The quantity is 1000 units. (Hint: Refer to Chapters 9 and 11.)

23.2. Prepare an operations plan for Figure 23.1B. Let the raw material be 0.187 + 0.005 − 0.000-in. OD free cutting brass. The quantity is 10,000 units.

23.3. Prepare an operations plan for the pinion, Figure 23.2, but with the following changes. (Hints: Your instructor will assign a subpart. Many of the operations are similar to the existing example. Pay attention to the changes.)

a. Let the 18.750-in. dimensions be 8.750 in., the 1.100-in. dimension be 1.125 in., and no holes.

b. Let the 4.75-in. flat dimension be 8.00 in., and the raw material be 2-in. OD instead of 1¾ in.-OD.

c. Let the 4-in. plain cutter be 3 in. in width, and operation number 10 with high-speed tool material instead of tungsten carbide tool material.

23.4. Prepare an operations plan for Figure 23.7, but let the material be gray cast iron with quantity of 25,000 units per year for the next 5 years. What tooling is standard versus special design tooling? (Hints: Determine a plan that considers the sequence number, equipment, process

description, and patterns or tooling. Assume that the raw material starts with a foundry operations and continues to machine shop work.)

23.5. Prepare an operations plan for Figure P23.5. Also sketch the piece on engineering computation paper. Keep the sketch to scale and indicate the successive metal removal with different colored pens or pencils. Your instructor will indicate the quantity. (Hints: Include the type of tools that are necessary in the operations plan. Use a variety of colored pencils or ball-point pens to identify the sequence of metal removal. An initial rough pass may be red, which would be the outside surface. This would be followed by a black finish pass, which would not be as large dimensionally as the rough pass. You may also use various

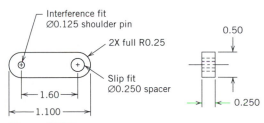

Figure P23.5

cross hatching symbols to indicate rough or finish passes, and the like, where they may be necessary. As processing continues, not all dimensions or views are given, so it is necessary to scale some dimensions for additional information. The initial stock diameter influences the sequence, number, and type of machining passes.)

a. Complete the above instructions for hot-rolled steel bar stock that is 1.75 by 0.5 in. and 0.25 in. thick.

b. Complete the above instructions for cold-rolled sheet stock that is 24 in. wide by 36 in. long and 0.25 in. thick.

23.6. Prepare an operations process plan for Figure P23.6. (Hints: Construct a plan that considers the sequence number, equipment, process description, and tooling that may be perishable or special. Tolerances for 0.XXX in. are ± 0.005, and 0.XX are ± 0.01. Use the information given

throughout the book for selection of equipment and advice for processing.)

a. Consider quantity variations of 25 and 25,000 per year for 10-gage (0.135-in.), AISI 1018 steel. (Hints: Begin by specifying initial stock conditions, such as individual blanks, or strip, or sheet, or perhaps coil stock, and then proceed to develop the process plan. Be sure you indicate the dimensions of the initial stock, and as the operations continue, be aware of the dimensional effects on the stock. As you proceed, it is important to indicate these material alterations. It is not necessary to sketch and design any special tooling, but indicate what it is.)

b. Repeat part a., but let the stock thickness be 0.0135-in. 3004-H19 aluminum.

c. Repeat part a. for a thermoplastic material for a quantity of 25,000 and 25 million units. Is there as much processing variation and complexities as you found in part a. in the processing of metal components?

23.7. Prepare an operations plan for Figure P23.7. Determine plans that consider the sequence number, equipment, process description, and tooling. Visit your school shop, if one is available, and identify the equipment and tools that are found there for your operations plan. If a school machine shop is unavailable, consider visiting a shop that may be nearby. Determine an appropriate quantity for your school "factory" or neighboring shop. (Hints: Before you start, realize that three operation plans are necessary. There are many possible routes, but only a few are practical. Tolerances for 0.XXX are ± 0.005, and 0.XX are ± 0.01.)

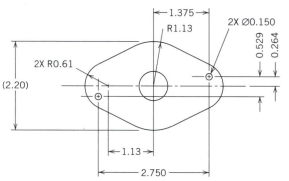

Figure P23.6

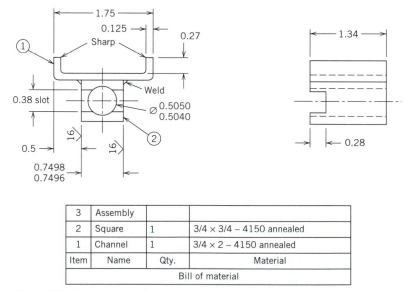

3	Assembly		
2	Square	1	3/4 × 3/4 – 4150 annealed
1	Channel	1	3/4 × 2 – 4150 annealed
Item	Name	Qty.	Material
Bill of material			

Figure P23.7

23.8. Prepare an operations process plan for Figure P23.8. Material is 11-gage, 0.1196-in. thick, low-carbon sheet, supplied in sheet stock 4 ft by 8 ft.

a. Quantity = 125 units.

b. Quantity = 12,500 units. Suggest the raw material for this quantity.

large. Stock is 0.048-in., 1018 CRS, 4 by 8 ft-sheets. What does changing the orginal stock from sheet to strip, or even coil, do to your plan?

23.10 Prepare an operations processing plan for Figure P23.10. The annual quantity is 12,000 units with 1000 units per month. (Hints: Specify the dimensions of the initial

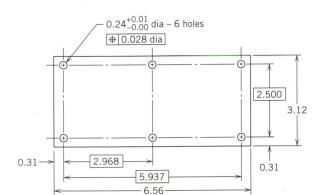

Figure P23.8

23.9. Determine an operations process plan for the manufacturing of the hinge bracket product, Figure P23.9, where this part is used for automobile doors. The quantity is very

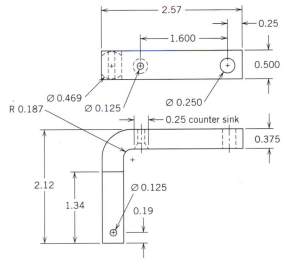

Figure P23.10

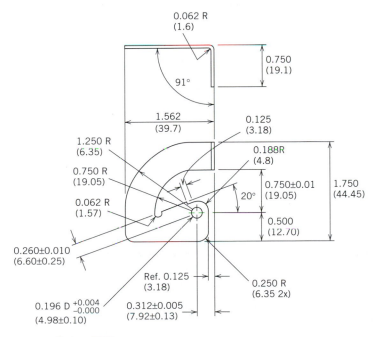

Part no: 8871
Title: Hinge bracket
Material: 0.048(18 ga.) C.R.S., 4 × 8-ft sheet

Figure P23.9

standard annealed bar stock, which is 4130 steel. Notice that one end is rectangular and the other is a partial round. Tolerances are 0.XXX ± 0.005 and 0.XX ± 0.01 in. and final material condition is Bhn 467.) For unspecified holes, the schedule is as follows:

0.250 dowel interference fit	0.2504/0.2500
0.125 shoulder pin slip fit	0.1252/0.1250
0.125 roll pin interference fit	0.1253/0.1250

MORE DIFFICULT PROBLEMS

23.11. Figure P23.11 is the dome light of a passenger car. Write an operations plan for the assembly of this dome lighting system, which is installed on the moving assembly of the car. The *kit* for this assembly consists of three sheet metal screws, bezel, and the lower and upper base. The lower and upper base are preassembled units. The lamps are already inserted into the upper frame. (Hints: Be descriptive in the writing of the steps for the assembly, remembering that assembly process planning tends to be less specific in its description. One installer is doing this work.)

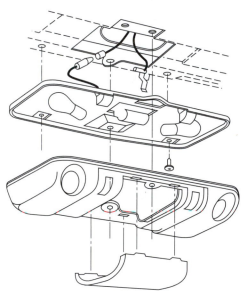

Figure P23.11

23.12. A steel spacer, Figure P23.12, is produced using "world class manufacturing" methods. A single-product company provides international markets with mass-produced quantities. Chemical analysis of the AISI C 1050 steel is C 0.48/0.55%; Mn 0.60/0.90%; P 0.04% Max; and S 0.05% Max. Stock is purchased in the hot-rolled annealed soft condition for machining. The desired post-hardened annealed yield strength is 97,000 psi and 293 Bhn.

Indicate the type of the raw stock, such as coil, bar, or sheet, and its purchased dimensions ready for your operations. Material selection is compatible with the preparation of an operations processing plan where you specify the operation number, equipment (including hardening equipment), description, standard tools and materials, rpms, and other operating conditions of the equipment to satisfy the finished material property specification. Refer to Figure 17.11 for the physical properties of C 1050 steel, and notice the heat-treating conditions.

Sketch a plant layout for your manufacturing system. Dimensions are unnecessary, but the plant is probably not much larger than 1000 ft². End with a discussion of the merits of this "world class manufacturing."

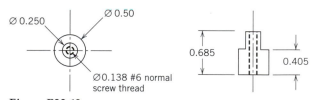

Figure P23.12

PRACTICAL APPLICATION

Purchase a fingernail clipper product or other simple assembly. Disassemble this product and "reverse engineer" the assembly procedure that is used. Visualize and describe an automatic machine that starts with material and ends with a product. Conclude with an operations process sheet for the assembly of the product. (Hints: You have choices; for example, the material may be cold-rolled flat stock, that is not "chromed" or otherwise finished, or the parts may have been fabricated previously and then welded or assembled. By examination of the pieces as a "detective," discover the assembly procedure.)

CASE STUDY: SUPER SNAP RING

Oil Head, Inc., a large transnational company specializing in manufacturing and installing long distance pipe lines, has developed a new scheme of attaching pipe ends together. The notion is that a snap ring, except much larger than small snap rings that can be bought at the hardware store, can be used to attach the bell ends of seamless steel pipe together. It is followed by tack and seam welding around the bell ends. The snap ring is an important part of the field assembly. Field welding is independent of the manufacture of the snap ring. Welding occurs in harsh and remote regions of the world.

The ring has two drilled 0.405/0.403-in.-holes, and a cross hole of 0.251/0.250 in. A simplified sketch is given as Figure C23.1. The material is AISI 8640 alloy and is received as annealed 10-in. bar stock, which is 12 ft long. On conclusion of the metal removal operations, a hardening operation is necessary.

Your job is to write an operations process sheet for an annual order of 18,000 units. This market opportunity is expected to be available for the next 10 years. What major changes in design and material selection will allow for improvement in the material yield? (Hints: Especially focus on the sequence, machine and process identifications, and material removal statements. Because not all dimensions are shown, roughly scale for missing dimensions. Use other chapters to aid in the selection of equipment. Try alternate schemes in the selection of equipment. Recognize that the *material yield* is going to be pretty bad if turning is a principal method.)

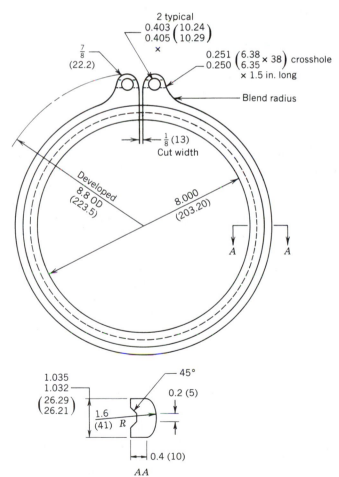

Figure C23.1
Case study.

CHAPTER 24

GEOMETRIC DIMENSIONING

AND TOLERANCING

Geometric dimensioning and tolerancing (*GDT*) specifies engineering design requirements for function and relationships of part features. GDT encourages the economic production of these features. Design and production systems, complexity, computerization, and global manufacturing have made exacting engineering drawing requirements mandatory. Functional gaging, tools, part dimensions, and manufacturing benefit with GDT. The study of GDT is important, because it is the communication glue among design, manufacturing processes, and quality.

Manufacturing and engineering systems require a language that is understandable, otherwise, it is not coherent and usable. A technical language is defined by a "standard," and one that is widely used is the American National Standards Institute (ANSI) known as *dimensioning and tolerancing*. Our purpose is to harmonize GDT with manufacturing processes. You may have been exposed to GDT in a CAD or drafting class. This chapter gives manufacturing understanding to these symbols.

24.1 DIMENSION AND TOLERANCE

Size and location of a *feature* are determined by a *dimension*. Dimensions are seen in engineering drawings, which represent a number that is approximated in actual manufacturing. A shaft may have a *nominal size* of 2½ in. (63.5 mm), but the exact decimal equivalent measurement is unobtainable in manufacturing. Although fractional dimensions are found in older drawings, decimal equivalents are preferred practice. Typically, it is unlikely that manufacturing is able to produce parts to 2.500 in. because there is natural variation in manufacturing.

Tolerance is the amount of variation permitted in the dimension. All dimensions have tolerance, either specific or general (which is stated in the title box). A dimension is a joint number, which includes a basic dimension value with its tolerance. A tolerance considers functional engineering and manufacturing requirements. For example, a dimension could be 2.500 ± 0.005, or 2.5000 ± 0.0005. The dimension given in thousandths or tenth-thousandths implies lesser or greater accuracy.

Tolerance is the range of allowable dimensions, knowing that a dimension is never exact, such as 2.50000 . . . in. A tolerance is similar to the term "significant digits." It provides the machinist with a cutoff number that is achievable. Tolerances may be "tight" or have too many significant digits. With a tight tolerance, cost rapidly increases.

Dimensions are given as limit, unilateral, or bilateral. An example of limit dimensioning is $\frac{2.505}{2.495}$. A *limit dimension* has the larger value on top, and the fractional line is omitted between the two values. Limit methods of dimensioning are preferred practice. *Bilateral dimensioning* form is 2.500 ± 0.005, where both tolerance values are equal in this case. Bilateral practice does not require both tolerance values to be the same. A special case is unilateral, such as $2.495 \, ^{+0.010}_{-0.000}$, where the lower value is zero. *Unilateral*

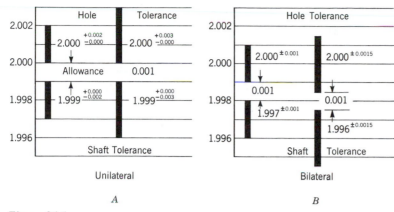

Figure 24.1
Tolerance and allowance. *A,* Unilateral application. *B,* Bilateral application. Dimensions are representative of either English or metric units.

tolerance means that any variation is made in only one direction from the nominal dimension.

Manufacturing deals with fitting and mating parts, a shaft in a hole, for example. *Allowance,* which is sometimes confused with tolerance, is defined differently. It is the minimum clearance space intended between mating parts and represents the condition of tightest possible fit. Notice Figure 24.1*A,* where unilateral tolerances are used. For the two cases in *A,* the allowance is 0.001. Similarly in Figure 24.1*B,* the use of bilateral tolerances shows an allowance of 0.001. An advantage of the unilateral tolerance permits changing the tolerance while still retaining the same allowance. A force fit or an interference fit between mating parts requires the tolerance to have a zero or negative allowance.

24.2 SYMBOLS

Geometric dimensioning and tolerancing symbols are given by Figure 24.2. These symbols replace many of the notes on engineering drawings. Advantages of symbols over notes are given as

1. *Uniform and defined meaning.* A note may be stated inconsistently.
2. *Compactness.*
3. *An international language.* Notes may require translation between languages.
4. *Manufacturing clarity.*

Some of these symbols are seen in Figure 24.3. Study this machined part, where dimensions are in millimeters.

Maximum material condition (MMC) or Ⓜ deals with hole and shaft diameters and similar features. Ⓜ enclosed in the circle is the drawing symbol, although MMC is used in discussion. A hole that is given as 2.500 ± 0.005 in. ranges from a maximum hole size of 2.505 to a minimum diameter of 2.495 in. The material is "maximum," however, when the diameter is smallest, or 2.495 in. The hole at its low limit contains more material than its high limit.

Thinking for MMC is opposite for shafts. For example, if the shaft diameter is 0.750 ± 0.004 in., the high limit of the tolerance, when applied to the shaft, gives 0.754 in., which is the MMC for the shaft, that having more material than the low limit.

SYMBOL	CHARACTERISTIC
▱	Flatness
—	Straightness
○	Circularity (roundness)
�	Cylindricity
⌒	Profile of a line
⌓	Profile of a surface
⊥	Perpendicularity (squareness)
∠	Angularity
//	Parallelism
↗	Circular runout
↗↗	Total runout
⊕	Position
◎	Concentricity
═	Symmetry
Ⓜ	Maximum material condition MMC regardless of feature size RFS
Ⓛ	Least material condition LMC
⌀	Diametrical (cylindrical) tol zone or feature
60.5	Basic, or exact, dimension
—A—	Datum feature symbol
⊕ ⌀0.2Ⓜ A	Feature control frame

Figure 24.2
Geometric characteristics and terms.

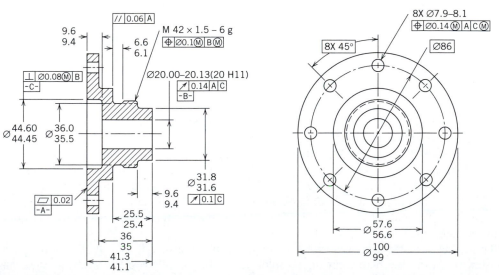

Figure 24.3
Sample GRT drawing.

Generally, MMC allows greater possible tolerances. It aids interchangeability and permits functional gaging techniques. The MMC implies worst condition or critical size, and relates mating part. The MMC principle applies to features of size, such as a hole, slot, or pin with an axis or center plane.

Least material condition (*LMC*) or the drawing symbol Ⓛ is the condition in which a feature of size contains the least amount of material within the stated limits of size. It is the maximum hole diameter or minimum shaft diameter.

Regardless of feature size (*RFS*) indicates that geometric tolerance or datum reference applies at any increment of size of the feature within its size tolerance. It indicates tolerances apply to a geometric feature regardless of size, from MMC to LMC. RFS permits no additional positional, form, or orientation tolerance. The RFS applies only to features of size, such as a hole, slot, pin with an axis or center plane. The RFS is implied, unless otherwise specified under MMC rules.

The term *basic* is a numerical value and dimension that describe the exact size, profile, orientation, or location of a feature or datum. It is the dimensional value that requires a tolerance, giving the permissible departure from the exact basic value. The *basic dimension* is enclosed in a feature control frame. (See Figures 24.2 and 24.3.) It is emphasized that not all dimensions are enclosed in frames. Only the important ones are framed and that depends on the engineering design requirements.

A *datum* is the theoretically exact point, plane, or axis that is the origin from which the location or geometric characteristics of features of a component are established. Each datum is featured in a frame and is identified by capital alphabetic letter. A workpiece shown with datums *D*, *E*, and *F* is given in Figure 24.4. The first datum plane *D* requires that the workpiece have a minimum of three locating points, which defines a plane surface or the top surface of a machine tool table. The secondary datum *E* requires two points

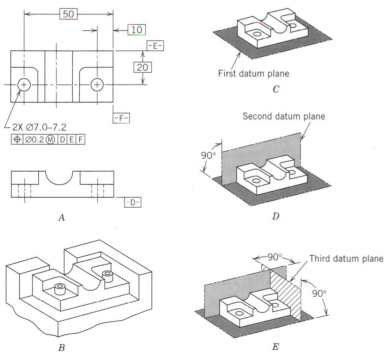

Figure 24.4
A, Part design. *B*, Plane surfaces. *C*, *D*, and *E*, Datum planes.

of contact, whereas the third datum *F* needs one point of contact. Datum planes are usually associated with manufacturing equipment, such as machine tables or with locating pins. A three-two-one location principle of datums is closely related to tool design, which is studied in Chapter 25.

For a *tolerance of position*, the datum reference letter is followed by the appropriate modifying symbol in the feature control frame. For all other geometric symbols, RFS is implied, unless otherwise specified. Where a datum feature of size is applied on an RFS basis, the datum is established by physical contact between the feature surface or surfaces and surfaces of the processing equipment. Machine elements, which are variable in size (chuck, mandrel, or centers), are used to simulate a true geometric counterpart of the feature to establish the datum.

Where a datum feature of size is applied on an MMC basis, machine and gaging elements in machine tools and processing equipment, which remain constant in size, may simulate a true geometric counterpart of the feature and establish the datum. In this case, the size of the simulated datum is found by the specified limit of size of the datum feature or its virtual condition. The boundary generated by the collective effects of a specified MMC limit of size of a feature and any geometric tolerances is *virtual condition*. Figure 24.5 shows the simulated datum, which is the surface of the machine tool or inspection devices, datum feature of the part, and the theoretical exact datum.

Measurements cannot be made from the theoretical or exact datum. The datum is assumed to be simulated by the manufacturing or inspection equipment. For example, surface plates or machine tool tables are assumed to be of sufficient quality, as compared to the part, and they are used to simulate the datums. For an external cylinder (shaft), the datum axis is the axis of the smallest circumscribed cylinder that contacts the cylindrical feature of the part. That is, the largest diameter of the part will make contact with the smallest contacting cylinder of the machine element that holds the part. For an internal cylinder (hole) the datum axis is the axis of the largest inscribed cylinder that contacts the inside of the hole. That is, the smallest diameter of the hole will make contact with the largest cylinder of the machine element inserted into the hole.

The major features of the part establish the *basic coordinate system*. These part features are defined in relation to the datums, which are the mutually perpendicular reference planes. The planes are compatible to the computer numerically controlled planes used by the machine tools and the manufacturing systems. Machine tool tables are used frequently as the plane surface for the three-point requirement.

A feature is the entity of the part, such as a surface, hole, relief, slot, or thread. Features may be individual or related. The *feature control frame* consists of a rectangular box geometric symbols, datum references, tolerance, and MMC, for example. Each frame is typically different and depends on the nature of the part. Feature control frames are shown in Figures 24.2. 24.3, 24.4, and 24.9, for example.

The GDT symbol for the diameter (on some drawings the symbol DIA is used or ID

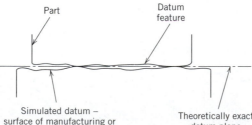

Part   Datum feature

Simulated datum – surface of manufacturing or verification equipment

Theoretically exact datum plane

Figure 24.5
Datum feature, simulated datum, and theoretical datum.

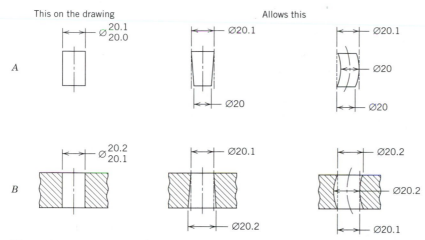

Figure 24.6
Extreme variations of form allowed by a size tolerance. *A*, Shaft. *B*, Hole.

or OD for internal and outside diameter) of a cylindrical feature or tolerance zone is ϕ. The ϕ symbol precedes the numerical value.

There are five types of geometric characteristics. A *form* tolerance gives the allowable and maximum variation for an actual surface or feature from the specified form on the drawing. A *profile* tolerance gives the maximum variation of an actual surface or feature from the desired form or datum, as indicated on the drawing. Further, *orientation* states how far an actual surface or feature is permitted to vary relative to a datum. *Runout* is the permitted variation of the actual surface from the desired form during full 360° rotation of the part on a datum axis. A *location* tolerance gives the maximum variation of an actual size feature from the perfect location implied by the drawing.

24.3 APPLICATIONS

There are limits of size that a dimension indicates for manufacturing. In Figure 24.6, the tolerance of size of an individual feature prescribes the extent of the variations in its geometric form. An envelope is the true geometric form of either the pin (*A*) or the hole (*B*). The left-hand sketches are the drawing statements of the dimensions, while the manufacturing and inspection interpretations are shown.

Tolerancing of an angular surface with a combination of linear and angular dimensions is given by Figure 24.7. The surface controlled by the angular dimension may be anywhere within the tolerance zone with one restriction. The manufacturing process must keep the angle between 29° 30′ and 30° 30′.

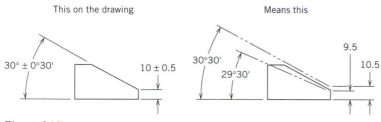

Figure 24.7
Tolerancing an angular surface using a combination of linear and angular dimensions.

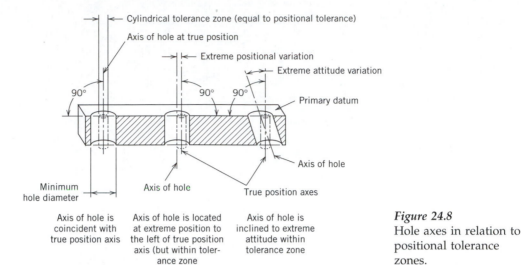

Axis of hole is coincident with true position axis

Axis of hole is located at extreme position to the left of true position axis (but within tolerance zone

Axis of hole is inclined to extreme attitude within tolerance zone

Figure 24.8
Hole axes in relation to positional tolerance zones.

Where a hole is at MMC (minimum diameter), its axis must fall within a cylindrical tolerance zone whose axis is located at true position. This is shown by Figure 24.8. The length of the tolerance zone is equal to the length of the feature, unless otherwise specified on the drawing.

An example for positional tolerancing for RFS for symmetry is given by Figure 24.9. The frame requires that the center plane of the slot lie between two parallel planes 0.8 mm apart, regardless of feature size, which are equal about the center plane of datum B.

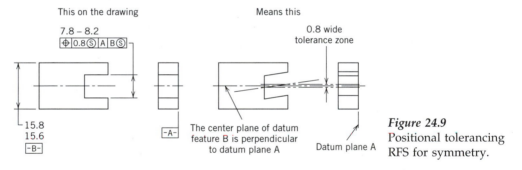

Figure 24.9
Positional tolerancing RFS for symmetry.

Concentricity tolerancing for coaxiality is given by the example of Figure 24.10. Concentricity is the condition where the axis of all cross-sectional elements of a surface of revolution are common to the axis of a datum feature. A concentricity tolerance specifies a cylindrical tolerance zone whose axis coincides with a datum axis and within which all cross-sectional axes of the feature being controlled must lie.

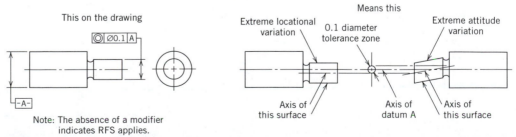

Figure 24.10
Concentricity tolerancing for coaxiality.

The design specifies *straightness* for manufacturing. Figure 24.11 shows the straightness symbol. Each longitudinal element of the surface must lie between two parallel lines (0.02 mm apart), where the two lines and the nominal axis of the part share a common plane. In addition, the feature must be within the specified limits of size and the boundary of perfect form at MMC.

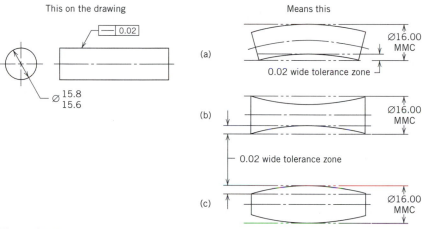

Figure 24.11
Specifying straightness of surface elements.

Flatness is specified by a parallelogram within the frame as shown by Figure 24.12. The surface must lie between two parallel planes 0.25 mm apart. In addition, the surface must be within the specified limits of size.

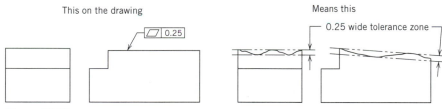

Figure 24.12
Specifying flatness.

Cylindricity or *roundness* is shown by Figure 24.13. The cylindrical surface must lie between two concentric cylinders, where the radius is 0.25 mm larger. In addition, the surface must be within the specified limits of size.

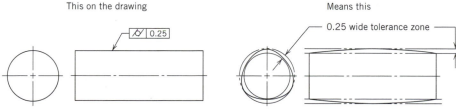

Figure 24.13
Specifying cylindricity.

Angularity or *taper* is a frequent manufacturing requirement. The drawing callout is shown by Figure 24.14, and refers to datum A. The surface must lie between two parallel planes 0.4 mm apart, which are inclined at 30° to datum plane A. In addition, the surface must be within the specified limits of size.

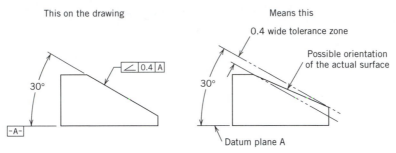

Figure 24.14
Specifying angularity for a plane surface.

Surfaces may require *parallelism*, and the engineering and manufacturing drawing is given by Figure 24.15. The meaning is shown. The surfaces must lie between two planes 0.12 mm apart, which are parallel to datum plane A. In addition, the surface must be within the specified limits of size.

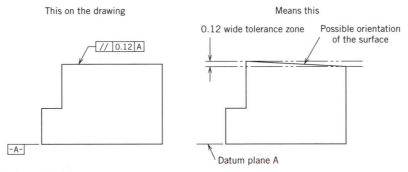

Figure 24.15
Specifying parallelism for a plane suface.

Perpendicularity specification is given in Figure 24.16. The surface must lie between two parallel planes 0.12 mm apart, which are perpendicular to datum plane A. In addition, the surface must be within the specified limits of size.

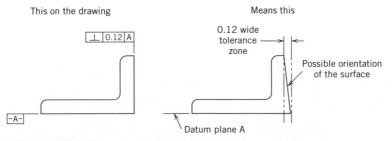

Figure 24.16
Specifying perpendicularity for a plane surface.

Circular runout can be either for a particular position or surface. In Figure 24.17, the requirement deals with single circular elements. The part is given a datum axis A. The part is rotated, perhaps using 60° conical centers at both ends. A dial indicator is mounted and the stylus is permitted to have a maximum movement of 0.02-mm *FIM* (*full indicator movement*). At any measuring position, each circular element of these surfaces must be within the specified runout tolerance (0.02-mm FIM) when the part is rotated 360° about the datum axis, with the indicator fixed in a position normal to the true geometric shape. This controls only the circular elements of the surfaces, not the total surfaces.

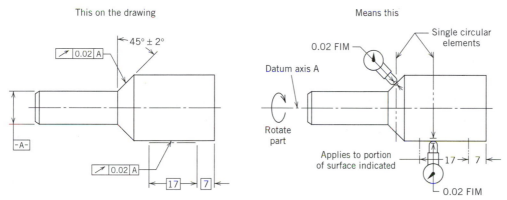

Figure 24.17
Specifying circular runout relative to datum diameter.

In Figure 24.18, the FIM deals with a tolerance zone. This deals with *total runout*. For manufacturing runout is a composite tolerance used to control the functional relationship of one or more features to a datum axis. The entire surface must lie within the specified runout tolerance zone (0.02-mm FIM) when the part is rotated 360° about the datum axis, with the indicator placed at every location along the surface in a position normal to the true geometric shape without reset of the indicator. In addition, the feature must be within the specified limits of size.

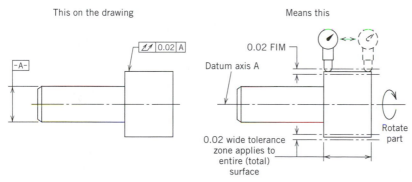

Figure 24.18
Specifying total runout relative to datum diameter.

Figure 24.19 is an example of an assembly where *angular orientation* is important. Notice the basic dimensions given the cover with respect to the key, body, and the bolt circle relationship between the cover and body.

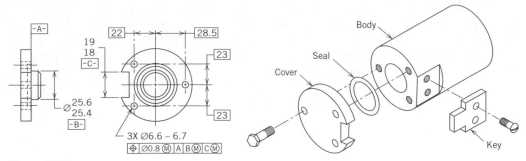

Figure 24.19
Part with angular orientation.

24.4 CAD AND CAM

Computer-aided design (CAD) and computer-aided manufacturing (CAM) are the systems that describe the part as a geometric model, and interactively adding manufacturing data, deliver this information to a manufacturing system or machine tool to complete the part. Geometric dimensioning and tolerancing is the practice that allows the dimensioning of the part into the manufacturing of the object.

The coordinate system is the same for the geometric model created by CAD and the CAM system that produces the part. The rectangular or Cartesian coordinates locate a point by its distance from each of two or three mutually perpendicular intersecting planes. Two-dimensional coordinates (in X and Y directions) locate a point on a plane. Three points in the coordinate XYZ direction locate a point in space. Once a point is defined, there is opportunity for commands to the machine tool, which execute the required relative motion between the cutting tool and the part. Linear, rotary, and composite machine tool motions are possible. Information for computer numerical control machining is given in Chapter 28.

24.5 APPROPRIATE TOLERANCES

When an engineer specifies dimensions for a part that is to be produced, it is necessary to include the tolerance along with the dimension. Tolerances should be specified from actual design requirements but with manufacturing capabilities clearly in mind. They should not be specified more accurately, even though the CAM equipment is capable of the greater accuracy. By knowing appropriate tolerances, an engineer is able to design a cost-effective part.

Tolerances may be expressed by general notes printed on the drawing immediately with the dimension such as 2.500 ± 0.005, or in the title box, which is applied to those dimensions. For example, if the dimension is 2.500 in., and there is no specified tolerance with the dimension, then the tolerance that is applied is based on the 0.xxx significant third-place digit, as given in the title box. An illustrative example of general title block tolerances for customary dimensions given in inch units is given as

2 places, 0.xx	± 0.010 in.
3 places, 0.xxx	± 0.005 in.
4 places, 0.xxxx	± 0.0005 in.

For a 2.240-in. dimension, the range of allowable dimensions is 2.235 to 2.245 in. Most dimensions on a drawing use the general title block tolerances.

Table 24.1 **Sample Tolerances for Traditional Machine Tools**

Machine	Condition	Tolerance
Mill	Minimum geometric	0.005 in.
	Surface roughness (side cut)	32 μin.
	Surface roughness (end cut)	63 μin.
	Straightness	0.005 in.
	Flatness	0.005 in.
	Parallelism (same side)	0.005 in.
	Parallelism (opposite side)	0.015 in.
	Angularity	0.1°
	Perpendicularity	0.015 in.
Lathe	Minimum geometric	0.002 in.
	Surface roughness	16 μin.
	Concentricity	0.005 in.
Grinder	Minimum geometric	0.001 in.
	Surface roughness	4 μin.
	Flatness (plate > 1 in. long)	0.002 in.
	Flatness (plate < 1 in. long)	0.001 in.

The tolerance may be applied directly to the dimension and appears between the dimensional arrows on the drawing. In this case the tolerance is not removed from the title block, but is given with the dimension.

Companies will determine appropriate tolerances that are suitable for their manufacturing equipment and product. An example of *effective tolerances* for a college or school shop, where the equipment is maintained, is given by Table 24.1. For example, the milling machine used for metal removal of parts that are not a surface of revolution will have a standard of accuracy of 0.001 in. Rigid machines may be capable of 0.0005 in. If deflection, heat expansion, nonrigid part, tool wear, inadequate tooling, and the like, are significant, the rule of thumb considers the appropriate tolerance as 0.005 in.

The lathe is the standard machine tool for surfaces of revolution. It has a standard accuracy of 0.001 in. Rigid or new CNC lathes can achieve an accuracy of 0.0005 in. Tighter tolerances are achieved more easily on the lathe than on the milling machine. But slender parts, interrupted cuts, hard metals, and other practical conditions do influence this tolerance. A minimum tolerance of 0.005 in. is frequently acceptable. Lathe *surface roughness* capability is superior to the milling machine, ranging from 250 to 16 μin. Concentricity tolerances greater than 0.005 in. are practical.

Grinders, as a family, may be rotational, planer, internal, or external. These machines are able to give tighter tolerances than either the milling machine or the lathe. Generally, accuracies of 0.0005 in. are achievable, although 0.001 in. is preferred. A surface roughness range of 63 to 4 μin. is possible. Generally, grinders are a more expensive way to remove metal as compared to turning or milling.

Figure 24.20 is a milled part with dimensions, surface roughness, and GDT symbols. Whereas these two views are considered sufficient for manufacture, each dimension is analyzed for overdimensioning. Next to each dimension is a number, which refers to an interpretation of the dimension. For example, the number "1" is placed next to 3.050 ± 0.001. Referring to Table 24.2, it is advisable to increase the tolerance to 0.005 in. or greater. Figure 24.20 is the drawing before interpretation and redimensioning. Table 24.2 gives the discussion regarding the interpretation. One needs to recall that this discussion deals with manufacturing capability, and functioning requirements for the product may

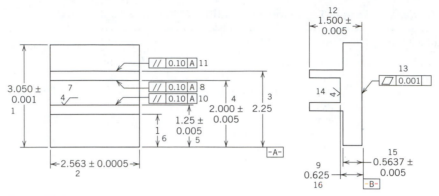

Figure 24.20
Appropriate tolerances.

not allow redimensioning. If the tolerances cannot be changed, some operations shift from milling to grinding.

The designer dimensions the part so that the geometric shape is completely defined and mathematically precise. Regular geometric profiles, such as ellipses, parabolas, and the like, may be defined on the drawing by mathematical formulas. Computer-aided manufacturing equipment can be programmed to generate these profiles by linear interpolation, that is, a series of short straight lines. These short lines have end points that are spaced closely, and the overall approximation becomes the desired profile with the profile tolerances. Thus, for certain CAM equipment, the profiles are handled by mathematical formulas and not coordinate dimensions, and the engineering drawing shows the formula rather than the dimensions. An exception to showing the formula might be for inspection requirements, especially if the tools and equipment do not have mathematical CNC capability.

A variety of coordinate systems are available. These *coordinate systems* are known as Cartesian (as generally used here with the discussion of the three mutually referenced planes), polar, spherical, or cylindrical. Usually, the same coordinate system is used on the drawing and the manufacturing equipment.

Table 24.2   **Interpretation of Milled Part**

Number	Remarks for Dimension
1.	Tolerance is too tight for mill, 0.005 in. or greater.
2.	Tolerance is too tight for mill, 0.005 in. or greater.
3.	A tolerance is necessary.
4.	Adequate.
5.	The tolerance and dimension should agree in decimals.
6.	A tolerance is necessary.
7.	The surface roughness is too tight; should be 32 μin.
8.	Parallism tolerance is too tight; should be 0.015 in. since it refers to the datum.
9.	No tolerance specified.
10.	Parallelism tolerance is too tight; should be 0.005 in. It is in addition to the other wall and parallel.
11.	Parallelism tolerance is too tight; 0.005 in. recommended.
12.	Tolerance is sufficient, although very tight. Could it be looser?
13.	Flatness tolerance is too tight; should be 0.005 in. since it is with datum.
14.	Too tight of a surface roughness, should be 63 μin.
15.	Too many significant digits in the dimension.

QUESTIONS

1. Give an explanation of the following terms:

GDT   Virtual condition
Feature   Concentricity
Dimension   Straightness
Tolerance   Flatness
Allowance   Angularity
MMC   Perpendicularity
LMC   Circular runout
RFS   Full indicator movement
Basic dimension   Coordinate systems
Datum

2. Discuss the advantage of GDT symbols over drawing notes.

3. Describe the practice of using the term "basic" or boxing the dimension as found on a drawing. What is the importance of the basic dimension for GDT?

4. List the types of symbols for form.

5. What is the difference between circular runout and total runout?

6. How is runout confirmed with a part?

7. How would you determine the typical tolerances for your school machine shop? Develop a plan that involves testing, metrology equipment, and types of samples.

8. List advantages of using datums for design and manufacturing.

9. Give the tolerance symbols for profile of a surface, circular runout, flatness, cylindricity, profile of a line, total runout, straightness, and position.

10. List orientation symbols.

11. What is the difference between profile of a surface and profile of a line?

12. How are design and manufacturing coordinated with GDT?

13. List coordinate systems.

14. How are datums assured in manufacturing?

PROBLEMS

24.1. Give the MMC for 0.963/0.964 in. if the dimension is a shaft or a hole.

24.2. Give the MMC for 1.025/1.023 in. if the dimension is a shaft or a hole. Give the LMC.

24.3. Sketch the following GDT applications:

a. A shaft ϕ 24.6/24.0 mm and a hole ϕ 25.0/25.6 mm. Indicate the MMC of the shaft and the hole.

b. A shaft of ϕ 7.4/7.3 mm and a hole of ϕ 7.6/7.5 mm. Indicate the MMC of the shaft and the hole. What is the maximum clearance and allowance between the shaft and hole?

24.4. Sketch the following design showing proper GDT symbols and dimensions for this layout: A ¼-in.-thick rectangular plate has an outline of 4.00 by 2.00 in. or (x dimension, y dimension). Coordinates for entities are given relative to the lower left-hand corner, which is the origin (0, 0). Positive measurements are above and to the right. Three 0.400-in. holes are located at centers given by (1, 1), (2, 1), and (3, 1). The holes have a + 0.010-in. size tolerance and a position tolerance diameter of 0.005 in. (Hints: It is not necessary to use drawing instruments or CAD systems.

Use 8½ by 11-in. engineering computation paper having 1-in. grids. Sketching implies pencil work, perhaps with the aid of a straight edge.)

24.5. Sketch the following design showing GDT symbols. A 1.000-in.-thick rectangular plate has an outline of 2.00 by 4.00 in. or (x dimension, y dimension). Coordinates for entities are given relative to the lower left-hand corner, which is the origin (0, 0). The datums are A, the top surface of the block, B or the y-axis left end, and C or the x-axis bottom. There are 3 holes, 0.250 ± 0.005 in. on true position with 0.010 in. diameter in relation to datums A, B, and C at MMC. Holes are located at (0.250 in., 0.375 in.), (1.000 in., 0.375 in.), and (0.250 in., 0.875 in.) where these centers are considered basic dimensions. Surface B is perpendicular to datum A within 0.001 in. Surface C is perpendicular to datum A within 0.001 in. and with datum B within 0.001 in. The top surface of the block is flat within 0.001 in. total for datum A. (Hint: Use 8½ by 11-in. engineering computation paper with 1-in. grids.)

24.6. Give appropriate dimensions and tolerances and GDT symbols for Figure P24.6. (Hints: You may choose to enlarge the figure with your solution, or copy and enlarge

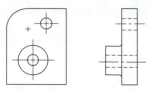

Figure P24.6

the figure and then do the layout of the dimensions. For reference, the left horizontal dimension is 2¼ in., and the boss diameter is 1⅛ in. Scale other dimensions from these reference dimensions.)

24.7. Sketch the design given by Figure P24.7 and give the meaning of the symbol in the sketch. How much can the hole vary?

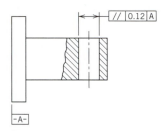

Figure P24.7

24.8. Figure P24.8 shows three diameters that are concentric to one another. Each of the dimensions, finish requirements, GDT symbols, and datums should be examined for purpose and level of tolerance. There are six values for checking, which are numbered on the sketch. Answer the question, "What is wrong with the tolerances?" Sketch the figure, and indicate appropriate tolerances, finishes, and GDT requirements. (Hint: Use engineering computation paper for the sketch. It is only necessary to show one view.)

24.9. Figure P24.9 is shown with only a datum symbol. Sketch what it means. (Hints: Assume that the object is a cylinder. Give it a diameter and length. Its manufacture and actual surface are rough, and over its length there is bowlike variation. Sketch a part of any diameter and length, but convenient for your engineering paper. Indicate dimen-

Figure P24.9

sions that are the smallest circumscribed cylinder for the part. Give the part solid lines, and give the enclosing surfaces dashed lines. The part is considered to be at RFS, since there is no modifying feature control symbol.)

24.10. Refer to Figure 24.9. Sketch the key with dimensions and provide symbols.

24.11. Sketch the meaning of Figure P24.11. Indicate inspection methods that will ensure the GDT requirement. (Hints: Each surface must lie between two common parallel planes 0.08 mm apart. In addition, both surfaces must lie between the specified limits of size. Repeat the sketch on paper and indicate the tolerance zone.)

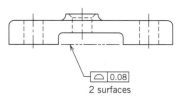

Figure P24.11

24.12. Sketch the meaning of Figure P24.12. Indicate inspection methods that will ensure the GDT requirements.

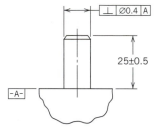

Figure P24.12

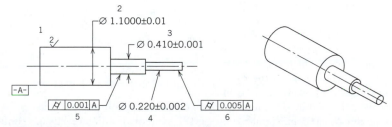

Figure P24.8

MORE DIFFICULT PROBLEMS

24.13. Sketch the meaning of Figure P24.13. Indicate inspection methods and tooling that will achieve the GDT requirements. (Hints: Create a table of the feature size and the diameter tolerance zone. Notice the requirement of 0 tolerance when the feature size of the hole is 50-mm MMC. However, when the feature size increases to 50.16 mm, each increment of hole diameter allows the diameter of the tolerance zone to increase. When the hole feature departs from MMC, the perpendicularity tolerance is equal to the departure from MMC.)

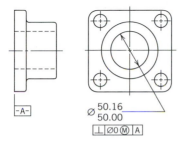

Figure P24.13

24.14. Sketch the meaning of Figure P24.14. The inspection devices are a dial indicator, surface plate, and centers that allow 360° rotation of the part. (Hints: There are four circular runout GDT symbols that need to be checked. Show the dial indicator as a simple plunger and a connecting circle representing the dial face.)

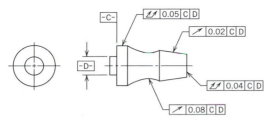

Figure P24.14

24.15. Figure P24.15 is manufactured from initial round stock, 1.25-in. OD; however, the concern is the final operation of total part inspection. Determine the inspection methods and tools for the dimensions, surface finishes, and GDT requirements. Assume small quantity. (Hints: Consider only the dimensions that are shown. Not all di-

mensions are given. If necessary, you may scale for missing dimensions to aid your inspection design.)

24.16. Bar stock, initially 1.5 in.^{2} by 16 ft long, is turned, milled, and deburred to the design shown by Figure P24.16. Reconsider the dimensions and tolerances by changing some of the numbers and callouts. Then for a large quantity requirement, determine the operations plan. Include choices for equipment, small perishable tools that will be used, and indicate the means of ensuring the dimensions and tolerances. (Hint: Assume that the inspection follows each operation.)

24.17. Do "reverse engineering" on a part with emphasis on determining the geometric characteristic and datums. Notice if there are machining or manufacturing marks that describe the final operations. Is there finishing or colorizing? (Hints: In reverse engineering, the emphasis is working back to the original material, and finding the operations, equipment, and tools that manufactured the part. Your instructor will provide additional details and provide an actual part for reverse engineering.)

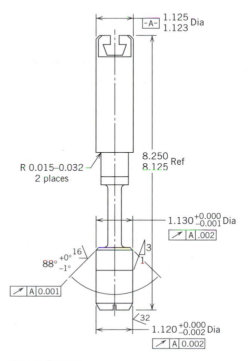

Figure P24.15

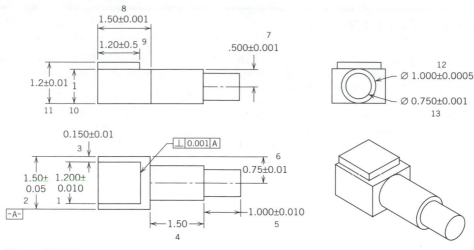

Figure P24.16

PRACTICAL APPLICATION

Visit a machine shop (either school or industrial) and study the characteristics of a lathe, milling machine, drill press, or some other machine tool. Select one machine and determine the types of datums that the equipment is capable of achieving. Sketch the machine tool by giving a front elevational view. Indicate the means of holding the part (chuck, collet, etc.) and show the relative planes of position for datums. Your instructor will give other instructions.

CASE STUDY: MINIMUM COST TOLERANCES FOR GEAR TRAIN

Lyell Production makes gear trains for industry. Luke, the designer, presents Figure C24.1A to Clayton, a metrologist on the Design for Manufacturing review team. The important components and their critical functional dimensions and tolerances are as follows:

Part Name	Critical Dimension	Basic Dimension (in.)	Design Tolerance (in.)
Spacer	A	0.250	± 0.0010
Spur gear	B	1.000	± 0.0010
Bevel gear	C	3.250	± 0.0010
	TOTAL:	4.500	± 0.0030

Design criteria dictate that the subassembly length ≤ 4.500 ± 0.0030 in. Luke is confident that if three maximum (or minimum) length parts happen to come together in the same assembly, the assembly will perform satisfactorily.

Clayton, after seeing the design, asks engineering to reconsider increasing the tolerances.

"All right," says Luke, "but don't increase the cost or make rejects."

Clayton reasons that a model given by

$$T_{\text{sum}} \geq 0.0030 \text{ in.} \geq [(T_s)^2 + (T_{\text{sg}})^2 + (T_{\text{bg}})^2]^{1/2}$$

will serve. For this model to work, it is necessary to achieve random selection of the components where a variety of actual tolerances are butted end-to-end. T minimizes the cost. Clayton has a cost curve that gives processing method against length-tolerance cost. That is shown as Figure C24.1B. As a first approximation, he tries the following:

Part Name	Machine Tool	Production Method	Trial Tolerance	Trial Cost ($)
Spacer, A	Turret lathe	End face	± 0.0030	0.043
Spur gear, B	Grinder	Flat end grind	± 0.0010	0.12
Bevel gear, C	Mill	End face	± 0.0030	0.10
		TOTAL:	± 0.0043	0.263

For an approximate cost of $0.263 for the three tolerances, the model gives a "stackup" tolerance of ± 0.0070 in., which is more than the design requirement of ± 0.0030. Clayton's model gives $(0.0003^2 + 0.001^2 + 0.003^2)^{1/2} = 0.0044$

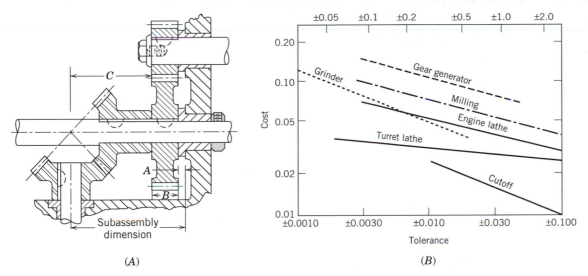

Figure C24.1
Case study. *A*, Gear subassembly. *B*, Sample cost data for several processes for producing a
length tolerance.

in., which is more than 0.0030 in. However, if the part
tolerance of 0.0010 in. and the same machine tools are
used, the cost is $0.303 (= 0.043 + 0.12 + 0.14).

Clayton, however, can do better. Using trial and error,
try several processing combinations with Figure C24.1*B*

and find a lower cost while meeting the design criteria for
a mixed-part assembly. What GDT symbols and requirements would you indicate to Luke for these components?
Is there a downside to Clayton's plan if the three parts are
not selected randomly?

CHAPTER 25
TOOL DESIGN

Tools are important for efficient manufacturing systems. In this chapter, tools are discussed that are designed rather than tools that are purchased from catalogs. Examples of *ordinary tools* are drills, milling cutters, standard diameter punch and die sets, vises, and support equipment, such as a rotary table for a milling machine.

Tools are sometimes defined as perishable or nonperishable. *Perishable tools* have a short life, such as drills, reamers, turning cutters, and broaches. *Nonperishable tools* have much longer life and they may be rebuilt back to the original design. Production quantities are large, which justifies their high cost. For example, plastic molds have quantities that range from thousands to millions.

Designed tools convert general-purpose machine tools into specific product equipment. For example, a C-frame punch press, if given a particular die design, becomes very efficient in the production of parts, and the tool will replace many single operations, if only standard tools were available.

The tools must be available before production is started. Almost all production processes use designed tools. The cost of tools is substantial, but they are worth it because they improve the productivity of the operation. Tools reduce the cost of the operation, increase production, ensure high accuracy, and provide for interchangeability. These tools allow less expensive equipment, and reduce the cost of labor during the operation, as lower skilled machinists are employed and the operations are quicker.

Tool design is handled mostly with computer-aided design (CAD). There are micro-computer software programs that are useful. With 3-dimensional solid modeling and finite element analysis, tools are designed using the part drawing as it appears on the CAD monitor. Whereas that practice is critical to businesses that design parts or tools, for understanding the focus is not on current practice but rather on the basics of tool design.

25.1 PRACTICES

Tool design requires that the part design, equipment, operations plan, inspection requirements, and trained and efficient workers be coordinated. The part design is usually concluded before the tool design is started. Sometimes a tool design leads to improve part design, or it may find mistakes in the part design.

Designed tools are unique. A fixture holding an automobile panel will be different from another fixture holding a different type of panel. Also, only one tool will be made, unless duplicate tools are required for greater production quantities.

The tool design often will overlay the part design on the drawing or the computer monitor. In CAD practices, the part design is imported to tool design as a "pattern design" and becomes a feature to the tool design. The part is then recolored on the computer monitor. The part design may be in red phantom line, and the tool design is a green line on the monitor screen. This double coloring improves the clarity and design of the tool. It distinguishes the part from the tool. If the part is an assembly, and the tool is an assembly device, the parts may be in different colors. Sequencing of the parts of the assembly in the tool design can be followed.

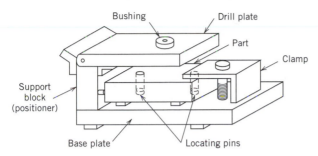

Figure 25.1
Basic components of a fixture.

Special tool designs are constructed by qualified *tool-and-die journeymen*. These journeymen are capable of using precise equipment that builds the tools to tighter tolerances than the part, otherwise the tools are unable to produce the part to the geometric dimensioning and tolerancing requirements. Special grades of tool material are required, and heat treating for hardness is helpful to the life and quality of the tool. Tools are constructed of standard and special components. Typically, tools are constructed of a frame, blocks, locators, clamps, bushings, set blocks, dowels, fasteners, springs, handles, motors, and electric/electronic controls.

The tool will have a frame or base that holds the other components. The base may be a simple plate with a regular grid of tapped holes for holding the other components, or tubular steel may be welded together as the frame. Struts, ribs, and gussets may be employed to reinforce the frame.

Figure 25.1 shows basic components of a fixture. It includes a base plate, locating pins, clamp, support block, and bushing.

There are many types of locators: pin, block, edge block, nest, and V block. The locators can be like a pin, or they may be integral to the plate, such as a boss or surface acting as a locator. Figure 25.2 shows some pin locators for positioning parts. The *diamond pin* has a full diameter in one direction, but in a 90° direction, the dimension across the pin is less. The pins are usually pressed into a block or base plate or a slip fit for repeated removal and insertion of the pin is found.

Uses for *locators* are frequent. If two or more holes are drilled in a flat plate, a workpiece can be positioned with *locating pins* in the next operation. Presswork tooling uses locators for consecutive punching of a blank from strip stock. Inspection tools will set the object on pins or a flat plate and take relative measurements off the locators.

Dowels give location accuracy to the position of one component relative to another, whereas fasteners hold components together. Fasteners, along with clamps, are used, obviously, to hold the workpiece to the tool.

A workpiece, theoretically suspended and free in space, is free to move in any direction. It is said to have 12 modes or *degrees of freedom*. It may move in either of two opposed directions of the three mutually perpendicular axes, *X*, *Y*, and *Z*, and may rotate in either of two opposed directions around each axis, clockwise and counterclockwise. Each direction of movement is considered one degree of freedom. To completely locate a workpiece, it must be restricted against movement in these 12 degrees of freedom, except

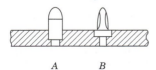 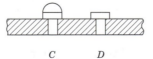

Figure 25.2
Pin locators. *A*, Straight bullet.
B, Diamond. *C*, Shoulder bullet.
D, Roller.

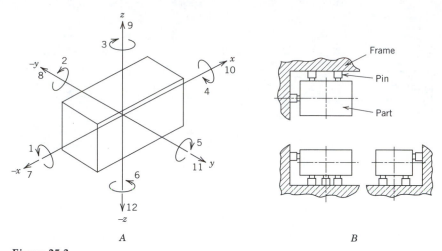

Figure 25.3
A, Translational and rotational types of movement of part in free space.
B, 3-2-1 principle of location. (Not all hidden pins shown.)

those required by the operation. When this condition is satisfied, the part is confined in the workholder. Notice Figure 25.3A.

Geometric dimensioning and tolerancing practices establish *datums*, usually in three planes that are in the X, Y, Z axis. Datums may be planes, lines, points, cylinders, and axis of cylinders. The datum is assumed to be exact for purposes of reference to other features. Location for tool design begins with this relationship. The primary datum, assumed to be the horizontal, is located by three pins, or three isolated flat surfaces, or by a flat plate, which is an infinite number of pins. The pins must be solid and stable, otherwise floating will prevent interchangeability. The second datum surface, which is perpendicular to the horizontal plane, is located by two pins (which gives a line). The third datum surface is controlled by one pin (which gives a point). This is called the *3-2-1 principle of location*. No more points should be used in a single plane, unless they serve a useful purpose. Generally, the objective is to avoid redundant locators. *Locational tolerance* of the pins on the tool design is 10% to 50% of the part location tolerance. Figure 25.3B shows the 3-2-1 pin principle.

Locators usually do not hold the part and clamps become necessary. Clamps come in a variety of types, sizes, and costs. The *clamp* is the mechanical means to apply force to the part to hold it against the locators. The clamp could be as simple as a bolt that is threaded through a hole in the part and tightens the part against the frame. Figure 25.4 shows typical clamps.

Clamping devices must be positive, and the design should *foolproof* the part, meaning that there is only one certain way of loading the part. There should be no ambiguity about the location of the part in the tool.

Once the part is positioned against locators and clamped, and the machine tool and the cutting tool are ready, the tool is guided into the part. A *bushing* guides the tool, perhaps a drill. A bushing is a hardened steel tube, and there are many varieties. The inside diameter of the bushing has a tight tolerance to guide the tool, typically a drill or reamer. Fasteners hold the tool parts together and may include screws, bolts, nuts, dowels, roll pins, keys, and adhesives.

Set blocks are used for positioning the tool and the part together in a machine tool. Figure 25.5 is an example where a set block and a shim positions a tool bit for relative

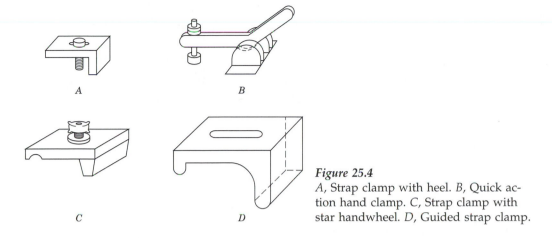

Figure 25.4
A, Strap clamp with heel. *B*, Quick action hand clamp. *C*, Strap clamp with star handwheel. *D*, Guided strap clamp.

dimensions. The position of the tool is monitored electronically and becomes a stationary point in numerical control operations.

25.2 WORKHOLDING

Many workpieces machined on lathes can be held quickly with chucks and collets. These general-purpose workholders are the simplest and are universal, and hold symmetrical workpieces, such as squares, rectangles, cylinders, hexagons, and similar regular-shaped parts. These standard workholders are ordered from catalogs.

However, a tool designer is sometimes asked to design special fixtures that mount irregularly shaped workpieces to a *faceplate*. The advantage of special fixtures for turret or other production lathes are savings of operating time, uniformity of quality from improved centering and locating, and the possibility of heavier cuts, because the workpiece is bolted more rigidly to the faceplate.

Chuck jaws attach to the geometry of the part. Notice Figure 25.6*A*, where a *pointed-pin jaw* holds a tapered surface. Figure 25.6*B* shows a setup where the end of the workpiece is supported by a pilot insert. Without the pilot, the workpiece will deflect under cutting load, and this may cause problems for accurate tolerances. Long flanged workpieces can be given additional support with clamps, as shown by Figure 25.6*C*. The clamps may be quick action or screw type.

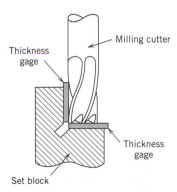

Thickness gage

Milling cutter

Thickness gage

Set block

Figure 25.5
Set block to position tool for relative dimensions.

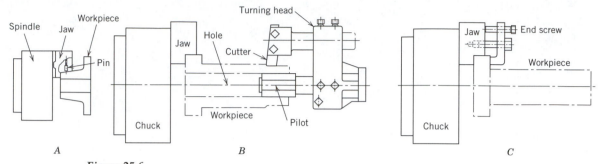

Figure 25.6
Fixtures. *A*, Pointed-pin jaw. *B*, Pilot to support work during turning. *C*, Holding clamps used with jaws.

Workholding can be with a *nest*. Notice Figure 25.7, which shows a part and its nest. Notice that removal of the part may be a problem, unless there is provision for ejection. In Figure 25.8, nest and edge blocks locate flat parts.

25.3 FIXTURES AND JIGS

Fixtures hold parts. The pliers can be considered a fixture. Even the hand holding a part for some simple manufacturing operation is behaving as a fixture. The hand is effective, but it is unsafe, inaccurate, and not as strong or as durable as a mechanical fixture. A plain fixture may be a vise having special jaws that are fitted or molded to the part. Complex fixtures are designed to hold, locate, check, and assemble. These fixtures are themselves constructed from many simple components.

Fixtures locate materials in a specific place and position, and hold them securely in a predesignated position under any strain that may result from the manufacturing operation. It needs to be simple and quick, durable, easy to maintain, safe for the operator, and ergonomically suited to the human.

The fixture is normally attached to the machine tool. In unusual cases, however, the fixture may be attached to the part, for example, bulkheads for the production of ships. The fixture may be designed for handling materials from one point of operation to a second point in the production process. Their primary purpose is to reduce the cost of the operation and to allow complex-shaped and often heavy parts to be machined by being

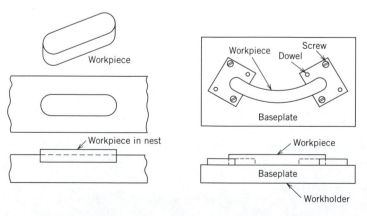

Figure 25.7
Workholding nests.

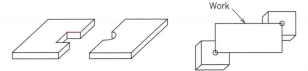

Work

Figure 25.8
Nest and edge blocks locate flat parts.

held rigidly. Fixtures are commonly used on lathes, milling machines, boring equipment, shapers, planers, and special machine tools.

Fixtures are made from gray cast-iron, steel, or aluminum plate by welding or bolting. They are bolted, clamped, or "set" with a low-temperature melting alloy to the machine. A fixture has locating pins or machined blocks against which the workpiece is held tightly by clamping or bolting. To ensure interchangeability and durability, locators are hardened steel. Many fixtures are massive because, like the machine frame, they must withstand large dynamic forces. Because fixtures are between the workpiece and the machine, their rigidity and the rigidity of their attachment to the machine are important. The fixture may be "duplex," which allows the loading and unloading of one side of a fixture while the machining operation is taking place on a part clamped to the other side.

Jigs are fixtures used in production drilling, tapping, boring, and reaming operations. If it is called a boring jig, then the tool is used for boring operations. If the jigs are small and light, they may be handheld, but if they are heavy, then they may be fixed to the table. The function of a *jig* is to hold the part like a fixture and guide the tool. Technically, the sketch in Figure 25.1 is a jig because it locates the drilling of a hole.

Figure 25.9 shows the components arranged for a jig. The drill plate is hinged, allowing for vertical removal of the part. Positioning is positive against locating pins and clamping and chip clearance are provided.

A plate or channel jig is shown in Figure 25.10. The tool designer is guided by the part design. Initially, observe the four holes, which are not particularly tight in tolerance, as a fractional dimension is given. However, the location is to thousandths (which is modified by title block tolerances, which are not shown). One pin is a datum point for the y coordinate, and the operator will load the ½-in.-thick piece butted against the pin for location. A thumb screw tightens the part against the wall and holds the part against the drilling thrust. Four shoulder bushings will then guide the drill. There is chip clearance above the part in the jig. The locations for the bushings and pin are not shown, that being left as an exercise for the student. Remember that the part as shown by Figure 25.10 could be manufactured by a computer numerically controlled (CNC) drilling machine able to move the table point-to-point with sufficient tolerance that a designed tool may not be required.

The classification of jigs is often based on their appearance. The channel jig of Figure 25.11 is one example. Figure 25.11 is a box type of jig with open sides. The *box jig* is

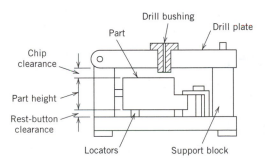

Figure 25.9
Support blocks position workpieces in fixtures, and drill plates position and guide the tools.

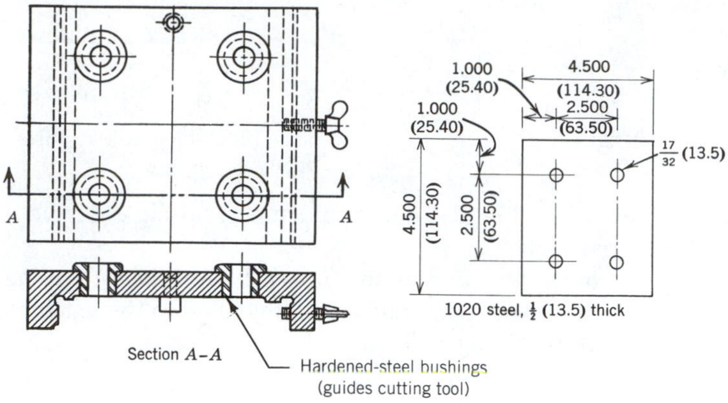

1.000
(25.40)

4.500
(114.30)

1.000
(25.40)

2.500
(63.50)

$\frac{17}{32}$ (13.5)

4.500
(114.30)

2.500
(63.50)

1020 steel, $\frac{1}{2}$ (13.5) thick

Section A–A

Hardened-steel bushings
(guides cutting tool)

Figure 25.10
Plate-drilling jig.

designed for drilling two sides of a block. Another example of a drill jig, shown by Figure 25.12, is a table-type jig that will be moved around on the top of the drill press table. This is another example of an operation that could be handled by a CNC machine, thus not requiring a jig. There are also sandwich, leaf, pump, and indexing jigs.

Notice that chip clearance is provided under the bushings to allow for their escape. It is poor practice to require that chips leave through the bushing. However, occasions are sometimes necessary that the chips exit through the drill flutes. It seems that the bushing is an inescapable part of a jig. There are many kinds of bushings: slip, removable, locking, drill and ream, and catalogs can be checked for sizes, types, and features. Figure 25.13 is the typical pressed bushing.

Jigs are not only limited to drilling, but also are used for tapping, counterboring, and reaming operations.

25.4 PRESSWORKING TOOLS

Chapter 16 discussed pressworking equipment, terms, and punching, blanking, bending, forming, and drawing operations, and the methods for efficient strip layout and high yield. Those operations are possible with standard or specially designed tooling. If the design requires unique tooling, it becomes necessary to design the tool. Like fixtures and

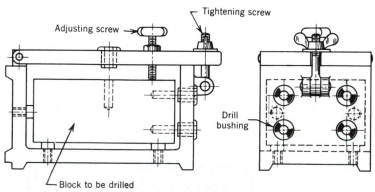

Adjusting screw

Tightening screw

Drill bushing

Block to be drilled

Figure 25.11
Box-type jig for drilling two sides of a block.

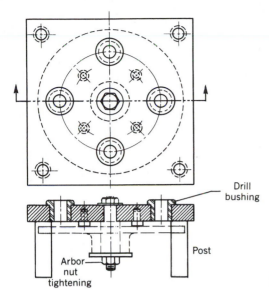

Figure 25.12
Table-type jig for drilling flange holes.

jigs, it may be possible to purchase many components for presswork tooling, and perhaps only grind a punch and its mating die to a size. Assembly of the tooling components, such as shoes, strippers, springs, and die blocks, follows the design and concludes the construction phase.

Whereas there are a variety of presswork operations, some of which are complex, they can be reduced to simple fundamentals. The three orthographic view of Figure 25.14 shows a plain blanking tool.

The stock material is called *strip*. During the working stroke, the punch penetrates the material, and on the return stroke by the press ram, the material is lifted with the punch, but it is removed from the punch by a *stripper*. Another idea for a stripper is a simple hook that snags the part on an upstroke of the ram. The *stop pin* is a gage and it sets the *advance* of the strip stock within the punch and die. The strip stock also is butted against the backstop, behaving as a datum location for the center of the blank. The good part escapes through the die opening, which is tapered to permit its ejection. The *waste*, or material lost because of the blanking operation, is a skeleton, which is recovered as salvaged material. The skeleton can be sold only for its salvage value. The blanking operation, involving a punch and die, produces a *blank*, which is a metal shape ready for a subsequent press operation.

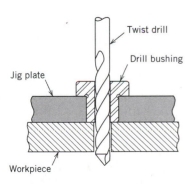

Figure 25.13
Drill bushing with no space for chip removal.

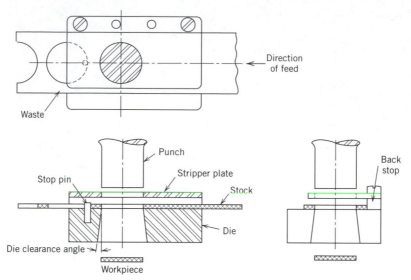

Figure 25.14
Blanking punch and die.

Piercing is closely related to blanking; however, it consists of the punching of holes. Piercing differs from blanking in that the punching results in waste, rather than in good parts. In effect, it produces the hole, as in a washer, where the hole is pierced, but the OD is blanked. Piercing is nearly always accompanied by a blanking operation, either before, after, or at the same time as the piercing.

Blanking and piercing require differing clearances between the punch and the die. The finding of these clearances differs, and that discussion is found in Chapter 16.

Figure 25.15 is an exploded view of a *piercing die* for punching a hole. Some dimensions are given for this part, which can be held in the hand. The only items that might be

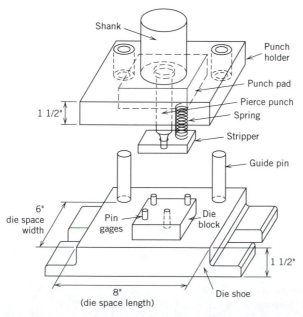

Figure 25.15
Piercing die.

specially ground is the punch and die. The other parts are standard and are purchased from catalogs; however, the components are assembled. The upper die half, or upper shoe, is attached to the ram of the punch press. There are holes for guide pins, and the press movement, both up and down, does not disengage the pins from the holes, thus keeping alignment between the upper and lower half of the tool. A pierce punch, stripper, and spring are attached to the upper shoe. The lower shoe is attached to the bolster plate of the punch press. Screw hold down slots can be seen. The die consists of the die block and locating pins. There is a hole in the die block and the die shoe, allowing the waste or hole of the washer to escape through the bottom.

A blanking punch and die for stamping stainless steel razor blades is shown as Figure 25.16. Material for the die parts is cemented carbide because of severe service. When a punch becomes dull, usually after 150,000 punches of steel, it is reground. Small high-speed presses use these tools and operate at 300 spm. After the press operation, the workpiece material is recoiled and hardened. Grinding, honing, and stropping follow. The razor blades are snapped apart, inspected by methods of metrology, and packaged.

Forming tools make changes in the shape of a metal piece and do not intentionally reduce the metal thickness. Their purpose is to form bends. Typically, forming tools do not pierce, blank, or generally otherwise cut stock. Forming tools are different than bending tools, even though there is a similarity. Bending tools normally form the bend in one direction, whereas forming can be one or more directions.

A forming die, designed to bend a flat strip of steel into a U-shape, is shown in Figure 25.17. As the punch descends and forms the piece the *knockout plate* is pressed down, compressing the spring at the bottom of the die. When the punch moves up during the return stroke of the press ram, the knockout plate forces the work out of the die with

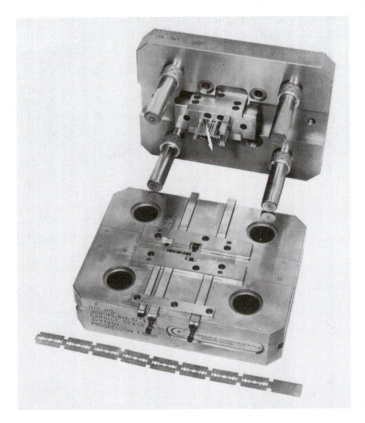

Figure 25.16
Punch and die for razor blades.

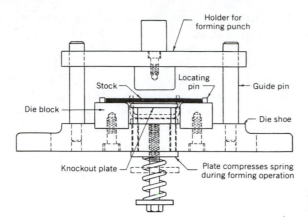

Figure 25.17
Forming punch and die.

the aid of a spring. This arrangement is necessary, because in most forming operations the formed metal hangs up against the die wall, making removal difficult. Parts that might stick to the punch are removed by a *knockout pin* that is engaged on the upstroke.

Drawing dies are important in the drawing of deep cavity parts. There is severe plastic flow of the thin metal strips or blanks. Review Chapter 16, which describes presses that are able to move two slides, one that first controls the blankholder ring and the second, which drives the punch relatively slowly into the stock, thus causing metal flow or ironing. Hydraulic presses or mechanical presses having toggle or linkage arm control are used.

The simplest draw die will have only a punch and die. This type of die forms the cup with low ratios of drawing of the blank to the cup diameter. In Figure 25.18*A*, a precut blank is placed in a recess on top of the die. The punch pushes the blank into the die, using a spring-loaded pressure pad to counteract the metal flow. The cup either drops through the die or is stripped from the punch by the pressure pad. If the spring pressure is increased, the greater the pressure on the blank, thereby limiting the depth of the draw. This type of die is used with light gage stock and shallow draws.

Figure 25.18*B* is a sectional view of a drawing die having an *inverted punch*. The punch is stationary and is mounted on the bed of the press. As the die descends, the blank is

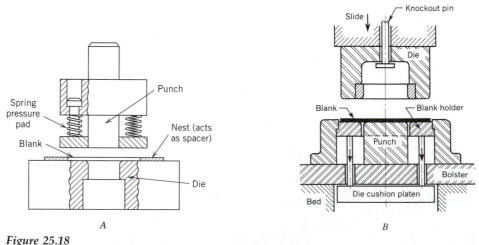

Figure 25.18
A, Draw die with spring pressure pad. *B*, Drawing operation using an inverted punch.

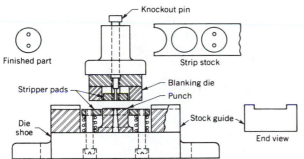

Figure 25.19
Compound punch and die.

touched and held. Then the die continues downward, and the blankholding ring maintains contact with the blank during drawing. The die cushion controls the holder pressure, and it can be increased or reduced to avoid various deep drawing problems.

The drawing operation may be done in stages. With the popular beverage 12-oz container, the first operation is *cupping*, which does not change the thickness of the material. Then the cup is *redrawn* and *ironed*, where severe metal flow is contained within ironing dies.

Dies that perform only one operation with each stroke of the ram are called simple dies. *Compound dies* combine two or more operations at one station, as shown in Figure 25.19. Strip stock is fed to the die, two holes are pierced, and the piece is blanked on each stoke of the ram. When the operations are not similar, as in the case of a blanking and forming operation, the dies are called *combination dies*.

A *progressive die* set performs two or more operations simultaneously, but at different stations. A punch-and-die set of this type is shown in Figure 25.20. As the strip enters the die, the small square hole is punched. The stock is advanced to the next station, where it is positioned by the pilot as the blanking punch descends to complete the part. This design is simpler than compound dies, because the respective operations are not jammed together. Regardless of the number of operations to be performed, the finished part is not separated from the strip until the last operation. A progressive die set that performs 15 operations on a can opener, completing one at each stroke, is shown in Figure 25.21. Production is rapid, but close tolerances are difficult to hold.

25.5 WELDING TOOLS

Workholding principles found with fixtures also apply to welding fixtures. However, the differences between machining and welding fixtures are locational tolerances and clamping methods. With welding fixtures, weight may be a problem. Many times a fixture is

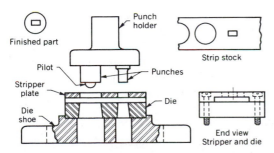

Figure 25.20
Progressive punch and die.

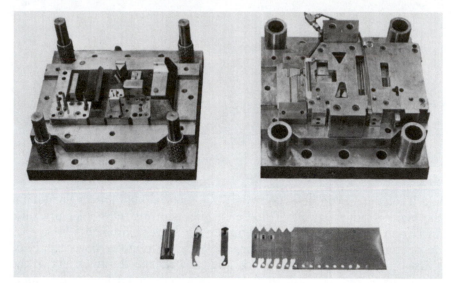

Figure 25.21
Progressive die set that performs 15 operations on can openers and completes one on each stroke.

assembled of welded sections, and these sections are usually positioned near the areas where the parts are to be joined. Whereas precision locators are found on machining fixtures, less precise angle clips, blocks, or similar elements are necessary as welding locators.

Toggle clamps are often used for welding fixtures. These clamps offer design flexibility, holding capacity, and loading speed. Toggle clamps, when released, clear the work area. Loading and unloading is simplified.

Other considerations for welding fixtures include:

1. Construct the fixture to have only one way for loading.
2. Design clamps and supports to prevent heat distortion.
3. Position locators and supports so that if heat distortion exists, the effect is to loosen the welded part, rather than tighten the part.
4. Design the fixture to encourage flat horizontal welding.
5. Include heat dissipation methods if excessive warpage and heat are expected.
6. Perform many welding operations before the operation is concluded.

25.6 MOLDING TOOLS

Molding operations are necessary for casting of molten metals, plastics, sands for the pattern and core of foundry operations, powder metallurgy, blow molding, and other special operations. Frequently, these molding operations are thought of as *net shape* processes because of their reduction of waste materials and extra operations that may result from machining, welding, assembly, or presswork. Intricate molds eliminate other fabrication operations.

If the engineering analysis of the product material allows plastic, zinc, or aluminum, and the like, and the quantities are sufficient, molding operations give significant cost reduction over machining operations. *Plastic molds* are extremely important because of the popularity of plastic parts. Many components are made from zinc or aluminum alloys, and the mold processes are commonplace.

In the case of plastic molding, the discussion of tool design considers thermosetting and thermoplastic plastic materials. In thermosetting processes, the material is cool as it enters a heated mold, whereas in the second, heated material is forced into a mold and then cooled. Thus, there are heat transfer considerations, electrical heating, and water cooling in the jacket that are factors. Warpage after cooling, parting plane of the die halves, and undercutting of parts that prevent the part release are a few design considerations.

Figure 25.22 is an example of a molding die design. It is viewed as the top sectional view, where there are two halves to the design.

In molds for plastic materials, ample draft and fillets should be provided to facilitate removal of the part from the mold. Ejector pins provided for this purpose should be located at points where the pin marks are not noticeable. Like metals, plastic materials shrink on cooling and allowances are required. Shrinkage varies, but a rule of thumb is 0.003 to 0.009 in./in.

Injection molds are made in two pieces, one half being fastened to the fixed platen and the other half to the movable platen. Contact between the halves is made on accurately ground surfaces or lands surrounding the mold cavities. Neither half telescopes with the other. The cavities should be centrally located with reference to the sprue hole in the fixed half, so as to obtain even distribution of material and pressure in the mold. For locating purposes, guide pins, which are similar to those employed on metal press dies, are fastened in the fixed half of the mold and enter hardened bushings in the movable part of the mold.

Mold cavity can extend into both halves of the mold. It is best to have the outside of the molded part in the fixed half, provided the shape is suitable for this plan. In the cooling process, the plastic material tends to shrink away from the cavity walls and is withdrawn from this half as the mold opens. It is retained on the cores of the movable half until the ejector mechanism operates.

Injection molds have cooling channels in both halves to permit the maintenance of a uniform temperature for chilling the molded part, because most materials fabricated by this process are thermoplastic. The material is forced into the mold from the heated cylinder under pressure ranging from 2 to 20 tons/in.² (30 to 275 MPa) and is ejected at a temperature of approximately 125°F (50°C). Ejection of the part occurs as the mold opens by either ejector pins or stripper plates.

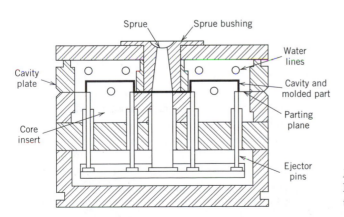

Figure 25.22
Molding die design. Top sectional view.

Any cores required in injection molding are placed on the movable half of the mold. The normal shrinkage of the molded part tends to cling to the cores, causing it to withdraw freely from the stationary half as the mold opens. Vents to permit the escape of entrapped air are extremely small and are located to encourage quick escape of the air.

25.7 INSPECTION GAGES

Production *gages* measure fixed shape or size in production work. They represent a standard for comparing manufactured parts. Application is limited to one or several dimensions on a part, since adjustments of a gage are not normally done during a production cycle. Inspecting a part requires a minimum of time with gages as compared to *adjustable inspection tools*, such as micrometers, which are not gages. The operators do the gaging while the equipment is in operation, thus causing no production time delay. Gages determine whether the part is manufactured to the design tolerance. The gage does not indicate a specific dimension like a micrometer. It only indicates a *go, no-go* choice.

Inspection gages are used by inspectors in the final acceptance of the product. They ensure that the product is made to the tolerance specification on the design. Working or manufacturing gages are made to slightly smaller tolerances than the inspection gages. This keeps the part size near the center of the limit tolerance. Parts manufactured around limit sizes will still pass the inspector's gage.

The success of a gage is measured by its accuracy and service life, which, in turn, depends on the workmanship and materials. Gages are subject to abrasive wear during service, and the selection of material is important. High-carbon and alloy tool steels are the principal materials. Glass is a rarer substitution.

Steel materials can be machined accurately and heat treated for additional hardness, if necessary. Heat-treating operations increase hardness and abrasive resistance. However, steel gages are subject to distortion because of the heat-treating operation.

Low-carbon steel materials have limited hardness opportunity and are not always suitable. However, lack of hardness for low-carbon steels is overcome by chrome plating the surface or in placing cemented carbides at points of wear. Chrome plating permits the choice of steel having inert qualities, since wear resistance is obtained with a hard chromium surface. Chrome resurfacing is used in reclaiming worn gages. However, cemented tungsten carbides attached to metal holders by brazing provides the hardest wearing surface.

Snap Gages

A *snap gage,* used in the measurement of plain external dimensions, consists of a U-shaped frame having jaws equipped with suitable gaging surfaces. The distance between these jaws is the dimension being measured. A *plain gage* has two parallel jaws or anvils that are produced to a standard dimension and are not typically adjusted during their use. This gage can be replaced by adjustable gages for changing tolerance settings or adjusting to wear. Most gages are provided with the go and no-go feature in a single jaw. These designs are both satisfactory and rapid.

The general design shown in Figure 25.23 is selected because it incorporates most of the advantages of similar gages. It is light in weight, sufficiently rigid, easy to adjust, provided with suitable locking means, and is designed to permit interchangeability with many parts.

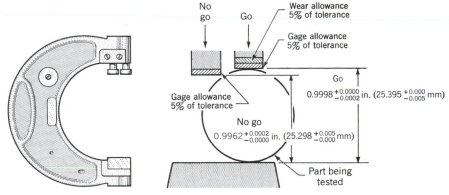

Figure 25.23
Snap gage set for inspecting a dimension of 1.000 $^{+0.000}_{-0.004}$ in. (25.40 $^{+0.00}_{-0.10}$ mm).

The tolerance for the settings as shown in Figure 25.23 must account for the total gage allowance, which is customarily taken as 10% of the tolerance of the part; that is, 5% of the part tolerance for each button and the wear allowance, which is 5% of the part tolerance. The allowance for wear is usually made only for "go" gages, since "no-go" gages have little wear.

The usual practice is to allocate both gage tolerance and wear allowance entirely within the tolerance limits of the part to be inspected. An occasional part may be rejected, even though the part may have been made within its tolerance.

Figure 25.23 indicates the appropriate sizes for the "go" and "no-go" dimensions when the gage is set to measure $^{+0.000}_{-0.004}$ in. (25.40 $^{+0.00}_{-0.10}$ mm). The total tolerance of the part is 0.004 in. and 10% is 0.0004 in. Five percent, or 0.0002, is subtracted from the high side, or 1.0000 − 0.0002, leaving 0.9998 in. The tolerance is applied negatively, or 0.9998 $^{+0.0000}_{-0.0002}$ to the high end. The low end, 0.996 in. of the part, has its tolerance applied positively, or 0.9962 $^{+0.0002}_{-0.0000}$. Thus, if the part lies between 0.9998 or 0.9962 in., the diameter is acceptable. If the diameter is 0.9999 or 0.9961 in., the diameter is unacceptable, even though the diameter is within the design limits. This is the slight risk that is paid for this snap gage inspection of the limit dimension.

Plug Gages

A plain *plug gage* is an accurate cylinder used as an internal gage for the size control of holes. One plug gage is used for one hole. The gage is provided with a handle for holding and is made in various styles. These gages may be either single or double ended. Double-ended plain gages have "go" and "no-go" members assembled on opposite ends, whereas progressive gages have both gaging sections combined on one end.

Plug gages check hole locations, distances between holes, and hole and slot sizes. To check a hole, the smaller plug gage is first inserted into the hole. Since the size of this gage is matched to the smallest size of the hole, the gage should enter. If the gage is too large, the hole is too small, and the hole is defective. If the small end enters the hole, and the larger end does not enter, then the hole diameter is okay. If the larger end enters the hole, the hole is too large, and the hole diameter is rejected.

The *allowance* for manufacturing snap gages and the allowance of the part to be inspected must be considered in the design. Figure 25.24 shows a "go" and "no-go" gage with the appropriate dimensions for checking a hole size of 0.750 $^{+0.000}_{-0.004}$ in. (19.05 $^{+0.00}_{-0.10}$ mm).

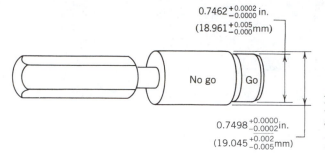

$0.7462^{+0.0002}_{-0.0000}$ in.

$(18.961^{+0.005}_{-0.000}mm)$

No go | Go

$0.7498^{+0.0000}_{-0.0002}$in.

$(19.045^{+0.002}_{-0.005}mm)$

Figure 25.24
Plug gage dimensioned for checking a hole size of 0.750 $^{+0.000}_{-0.004}$ in. (19.05 $^{+0.00}_{-0.10}$ mm).

Other gages include ring, taper, thread, and thickness. *Ring gages*, for outside diameters are used in pairs, a "go" and "no-go." *Taper gages* are not dimensional gages but rather a means of checking in terms of degrees. Their use is a matter more of fitting rather than measuring. A thickness or feeler gage consists of a number of thin blades and is used in checking clearances and for gaging in narrow places.

A ring gage is a set of two that is used to inspect diameters. There will be a go and a no-go ring. These gages look very much like round washers. There can be cases where a set of ring gages will accept an external diameter, whereas a snap gage will properly reject the diameter. Notice Figure 25.25, where an ellipse, which represents the perfect round outside diameter, is accepted by the ring gage in sketches *A* and *B*, but is rejected by the snap gage in sketches *C* and *D*.

Two-Hole Pin Gage

The principles of GDT and gaging can be combined to check two holes. Notice Figure 25.26*A*, which is the design for the two-hole pattern. The two holes, which are positioned 2.000 in. apart as a *basic dimension*, have limits of 0.500 and 0.502 in., with a tolerance of 0.002 in., and are located at *true position* within a diameter of 0.004 in. The gage pin diameter is calculated to be 0.4960 in. (the smallest hole size minus the true position tolerance), as Figure 25.26*B* shows. This means that a gage with two pins with diameters of 0.4960 in. spaced exactly 2 in. apart could be used to check the diameters and the spacing of the holes at MMC, the most critical size.

If the pins can be inserted into the holes, the holes are properly sized and located. In Figure 25.26*B*, when the two holes are located at true position at maximum material

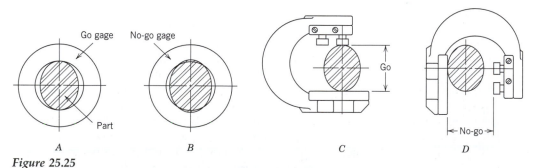

Go gage   No-go gage

Go

Part

No-go

A   B   C   D

Figure 25.25
A, Go-ring gage slips over diameter. *B*, No-go ring gage does not slip over shaft and diameter is accepted. *C*, Snap go-gage accepts diameter. *D*, Snap no-go gage accepts diameter when part rotated 90°.

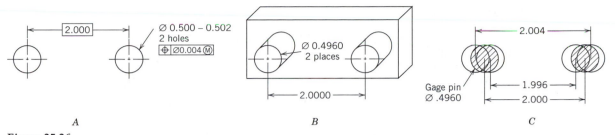

Figure 25.26
A, Design of hole pattern using geometric dimensions. B, Two-pin gage with φ 0.4960 pins spaced at 2.0000 in. apart. C, Allowable location of holes when holes are at φ 0.502 in.

condition, they may be gaged with two pins, which are 0.496 in. in diameter. In Figure 25.26C, the φ 0.004 in. locational tolerance allows the hole to hole centers to vary from 2.004 to 1.996 in. and be acceptable.

When the holes are not at MMC, that is, when they are larger than their minimum size, the pin gage will allow a greater range of variation. When the holes are at their maximum size of 0.502 in., the holes can be located as close as 1.994 in. from center to center, or as far apart as 2.006 in. from center to center.

Obviously, the gage pins cannot be made exactly to 0.4960 in. They are given a *gagemaker's tolerance* of 0.0002 in. and the pin diameter range then becomes 0.4960 to 0.4958 in.

Gaging Designs

In most cases, inspection gaging is electronic, optical, and mechanical in design. Notice Figure 25.27A, which applies the principle of *autocollimator* and the *light beam*. The light is aligned with the datum plane and turned at an exact right angle with an optical square called a pentaprism. This allows the measurement of perpendicularity by observing the reflection from a mirror moved along the second plane. The autocollimator, optical square, and reflector allow large measurements for length, levelness, or perpendicularity. A surface plate is frequently used for measurements and gaging of various kinds. In Figure 25.27B, a special fixture is designed using planes and dial indicators. The part surfaces are searched by sliding indicator gages. Parallelism, squareness, and angularity are checked.

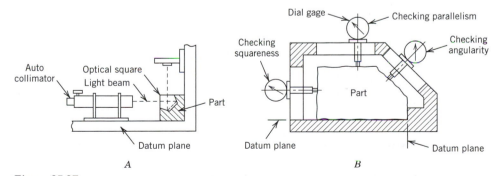

Figure 25.27
A, Auto collimotor with pentaprism for checking perpendicularity. B, Fixture with dial gages for checking perpendicularity, parallelism, and angularity.

25.8 STEREOLITHOGRAPHY

The *stereolithography* application is known as desktop manufacturing, tool-less manufacturing, rapid prototyping, and its purpose is to convert a three-dimensional solid or surfaced CAD model into a plastic part in the matter of hours or days without tooling, machining, or molding. It reduces the product/tool design-to-build time cycle and is used to improve designs. Complex models with intricate details are created as easily as simple parts.

The stereolithography process starts with a CAD model, which is processed to create a file that is interpreted by the "slice" computer software. The data are processed mathematically to give a series of horizontal cross sections. Next, a computer-controlled HeCd laser generates an ultraviolet beam, which travels across the surface of a vat of a photocurable liquid polymer. The laser traces the path of the cross section, changing the liquid into a solid until a solid layer is formed. This newly formed layer is then lowered into the vat by an elevator, while a recoating and leveling system establishes the thickness for the next layer. Successive cross sections are built up layer by layer, one on top of another, each adhering to the last until a three-dimensional part or tool is formed. It may be hand-sanded to smooth the surface. Then, the part is ready for examination and subsequent reengineering.

QUESTIONS

1. Give an explanation of the following terms:

Nonperishable tools
Tool-and-die journeymen
Locating pins
Degrees of freedom
Bushing
Set blocks
Nest
Fixtures

Box jig
Knockout pin
Combination dies
Progressive die
Injection molds
Snap gage
Plug gage
Stereolithography

2. Using word-processor software, describe a classification for tool design.

3. Cite advantages for tools.

4. Why do we overlay a tool design on the part design?

5. Give the purpose for locators. Describe the 3-2-1 location principle.

6. When is design tooling justified? Compare to ordinary tooling.

7. How does a jig differ from a fixture?

8. How does a blanking die differ from a piercing die?

9. Describe the sequence of the forming die set.

10. What is the purpose of the "redraw" in the deep drawing of shells?

11. Discuss the differences between welding and work-holding fixtures.

12. Indicate the component parts of a drawing die and give their role in the drawing process.

13. Describe production examples that use snap and plug gages.

14. Why are bushings used in aluminum sandwich jigs?

15. Reconfirm the calculations for Figure 25.26.

PROBLEMS

General remarks: These problems can be done with a CAD software program, with pad and pencil, or the school may have access to a special tool design package. Your instructor will give additional details for the following problem.

25.1. Sketch various locators using engineering paper.

25.2. Sketch an elevational view of several types of bushings.

25.3. Provide a dimensioned sketch of two diamond locating pins that locate a part that is 1 in. thick, 4 in. wide, and 6 in. long. There are two 0.500/0.501 in. holes irregularly spaced.

25.4. A 2-in. cube is located by the 3-2-1 principle. Sketch the object in three views with locator pins and indicate the principal datums. Add a clamping method.

25.5. A steel object, irregular in size, 1 in. thick, 4 in. wide, and 6 in. long, is located using the 3-2-1 principle. Sketch the object in three-view projection with locating pins and indicate the principal datums with GDT symbols. Dimension the size and location of the locating pins. (Hint: Assume plate backing for the locators.)

25.6. Examine an object on your desk, such as a stapler, phone, computer, or cup. Sketch a nest to hold the object. (Hint: Assume that the nest is mounted to T-slots on a machine tool table.)

25.7. Examine an object on your desk. List the necessary tools that you believe are necessary for its large quantity production.

25.8. A cross-sectional view of a machine-tool "way" is given as Figure P25.8. Ways are heavy sections that guide and support the major elements of machine tools as they move. Lengths of these steel ways vary according to the machine tool. Select components, and sketch and dimension a fixture to hold one of the following choices: 12 in., 4 ft, or 12 ft. Operations, such as milling, drilling, grinding, flame hardening, and the like, are going to be performed on the top surface. (Hints: The requirement to "sketch" implies pencil, but not a straight edge, and the use of engineering computation paper. The sketch is confined to a single sheet. Although a sketch is drawn with care, it is not an engineering drawing done with instruments, or the use of a CAD software package. Repeat the "way" as a first step on your sketch. Do the object in some other color than pencil. Your instructor will give additional instructions.)

25.9. A fixture is needed for Figure P25.9. The figure is given with partial dimensions and other dimensions may be scaled, as required for the fixture design. The steel workpiece is manufactured from 10-ft solid stock, 3.313-in. and 1.563-in. bar stock. The figure represents the final shape. Write an operations processing plan, specify the tooling, and design a fixture that will satisfy as many operations as possible.

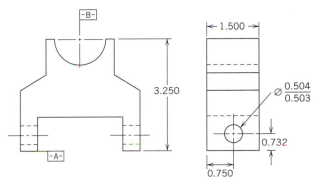

Figure P25.9

25.10. Design a drill jig for Figure P25.10. All work is completed except for the operations on the holes.

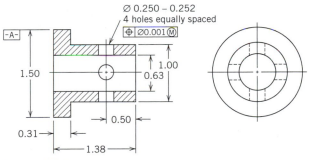

Figure P25.10

25.11. Design a drill jig for Figure P25.11. Prepare a simple bill of material. (Hints: Show dimensions for the tool design. Indicate the material for the body of the tool. Specify the components by name. Enlarge and duplicate Figure P25.11 for your design. Then build the tool design around it. Assume that for "thousandths dimensions," the standard title block tolerances are ± 0.005 in. Use a tool design tolerance of 10% of the part tolerance. If assigned by the instructor, do the work with a CAD package.)

25.12. Study a single-hole or three-hole paper punch that is readily available. Is the tool doing piercing or blanking? Is there "shear" on the punch? What can you say about

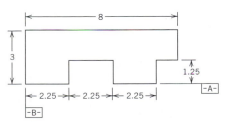

Figure P25.8

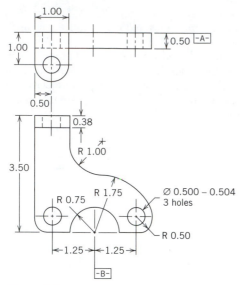

Figure P25.11

	Hole Diameter (in.)	Stock Thickness (in.)	Strip Width (in.)
A.	1.000	18 ga (0.0478)	1.191
B.	1.500	16 (0.0598)	1.75
C.	2.000 ± 0.003	14 (0.0747)	2.3

25.14. Study Figure 25.15, an isometric of a piercing die. Now an OD washer-blanking operation is followed by a piercing operation that punches a 0.5-in. square in the center of the washer. Sketch an elevational view of a piercing die, and label the parts of the tool. Also dimension the punch and die. Choose one of the following:

	Blank Diameter (in.)	Stock Thickness (in.)
A.	1.000	18 ga (0.0478)
B.	1.500	16 (0.0598)
C.	2.000 ± 0.003	14 (0.0747)

25.15. Design an elevational view of a blanking die for Figure P25.15. Determine the advance and developed length of the workpiece. Material is aluminum 5052-H34, 0.012 in. thick. Identify the parts of the tool. Separately, apart from the elevational view, sketch and dimension the punch and die. (Hints: Let the width of the strip stock equal developed part width plus two thicknesses, and the part-to-part separation is three stock thicknesses. Refer to Chapter 16 for information on the development of clearances for blanking.)

the clearance between the punch and die? Is a stripper used? What about paper location within the tool? Prepare your answer using a word-processing software package.

25.13. Design and dimension the plain blanking tool, Figure 25.14 for either A, B, or C, as specified below. Material is hot-rolled pickled-and-oiled steel strips. (Hints: Enlarge the figure by photocopying, trim any waste, and provide a copy of the figure. Then add dimensions. Refer to Chapter 16 for clearances between the punch and die.)

25.16. Design a piercing die for Figure P25.15. Show an elevational view, label the major parts, and give dimensions for the punch and die block and holes.

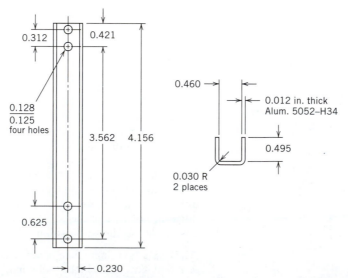

Figure P25.15

25.17. Design a forming die for Figure P25.15. Sketch an elevational view of the forming punch and die only. (Hint: The material is previously blanked, pierced, and now is formed. The workpiece is flat before the operation of forming.)

25.18. Design a forming die for Figure P25.18. The processing plan indicates that 40,000 units are produced. The soft brass material is previously blanked and pierced. The part is hardened following the forming operation.

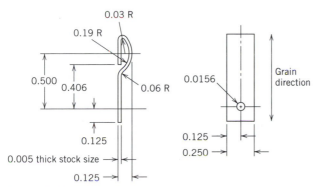

Figure P25.18

25.19. An aluminum 3004H-19 circular blank is 5.700 in. in diameter and is 0.0135 in. thick. Sketch a drawing punch and die that provides the cup. Notice that there is no change in thickness. Dimensions of the cup are given in Figure P25.19.

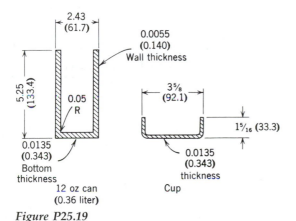

Figure P25.19

25.20. The cupping operation for Figure P25.19 is followed by drawing and ironing operation. The 12-oz beverage container is shown in Figure P25.19. After drawing and ironing, there is a ragged edge on the top that is later trimmed. Sketch the drawing and ironing punches and dies that give the drawn can. (Hints: Remember to provide the drawing and ironing corner radii. Give the dimensions for the designs. Assume a two-stage operation.)

25.21. Notice Figure 25.19, a compound punch and die. Sketch an elevational view of the tool, paying attention to the punch and die. The strip stock is 3.500 in. wide, ¼ in. thick, and indefinite length. Material is low-carbon steel annealed. The circular blank is ϕ 3.000 in., and the two internal holes are ϕ 0.197 in.

25.22. Notice Figure 25.20, a progressive punch and die. Sketch an elevational view of the tool, giving attention to the punch and die and the component parts of the tool. Label the parts of the tool. The strip stock is 3.500 in. wide and ¼-in.-thick. Material is low-carbon steel annealed. The circular blank is ϕ 3.000 in., and the slot is ½- by 1-in. long.

25.23. Design a welding jig for Figure P25.23. Two sheet metal parts are spotwelded together to form an assembly. Spotwelding is along the bottom lip and the two inner lips of the inside part. (Hints: Dimensions are unnecessary. For the sake of scale, the length of the bottom is 6 in. Keep the weight of the tool in mind.)

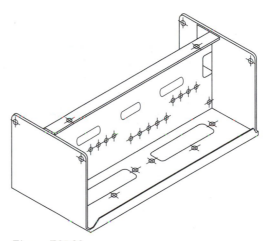

Figure P25.23

25.24. Design a welding fixture for Figure 23.7. (Hints: Duplicate the design and enlarge it with photocopying techniques. Sketch the welding fixture within the three views.)

25.25. Sketch a top view of injection plastic molding die for Figure 18.11. Notice that there are two pieces in the mold.

25.26. Examine Figure C18.1, where three layouts are given for the distribution of plastic to a component. Compare the molding designs and the opportunities and problems of designing, and producing molds for the three sketches.

25.27. Design snap gages for the following problems, as selected by the instructor. The length of the diameter is 2 in. (Hints: Sketch the C-frame and the buttons and provide key dimensions.)

	Diameter	Gage Allowance (%)	Toolmaker's Allowance (in.)
A.	1.000 ± 0.002	10	0.0004
B.	0.9873 ± 0.0005	10	0.0002
C.	$5.6942 \, {}^{+0.0002}_{-0.0000}$	10	0.0001

25.28 Design plug gages for the following problems, as selected by the instructor. The length of the diameter is 1.5 in. (Hints: Sketch the handle with the cylinders and provide dimensions.)

	Diameter	Gage Allowance (%)	Toolmaker's Allowance (in.)
A.	1.000 ± 0.002	10	0.0004
B.	0.9873 ± 0.0005	10	0.0002
C.	$5.6942 \, {}^{+0.0002}_{-0.0000}$	10	0.0001

25.29. Devise and sketch a gage to check Figure P25.29. (Hints: The radii are the significant dimensions. Measurements are millimeters.)

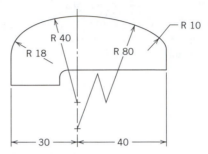

Figure P25.29

25.30. Develop gaging for Figure P25.30, where the hole and slot are checked by a plug gage. Design two two-pin gages to check the location of the holes and slots.

MORE DIFFICULT PROBLEMS

25.31. A sheet metal part is shown by Figure P25.31. Material is annealed cold-rolled steel and is provided in strips 44 mm wide by an indefinite length.

a. Find the preprocessed length of the part if it is flat. What is the advance of the part for a blanking die?

b. Sketch, using a straight edge, an elevation view of a blanking die.

c. Sketch a piercing die elevation view.

d. Sketch a forming die.

e. Find the yield. (Hints: Form a team of classmates for these requirements. Bend radius is equal to the thickness of the material. Distance between successive parts is three stock thickness. The die shoes are 150 by 150 mm. Proportionally sketch the other parts of the tooling design. It is unnecessary to dimension the sketches, but you should dimension the blanking and piercing punch and die. Refer to Chapter 16 for clearances. Label the components of the dies.)

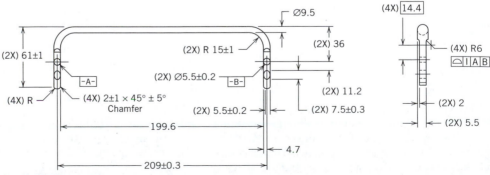

Figure P25.30

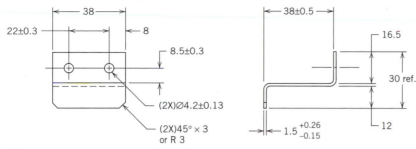

Figure P25.31

25.32. Figure P25.32 is a forged 0.5-lb stainless steel (17.4% chromium precipitation hardened material) crank arm. Annual production quantities are 240,000 units. The part is considered troublesome, and while quality control techniques are in place, QC uses direct reading micrometers for critical dimensions. Consider the process and the operations. The part starts out as bar stock, is forged, and during the sequence there is a heat-treating operation, in addition to about 20 other operations. Form a team of classmates for this problem. Determine a schedule of the dimensions that are necessary for gaging. What are the critical, major, and minor dimensions? Determine the gages. Should these gages be production or inspection gages? What approximate sample size should be inspected? How will the production system avoid final inspection and gaging of the part? (Hints: Your teacher will give instructions for forming the team. Not all dimensions are shown, but use a scale and estimate other noncritical dimensions, if necessary. Let the tolerance on any gages be 10% of the part tolerance. In determining the gages, the following are options: Develop a schedule of dimensions for a standard gage and sketch what the gage will be, or use some CAD system and determine a typical design. Indicate the materials for the gage. Examine catalogs of tooling components for information.)

25.33. A spacer block, shown in Figure P25.33, is made of AISI 1020 cold-rolled steel. The workpiece is cut to rough size and ground to the measurements given on the figure. A jig is necessary for drilling and reaming the two holes, and the third hole is subsequently milled to give the radius. Sketch a plate jig that will provide the accuracy. (Hints: The jig tolerances are 20% of the workpiece tolerances. Use ground pins for location. Include a bill of material for the jig items. Sketch the workpiece in a different color than the jig. The instructor will select one of the two problems. Choose either customary or metric dimension for the design.)

		x	y	ID
A	in.	3.018	1.875	0.305
		3.006	1.865	0.304
	mm	76.65	47.63	7.75
		76.40	47.37	7.72
B	in.	4.268	3.125	0.2501
		4.256	3.115	0.2503
	mm	108.41	79.120	6.35
		108.10	79.375	6.36

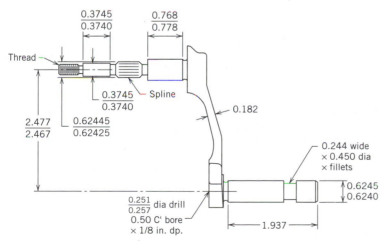

Figure P25.32

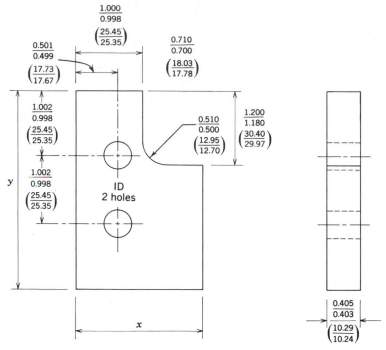

Figure P25.33

PRACTICAL APPLICATION

Conduct a study of tool design practice by calling and interviewing tool designers. Before you start this activity, prepare a list of questions and objectives for your survey.

CASE STUDY: OIL FILTER WRENCH

Hirsh Precision Products Company is contracted for lot of an oil filter wrench. Annual production quantities are high for this consumer product. Figure P23.7 is an assembly for this novel and innovative design. You, as the process and tool planner, are to develop the manufacturing design package for the part.

Steve Hirsh tells you, "We need to be very bright in the way in which we make this part. Our profit margin is small, and if there are production problems, we will end up losing money. See what you can do in preparing a manufacturing plan. Get back to me tomorrow."

Steve has in mind that you will prepare an operations process plan using the equipment with which you are familiar. Tooling, gaging, equipment, operations sequence, description, inspection and metrology equipment, and dimensions are stated on the ops plan. Item 1 starts out as ⅜ by 2-in. annealed AISI 4150 steel bar stock, and it is in 10-ft lengths. Item 2 is ¾ by ¾-in. AISI 4150 steel bar stock by 8 ft long. The two items are welded together in a welding fixture after they are machined. There are three operations process plans. Steve expects that you will prepare sketches of the tooling that are used with the equipment. "Plug and snap gages are appropriate, too," Steve says.

CHAPTER 26

METROLOGY AND TESTING

Simply stated, *metrology* is the science of measurement. It involves inspectors, technologists, operators, and engineers. Everything that has to do with measurement, be it designing, conducting, or analyzing the results of a test, exists within the metrology realm. Metrology activities cover the range from the abstract, comparing statistical methods, for example, to the practical, such as deciding what test to run or which scale of a ruler to read.

The effects of the science of measurement are seen everywhere, allowing people to plan their lives and make commercial exchanges with confidence. For example, most people assume that the clocks in their homes and the clocks in their workplaces all display approximately the same time. A pound of hamburger purchased at one grocery store will contain the same quantity of food as the same amount purchased at a store across town. Also, a screw purchased from Company A will fit into a hole made by a drill purchased from Company B, assuming they are specified to the same size.

26.1 FUNDAMENTALS OF METROLOGY

Working in and around metrology, there are necessary terms one needs to know; Table 26.1 defines a few of them.

The basic principles of metrology are simple. Metrology involves measurements and comparisons of a measurement standard to either another measurement standard or to a device of unknown accuracy. It involves the design of tests and methods by which the measurements and comparisons are made, and analysis of the results of the tests. The ultimate references for all measurement quantities (length, current, time, etc.) have been agreed on by international treaties.

Since measurement is necessarily a science of comparisons, instruments must be calibrated. *Calibration* is the comparing of a measurement device (an unknown) against an equal or better standard. A standard in a measurement is considered the *reference*; it is the one in the comparison taken to be the more correct of the two.

One calibrates to find out how far the unknown is from the standard. In this way, the instrument's *accuracy* and *absolute error* are evaluated as

$$\text{Absolute error} = \varepsilon = \text{true value} - \text{read value} \tag{26.1}$$

and

$$\text{Accuracy} = 1 - \frac{|\varepsilon|}{\text{true value}} \tag{26.2}$$

If several measurements are taken, then the arithmetic average is normally calculated.

There are five main types of metrology laboratories, and each has a different function. *Primary laboratories* are where new research into better methods of measuring is conducted. The calibration of primary and secondary standards is done here. Secondary and working standards are calibrated in *secondary laboratories*. *Research laboratories* are

Table 26.1 **Definition of Some Common Metrology Terms**

Accuracy[a]**:** The ability of an instrument to make measurements with small uncertainty.

Calibration: A set of operations, performed in accordance with a definite documented procedure, that compares the measurements performed by an instrument to those made by a more accurate instrument or standard, for the purpose of detecting and reporting, or eliminating by adjustment, errors in the instrument tested.

Calibration laboratory: A work space, provided with test equipment, controlled environment and trained personnel, established for the purpose of maintaining proper operation and accuracy of measuring and test equipment.

Error: The difference between the measured value and the true value of the object of a measurement.

Measurement: A set of operations performed on a physical object or system according to an established documented procedure to determine a physical property.

Precision[a]**:** The repeatability of the measurement process or how well identically performed measurements agree.

Readability[a]**:** A device is readable if the measurements are converted to a readily ascertained number. A vernier on a micrometer makes the instrument more readable.

Sensitivity[a]**:** The ability to detect differences in a quantity being measured.

Standard: Object, artifact, instrument, system, or experiment that provides a physical quantity which serves as the basis for measurements of the quantity.

Standards laboratory: A work space, provided with equipment and standards, a properly controlled environment, and trained personnel, established for the purpose of maintaining traceability of standards and measuring equipment used by the organization it supports.

Tolerance: In metrology, the limits of the range of values (the uncertainty) that apply to a properly functioning measuring instrument.

Traceability: Unbroken chain of comparisons from the measurement being made to a recognized national, legal standard.

Uncertainty: An estimate of the possible error in a measurement. More precisely, an estimate of the range of values that contains the true value of a measured quantity. Uncertainty is usually reported in terms of the probability that the true value lies within a stated range of values.

Verification: The set of operations that ensures that specified requirements have been met, or leads to a decision to perform adjustments, repair, downgrade performance, or remove from use.

[a] Sensitivity and readability are associated with the device, whereas precision and accuracy are associated with the measuring process.

where some of the most abstract projects (such as measuring the current of a single electron) are done. *Calibration laboratories*, which tend to be oriented toward the production of higher-volume calibrations, utilize standards supported by secondary or primary laboratories. *Mobile laboratories* are equipped with metrology standards that can be moved to a variety of locations.

The process of making a measurement, by either of these laboratories, is only a part of calibration. After the measurement is finished, all the information about the measurement standard and the test instrument is placed together with the test data. The information in this test record supports the *traceability* of the test.

A traceable calibration is achieved when each instrument and standard, in a hierarchy stretching back to the national standard, is itself properly calibrated, and the results properly documented. The documentation provides the information needed to show that all the calibrations in the chain of calibrations are properly performed. If one were to follow the paper trail up from the calibrated unit through the secondary and primary standards, it would eventually end at the record of an experiment made to establish the quantity in question in terms of one or more of the seven basic measurement units.

A system of measurements, such as *Le Système International d'Unités* (*SI*), satisfies certain concepts. For each quantity of length, mass, time, or temperature, a unit is necessary. The SI system is a decimal system composed of six base units, two supplementary units, and additional derived units. (The units and conversions are provided in the front and back end papers.) In SI, the quantities of length, mass, and time have the units meter, kilogram, and second.

Every unit of measurement has a definition, a realization, and a representation. The *definition* is the ideal, and it is usually a member of the SI. The *realization* is achieved by means of an experiment whose result matches the definition as closely as possible. When the realization is obtained, a national laboratory stores its value as a *representation* of the unit. The national laboratory uses its representation as a master standard to which other representations are compared, as described in this chapter.

Ideally, any manufacturer adopting a defined standard for their measurements of a quantity would expect that everyone considers the definition of the standard fixed and correct. Furthermore, the standard value must be identical at all times and at all places, under all environmental conditions.

EXAMPLE 26.1

A local manufacturer sends a pneumatic gage to a calibration laboratory. In the calibration process, the input value (known pressure) is a controlled independent variable, whereas the measured output value (dial displacement) becomes the dependent variable of calibration.

By application of a range of known values for the input and observation of the output, a direct *calibration curve* is developed for the measurement instrument. The curve in Figure 26.1 is obtained. If the expected true values are represented by the exponential curve, then the average absolute error and the average accuracy of the instrument are 0.015 and 0.97 (or 97%), respectively.

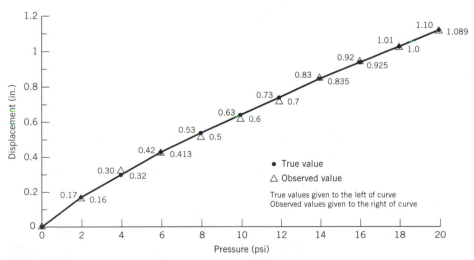

Figure 26.1
Calibration curve for a pneumatic gage.

26.2 LINEAR MEASUREMENTS

The meter is defined as the length of 1,650,763.73 wavelengths of radiation of the atom of Krypton 86, the orange-red line. This definition points out the accuracy of metrology but does not provide a standard located in the plant. Measuring length accurately can be accomplished only through comparison with a standard, preferably one that is traceable to a primary standard.

A standard used for comparison to other measurement devices is the precision gage block set.

Precision Gage Blocks

Gage blocks are square, round, or rectangular in shape, with two parallel sides very accurately lapped to size. In order of hardness and cost the blocks can be made from tool steel, chrome plate steel, stainless, chrome carbide, or tungsten carbide. Tungsten carbide is the hardest and most expensive. Laboratory sets may be obtained with a guaranteed accuracy within two-millionths of an inch per block. These blocks are used mainly for reference in setting gages; for accurate measurements in tool, gage, and die manufacturing; and as master laboratory standards for control of measurements in manufacturing. Their accuracy is valid only at 68°F (20°C). With a standard set of 81 blocks it is possible to obtain practically any dimension in increments of 0.0001 in. for 0.100 to more than 10 in. by combining those of the proper size. Micrometers and vernier equipped instruments are used to check tolerance to within one-thousandth of an in., but gage blocks are usually necessary when tolerances are in the ten-thousandths. If millionths are to be measured, a constant temperature laboratory with optical or electronic equipment for gage block calibrations and comparisons is necessary. Special *angle gage blocks* are available with accuracies compatible to the precision gage blocks. A set of 16 angle blocks will permit the measuring of most any angle to an accuracy of one second.

In the gage block set of 81 blocks (plus two wear blocks) dimensions are as follows:

9 blocks with 0.0001-in. increment from 0.1001 to 0.1009

49 blocks with 0.001-in. increment from 0.101 to 0.149

19 blocks with 0.050-in. increment from 0.050 to 0.950

4 blocks with 1-in. increment from 1 to 4

2 tungsten steel blocks, each 0.050 in. thick

If a dimensional standard of 3.6083 in. is desired, the procedure for arriving at a suitable combination is determined by successive subtraction, as shown in Table 26.2. This stack of blocks is shown in Figure 26.2.

Vernier Caliper

Figure 26.3 shows a vernier caliper that can be used for determining inside and outside measurements over a wide range of dimensions. The vernier consists of a main scale graduated in inches and an auxiliary scale having 25 divisions. Each inch on the main scale is divided into tenths and each tenth into four divisions, so that in all there are 40 divisions (each 0.025 in.) to the inch.

Vernier and micrometer calipers also are available in metric units. The 25 divisions on the auxiliary or sliding scale correspond to the length of 24 divisions on the main scale and are equal to 24/40 or 24/1000 in., which is 1/1000 in. less than a division on the main scale. Hence, if the two scales were on zero readings, the first two lines would be 0.001 in. apart, the tenth lines 0.010 in. apart, and so on.

Table 26.2   Successive Subtraction Procedure for Using Gage Blocks

Procedural Step		Block Used
Desired dimension	= 3.6083	
Ten-thousandth's place	= 0.1003	0.1003
Remainder	= 3.508	
Thousandth's place	= 0.108	0.108
Remainder	= 3.40	
Hundredth's place	= 0.30	0.30
Remainder	= 3.10	
Two wear blocks	= 0.10	0.10
Remainder	= 3.0	
Units place	= 3.0	3.0
Remainder	= 0	
TOTAL:		3.6083

The reading on the main scale is first observed and converted into thousandths and to this figure is added the reading on the vernier. The vernier reading is obtained by noting which line coincides with a line on the main scale. If it is the fifteenth line, 0.015 in. is added to the main scale. These scales are shown in detail in the enlarged view of the vernier scale in Figure 26.3. As shown, the vernier reads exactly 0.400 in.

Outside measurements are taken with the work between the jaws and inside measurements with the work over the ends of the two jaws. This method of measurement is not so rapid as a micrometer but has the advantage of a wider range with equal accuracy. The principle of the vernier also is used on protractors for angular measurement.

A *divider* is similar in construction to a caliper, except that both legs are straight with sharp, hardened points at the end. This tool is used for transferring dimensions, scribing circles, and doing general layout work.

Micrometer

The *micrometer*, Figure 26.4, is used for quick accurate measurements to the thousandth part of an inch. Micrometers are available, which read in metric units. This tool requires an accurate screw thread in obtaining a measurement. The screw is attached to a spindle and is turned by a thimble at the end. The barrel, which is attached to the frame, acts

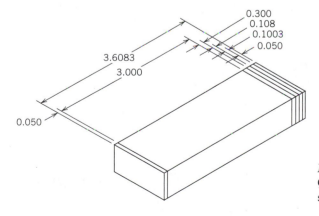

Figure 26.2
Gage blocks assembled to a dimension of 3.6083 in.

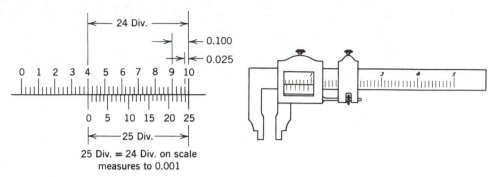

Figure 26.3
Vernier scale.

as a nut to engage the screw threads, which are made accurately with a pitch of 40 threads per inch. Each revolution of the thimble advances the screw $\frac{1}{40}$ of an inch or 0.025 in. The outside of the barrel is graduated in 40 divisions, and any movement of the thimble down the barrel can be read next to its beveled end. When the spindle is in contact with the anvil on a 1-in. micrometer, the zero readings on barrel and thimble should coincide. The graduations on a metric micrometer caliper in the 0- to 25-mm range are 0.002 mm with a vernier reading.

The scale on the barrel and thimble edge are best understood by reference to the view of Figure 26.4. On the beveled edge of the thimble are 25 divisions, each division representing 0.001 in. To read the micrometer the division on the thimble coinciding with the line on the barrel is added to the number of exposed divisions on the barrel converted into thousandths. Thus, the reading shown in Figure 26.4 is made up of 0.200 plus 0.025 on the barrel, or 0.225 in., to which is added 0.016 on the thimble to give a total reading of 0.241.

Since most micrometers read only more than a 1-in. range to cover a wide range of dimensions, several micrometers or different length spindles are employed. The micrometer principle of measurement also is applied to inside measurements, depth reading, and the measurements of screw threads. Micrometers are available with optical and electronic readouts.

Optical Instruments

Because of their extreme accuracy and ability to measure parts without pressure of contact, numerous optical instruments have been devised for inspecting and measuring. A microscope for toolroom work, for example, can be used to enlarge an object's view

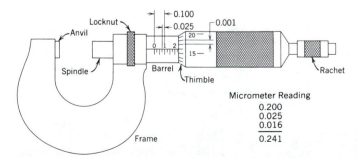

Figure 26.4
One-inch micrometer caliper.

without reversing the image as ordinary microscopes do. The instrument has micrometer screws that operate in either direction and read to an accuracy of 1 part in 10,000.

Optical gages can be used to measure heights to 1 part in 10,000 in the production shop or gage laboratory. The gage is equipped with interchangeable spindles to accommodate dimensions to 9 in. (230 mm). The spindle is lowered by an electric motor until it contacts the workpiece with a predetermined load between 7 and 10 oz (2 to 3 N). The actual dimension can be read on an illuminated, knob-adjusted vernier scale. Used like a micrometer or visual gage, optical gages are faster than a manual micrometer.

Laser

Helium neon *gas lasers* are popular in inspection and in the assembly of large machines because they are the only method of providing a visible straight line. The bright red beam does not sag or bend. There is no other inspection system with this accuracy over large distances. Lasers are used in production shops and in inspection laboratories for checking straightness, flatness, squareness, and levelness.

A laser is a device capable of generating coherent electromagnetic radiation at wavelengths shorter than those of the microwave. The gas laser contains a gas whose atoms or molecules are raised to higher energy levels in the gas. The material portion of the laser consists of a long cylindrical tube containing the gaseous medium, a means for exciting a discharge in the medium, and a pair of mirrors facing each other, which constitute the resonator for the laser energy. Generally, this laser may be summarized as a gaseous medium excited by electric discharge and containing within it a closed optical path in which optical energy can be contained for relatively long periods of time. Figure 26.5 is a sketch of the essentials of the laser gun.

The most accurate measurement and calibration techniques involve a process called *interferometry*. This method is simply measuring distances making use of a known coherent source (most often a helium–neon laser). Laser light from a coherent source is divided into two beams, as illustrated in Figure 26.6. A light from a mirror mounted on one surface interferes with reflected light from a second surface. One mirror remains in a fixed position, while the other is moved to cause alternately constructive and destructive interference. This alternation produces variations of wave intensity that can be detected by a photoelectric detector and counted with a digital counter.

This method (accurate to two millionths of the desired distance) can measure lengths from 0.2 mm to 120 m and is used primarily for testing and measuring machine tools, precision measurement, comparisons with standards, and calibration.

The *laser interferometer* consists of three parts: a power supply, a combination laser and interferometer, and a retroflector. Simply stated, beam splitters send one-half the

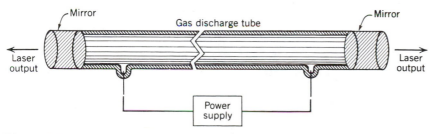

Figure 26.5
Essentials of a gas laser.

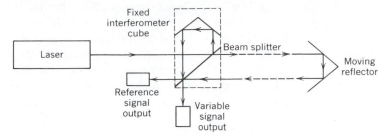

Figure 26.6
Principle of operation
of laser interferometer.
(Adapted from Harry,
*Industrial Lasers and
Their Applications.*)

laser light to the retroflector and the other half is directed to a photo detector. The light going to the retroflector is reflected back to the interferometer and the fringes created by the interference of the two sources of light are a measure of distance. Such units can be equipped with digital or graphic readout. So fast is the interpretation of distance that the retroflector may move at rates up to 200 in./min (0.1 m/s).

26.3 ANGULAR MEASUREMENTS

Angular measurement is standardized by the *radian*. This is the unit of measure of a plane angle with its vertex at the center of a circle and subtended by an arc equal in length to the radius. Usually common angular measuring instruments read degrees directly from a circular scale scribed on the dial or circumference. There also are devices that require the aid of other measuring instruments and calculations to obtain the result.

The plain or universal *bevel protractor* measures directly in degrees and is adapted to all classes of work in which angles are to be laid out or established.

Sine Bar

A *sine bar* is a simple device used either for accurately measuring angles or for locating work to a given angle. Mounted on the center line are two buttons of the same diameter at a known distance apart. For purposes of accurate measurement, the bar must be used in connection with a true surface.

The operation of the sine bar is based on the trigonometric relationship that the sine of an angle is equal to the opposite side divided by the hypotenuse. Measurement of the unknown side is accomplished by a height gage or precision blocks.

Figure 26.7 shows a sine bar set to check the angle on the end of a machined part. In this case

$$\sin \Theta = \frac{(h_1 - h_2)}{L} \tag{26.3}$$

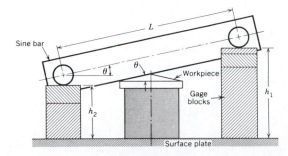

Figure 26.7
Sine bar setup on gage blocks for measuring an angle on workpiece.

where L is a known distance. The heights h_1 and h_2 are built up to correct heights in corresponding units of linear dimension with precision gage blocks and their difference in elevation over L gives the sine of the angle Θ.

26.4 SURFACE MEASUREMENTS

Surface checking instruments are for finding a measure of the *accuracy of a surface* or the condition of a finish. Much of this work is done on a flat, which is an accurately machined casting or lapped granite block known as a *surface plate*. It is the base on which parts are laid out and checked with the aid of other measuring tools. These plates are made carefully and should be accurate to within 0.0001-in. (0.003-mm) flatness from the mean plane to any point on the surface. Small plates, known as *toolmakers' flats*, are lapped to a greater degree of accuracy. Their field of application is limited to small parts and they are normally used with precision gage blocks.

Surface Gage

The *surface gage*, shown in Figure 26.8, checks the accuracy or parallelism of surfaces. It also transfers measurements in layout work by scribing lines on a vertical surface. When in use, it is set in an approximate position and locked. The spindle can be finely adjusted by turning the knurled nut that controls the rocking bracket. When used with the scriber, it is a line measuring or locating instrument. If the scriber is replaced by a dial indicator or a transducer, it becomes a precision instrument for checking surfaces.

Optical Flat

Measurements to the millionth part of an inch (approximately 25 μm) are made by interferometry, the science of measuring with light waves. Measurements by this principle

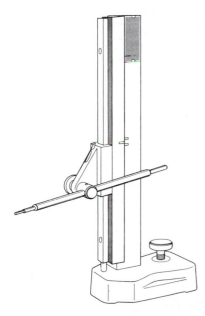

Figure 26.8
Height transfer (surface) gage.

are made with small instruments known as *optical flats*. They are flat lenses with very accurately polished surfaces that have light-transmitting quality. Optical flats are usually made from natural quartz because of its hardness, low coefficient of expansion, and resistance to corrosion. Optical flats are available from 1 to 12 in. (25 to 300 mm) in diameter with a thickness about ⅛ the diameter. It is not necessary that the two surfaces of a flat be absolutely parallel.

One of the common uses for optical flats is the testing of plane surfaces. The optical flat is placed on the flat surface to be tested, and light is reflected both from the optical flat and the surface being tested through the very thin layer of air between the two surfaces. When the light waves are in phase, there is a light band; when they are out of phase, a dark band is created. If the thickness of the air layer measures one-half a wavelength of light or more, an interference effect occurs. The interference between the rays reflected from the bottom of the flat and from the top of the work causes dark bands, called newton's rings, to appear.

If the surface is irregular, the appearance is similar to a contour map. The position and number of lines show the location and extent of the irregularities. When the bands are straight, evenly spaced, and parallel to the line of contact, as shown in Figure 26.9, the surface is perfectly flat. If the bands were straight but not evenly spaced or if the bands curved, the surface would not be flat. If the wavelength of the light source is known, any deviation from this pattern indicates an error in the surface, the amount of which can be measured. A monochromatic light of one wavelength, such as fluorescent helium is used for the bands to be sharply defined. Each fringe or band in such a light indicates a difference in height of 0.0000116 in. (295 nm), the one-half wavelength of helium.

The fringe patterns are interpreted differently according to whether the optical flat is pressed hard against the surface or slanted slightly with a wedge-shaped air space between the block and the workpiece, as shown in Figure 26.10*A* and *B*. Because contact fringes are more difficult to interpret, the wedge method is used whenever possible. Figure 26.10 shows two parts being inspected, one concave and one convex. The number of bands or fringes that appear depend on the wedge thickness, but the curvature is the measure of surface flatness. Figure 26.10*A* shows a condition in which each band curves an amount slightly more than two band intervals. The 2.2 bands of curvature indicate that the workpiece is 2.2 by 0.0000116 in. = 0.0000255 in. = 25 μin. (625 nm) high in the center, since the bands curve toward the thin part of the wedge. In Figure 26.10*B*, the bands curve away from the thin side of the wedge so that the surface of the workpiece is concave. The four-band curvature means that the surface is 46 μin. (1150 nm) lower in the center than on the edges.

Figure 26.9
Straight interference fringes shown by this optical flat indicate a flat gage block.

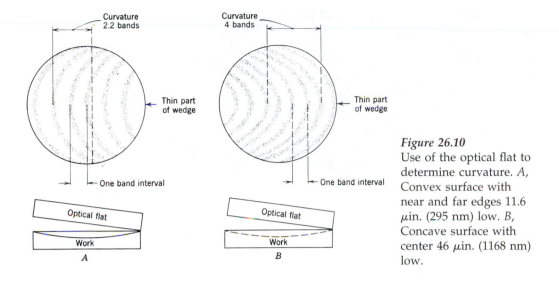

Figure 26.10
Use of the optical flat to determine curvature. *A*, Convex surface with near and far edges 11.6 μin. (295 nm) low. *B*, Concave surface with center 46 μin. (1168 nm) low.

Surface Roughness

Several devices have been developed to measure surface roughness. The simplest procedure is a visual comparison with an established standard. Other methods include microscopic comparison, direct measurement of scratch depth by light interference, and the measurement of the magnified shadows cast by scratches on a surface or even touch comparison of tactile standards. The usual procedure is to employ a diamond stylus to trace over the surface being investigated and to record a magnified profile of the irregularities.

To measure roughness and other surface characteristics, a standard was developed by the American Standards Association (ASA B46.1-1962), which deals with such surface irregularities as height, width, and direction of the surface pattern. These surface irregularities, as well as the symbols for specifying *surface roughness* on a drawing, are shown in Figure 26.11.

An instrument used in making surface roughness measurements is shown in Figure 26.12. This is a direct reading instrument that measures the number of roughness peaks per inch above a preselected height by passing a fine tracing point over the surface. The unit consists of a tracer, which converts the vertical movements of the tracing point into a small fluctuating voltage that is related to the height of the surface irregularity, a motor-driven device (pilotor) for operating the tracer, and the amplimeter. The amplimeter receives the voltage from the tracer, amplifies, and integrates it so that it may be read

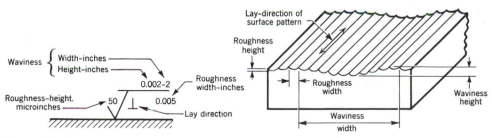

Figure 26.11
Surface characteristics and symbols for indicating their maximum values.

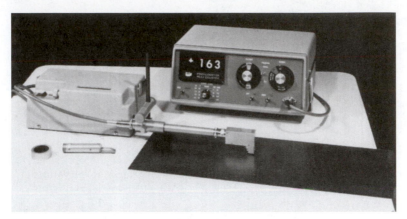

Figure 26.12
Surface gage including transducer tracer, amplifier, and indicator for
measuring surface roughness.

as digital values or shown on a strip chart recorder. The process is a continuous one,
and the instrument shows the variation in average roughness from a reference line, as
illustrated in the magnified profile of a surface in Figure 26.13. Readings may be either
arithmetical (AA) or *root mean square* (RMS) average deviation height from the reference
line CD. The difference in result of the two methods of calculation is indicated in the
example in Figure 26.13. The instrument may be operated either manually or mechanically, and readings can be taken on both plain and curved surfaces.

Surfaces with the same average roughness height can be very different, since the height
and number of the peaks and valleys and the roughness width can be dissimilar. The
unit shown in Figure 26.12 also will determine the number of peaks per inch above a
preselected height so that surfaces can be identified more thoroughly and inspected.
Surface roughnesses available by common production methods are indicated in Figure
26.14.

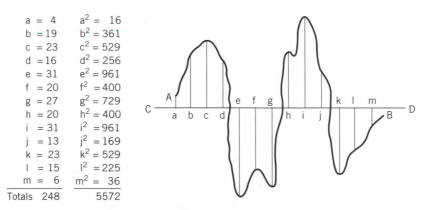

a = 4	a^2 = 16
b = 19	b^2 = 361
c = 23	c^2 = 529
d = 16	d^2 = 256
e = 31	e^2 = 961
f = 20	f^2 = 400
g = 27	g^2 = 729
h = 20	h^2 = 400
i = 31	i^2 = 961
j = 13	j^2 = 169
k = 23	k^2 = 529
l = 15	l^2 = 225
m = 6	m^2 = 36
Totals 248	5572

Arithmetical average = $\dfrac{248}{13}$ = 19.1 microinches (485 nm)

Root-mean-square average = $\sqrt{\dfrac{5572}{13}}$ = 20.7 microinches rms (525)

Figure 26.13
Relationship between arithmetic average and root-mean-square values
used in determining surface roughness.

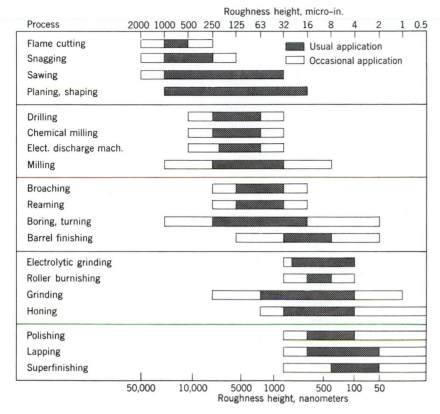

Figure 26.14
Surface roughness available from common production methods.

26.5 ELECTRICAL MEASUREMENTS

Many parameters can be measured in electrical systems. Resistance, impedance, capacitance, inductance, and electromagnetic fields are a few examples. The ampere is the base unit for electricity in the International System of units. The SI definition of the volt is derived from the ampere and the mechanical unit of the watt.

Current, Resistance, and Voltage

The SI links the mechanical and electrical units by defining the *ampere* as "that constant current which, if maintained in two straight parallel conductors of negligible cross section and placed 1 meter apart in a vacuum, would produce between these conductors a force equal to 2 by 10^{-7} newton per meter of length."

Because the standard ampere is very difficult to realize, the uncertainty of its realization is large, about 15 parts per million (ppm). Also, it is difficult to maintain for more than a few minutes.

Figure 26.15 illustrates the process by which the ampere is realized. In this simplified drawing, the magnetic fields of the coils on the left side of the balance arm produce an attractive force proportional to the current flowing and the number of turns in each coil. The force is balanced by the weight on the right side of the balance arm.

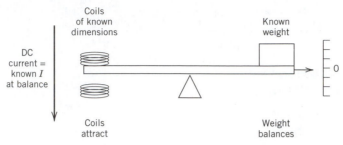

Electrical units unified with mechanical units

Figure 26.15
Realization of the ampere.

However, what causes the force? According to physics, the force is produced by the electromagnetic field associated with moving electrons. In these terms, 1 ampere is equal to the flow of 1 coulomb of electrons per second past a given point in an electrical circuit.

Because of the difficulty of realizing the ampere, there is no standard representation for it. Instead, it is derived locally from the ratio of the volt and the ohm according to Ohm's law, $I = V/R$. The volt and the ohm are maintained in standards both in a local laboratory and at the national level.

On the practical level, working standards for resistance take the form of certified standard resistors that are used for comparison in the calibration of resistance-measuring devices. The practical potential standard makes use of a standard cell consisting of a saturated solution of cadmium sulfate.

The *volt* is defined in terms of the SI base electrical unit, the ampere, as "the difference of electrical potential between two points of a conducting wire carrying a constant current of 1 ampere, when the power dissipated between these points is equal to 1 watt" (1 watt = 1 joule per second). Based on mechanical units, then

$$V = W/A \tag{26.4}$$
$$W = J/s \tag{26.5}$$
$$J = Nm \tag{26.6}$$
$$N = kgm/s^2 \tag{26.7}$$

where

W = Watt (SI derived unit for power)
A = Ampere (SI base unit for electricity)
J = Joule (SI derived unit for work/energy)
s = Second (SI base unit for time)
kg = Kilogram (SI base unit for mass)
m = Meter (SI base unit for length)

26.6 GAGES AND OTHER MEASUREMENTS

A significant amount of new techniques are introduced constantly to measure systems and processes' performance. Some of these techniques are summarized in Table 26.3.

For controlling dimensions in the manufacture of all products, various types of *standard gages* are used. Gages are used to measure fixed shape or size in production work. They represent a standard for comparing manufactured parts. Their use is limited to one or several dimensions on a part to determine whether the part has been made to the design tolerance.

Table 26.3 **Other Measurements—Summary Table**

Measurement Type	Parameters	Some Instruments and Calibration
Mechanical	Flow rate	Obstruction meters: orifice plate, nozzle, venturi
	Mass	Analytical balances
	Force, stress, and strain	Load cells, proving rings, strain gages
	Vibrations	Piezoelectric accelerometers, seismic transducers
		Accelerometers are normally calibrated by subjecting them to half-sinewave pulses with peak amplitudes of 50 to 5000 g and pulse widths from 0.2 to 40 ms.
	Sound	Sound-level meters, comprising a microphone, attenuator, amplifier, frequency-weighting network, and a recorder. Microphones are calibrated by pressure and free-field responses. Calibration is reported in terms of open-circuit voltage per unit of sound pressure applied uniformly to the diaphragm.
	Ultrasonics	Aluminum, titanium, and steel reference blocks are used and calibrated using ASTM E 127 or E 428.
Thermodynamic	Temperature	Thermometers: liquid-in-glass, bimetallic. Thermistors. Resistance temperature detectors (RTD). Calibration is done by placing the thermometer in a constant temperature bath along with a NIST-calibrated standard platinum-resistant thermometer.
	Pressure	Barometer, manometer, deadweight testers
	Humidity	Hygrometers: dew-point, electric, electrolytic, aspirated, pneumatic
	Heat Flux	Heat flux gages are calibrated in an infrared radiation field at flux levels to 4 W/cm^2. The radiant field is produced by a stable infrared electric heater. A reference radiometer is used to establish the flux level at the measurement point within the radiative field.
Radiation	Optical Radiation Photometry	Luminous intensity lamps, spectral transmitance filters, spectral radiance lamps, radiometric detectors
	Spectrophotometry Radiometry	Special test of radiometric detectors, generally used in the ultraviolet spectrum, are performed to determine linearity, quantum efficiency, and uniformity.
	Ionizing radiation Gamma and x-rays	Radioactive solutions, dosimeters (passive instruments), and remmeters (active instruments)
	Alpha and beta particle emission Neutron emissions	For radioactive solution calibration the uncertainty, varying from 0.7% to 2.5%, depends on the activity level and chemical form. Calibration for instruments (X- and gamma-ray measuring in particular) is done by a substitution method in a ray beam at a point where the rate has been determined by means of a standard free-air ionization chamber.
Time and frequency	Time	Digital and atomic clocks
	Oscillations	Precision oscillators are used and can be calibrated in frequency range from 1 to 100 MHz. Reference calibration accuracy is nominally that of the NIST primary frequency standard, that is 8 by 10^{-14}. The accuracy level transferable to the oscillator depends on the stability and noise properties of the oscillator.

According to their function, gages can be classified as inspection or as manufacturing gages. *Inspection gages* are used by inspectors in the final acceptance of the product to determine if the product is made in accordance with the tolerance specification of the design. *Working* or *manufacturing gages* frequently are made to slightly smaller tolerances than the inspection gages, to keep the size near the center of the limit tolerance. The assumption is that parts manufactured around limit sizes will pass the inspector's gage.

In this text, snap and plug gages (go, no-go gages) are considered tools of special application, so they are discussed in Chapter 25, Tool Design.

Dial Indicator

A *dial indicator* is composed of a graduated dial, spindle, pointer, and a satisfactory means for supporting or clamping it firmly. Most indicators have a spindle equal to 2½ revolutions of the hand. Between the test point and the hand is interposed an accurate multiplying mechanism, which magnifies on the dial any movement of the point. This tool may be considered either a measuring device or a gage. As a measuring device, it measures inaccuracies in alignment, eccentricity, and deviations on surfaces supposed to be parallel. In gaging work, dial indicators give a direct reading of tolerance variations from the exact size.

Projecting Comparators

Projecting comparators are designed on the same principle as a projection lantern. Figure 26.16 represents two types of such an instrument. In Figure 26.16*A*, an object is placed before a light source and the shadow of the profile is projected on the screen at some enlarged scale. In Figure 26.16*B*, a projecting portable camera lens is used. Usual magnifications run from × 5 to × 100, with an accuracy of ± 0.05% for contour and ± 0.01% for surface readings.

A   *B*

Figure 26.16
A, Horizontal comparator. *B*, Comparator with a portable projecting camera.

The outline of the object is reflected to the screen facilitating the inspection of small parts such as needles, saw teeth, threads, forming tools, taps, and gear teeth. Since comparators check work to definite tolerances, they are useful for studying wear on tools or distortion caused by heat treatment.

Pneumatic, Electric, and Electronic Gaging

Because of speed, accuracy, and adaptability to automatic inspection, pneumatic, electric, and electronic gaging are used as inspection and production devices. These gages are used for checking other gages and dimensional standards, piece by piece inspection, automatic inspection, and machine control.

Pneumatic gaging employs compressed air and measures back pressure of the air as it exits from the gage by metering its flow. An air spindle, shown in Figure 26.17, has two small diametrically opposed holes for air flow. The amount of air flow is controlled by the size of the annulus space between the air spindle and the work. This change in flow is registered on the dial, which is calibrated to read in fractions of a thousandth of an inch (0.02 mm). The relationship between the rate of flow and the size of hole is true only for small clearances, and the maximum range is around 0.003 in. (0.08 mm). High amplification permits reading in fractions of a tenth of a thousandth (0.002 mm).

Pneumatic gages can check internal and external dimensions, and multiple checking of several dimensions can be done simultaneously. Air gages will reveal hole or shaft taper, out-of-roundness, and tool gouges, which are difficult to detect with a plug gage. Pneumatic gages have the following advantages: speed and simple operation, accurate to about 0.0005 in. (0.013 mm), relatively inexpensive, do not scratch even the finest or softest finish, low gage wear, and require minimum skill. Dimension can be magnified and displayed. Many dimensions can be read at once, which facilitates inspection and selective assembly. These gages can be fitted with electric or electronic gages and amplifiers.

Electric Gaging

Two types of electric gaging employ either sensitive microswitches (contact gages) or a dial indicator with two limit switches. Microswitch-type units are usually employed for inspection of very large parts that have tolerances of at least 0.002 in. (0.05 mm). Large die castings, for example, can have many dimensions inspected simultaneously by deploying microswitches previously set with gage blocks on a surface plate. Each dimension to be checked is provided with pairs of red and green lights that indicate whether the dimensions meet the specified tolerance. These gages are designed for simultaneous checking of

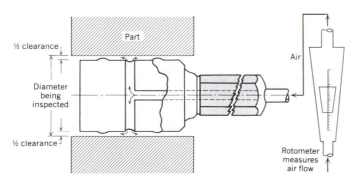

Figure 26.17
Schematic drawing of a spindle used to measure internal diameters.

several dimensions. Master parts and their dimensions are employed in setting the gage to the correct limits.

The dial indicator type of electric gage has two switching limits. One limit is set at the upper and the other at the lower tolerance limit. Production work can be divided into three categories. Parts too large, too small, and within tolerance can be electrically shuttled if a control circuit is incorporated. This type of gage is fast, accurate to 0.0001 in. (0.003 mm), and can be used in automatic inspection and control.

Electronic Gaging

Electronic gaging systems are popular in production, inspection, and control because, unlike mechanical or pneumatic gaging, they require less response time and have improved transducer linearity over the range of measurement. Accuracies to 0.00005 in. (0.0001 mm) are reported.

Most measurement systems are composed of a generalized three-stage arrangement, as shown in Figure 26.18. The first stage is the detector transducer followed by an intermediate modifying stage and a terminating stage. The purpose of the detector transducer is to sense the input signal while being insensitive to other inputs. A strain gage should be sensitive to strain but unaffected by temperature. The second stage modifies or sometimes amplifies the transduced information to have it acceptable to the third stage. It may amplify power to drive the terminating device. The terminating stage provides the information in a form understandable to one of the human senses or to a controller, minicomputer, digital readout, or recorder.

The first contact that a measuring system has with the quantity to be measured is through the input information accepted by the transducer. The transducer must be changed by the quantity to be measured. Figure 26.19*A* describes a variable inductance transducer or more commonly, the *linear variable differential transformer* (*LVDT*). The LVDT provides an a-c voltage output proportional to the displacement of a core passing through the windings. It is a mutual inductive device using three coils. The LVDT has advantages over other transducers. It converts length displacement into a proportional electric voltage and it cannot be overloaded mechanically, since the core is separable from the remainder of the device and is relatively insensitive to temperature.

Figure 26.19*B* is a resistance strain gage bonded to a surface. Once the block is loaded, say by a compressive force, the force indirectly strains the wires of the gage. The elongation of the gage wires reduces their diameter and a longer length results in increased resistance. Figure 26.19*C* is a self-inductance transducer, in which the length displacement changes, thereby changing inductance.

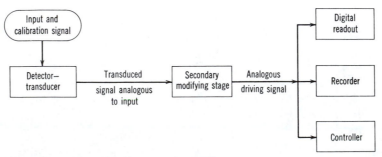

Figure 26.18
Three stages of a generalized measurement signal.

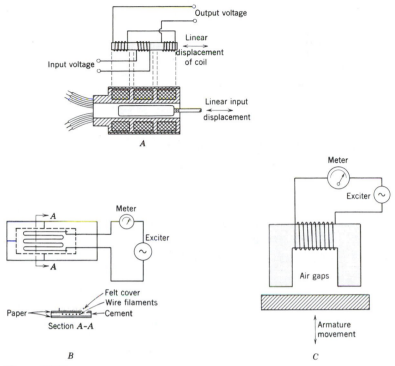

Figure 26.19
Schematics of transducers. *A*, Variable linear differential transformer
with a typical section. *B*, Bonded-wire strain gage and a typical section.
C, Self-inductance transducer where the length of the air gap changes
the pickup output.

The electronic gage like the electric type can be "tailored to fit" large unusually
shaped parts. The advantages include excellent accuracy and sensitivity, fast readings,
zero adjustments, high magnification, and versatility.

Machine Gaging

Electronic measurement devices can be mounted on machines to monitor the dimensions
dynamically. Gaging can control the amount of metal removed, for example. Consider
a dynamic balancing machine integrated into a production line where engines are being
produced continuously. The crank shaft, which is sensitive to imbalance, requires a
monitoring activity to determine the amount of metal to remove to satisfy performance
specification for vibration control. Once the nominal dimension is achieved, a control
system produces a feedback signal to stop machining.

Figure 26.20 is a unit adaptable to continuous monitoring of ODs. In this simple device
hydraulic retractable fingers move the gage out of the way during the load and unload
cycle. These transducer–monitor instruments are used in several ways for machine gaging,
and the example of the dynamic balancing machine for the rotary crankshaft will have
a feedback computer control system capable of continuously interpreting results and
making adjustments in the metal removal.

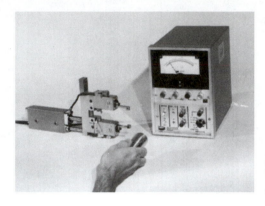

Figure 26.20
LVDT and controlled fingers for on-line gaging
and control of ODs.

Roundness and geometric gages can be mounted on machines to measure roundness, concentricity, and alignment of inside and outside diameters, squareness, and flatness. Figure 26.21 shows two kinds of faults with bar stock, roughness, and ovality, which are measured with a roundness meter.

Other standard gages include ring, taper, thread, and thickness. *Ring gages*, for outside diameters are used in pairs, as a "go" and "no-go." *Taper gages* are not dimensional gages but rather a means of checking in terms of degrees. Their use is a matter more of fitting rather than measuring. A *thickness* or *feeler gage* consists of a number of thin blades and is used in checking clearances and for gaging in narrow places.

Automatic Inspection Machines

Coordinate measuring machines (CMMs) are some of the most popular automatic inspection machines in industry today. They consist of a contact probe and a means of positioning it in a three-dimensional space relative to the surface of the part. Figure 26.22 shows a CMM of the gantry construction type. This type of machine provides a decimal readout of a probe position. The part is loaded and leveled on the gaging table, and a zero point is located by a reference point. As the probe is moved about, the readout console displays the x, y, and z coordinate positions from the reference point. Bore location and size can

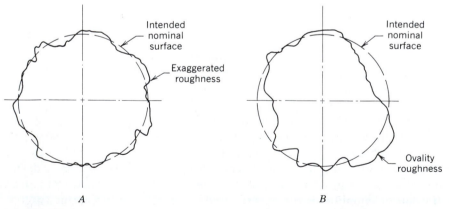

Figure 26.21
A, Transverse variations in roughness. *B*, Ovality variations in roughness.

Figure 26.22
Coordinate measuring ma-
chine.

be determined. Movements in all axes are measured by a reading head as it travels over a steel grating having 1000 lines/in. (40 lines/mm). A corresponding segment of grating mounted on the reading head creates a *moire fringe pattern* as it passes over the grating. As the fringe patterns are counted, output signals from the head provide a digital readout of movement and position. Accuracies of 0.001 in. (0.03 mm) are common.

26.7 NONDESTRUCTIVE INSPECTION

The purpose of nondestructive testing (NDT) is simply to determine flaws or defects without damage to the object. Preservation of the part and economics are the motivating considerations. When any of these tests is conducted, measurement instruments are used, so their calibration is fundamental to minimize any instrument error. Some NDT methods are discussed here and methods of evaluating uncertainty will follow.

Hardness Measurements

It is common practice to have hardness specified by engineering for heat-treated parts. Since most parts are near their final stage of production when hardness is verified, equipment is selected so that minimum impression and distortion of the workpiece are caused. In Chapter 2, the methods of determining material hardness are discussed. Figure 26.23 is a photograph of a portable, handheld penetration hardness tester suitable for determining the hardness of mild steel and nonferrous alloys. Most hardness testing units are semiportable and require movement of the machine to the production area or else the parts must be taken to the hardness tester.

Figure 26.23
Portable hardness tester.

Magnetic Particle Inspection

In *magnetic particle inspection* an intense magnetic field is set up in the part to be inspected. Cracks, voids, and material discontinuities cause the lines of magnetic flux to be distorted and they break through the surface, as shown in Figure 26.24. Ferromagnetic powders applied to the part build up at the point where the defect occurs. This method is used to indicate surface imperfections in any material that can be magnetized. Fluorescent magnetic particles that glow in ultraviolet light can be used to intensify the effect.

Figure 26.25 indicates the way in which cracks in a truck axle king pin are revealed using this technique. Trade terms that describe the process are "*Magnaflux*" for magnetic particle inspection and "Magnaglow" for fluorescent particle inspection.

In some alloys hardness measurements cannot be correlated with strength, but certain magnetic properties do correlate with strength and other physical characteristics of the metal. Testing of magnetic properties is gaining importance.

Radiographic Inspection

Radiographic inspection is accomplished by exposing a part to either X-rays, gamma rays, or radioisotopes and viewing the image created by the radiation on a fluoroscope

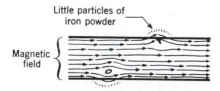

Figure 26.24
Principle of magnetic particle inspection.

Crank pin as it
appears in visual
inspection.

Cracks revealed
with magnetic flux
inspection —
Magnaflux process.

Cracks revealed
with fluorescent
particles —
Magnaglow process.

Figure 26.25
Cracks in crank pin revealed with magnaflux and magnaglow inspection.

or film. To examine a piece of steel 5 in. (127 mm) thick with X-rays requires a machine of more than 1000 kVA capacity. X-rays are very sensitive and can be used to inspect any thickness of almost any material. Because of the powerful nature of gamma rays, much radiographic work is done with radium or "cobalt 60" sources. The principal advantages of gamma-ray inspection are the low cost and portability of the source. Electrical power and cooling water are unnecessary. Special techniques shield the capsule of radioactive material when it is not in use. *Radiography* is employed to examine for internal defects and to check for alignment and operation of assembled parts.

Gamma rays may be used to inspect a casting or welds. Flaws that may be found by radiography in welds include porosity, slag, incomplete fusion, and undercutting.

Fluorescent Penetrants

Various *fluorescent penetrants* can find surface defects in almost any material. The penetrants are normally oil-based and may be applied by dipping, spraying, or brushing. The penetrant is later washed off the surface and a powder is applied to absorb the penetrant remaining in cracks and voids. The part is then examined under special light and the colored powder reveals the flaws.

Ultrasonic Testing

In *ultrasonic testing* a high-frequency vibration or supra-audible signal is directed into the part to be tested. A quartz crystal that changes electrical signals to ultrasonic inaudible sound waves is pressed against the part. When the sound waves reach the other side of the part or reach a discontinuity, they are reflected back and the crystal generates a signal upon receiving them. A cathode ray tube measures the time lag between the initial signals and the returning ones; hence, metal thicknesses or distances to discontinuities may be measured with precision, and the metallurgical characteristics may be monitored with an unusual degree of precision in some alloys.

Eddy Current Testing

This NDT method is useful for flaw detection, sorting by metallurgical properties such as hardness, and thickness measurement. This method induces an eddy current from a coil adjacent to the surface. Discontinuities in the part change the amplitude and direction of flow of the induced current. The changes of magnitude and phase difference can be used to sort parts according to alloy, temper, and other metallurgical properties.

26.8 STATISTICS AND UNCERTAINTY

By using statistical tools in production, the metrologist can establish a common ground for understanding, a common ground largely missing in the past.

Statistical process control (SPC) was designed as a statistical aid to the control of processes. It is not itself a controlling element, but provides evidence of the state of control of the process. It can be applied to any repeated operation that has a measurable output, including measurement processes. The simplest application of SPC in the metrology laboratory is in controlling the uncertainty obtained in operating measuring systems.

Controlling Measurement Uncertainties

Figure 26.26 illustrates the three parameters of interest to a metrologist: *error*, the difference between the measured value and the true value; *uncertainty*, the range of values that will contain the true value; and *offset*, the difference between a target (nominal) value and the actual value.

Offset is calculated by means of an experiment to determine the absolute value of a parameter, the simplest such experiment being a calibration against a superior standard.

Uncertainty is determined by an uncertainty analysis that takes into consideration the effects of systematic and random errors in all the processes that lead to the assignment of a value to a measurement result. Both standardization and uncertainty analyses are relatively difficult and costly operations that should not be undertaken more often than necessary. It is most cost effective to perform these analyses once, and then to use process control techniques to provide evidence that the system used is the system analyzed; that is, nothing has changed.

A process can change in two fundamentally different ways: The setpoint can change, or the variability of its processes can change. A *change in setpoint* corresponds to a change in the calibration of the system. A *change in variability* indicates that the uncertainty analysis is no longer valid, since the assumptions on which it was based have changed. Either change could lead to measurement results outside the range of results expected, based on the uncertainty stated. Chapter 27 discusses SPC applications further.

Distributions Versus Test Uncertainty Ratios

The immediate purpose of calibrating inspection and test equipment is to gain confidence that a unit under test (UUT) is able to make measurements within its specifications. *Test*

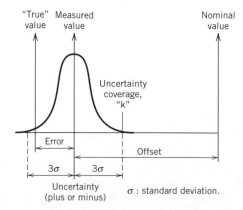

Figure 26.26
Error, uncertainty, and offset.

uncertainty ratio (TUR) is one approach to evaluate a UUT's uncertainty. It is simply the specified uncertainty of the test instrument divided by the uncertainty of the calibrating instrument.

EXAMPLE 26.2

Suppose one wants to check a differential-method measurement device's (DMM's) performance by using a multifunction calibrator (MFC). The DMM's 10-V range specifications are: $\pm$ 20 ppm of reading, + 1.6 ppm of range. The calibration of this DMM is checked by applying a known 10 V from an MFC to its input. If exactly 10 V is applied to this DMM, any reading within $\pm$ 216 μV of 10 V will be within specifications.

Calibration laboratories generally do not have access to exact values of stimulus to test DMMs or other equipment. Instead, they must use commercially available calibrators such as MFCs, which also have uncertainty specifications. Assume that data for a commercial MFC specify that the outputs on its 10-V range are within $\pm$ 5 ppm of output, $\pm$ 4 μV of the true value. When this specification is converted to units, the uncertainty in the MFC's 10-V output is $\pm$ 54 μ or $\pm$ 5.4 ppm. This is its 99% confidence level (or 2.58 sigma value), which corresponds to $1\sigma = 5.4/2.6 = 2.08$ ppm.

If the vendors of the DMM and MFC in this example have based their respective specifications on a normal distribution of uncertainty, and if the confidence interval for the specifications is the same, the TUR between the DMM and MFC is exactly 4:1. (According to MIL-STD-45662A and other similar calibration standards, this is an acceptable TUR.)

However, a reliable TUR is obtained only when the specifications for the UUT and the commercial calibration standard are correlated according to their respective uncertainty distributions and confidence intervals.

Suppose, for example, that the DMM specification is $\pm$ 15 ppm of input $\pm$ 20 μV (which corresponds to a 2σ confidence interval). When 10 V is applied to this DMM, readings within $\pm$ 170 μV of 10 V will be within specifications. If the calibrator uncertainty of $\pm$ 54 μV is divided into this value, an apparent TUR of 3.148 is obtained. However, the DMM specifications have a 2σ confidence interval, whereas the calibrator's specifications have a 2.58σ confidence interval. The difference in the confidence intervals must be taken into consideration to compute the true TUR.

When the specifications for the calibrator and the DMM are based on a normal distribution of uncertainty, the true TUR is computed as

$$TUR_{true} = TUR_{apparent}\, K_{cal}/K_{DMM} \qquad (26.8)$$

where

$$TUR_{apparent} = \text{Ratio of published specifications}$$
$$K_{cal} = \text{Coverage factor for the calibrator specifications}$$
$$K_{DMM} = \text{Coverage factor for the UUT specifications}$$

For the example under discussion, the true TUR is:

$$TUR_{true} = 3.148(2.58/2.0) = 4.06$$

Clearly, one cannot simply divide the smaller number into the bigger number and decide that the TUR is 4:1 when the quotient of the specification ratios is $\geq$ 4.

QUESTIONS

1. Give an explanation of the following terms:

Metrology   Standard
Gage blocks   Inspection
Micrometer   Electronic gaging
Interferometry   LVTD
Sine bar   Magnaflux
Surface plate   Radiography
Surface roughness   Test uncertainty ratio

2. What instruments can be used to measure humidity?

3. What is the difference between a standards laboratory and a calibration laboratory?

4. State your understanding of error and uncertainty. Are they the same?

5. What do you understand by the terms "calibration" and "traceability"?

6. What is a calibration curve and what can it be used for?

7. What instrument would you use to check tolerances to within one-thousandth, one ten-thousandth, or one-millionth of an inch?

8. Name at least three instruments you can use to measure flow rate.

9. It is said that sensitivity and readability are associated with the device, whereas precision and accuracy are associated with the measuring process. Explain why.

10. Write a method for certifying the accuracy of a micrometer.

11. Explain the difference between inspection and manufacturing gages.

12. Explain the three components of a standard unit of measurement: definition, realization, and representation.

13. How are error, offset, and uncertainty determined in metrology?

14. You are interested in measuring the sound level in your shop. What instruments could you use? How are they calibrated?

15. Explain how a laser works and how it is used in inspection of large machinery.

16. How is surface roughness measured?

17. How could a laser determine if a machine were set so that its bed was horizontal?

18. Explain the statement "surfaces with the same average roughness height can be very different."

19. Explain the process by which the ampere is realized.

20. How would you measure the lead of a small screw using a toolmaker's microscope?

21. Indicate ways of ensuring long linear measurements, 3 or more meters, that are correct to 1 part in 100.

22. Define, sketch, or describe the differences among a surface gage, profilometer, toolmaker's flat, optical flat, and a surface plate.

23. List advantages of the principal types of nondestructive tests.

24. Give the advantages of pneumatic, electric, and electronic gages.

25. Why is the electronic gage preferred over the electric gage?

26. Laser light has unique advantages for inspection. What are they?

27. What general type of dimensional inspection equipment do you recommend for "mass production" inspection of bottle caps, gears for a watch, large forging, extruding tube, and plastic steering wheels?

PROBLEMS

26.1. Sketch a vernier caliper reading of 0.702 in.

26.2. Sketch the barrel of a 4-in. (100-mm) micrometer that is reading 3.762 in. (95.56 mm).

26.3. Sketch a micrometer barrel showing 0.862 in. (21.89 mm). Also show in metric dimension.

26.4. Sketch a ground scissor blade and give inspection procedure for its acceptance or rejection.

26.5. Sketch the fringes on an optical flat if the surface of the part being inspected is high in the center and low on all edges; if the part has a "valley" going one direction down the center and a "hill" going perpendicular to it; if the part is perfectly level.

26.6. Calculate the resolution of a laser beam in metric units. How much does this amount to in a 5-m length?

26.7. Find the average and root-mean-square for the following measurements: a = 0.0042, b = 0.0043, c = 0.0039, d = 0.0047, and e = 0.0045.

26.8. Find the gage blocks for a 0.3125-in. dimension and for a 4.0245-in. dimension.

26.9. Determine gage block set for a 3.5687-in. dimension using the fewest number of blocks. Repeat for 2.178 in.

MORE DIFFICULT PROBLEMS

26.10. Referring to Figure 26.14, estimate the surface roughness of the following: scissor blade, window glass, carburetor die casting, airplane wing, master gage block, surface plate, structural steel, paper clip, drawing triangle, and the bottom of a cast-iron frying pan.

26.11. Examining Figure 26.10, what is the irregularity if the fringes are made by using monochromatic light that has a wavelength of 0.000017 in. (432 nm)?

26.12. A 3-in. gage block is set up under one end of a 10-in. sine bar. What height would have to be used on the other end to check an angle of 60 degrees?

a. For 48 degrees?

b. Repeat for 30 degrees.

c. Repeat a. if L = 20 in.

26.13. A surface is measured and the following deviations are noted. Calculate the arithmetic and root-mean-square average roughness.

a = 0.0039	g = 0.0049	m = 0.0040
b = 0.0045	h = 0.0058	n = 0.0031
c = 0.0057	i = 0.0061	o = 0.0042
d = 0.0046	j = 0.0052	p = 0.0045
e = 0.0030	k = 0.0045	
f = 0.0035	l = 0.0053	

26.14. Determine the out-of-flatness constant error for the following calibration data of a micrometer caliper. Using a 0.2500-in. ball gage, five different locations on the anvil

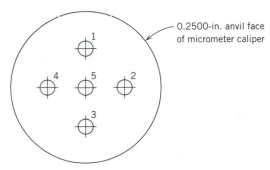

Figure P26.14

and spindle face were tested. The locations are shown in Figure P26.14. The average measurement is determined from a series of readings made of each location.

Ball Gage Size		Position				
		1	2	3	4	5
0.2500	Reading	0.2501	0.2500	0.2500	0.2501	0.2501
0.3125	Reading	0.3125	0.3122	0.3122	0.3122	0.3125

a. Give the out-of-flatness for the 0.2500-in. ball.

b. Parallelism for the anvil and spindle face can be found using a second 0.3125-in. ball gage, which is placed in the same positions, and the data are recorded as above. The 0.3125-in. dimension results in a 180-degree difference of the angular position of the spindle. Find the average difference to indicate any nonparallelism.

PRACTICAL APPLICATION

Test and measurement equipment must be calibrated regularly. Identify a nearby manufacturer and set up a visit to discuss calibration practices. Design a set of questions before your visit. Make sure that your questions address issues such as frequency of calibration and how calibrated equipment is identified. What is done if a particular gage is found out of tolerance during calibration? Are records and certifications kept current? What calibration standards are observed?

CASE STUDY: THE SWING FREQUENCY

Swing & Don't Fall, Inc., is conducting tests to design a new swing set model containing composite materials. Nina Molinaro, the company's owner, calls your supervisor and manufacturing representative in the design team, Tina Fitch, and says:

"Ms. Fitch, as a dedicated manufacturing practitioner, and as someone who has raised several children, you might have some ideas for the design of our new swing set."

"Yes," Tina says. "I talked to the design department already and gave them a few pointers. . . ."

"Great!" Nina interrupts. "And I have a specific task for you. I want you to evaluate the expected number of oscillations per minute that the swing set will have. And since we need to keep good records, I need a report stating the accuracy and absolute error of your estimate."

"Okay . . . and when do you need this report?" Tina asks.

"Tomorrow morning will be just fine. See you then!"

Tina leaves Ms. Molinaro's office a little concerned, and stops to see you "The Experimenter." After she explains the assignment, you say, "Well, Tina, the basic swing system may be simulated by the motion of a simple pendulum. Let's use 10 strings of different lengths ranging from 5 to 20 in., attach a weight to one end and count the number of complete oscillations made in 30 s. Then we can plot length versus frequency, and it's done! That plot will represent the *static calibration curve* for your report."

"Oh, thanks!" Tina says.

You continue, "As you know, the frequency of a simple pendulum in free motion is given by

$$f = (2\pi\sqrt{L/g})^{-1}$$

where

f = Frequency, oscillations/s
L = Length of the string, ft (m)
g = Acceleration due to gravity, ft/s^2 (m/s^2)

Using this function as the expected true value, we find out the accuracy and the absolute error of our measuring system."

"Excellent!" Tina says with a sigh of relief.

You conclude proudly, "And to show Nina your dedication, just change some of the parameters of the experiment (i.e., weight, time, initial force, initial starting point of travel, etc.) and describe any deviations of the results."

"I'm impressed!" says Tina. "Go ahead and do everything that you said. Write the report describing the experimental process and stating all the findings. I'll review it first thing tomorrow morning. See you later Experimenter!"

Chapter 27
QUALITY SYSTEMS

International competition has changed the quality function since the 1980s. The challenge of the manufacturing enterprise is to minimize process variation to improve production as well as design and business practices. To accomplish these goals either *prevention* systems or *detection* methods are used. Prevention systems include statistical process control (SPC), capability studies, and other process analysis tools. Detection methods require inspection practices including the destructive and nondestructive tests explained previously, and reliability studies.

To document processes and the changes made to reduce variation, producers must implement reliable quality systems. Depending on their industry sector, such systems follow conventions, or *standards*, that serve as guidelines for process improvement. This chapter describes the basic components of a quality system, some guidelines given by ISO 9000, the main steps of a methodology to ensure quality of a design and quality of a process, and the prevention and detection techniques needed to support it.

27.1 QUALITY SYSTEMS AND PROCESS IMPROVEMENT

The goal of the manufacturing enterprise is to provide the best product at a lower price and with the lowest operating cost. Costs must be reduced without sacrificing the technical superiority of products. Such a challenge demands proper documentation of a quality system. Mil-Q-9858 and ISO 9000, the most stringent standards of the military and commercial industries, respectively, present a framework to make continuous improvement (CI) a reality.

ISO 9000, for example, is composed of a set of four standards that provides quality systems guidelines and requirements for organizations that design, develop, produce, install, and service their products.

The effort to achieve continuous improvement at every level cannot be a short-term program. It demands commitment to identify and eliminate the waste in all business systems that detracts from the ability to improve quality and productivity. A documented quality system provides the means to identify and correct inefficiencies. Figure 27.1 shows the documentation structure that ISO 9000 requires. The bottom line is document what you do and actually do what you document.

The Improvement Process

One of the first steps in the CI journey is to define two terms: quality and continuous *quality improvement*. Keeping the customer as the driver of all efforts, *quality* is defined as "the act to provide customers with products and services that consistently meet their needs and expectations." "Consistently" means that there is minimal process variation. It means that flow time is always the same, conformance to specifications is the same, product performance is always the same, and the like.

Keeping in mind this definition of quality, *continuous quality improvement*, then, is a systematic method of improving processes to better meet customer needs and expecta-

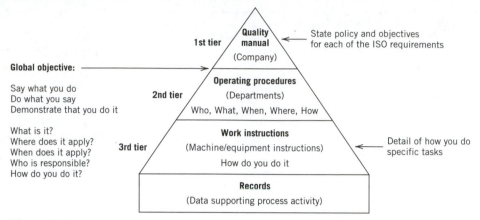

Figure 27.1
The ISO 9000 quality system documentation.

tions. The overall CI process must be straightforward, following the classic "Plan–Do–Check–Act" wheel that the late Dr. W. E. Deming proposed. *Plan*: Begin by setting goals and planning how to achieve them. *Do*: Continue implementing the plan. *Check*: Gather and analyze data to find out what happened. *Act*: After the analysis of results, act to improve the process.

Based on this basic philosophy, a CI approach is illustrated in Figure 27.2. The process is divided into five major blocks:

The first block is the *Planning and Education* (*P/E*) step, which is conducted by top management. Each department director is responsible for developing a quality policy

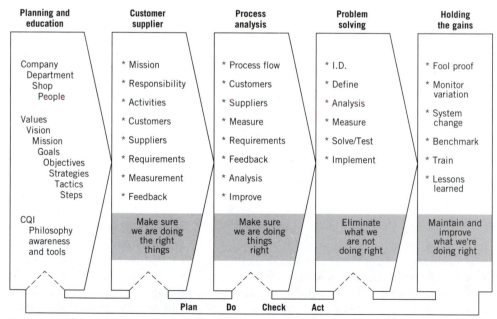

Figure 27.2
Continuous improvement methodology.

and a CI plan. Directors develop their plan with their own staffs and provide updates to the highest decision-making body.

The second block is the *Customer/Supplier (C/S) Analysis*. The main purpose of C/S analysis is to ensure that organizations have a basis for understanding and responding to both internal and external customer requirements and expectations. As a result of this analysis, each member of the organization understands how his or her particular function and task relates to meeting customer requirements. By conducting this analysis, organizations become more effective by making sure they are doing the right things to satisfy their customers.

At the end of the C/S analysis, organizations have developed and documented ongoing procedures and channels of communication with their respective customers and suppliers. They have a clear and documented understanding of their requirements, and they know their mission and quality policies.

The purpose of the third block, *Process Flow Analysis (PFA)*, is to document all the activities that take place in the business process so that nonvalue–added steps and bottlenecks are identified and then eliminated. This helps organizations become more efficient by making sure they are doing things right. As a result of PFA, organizations compile diagrammed and documented *flowcharts* that are used to isolate problem areas, identify nonvalue added steps, analyze C/S relationships, identify control and measurement points, and evaluate process efficiency, among other things. Figure 27.3 illustrates a simplified flowchart for a generic foundry process.

The purpose of the fourth block, *Problem Analysis and Improvements (PA&I)*, is to identify and to eliminate nonvalue–added activities within the business and manufacturing process, to improve quality and capability, and to better meet customer requirements. It forces a company to look at all processes, whether in the factory or in office environments, as a sequence of activities or events that have measurable outputs.

Problem analysis and improvements uses the flowchart analysis to identify problems and nonconforming characteristics of the process and to determine what data can be

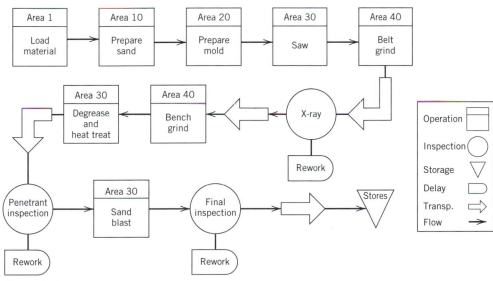

Figure 27.3
Simplified process flowchart.

collected and analyzed. After *root causes* of variation are identified, solutions are generated and prioritized to choose the best option for implementation. This step concludes with the evaluation and control step, which determines whether real gains are obtained.

The fifth block, *Holding the Gains*, must not be taken for granted. Many things, such as equipment deterioration and breakdown, material shortages and deficiencies, human error and backsliding, can weaken the gains realized by the changes made in the process. Faced with this possibility, good managers and manufacturing professionals do not just walk away once the "best" solution has been implemented. They provide a systematic means for holding the gains. Statistical process control charts and random sampling for nondestructive inspection are of paramount benefit to carry out this step.

To remain competitive, though, holding the gains cannot stop here. Organizations must seek new opportunities to improve their processes by further reducing variation. Correlation studies and statistical design of experiments are effective tools to identify new areas for improvement.

When solutions are implemented, different things in the process can change. Some suppliers may go away, some tasks may be eliminated, the organization's mission may need revision, equipment and plant layout changes may vary the process flow, and so on. These changes create the need to monitor process variation continuously. Some statistical tools to facilitate that effort are explained next.

27.2 PROCESS VARIATION

Variation is the inability to perform a task consistently according to a specification. Such dispersion is usually estimated by the *variance* (standard deviation squared), which is calculated in several ways depending on the type of distribution that the data follow.

Distribution	Variance Estimator, σ'^2
Normal	$\sum_{i,n} (x_i - \text{average})^2/n$, $n = $ sample size
Binomial	$(\text{Average}) \times (1 - \text{average})/n$
Poisson	Process average

In many factories, automated inspection and measurement systems present the opportunity for having information on each and every variable of interest for every item produced. The classical control charts developed in the 1920s successfully addressed the problem of the high cost of data collection and the need for operator intervention in process adjustment. However, the potential now exists for acquiring inexpensive measurements and for implementing a system of self-adjusting machines.

With tactile systems and end effectors, with vision systems and laser telemetry, and with voice recognition systems, it is common to find "intelligent" machines and controllers in large factories. The challenge is to select techniques (probabilistic, deterministic, or uncertainty-based) to exploit the computer-controlled environment and to reduce the variation of the outgoing product.

The goal, regardless of the tools used to control and monitor variation, is to identify and eliminate sources that cause it. Sources of variation include *process variability* (i.e., the inability of manufacturing to produce products that are completely identical), *product degradation* over its life cycle, *external factors* (e.g., variation in environmental or field conditions where the product is used), and *process degradation* or drift (resulting from changes in raw material input sources, new operators, etc.).

Dimensions in a manufactured part, then, vary as a result of some causes that are inevitable, because they are inherent to the process, or as a result of some others that are induced. The former are known as *chance* or *random* causes of variation, while the latter are known as *assignable* or *special* causes of variation. Examples of the first type are equipment accuracy, weather effects, and material properties. The second type include factors such as worn-out equipment, out-of-specification machine adjustments, improper tooling, material defects, and human error.

Monitoring Variation

One technique that has succeeded in understanding, monitoring, and controlling variation is *statistical process control* (*SPC*), which is a systematic method of tracking, predicting, and minimizing process variation. Used properly, SPC aids in cost reduction, quality and yields improvement, scrap and rework reduction, elimination of unnecessary inspections, exposure of design and specification problems, and isolation of sources of process variation.

Manufacturing professionals unfamiliar with SPC, however, usually find it difficult to decide the type of control chart to use. It must be understood that the purpose of a control chart is to distinguish between random fluctuations and the variation attributed to assignable causes. This is achieved with an appropriate choice of control limits calculated using probability laws. If the control limits are exceeded, the process is *out of (statistical) control*.

Knowing the purpose of control charts, the decision tree depicted in Figure 27.4, makes the selection of control charts easier. First, the type of data to be collected is decided. Data are either *variable* (measurable quality characteristics) or *attribute* (non-measurable but countable quality characteristics). Examples of variable data are values read from a micrometer. A go, no-go gage, however, gives attribute data, since it determines if a diameter is satisfactory or nonsatisfactory, without actually measuring it. A go, no-go gage is discussed in Chapter 25, Tool Design.

If the decision is to collect variable data, two control charts are always required, one to monitor "the process average" and another one to monitor "changes in the dispersion of the process." The control chart used to monitor the process average is normally the *x-bar chart*. The chart to monitor changes in dispersion, however, depends on the sample

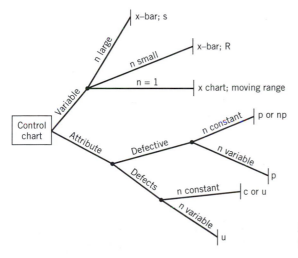

Figure 27.4
Control chart decision tree.

size, *n*. If *n* is large (more than 10 observations per sample), the *s chart* (sample's standard deviation) is preferred. If *n* is small (less than 10 observations per sample), the *R* (range) *chart* is the optimal choice. If only one observation per sample is possible, then an *x chart* (control chart for individual observations) and the *moving range chart* are the options.

However, if attribute data are collected, only one control chart is required. The decision that needs to be made is whether "defective parts" or "number of defects" are being counted. If the process is being monitored by collecting data on nonconforming (defective) parts, then the question is whether or not the sample size, *n*, is constant. If it is constant, then a *p* (fraction defective) *chart* or a *np* (number of defective units) *chart* would work. However, if the sample size varies, then only the *p* chart displaying variable control limits is appropriate.

If the attribute is monitored by collecting data on the number of defects (or nonconformances, or discrepancies), then the condition of the sample size determines the type of chart. If *n* is constant, *c* (number of defects per sample) or *u* (number of defects per unit) charts are used, but if *n* is variable, only the *u* chart is acceptable.

Notice that sample size plays a key role in the decision-making process, thus a sample must be chosen with impartiality but with the cost of data collection in mind. For most practical purposes, inspectors group data in subgroups of 4 to 10 parts and call this a *sample* even if the parts are not gathered at the same time.

27.3 CONTROL CHARTS FOR VARIABLE DATA

Two control charts are *always* needed when variable data are used: the *x*-bar chart to monitor changes in the average from sample to sample, and a second chart (*R*, *s*, or moving range) to monitor dispersion changes from sample to sample. These types of charts are constructed under the assumption that the data follow a normal distribution. This means that if all the measurements taken over a very long period of time are recorded and there are no assignable causes of variation, the frequency with which each value occurs can be plotted against such value giving a curve, similar to Figure 27.5.

This bell-shaped curve, known as a *normal curve*, has the remarkable property that 68.27% of the data fall within $\pm 1\sigma$, 95.45% within $\pm 2\sigma$, and 99.73% within $\pm 3\sigma$.

Using this assumption then, a control chart, such as that shown in Figure 27.6, is constructed by plotting the average dimension of a sample against time. The *upper control limits* (UCLs) and low control limits (LCLs) are drawn at a distance equal to $3\sigma_x$ above

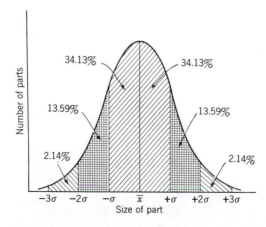

Figure 27.5
Normal distribution and percentage of parts that will fall within sigma limits.

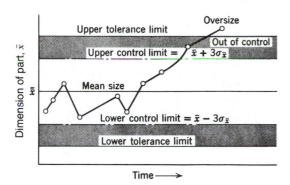

Figure 27.6
Control chart characteristics.

and $3\sigma_x$ below the mean dimension line. The value of σ_x (the standard deviation of the samples' averages) is estimated in different ways, as shown later. The control limits define a $6\sigma_x$ band that represents *random* or *normal variation*. Thus, UCL and LCL are set so that only 2700 pieces out of 10^6 (2700 ppm) are expected to be defective. As a reference, the tolerance limits are drawn. However, most SPC experts discourage this practice.

X-bar and R Charts

In most practical applications the *standard deviation of the process, σ,* is unknown. Therefore, if the sample size, n, is small (normally less than 10 observations per subgroup) control limits are calculated based on the *range, R,* of the dimensions inspected. R is simply the difference between the maximum and the minimum value in the sample

$$R = x_{max} - x_{min} \tag{27.1}$$

Once sufficient data, free of assignable causes of variation, are obtained, the two charts are constructed as follows:

1. Calculate range and average for each subgroup or sample.
2. Find the average of all ranges ($\overline{R}$).
3. Find the grand average ($\overline{\overline{x}}$) or average of sample averages.
4. Calculate control limits as

$$\text{For } x\text{-bar chart} \quad (\text{UCL}, \text{LCL}) = \overline{\overline{x}} \pm A_2\overline{R} \tag{27.2}$$

$$\text{For } R \text{ chart} \quad (\text{UCL}, \text{LCL}) = (D_4\overline{R}, D_3\overline{R}) \tag{27.3}$$

Where A_2, D_3, and D_4, are constants calculated by probability theory so that the 6σ band is provided. The value of the constants depends on the number of pieces in the sample and is given in Table 27.1.

5. Plot x-bar and R data in respective charts.
6. If all plots fall within the control limits, the charts can be used to monitor the variation of the process. Otherwise, the variation is not *random*, outlying points must be excluded (but plotted nevertheless) and control limits for both charts recalculated.

Once the control charts are established, data are recorded on them and they become a record of the variation of an inspected dimension over a period of time. Control limits are not recalculated unless there is an evident and identified change in the process. The data plotted should fall in random fashion between the control lines if all *assignable*

Table 27.1   Factors for 3σ Control Limits

Sample Size	c_2	A_1	B_1	B_2	A_2	D_3	D_4	d_2
2	0.564	3.76	0	1.84	1.88	0	3.28	1.128
3	0.724	2.39	0	1.86	1.02	0	2.57	1.693
4	0.798	1.88	0	1.01	0.73	0	2.28	2.059
5	0.841	1.60	0	1.76	0.58	0	2.11	2.326
10	0.923	1.03	0.28	1.58	0.31	0.22	1.78	3.078
12	0.936	0.93	0.35	1.54	0.27	0.28	1.72	3.258
$n > 12$	$\dfrac{4(n-1)}{4n-1}$	$\dfrac{3}{c_2\sqrt{n}}$	$1 - \dfrac{3}{2\sqrt{n}}$	$1 + \dfrac{3}{2\sqrt{n}}$	—	—	—	—

causes for variation are absent. When the data fall in this manner, the *process is under statistical control*, and it can be assumed that the part is being made correctly 99.73% of the time.

As long as the points fall between the control lines, no adjustments or changes in the process are necessary. If five to seven consecutive points fall on one side of the mean, the process should be checked. When points fall outside of the control lines, the cause must be located and corrected immediately. Observe Figure 27.6, where the dimension as measured moves above the upper control line and the upper tolerance limit. This dimension is out of control, and action would be taken to correct the problem.

EXAMPLE 27.1

To illustrate the manner in which control limits are calculated, assume that round pieces of stock, called gear blanks, have been inspected. The dimensions, which may be viewed as numbers representing either English or metric units, are grouped in sample sizes of three, and are shown in Table 27.2. The process was controlled carefully in that no assignable causes of errors were known to exist.

Table 27.2   Control Limit Calculations for Variable Control Charts

Sample Number	Dimensions			Average Size in Sample X-bar	Range R	Standard Deviation $s \times 10^{-3}$
	x_1	x_2	x_3			
1	2.495	2.501	2.499	2.498	0.006	2.49
2	2.501	2.500	2.496	2.499	0.005	2.16
3	2.501	2.495	2.498	2.498	0.006	2.45
4	2.497	2.500	2.503	2.500	0.006	2.45
5	2.497	2.503	2.501	2.500	0.006	2.52
6	2.502	2.500	2.498	2.500	0.004	1.63
7	2.499	2.499	2.496	2.498	0.003	1.41
8	2.500	2.503	2.505	2.503	0.005	2.16
9	2.500	2.497	2.499	2.498	0.003	1.41
10	2.499	2.503	2.501	2.501	0.004	1.63
11	2.503	2.497	2.501	2.500	0.006	2.52
		SUM:		27.495	0.054	22.83

After calculating the average and the range for each of the 11 samples, the grand average and the average range are

$$\bar{\bar{x}} = \frac{27.495}{11} = 2.500 \quad \text{and} \quad \bar{R} = \frac{0.054}{11} = 0.005$$

and using Equations (27.2) and (27.3), and Table 27.1, the control limits are

$$(\text{UCL}, \text{LCL})_x = 2.500 \pm 1.02\,(0.005) = [2.505, 2.495]$$
$$[\text{UCL}, \text{LCL}]_R = [(2.57 \times 0.005), 0] = [0.013, 0]$$

Figure 27.7 represents the control charts, and future data can be plotted on them to determine the variation of the process. Usually more data are used in determining control limits. This example indicates the method only.

X-bar and s Charts

When the sample size, n, is large, the sample's standard deviation must be used to evaluate variability. The student is encouraged to think about why this is the case.

The standard deviation is a measure of dispersion of a dimension about the average dimension. The standard deviation for each sample or subgroup, s_i, is found as follows:

1. Calculate the average size, $\overline{X}_i$, called x-bar, of the dimensions measured for each subgroup.

2. Calculate the standard deviation, s_i, for each subgroup by

$$s_i = \sqrt{\frac{(x_1 - \overline{X}_i)^2 + (x_2 - \overline{X}_i)^2 + \ldots + (x_n - \overline{X}_i)^2}{n}} \tag{27.4}$$

where

$$x_1, x_2, \ldots, x_n = \text{Individual dimensions in the } i^{\text{th}} \text{ sample}$$
$$\overline{X}_i = \text{Average dimension of the } i^{\text{th}} \text{ sample}$$
$$n = \text{Number of parts in each sample}$$

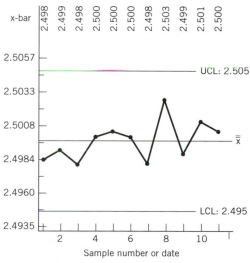

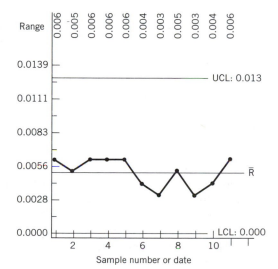

Figure 27.7
X-bar and R-charts for Example 27.1.

3. Calculate the average standard deviation, $\bar{\sigma}$ or sigma-bar, by using the number of subgroups, N,

$$\bar{\sigma} = (\Sigma s_i)/N \tag{27.5}$$

Now control limits for the two control charts are calculated as

For x-bar chart $(\text{UCL}, \text{LCL}) = \bar{\bar{x}} \pm A_1 \bar{\sigma}$ (27.6)

For s-chart $(\text{UCL}, \text{LCL}) = (B_2 \bar{\sigma}, B_1 \bar{\sigma})$ (27.7)

Where A_1, B_1, and B_2 are constants calculated by probability theory so that the 6σ band is provided. The value of the constants depends on the number of pieces in the sample and is given in Table 27.1.

As a *note* for the inquisitive student: Some authors define standard deviation in Equation (27.4) with $n - 1$ in the denominator. When this is the case, the respective constants for control limits are A_3, B_3, and B_4, whose values are tabulated in SPC handbooks. For all practical purposes in monitoring process variation, the effect is the same.

EXAMPLE 27.2

Use the same data of Example 27.1, but use the sample's standard deviation (instead of the range) as the measure for dispersion.

The standard deviation for each sample is calculated using Equation (27.4) and is given in Table 27.2. The average standard deviation is

$$\bar{\sigma} = \frac{0.02283}{11} = 0.00208 \text{ (say } 0.002)$$

The grand average was calculated as 2.500, so using A_1, B_1, and B_2 values for a sample size of 3, the control limits are

$$(\text{UCL}, \text{LCL})_{x\text{-bar}} = (2.500 \pm 2.39 \times 0.002) = (2.500 \pm 0.005)$$
$$(\text{UCL}, \text{LCL})_s = (1.86 \times 0.002, 0 \times 0.002) = (0.0037, 0)$$

Notice that the control limits for x-bar are the same as in Example 27.1. This demonstrates that for small sample sizes the use of R-bar or sigma-bar to monitor variation gives equivalent results.

Control Charts for Individual Parts

Selection of inspection parts is influenced by many factors. Sometimes, when automated inspection is used, every part manufactured is analyzed on-line. This is called *100% inspection*, and it normally slows production and increases costs. In other instances, destructive testing is required, or the inspection procedure is very time consuming, or repeated measurements differ only because the accuracy level of the test is low.

In all these circumstances, however, the sample size, n, is equal to 1, which makes "sample" calculations different. One approach uses the x and *moving range* charts. Another approach is known as the *short run* method. The student interested in these subjects is referred to the Bibliography.

27.4 CONTROL CHARTS FOR ATTRIBUTE DATA

There are four control charts of common use when *attribute data* are collected: the *p chart*, used when the fraction of defective parts is monitored; the *np chart,* used when the number of defective parts is the important parameter to observe; the *c chart*, used when the number of defects per sample is desired; and the *u chart*, used when the number of defects per unit is of interest.

The procedure to construct any of these charts is the same, so only the *p* chart is explained here. The only difference is in the estimation of the standard deviation. Data of fraction defective as well as data on number of defective parts follow a binomial distribution, whereas number of defects is modeled by a Poisson distribution. Therefore, the standard deviation must be estimated as explained in Section 27.2.

The *p* Chart

The characteristic being plotted on these charts is *p*, the fraction of defective parts found in a sample.

Sometimes *p* is called a *proportion*. This type of *quality control* analysis evaluates characteristics that are obtained on a "go" or "no-go" basis. Therefore the *p* chart provides an overall picture of quality, and the products are thus divided into two categories only, either acceptable or unacceptable. For this chart, the distinction is whether the lot of product is a conforming one or is defective. The lot is accepted on the basis of the *p* value; that is, if the *p* value falls within the control limits. Otherwise, the lot is rejected and returned to the supplier.

In the formation of *p* charts, the observations are classified into subgroups called samples. The sample proportion defective for the ith sample is

$$p_i = d_i/n_i \tag{27.8}$$

where

d_i = Number of defectives found in the ith sample
n_i = Pieces inspected, or size sample

If all samples are of equal size, *n*, then the overall average proportion of defectives, $\bar{p}$ (called *p*-bar) is the sum of sample proportion defectives divided by the number of samples, *N*. The central line on the *p* chart will be this *p*-bar value.

The standard deviation of the number of defectives, derived from probability mathematics for a binomial distribution, is

$$\sigma_P = \sqrt{\frac{p(1-p)}{n}} \tag{27.9}$$

where *n* is the *sample size* not to be confused with the number of samples, *N*, nor with the total number of parts sampled, $n \times N$.

Instead of operating in terms of the number of defectives, the control limits for the *p* chart are established in terms of the mean proportion of defectives, *p*-bar. The choice of 3-sigma control limits (three standard deviations from the central line *p*-bar) is an economic choice based on experience in the field that has proved satisfactory for detecting assignable causes of variation. Therefore, control limits are

$$(\text{UCL}, \text{LCL})_p = \bar{p} \pm 3\sigma_P \tag{27.10}$$

If a negative lower limit is obtained, the limit is replaced by zero. On a p chart plot the sample number along the x-axis direction and the proportion defective p in the y-axis direction.

EXAMPLE 27.3

A producer of silicon computer chips randomly selects 1000 chips per day ($n = 1000$) for inspection. The results of the inspection are to either accept the chip as okay or not okay. This type of inspection results in a p chart. Data for 20 days of inspection ($N = 20$) are in Table 27.3.

$$\text{Average proportion defective} = \bar{p} = 1025/(20 \times 1000) = 0.0513$$

The standard deviation for the p chart is found using Equation (27.9).

$$\sigma_P = \sqrt{\frac{0.0513(0.9488)}{1000}} = 0.0070$$

then,

$$\text{UCL} = 0.0513 + 3 \times 0.007 = 0.072$$
$$\text{LCL} = 0.0513 - 3 \times 0.007 = 0.030$$

Observe Figure 27.8, where the points are plotted on the p chart. The points fall within the upper and lower control limits, and the process is considered in control. The inspection will continue and when sample fraction defective points fall outside of the limits, immediate action is required.

Formulas to calculate control limits for the other three control charts for attribute data are

Chart	Control Limits
np	$(\text{UCL, LCL})_{np} = n\bar{p} \pm 3\sqrt{np(1-p)}$
c	$(\text{UCL, LCL})_c = \bar{c} \pm 3\sqrt{c}$
u	$(\text{UCL, LCL})_u = \bar{u} \pm 3\sqrt{u/n}$

Table 27.3 **Data for 20 Days of Inspection**

Day	Number of Defects	Proportion Defective	Day	Number of Defects	Proportion Defective
1	44	0.044	11	35	0.035
2	57	0.057	12	69	0.069
3	44	0.044	13	36	0.036
4	68	0.068	14	51	0.051
5	58	0.058	15	51	0.051
6	36	0.036	16	64	0.064
7	52	0.052	17	42	0.042
8	44	0.044	18	49	0.049
9	69	0.069	19	43	0.043
10	63	0.063	20	50	0.050
				TOTAL: 1025	

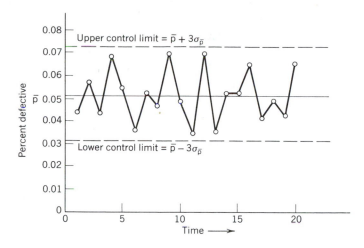

Figure 27.8
p-Chart of the fraction of good units to sample number of units.

27.5 PROCESS CAPABILITY ANALYSIS

Process capability is a measure of the natural behavior of the process after special causes of variation are eliminated. The basic purpose of process capability is to compare the "normal variation" of the process against the design tolerances to assess whether the process is able to meet those specifications.

Control charts and process variability are directly related. Control limits are based on sampling variability, and sample variability is a function of process variability. This relationship is seen in the way control limits are calculated.

$$\text{Control limits} = \text{Process mean} \pm 3(\text{process variability estimator})$$

Several vehicles exist to conduct a process-capability analysis. Probability plots and histograms, control charts limits, and design of experiments are the most common.

Process capability is usually expressed via indices and ratios that require that the data be distributed normally and that the process be in a state of statistical control. All the known indices relate the normal variation of the process to the engineering specifications. Table 27.4 records five process-capability ratios of common use. In the equations in Table

Table 27.4 **Common Process Capability Indices**

Index	Remarks
$C_p = \dfrac{\text{USL} - \text{LSL}}{NT}$	Measures the capability of the process if it is *centered* (i.e., $m = \mu$). It represents the process *potential*.
$\text{CPU} = \dfrac{\text{USL} - \mu}{NT/2}$	Considers unilateral tolerance when only upper specification limit is necessary.
$\text{CPL} = \dfrac{\mu - \text{LSL}}{NT/2}$	Considers unilateral tolerance when only lower specification limit is necessary.
$C_{pk} = \text{Min}\{\text{CPL}, \text{CPU}\}$	Measures the *actual* capability when the process is not centered (i.e., $m \neq \mu$).
$K = \dfrac{\lvert m - \mu \rvert}{(\text{USL} - \text{LSL})/2}$	Measures the difference between the actual capability and the potential of the process.

27.4, USL = upper specification limit, LSL = lower specification limit, m = midpoint of the specification range, μ = process average, which is estimated by $\bar{\bar{x}}$, and NT = natural tolerance of the process, which is equal to six process standard deviations (or 6σ).

In most manufacturing applications, the standard deviation of the process, σ, is unknown. To calculate a process capability index the most common approach is to estimate σ by the σ' value

$$\sigma' = \overline{R}/d_2 \tag{27.11}$$

where d_2, given in Table 27.1, is a constant that depends on the sample size, n.

EXAMPLE 27.4

Referring to Example 27.1, if the tolerances for the process are 2.500 ± 0.005, the process is *centered* (average of the process is equal to midpoint of the specifications). From Table 27.4 the process-capability ratio is

$$C_p = \frac{\text{USL} - \text{LSL}}{6\sigma}$$

where σ is estimated by σ' of Equation (27.11).

If C_p is at least 1, the maximum expected variation of the process (6σ) will be less than the given specification span and the process is said to be *capable*.

Since d_2 for $n = 3$ is 1.693, the estimate of the standard deviation in this example is equal to $(0.051/11)/1.693 = 0.0027$ and

$$C_p = \frac{0.010}{6(0.0027)} = 0.6173$$

This is a very low capability ratio, which means that the natural variation of the process (6σ) is much larger than the tolerance band given. Then, the process is *incapable* of ensuring 99.73% of good parts. Figure 27.9 shows the bilateral tolerance limits at 2.500 ± 0.005.

It is common practice to add some leeway between control and tolerance limits. However, in this example, tolerance limits would need to be set at 2.500 ± 0.0081 to make C_p equal to 1 and, therefore, guarantee that only 2700 ppm are expected to be defective. World-class manufacturers realize Cp's well above 2.

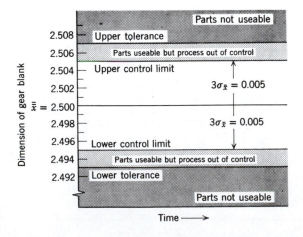

Figure 27.9
Control and tolerance limits for gear blanks. Tolerance limits are for reference and analysis only.

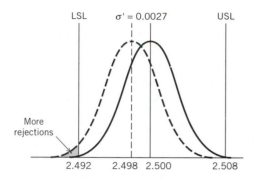

Figure 27.10
Effect on process capability because of a shift of the average of the process mean.

EXAMPLE 27.5

Assume that in Example 27.4, the process is biased (i.e., the dispersion is the same as before but the average has shifted). Let the new average of the process be 2.498.

In this case, the process capability ratio to be used is the $C_{pk} = \text{Min}\{\text{CPL, CPU}\}$, where

$$\text{CPU} = \frac{\text{USL} - 2.498}{3\sigma'} = \frac{2.505 - 2.498}{3(0.0027)} = 0.8642$$

$$\text{CPL} = \frac{2.498 - \text{LSL}}{3\sigma'} = \frac{2.498 - 2.495}{3(0.0027)} = 0.3704$$

Therefore, $C_{pk} = 0.3704$, which means that a 2-mil shift in the process average has slashed the capability of the process almost by half! Figure 27.10 illustrates the situation.

To reduce the "normal variation" of the process, more advanced tools, such as *Statistical Design of Experiments (DOE)*, are needed. DOE uses *Analysis of Variance (ANOVA)*, linear regression, and nonlinear regression analysis, which are beyond the level of this textbook.

27.6 STATISTICAL DESIGN OF EXPERIMENTS

Consider a generic manufacturing process, as illustrated in Figure 27.11. The process may be understood as a black box whose input is a set of adjustable and controllable

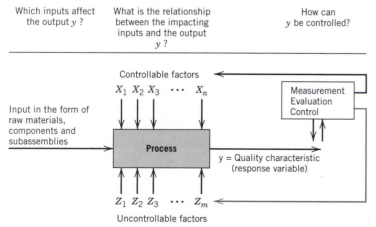

Figure 27.11
Process inputs and outputs.

factors x_1, x_2, . . . , x_n, and a set of uncontrollable factors z_1, z_2, . . . , z_m. These inputs interact with basic raw materials, processing equipment, and operators to manufacture a product that has some specifications or quality characteristics y_1, y_2, . . . , y_k.

The *controllable input variables* x_i include factors that can be manipulated and held to predetermined levels by the operator. Such variables potentially influence either the nominal value of the quality characteristic y_k or its normal variability. The *uncontrollable input variables*, however, cannot be manipulated easily, nor measured accurately and held to specific levels by the operator, which adds uncertainty to the input data.

The general subject of process control, then, is concerned with the stabilization and optimization of the ratio output/controllable–input (y/x), while the uncontrollable input z is interacting simultaneously with the process. In other words, the objective of *process control* is to ensure that quality characteristics, y_k, are consistent at every moment, for every product.

Statistical Design of Experiments is a special technique for studying the effects of controllable variables. Every process change should be treated as an experiment. A *designed experiment* is an orderly arrangement of these changes and/or variables to obtain the most information with a minimum of data (experiments). DOE methods run all the way from simple comparisons of machines, methods, materials, and the like through the more complicated *fractional factorial* experiments involving many variables and their interactions.

The following factors need to be considered in DOE:

- The manner in which experiments are arranged into production lots.
- The way of assigning various experimental conditions (treatments) to the units within the experimental plan.
- A plan (worked out beforehand) for analysis and interpretation of the results.

The field of DOE is a specialized knowledge. Manufacturing engineers are encouraged to work with a statistician in planning the designed experiment. What is covered next is a simple outline of the technique so that the student is conversant with the language and the methods. For a more complete understanding, the student is encouraged to consult the references.

Factorial Experimentation

In *factorial experimentation* the effects of a number of different factors (or variables) are investigated simultaneously. The advantages to including as many factors as possible in the same experiment are:

- Much greater efficiency can be obtained in a much smaller total number of experiments. When the experiments are on a full industrial scale, this may represent a very considerable reduction in their cost.
- Information is gained on the extent to which the factors interact. That is the way that the effect of one factor is affected by the other factors. The experiment will thus give a wider basis for any conclusions that may be reached.
- Information is also gained about the factors that are having no effect on the process.

EXAMPLE 27.6

Consider the following simple factorial experiment. Four parts, P_i, with the same characteristics are allocated randomly to three machines, M_i, on three shifts, S_i, to determine if the product variation is coming from the shifts, machines, or parts. Two factors, machines and shifts, are of equal interest, as is the possibility that they interact. The two factors are set at three different levels or conditions and all combinations are evaluated.

The number of experiment runs is given by

$$\text{Replications} * (\text{Number of levels})^{\text{number of factors}} \tag{27.12}$$

where *replications* mean the number of times each combination is repeated (four in this example). The combinations are

Machine	Shifts S_1	S_2	S_3	Sample Size
M_1	P_1	P_1	P_1	12
	P_2	P_2	P_2	
	P_3	P_3	P_3	
	P_4	P_4	P_4	
M_2	P_1	P_1	P_1	12
	P_2	P_2	P_2	
	P_3	P_3	P_3	
	P_4	P_4	P_4	
M_3	P_1	P_1	P_1	12
	P_2	P_2	P_2	
	P_3	P_3	P_3	
	P_4	P_4	P_4	
Sample size $N =$	12	12	12	36

After running each combination in random order, the dimension of interest is recorded for each part and data are analyzed. The analysis is accomplished by the ANOVA technique, which allows to determine if the factors interact as well as to define which factor has the higher impact on the output.

The use of factorial designs has now become widely accepted as an efficient way for carrying out experiments involving many different factors. However, one of the main difficulties with factorial designs is that the number of measurements required may be large and, in some cases, cost prohibitive. Another disadvantage is that in many experimental situations it is not practical to plan an entire experimental program in advance, but to make a few smaller experiments that serve as a guide to further experimentation. Factorial experiments requiring a reduced set of measurements are referred to as *fractional factorial designs* or fractional replicas.

Evolutionary Operations

Evolutionary operations (EVOP) is an ongoing method of using the regular operating process so that information on process variables are tested from a simple experimental design *while the production line is running*. To avoid changes that may affect the required characteristics of the product being produced, only small changes are made in the levels of the process variables. To determine the effects of these small changes, the changes are repeated a number of times.

EXAMPLE 27.7

A two-level, two-factor experiment with four replications is designed to determine the effect of temperature and quench, as well as their interactions, on a heat-treating process.

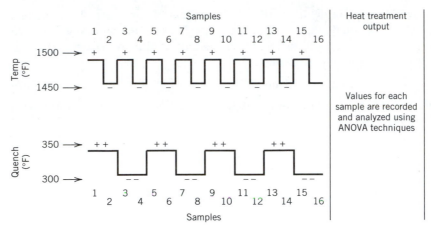

Figure 27.12
Two variable EVOP experiments.

The temperature levels are 1450°F and 1500°F, while the quench levels are 300°F and 350°F. The response variable is then analyzed by ANOVA techniques. Figure 27.12 illustrates the setup of each replication.

For a successful application of EVOP:

• Make sure the process is under statistical control first.
• Make small within-specifications changes in the process.
• Repeat the change and evaluate if the effect is consistent.
• Analyze the change in response.

27.7 RELIABILITY THEORY

Design of experiments, testing, and other process analysis tools allow one to determine the ability of a product, a part, or a system to perform its intended function; in other words, to determine its *reliability*.

The importance of understanding fundamental reliability concepts is underscored by the fact that process and/or product design are affected by the probability of failure. Therefore, the cost of manufacturing is impacted. For example, inadequate thermal design is currently one of the primary causes of poor reliability in electronic equipment; deficient operation sequence reduces production yield and increases product cost.

Reliability is measured in two ways: evaluating the probability that the product or system will function on any given moment, or estimating the probability that the product or system will function for a given length of time.

In the first case, reliability of the system is a function of its components and how they are interrelated. If the components must function sequentially and all of them must work for the system to function, then the reliability, R, of the system is

$$R = R_1 \times R_2 \times R_3 \times \ldots \times R_n \qquad (27.13)$$

where

$$R_i = \text{Probability that the } i^{th} \text{ component works}$$

Obviously, the total reliability decreases substantially with the number of components. One way of increasing it is by using *redundancy* in the design; in other words, to provide backup parts in the product, or backup machines in a process that can be used if one fails.

EXAMPLE 27.8

Consider the case of a cutter that works properly 90% of the time. Another cutter with a reliability of 80% can be used immediately if the original fails. What is the yield (reliability) of this cutting operation?

The probability that the original cutter functions is 0.9. Of the 10 out of 100 times (i.e., probability of failure = 1 − 0.9) that it does not work, 80% of the time the backup cutter will help. Therefore, the resulting reliability of the cutting operation is 0.9 + (1 − 0.9) × 0.8 or 98%.

When reliability is evaluated based on the probability that the system functions for a given length of time, the so-called *bathtub curve* (Figure 27.13) is normally used. It is assumed that early failures of the system (usually design failures) follow an exponential distribution, that during normal operation the system stabilizes resulting in few failures, and that after the expected product life failures increase again because of wearout.

In this approach, reliability is concerned with the probability that a product will last *at least* until time T and it is estimated by the area under the curve beyond T. Thus,

$$\text{Reliability} = e^{-T/\text{MTBF}} \tag{27.14}$$

where

$$e = \text{Natural logarithm, 2.7183} \ldots$$
$$T = \text{Length of service before failure}$$
$$\text{MTBF} = \text{Mean time between failures}$$

Mean time between failures (*MTBF*) is a parameter calculated by experimentation and tabulated in handbooks for a variety of electric and mechanical components. For manufacturing machinery and operations, it can be estimated by keeping good records of equipment maintenance.

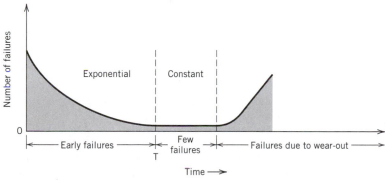

Figure 27.13
Expected product life; "bathtub curve" concept.

QUESTIONS

1. Give an explanation of the following terms:

Standards
Quality improvement
Quality
Sample
Upper control limits
Standard deviation
Assignable causes

Attribute data
Quality control
Process capability
Controllable input variables
Fractional factorial
Reliability
MTBF

2. State the difference between preventive and detection systems for quality control. Give examples of each.

3. Is 100% inspection impractical for most purposes? When is it necessary?

4. Discuss the pros and cons to the statement that quality does not add to product value.

5. What is the documentation level structure required by ISO 9000?

6. Describe what process flow analysis is and how to conduct it.

7. What is the difference between assignable and random causes of variation?

8. Give 10 examples of assignable and random causes of variation.

9. What percentage of the data would fall outside control limits naturally if $\pm 2\sigma$ were used as control limits?

10. What percentage of the data should fall within $\pm 1\sigma$ on the normal distribution curve? Within $\pm 2\sigma$?

11. Explain the setting of tolerance limits using a control chart.

12. What is the difference between potential and actual process capability?

13. Discuss the sample size and its effect on control chart construction and the use of a control chart for inspection.

14. Why do you need two control charts for variable data? Why do you need only one for attribute data?

15. When using control charts for variable data, why should you calculate s instead of R when the sample size is large?

16. What are some benefits and some limitations of statistical design of experiments?

17. What do you understand by evolutionary operations?

18. If three factors are considered critical in a process and four levels are of interest for each of them, how many experiments are needed?

19. What alternative approaches are there to operationally define reliability?

20. Explain the concept of bathtub curve in reliability.

PROBLEMS

27.1. A rectangular aluminum case is to be built and fastened to a predrilled base plate. The raw material for the case is sheet metal and the final product must be black anodized. Draw a flowchart that illustrates all the steps needed to build and assemble the product.

27.2. Find the average, the range, and the standard deviation for the following samples: (1.501, 1.504, 1.503), (1.501, 1.504, 1.503, 1.503), and (2.501, 2.503, 2.506, 2.506). Discuss the difference between the range and the standard deviation.

27.3. Find the upper and lower control limits for a process if the average is known to be 1.503, the standard deviation is 0.005, and sample size is 4. Reconstruct the limits for a sample size of 5.

27.4. Samples of five rods per hour are taken for diameter inspection. The average standard deviation is found to be 15 mils. What is the average range?

27.5. Calculate the standard deviation and the range of the following dimensions (variable data) if they are grouped in samples of size 3. Use horizontal groupings. Are there significant differences between the two measurements of variation? Calculate control limits and draw the control charts (x-bar and R) showing upper and lower control limits.

4.188	4.186	4.190
4.186	4.187	4.189
4.184	4.183	4.187
4.191	4.189	4.189
4.183	4.190	4.186
4.186	4.182	4.186
4.188	4.183	4.184
4.182	4.188	4.190
4.184	4.185	4.189
4.189	4.187	4.183
4.189	4.191	4.183
4.183	4.184	4.185

27.6. If after constructing the control charts of Problem 27.5 the following average sample dimensions and ranges are found by inspection, give your opinion as to whether the process is in control:

X-bar: 4.189   4.191   4.189   4.196   4.192   4.193
Range: 0.009   0.001   0.003   0.008   0.002   0.004

X-bar: 4.194   4.191
Range: 0.006   0.007

27.7. Tolerances for data given in Problem 27.5 are 4.185 ± 0.006. Are the tolerances being met? What is the capability of the process? What would you recommend to improve the capability of this process?

27.8. Five bags of a cleaning compound are weighted every hour. The average range is 3 oz and the overall average weight is 2.003 lb. If the nominal weight is 2 lb and the tolerance is $\pm$ 4 oz, what is the actual capability of this process?

27.9. The nominal thermal expansion of a phenolic glass-reinforced resin is 0.88×10^{-5} in./in.°F. This characteristic is verified by testing samples of 10 specimens per production shift. After 20 samples, the average expansion is found to be 0.89×10^{-5} in./in.°F and the average range is 4.5×10^{-7} in./in.°F. If the maximum thermal expansion allowed is 5% above nominal, what is the capability ratio of this process? What does this value mean to you?

27.10. Calculate control limits for a *p* chart if the number of total defective capacitors is 1350 for 20 batches of 1000 units each. Can you assess the capability of the process with this information? Explain your answer. Five of these capacitors are needed in a board assembly. If all of them must work for the assembly to function, what is the reliability of the board because of the capacitors?

27.11. Twenty-three rolls of coiled wire are examined and the number of defects per roll have been recorded. On the average, 3.6 defects per roll were found. What control chart do you recommend? What are the control limits for this process?

27.12. In a 30-day period, 1907 defectives were determined from equal-sized samples of 1000 units. Determine the upper and lower control limits for the control chart that you would recommend.

27.13. A process for manufacturing computer keyboards has a C_p equal to 1 and it is centered. Automated inspection is in place at a cost of $0.35 per unit and 100% inspection generally catches all defectives. Any defective keyboard must be replaced at a cost of $120. Is 100% inspection justified?

MORE DIFFICULT PROBLEMS

27.14. It is known that a process for making threaded fasteners has an average of 2 in. for the external diameter. If the process variability is normal and has a standard deviation of 0.1 in., determine control limits that will include 99.73% of sample averages. Assume a sample size of 3, 5, and 10. What effect does sample size have in the results?

27.15. Examine the information below. Assuming that each sample contains 100 observations, construct a control chart that will describe approximately 95.5% of the random variation in the process when it is in statistical control.

Sample	No. of Defects	Sample	No. of Defects	Sample	No. of Defects
1	16	8	9	15	8
2	12	9	10	16	10
3	10	10	13	17	10
4	10	11	8	18	13
5	12	12	9	19	11
6	14	13	11	20	12
7	12	14	10		

27.16. Tight-fit rings for a hydraulic system are produced by a press work process. The inside diameter of 20 O-ring samples is shown as follows.

Sample	Value 1	Value 2	Value 3
1	18.797	18.804	18.8
2	18.796	18.799	18.797
3	18.793	18.802	18.801
4	18.794	18.798	18.8
5	18.793	18.8	18.797
6	18.795	18.798	18.794
7	18.794	18.8	18.797
8	18.795	18.794	18.795
9	18.793	18.795	18.797
10	18.797	18.796	18.796
11	18.796	18.788	18.796
12	18.792	18.797	18.799
13	18.795	18.795	18.798
14	18.795	18.796	18.797
15	18.795	18.797	18.796
16	18.794	18.799	18.797
17	18.792	18.798	18.796
18	18.799	18.796	18.794
18	18.799	18.796	18.794
19	18.798	18.797	18.799
20	18.795	18.795	18.793

a. Develop the appropriate control chart(s) to monitor the variation of this process.

b. Assuming that the process follows a normal distribution and that the specifications are 18.800 ± 0.008, what is the *actual* capability of the process?

c. What is the *potential* capability of the process?

d. If the expected life of these rings follows an exponential distribution with an average MTBF of 45 months, what is the probability that they fail before 4 years of service?

27.17. The assembly shown in Figure P27.17 consists of Parts A, B, and C. The three parts are machined and it has been demonstrated that production is clearly under *statistical control*. If the needed length of the assembly is $12 + 0.012$ in., and the reliability of machining each part correctly is 0.98 for Part A, 0.94 for Part B, and 0.97 for Part C, what is the probability that the total assembly will not meet specifications? Now, assume that all parts are

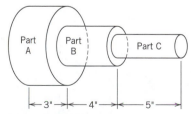

Figure P27.17

produced with equal precision. What are the maximum tolerances for the length of each of the three component parts?

a. Use the statistical method of tolerancing discussed in Chapter 24.

b. Use any conventional method of tolerancing that you know.

c. Compare the two methods and determine which one is more economical and yet meets the overall tolerance requirements. Justify your answer and state any assumptions you make.

27.18. Assume that the nominal value of the diameter of Part C in Figure P27.17 is 10 mm, with tolerance ± 0.10 mm. If the process average is at the target value (i.e., 10.00 mm), and if the standard deviation is known to be 0.05 mm:

a. What is the process capability index?

b. About what percentage of the process output will be acceptable?

c. Suppose that the process deteriorates such that the process average shifts $+ 1.5$ standard deviations. What is the *actual* process capability in this case?

27.19. Five power supplies per lot are inspected to monitor output voltage ripple, which is measured in mV. The lower ripple (noise) the better. Data for 10 lots are given as follows. Set up the initial control charts for the process.

Sample	Value 1	Value 2	Value 3
1	44.00	44.00	32.00
2	34.00	33.00	36.00
3	46.00	47.00	48.00
4	36.00	42.00	36.00
5	42.00	34.00	44.00
6	40.00	38.00	24.00
7	44.00	34.00	38.00
8	24.00	26.00	30.00
9	30.00	50.00	48.00
10	26.00	24.00	18.00

Sample	Value 4	Value 5
1	44.00	32.00
2	34.00	38.00
3	44.00	52.00
4	24.00	38.00
5	52.00	50.00
6	36.00	32.00
7	28.00	50.00
8	36.00	34.00
9	58.00	60.00
10	32.00	16.00

a. If the process is out of statistical control, recalculate the control limits for the given data (i.e., exclude samples out of control and recalculate the limits).

b. If the maximum allowable ripple is 250 mV, estimate the capability of this process.

c. Based on a. and b., what would you say about the quality of these power supplies?

27.20 Control charts are maintained on the shear strength of GMAW welds. After 25 samples of 16 welds each, the overall average is 422 psi and the average sample standard deviation is 30 psi. Assume that the process is in control and calculate the required control limits. If the specifications for these welds are 420 ± 20 psi, what is the capability of the process? What would you suggest to improve the capability of the process?

PRACTICAL APPLICATION

Identify a company in your vicinity that has implemented a continuous improvement methodology, that has earned ISO 9000 certification, or that has received a national quality award. Set up an interview with a company representative and learn about the challenges involved in pursuing continuous improvement. Write a report with your observations. (Hint: The local chapter of SME or of the American Society of Quality Control can assist you in contacting a suitable company.)

CASE STUDY: U CHART FOR CONTROL

U-Blow Thin Manufacturing uses high-speed machinery to cut flexible tubing to predetermined lengths and install terminals on the ends for pipe fittings.

The products are basically all the same: The tubing is ABS and the terminals are brass with an insulation coating. The length and diameter of the tubing and the shape of the terminals vary, but the types of defects are the same for all sizes and shapes.

The floor facilitator makes a tour of the machines and visually inspects the output of the machines that are running at that time. (Several machines perform this assembly operation but all of them do not run at the same time.)

One hundred pieces are inspected visually at each machine that is running, and the number of defects found in the sample is recorded. If the machine is not running, the record is left blank.

a. Without constructing control charts, rank the six machines from best to worst. Justify your answer.

b. Using the appropriate control chart(s), provide your ranking of the six machines. Again, justify your answer and propose a maintenance program.

c. What advantages or disadvantages do you find when using the control charts to evaluate the performance of these six machines?

Inspection Round	1	2	3	4	5	6	7	8
Machine 1	0	0	0	0	—	6	0	0
Machine 2	—	—	1	0	—	0	—	0
Machine 3	6	6	—	—	—	4	—	—
Machine 4	0	1	0	0	5	—	—	0
Machine 5	—	—	—	—	0	3	—	0
Machine 6	1	0	0	1	—	0	—	0

Inspection Round	9	10	11	12	13	14	15	16
Machine 1	2	1	1	—	0	—	—	—
Machine 2	0	0	0	2	0	0	0	0
Machine 3	—	—	—	—	0	3	1	2
Machine 4	0	0	0	1	0	—	2	0
Machine 5	0	0	0	0	0	4	1	5
Machine 6	0	0	0	0	0	0	1	1

Inspection Round	17	18	19	20	21	22	23	24
Machine 1	—	—	—	—	—	4	12	8
Machine 2	0	0	0	0	2	0	0	0
Machine 3	—	—	6	1	7	8	2	1
Machine 4	0	0	0	0	1	0	0	0
Machine 5	5	0	0	0	0	0	0	0
Machine 6	0	0	0	1	—	—	2	1

Inspection Round	25	26	27	28	29	30
Machine 1	—	—	—	—	—	—
Machine 2	0	0	0	1	0	—
Machine 3	3	1	1	3	—	—
Machine 4	2	0	0	0	—	—
Machine 5	1	0	0	0	0	0
Machine 6	2	2	1	1	0	—

CHAPTER 28

COMPUTER NUMERICAL CONTROL SYSTEMS

Numerical control (*NC*) refers to the operation of machine tools from numerical data. Data are stored on paper, magnetic tape, computer storage disks, or direct computer information. A historical example of using instructions punched on paper tape is the player piano. Notes to be played (instructions) are defined as a series of holes on a piano roll (punched paper tape), then sensed by the piano (using a pneumatic system powered by a foot-operated bellows), which plays the notes (executes the instructions).

Because mathematical information is used, the concept is called numerical control or NC. Numerical control is the operation of machine tools and other processing machines by a series of coded instructions. With a built-in computer supporting the machine tool functions, the system is known as *computer numerical control* or *CNC*.

Computer numerical control is not a machining method; it is a means for machine control that anyone involved with manufacturing must understand. In this chapter the fundamentals of numerical control are presented, the key concepts that engineers and technologists must understand about designing for CNC equipment are highlighted, basic programming techniques are illustrated, and the impact of artificial intelligence in machine control is discussed. There is much more to CNC, and the student is encouraged to consult specific books on the subject for greater information.

28.1 TYPES OF CNC SYSTEMS

The most basic function of any CNC machine is automatic, precise, and consistent motion control. This makes the relative positioning of the tool to the workpiece the most important instruction of a CNC program. The program is an *organized list of commands* used repeatedly to obtain identical results.

The way in which the CNC program is used to interact with the machine tool defines the type of CNC system. Systems using manual programming and punched paper or magnetic tape to store the program are known as *numerical control machines*. Equipment having a host computer controlling one or more machine tools are known as CNC systems. Either type may have a *dial control* feature, which is the ability to dial directly each axis dimension for the workpiece.

Computer numerical control systems use a dedicated program to perform NC functions in accordance with control commands stored in computer memory. The computer provides basic computing capacity and data buffering as a part of the control unit. Part programs are entered either manually to the tape reader, or interactively using CAM (computer-aided manufacturing) software. In the manual case, however, the tape reader is not used for subsequent parts, as the computer does the directing. CNC is also known as "soft wired," implying that the program can be changed along with built-in control features. In addition, the computer is used as a terminal to accept information from another computer or telephone data.

If the host computer is external to the machine tool and commands several machines as well as other NC devices, the system is a *direct numerical control* (DNC) system. If a microprocessor that interacts with the host computer is used as a resident controller in the machine tool, the equipment is a *distributed numerical control* (DNC) system.

In DNC, one or more NC machine tools is connected to a common computer memory to receive "on demand" or real time, distribution of data. DNC, which most people associate with *distributed* rather than with *direct* NC, may also comprise a management information retrieval package, where information is returned to the central computer and a variety of reports and actions are presented for management interpretation. The DNC system includes telecommunication lines or other methods of transmitting the instructions. DNC overcomes the expensive nature of large computers dedicated to one machine tool and allows storage of extremely long programs in a distributed fashion.

NC and CNC offer economic advantages in moderate production and job-lot industries, including the following: (1) The amount of nonproductive time is reduced and "chip time" is increased. (2) The number of jigs and fixture, particularly those used to define positioning, is reduced because the tape does the positioning. (3) A complex component may be machined almost as easily as a simple one once a tape is prepared. (4) There is usually a reduction in machining setup and cycle times, although the machining requirements for cutting are identical to those machines that have no NC. Scheduled throughput time is reduced. (5) It is adaptable to short runs, as compared to special-purpose production machines. (6) There are fewer rejects since reliability and quality are consistent. (7) The program may be changed to allow for machining modifications. (8) Inspection costs are usually reduced.

However, some disadvantages to NC and CNC machines include: (1) The capital cost for buying a machine is high, and sometimes the marginal investment saving over non–NC general-purpose machines is not warranted. (2) There is a loss in machine flexibility if a tape or the control malfunctions. (3) Control systems are relatively expensive. (4) Maintenance costs increase because of the sophistication of the control systems.

28.2 EVOLUTION OF CNC MACHINE TOOLS

The first successful application of an NC machine tool is the John Parsons' machine developed by MIT in 1952. Since then, mostly lathes and milling machines have incorporated NC techniques. With the developments of DNC, more industrial machines are becoming CNC devices.

Early NC design placed control units on existing machine tool structures to accomplish numerical control. As experience was gained, it was apparent that NC machines were more efficient in the overall operation than conventional machines. Inspection, pipe bending, flame cutting, wire wrapping, circuit board stuffing of electronic chips, laser cutting of fabric, drafting machines, and production processes have proven applications. In some instances, the added controls cost more than the basic machine tool, but modern solid-state circuitry has provided more reliable control at lower cost than previous electronic technology.

CNC machine tools incorporate many advantages, such as programmed optimization of cutting speeds and feeds, work positioning, tool selection, chip disposal, and accuracy and repeatability. This last advantage is an important feature to evaluate CNC equipment. *Accuracy* is the ability to position the machine table at a desired location.

Repeatability is the ability of the control system to return to a given location that was previously programmed into the controller. An approach to estimate accuracy and

repeatability is

$$A = \overline{X} \pm R \qquad (28.1)$$

where

A = Accuracy of the CNC equipment
R = Repeatability = $\pm 3\sigma$
$\overline{X}$ = The average error from the programmed point
σ = 1 standard deviation (see Chapter 27)

Notice that accuracy is based on the error that the CNC equipment makes. These errors are corrected by features, such as leadscrew compensation, axis alignment, and self-adaptive control.

The adaptation of numerical control has altered existing designs to the point that CNC machine tools have their own characteristics separate from the machine tools described in other sections of this textbook. For example, modifications to the turret lathe have resulted with a turret slanted on the backside rather than placed on the horizontal. A greater number of tools can be mounted on the turret as a result of the structural adaptation. This can be seen in Figure 28.1.

Development of the *machining center* with *tool storage* resulted from NC. Figure 28.2 shows an NC machining center with a storage of 24 tools in a magazine. Each tool can be selected and used as programmed. These machining centers can do almost all types of machining such as milling, drilling, boring, facing, spotting, and counter boring. Some machining operations can be programmed to occur simultaneously. The NC program selects and returns cutting tools to and from the storage magazine, if equipped, as well as inserts them into a spindle. Parts can be loaded and moved between pallets, manipulated by rotation, and inspected after the work is finished. Robotic operation is possible, also being accomplished by numerical control.

28.3 TYPES OF CONTROLLERS

Machine tools, in general, are controlled by one of four types of control programs:

1. *Sequential control,* which is based on electromechanical devices to control machine's motion. Stepping-drum and perforated paper-tape programmers are the most common controllers.

Figure 28.1
NC controlled slant-bed turret combination turning-chucking lathe.

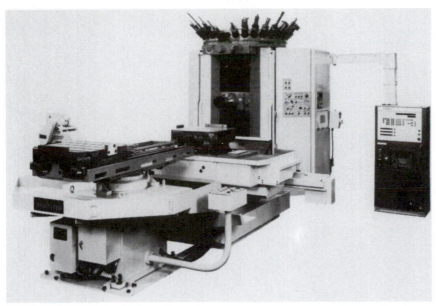

Figure 28.2
NC machine with tool carrousel on top of column and pallet for part loading in front.

2. *Programmable logic control* (PLC) is a solid-state device with a central processing
unit interacting with input and output devices to monitor motion.

3. *Automatic adaptive control* (ACC), which continuously identifies on-line performance of an operation, compares it with the expected value and automatically adjusts
one or more parameters to improve the process.

4. *Numerical control* uses prerecorded written symbolic instruction.

Depending on the type of feedback, controllers are either open or closed loop. This
classification, although applicable to all controllers, is primarily used for NC and CNC
equipment.

Open loop control is defined as a system where the output or other system variables
have no effect, or *feedback* on the control of the input. In the open-loop system an
operational device, such as a machine slide, is instructed to move to a certain location,
but whether or not the slide reaches the predetermined location is not ascertained by
the control unit. Figure 28.3 illustrates an open-loop, two-axis system. The axis coordinates
are in the X–Y plane, and a third axis, Z, is possible. The input media is scanned in a
unit called a reader. Discrete signals feed into the control unit, and instructions proceed
to the stepping motor drive unit. One *pulse* of the motor drives *one step* (or fraction of
a revolution) of the leadscrew. Each machine slide or movement that is to be controlled
has its own *stepping motor* and drive. The stepping motor is usually electric, but hydraulic
units are sometimes found. The drive to the machine element may be conventional
leadscrews, ball-bearing screws, or pinion and rack arrangements.

Open-loop control is simple and less costly, but it is not as accurate as closed-
loop control systems. The main variable to address in the open-loop control deal with
the stepping motor; step angle, angle of rotation, and rotational speed of the motor
need attention. The following example illustrates the importance of these three parameters.

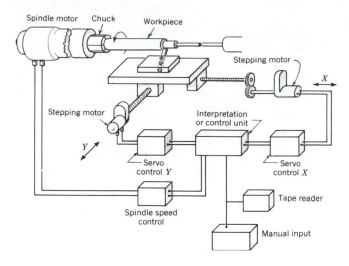

Figure 28.3
Two-axis, open-loop numerical control system.

EXAMPLE 28.1

In Figure 28.3, the shaft connected to the x-axis leadscrew has step angle of 1.8°. If the pitch of the leadscrew is 0.118 in. (3 mm) how closely can the table's x position be controlled?

The number of step angles represents the number of positions at which the motor is controlled. In other words, 1 revolution of the motor (360°) divided by the step angle (1.8°) determines that the motor can be controlled at 200 points. In addition, 1 revolution of the motor represents a 0.118-in. (leadscrew's pitch) advance of the machine table. If the motor is controlled at 200 times per revolution, the machine table is controlled at the same frequency. Therefore, 0.118 in./200 determines that the table is controlled at 0.0006-in. (0.015-mm) increments.

With this information, the manufacturing engineer can determine things like *pulse frequency*, f_p, and the required rotational speed, N, of the motor for a specific feeding speed. Assume, for example, that the machine table needs to move 3.94 in./min (100 mm/min). This speed is accomplished only if the leadscrew rotates at 33.3 rpm (speed/pitch). Since the motor shaft and the leadscrew are connected, their rotational speeds are the same. The pulse frequency, f_p, given in pulses per second, is simply

$$f_p = 60 \times N \times (\text{pulses per rev.})$$
$$= 60 \times 33.3 \times 200 = 111.11 \text{ pulses/s} \tag{28.2}$$

A *closed-loop control* system for a single axis control is given in Figure 28.4. The machine motion as actuated by the *servomotors* is recorded or monitored by a feedback unit that may be electronic, mechanical, or optical. It may be a *transducing* device that indicates the position the machine table, slide, or tool has reached in response to the tape command. The feedback unit transmits position signals through the feedback signal circuit to the control unit where the signals are compared continuously with program signals. The command signal is fed through an amplifier to actuate the drive motor until the difference between the command signal and the actual slide position reaches zero error. When the error signal is zero or null in a closed-loop system, the machine move-

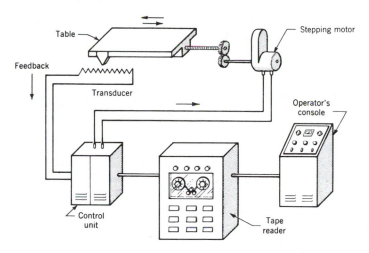

Figure 28.4
One-axis, closed-loop numerical control system.

ments are at the exact position commanded. Most NC systems are closed loop. The control unit may initiate one or more of the following actions:

1. Record the accuracy of the command, that is, the position of the tool or the selection of the component.
2. Automatically compensate for error and create an additional requirement for the tool to move to a new position
3. Stop the motion when the input and feedback signals are the same.

Figure 28.4 illustrates this principle for one axis, but it is also pertinent to three-axis machines.

28.4 CNC OPERATIONAL SEQUENCE

Computer numerical control starts with the *parts programmer* who, after studying the engineering drawing, visualizes the operations to machine the workpiece. These instructions, commonly called a program, are prepared before the part is manufactured and consist of a sequence of symbolic codes that specify the desired action of the tool workpiece and machine. Even in computer-aided design and manufacturing this interpretation is necessary. The engineering drawing of the workpiece is examined, and processes are selected to transform the raw material to the finished part that meets the dimensions, tolerances, and specifications. This process planning is concerned with the preparation of an operations sheet or a route sheet or traveler. These different titles describe the procedure or the sequence of the operations, and it lists the machines, tools, and operational costs. The particular order is important. Once the operations are known, those that pertain to numerical control are further engineered in that detail sequences are selected. Chapter 23 discusses operations planning.

A program is prepared by listing codes that define the sequence. A parts programmer is trained in manufacturing processes, is knowledgeable of the steps required to machine a part, and documents these steps in a special format. There are two ways to program for NC, by *manual* or *computer-assisted* part programming. The parts programmer must understand the *processor language* used by the computer and the NC machine.

If manual programming is required, the machining instructions are listed on a form called a part program *manuscript.* This manuscript gives instructions for the cutter and workpiece, which must be positioned relative to each other for the path instructions to machine the design. Computer-assisted part programming, however, does much of the calculation and translates brief instructions into a detailed instruction and coded language for the control tape. Complex geometries, many common hole centers, and symmetry of surface treatment can be programmed simply under computer assistance, which saves programmer time.

Tape preparation is next, as the program is "typed" onto a tape or perforated paper. If the programming is manual, a 1-in. (25-mm) wide perforated tape is prepared from the part manuscript on a printer equipped with a punch device capable of punching holes along the length of the tape. If the computer is used, the internal memory interprets the programming steps, does the calculations to provide a listing of the NC steps, and prepares the tape. Some tapes contain electronic or magnetic signals; other systems use disks or direct computer inputs. In summary, input translation takes place.

Verification is the next step as the tape is run though a computer, and a plotter will simulate the movements of the tool and graphically display the final paper part often in a two-dimensional layout describing the final part dimensions. This verification uncovers major mistakes.

The final step is production using the NC tape or any other CNC input media, which involves ordering special tooling, fixtures, and scheduling the job. A machine operator loads the tape onto a tape reader that is part of the *machine control unit*, often called a MCU. This *post-processing* step converts coded instructions into machine tool actions. The media that the MCU can sense may be perforated tape, magnetic tape, floppy disks, or direct computer signals from other computers or satellites. *Perforated paper tape* is used in this text because it is easier to illustrate, but the concepts are the same whatever the input. The student is advised that CNC, where the computer is used, is more frequent. However, the general steps are illustrated in Figure 28.5.

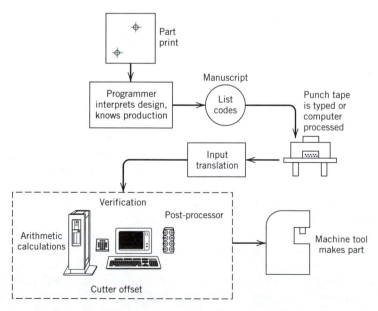

Figure 28.5
Flowchart of numerical control steps.

28.5 RECTANGULAR COORDINATES

To program the CNC processing equipment, it is necessary to establish a standard axis system. *Rectangular coordinates* or *Cartesian coordinates* is the most common system used to define a point in space. The advantage is that distances between points are equal; for example, between 2 and 3 is the same distance as between 8 and 9. Through this coordinate system a point in space is described in mathematical terms from any other point along three mutually perpendicular axes. Each machine tool has a standard coordinate axis system, which allows the parts programmer to define unambiguously the sequence of operations and movements of the machine tool, cutting tool, and part. Machine tool construction is based on two or three perpendicular axes of motion and an axis of rotation.

Generally, the Z axis of motion is parallel to the principal spindle of the machine, whereas the X axis of motion is horizontal and parallel to the workholding surface. Once the X axis is oriented, the remaining planes fall into place. The Y axis of motion is perpendicular to both X and Z. The location of the router in Figure 28.6 is $X = -2$, $Y = +3$, and $Z = +1$. The axes designations for typical machine tools are given by Figure 28.7. A similar program of notation exists for rotary motion. Separate axes motions may include tilt and swivel of gimballed heads. Milling an elliptical part with sloping walls may use five axes of the machine simultaneously.

Once the coordinate axis is known, the parts programmer may have the option of specifying the tool position relative to the origin of the coordinate axes. NC machines may specify the zero point as a *fixed zero* or *floating point*. In fixed zero the origin is always located at the same point on the machine table and is the lower left-hand corner. Locations are defined by positive X and Y coordinates. A floating point allows the zero point to be set at any position on the machine table. The workpiece may be symmetrical and the zero point would be at the center of symmetry. The floating point method is the more common.

Absolute dimensions always start from a fixed zero reference point and span the distance. For NC the tool locations are defined in relation to the zero point. An *incremental*

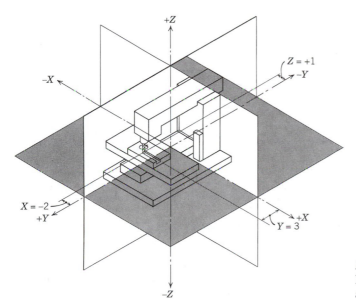

Figure 28.6
Principal X, Y, and Z axes for a milling machine.

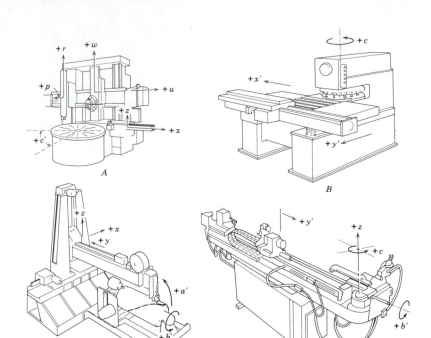

Figure 28.7
Principal numerical control machine axes. *A*, Vertical turret lathe or vertical boring mill. *B*, Turret punch press. *C*, Welding machine. *D*, Right-hand tube bender.

dimension always starts from a prior location that is not a zero reference. Figure 28.8 is a part requiring two holes to be drilled. The holes are labeled 1 and 2 and their coordinate axes numbers are given. For absolute dimensioning, the tool will move to hole 1 with dimensions $x = 4$ and $y = 3$, and after drilling, the tool will move to hole 2 with dimensions $x = 7$ and $y = 7$. For incremental positioning, the tool will move from hole position 1 to 2 along the x axis 3 units ($\Delta x = +3 = 7 - 4$) and 4 units along the y axis ($\Delta y = +4 = 7 - 3$).

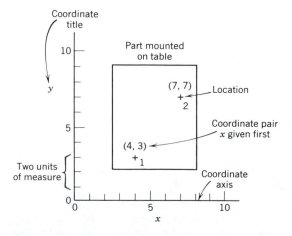

Figure 28.8
Example of absolute and incremental locations.

28.6 PROGRAM FORMATTING AND CODING

The basic coding and formatting is very similar for any form of input media. The *punched tape* was an early form of standardized input, and it is used in this textbook to explain the coding of an NC program. Material for the tape varies from punched or perforated paper to mylar-reinforced paper, mylar-coated aluminum, or certain plastics materials.

Although various tape formats are available, emphasis continues to be with those NC systems manufactured to standards[1] that define variable block tape formats for positioning and contouring controls. The tape is 1 in. (25.4 mm) in width and has eight channels. Holes are punched in the channels in code patterns. A tape reader senses the hole pattern by photoelectric cells, fingers, brushes, or a vacuum method.

During part production, the tape is fed through the tape reader once for each work-piece. While the machine is performing one machining sequence, the tape reader is feeding the next instruction into the controller's data buffer. Machine operation is more efficient in this manner, as the machine tool is not waiting for the next instruction to be fed into its active memory. After the last instruction is recorded into the data buffer, the tape is rewound ready for the next part.

Figure 28.9 illustrates the EIA standard coding for 1-in. (24.5-mm) wide, eight-channel punched tape for the *binary coded decimal* (BCD) system. In addition, an ASCII standard is used widely. The coding of a tape is by the absence or presence of a hole. This absence or presence of a hole is binary, that is, two. The base number 2 system can represent any number in the more familiar 10 or decimal system. The binary code uses only 0 or 1. The 0 or 1 is referred to as a bit. Computers operate on a form of binary arithmetic, where a number can be expressed by a combination of "on" and "off" circuits. This concept is suitable for NC and CNC, since a number may be expressed by a hole or no-hole in a tape or by a 0 or a 1 in machine language. Also, letters can be expressed by a combination of binary bits. The meaning of successive digits in the binary system is based on the number 2 raised to successive powers.

This system is used in almost all NC operations. There are eight designated channels and one line of sprocket holes on the tape. When numerical data are the input, channel numbers 1, 2, 3, 4, and 6 are employed. The first four channels represent the numbers 1, 2, 4, and 8, which are powers of 2; that is, $(2)^0 = 1$, $(2)^1 = 2$, $(2)^2 = 4$, $(2)^3 = 8$. Hence, referring Figure 28.9, the number 7 is read into the tape by punching holes in the first, second, and third channels, which total $1 + 2 + 4 = 7$. To indicate the number 5, a hole is punched in channels number 1 and number 3. The tape reader makes elementary checks on the accuracy with which the tape has been punched. This is called a parity check. There must be an odd number of holes in each row or the tape reader and, hence, the machine stops. Therefore, each time a command calls for an even number of holes to be punched, an additional one must be punched in channel 5, the parity check channel. The sprocket drive holes are not considered in the parity check.

Figure 28.10 is a schematic of the way a short strip of tape might appear for the simplified program shown. All numbers are usually depicted with either five or six digits, and the first two refer to whole inches (mm or cm) and the latter ones to the decimal fraction.

The coding in Figures 28.9 and 28.10 will not be the same necessarily for different manufacturers' products. Such things as "coolant on" and "coolant off" are regularly punched into the tape and meet machine tool builder's specifications. Various commands

[1] Electronics Industries Association (EIA) RS 273-A and RS 274-B for positioning and contouring.

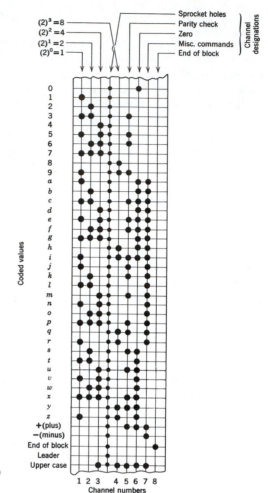

Figure 28.9
BCD tape.

may be coded as illustrated, although programming of a given tape is characteristic of a particular machine tool and control arrangement, and universally used standards for all commands are not available.

Tape Formats

The organization of words within blocks is called the *tape format*. The most common formats are the *tab sequential* format and the *word address* format.

Figure 28.11 illustrates an example of the tab sequential format. It is a simple positioning program to EIA RS 273 standards for drilling four holes. This standard requires that a *tab code* (holes in track numbers 2, 3, 4, 5, and 6 in a traverse row) precede each command block and that the instruction for the *x* coordinate always precede that for the *y* coordinate.

Almost all modern CNC controls use the *word address* format. In this format, the CNC program is made up of sentencelike commands. Each command is made up of a CNC word, each of which has a letter address (also known as the code) and a numerical value. The letter address tells the control the kind of word and the numerical value tells

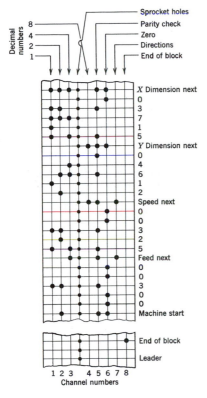

Simplified program
X = 03.715
Y = 04.612
Speed = 00.325 rpm
Feed = 00.300 in./rev
Start machine
End of block

Figure 28.10
BCD program.

EXAMPLE 28.2

This program drills two holes in a workpiece on a CNC machining center:

Command	Meaning
PROGRAM O0001	Program number.
N005 G54 G90 S499 M03	Select coordinate system, absolute mode, and turn spindle on CW at 400 rpm.
N010 G00 X1. Y1.	Rapid traverse to *XY* location of first hole.
N015 G43 H01 Z.1 M08	Instate tool length compensation, rapid traverse in *Z* to clearance position above surface to drill, turn on coolant.
N020 G01 Z-1.25 F3.5	Feed into first hole at 3.5 in./min.
N025 G00 Z.1	Rapid back out of hole.
N030 X2.	Rapid to second hole.
N035 G01 Z-1.25	Feed into second hole.
N040 G00 Z.1 M09	Rapid out of second hole, turn off coolant.
N045 G91 G28 Z0	Return to reference position in *Z*.
N050 M30	End of program.

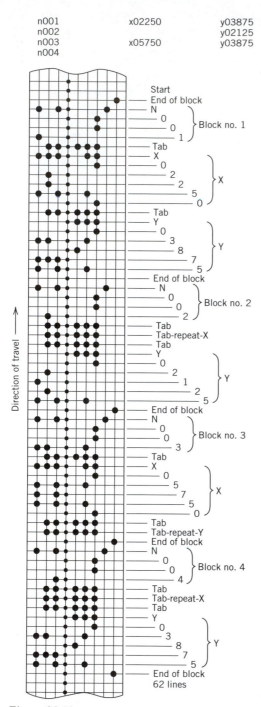

n001	x02250	y03875
n002		y02125
n003	x05750	y03875
n004		

Direction of travel →

Start
End of block
N
0
0 } Block no. 1
1
Tab
X
0
2
2 } X
5
0
Tab
Y
0
3
8 } Y
7
5
End of block
N
0
0 } Block no. 2
2
Tab
Tab-repeat-X
Tab
Y
0
2
1 } Y
2
5
End of block
N
0
0 } Block no. 3
3
Tab
X
0
5
7 } X
5
0
Tab
Tab-repeat-Y
End of block
N
0
0 } Block no. 4
4
Tab
Tab-repeat-X
Tab
Y
0
3
8 } Y
7
5
End of block
62 lines

4.750 (120.65)
1.250 (31.75)
Y
4" (101.6)
2.875 (73.03)

1 4
2 3

1.125
(28.58)

1"
(25.4)

Origin
1"
(25.4)
6" (152.4)
X

(Metric measurements are given in parentheses)

Figure 28.11
EIA RS 273 tape format for drilling four holes. Channel No. 1 left side of tape.

Table 28.1   **Common Letter Address Specifications**

Type[a]	Function
N-word	Sequence number; used for line identification
O-word	Program number; used for program identification
G-word	Preparatory function (see below)
M-word	Miscellaneous/auxiliary function (see below)
F-word	Feed rate designation, (in./min)
S-word	Spindle speed designation, (rpm)
T-word	Tool designation
Z-word	Z-axis designation (similar for X and Y axes)
H-word	Tool length offset designation

G-word examples:

G00	Prepare for a point-to-point operation
G01	Linear interpolation in contouring systems
G03	Circular interpolation, counterclockwise

M-word examples:

M00	Stops machine; operator must restart
M03	Starts spindle in clockwise direction
M14	Start spindle in CCW direction and turn coolant off

[a] Only two letters of the alphabet, H and L, are unassigned.

the control the value of the word. CNC control manufacturers do vary with regard to how they determine the word names. Table 28.1 presents a brief list of some of the word types and their common letter address specifications.

28.7 TYPES OF PROGRAMMING AND INTERPOLATION

Numerical control programming is often segregated into *point-to-point* or *continuous path*. Although many NC controls have capability in both methods and distinctions are obscured, the concepts are sufficiently different for learning purposes.

The point-to-point (PTP) or positioning method is characterized by punching, spot welding, or drilling machines. It is used extensively on machines that can move in one direction only. Point-to-point locates the working spindle or workpiece in a specific relative position, and the tool operates either by tape instruction or manually. The tool does not contact the work when moving between coordinate positions. For example, the holes for Figure 28.8 would be drilled in successive stops. Whether the tool moves along the *x* axis first and then the *y* axis is immaterial to PTP. Actually, some machines move simultaneously along both axes. Many positioning control systems use the eight-channel, 1-in. (25.4-mm) wide perforated tape with standard codes and formats. Some NC point-to-point machines have only the *x* and *y* axes controlled, whereas others will be programmed on three or more axes as well as have tool selection, feed, speed, spindle rotation, coolant flow, and other functions controlled.

In some cases, the PTP method is programmed to machine a *straight line* and a *contour*. To machine the surface *FG* in Figure 28.12*A*, the tool is positioned by tape instruction at nine different *x–y* locations. If 18 different positions were programmed, the actual

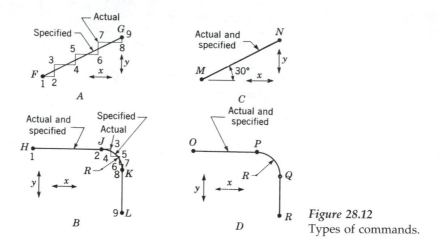

Figure 28.12
Types of commands.

machine path is more accurate. For Figure 28.12B, 9 program steps are shown, but it is usually necessary to have 100 programmed commands per quadrant. For these operations the tape length may become excessive in machining complex contours involving point-to-point commands.

Some machines are basically positioning ones but do have some contouring features such as straight-line milling, as shown in Figure 28.12C. Turret lathes and turret drills are machines that are positioning ones with some contouring ability.

In *continuous path programming* (CPP) the cutting tool contacts the workpiece as coordinate movements take place, as shown in Figure 28.13. *Contouring operations* include milling, turning, and flame cutting. Contouring differs in movement between program points. An *interpolation* routine differentiates CPP from point-to-point programming. The problem is to provide control for the tool continuously, which requires frequent changes in two or more axes simultaneously. During this movement, the tool touches the workpiece.

There are several interpolation methods used to connect defined coordinate points. The most common are linear, circular, and parabolic.

In *linear interpolation*, the machine shape is the result of a series of straight-line machining moves programmed in sufficient quantity to give an acceptable comparison between the drawing contours and the finished shape. Linear interpolation allows movement of two or more axes of the cutter at the same time. If a circle is machined, perhaps several thousand finite and discrete points connected by straight lines would be required.

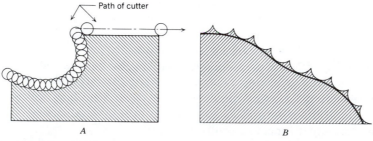

Figure 28.13
A, Cutter path for continuous programming. B, Cusps resulting in nonparallel axes machining.

Inasmuch as the points appear on the program tape, it can be seen that the tape may become lengthy.

In *circular interpolation*, the programming for a circle would be the end points of the arc, the radius and center, and the direction of the cutter, as shown by Figure 28.12*D*. The circular interpolator in most machine control units is a computer component that breaks up the span into the smallest straight-line *resolution units* available in the control 0.0001 in. to 0.0002 in. (0.003 to 0.005 mm). The control computes and generates the controlling signal for the tool. For 1000 blocks of linear interpolation information on a tape, only 5 blocks are necessary with circular information. *Parabolic* interpolation has application in free-form designs, such as molds or die sculpturing.

In continuous programming feed rates, tool geometry and offset, depth of cut, and materials are entered into the program logic. When a surface not parallel to one of the machine axes is being machined, a smooth surface is usually unobtainable, as shown in Figure 28.13*B*. The choice of cutter, ball-nosed, or square-end mill, does not eliminate the problem. There may be cusps that are ground away in a later operation. The deepest part of the cut represents the finished dimension.

28.8 HIGH-LEVEL LANGUAGES

Computer languages, such as *APT* (Automatic Programmed Tools, MIT, 1959), AUTO-SPOT (Automatic System for Positioning Tools, IBM, 1962), and others, are *high-level languages* especially designed for numerical control. A postprocessor is required to convert the program into the specific language that a particular CNC machine tool can understand. The final program used by the machine tool will be identical, whether it is prepared in machine language, by a CAM system, or by the computer in a high-level programming language. Different machine tools, however, require their information differently so there are as many postprocessors as CNC manufacturers.

Computer-assisted NC programming, in particular the popular APT language, is the de facto high-level programming language for numerical control of machine tools. Space does not permit its full development but a brief overview follows.

Most APT statements are composed of two parts separated by a slash. A statement, such as "GORGT/TO, LIN2, PAST, LIN3," directs the tool to move right from its present position until its periphery is tangent to LIN2, and its center is past LIN3 on the farside. Normally, a parts programmer will define the lines, points, circle, and the like to describe the part near the beginning of the program prior to the first motion statement.

There are four types of statements in the APT language (see Table 28.2). The *geometry* statements define the elements that comprise the part. The *motion* statements define the required tool path. The *postprocessor* statements specify feeds, speeds, and other machine-specific features; they vary among control systems. The *auxiliary* statements describe tolerances, identify tools and parts, and specify other miscellaneous functions.

In the geometry statement, the *symbol* is a six-character identifier for a geometric element. The geometry type is an APT vocabulary word such as POINT, LINE, PLANE, or CIRCLE. The *descriptive data* is an expression identifying the element precisely and uniquely. A line, for example, can be defined by two points or by the intersection of two planes, whereas a plane may be defined by three points or as being parallel to another plane. Spatial geometry ability is very helpful here.

In the motion statement, the *motion command* is an APT word that tells the machine what to do. The *descriptive data* tells the machine where to go. First, the starting point must be defined, so a FROM/ command is given. Then, since the moving geometry of a

Table 28.2   **The Statements of the APT Language**

Type	General Form	Examples
Geometry	Symbol = geometry type/descriptive data	P1 = POINT/3.0, 5.0, 0.0 P4 = POINT/(3*sin α), 0, (3**2) L1 = LINE/P1, P4 PL1 = PLANE/P2, P3, P4 C1 = CIRCLE/CENTER, P1, RADIUS, 4.0
Motion	Motion command/descriptive data	FROM/P4 GOTO/P1
Postprocessor	APT word/machine type, processor type	MACHIN/MILL,1; TURRET/T30; FEDRAT/2.3; SPINDL/rpm; RAPID; END
Auxiliary	Auxiliary word/descriptive data	OUTTOL/0.005 (allowable tolerance between the outside of a curved surface and any straight line used to approximate the curve) INTOL/0.001 (defines maximum inside tolerance)

cutter must correspond to the engineering drawing requirements, commands are given, depending on the type of motion (PTP or CPP).

For PTP the only motion commands are GOTO/ and GODLTA/, meaning go directly and go incrementally to the location specified in the descriptive data portion of the statement. For CPP six self-explanatory motion commands exist: GOLFT/, GOFWD/, GOUP/, GORGT/, GOBACK/, and GODOWN/. The descriptive data portion for CPP motions requires *modifying words* such as TO, ON, PAST, and TANTO. An example was given previously.

In the postprocessor statement APT words may or may not be followed by a slash. If a slash is used, descriptive data are required. For example, the APT word COOLNT/ must be followed by ON, OFF, FLOOD, or MIST to control the coolant status.

In the auxiliary statement APT words may or may not be followed by a slash. If a slash is used, descriptive data are required. A few common auxiliary statements are: PARTNO (typed in columns 1–6 identifies the workpart), FINI (indicates the end of the APT program), and CUTTER/0.500 (which defines the diameter of the cutter in inches). As a final observation, every time a cutter is defined the tool path must be offset by half the cutter's diameter.

28.9 EMERGENT CONTROL METHODS

As a rule, the programmer establishes feeds and speeds. If during the machine operation some unforeseen problem such as hardspots or worn or broken tools occurs, an *adaptive control* is employed to slow or stop the machine. Conversely, adaptive control senses machining conditions and can increase speeds or feeds as the situation dictates. The sensed variables include torque, heat, deflection, or vibration.

Many adaptive control techniques exist: proportional, PI (proportional–integrating), and PID (proportional–integrating–differentiating) are methods based on linear adaptation. Among the most current developments, nonlinear adaptation such as the *fuzzy logic* and the *artificial neural network* (*ANN*) schemes are the front runners.

Fuzzy logic control (*FLC*) bases its algorithms in the concepts of fuzzy logic that allow manipulation of linguistic variables, rather than numerical variables. Essentially, FLC is the means by which linguistic rules, which qualitatively express the control strategy, are interpreted and implemented on a computer.

For example, the forces on a cutting tool depend on a number of considerations. To keep them low the linguistic rule "IF chip size is large, THEN increase the side cutting edge angle by 15%" is coded into the control program. Such a rule, however, is open to interpretation because the qualifier "large" for the variable "chip size" may have different numerical values for different materials. In FLC, both inputs and outputs are not single-valued and describe a *fuzzy set*; in other words, a set with vague (fuzzy) boundaries, whose elements have an assigned level of membership.

ANN controllers base their algorithms in the power of interconnection. An ANN consists of many nonlinear computational elements (*nodes*) operating in parallel and arranged in patterns resembling those of biological neurons. Nodes are interconnected and weights are assigned to each link. These weights are adapted numerous times during the network's learning process.

Input nodes get process data from sensors. The input values are then propagated through the links to the other regions of the network. As they propagate and arrive to common nodes, they are combined and changed according to the computational rules of the links and the nodes through which they pass. *Output nodes* provide numeric outputs of the neural network that are used to modify the performance of the machine. Unlike FLC, ANN takes an input numeric pattern and provides an output numeric pattern.

These new developments open up a variety of applications. For example, nonlinear systems such as in process, tension, and position control; systems with gross input deviations or insufficient input resolution; difficult-to-control systems that require human intuition and judgment; systems that require adaptive processing to overcome changing environmental conditions; and processes that must balance multiple inputs or conflicting constraints.

QUESTIONS

1. Give an explanation of the following terms:

Numerical control machine
Machining center
Tool storage
Open-loop control
Closed-loop control
Programmer
Processor language
Machine control unit
Rectangular coordinates

Punched tape
Binary coded decimal
Tape format
Point-to-point
Continuous path
Interpolation
APT
High-level languages
Fuzzy logic control

2. What is meant by feedback?

3. How does a player piano tape differ from numerical control tape?

4. Make a schematic diagram of a two axis, closed-loop control system.

5. How does a machining center differ from a numerically controlled turret lathe?

6. List the media for NC. Which is most common?

7. What is more important in CNC equipment, accuracy or repeatability? Justify your answer.

8. Name some disadvantages of NC and CNC systems.

9. What is the purpose of the parity check?

10. Explain your interpretation of DNC.

11. Write a paragraph about the machine control unit.

12. What are the characteristics of a stepping motor?

13. What types of operations are best suited for point-to-point commands?

14. Why is a coordinate system necessary for numerical control?

15. What is the importance of fixed zero and floating point?

16. What is adaptive control and what is its purpose?

17. Why does the first NC machine tool located in a plant have disproportionate costs associated with it?

18. Explain how the word address tape format works.

19. What is linear interpolation? How does it work?

20. What are the main types of statements in APT? What is the purpose of each?

21. What is the basis of an ANN controller?

22. Write a fuzzy rule to control a car driver's behavior while driving near an elementary school.

PROBLEMS

28.1. Show the coordinate location in a three-dimensional place of $x = 5$, $y = 3$, and $z = 4$.

28.2. Sketch the following machines and show the X, Y, and Z axes for each: drill press, lathes, and milling machine.

28.3. Design a numerical system that can be used to remove chips from a NC turret lathe.

28.4. Make a schematic of tape and, with the binary coded decimal system, show the numbers 0 to 9, end block, and six miscellaneous operations identified by the first six letters of the alphabet.

28.5. Using the EIA standard program on a tape, facsimile the following: "Manufacturing Processes," "The Year of 2421," "6 + 7 = 13."

28.6. The following sample was taken from $\frac{7}{16}$ in.-diameter motion control couplings made by a CNC machine tool.

0.4375, 0.4374, 0.4376, 0.4377, 0.4373, 0.4373, 0.4375, 0.4372, 0.4377, 0.4374

Determine the repeatability of the machine. If the CNC manufacturers claim an accuracy of 0.1%, what would you say about their claim?

28.7. A stepping motor is connected to a leadscrew of 0.13-in. pitch. Determine the size of each step angle in the motor, the rotational speed of the motor, and the linear travel rate of the table if the pulse frequency is 200 pulses/s and the number of step angles required is 245 pulses/rev.

28.8. Prepare a schematic of BCD tape for the following: $X = 30.720$, $Y = 2.000$, $Z = 0.500$, feed = 0.025 in./min, coolant on, coolant off, end block. Repeat using the word address format.

28.9. Prepare a schematic of the binary coded decimal tape for the following: $x = 22.750$, $y = 1.250$, $z = 2.117$, feed = 0.015 in./min, coolant on, coolant off, end block. Repeat using the word address format.

28.10. Prepare a tape schematic for the following: $X = 15.005$, $Y = 0.005$, $Z = 1.275$, feed = 0.009 ipr, coolant on, coolant off, end of block. Repeat using the word address format.

MORE DIFFICULT PROBLEMS

28.11. A two-flute, 0.5 in.-diameter cutter is used to cut out and rough shape the bracket shown in Figure P28.11. Write an APT program to produce the part if the cutting speed is 75 ft/min and the feed rate, f_r, is 2 mil/cutter rev. (Hint: Number of turns, $N = 12 V/\pi D$; feed, ipm = $N \times f_r \times$ teeth.)

28.12. Explain each line of the following program and sketch the workpiece contour. (Hint: See Table 28.1 and Example 28.2.)

PROGRAM O0002

N005 G54 G90 S350 M03

N010 G00 X-.625 Y-.25

N015 G43 H01 Z-.25

N020 G01 X5.25 F3.5

N040 G01 X.75

N045 G03 X-.25 Y3.25 R1.0

N050 G01 Y.75

N055 G03 X.75 Y-.25 R1.0

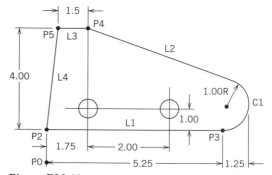

Figure P28.11

N025 G03 x6.25 y.75 r1.0   N060 G00 z.1

N030 G01 y3.25   N065 G91 G28 z0

N035 G03 x5.25 y4.25 r1.0   N070 m30

28.13. Repeat Problem 28.11, but use the letter address format to write the NC program.

28.14. Write an NC program to manufacture the brake mount of Figure P28.14. The raw material is ⅛-in. steel plate, all holes are through, all fillets are ⅛ radius. Choose a proper diameter for the cutting and drilling tools. Recommend and use adequate cutting speeds (V) and feed rates (f_r). (Hint: Review Chapters 8 and 9; number of turns, $N = 12\ V/\pi\ D$; feed, ipm $= N \times f_r \times$ teeth.)

28.15. Repeat Problem 28.14, but use the letter address format to write the NC program.

28.16. Study Figure C28.1 and then prepare an APT program to rough cut to size and to drill the holes shown. Make one change, though: Assume that the four corners

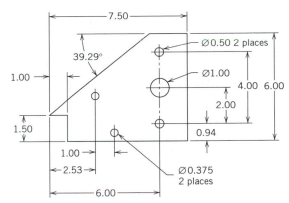

Figure P28.14

of the plate are *not* square but rounded with a 1-in. radius. The end-milling cutter is a two-flute, 0.5-in. diameter tool. Let the drilling feed and the cutting feed be 0.025 in./min and the speed = 325 rpm. (Hint: Number of turns = 12 V/π D.)

PRACTICAL APPLICATION

Visit a local machine shop and learn what techniques are used for CNC part programming. Identify how programs are downloaded, what kind of machine tools are being programmed, and what kind of systems the shop is planning for. Ask if part programs are generated off-line (with CAD/ CAM systems), at the machine, or manually (G-Code). Identify the type of computer that supports these functions. If the shops in your area do not use CNC, identify a machine shop that would like to implement CNC and discuss the same questions with them.

CASE STUDY: DRILLED PLATE

"Well, what do you think about the new request for a quotation from Baltimore Machine Tool for machining their drilled plates?" Ralph Digiamco, president of Ace Tool and Die, a company that subcontracts work for larger

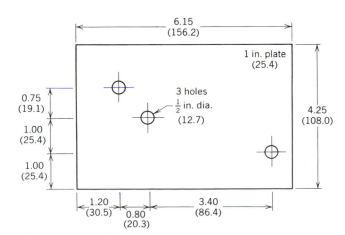

Figure C28.1
Case study.

companies, was addressing his production engineer, Bob Quinn.

"I don't know if we can do it," answered Bob, looking at the drawing. "We can flame cut the blank from plate stock oversize and then mill the external dimensions. Perhaps while it is on the miller we can drill the holes, but the drilled plate may not meet quality specs. There probably won't be more than a few that meet quality control specs."

Ralph nodded in agreement. "It will be definitely too expensive to do it that way. What if we use the NC to drill the holes instead of the vertical miller?"

"A good idea," said Bob, sitting up in his chair. "That way we would only have to change the NC tape to get a new hole layout for other jobs that might come in. I'll look into this and get back to you in a few days."

Help Bob with his program (see Figure C28.1) using the following hints: Assume that the outline has been completed by the milling machine. Use the axis system suggested by the drawing where the origin is defined in the lower left-hand corner of the part. The end-milling cutter is a two-flute, 0.5-in. diameter tool. Let the drilling feed = 0.025 in./min, speed = 325 rpm, and use start-and-stop machine and end of block (EOB). Prepare a schematic of the tape for this job.

CHAPTER 29

PROCESS AUTOMATION

With the development of computers and electronic communication equipment a variety of technologies have evolved to improve the efficiency of manufacturing processes and systems. Such developments have been linked by what has become known as *computer integrated manufacturing*, or CIM. It is impossible to discuss all these techniques in one chapter, so a summary and definitions are given in Table 29.1. Students are encouraged to review the references to expand their understanding of the most common computer-aided technologies in use.

This chapter, then, concentrates on one segment of the big picture of CIM: *process automation*. Such a discussion, however, requires that related subjects such as simulation, material handling, robotics, group technology, and flexible systems be presented. The chapter concludes by contrasting automation to other production systems of common application in manufacturing.

29.1 SIMULATION

Simulation is the process of developing mathematical models of a real system to evaluate the impact that some parameters have on a given output. If the relationships in the model are simple, *analytical solutions* are developed to find exact answers. Otherwise, *numerical methods* to find approximate solutions are used.

As an example of an analytical solution, consider jobs arriving sequentially at a single *server*. Assume the server is a turret lathe that processes the job immediately on arrival if it is idle. Otherwise, jobs join the end of a *queue*. When the server finishes processing one job, it begins processing the first job in the queue until no jobs are left.

It is possible, then, to develop a model to calculate the average time, t_s, that a job spends in the system, by evaluating the rate at which jobs arrive at the server, and the rate at which the machine processes jobs. Under the common assumption that waiting time follows an exponential distribution, the long-run average time, t_s, that a job spends in the system (i.e., in queue, plus being machined) is

$$t_s = 1/(r_p - r_a) \qquad (29.1)$$

where

$$r_p = \text{Rate at which server processes job, job/min}$$
$$r_a = \text{Rate at which jobs arrive at server, job/min}$$

Most manufacturing processes are too complex to allow exact analytical solutions. Therefore, a computer is used to apply numerical methods to simulate a process' behavior over a time period of interest. Figure 29.1 is a picture of the screen of a popular simulation program using numerical methods to simulate a manufacturing process. Depending on the computer program, one can concentrate on a type of simulation known as *discrete*, on a type known as *dynamic*, or on a type known as *continuous*.

Discrete event simulation is *stochastic* in nature, which means that random samples from probability distributions are used to derive the model through time. *Dynamic*

Table 29.1   **Glossary of Some Common CIM Terms and Technologies**

AGV: Automated guided vehicles. They are self-controlled and follow specified paths in a plant floor to move materials, tools, and other items.

AS/RS: Automated storage and retrieval system. Computer-controlled, high-density rack system for rapid storage and retrieval of parts and tools.

CACE: Computer-aided cost estimating. Knowledge-based system used for manufacturing cost estimating.

CAD: Computer-aided design. Broad term that denotes the use of computer systems in designing components and end products. It encompasses modeling, drafting, and analysis of the design.

CAE: Computer-aided engineering. The use of computer systems to support essential functions to engineering a product.

CAM: Computer-aided manufacturing. The use of computers to aid management, control, and operation of a manufacturing plant. It includes tasks such as planning, part programming, robotics, material handling, tool management, fixturing, and automated inspection.

CAMAC: Computer-assisted measurement and control. The use of computer systems in measurement and control of processes.

CAPP: Computer-aided process planning. Use of computer systems to select processes and parameters needed to manufacture a product. CAPP is the bridge between CAD and CAM.

DSS: Decision support system. Artificial intelligence-based system that aids manufacturing personnel in decision making.

IGES: Initial graphics exchange specification. A standard developed to aid in the exchange of design data between CAD systems of different makes.

LAN: Local area network. Data communication network that interconnects workstations, computers, machine tools, robots, and other digital devices in a CIM environment.

MAP: Manufacturing automation protocol. Communication standard specifically tailored for the needs of automation in manufacturing.

MRP II: Manufacturing resources planning. Planning of materials, machines, and operators required to build a product. It integrates sales forecasts with capacity requirement planning (CRP), master production schedule (MPS), and the bill of materials (BOM).

TOP: Technical office protocol. Specification that facilitates the integration of systems in the technical office so the business side of manufacturing can be integrated with the production floor.

VLSI: Very large-scale integration. Technology in which a huge number of circuit components are integrated in one chip so that data transfer and analysis are more efficient.

simulation is time-based and requires that system performance with respect to time be taken into account. *Continuous process simulation* focuses on processes in which no discrete events occur over time. Common examples are startup, steady-state operation, and shutdown events in continuous processes, such as those in the chemical industry, heat exchangers, and pumps.

29.2 AUTOMATION

Automation involves automatic handling between machines and continuous automatic processing at the machines. The elements *continuous* and *automatic* are necessary to separate automation from other production systems, such as mechanization. Automation exists only when a group of related operations are tied together mechanically, electronically, or with the assistance of computers.

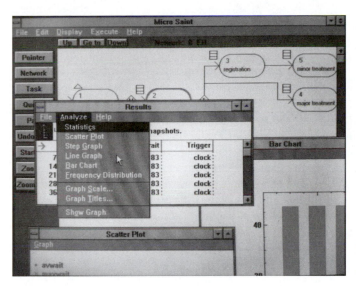

Figure 29.1
Typical simulation software
display.

Automation is not a new technology. Food and beverage processing, petroleum and chemical industries, and telephone services have been "automated" for decades. Grain milling, for example, is a processing operation that approached complete automation about two centuries ago. Grain was fed into a flour mill by means of a bucket conveyor, and water power moved it over a series of endless belt and screw conveyors through coarse and fine grinding operations until flour emerged.

The word "automation" was coined at Ford Motor Company in 1945 to describe "a logical development" in technical progress where *automatic handling* between machines is combined with *continuous processing* at machines. This implies that combining two or more automatic operations on a standard machine (such as found with automatic bar machines, vertical turret lathes, and others) does not constitute an automated system. Machines are considered automated only when they are mechanically joined for continuous automatic handling and processing. If an automatic bar machine is connected mechanically to a material feed or to the conveyor that advances parts, it is called an automated system.

Purposes for Automation

The original objective of automation was the reduction in direct labor costs. Other reasons, however, are now more prominent. Companies use automation to obtain *uniform quality* with machines instead of employing the variable skills of labor. Operator fatigue, boredom, or other human frailties are reduced.

Safety is another reason for automation. Automation reduces industrial accidents, which is important for positive employee morale and good operating practice.

Shop efficiency is improved as loaders, unloaders, feeders, automatic inspection, and other automation devices require efficient distribution of parts from one machine to another. This smoothes out the delivery of parts between machine operations.

Automation improves efficiency by segregating short runs from high-volume production. Flip-flop between short and long runs results in extended setups; time is spent changing over from one to the other, and it is difficult to maintain systematic work scheduling. Automated setups, of course, must be left that way within practical limits.

Keeping long-run jobs progressing through one line and low-volume work through another is a planning necessity to maintain the economic efficiency.

Another reason is the *use of standard tools* instead of specially designed tooling, which minimizes tool change time. Tools are located in carousels or in AS/RS (automatic storage and retrieval) systems and are retrieved automatically or under the control of robots or attending operators. There is a specific place for each tool, and each is identified and computer keyed to the position of the tool in the machine. Moreover, each tool in the bin is *preset*, so that it can replace a worn tool in the machine without adjustments. A clock or counter is connected to the tool on the machine and indicates when a predetermined number of parts are machined. Setting these clocks, of course, is based on the known life of a tool. When any particular tool has machined a predetermined number of pieces, the machine automatically stops, and the preset tool on the board is installed. Out-of-tolerance parts caused by worn tooling are minimized and fewer tools are broken.

Part and Process Design

A basic principle in automation is that the design of the part and the design of the process should be related as closely as possible. Close cooperation between product and equipment designers is essential, not only to avoid excessive costs in automated machines and to ensure maximum efficiency in processing but also to provide the highest degree of flexibility. Without some idea of what future changes may be made in a part, the equipment builder cannot provide for them. For example, the power source, the type of transfer equipment, and the orientation of parts need to be defined.

1. *Power sources.* The basic machine movements are linear, as in the travel of a drill or rotary, such as an indexing table. The power for these movements are electrical, pneumatic, or hydraulic. The selection of the power source depends on availability, relative cost, amount of power, space, and speed requirements.

2. *Transfer equipment.* Mechanical loading and transfer devices are used to move components of varied geometry from machine to machine. Special jaws grip the part, lift, move, and turn it on arms, and place it into the new work position. Travel distance, direction, sequence, and speed are controlled mechanically, electromechanically, or with fluidic controls. Robots and computer control are used. Dead stops, *mechanical arms*, or *iron hands* can be reprogrammed, but not very readily. However, robotic manipulators overcome the inflexibility of mechanical manipulators. Figure 29.2 illustrates transfer equipment.

3. *Parts orientation.* A step in the assembly process is the orientation of parts. This work can be done by operators who will pick, orient, and place the part in another location. The parts may be jumbled from a previous operation or they may be stacked horizontally, vertically, in order, random, or arranged by size. The choice here is almost limitless. Parts orientation work is popular, especially in low-volume work, where direct labor employees arrange the parts. As volume increases, it pays to consider other ways. Numerous mechanical, pneumatic, vibratory, and ingenious devices exist to orient parts.

In some systems, a robot or a part-feeding system is used to orient an individual part and present the component consistently for assembly. Parts are manually oriented either directly into a feeding system on the robot arm or into a magazine tray from which the robot removes them. In contrast, a computerized visual detection system can orient and present the part without the need for manual labor.

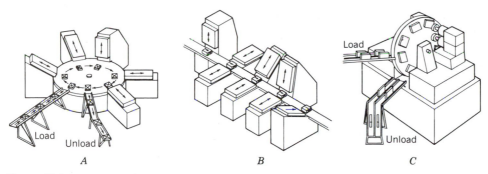

Figure 29.2
Transfer machines. *A*, Circular table. *B*, In-line. *C*, Drum.

A computerized *visual feeding* system is composed of a feeding and an orientation section. The feeding section consists of a bowl and track feeding unit that separates the disoriented parts and passes them along to the orientation section. One method for separation is vibratory motion, which first reduces the pile of parts to a single layer, and then aligns them into a single row. At this point, single parts are removed from the row and fed for orientation.

Take, for example, the part shown in Figure 29.3. It must be checked to see if it is inverted and turned so that all pieces face the same direction. Determining which side of the part is up is first. Observe Figure 29.4*A*. The isolated part is fed into a holding unit and held between a horizontal light source and a light-detection sensor. The sensor reads the position of the part and the computer determines if any orientation is necessary. If the component is positioned wrong side up, it is vertically rotated 180°. The part is returned to the orientation sensing unit, where it is checked using vertical light sensors, as shown by Figure 29.5. If part orientation is wrong, it is rotated horizontally 180°.

Bowl feeders, Figure 29.6, are popular for small parts, including nuts, screws, shafts, pins, washers, and the like. Some devices allow vibration from an electromagnet supported on the base. When components are placed in the bowl, the vibratory motion causes them to climb up a track to the outlet at the top of the bowl. Another device (Figure 29.6*B*), is a centerboard feeder. Basically, this consists of a hopper for the random dumping of parts. A blade that has a slot in its upper edge is periodically pushed through the parts and catches some in its slot. When the blade is in its highest position, parts run down a chute and are fed to the assembly below.

Individual Machine Automation

Automatic screw machines, multispindle automatic lathes, and gear-cutting machines are examples of individual machine mechanization. Magazine-type feeders are used to supply lathes with bar stock, and gear blanks to the gear-cutting machines, to make them fully

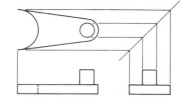

Figure 29.3
Design of part for orientation.

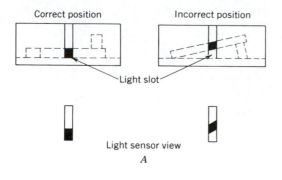

Light slot

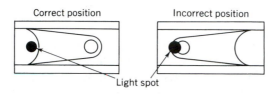

Light sensor view

A

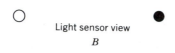

Light spot

Light sensor view

B

Figure 29.4
Vision-determining system. *A*, Inversion from side view. *B*, Direction from top view.

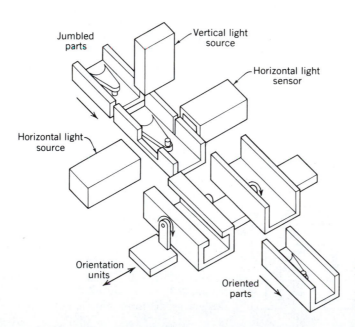

Jumbled parts

Vertical light source

Horizontal light sensor

Horizontal light source

Orientation units

Oriented parts

Figure 29.5
Use of light sources and computer choices for part orientation.

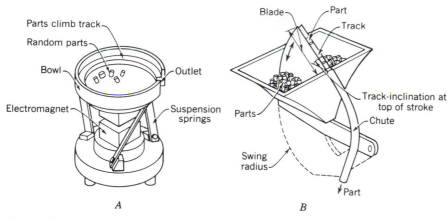

Figure 29.6
Bowl feeders. *A*, Vibratory. *B*, Centerboard.

automatic. If they do not have automatic stock loading and unloading, they are classified as semiautomatic.

The razor blade assembly is an example of individual machine automation. Figure 29.7 shows a spider that has received prior subassembly. Of the seven components in the complete assembly, three (the spider and two half-caps) make up a half-cap subassembly. These are put together on an in-line subassembly machine at a rate of 60 per minute. Spiders are fed into a chute, vibrating feeders supply half-caps to each side of the spider, half-caps are sprung into place, and subassemblies are chuted into trays, which serve as oriented storage magazines for the next machine assembly.

Complete razors, Figure 29.8, are assembled on the second machine at a rate of 30 razors per minute. This machine uses rotary indexing with work fixtures located on the perimeter. Vibratory bowl feeders feed guard extensions, outer tubes, inner tubes, and screws. The spider subassembly was previously completed and is now joined to the other units. This machine is an example of a *specific product machine automation*. This automation equipment is custom designed for a particular product.

Assembly Considerations

Assembly is fitting together individual parts to produce a product. The parts may first be combined into mini-assemblies, then subassemblies, and finally, perhaps, final assembly. Figure 29.9 is a sketch of a highly flexible assembly line, in which the parts are fitted

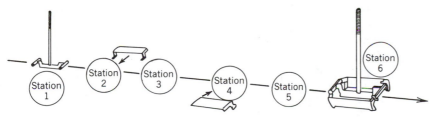

Figure 29.7
Sequence of half-cap assembly. 1, Load spiders onto track, escapes into fixture; 2 and 4, hopper feed of half-cap into spider; 3 and 5, snap half-cap onto spider; and 6, unload half-cap subassembly into magazine trays for final assembly.

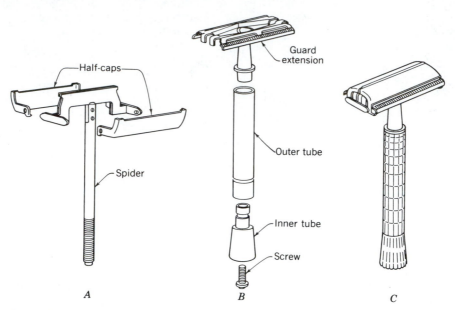

Half-caps

Spider

Guard extension

Outer tube

Inner tube

Screw

A

B

C

Figure 29.8
Two assembly machines complete this Gillette safety razor. *A*, Half-cap subassembly. *B*, Final assembly. *C*, Final product.

manually. This line is a strong candidate for automation, flexible cells, or cellular assembly if economic conditions warrant. Robots also may assist the assembly process (Figure 29.10).

In all instances assembly may be performed *in line*; that is, along a conveyor on which parts move or on pallets that ensure accurate positioning. Alternatively, the assembly may be of the rotary kind with a carousel carrying the unit from station to station.

Assembly Lines and Materials Moving. In assembling a simple or complex product, a first step is breaking down the assembly into smaller steps. This facilitates material handling by ensuring that parts are supplied in proper place and sequence. Materials handling is achieved by many methods. Notice Figure 29.11, which shows a bay crane, overhead track crane, magnetic chuck on a moveable arm, motorized truck, and other simple conveyances. Installing automatic or semiautomatic handling equipment between

Figure 29.9
Labor-intensive conveyor assembly.

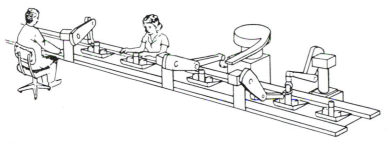

Figure 29.10
Robot-assisted placement of parts alongside manual assembly.

machines already on line is successful and permits easy introduction of automation into existing production systems. The description of some material moving systems follows.

1. *Towline or wire guidance.* In this system workpieces are attached to pallet fixtures or platforms, which are carried on carts towed by a chain located beneath the floor. The pallet fixture is designed so that it may be moved conveniently and clamped at successive machines in manufacturing cells. The advantage of this method is that the part is accurately located in the pallet, and it is correctly positioned for each machining or assembly operation.

With the *wire guidance system*, carts can move along a path determined by wire embedded in the floor. A cart picks up a finished palletized workpiece from the machining center and delivers it to an unload station elsewhere in the system.

2. *Roller conveyor.* A conveyor consisting of rotating rollers may be used throughout the factory. The conveyor transports palletized workpieces or parts that are moving at constant speed between the manufacturing cells. When a workpiece approaches the required cell, it is picked up by the robot or routed to the cell via a cross-roller conveyor. The rollers are powered either by a chain drive or by a moving belt, which provides the rotation of the rollers by friction. Figure 29.12 is an example of a gravity roller conveyor and a belt, power-driven, portable conveyor.

3. *Belt conveyor.* In this materials-moving system, either a steel belt or a chain driven by pulleys transports the parts. This system operates by three different methods. In *continuous transfer*, the workpieces are moving continuously and the processing is either performed during the motion or the cell's robot picks up the workpiece when it approaches the cell.

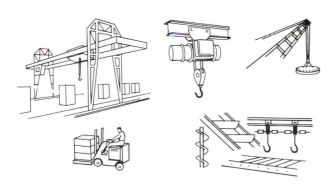

Figure 29.11
Material-handling methods.

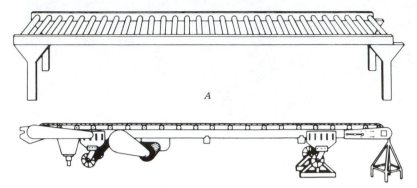

Figure 29.12
A, Roller conveyor. *B*, Belt, power-driven, and portable conveyor.

Synchronous transfer is used mainly in automatic assembly lines. The assembly stations are located with the same distance between them, and the parts to be assembled are positioned at equal distance along the conveyor. In each station a few parts are assembled by a robot or automatic device with fixed motions. The conveyor is of an indexing type; namely, it moves a short distance and stops when the product is in the station, and, subsequently, the assembly takes place simultaneously in all stations. This method is applied where station cycle times are almost equal.

Power and free material handling allows each workpiece to move independently to the next manufacturing cell for processing.

Adaptable Programming Assembly. Batch assembly is different from high-volume assembly, as described earlier by the razor example. In that case parts and subassemblies are limited and a large quantity is necessary. Outside of consumer products, however, high volume is not always characteristic. *Batch production* is more the norm, covering about 75% of all production. Batch production is labor intensive and is not easily automated. The batch sizes can range from several to hundreds of the same product.

Figure 29.13 is an assembly line to build end bell subassemblies, rotors, and stators, and represents an example of adaptable programming assembly. This project, engineered by Westinghouse and Unimation, uses fractional horsepower motors, a product Westinghouse sells by the millions and manufactures in batches of a few hundred. These motors come in 8 classes and 450 styles. A representative batch has a quantity of 600. An assembly line averages 13 style changeovers per day. Assuming an hour per changeover, there is as much time spent in setup as in cycle production.

The assembly process begins at the top center of the picture and proceeds counterclockwise. Vision is used with the robots, and transmitted infrared light is the main source. One of the first stations determines the bell end orientation, then grips it, rotates it, and places it on a special pallet that preserves the orientation through the entire assembly. A gripper design is a three-stage device using different holding techniques for various pieces it might hold.

Image processing algorithms aid the interface between the camera and computer, which necessitates preprocessing hardware to increase speed. Some vision systems are binary, that is, black and white only, without shades of gray. Variations in the reflectivity of the painted surfaces precluded the use of reflected light. Instead, the choice of infrared light source minimizes problems resulting from ambient light. The vision algorithm provides information about end bell characteristics, such as total dark area, number, location,

Figure 29.13
Adaptable programmable assembly system (APAS) facility.

and size of holes, and compares that to the library of parts. The computer instructs the robot manipulator to accept or reject and dispose in a reject bin if necessary. Other *pick-and-place robots* add a felt wick and a plastic cage. Lubricants are added later to the parts.

The system also incorporates *hard automation*, such as automatic screwdrivers fed by vibratory screw feeders. In this experimental system, the automatic screw driving was a problem area because of misalignment, wrong screws, bad threads, and bad holes. If the station had been converted to robots, these problems could have been detected by sensors.

29.3 ROBOTS

The Robot Institute of America defines the *industrial robot* as "a reprogrammable multifunctional manipulator designed to move materials, parts, tools, or other specialized devices through variable programmed motions for the performance of a variety of tasks." It is understood that an industrial robot must include an end effector, factory work, and stand-alone operation.

Robots are used in light manufacturing, casting and foundry, automotive, electrical, heavy manufacturing, and aerospace industries. Labor-intensive operations, where end effectors and sensing equipment are necessary to replace people, is a growing field of application. In the sense of operations for manufacturing processes, robots are used for palletizing, spraying, grinding, welding, deburring, searching, machine load and unload, packaging, assembly, tool carrying, and line tracking. A robot-welding cell is shown by Figure 29.14, which includes the most relevant cell components. In the forging and

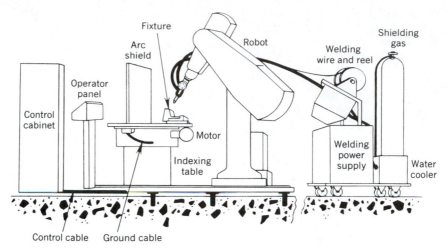

Figure 29.14
Welding cell using robot.

foundry area alone robots are used for upsetting, die forging, press forging, heat treating, loading and unloading ovens and furnaces, and for flame cutting.

Components

Robots consist of three major components, namely, the manipulator, controller, and power supply. The structure of a robot manipulator, in general, is composed of a *main frame* and a *wrist* at its end. The *manipulator* is a series of mechanical linkages and joints capable of motions in various directions to perform work. The manipulator simulates the movement of the human arm, wrist, hand, and fingers. The main frame is called the *arm*, and the most distal group of joints affecting the motion of the end effector is called the wrist. The *end effector* can be a welding head, spray gun, machining tool, or a gripper containing on–off jaws depending on the specific application of the robot.

The components permit the following classification of industrial robots:

1. *Manual manipulator,* which is worked by an operator.

2. *Fixed-sequence robot.* This manipulator performs repetitive operations according to a predetermined pattern. The set of information cannot be easily changed.

3. *Variable-sequence robot.* A manipulator that repetitively performs operations according to a predetermined set of rules whose set of information can be changed easily.

4. *Playback robot.* From memory, this manipulator produces operations originally executed under human control. A human operator initially operates the robot to establish the rules. All the information relevant to the operations, sequence, conditions, and positions is loaded into memory. This information is recalled when required, hence, the term "playback" robot. The operations are executed automatically from memory. As an example, consider an arc-welding robot that is roughly programmed to weld automobile body sections together. At the time of the production, it will be fine-tuned by operators who may hand guide it through the motions that are necessary for the compound surfaces of the body. These robots are sometimes called "teachable."

5. *NC robots.* Using media such as a punched tape and computer, this manipulator

performs a given task according to the sequence, conditions, and position as commanded via numerical data.

6. *Intelligent robot.* This robot uses vision and/or touch to notice changes in the work environment or condition. Using its own decision-making ability, the robot proceeds with its operation.

The *controller* is considered the "brain" of the robot. It stores data and directs the movement of the manipulator. Controllers can be simple or complex, but a typical controller permits storage and execution of more than one program. The controller first initiates and terminates motions of the manipulator in a desired sequence and at specific points. Second, it stores position and sequences data in memory; third, it interfaces with the manufacturing operation. *Feedback* is often a part of the controller. The controllers range from simple step sequencers through pneumatic logic circuits, electrical and electronic sequencers, microprocessors, and minicomputers. The controller may be installed physically in the manipulator or have a special cabinet.

Robots may be further classified as nonservo or servo-controlled devices. *Nonservo robots* are often referred to as end point, pick-and-place, *bang bang*, or *limited-sequence* robots. Nonservomechanism robots have directional control valves that are either fully opened or closed, thus limiting program and positioning capacity. Their arms travel only at one speed and can stop only at the end point of their axes. Significant features include relatively high speed, good repeatability to within 0.25 mm, limited flexibility in terms of program capacity, simple operation and programming, low maintenance, and comparatively low cost. These nonservo robots are hydraulic or pneumatically powered. Nonservo machines are used on high speed and precise tasks.

Servo-controlled robots have one or more servomechanism or motors that allow the arm and gripper to change direction in mid-air without having to trip a mechanical switch. They can vary speed at any point in the work envelope. The features of a servo robot include the following: ability to move heavy loads in a controlled fashion, maximum flexibility to program the manipulator to any position within the limits of travel, and more than one program may be stored and executed from memory. The end-of-arm positioning has a positioning accuracy of 1.5 mm and a repeatability of ± 1.5 mm. Figure 29.15 is a six-axis industrial robot using electrohydraulic servo-control systems.

The *end effector* is another key component of industrial robots. It is analogous to the human hand and is sometimes called the *gripper* or *end-of-arm tooling.* Various factors determine the end effector use. For example, type of power, floor layout and work envelope size, work environment, part configuration, characteristic of the robot (payload, accuracy, reach, construction), part fixturing, cycle time, and maintenance. Cost, of course, is an important consideration. Figure 29.16 illustrates end effectors.

Figure 29.16*A* is a clamp or crimp. Stud welding is possible with the tool shown in Figure 29.16*B*, and spare studs are fed down a tube. Torches for welding or heating are possibilities. A ladle for the hot and dirty job of pouring molten metals is another application. Spot-welding guns, pneumatic nut runners, drills, impact wrenches, and tool changing are favorite tools.

Paths and Coordinate Systems

Servo-controlled robots are further classified as *point to point* or *continuous path.* In many respects this classification is similar to computer numerical control systems, as discussed in Chapter 28. Point to point is used for tool and part handling. The path through which the various members of the manipulator move when traveling from point

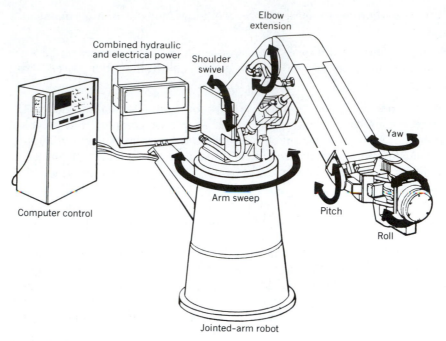

Figure 29.15
Jointed-arm robot.

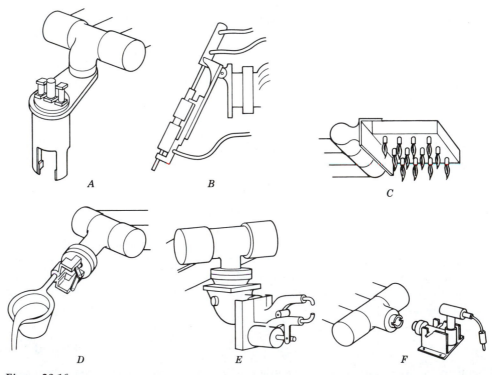

Figure 29.16
End effectors. *A*, Clamping, crimping, or nut running. *B*, Stud welding. *C*, Heating with flame torch. *D*, Pouring molten metal. *E*, Spot welding. *F*, Tool changing.

to point is not directly programmed. Instead, end points are indicated. In point to point, the servo-controlled robot is "taught" to perform its job one point at a time. A human operator positions the robot hand at a particular point in space and instructs the robot to store that position in memory. This procedure is repeated until the robot has stored in its memory the complete sequence it will be expected to perform.

With continuous path servo-controlled robots, feedback and positioning requirements are more important. To "teach" a continuous path robot its task, a human operator physically moves the robot manipulator through whatever series of motions it is expected to perform. These "learned" motions are stored in the controller for later playback.

Three common programming methods exist for teaching a robot specific motion paths: manual, lead-through, and walk-through programming. *Manual* refers to off-line method for programming either a mechanical, pneumatic, or electrical memory. It involves setting the limit switches on each axis by presetting cams on a rotating step drum, connecting air logic tubing, or prewiring appropriate connections. *Lead through* consists of maneuvering the robot arm from one path point to the next by means of a control console. *Walk-through programming* involves physically guiding the robot arm through the desired motions.

Power to the manipulator actuator is by electric, hydraulic, or pneumatic means. A robot with hydraulic power usually consists of a motor-driven pump, filter, reservoir, and heat exchanger. Remote air compressors provide the power for "air logic" robots, and this compressor may serve other requirements.

Coordinate Systems. Robots are structured in four coordinate systems: Cartesian, cylindrical, jointed spherical, and spherical. These axes of motion, sometimes called *degrees of freedom*, refer to the separate motions a robot has in its manipulator, wrist, and base.

Robots having *Cartesian coordinate motions* travel in right angle lines to each other, as shown by Figure 29.17A. There are no radial motions. The profile of the robot's work envelope is rectangular in shape. Also called "rectangular robots," these robots tend to have greater accuracy and repeatability than other types, especially for heavier loads.

A *coordinate robot*, Figure 29.17B, has a horizontal shaft that goes in and out and rides up and down on a vertical shaft, which rotates about the shaft. These two members can rotate as a unit on a base.

The *jointed spherical coordinate robot*, also called "jointed arm," can perform similar actions to a human's shoulder, arm, and elbow arrangement. In Figure 29.17C, the work envelope approximates a portion of a sphere.

A *spherical coordinate robot*, Figure 29.17D, moves so that the work envelope forms the outline of a sphere. A robot with a cylindrical coordinate system moves so that the work forms the outline of a cylinder. The spherical coordinate robot, sometimes called "polar," has a configuration similar to a tank turret. The arm moves in and out, and is raised and lowered through an arc while rotating about the base.

Robots have differing work envelopes. The *envelope* is defined as the area in space that the robot can touch with the mounting plate on the end of its arm. Note Figure 29.18, in which the top and elevation views are given. Figure 29.18A is the work envelope for the jointed arm, while Figure 29.18B provides the diagram for the spherical coordinate robot.

The wrist, which is the unit mounted on the end of the robot's arm, has a tool or gripper installed. The wrist has three basic motions. *Yaw* refers to rotation in a horizontal plane through the arm. *Pitch* refers to the rotation in a vertical plane through the arm, and *roll* is rotation in a plane perpendicular to the end of the arm.

These six basic motions, or degrees of freedom, provide the capacity to move the end

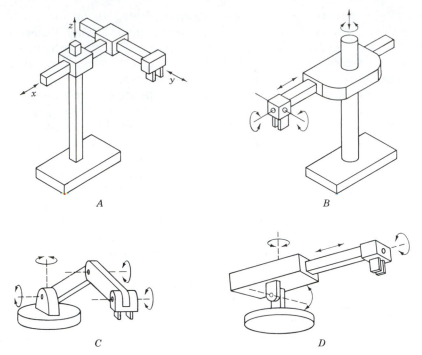

Figure 29.17
Coordinate systems. *A*, Cartesian. *B*, Cylindrical. *C*, Jointed arm. *D*, Spherical.

effector through a sequence, as shown by Figure 29.19. Not all robots are equipped with six degrees, because it may not be economically wise.

Robotic Sensors

Sensors are used to give robots more humanlike capabilities to perform a task. These capabilities include vision and hand–eye coordination, touch, and hearing. The types of sensors used in robotics fall in the following three categories: vision, tactile, and voice sensors.

Applications for sensors are several. They may be used in support and direction for robots and in confirming assembly operations, inspection, simple fixturing requirements, and providing continuous production monitoring.

Vision Sensors. Robot *vision* uses a video camera, light source, and a computer programmed to process image data. The camera is mounted either on the robot or in a fixed position above it so that its field of vision includes the robot's work volume. The computer software enables the vision system to sense the presence of an object and its position and orientations. Vision capability enables the robot to retrieve parts that are randomly oriented on a conveyor and to recognize particular parts that are intermixed with other objects. It also permits visual inspection tasks and assembly operations requiring alignment.

A simple sensor may be a *photoelectric switch*. Using light rays, it determines if a characteristic of an object is present or absent. Door opening is a popular application. Another more complicated application locates a hole and allows a fastening head to position on the hole.

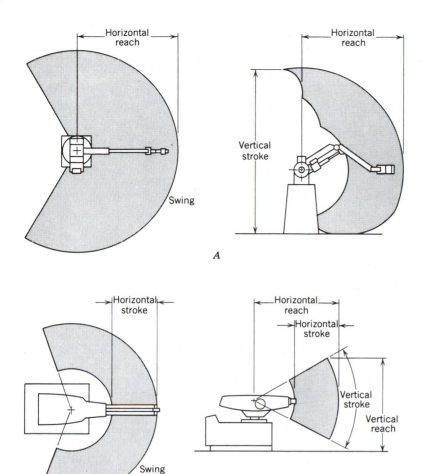

Figure 29.18
Work envelopes. *A*, For jointed arm. *B*, For spherical coordinate robot.

Machine vision is a noncontact optical technique used for a variety of automatic tasks of stationary or moving objects in two or three dimensions. It uses a high-resolution video camera, laser or incandescent light source, microcomputers, and image-processing hardware and software to determine defined product features. The process of machine vision requires that as an image is received from the part or product, it be converted to a usable form for processing by the computer. It is compared to a nominal part or standard set of information, and for inspection work a decision is made about the part regarding its acceptability.

The camera uses electronic circuitry to determine the image, so film is not needed. The circuitry is composed of an array of light-sensitive devices, referred to as *pixels* or picture elements. Each pixel produces a signal proportional to the light striking it during an image capture cycle. Cameras are either "line" or "area." Pixels are arranged in a one-dimensional array for a line camera, but the area camera is two-dimensional, which is illustrated in Figure 29.20*A*.

Both cameras may be equipped with different lensing arrangements to provide for a

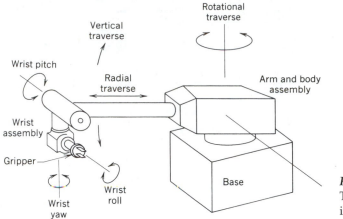

Figure 29.19
Typical six degrees of freedom in robot motion.

field of view and spatial relationships, especially true for inspection tasks. Notice Figure 29.20*B*. Resolution for line cameras varies from 64 to 4096 pixels. Area cameras vary from 64 by 64 to 240 by 320 pixels.

Lighting for vision is important. Either lasers, fluorescent or incandescent lighting, or a combination of these is arranged to create spots, shadows, narrow lines, crosshairs, parallel lines, off-angle projections, or front or back illuminations to light the product. Figure 29.20*C* shows the arrangements for lighting. These arrangements cause high contrast to make the information of interest appear as a dark or light image as compared to the rest of the information in the camera's field of vision.

There are four types of systems in the vision of parts: range, contour, surface, and feature. The range system measures the distance that a part surface is from the face of the camera or datum surface. The method is a triangulation, in which a spot of low-power laser is projected at the surface at a known angle. If the spot's image strikes the camera single-dimension array to the left or right of the normal position, the part is closer or further from the camera. The range sensor method is used to measure surface

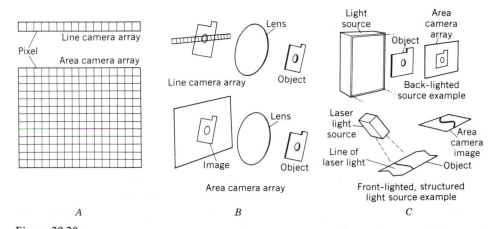

Figure 29.20
A, Electronic circuitry of vision is composed of pixels of light-sensitive devices. *B*, Field of view and lensing. *C*, Light sources create spots, shadows, and contrast light and dark regions.

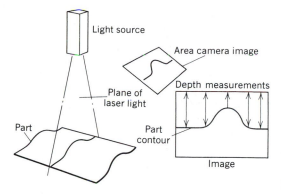

Figure 29.21
Coordinates for hundreds of two-dimensional points on the pixel array that are representative of the surface contour.

position, cavity and hole depths, or to detect part features. An area camera uses triangulation methods to define contour, as shown in Figure 29.21.

Tactile Sensors. Tactile sensors provide the robot with the capability to respond to contact forces between itself and other objects within its work volume. Figure 29.22 is a robot gripper with tactile touch for assembly work and movement of parts.

The human touch feels various sensations such as shape, pressure, temperature, and texture. Immediately, it is apparent if the object is hard, round, or pointed. The material often is identifiable by touch. Robot operations for painting, welding, drilling, and basic part handling, usually do not require touch perception, but other automatic manipulators do require tactile sensing. For example, Figure 29.23 shows a tactile sensor picking up the outline of a wrench end.

Diverse and complicated assembly requires tactile sensing, not simply touch, which means more than knowing that touch has occurred. A comprehensive ability to grope and identify shape, surface features, texture, force, and slippage are important requirements. Tactile sensing or *taction* means continuous variable measurement of forces in a large surface area. *Touch* usually means simple contact or force at a single point. Diverse approaches exist, for example:

1. *Strain gages.* Transducers that monitor the reactions at the supports to a force acting anywhere on the end-of-arm platform.

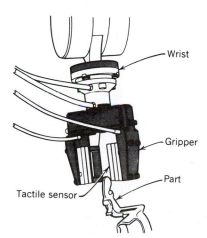

Figure 29.22
Robot gripper with tactile sensitivity.

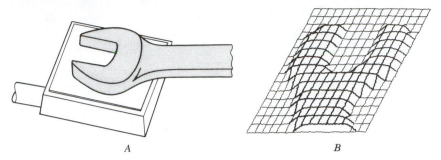

Figure 29.23
A, Tactile sensor. *B*, Outline displayed by sensor of wrench end.

2. *Silicon sensors*. Three-dimensional mechanical devices etched in silicon using specialized integrated processing techniques. The sensor measures only perpendicular forces, as it does not measure slip or shear forces. It is shaped like a box, and the forces transmitted are transduced at the silicon diaphragm. A protective layer of hard plastic 250 μm thick covers the chip, and an elastomeric cover rests on that. Notice the elements of Figure 29.24.

3. *Optical fiber-based sensor*. Fiber optic technology is displayed as a 16 by 16 array, roughly 38 mm². There are 256 individual sensing points with approximately 2.5-mm spacing. Normal forces at the surface result in deflection of the light being coupled into the 256 receiving fibers, yielding a corresponding reduction in the electrical output of the light detectors. This information is processed so that shape, location, size, or weight are found in real time.

4. *Conductive elastomer*. This material measures the change of resistance based on the surface area contact. The outer layer is a conductive surface that seals the sensor from the environment. Beneath this surface are conductive V-shaped rows that change resistance when depressed. These rows have *piezoelectric properties*. A disadvantage to piezoelectric materials is that response to pressure is transient.

Voice Sensors. Voice programming is defined as the oral communication of commands to the robot or other machine. Voice and sound recognition permit the system to recognize the symptoms of equipment malfunction, such as air leaks, abnormal vibrations, or broken drill bits. Talking robots warn factory personnel of machine breakdowns, rejected parts, and even low inventory levels.

The integration of the robot into individual manufacturing processes requires systems engineering. While robots are available off-the-shelf, a robotic system is not. It must be

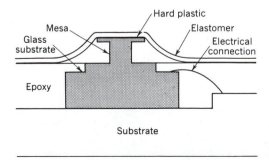

Figure 29.24
Single-element tactile sensor made by micromachining technology allowing mesa, diaphragm, and connections to be made from a silicon wafer.

carefully planned and customized to each application. In a typical application the robot represents about 30%–40% of total system cost. Engineering and peripheral equipment might comprise 60%–70% of the total cost. Yet, these nonrobot costs may yield a greater return on investment than the robot itself.

29.4 GROUP TECHNOLOGY

Group technology (GT) is a management philosophy based on the recognition that similarities exist in design and manufacture of discrete parts. In "family of parts manufacturing," group technology achieves advantages on the basis of these similarities. Similar parts are arranged into part families. For example, a plant producing many part numbers of shafts, chassis, and castings will group the parts into families of their physical features, such as shafts. Each family possesses similar design and manufacturing characteristics. Efficiencies result from reduced setup times, lower in-process inventories, better scheduling, streamlined material flow, improved quality, improved tool control, and the use of standardized process plans. In some plants where GT has been implemented, the production equipment is arranged into machine groups or cells to facilitate work flow and parts handling. In product design there are also advantages obtained by grouping parts into families. A parts classification and coding system is required in a design retrieval system. Group technology is a prerequisite for computer-integrated manufacturing.

Parts classification and coding are concerned with identifying the similarities among parts and relating these similarities to a coding or a number system. Part similarities are of two types: *design attributes* (such as geometric shape and size), and *manufacturing attributes* (the sequence of processing steps required to make the part). It should be noted that GT is not a science with precise formulas but, rather, is a tool to be developed in each plant.

The coding developed for GT is used in computer-aided process planning (CAPP), which involves the computer generation of an operations sheet or route sheet to manufacture the part. The operations sheet is discussed in Chapter 23.

Part Families

A *part family* is a collection of parts that are similar either because of geometric shape and size or because similar processing steps are required in their manufacture. The parts within a family are different, but are close enough to merit identification as members of the part family.

Figure 29.25 illustrates a collection of similar shafts. The parts, used as drill ends for mining drills, have many similar details, materials, and dimensions. These parts are grouped sequentially for manufacturing and, thus, reduce setup, cycle time, inventory, and improve shop efficiency.

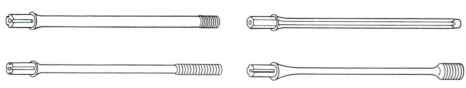

Figure 29.25
Sample of grouped parts.

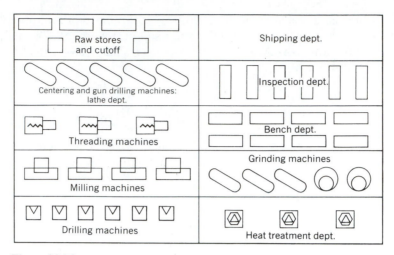

Figure 29.26
Typical plant layout by machine-type classification.

The obstacle in changing over to GT from a traditional production shop is the problem of grouping parts into families. Methods to solve this problem are visual inspection, production flow analysis, and parts classification and coding system.

Grouping parts into families involves an examination of the individual design and/or manufacturing characteristics of each part. The attributes of the part are identified uniquely by means of an alphanumeric code number. This classification and coding may be carried out on the entire list of active parts of the firm, or a sampling process may be used to establish the part families.

Group Technology for Cell Layout

One variation of GT includes the concept of GT machine cells, groups of machines arranged to produce similar part families. A cellular arrangement of production equipment achieves an efficient work flow within the *cell*. Labor and machine specialization

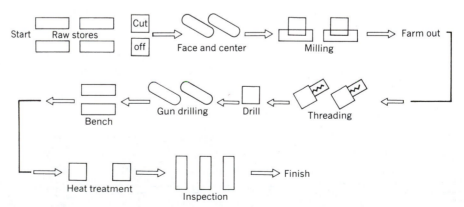

Figure 29.27
Group technology machining center layout.

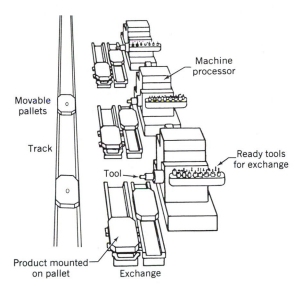

Movable pallets

Track

Tool

Machine processor

Ready tools for exchange

Product mounted on pallet

Exchange

Figure 29.28
In-line manufacturing cell with movable pallets.

for the particular part families produced by the cell are possible, raising the total productivity of the cell.

Figure 29.26 is a typical plant layout by *machine type* classification. If a part is moved between these departments, the shop path would appear as spaghetti or as a random walk. The logic to this type of layout is that the skills to operate the equipment are clustered and the operators can, perhaps, interchange among the equipment. A group technology layout for common parts is given in Figure 29.27. This pattern reduces in-process storage, the distance parts are being moved, and indirect labor costs of handling.

One limitation of the *flow line layout* requires that all parts in the family be processed through the machines in the same sequence. While some processing steps can be ignored, it is necessary that the flow of work through the system be unidirectional. Reversal of work flow is accommodated in the more flexible group machine layout, but not conveniently in the flow line configuration, as shown by Figure 29.28.

29.5 FLEXIBLE MANUFACTURING SYSTEMS

Flexible manufacturing systems (FMS) are arrangements of individual workstations, cells, machining cells, and robots under the control of a computer. Workpieces are mounted on pallets, which move through the system transferred by towlines, conveyors, or drag chains. Operator interdiction is discouraged by FMS. As jobs are changed, the computer is reprogrammed to handle new requirements.

Flexible manufacturing systems are closely related to cellular systems. Equipment and manufacturing cells are located along the material transfer highway. Different parts move on the conveyor and generally the quantity is small. The workpieces are complex, however, and can require complicated manufacturing steps. The parts along the highway will not need all the equipment or cells. Production of the various parts requires processing by different combinations of manufacturing, but FMS is versatile and can perform different operations on a variety of products.

Often a FMS machine can perform many processing steps. The process begins with a robot or operator loading or unloading a CNC machine in FMS. Notice Figure 29.29.

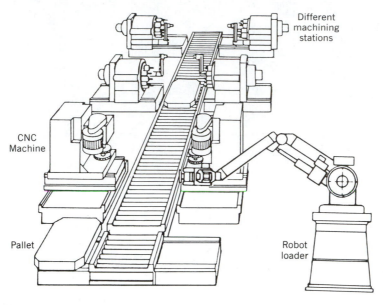

Different machining stations

CNC Machine

Pallet

Robot loader

Figure 29.29
Flexible manufacturing system with robot-handling system.

After processing in FMS, the robot returns the semifinished or finished part to the conveyor.

Pallet transfer to and from machines is done by *automatically guided vehicles* (AGVs). These carriers are often rated by the tonnage capacity they are able to move. Guidance of the vehicles is by several methods. A hidden wire, tracks, or electronic targeting may be used to guide the vehicles.

The moving pallets play synthesized tunes through loud speakers to warn pedestrians as they move about the factory floor, and tunes are changed from time to time. These unmanned pallets can move the part to a central area where special work is conducted. Some loading stations prepare work for the night shift, and these are inventoried for accessibility on the pallets by robots. The fixtures aboard the pallet are designed for a specific class of parts such as prismatic, shafts, or plate, for example. Automatic-loading palletizers are used to mount and unmount the part on the vehicle.

Not only are parts moved but also tools are moved between the stations if a part requires tooling that is not available on a specific machine. The robots mount the tool into the carriage, where it is available for the part at the time it is needed. The robot arms are able to reach the tool and extend it the required distance.

For *unmanned machines* or factories it is necessary to control chips, as cleanliness is important. Various means are used to keep pallets clean. Designs are available to dump the pallets, blow off the chips, or use vacuum systems to suck up even the smallest chips. Chips are often conveyed in underfloor tubes, where they are gathered eventually in a large cyclone for collection, cleaning if economical, and compacting and recycling, if possible.

Tool breakage for the unmanned machines is controlled by several means. Probes on each machine check for tool breakage, and spare tools are contained in the magazine if necessary. In some cases acoustic emission monitoring checks for broken tools. When tools need to be changed, they are sometimes supplied from a central location, where they have been preset. Changing may be automatic and follow tool life criteria rather than a breakage alarm. This tool life is programmed, and then maintenance follows the tool life requirements as instructed by the central computer. If the tool breaks while machining, the tool and part are removed for later inspection. The operation continues, however, with replacement of a new part on the machine.

FMS is integrated with computer-aided design (CAD) and manufacturing. CAD, for example, limits the number of tools to a preset number, such that the factory does not store more than a specific number. Another approach finds the number of tools and then reduces that number by cost control methods. Standardization of tools, their kind and quantity, and specifications are a natural development of FMS.

The economic objective of FMS is to approach the efficiency of mass production for *batch production*. Batch production exists whenever the number of parts manufactured occurs in lots ranging from several units to many. Annual demand may require several lots or batches. There is a setup and a teardown for the general-purpose equipment, which prepares for the design, then readies the equipment for the next lot. *Mass production* occurs when high annual volumes of production utilize special-purpose equipment. Setup approaches a zero time per unit for mass production quantities.

In some FMS experiences, many of the parts are *nonrecurring*, that is, parts that are done only once and there is little chance of repeated production. However, many companies report paybacks of about 4 years. Payback is usually defined to be gross investment divided by net annual savings or

$$\text{Years payback} = \frac{\text{Investment in manufacturing system}}{\text{Net annual savings}} \qquad (29.2)$$

This is a crude measure of desirability and is used because of its ease of understanding. Engineering economic methods that find the rate of return or net present worth are preferred, and that subject is deferred to those textbooks.

29.6 OTHER PRODUCTION SYSTEMS

A *production system* includes only converting equipment, such as machines, processes, benches, and tools; therefore, it is more limiting than a manufacturing system. Production can be classified according to the conceptual model shown in Figure 29.30, in which product variety and quantity suggest differing modes of production systems. According to this model, production systems are divided into six categories: workstations, cells, flexible workstations, mechanization, automation, and continuous flow processes.

Workstation

This production design is the most primitive production system. Chapters 7 through 11 consider individual *workstations*. It is characterized by individual bench, machine, or discrete processes. Craft work, single stations, or numerical control machine tools working independently are typical examples.

Bench work or manual assembly is an example of a simple workstation. Manual assembly is still an important application for small quantity production. "Ones and twos" are not unknown quantities in prototype work, and these circumstances dictate human assembly. These workstations may have little to no numerical control or computer involvement. With larger-production quantities there is a danger of human error in production, and costs of labor can be significant. Numerical controlled machine tools then become the workstation. Because of the low overall efficiency of single workstations, there are numerous attempts to connect them to a system.

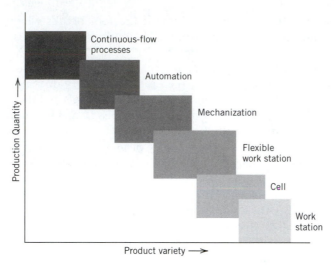

Figure 29.30
Concepts of production systems
for product variety and quantity.

Cell Production

This is the simplest organized production effort composed of two or more workstations. It may or may not be computer-controlled or robot-assisted. Cell production can be thought of as "garage style" production located in a bigger plant. While the comparison is not always apt, it suggests several characteristics. It is the next higher level of manufacturing workstations as compared to a single workstation. Cell production may involve several machines connected by a conveyor system. If it is a specialized cell, it will make only certain classes of parts. In some situations, cell design may approach the efficiency of an automated transfer line.

Cell production encourages the following benefits: for simple product design fewer and more straightforward tooling and setups, shorter material handling distances, and less complicated production and inventory control. Improved process planning procedures are possible.

Unlike "garage style" production, there is often a high technology core within the cell. Several CNC machine tools are arranged within the reaching arm radius of a robot. A *robot slave* is responsible for part handling. The cell computer supervises and coordinates the various operations, which makes it compatible to the direct numerical control (DNC) computer. In developmental experiences, the core runs 24 hours continuously and requires workers' participation only during the day. The day shift loads the computer program, enabling the production scheduling, and may do secondary or noncell operations, such as heat treatment, mounting part on special pallets, unloading pallets, and cleaning and removal of chips.

During a part-processing activity by the machine in a cell, the robot can perform housekeeping functions such as chip removal, staging tools in the tool changer, and inspection of tools for breakage or excessive wear. A more sophisticated robot will alter its functions, depending on the outcome of these tests or the presence of any alert situations during this housekeeping. For example, if machine control detects tool breakage during a machine operation, the robot quits the routine tasks and corrects the problem or initiates an emergency procedure for the total system.

Cell Layout. As cells are a higher order over a single machine or processing unit, cell or plant layout becomes a consideration. *Plant layout* is the physical arrangement of all

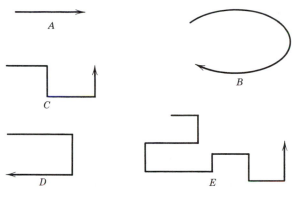

Figure 29.31
Basic layout plans. *A*, Straight line. *B*,
Circular. *C*, S-shaped. *D*, U-shaped.
E, Random.

facilities within the factory. Like all planning acts, plant layout decisions depend on product design and specifications, production volume, manufacturing operations for the product, and assembly sequence for the product. The best layout will attempt to use the least space consistent with safety, comfort, and product manufacture. It will consider operations, inspections, delays, transportation (material handling), and storage. Figure 29.31 provides basic plant and *cellular layouts*, and uses one of the following patterns: *straight line*, *S-shaped*, *U-shaped*, *circular*, and *random angle*. The type employed depends on many factors, and the existing shape of the building is a prominent one.

Cells need not be in a circle, as a variety of layouts are possible. Nor do cell layouts require that the cells be far from each other. The cells may be stationed along an in-line material transfer system, such as a conveyor. Raw and intermediate finished parts move along the conveyor. A "ready for workpiece" signal from the control unit of the first machine in a manufacturing cell instructs the robot to look for the required workpiece on the conveyor. The robot picks up the workpiece, loads it onto the machine, and sends a signal to the machine control to begin its operations on the workpiece. A "part finished" signal from the last machine tool to the robot requests that the completed part be unloaded and transferred to the outgoing conveyor. The cycle would then be repeated.

A machine cell is able to provide machining operations. The machines, fixtures, and tools are arranged for efficient flow of work parts through the cell. A single machine approach can be used for work parts whose design allows them to be made using one type of process, such as a turning or milling machine center. The group machine layout is a cell design in which several machines are used together with no provision for conveyorized parts movement between the machines. Manual operations may be substituted for the robot-assisted conveyors.

Planning for production in a manufacturing cell is different than for a single machine tool. In a manufacturing cell, only one machine can operate with a cutting speed derived by the conventional minimum production time criterion or with an adaptive control system, which maximizes the material removal rate. Such a cell drifts down to the slowest workstation in the production process.

Flexible Workstations

A flexible workstation (FWS) usually consists of a CNC machining center and one or two special machines, which are linked with a tool store through an industrial robot. This type of production system is a kind of automation that fills the gap between FMSs and individual stand-alone CNC machine tools. Volume of production has less effect, and the use of the computer is characteristic.

The flexibility inherent in the design of a FWS allows for high productivity when parts come in small lots and in a variety of part configurations. Lots of 30 to 400 pieces of 20 to 300 different shapes are common. Examples include low quantities of engine blocks.

Mechanization

Mechanization is a well-developed production system. For example, automatic screw machines (which may be cam and drum-driven) have been available for a century. Transfer dies in press work is also a highly developed form of mechanization.

Traditionally, this type of production system has not involved the computer or adaptive controls, although sensors, electronic and electrical controls are common. Significant mechanization is found in dedicated production of large quantities of one product with little model variation. Examples include high volume and automobile parts.

Many examples of small batch mechanization are found in job shops, die-making shops, and in the aerospace and machine tool industries. The accuracy of the final product depends not only on the machine but also on the skill of the operator who has to set up the workpiece. The task is facilitated by *jigs*, which hold the workpiece and incorporate a guide for the tool, and *fixtures*, which firmly hold one or more workpieces in the correct position in relation to the machine tool bed. Machining or processing then proceeds. Machining operations are discussed in Chapter 11.

The use of several tools in a machine to perform several operations (either simultaneously or in sequence) saves time over individual machining elements and is a form of mechanization. Mechanization requires automatic workpiece handling and tooling. Handling will sequence a workpiece through a work cycle, perhaps "index and travel" the part from one location to another. Clamping and locating of the workpiece occur at the workstation, and machining follows.

A variety of design configurations is possible. The movement of the workpiece may be circular around the machine or linear along the machine if one machine is used. Circular movement uses less floor area. Straight-line type machines allow addition and subtraction of stations, thus facilitating interchangeability and continuous chip or waste removal. Indexing is most suited for drilling, tapping, and turning operations. In constant travel mechanization the workpiece is advanced with indexing in either a circular or straight-line path, and locating and clamping the workpiece is required only once. Constant travel mechanization is preferred for milling, broaching, and grinding operations. Observe Figure 29.2 for three types of mechanization.

Some assembly operations lend themselves to simple mechanical methods. Thus, screws or bolts can be driven and parts placed, crimped, or riveted with mechanical devices. Complicated machines are available, for example, to manufacture completely and surgically hospital syringes in large daily volume. The cost is reduced, whereas productivity and consistency of product increases, but only if the reliability of mechanization can be ensured. The cost of down time repairs can cancel all savings. In-line inspection to pinpoint and remove imperfect assemblies, either during or at the end of the assembly operation, is an important element for complicated mechanization operations. Many of the elements used in automatic assembly are the same as in mechanized production and workpiece handling, and may be mechanical, electromechanical, fluidic, or numerically or computer controlled.

Automation

Automation was discussed in detail in Section 29.2. Examples of this type of production system include transfer lines, which may or may not be computer controlled. More recent automation includes robots, which are used in many processes such as arc welding and parts handling.

Robots will continue to develop in the decades ahead to support automation systems. A Multi-Armed Assembly Robot, nicknamed the "*troikabot*," performs a variety of complicated assembly tasks at a single workstation. Figure 29.32 is a CAM-generated solid model showing the work cell. This model has not been completely realized, but research and development are continuing. For example, the plans provide for three interacting six degree of freedom arms to perform three-dimensional assembly tasks without fixtures. These assembly tasks would include parts mating, insertion, and use of gravity and friction forces. The robots use vision, electrical grippers, and a sensor mitt that is acquired and used by the arms.

The goal of the troikabot is a practical work cell extending robots to assembly processes. The robot recognizes that the "factory of the future" will be closely integrated such that order taking, CAD, CAM, inventory control, assembly, inspection, packaging and delivery will be automatic. Second, the factory of the future will have the flexibility to manufacture a variety of products and be capable of changing products from time to time without major changes in equipment or arrangement. Finally, the factory will have the ability to plan and perform complex assembly tasks. The troikabot is a research and development effort to anticipate a comprehensive work cell composed of robots.

The troikabot is a highly integrated work cell that has three interactive robot manipulators, vision, multiple interactive sensors, and advanced electromechanical components, such as sensor wrists, electronic grippers, and end-effector mitts. The grippers acquire these mitts and thereby extend their ability. The troikabot is programmed off-line and uses a hierarchical *real-time control system*. Three arms were chosen to permit free space assembly with simple grippers. One robot holds the subassembly of the components, while a second robot acquires a second component and inserts it into the subassembly. A third robot is required to clamp or regrip the subassembly so that the other robot can release its grippers without the danger of the subassembly falling apart.

Figure 29.32
CAM-generated solid model depicting the troikabot assembly work cell.

Continuous Flow Processes

Examples include production of bulk product, such as chemical plants and oil refineries. Features are: flow process from beginning to end, sensor technology available to measure important process variables, use of sophisticated control and optimization strategies, and full computer control.

29.7 ECONOMIC CONSIDERATIONS

Competition in high-quantity production requires the greatest productive efficiency. Automation has traditionally concentrated on the manufacturing operation and its equipment for high volume. Today, however, computer-aided design and computer-aided manufacturing (CAD/*CAM*) includes the manufacturing operations, design, and planning functions that precede the actual production of the product; thus the need for computer-integrated manufacturing systems. Although high volume was the first recipient of this technology, mini, personal, and mainframe computers are involved with cell and flexible workstations.

In planning for CAM, it is necessary to outline the essential requirements that must be met. *Designing the product* with CAM and robots concentrates on part geometry for easier automatic handling, feeding, locating, holding, loading, and unloading, and many management functions. In addition, the basic components of CAM are described in detail and are called specifications. These include both special and commercially available feeders, selectors, loaders, unloaders, transfer devices, indexing equipment, vision systems, and numerous others.

Is CAM really necessary? If a manufacturer hopes to stay in business over the long term, the firm must steadily improve manufacturing methods. Whenever a manufacturer can produce a comparable or better product at lower cost than a competitor, the firm will prosper. The selection of the CAM system depends on planning and economics.

If a company automates one of its assembly lines, the savings in assembly time must pay for the expense of the new machinery. By comparing the efficiency of a given process with the expected efficiency realized with automation, it is possible to examine the benefits.

To develop an efficiency index, operations are divided into primary and secondary assembly operations. *Primary operations* are those that add value to the assembly, such as painting, fitting parts to the assembly, and packaging. All other necessary operations, such as rotating and moving, that do not add value to the assembly are secondary. Using primary and secondary assembly operations, an overall efficiency for an assembly process is rationalized as follows

$$E = [P/(P + S)] \times 100 \tag{29.3}$$

where

E = The *process efficiency*, %
P = Total time to complete primary operations
S = Total time to complete secondary operations

In the economic evaluation step, a number of strategies are possible. Consider the situation where there is a minimum product variety but a large quantity of output. Automation may be a choice but full automation may not be economical. Instead, *partial automation*, with the addition of automatic devices until *full automation* or until the point of diminishing economic returns has been reached, is preferred. It is possible to describe levels of automation that range from an automatic fixture at a single machine, or mechanization, to full automation. Part variety may be a dominant requirement and with low volume, flexible manufacturing systems may be the choice.

The choice of the production system depends on many factors. Very important, however, is the objective of *profit maximization* or *cost minimization*. These ideas are discussed further in Chapters 23 and 31, but the motive of profit is strong in the selection of a production system.

Planning at this stage includes the economic considerations of the design, quantity, equipment, plant layout of equipment, computers, programming tool change time, maintenance, load and unload cycles, and other normal production interruptions. Cycle time per station is required, type and number of machines employed, and the cost of both direct and indirect labor for both automated and conventional methods are among the considerations made.

QUESTIONS

1. Give an explanation of the following terms:

Analytical solutions	Pixels
Numerical methods	Conductive elastomer
Discrete event simulation	Part family
Stochastic	Cell
Automation	Pallet transfer
Bowl feeders	Workstations
Wire guidance system	S-shaped
Synchronous transfer	Mechanization
Pick-and-place robots	Troikabot
Manipulator	Real-time control system
Variable-sequence robot	CAM
Nonservo robots	Process efficiency
Walk-through programming	Partial automation

2. You are required to estimate the average time spent by a job in a system composed of a single server. What data do you need to do your job?

3. Explain the difference among discrete, dynamic, and continuous simulation.

4. In the general scheme of computerized visual feeding system, what steps follow separation and holding of a single product component?

5. Referring to robotic sensors, what is the difference between taction and touch?

6. What are the advantages and disadvantages of individual workstations?

7. If you were designing automation equipment, would you stress primary or secondary operations? Can automation be conducted without secondary operations?

8. State the classification of cost for the following items: Machine operator, raw material ending up in the consumer product, robot, technician, machine operator, and engineer.

9. What products are customarily produced with continuous flow products processes?

10. Describe various layouts for cell production.

11. List various ways of moving the following materials: thousands of electronic wafers, several printed circuit boards, cast iron automobile engines, turbine shroud, fractional horsepower motors, and light bulbs.

12. How does mechanization differ from automation?

13. Develop a list of practices for preventive maintenance on automation equipment producing razors.

14. Describe various methods to orient parts.

15. What are the possible ways for parts to be disoriented?

16. Sketch an optical means of ensuring the orientation of a toothbrush before it is packaged.

17. What is the purpose of image-processing algorithms in vision orientation systems?

18. Describe the types of vision systems.

19. For vision application, is it important to have bright surfaces, contrast, ambient light, or no outside light?

20. What are some difficulties in assembly of products?

21. Define a robot and list several industrial applications.

22. What is a manipulator, wrist, controller, end effector, and control system for a robot?

23. Contrast the coordinate systems used for robots.

24. List the main differences between a nonservo and a servo-type robot.

25. What function does an error signal have in control of robots?

26. Write a paragraph description for handling parts, such as printed circuit boards, automobile wheel hubs, Christmas tree ornaments, fruit, survey stakes, and clothing. Consider tactile sensors and their role in the handling. Assume a large volume of product to be handled.

27. Write a newspaper article of 250 words describing the troikabot. Indicate its functions, purpose, and role in production.

28. Describe various methods of assigning group technology code numbers for parts that are washerlike, except they differ in thicknesses and both inside and outside diameters. Also compensate for material. Use a four-digit number, then a five-digit number.

PROBLEMS

29.1. You are required to perform an assembly process for less than 500 units. Would you choose a magazine loading system or a robot arm loading system?

29.2. List the manufacturing system you might adopt for the production of a consumer item where production is

a. 2 million units

b. 200,000 units

c. 20,000 units

d. 2,000 units

e. 200 units

29.3. Jobs arrive at a press punch at a rate of eight jobs/ h. If each job is processed in approximately 7 min, and waiting time is exponentially distributed, estimate the long-run average time that a job spends in the system.

29.4. An automation cycle consists of seven operations using 17, 11, 14, 18, 21, 16, and 23 s. Indexing time between the stations is 14 s. If there are no means of providing for inventory between the stations, what is the process efficiency? Which operation is limiting in this in-line sequence?

29.5. An automation system is suggested as a replacement for a manual system. The first cost for the automatic equipment is $5 million, and annual net saving is anticipated to be $725,000. What is the payback?

29.6. The total budget for the automation system is $5 million. What amounts can be expected for robot equipment? For engineering and peripheral equipment?

29.7. List the steps for mechanizing the production of a wire paper clip with the following specifications: three U-bends, 0.037-in. wire, prebent length 3.675 in., and ductile cold-rolled AISI 1008 steel.

29.8. Material-handling equipment is expected to cost $80,000 installed. Annual net saving is $18,000. What is the payback?

29.9. Automation equipment produces bulletlike containers and the four steps require 14 s each. There are five handling steps each 8 s in duration. What is the process efficiency?

MORE DIFFICULT PROBLEMS

29.10. Production schedule demands that 14 jobs be manually machined by a single turret lathe during an 8-h. shift. If jobs arrive at a rate of 10 per h, and each job is processed in approximately 5 min,

a. Can this single-server system handle the requirements? Assume that waiting time is exponentially distributed and that the server is 100% efficient.

b. If the machine is 2% less efficient every month, after how long would you need to adjust the system? How would you adjust it? Justify your answers.

PRACTICAL APPLICATION

Review a manufacturing magazine (SME's or Modern Machine Shop, for example) and identify two CAD/CAM systems. Write to the makers of the systems to request product literature and then evaluate the two systems. Look into the programming functions that are supported, the graphic exchange standards used, the hardware platforms supported, and the postprocessors available. Document your findings in a report.

CASE STUDY: THE ROUND PLATE COMPANY

The Round Plate Company specializes in round plates. This company has standardized its product into two families shown as A and B in Figure C29.1. These plates are used for many applications, such as bearings plates, post rests, flanges, bases, and any engineered design that starts with round bar stock in 12- and 10-in. OD and ends with these general shapes. Operations are limited to turning, drilling, tapping, reaming, step milling, and grinding.

Mr. Sander Friedman, owner of the shop, tells any potential customer, "I can win any bid starting with standard bar stock in two sizes and concluding with no more features than shown on the sketches." He has been talking to you about investing in his company.

A registered and legal financial prospectus about Round Plate suggests that the capitalization using shareholder's investment will result in a modern factory of the future designed for vertical market penetration in the round plate business. The prospectus gives various details about the product and factory. A summary is: A sketch is given of the product, shown by Figure C29.1. Dimensions on the sketches are maximum, and other dimensions may be roughly scaled to approximate missing dimensions. These maximum dimensions are useful as limitations for standard pallets, fixtures, grippers, and robots required in manipulation and loading of machines. Bolt circles are an important part of the company's operation.

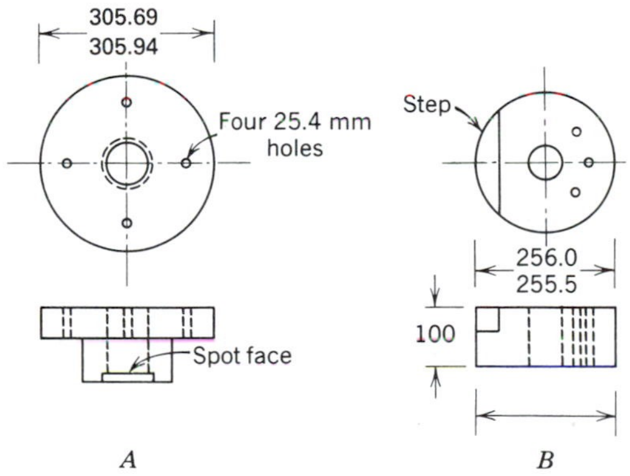

Round Plate's operation is a garage-style cell with CNC machines, robotic operation for loading, and nighttime supervision and inspection. According to the prospectus, Round Plate will employ the president, maintenance operator, toolroom grinder, computer programmer, manufacturing engineer, robot maintainer and teacher, accountant and bill collector, and finally a materials handler for shipping, raw stock receiving, and chip and waste removal.

"We will operate the night shift without anybody," boasts Friedman. "Even the lights are out," he says. "This is a good time to buy shares." He looks at you expectantly.

A pause and then you reply. "Well, everything sounds pretty good, and I do like your idea. But I don't buy into businesses without first sleeping on it. I'll see you tomorrow and we will go over some details."

That night you begin to ponder the offer. You mutter, "What are the critical requirements?" Continue your questioning attitude and develop a series of questions for Mr. Friedman. Develop your comments along these general groupings: (1) systems engineering; (2) equipment and ancillary support in tooling, material movement, controls, and vision; (3) part family and group technology; and (4) marketing strategy and business factors. Prepare these questions in a report.

Figure C29.1
Maximum dimensions (in mm) and general part family configuration for Round Plate Company's product. A, Standard plates. B, One-step plates.

CHAPTER 30
OPERATOR-MACHINE SYSTEMS

Interest in various aspects of operator-machine systems has increased recently in many areas, including education, medicine, computing, and manufacturing. The interaction between operators and machines, the interface between workers and equipment, the understanding of manufacturing as an interactive and evolving system, the impact of ergonomics and safety regulations, as well as the revolution that intelligent systems bring into the manufacturing enterprise, are matters addressed by this subject.

The price of a workstation or a hand tool, for example, is only a very modest share of the total production cost. One has to make allowance for several other factors, such as high absence because of sickness, unnecessary disruptions, and rejects. With safety practices and ergonomically designed systems, many of these problems can be minimized and, thus, lower manufacturing costs. The purpose of this chapter is to give the student an understanding of how hand tools and workstations can be designed to avoid unnecessary strain on the operator. Important safety and ergonomics issues needed for proper system design are addressed. The chapter concludes with a discussion of the impact that artificial intelligence techniques have in the human-machine interaction.

30.1 OPERATOR-MACHINE SYSTEMS STRUCTURE

Figure 30.1 illustrates a simplified scheme of an operator-machine system. Here, the term "machine" is used generically to refer to tools, workstations, and other hardware.

Machines are designed to achieve certain goals performing at some given specification levels. From the specifications a functional description in the form of block diagrams can be developed. Finally, the physical form of the tool, workstation, or machine components is defined. Three levels of the description of a machine can be distinguished:

1. A *symbolic* representation describing the purpose and goals for which the machine is constructed. Such representation is traditionally done in natural language.
2. A *functional* representation, which describes the functional layout of the machine.
3. A *physical* representation of parts and components of the machine.

Furthermore, the interaction between an operator and a machine takes place at three different activity levels:

1. *Skill-based activities*, which are carried out almost subconsciously and automatically by the natural "sensory motor subroutines" of the operator. Skillful workers react to the signals and the physical state of the machine.
2. *Rule-based activities* cover all situations that have been considered in the design process. These interactions are driven by "stored rules" that have been learned during the training of the operator or in former experiences. Reactions at this interface level are evoked by clear signs or rules of the type: "If machine overheats, then lower frequency."
3. *Knowledge-based activities* deal with problem-solving and planning. In case of unexpected events, such as system failure, no rules may exist. Operators must rely on

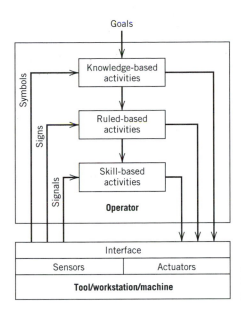

Figure 30.1
Simplified description of a conventional operator-machine system.

their knowledge about the goal to be achieved by the machine, about the functional layout, and about the physical form of the machine. This knowledge and the concepts the operator develops are known as symbols, which require sophisticated techniques for representation.

At the core of these systems is the operator and its environment. The interaction between them and the impact of tools, machines, and processes on the worker is the study of safety and ergonomics.

30.2 ERGONOMICS

In today's competitive market, the drive to achieve high productivity with high-quality production is strong. Many companies now realize the important contribution that ergonomic workstations can make toward achieving this goal.

Ergonomics is the study of the interaction between the worker and the process at the workplace. When discussing tool design, only one aspect of the working environment is being considered. However, it can be said that in a given workstation an *ergonomic tool* is one that allows people to do their appointed tasks in the most efficient way. Workstations and working practices must be designed to avoid exposing the operator to unnecessary physical load, noise, vibration, and dust. Hand tools are important because they form the direct link between the worker and the process.

Cumulative Trauma Disorders

Work injuries to the musculoskeletal system are very common. More workers are disabled from such problems than from any other category of disorder. It is estimated that the average employed person loses nearly 2 days of work each year because of this type of injury. *Cumulative trauma disorders* (*CTDs*), also known as repetitive stress injuries, are work-related musculoskeletal impairments caused by repetitive and straining motions. *Cumulative* indicates that these injuries develop gradually, *trauma* signifies bodily injury from mechanical stresses, and *disorders* refers to physical ailments or abnormal conditions.

The human body has great recuperative powers if a sufficient interval of rest time is allowed between episodes of high usage. When the recovery time is insufficient, and when high repetition is combined with forceful and awkward postures, the worker is at risk of developing a CTD. The primary signs of CTDs include loss of strength, numbness, inflammation of tendons, soreness, achiness in neck, shoulders, arms, and hands.

The Working Arm

Motion and leverage for the arms are provided by ligaments and tendons at three major joints: wrist, elbow, and shoulder. Many muscles, tendons, ligaments, and nerves form an exceedingly complex system of pulleys and canals that move the 32 bones that make up the arm and hand. Any of these components is at risk of presenting a CTD. A few of the most common disorders are described in Table 30.1.

Figure 30.2 highlights some of the main components of the working arm. Muscles, which are not usually attached to bones but to tendons, are composed of thousands of

Table 30.1   **Definition of the Most Common Cumulative Trauma Disorders**

Code	CTD Name	Description
TO	Thoracic outlet	General term for compression of the nerves and blood vessels between the neck and shoulder.
CTS	Carpal tunnel syndrome	The tendons for flexing the fingers, the median nerve and blood vessels pass through the carpal tunnel. If any of the tendon sheaths become swollen, the median nerve may be pinched.
UNE	Ulnar nerve entrapment	The ulnar nerve runs along the ulna (the bone on the little finger side of the lower arm). Sustained elbow flexion with pressure on the ulnar groove interferes with normal sensations felt on the palm and back hand areas.
TW/S	Tendonitis (of the wrist and/or of the shoulder)	It is a form of tendon inflammation that occurs when a muscle/tendon unit is repeatedly tensed. With further exertion, some of the fibers can fray or tear apart. Tendon becomes thickened, bumpy, and irregular. Tendons without sheaths, such as in the shoulder, may cause the injured area to calcify.
TS	Tenosynovitis	Extreme repetition (more than 1500 movements per min) stimulates the synovial sheath to produce excessive synovial fluid. The fluid accumulates and the sheath becomes swollen and painful.
DQ	De Quervain's	Specific type of tenosynovitis that affects the tendons on the side of the wrist and at the base of the thumb.
TN	Tension neck	The three major nerves that feed into the arm and hand (ulnar, median, and radial) originate in the neck passing under the clavicle. Prolonged restricted postures and heavy loads on shoulders press these nerves resulting in numbness and tingling.
PT	Pronator teres	Inflammation of the tendon attached to the muscle of the forearm that turns the hand with the palm down.
EC	Epicondylitis	Also known as golfer's elbow. It is an irritation of the tendon attachments of the finger flexor muscles on the inside of the elbow.

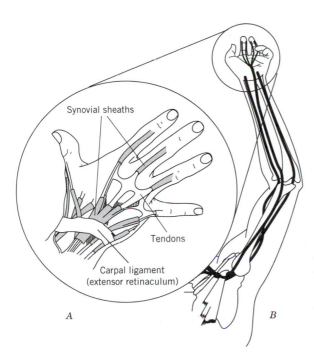

Figure 30.2
A, A pictorial view of the finger tendons, their sheaths, and the carpal ligament in the hand. *B*, Pathway traced by three major nerves that originate in the neck and lead into the arm and hand: ulnar, median, and radial.

tiny fibers, all running in the same direction. *Muscles* are red because they are filled with many blood vessels that supply oxygen and nutrients, and carry away carbon dioxide and waste materials. *Tendons* are made of tough, ropelike material that is smooth, white, and shiny. Tendons do not stretch; they just transfer forces and movements from the muscles to the bones. When tendons tear, a *strain* takes place. *Ligaments* are strong, ropelike fibers that connect one bone to another to form a joint. Their function is to bind the bones and limit the range of joint motion. When they tear, a *sprain* takes place.

Jobs and CTD

Upper-extremity CTDs are associated with work activity in a variety of industries. Table 30.2 lists a sample of manufacturing jobs describing the occupational factors that lead to some common CTDs. The majority of the occupational factors in Table 30.2 are categorized as involving one or more of the following components: awkward *postures* of the wrist or shoulders, excessive manual *force*, and high rates of manual *repetition*. For example, even mechanical stresses on hand tissue, a common CTD in heavy manual work, can be traced to combinations of forceful gripping and repetitive pounding with hands.

With the proliferation of assembly-line techniques, the ever-increasing production time, and the widespread use of vibrating and air-powered tools, CTDs are common in industrial life. The need for ergonomically designed tools and workstations is very crucial for successful manufacturing systems.

30.3 DESIGNING ERGONOMIC TOOLS

All tools should be designed to have a high power-to-weight ratio. Heavy tools not only put a higher load on the wrist but they also make precision tasks more difficult. This section looks at the design of the different handles and explains their applications.

Table 30.2   **Some Manufacturing Jobs and Their Occupational Risk Factors**

Type of Job (Disorder Code[a])	Occupational Factors
Buffing/grinding (TS, TO, CTS, DQ, PT, TW/S)	Repetitive wrist motions, prolonged flexed shoulders, vibration, forceful ulnar deviation, repetitive forearm turns.
Punch press (TW/S)	Repetitive forceful wrist extension/flexion, repetitive shoulder flexion, forearm turns.
Overhead assembly (DQ, TO, TW/S)	Repetitive ulnar deviation in pushing controls. Sustained hyperextension of arms. Hands above shoulders.
Belt-conveyor assembly (TW/S, CTS, TO)	Arms extended or flexed more than 60°. Repetitive, forceful wrist motions.
Typing, keypunch (TN, TO, CTS)	Static, restricted posture, arms extended/flexed, high-speed finger movement, palmar base pressure, ulnar deviation.
Small parts assembly (DQ, CTS, TN, TO, TW/S, EC)	Repetitive wrist flexion/extension, palmar base pressure. Prolonged restricted posture, forceful ulnar deviation and thumb pressure. Repetitive wrist motion, forceful wrist extension and turns.
Bench work (UNE)	Sustained elbow flexion with pressure on ulnar groove.
Packing (TW/S, TN, CTS, DQ)	Prolonged load on shoulders, repetitive wrist motions, overexertion, forceful ulnar deviation.
Stockroom, shipping (TO, TW/S)	Reaching overhead. Prolonged load on shoulder in unnatural position.
Materials handling (TO, TW/S)	Carrying heavy load on shoulders.

[a] Refer to Table 30.1.

Static Forces

Handles and Grips. The handle is the interface between the worker and the tool. All forces associated with the task are transmitted between the hand and the handle. An ergonomic handle design is vital in producing a tool that places the minimum strain on the operator. The optimum position for transmitting forces between hand and arm is to ensure the wrist is kept straight. In addition, it is important to reduce, as far as possible, the twisting and bending forces applied to the wrist.

Bow Handle. The bow handle is used to transmit high-feed forces without causing a torque in the wrist. To achieve this the handle should be designed to ensure the feed force is transmitted through the centerline of arm and wrist. The natural angle of the hand grip is independent of hand size. This angle is approximately 70°. By making the handle to this angle, the muscles stabilizing the wrist are in their optimal position. The limitation of this design is that the weight of the tool is in front of the hand. If used as a single-handed tool, it would put a high bending force on the wrist, which is undesirable. This handle is ideal for chipping hammers and large rivet hammers. See Figure 30.3.

Pistol Handles. The pistol handle is used to shorten the length of the tool, reducing the bending load on the wrist. Keeping the distance between hand and workpiece as short as possible allows precision tasks to be performed efficiently. For tasks requiring high-feed forces the handle needs to be angled at 70°. Where only twisting forces occur at the center of the tools, weight should be on top of the handle. This lowers the bending

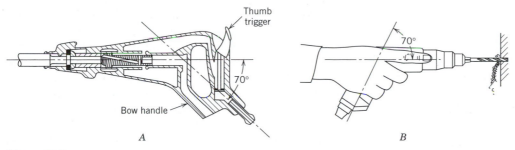

Figure 30.3
A, Chipping hammers. *B,* Pistol handles.

force on the wrist. A properly designed handle makes the work easier in that the feed forces and/or resist twisting forces are transmitted gently to the hand. See Figure 30.3.

Straight Handles. The diameter of a handle that when held allows the strongest grip, and causes the minimum strain in the hand, has been found to be a 1.5-in. (38-mm) handle for men and 1.34-in. (34-mm) handle for women. Consideration to the probable sex of the worker should be given when designing a tool. The next aspect to consider is whether the tool produces a twisting torque in the hand (screwdrivers and drills) or produces a levering torque on the wrist and arm (grinders and angle nutrunners).

Screwdrivers and Drills. For tools used in light assembly work a diameter of 1.34 in. (34 mm) is desired. This helps workers with a light grip. When grip tends to be stronger, workers are less affected by a small deviation from their optimum diameter. For applications, such as drilling, where workers with strong grip dominate, a 1.5-in. (38-mm) handle diameter should be used. By keeping the optimal diameter and by making the handle from a high friction surface, which also allows ventilation of the grip, higher torques can be taken comfortably by the worker. A further improvement is gained by tapering the handle to allow all fingers to grip equally.

For ease of movement and to minimize loading of the wrist, the tool should be kept short with the center of weight where the tool is gripped. Keeping the distance between hand and workpiece as short as possible speeds precision tasks.

Grinders and Angle Nutrunners. In the case of angle nutrunners, particularly on tools with a high torque capacity it is important to ensure a secure grip on the tool. As these tools are mainly used by workers with a strong grip, a handle diameter of 1.5 in. (38 mm) is most suitable.

With grinders the forces on the hand and wrist are lower than with other tools, but they are used over longer time periods. The comfort level for workers can be improved by making the handle from an insulating material, protecting the hand from the cold temperature of the expanding exhaust air. Vertical grinders that require two-handed operation can benefit from adjustable handles. This allows the worker to adjust the grip to suit the task.

Finger Triggers. Finger triggers are used mainly on pistol handle tools where the trigger can be operated without affecting the grip on the handle. The trigger should be designed to ensure that it is pressed with the middle part of the finger. If the tip of the finger is used, tendon damage can result. This type of trigger gives very precise control over the speed of the tool. To give the worker the right "feel" when using the tool, the trigger should be designed so that to increase speed the pressure on the trigger must be increased. Finger triggers are used for frequent operations, precision tasks, and where the tool needs to be positioned before it is started. See Figure 30.3. The same figure illustrates *thumb triggers*, which can be operated while maintaining a firm grip on the tool. It is particularly suitable for tools, such as chipping hammers, where high feed forces are used. In this application the trigger operation becomes part of the feed force.

Other Triggers.   Figure 30.4 illustrates other triggers. The operation of *strip triggers* is achieved with low trigger forces. All fingers are used, minimizing strain in the hand. With this type of trigger, the grip on the tool is unstable. However, it is ideal for nonprecision repetitive functions, as found in blow gun applications. *Push start* triggers are used primarily on straight screwdrivers, where it is ideal for repetitive high-volume operations. The trigger is activated by putting a feed force on the tool. A stable grip is maintained on the handle at all times. With the *lever start* the trigger operation becomes part of the grip on the handle, giving great stability. It is ideal for long-cycle operations and for tasks that place high forces on the tool. The lever start enables the tool to be positioned prior to start-up, and a safety device can be built into the trigger without affecting its operation. The lever start triggers are used on grinders, nutrunners, drills, and screwdrivers.

Torque Reaction in Fastening Tools

Torque reaction is most apparent in fastening tools. It is caused by the turning force, required to tighten a screw or bolt, and applying an equal and opposite force on the handle of the tool. Figure 30.5 shows torque applied and torque reactions in the handle for four different tools.

Stall Tools.   With this type of tool the fastener is connected directly to the motor through a gear train. All the torque applied to the fastener must be reacted by the operator holding the handle of the tool. For certain applications a ratchet clutch can be put into

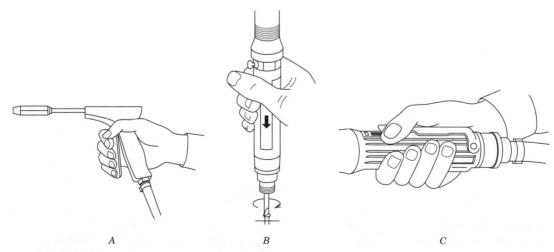

A   *B*   *C*

Figure 30.4
Triggers. *A*, Strip trigger. *B*, Push start. *C*, Lever start.

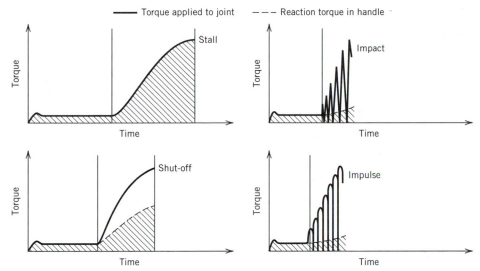

Figure 30.5
Graphs showing torque applied to a fastener and torque reaction in handle generated by the different tools on the same joint.

the drive line; this has little effect on the torque reaction. These tools need to be relatively large and heavy. The maximum practical torque that can be transmitted without a reaction bar is around 73.8 lb-ft (100 Nm).

Impact Wrenches. The tightening mechanism consists of a rotary hammer and an anvil on which the fastener is located. During run down the hammer and anvil rotate as one. When resistance is met, the anvil is stopped. However, the rotary hammer continues to be turned by the air motor. Further torque is applied to the fastener when the rotary hammer hits the anvil. The only torque reaction occurring in the handle is that caused in accelerating the hammer up to speed. These are very fast tools and can be used to very high torques (5000 Nm +) without a reaction bar.

Impulse Tools. The impulse tool works on the same principle as the impact wrench, but the mechanical impact mechanism is replaced by a hydraulic unit. These are compact tools that can achieve high torques (500 Nm) without reaction bars.

Shutoff Tools. These tools have a clutch mechanism that shuts off the air supply to the motor at a preset torque level, which is built into the drive line. With a fast-acting clutch the inherent inertia in the tool takes a large percentage of the reaction force (on most joint conditions). Although not as compact as the impact/impulse tools, they are relatively small and fast compared to a stall tool.

Balancers. Where possible, tools should be fixed to balancers to reduce the strain on the worker, to prevent damage to the tool and to keep the work area tidy. There are two types of balancers: standard spring balancer and constant tension balancer.

The *standard spring balancer* applies a force on the tool that retracts the tool to a standby position. This force increases the further the wire is extended. Every time the tool is used, the worker has to overcome this load. The *constant tension balancer* applies a constant force on the tool regardless of wire extension, which reduces the load on the worker. A drawback with this balancer is that the tool has to be consciously moved to the standby position.

Sound

Sound is the transmission of mechanical waves in matter. This process transmits energy through the media, and the *intensity of sound waves* passing can be estimated as

$$I = \text{Intensity} = \text{Power/Area} \qquad (30.1)$$

The ear's ability to detect variations in air pressure determines the audible range, which is expressed in terms of *relative intensity* or *power level* and is measured in units called decibels (dB).

$$\text{Relative intensity (dB)} = 10 \log(I/I_o) \qquad (30.2)$$

where

I = Intensity level of interest, in W/m^2
I_o = Intensity of sound at the threshold of human hearing = 10^{-12} W/m^2

EXAMPLE 30.1

To find the intensity in W/m^2 of a 50 dB sound, one uses Equation (30.2)

$$50 = 10 \log(I/10^{-12})$$

solving for I, the intensity of interest corresponds to 10^{-7} W/m^2.

The sensitivity to sound is dependent on the frequency, also known as *pitch*, of pressure variation. When measuring sound levels, a filter simulating the characteristic of the ear is used. This filter is called the A filter. As shown in Figure 30.6, noise is usually described in dB(A), indicating the filter has been used in sound measurement.

Continued exposure to high sound levels destroys the sound sensitive hair cells in the inner ear, which reduces hearing ability. To a certain extent the ear can protect itself against loud noise by dampening the movement of the bones in the middle ear. As this mechanism takes about 0.04 of a second to activate, impulse noises from tools, such as impact wrenches and rivet guns, can be particularly damaging. As hearing is one of the brain's main senses, even relatively low levels of noise can be distracting. This can lead to loss of concentration and fatigue.

Health and Safety Regulations. The health and safety regulations are most concerned with overall noise levels in the workplace. Although brief exposure to very high noise levels can cause pain and a permanent loss of hearing, it is prolonged exposure to relatively high noise levels that affects the majority of workers. Maximum noise levels defined by the U.S. Occupational Safety and Health Act (*OSHA*) for selected machinery are tabulated in Figure 30.6.

Most countries have similar legislation on noise levels. For a 40-hour week the average noise level must not exceed 85 dB(A) and the maximum noise level must not exceed 115 dB(A) at any time. The average noise level can be defined as that value of continuous

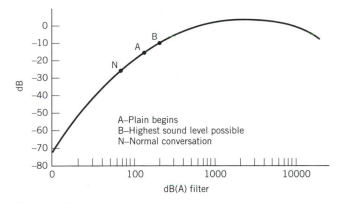

Machine type	dB(A)/hr/day
Riveting	125
Lathes and grinders	110
Punch press	106
Screw machines	104
Milling machines	95

Permissible noise exposures (OSHA)

Figure 30.6
Relative intensity of sound curve and OSHA limits for selected machines.

noise that produces the same sound energy as the varying noise levels produced through the working period.

Legislation also states a *halving level*, which is an average level of noise that reduces the allowable exposure time by half. In the case of some industrialized countries the "halving level" is 88 dB(A). This is a good indication of how significant even a small increase in noise level can be and why tools should be made as quiet as possible.

Sources of Noise in Pneumatic Tools. Most pneumatic tool manufacturers use the *Pneurop Cagi* test method to measure the noise levels of their product. This method involves measuring the noise level (in an anechoic chamber) at 5 points around the tool. From this, the noise level for the tool is obtained. There are three main sources of noise associated with pneumatic tools: process, airflow, and vibration-induced noise.

1. Process noise is caused by contact between machine and workpiece. With operations such as grinding, riveting and chipping, it is very difficult to reduce process noise levels. In many cases, the best solution is to isolate these processes from the rest of the factory and ensure that the workers in that area wear ear protection.

A fastening tool that produces high process noise levels is the impact wrench. Despite this, these tools remain popular because of their high torque capacity and low reaction force. A solution to the noise problem has been found in the hydraulic impulse tool. As shown in Figure 30.7, the mechanical impact mechanism has been replaced with a hydraulic impulse unit, which generates a smoother torque pulse and causes much lower process noise levels.

2. Airflow noise can be either "pulsating air flow" (caused when a vane passes an exhaust port and the trapped volume of air expands into the exhaust) or "aerodynamic noise" (generated when air at high velocity passes through a hole). A simple but a very effective method of reducing air flow noise is to pipe the exhaust air away from the working area. To reduce noise levels further a silencer can be put into the exhaust pipe.

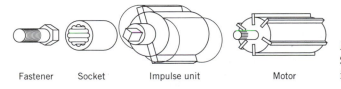

Fastener   Socket   Impulse unit   Motor

Figure 30.7
Simplified layout of a hydraulic impulse tool.

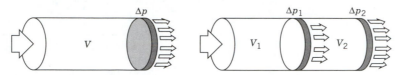

Figure 30.8
Air flow noise reduction.

To reach very low noise levels a two-stage silencer can be used, as shown in Figure 30.8. A large back pressure (which smoothes pressure pulses) is achieved by reducing flow area. However, this produces a high air flow velocity at exit. To reduce this, the air is allowed to expand into a second chamber (thereby reducing pressure) before being exhausted to the atmosphere through diffusing holes.

3. Vibration-induced noise occurs when the pulsating air flow sets up a vibration in the tool casing. The movement of the casing acts like a speaker and creates sound energy. There are two methods of reducing this noise. The first method used on conventional metal-bodied tools is to isolate the handle from the motor drive line by means of rubber suspension. The second method is to use materials such as nylon reinforced plastics, which are nonresonating.

Vibrations in Hand Tools

Vibration is a means of describing how a body is accelerated by oscillating external or internal forces. These oscillating forces have different frequencies, depending on their source. Exposure to excessive levels of vibration can cause vascular injury, can damage nerve cells, and can cause damage to the bones at the joints.

Vascular injury is where the arteries in the fingers thicken, reducing flow area for the blood. The condition is irreversible. This injury becomes apparent under cold temperatures where the body reduces blood supply to the extremities. Because of the restricted flow path, the blood cannot pass through the affected areas, and sensation is lost. The affected fingers appear white; this phenomena is called vibration-induced *white finger*. Nerve cells can be damaged by vibration and is associated with loss of sensitivity in the fingers. Although not permanent, this can affect the efficiency of the worker. Bone damage is caused by using tools that require high-feed forces. The vibration is passed through the hand and arm, causing wear and even fracturing at the joints.

Health and Safety Regulations. It is difficult to quantify the time exposure, the amplitude, and the frequency of vibration that would cause the above injuries. The most in-depth analysis of vibration-induced injuries is the ISO paper 5349 of the International Standards Organization (*ISO*). However, the study can be used as a guide only because, since many factors influence the effect of vibration, it uses very narrow parameters to determine acceptable vibration levels and exposure time.

Reducing Vibration in Hand Tools

There are many tools that generate vibrations in their operation. In this section, the methods of minimizing the effect on the operator are discussed.

Impact and Impulse Wrenches. The vibrations in the handle of these tools is small; however, incorrect use can cause the worker to experience high vibration levels. The main source of vibration is from the socket. To reduce vibration the tool and socket should be kept in line, a worn-out socket should not be used, and the worker should never hold the socket.

Percussive Tools. With all percussive tools there is a potential vibration source from the movement of the piston and from the changing air pressure that powers it. In a conventional percussive tool, this vibration is transferred to the handle. A high level of vibration is generated in the chisel, which is caused when the piston strikes the chisel. This vibration cannot be removed; therefore, the chisel should never be held.

Chipping Hammers. Vibration levels in the handle of a chipping hammer can be reduced by using a "differential piston" percussive mechanism. With this design the air pressure acting on the handle remains almost constant and, hence, the vibrations in the handle are smaller.

Riveting Hammer. For this application very precise control over tool speed and feed force is required. A successful design uses a conventional percussive mechanism, which is isolated from the handle of the tool. The movement of the percussive mechanism is controlled by an air cushion between it and the handle. Placing a feed force on the handle of the tool increases pressure in the air cushion, which increases the force on the percussive mechanism. This way precise tool control is maintained and the vibrations in the percussive mechanism are not transferred to the handle.

Scaling Hammer. For scaling a percussive tool is not required. By fixing the chisel to the piston and by allowing the cylinder to move in the opposite direction, the tool can be designed so that the forces caused in the movement of these two masses approximately cancels each other out. Therefore, the vibration levels generated in the handle of the tool are low.

Grinders. As grinders tend to be used over long periods, it is important to reduce vibration levels as much as possible. The greatest source of vibration is in the cutting wheel. It is important to always use a well-balanced wheel.

Die Grinders. Vibrations are caused by out-of-balance cutting burrs and by the grinding process itself. A method of reducing vibration in the handle is to mount the motor and drive line in a weak suspension, which in this case is rubber elements. The vibrations are absorbed in the suspension and are not transmitted to the handle.

Angle Grinders. The vibrations have the same source as the die grinder, but because of the angle gear a weak suspension cannot be used at the front of the drive line. In this case a tuned support handle can be used. The pin (through which the handle is fixed to the tool) is tuned to the mass of the handle so that vibrations in the tool do not excite the handle.

Vertical Grinders. With vertical grinders one method of reducing vibration levels is to fix the handles to the tool body through rubber elements. These rubber elements, as in the die grinder, act as a weak suspension and do not pass all vibrations generated in the tool on to the handles.

Dust Control

The dust from many materials is inert and causes no permanent damage, but it can make working conditions difficult and unpleasant, which can affect productivity. The dust from other materials, however, is harmful and diseases, such as cancer, silicoses, and asbestos and toxic poisoning can result. The size of dust particles that are of most concern are those below 5 microns (5×10^{-6} m). Particles of this size can penetrate the body's defenses and reach the oxygen transfer areas in the lungs.

Health and Safety Legislation. Every country has its own national threshold values of allowable dust levels for different materials. These levels and new materials are constantly under review. There are methods to ensure that the worker is not exposed to dust when using hand tools. The workbench can have extraction openings connected to a strong vacuum source, a movable extraction hood can be used, and the tool itself can be fitted with an extraction system. In extreme cases, the worker can wear breathing apparatus as an additional protection.

Of the dust collection methods, dust extraction on the tool is most efficient, as collection is at the point of dust creation. This not only ensures very efficient extraction but also allows a relatively low power vacuum source to be used.

Extraction systems for some selected tools are:

1. *Disc sanders.* On all rotating cutting tools, heavier particles are propelled at a tangent from the disc; as the particle size reduces, the dust tends to follow the rotation of the disc. It is the smaller particles that are most harmful. To remove the lighter particles efficiently the extraction should conform to the periphery of the disc as much as possible.

There are two basic systems used on disc sanders: the rotating and the fixed hood types. The rotating hood, as the name suggests, rotates with the disc and is flexible so it can bend with the disc. This hood sits on spacers over the disc, so the suction gap is constant. The hood is available in two forms. The first type has an overlapping lip that traps the heavier particles as well as extracting smaller ones. The second type has an open lip that allows sanding into corners. The fixed hood type has the extraction hood fixed to the tool. This has a plastic or a brush edge to trap heavier particles. There is a cut out on the front of the hood that allows sanding into corners and gives good visibility. This design has improved suction in the area that dust is formed.

2. *Orbital and random orbital sanders.* On an orbital sander, dust extraction is made through holes in the pad. As these tools tend to be used on flat surfaces, the dust is kept under the pad. Random orbital sanders have suction around the periphery of the disc. On larger diameters discs holes are put in the disc to extract dust from under the disc.

3. *Grinders. Depressed center grinders* use the same principle for grinding as on the fixed hood for sanders. *Straight grinders* are equipped with a shield placed in front of the wheel. This traps the dust, which can then be extracted. In *die grinding* the burr is enclosed by the hood, apart from a cut out to give access to the workpiece. This gives dust extraction over the largest possible area. *Diamond-tipped grinding* is illustrated in Figure 30.9*A*. In this case the dust is extracted through the center of the drum.

4. *Other machines.* A flexible hood is used with a *drilling machine*, which totally encloses the drill bit during process, as shown in Figure 30.9*B*. It is designed so that the hood can be pulled back to help locate the drill. It is almost impossible to offer a high efficiency extraction system on a *chipping hammer*, because of the difficulty in getting an extraction hood near the point of dust generation. Also, because of the different sizes

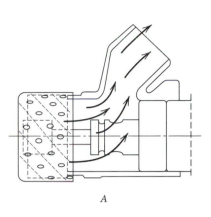

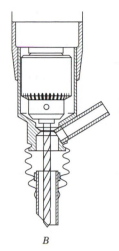

A

B

Figure 30.9
Dust extraction. *A*, Dia-
mond-tipped grinding. *B*,
Drilling machine.

of pieces removed, an extraction hood can easily be blocked. The best compromise is to use a flexible rubber hose through which the chisel protrudes. This gives the flexibility of positioning the hood in the best position.

An alternate approach to on-the-tool dust collection is the vacuum source system. There are basically three choices of vacuum source: self-generating vacuum source, portable vacuum source, and central vacuum source.

Tools with a *self-generating vacuum* have the lowest set-up cost and offer great flexibility. However, the extraction strength is low compared with the other methods. *Portable units* provide a strong vacuum for one or two tools. They are ideal for small workshops and for flexible working conditions. The initial set-up cost is low, but the units take up work space and the vacuum unit must be kept close to the tool. A *central vacuum source* is used where there are fixed workstations. This can provide a very strong vacuum and the dust is easily collected. The system keeps the work area clear. However, the initial set-up cost is high.

For efficient dust extraction the designed flow rate through the extraction hood must be maintained. For hand tools, this is usually between 1.47 and 2.45 ft³/s (150 to 250 m³/h). When selecting the vacuum source, the number of tools that will be used, and their required extraction rate, must be known. In addition, the losses in the pipes and hoses must be calculated.

EXAMPLE 30.2

Determining Required Vacuum Pressure

Assume a single tool that requires a flow of 1.96 ft³/s (200 m³/h) for dust extraction. The tool is connected to a vacuum source through a 16-ft, 1½-in. diameter pipe. A 1½-in. pipe will cause a pressure drop of approximately 35 psi (see Problem 30.8). In addition, there will be a further pressure drop generated in the extraction hood, couplings, filters, and the like. This will be around 9.6 psi. Therefore, for this application the vacuum source must maintain an extraction rate of 1.96 ft³/s (200 m³/h) at 44.6 psi.

30.4 REDESIGNING WORKSTATIONS

In addition to the noise, dust, and vibration issues, two more factors must be considered when designing workstations: illumination levels and carpal tunnel syndrome prevention.

Illumination affects worker performance. If it is inadequate, fatigue or damage to eyesight can occur. The quantity of light with which a work surface is illuminated is measured in *lux*. The following are recommended ranges of illumination: working areas where visual tasks are occasionally performed, 100–200 lux; rough assembly work, 200–500 lux; detailed inspection work, 1000–2000 lux; circuit board repair, 10,000–20,000 lux.

The majority of the guidelines that follow are intended for achieving ergonomic control over cumulative trauma. They have one or more of the following three objectives:

1. Reduction of extreme joint movement
2. Reduction of excessive force levels
3. Reduction of highly repetitive and stereotyped movements

What follows is a discussion of the principles and recommendations of ergonomic experts on the prevention of CTDs.

Reduction of Extreme Joint Movement

Since excessive stress on joints and tendons is a principal cause of CTDs, motions that are performed often should be kept well within the range of motion of that joint. Work activities should ideally be performed with the joints at about the midpoint of their range of movement. When force is being applied by the hand, the wrist should be kept straight and the elbow bent at a right angle. All side-to-side deviations of the wrist should be avoided. The hands should be kept in line with the forearms.

At least three methods exist for reducing deviations of the wrist: (1) altering the tool or controls (e.g., bending the tool or handle instead of the wrist), (2) moving the part (e.g., rotating the part in front of the worker so the wrist can be straight), and (3) moving the worker (e.g., changing the position of the worker in relation to the part). Figure 30.10 illustrates proper wrist postures used with various hand tools and workstation layouts.

Reduction of Excessive Force

Tasks requiring prolonged and excessive muscle contractions to maintain a posture (or to assemble a part or hold an ill-fitting tool) contribute to the development of CTDs.

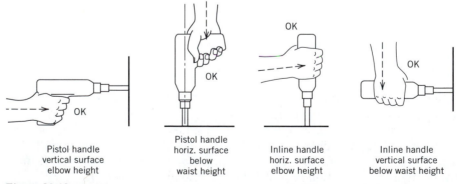

| Pistol handle vertical surface elbow height | Pistol handle horiz. surface below waist height | Inline handle horiz. surface elbow height | Inline handle vertical surface below waist height |

Figure 30.10
Example of wrist postures used with various hand tools and workstation layouts.

Jobs, therefore, should not require workers to exert more than 30% of their maximum force for a particular muscle, in a prolonged or repetitive way. All muscular contractions, including occasional ones, in excess of 50% of the maximum should be avoided.

Exertion level or fatigue also can be minimized by considering the relationship between load and duration. Simply decreasing the required effort (load) by as little as 10% will allow work to continue at a constant level for a period that exceeds the original duration by a factor of five or six. In essence, work fatigue is influenced more by load than duration. The ratio of load to duration may be as high as $10:1$, respectively.

Available charts and tables provide a general range of strength for specific muscle groups in men and women of different age groups and worker stature. Three general approaches to controlling job forces are: (1) reducing the force required (use weaker springs in triggers, power with motors rather than muscles, use jigs and clamps instead of hands to grip parts), (2) spreading the force (use trigger levers rather than single-finger push buttons and allow the worker to alternate hands), and (3) getting better mechanical advantage (use stronger muscle groups and use tools with longer handles).

Reduction of Highly Repetitive Movements

Since highly repetitive and stereotyped manual movements contribute to CTDs, potentially aggravating production and design factors must be identified and altered to reduce the repetitive levels of a work cycle. Countermeasures include limiting the duration of continuous work or restructuring of work methods.

Several approaches may be taken to reduce rates of repetition: (1) *Task enlargement.* Restructure jobs so that each worker has a larger and more varied number of tasks to perform. This must include a corresponding increase in job-cycle time. (2) *Mechanization.* The use of special tools with ratchet devices or power drivers can reduce stressful repetition. (3) *Automation.* Repetitive tasks are performed best by a machine. To be cost-effective, this must typically involve a high-volume, long-term production process.

General Guidelines for Workstations

Improvements in workplace design are usually the result of trial and error. This section provides some guidelines based on ergonomic principles for planning the layout or modifications of workstations to minimize the stressful effects of repetitive and static work patterns.

1. Design workstations to accommodate different people. An ergonomic workstation should accommodate a vast majority of the people who work on a given job and not merely the average. A workstation that is adjustable and is either designed or selected to fit a specific task should be experienced as relatively comfortable by 90% to 95% of the worker population. The work space also should be large enough to accommodate the full range of required movements. For the seated operator, certain space requirements should be observed. Minimum requirements are given in Figures 30.11 and 30.12.

Data for design decisions can come from a variety of sources: (1) *Anthropometric tables* offer extensive listings of the size measurements and proportions of the human body as distributed in the adult population and are especially helpful at the drawing board stage of a new workstation. (2) *Fitting trials* use actual workers in an inexpensive cardboard mock-up of a possible workstation. This mock-up trial allows for adjustments of key dimensions for each subject so that an optimum final design may be made.

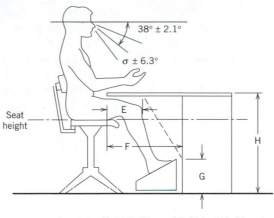

Leg room: 1) depth (E)   17–20 in. (43–51 cm)
          2) depth (F)   20–26 in. (51–66 cm)
          2) footroom (G)   1–10 in. (3–25 cm)
Work surface height: 3) (H) 25–32 in. (64–81 cm)

Figure 30.11
Work space design specifications for seated worker. Side view.

(3) *Fatigue measurements* in the laboratory and factory involve using scientific instruments to measure muscle activity and force.

2. Permit several different working positions. To avoid static loading of muscles, good workstation design should permit the worker to adopt several different but equally healthy and safe postures that still allow performance of the job. Ideally, a worker should be able to choose either a sitting or standing position. Arm rests and foot rests should be supplied when appropriate. Sometimes the shape of the tool handle may have to be changed to maintain a neutral wrist position when the worker assumes different positions.

3. Design should start from the working point, where the hands spend most of their time. Frequent work should be kept within the area that can be conveniently reached

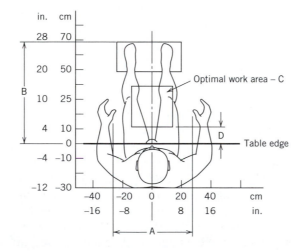

Leg room: 1) width (A)   Minimum of 20 in. (50 cm)
          2) depth (B)   26 in. (65 cm) if limited
Optimum work area: 3) (C)   Approximately 10 × 10 in. (25 × 25 cm)
          4) (D)   4 in. (10 cm) from edge of work surface

Figure 30.12
Work space design specifications for seated worker. Top view.

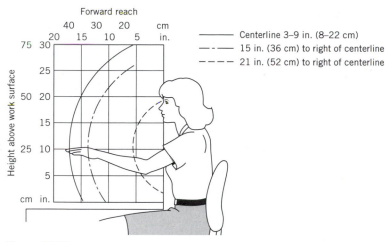

Figure 30.13
Graphic aid for estimating reach distances for sitting or standing
(average worker).

by the sweep of the forearm, with the upper arm hanging in a natural position at the side of the trunk. The area that can be reached by extending the arm from the shoulder will vary greatly, depending on the size of the worker.

If the worker has to use the hand to grip an object, the maximum reach of the arm should be reduced by 2 in. (5 cm) or more. An occasional stretch to reach beyond this range is permissible, since the momentary effect on shoulders and trunk is transient. However, if this becomes a permanent part of a job, forearm or shoulder fatigue may be experienced. In general, the design should allow the worker to maintain an upright and forward-facing posture during work. Figure 30.13 provides a graph layout of average reach distances for wrist, thumb, and fingertips for a sitting position.

4. Place controls, tools, and materials between shoulder and waist height to be easily reached and manipulated. Avoid reaching above shoulder level or behind the body. All reaching should be below and in front of the shoulder. There should be a definite and fixed space for all tools and materials. Their location should be close to the point of use, and they should be arranged so that they permit the best sequence of motions.

30.5 JOB ANALYSIS

Job analysis is useful to evaluate improvements in operator-machine systems, to redesign tools and workstations, to identify safety needs, and to define sources of CTDs. One approach to job analysis is the *work-methods analysis*. To conduct a work-methods analysis, the job is described as a set of tasks, each task is defined in terms of a series of steps or elements (*Therbligs*), and each element is described as fundamental movements or acts required to perform a job. Tasks are determined from available job descriptions, whereas elements are determined by observation. A *fundamental cycle* is defined, then, as a work cycle that has a sequence of Therbligs that repeat themselves within the cycle. Historically, the Therbligs were proposed by Frank and Lillian Gilbreth. A summary of them is in Table 30.3.

Table 30.3   **The Therbligs: Gilbreth's Work Element Definition**

Element	Description
Assemble	Joining together two or more objects.
Delay—unavoidable	Interrupting activity because of a factor beyond one's control.
Delay—avoidable	Interrupting activity because of a factor under one's control.
Disassemble	Separating two or more objects.
Grasp	Touching or gripping an object with the hand.
Hold	Exerting force to hold an object on a fixed location.
Inspect	Examining an object by sight, sound, touch, etc.
Move	Movement of some object from one location to another.
Position	Moving an object in a desired orientation.
Plan	Perform mental process that precedes movement.
Reach	Moving of the hand to some object or location.
Rest	Interrupting work activity to overcome fatigue.
Search	Looking for something with the eyes or hand.
Select	Locating one object that is mixed with others.
Use	Manipulating a tool or device with the hand.

EXAMPLE 30.3

Assume a worker is to sand fiberglass parts. Parts come in to the left-hand side of the work bench, are hand sanded, and stored to the right-hand side of the bench when finished. According to Table 30.3, the elements required to sand the edge of the fiberglass part are as follows:

Left Hand	Right Hand
Reach for part	Hold sander
Grasp part	Hold sander
Move part	Hold sander
Position part	Hold sander
Hold part	Hold sander to part
Hold part	Move sander over surface five times
Move to pallet	Hold sander
Position on pallet	Hold sander
Release part	Hold sander

After elements are defined and documented, completion time for each is evaluated, and the work content is analyzed with respect to potential biomechanical risk factors:

Element	Biomechanical Risks
Grasp, move, and hold	Pinching part. Stress concentrations from sharp edges of part.
Hold and move sander	Pinching sander. Sharp edges on sander handle.
Move sander	Sustained static effort to support sander and airline. Inward rotation of forearm with wrist flexion. Pinching sander. Sharp edges on handle. Vibration. Cold exhaust air on fingers.

Although it is difficult to define repetition levels that can be labeled as always harmful, jobs that have a cycle time of less than 30 s and a fundamental cycle that exceeds 50% of the total cycle time, should be considered as posing a risk for CTDs.

Job Safety

Many methods of safety are considered in manufacturing systems. Several have been addressed in this chapter. While evaluating job safety, the following must be considered: use of safety glasses, avoid wearing loose clothing, use of warning signs and labels, continuous training, and *point-of-operation guards,* such as the ones described as follows.

Figure 30.14*A* shows a machine guard that surrounds the tool working area of a punch press. Before the machine cycle starts, a *gate barrier* device must be closed, forcing operators to keep hands or arms away from danger. Figure 30.14*B* illustrates another safety measure, a sensing device that normally uses a photoelectric *light curtain* or a radio frequency sensing field. These light curtains are installed to prevent or stop normal stroking of the press or operation of the machine if the operator's hands are placed in or near the point of operation.

Pull-back devices, as shown in Figure 30.14*C*, are designed with built-in wristlets for each operator's hand and are attached to a pulling cable. In this way, reaching into the point of operation becomes impossible. Other popular safety devices, illustrated in Figure

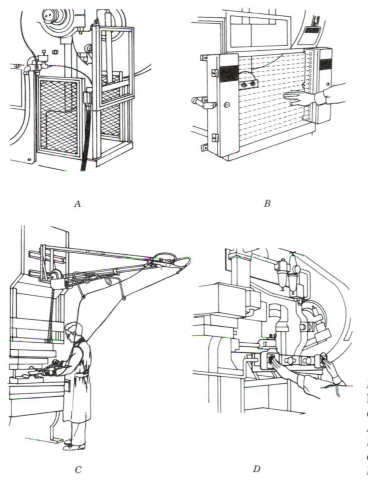

A

B

C

D

Figure 30.14
Methods of point-of-operation protection.
A, Gate barrier.
B, Photoelectric light curtain. *C*, Pull-back.
D, Two-hand controls.

30.14D, are the two-hand buttons or palm buttons and trips that require both trips to be operated jointly before the machine cycle can begin.

Economic Issues

When a company is in the process of deciding on safety and ergonomic equipment redesign or purchase, it needs to know what the *return on investment* will be. The payback of such equipment comes from three sources: (1) increase in efficiency, (2) decrease in lost time and workers' compensation costs, and (3) minimal OSHA fines.

Increase in *efficiency* comes from many factors. If the time to move an object from point A to point B can be decreased by 10%, the handling costs are cut by 10%. If a worker gets tired after 6 hours of labor instead of 4, an added benefit of higher productivity is earned. If a worker is off with an injury, the replacement worker is less productive and adds direct costs.

The cost of *lost time and medical expenses* can be substantial. Recent data from the National Safety Council show that an average lost time injury costs $20,000. Finger and hand injuries total 26% of reported cases. Back injuries constitute 22% of the claims and account for 33% of the medical expenses. If lost time accidents reach to $50 billion a year for all American manufacturers, these costs are extenuating.

Fines assessed by OSHA are for violations of the "general duty clause," which requires all manufacturers to provide a safe workplace. Fines of more than $1 million to individual companies have been assessed. The impact can be illustrated better with an example.

EXAMPLE 30.4

What percentage of savings in labor is needed to pay for a $2000 piece of equipment in 1 year? Assume the cost of labor and burden at $20 per hour.

The number of hours to pay for equipment is given by the relationship

$$\frac{\text{Equipment cost}}{\text{Labor per hour}} \tag{30.3}$$

which, in this case, would be $2000/$20/h, or 100 h.

The percent increase to pay for equipment in 1 year, then, can be calculated as

$$\frac{\text{Number of hours to pay for equipment}}{\text{Average hours worked per year}} \times 100 \tag{30.4}$$

which gives 100/2080, or 4.8%.

If efficiency can be increased 4.8%, the $2000 purchase can be paid for in 1 year. Additional savings are realized by reducing lost time and medical expenses or perhaps OSHA violations.

30.6 SYSTEMS TO MEASURE INJURY FREQUENCY

Any system that is used to measure the frequency of occupational injuries should have the following attributes:

1. The system should indicate the rate at which the disorders occur in a specified group of workers.

2. The system should be able to compare the rates of disorders between two groups of workers.

3. The system should be able to identify areas (e.g., plants, departments, section, job titles) in which injuries are emerging or unacceptably high.

4. The system should be able to measure progress (or lack of it) in efforts to control disorders.

Three pieces of information are needed to compute a meaningful frequency rate: the number of workers in a specified group that experience a disorder, the total number of workers in the specified group, and the period of time in which the disorders occurred.

EXAMPLE 30.5

A plant that manufactures computer circuit boards employed 320 assembly line workers on July 1, last year. On that date, all these workers were given a physical examination to detect cases of carpal tunnel syndrome. Ten cases were found during the examinations. The *prevalence rate* (i.e., the frequency of a disorder or the proportion of a population that experiences it at a specified point in time) for a population is computed as

$$\text{Prevalence rate} = \text{Number of injuries/Number of workers} \tag{30.5}$$

which in this example is 10/320, or 0.0313.

The *incidence* of a disorder is the number of new cases that come into being during a specific period of time. For occupational disorders, time is usually measured in terms of exposure hours to a job, in addition to calendar time. The *incidence rate* is the number of new cases in a population per specified unit of exposure time. Since the passage of the OSHA, most industries compute incidence rates as the number of cases of a disorder per 200,000 hours as the standard unit of exposure time.

EXAMPLE 30.6

Illustration of the OSHA Method

In the previous example, the circuit board manufacturer employed an average of 225 full-time employees during the calendar year. (A full-time employee works 40 hours per week, 50 weeks per year, for an exposure time of 2000 hours per year.) During that year, eight new cases of *carpal tunnel syndrome* (CTS) were diagnosed among assembly line workers. The OSHA-computed incidence rate of CTS for the year can be computed using the formula:

$$\text{Incidence rate} = \frac{\text{Cases} \times 200,000}{\text{Hours worked}} \tag{30.6}$$

Substituting plant data into Equation (30.7), the incidence rate of CTS is found to be 3.56. This may be considered a moderate incidence level and may serve as an early indicator of an emerging problem. An ergonomic hazard analysis should be conducted to determine if the incidence rate represents a trend requiring some immediate intervention or only a temporary condition that is self-limiting.

Severity statistics attempt to describe the seriousness of a disorder in terms of its cost to an employer. Severity is not a measure of disease frequency. A commonly used measure is the *American National Standards Institute* (*ANSI*) severity rate, defined as the number of workdays lost during a calendar year per million hours of work time. Other commonly used severity measures include: (1) number of days lost per employee per year or number of days lost per 100 employees per year, (2) workers' compensation cost per employee per year or per 100 employees per year, (3) number of days lost per case of a given type of disorder or injury, and (4) workers' compensation cost per case of a given type of disorder or injury.

Because of the lack of standardization in severity statistics, it is often impossible to compare the experience of one department of the plant with that of another. However, *expected number of incidents* and chi-square statistics can be estimated for planning and prevention purposes. See Problems 30.6 and 30.10.

Statistical Method for Identifying Problem Areas

To determine whether differences exist in the incidence rates of a particular injury in several departments of the same plant, a chi-square statistic can be calculated, according to the following formula:

$$\chi^2_{m-1} = \sum_{i=1}^{m} \frac{(E_i - O_i)^2}{E_i} \tag{30.7}$$

where

χ^2 = Test statistic with $(m - 1)$ degrees of freedom
m = Number of groups being compared
E_i = Expected number of incidents in group i
O_i = Observed number of incidents in group i
$i = 1, 2, 3, \ldots m$

The value of E_i is computed using the following equation

$$E_i = \frac{H_i \times O_T}{H_T} \tag{30.8}$$

where

H_i = Hours of exposure in group i
H_T = Total hours of exposure for all groups
O_T = Total observed incidents for all groups

A restriction on this test is that the expected number of incidents in each group (E_i) must be at least five. Unfortunately, this requirement is not met in many settings because population sizes are smaller. In situations like these, it is often necessary to make decisions without conclusive results of statistical tests. Problem 30.10 illustrates the use of the chi-square statistic.

30.7 IMPACT OF INTELLIGENT SYSTEMS

The inventions of automation and artificial intelligence have brought about a technological revolution in the operator-machine systems relationship. These technologies have made it possible to reduce safety risks and to implement cognitive capabilities, such as speech understanding, logic deduction, picture understanding, reasoning about expert knowledge, problem solving, and decision making in machines.

With *automation* most of the physical human work is transferred to machines. Optimal controllers are developed to take over skill-based activities. The human role shifts from the doer to that of a monitor and problem-solver. Awkward postures of the human body, excessive manual force, and high rates of manual repetition are less impacting.

Artificial intelligence (AI) is the science of making machines do things that would require intelligence if done by humans. Techniques such as expert systems, fuzzy logic, and artificial neural networks, described in Chapter 29 of this textbook, are some examples. AI technology offers the potential to transfer rule-based and knowledge-based functions from humans to machines and computers. Examples of these functions are: monitoring the state of the machine and the environment; if operational conditions of the machine change, self-adapting mechanisms enter to modify goals; when goals change, a set of suitable plans is developed to react, and alternative plans are evaluated; after simulating the effects of alternate plans, the optimal solution is implemented.

A description of an *intelligent operator-machine system* is outlined in Figure 30.15. Several points are key to a successful realization of such systems.

1. Reliable symbolic representation schemes must be developed.
2. Appropriate levels of operator alertness must exist.
3. When the system works in the automated mode, the operator must stay in the loop to minimize reaction time for take-over in case of system failure.
4. The operator must be able to intervene at all three levels of interaction (skill-, rule-, and knowledge-base) to serve as a backup in case of failure of the automation system.
5. The display and interface systems must be laid out to ease communication between worker and machine on the three levels of interaction.

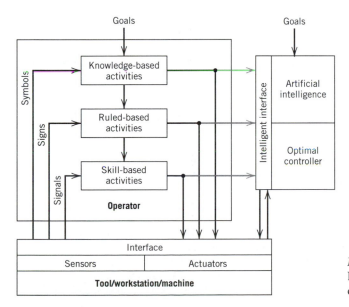

Figure 30.15
Description of an intelligent operator-machine system.

QUESTIONS

1. Define the following terms:

Skill-based activities   Anthropometric tables
Ergonomics   Therbligs
CTDs   Prevalence rate
Intensity of sound   Carpal tunnel syndome
Relative intensity   Severity rate
Pitch   Artificial intelligence
OSHA   Intelligent operator-machine
ISO   system
Lux

2. What are the components involved in the majority of occupational factors involving CTDs?

3. Name at least four design principles to avoid injuries caused by static forces.

4. If you want to minimize the torque applied to a joint, which fastening tool would you recommend: an impact wrench or an impulse tool?

5. Describe your understanding of a conventional operator-machine system. What is the difference with an "intelligent" operator-machine system?

6. Describe a test method to measure the noise level of a pneumatic tool.

7. Describe some methods of point-of-operation guards of machinery.

8. What is the "halving level" approach as it relates to noise levels?

9. Describe three sources of noise associated with pneumatic tools.

10. Name at least four design principles to avoid injuries caused by vibration in hand tools.

11. Describe a method to control dust when disc sanders are in use.

12. What is the dust extraction flow rate range required for hand tools?

13. What is the illumination range for inspection of a densely populated electronic board?

14. Describe three methods to reduce extreme joint movement.

15. Name three methods to control job forces.

16. Discuss some ways to minimize highly repetitive movements.

17. Describe four fundamental principles to design ergonomic workstations.

18. What is a fundamental cycle when the work-methods analysis technique is used?

19. After investing in safety equipment, from which sources do you think the return on investment can come?

20. What attributes should a system to measure injury frequency have?

21. Discuss some key points for a successful intelligent operator-machine system.

PROBLEMS

30.1. Find the intensity in W/m^2 of a jackhammer and a jet aircraft if their relative intensity is measured as 100 dB and 120 dB, respectively.

30.2. A 3 by 4-ft sheet of plastic is in front of a 50 kW electric motor that emits sound at 85 dB. What power is being transferred to the sheet?

30.3. EZ Products, Inc.'s direct labor costs are $700,000 per month and employs 300 operators. It is estimated that efficiency can be increased by 6% if equipment improvements are made. How much can the company invest in capital equipment if the payback period expected is 18 months?

30.4. Study Figure 30.13. Redraw the figure using your own physical measurements.

30.5. In Figure P30.5, explain why these are dangerous wrist postures. Redesign the tool or the posture to prevent injuries.

30.6. To determine whether an injury problem exists on a given job or in a given area, it is necessary to collect the data required to compute job- or area-specific incidence rates. Ergonot, Inc., employed an average of 1450 workers last year in three different departments. A review of the OSHA log disclosed that there were 36 new cases of tendi-

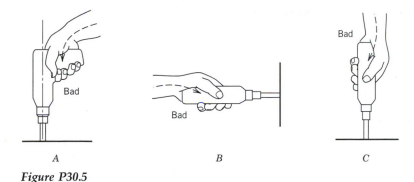

A   *B*   *C*

Figure P30.5

nitis in the plant during the year. Human resources records show the following for each department: Area A averaged 450 workers and had 22 tendinitis cases, Area B had 4 cases with 300 employees, and Area C had 10 cases with 700 workers. Determine the incidence rate and the expected number of cases per department. Compare and comment.

MORE DIFFICULT PROBLEMS

30.7. Wind blows through a 5-in. diameter wheel that has seven slots in it. The wheel is mounted in a high-speed conveyor that could travel at 50 ft/min. What is the pitch (in Hz) of the sound produced? If the minimum frequency the human ear can perceive is approximately 20 Hz, would the sound of the wind disturb operators of the conveyor? (Hint: Velocity = angular velocity × radius. Angular velocity = $2\pi f$).

30.8. A single tool requires a flow of 200 m³/h for dust extraction. Because of plant layout, the vacuum source is located 10 m away. Using the approximate data given in Figure P30.8, suggest at least three suitable combinations to connect a hose to this tool. If there is a further pressure drop generated in the extraction hood, couplings, filters, and the like of around 9 psi, what is the total pressure drop required by the extraction system? Which of the three alternatives would you recommend? Why?

30.9. Study Figure 30.13. Construct a similar chart to determine the reach distance for a standing position using your own physical measurements.

30.10. Referring to Problem 30.6, determine if differences exist in the incidence rates of tendinitis for the three departments of Ergonot, Inc. What statistical method would you recommend? Why? After conducting the test, what would you recommend Ergonot, Inc., should do? (Hint: At a probability level of 1%, a chi-square test statistic of 10 or less means that it is highly probable that the recorded incident rates for the three departments are equivalent.)

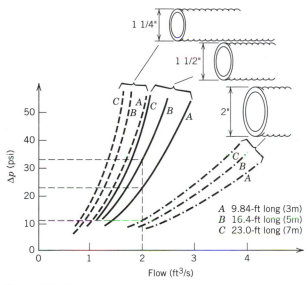

Figure P30.8

PRACTICAL APPLICATION

Contact the safety officer of a local company. Explain to them your interest in ergonomic practices and solicit a visit of their facilities. Once there, ask them to describe the jobs performed in the company and any CTDs associated with them. Identify what occupational factors are associated with those jobs and any measures taken to minimize the risks. Write a report detailing your observations and make recommendations to improve safety and ergonomic conditions.

CASE STUDY: CTD COSTS AND FREQUENCY IN ELECTRONIC PARTS MANUFACTURING

Magnapart is a medium-size facility that manufactures small electronic parts for circuit boards and electronic consumer products. Depending on the demand, hourly employment varies. Most workers live in rural communities near the plant. Demographics distribution is normally males 40%, females 60%, and ages between 20 and 40 years old. Virtually all the production jobs in the plant are sedentary and consist of repetitive hand tasks such as inspecting parts, loading and unloading parts to racks, jigs, fixtures, and machines, operating machine tools, and using hand tools, such as wire cutters. These jobs are performed while standing or sitting at a workbench. Most employees are paid under an incentive system that awards bonuses for each unit processed above a base rate.

Management has become concerned about CTDs in the plant (see table below) because the medical department reported an increase in lost-time cases. The sum of medical costs and disability payments last year averaged approximately $2500 per case and there was an OSHA fine of $350,000. Top management knows that if something is not done, a more stringent fine is coming this year. They are aware of your understanding of ergonomic issues and how they impact the manufacturing system, so they have called you to study the situation and recommend some changes. Your analysis must be well-documented and should include at least the following: potential causes of the disorder, incidence rate of each disorder per department and for the entire plant, the expected number of cases per department, workstation(s) design proposal, and the economic analysis to justify the implementation of your recommendations.

Magnapart CTDs records for last year are as follows:

		Type of Disorder		
Department/Task	Hours Worked	Nerve No. of Cases	Tendons No. of Cases	Other No. of Cases
Lead preparation	58,000	0	0	5
Solderability testing	10,300	0	1	1
Subassembly area	115,700	1	1	9
Final assembly	50,000	0	1	6
Product testing	30,200	1	0	1
Final inspection	5,000	0	0	1
Stock room	11,500	1	0	1
Packing/shipping	12,000	0	1	2

CHAPTER 31

COST ESTIMATING

The purpose of cost estimating is to find the cost of the manufacturing operations and to assist in setting the price for the product. This important activity, which is concurrent to engineering design, precedes production, and, thus, is forecasting future costs. Cost estimating and process planning are prominent activities in the manufacturing system. Using modern cost-estimating methods, manufacturing businesses are able to respond to *electronic commercing*. Much of cost estimating is done with computer programs, which seamlessly work with the computer-aided design of the part.

Cost estimates include the cost of labor, material, machine costs, overhead, and profit. The important objective of the manufacturing enterprise is to return a *profit* to the owners of the firm, satisfy customer needs, and provide worth of employment to the employees. Profit pays taxes, buys new equipment, and declares dividends to the company owners.

This chapter reviews the important classical metal cutting approach to cost estimating, and the student needs to appreciate this rich heritage. It concludes with a modern approach to cost estimating.

31.1 CLASSICAL METAL CUTTING COST ANALYSIS

Involved in a metal cutting operation cycle is load work (LW), advance tool (AT), machining, retract tool (RT), and unload work (UW). These elements are shown in Figure 31.1A and are labeled "one work cycle." After a number of parts are machined, tool maintenance is required, which includes removing the tool and replacing or regrinding the tool point and reinserting the tool ready for metal cutting; this is shown by Figure 31.1B. The classical mathematical model used most frequently in the study of machining estimating describes the cost of a single-point-tool rough-turning operation. *Operation unit cost* is composed of handling, machining, tool changing, and tool cost.

Handling

Handling time is the minutes to load and unload the workpiece from the machine. It can include the time to advance and retract the tool from the cut and the occasional dimensional inspection of the part. Independent of cutting speed, V_c, handling is a constant for a specified part and machine. It is defined as

$$\text{Handling cost} = C_o t_h \tag{31.1}$$

where

$$C_o = \text{Direct labor wage, dollars/min}$$
$$t_h = \text{Time for handling, min}$$

Figure 31.2A is the constant *handling cost* plotted against cutting speed. As it is independent of cutting velocity, V_c, it appears as a straight line. C_o does not include overhead costs.

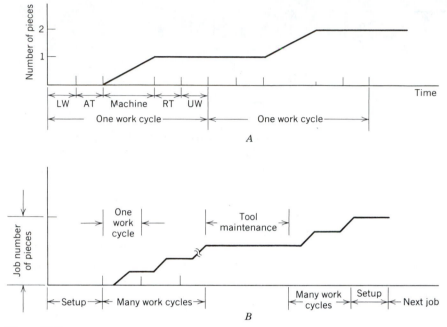

Figure 31.1
Operation items including setup, cycle, and tool maintenance.

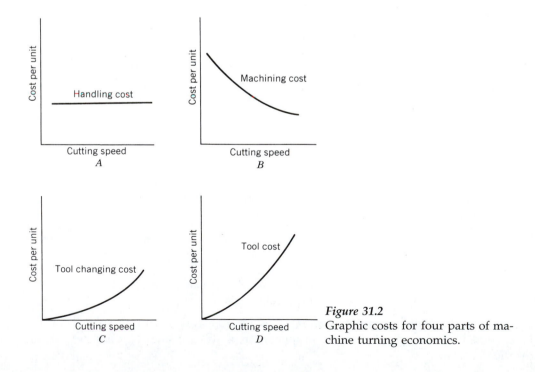

Figure 31.2
Graphic costs for four parts of machine turning economics.

Machining

Machining time is the time that the tool is actually in the feed mode or cutting and removing chips. This has been discussed in Chapter 11, but is studied again.

$$t_\mathrm{m} = \frac{L\pi D}{12 V_\mathrm{c} f} \tag{31.2}$$

and

$$\text{machining cost} = C_\mathrm{o} t_\mathrm{m} \tag{31.3}$$

where

t_m = Machining time, min
L = Length of cut for metal removal, in.
D = Diameter, in.
V_c = Cutting speed, ft/min
f = Feed rate, in./rev
N = Rotary cutting speed = $\dfrac{12 V_\mathrm{c}}{D}$, rpm

The *cutting feed rate*, f, ipr, is substantially less than *rapid travel velocity* of various machine elements. Each material has special turning and milling cutting speeds and feeds as determined by testing or experience. Values are different for roughing or finish. A roughing pass removes more material but does not satisfy dimensional and surface finish requirements. The material removal rate is expressed as min/in. (min/mm) for drilling and tapping. Table 11.1 is an example of machining speeds and feeds for high-speed steel and tungsten carbide tool material. Table 31.1 is an abbreviated set. As cutting velocity, V_c, increases, machining unit cost decreases. This is seen in Figure 31.2*B*.

The *length of the machining cut* depends on the type of machining. In turning, the length that is being machined in a lathe is at least equal to the final drawing dimension and is usually greater because of additional stock for roughing or finishing. In a lathe facing element, the length of cut is from the center of the bar stock circle to the outside diameter (or the stock OD), and the facing length is at least equal to one-half the diameter.

For all types of machining elements, the length of the cut includes safety length, approach, design length, and overtravel. This is discussed in Chapter 11, Section 11.3.

The diameter, D, of Equation (31.2) is either the workpiece or the tool. When a lathe-turning operation is visualized for cost estimating, the diameter, D, is the largest unmachined bar stock dimension. In turning the periphery, it is the maximum diameter resulting from the raw stock or the previous machining element. If the cut is to finish turn the diameter, D, for purposes of Equation (31.2), the rough cut is the outside dimension.

For milling and drilling, the diameter, D, is the cutting tool diameter. A rotary milling cutter 4 in. (100 mm) in diameter is an example. The cutting speed velocity, V_c, has the dimensions of ft/min (m/s). Its value depends on many factors, and production time data will consider these effects. Table 31.1 shows a small sampling of speeds and feeds for metal cutting, which are used for the examples and problems.

Tool Life

Cutting tools become dull as they continue to machine. Once dull, they are replaced by new tools or they are removed, reground, and reinserted in the toolholder. Empirical studies can relate tool life to cutting velocity V for a specified tool and workpiece material.

Table 31.1 **Machining Speeds and Feeds**

| | Turning and Facing (ft/min, ipr) | | | | | |
| | High-Speed Steel | | | Tungsten Carbide | | |
Material	Rough		Finish		Rough		Finish	
Stainless steel	150	0.015	160	0.007	350	0.015	350	0.007
Medium-carbon steel	190	0.015	125	0.007	325	0.020	400	0.007
Gray cast iron	145	0.015	185	0.007	500	0.020	675	0.010

| | High-Speed Steel | | | High-Speed Steel | | |
| | Slab Milling | | | Slot 1 in. | | |
Material	Rough		Finish		Rough		Finish	
Stainless steel	140	0.006	210	0.005	85	0.002	95	0.0015
Medium-carbon steel	170	0.008	225	0.006	85	0.0025	95	0.002
Gray cast iron	200	0.012	250	0.010	85	0.004	95	0.003

Drill Diameter[a]	Stainless Steel	Medium-Carbon Steel	Gray Cast Iron
¼	0.55	0.20	0.20
⁵⁄₁₆	0.61	0.23	0.23
⅜	0.65	0.25	0.25

Threads/in.[a]	Steel	Stainless Steel
32	0.18	0.33
20	0.15	0.30
16	0.19	0.33
10	0.32	0.48

[a] Times are min/in. for power-feed drilling; threading times are min/in.

Two popular tool materials are high-speed steel (HSS) and tungsten carbide. Most studies of tool life are based on *Taylor's tool life* cutting speed equation. This is discussed in Chapter 11.

$$VT^n = K \tag{31.4}$$

where

T = Average tool life, min/cutting edge or tool point
n, K = Empirical constants resulting from statistical analysis and field studies, and $0 < n \le 1, K > 0$

The average tool life T is found as

$$T = \left(\frac{K}{V}\right)^{1/n} \tag{31.5}$$

Table 31.2 gives a small set of tool life data for three workpieces and two tool materials.

If a tool life equation is $VT^{0.16} = 400$, one can find either V or T, given the other variable. If $V = 200$ ft/min, one can expect an average of 76 min of machining for a tungsten carbide replaceable insert, before the tool is indexed to a new corner. If the

Table 31.2 **Taylor's Tool Life Parameters**

	High-Speed Steel		Tungsten Carbide	
Material	K	n	K	n
Stainless steel	170	0.08	400	0.16
Medium-carbon steel	190	0.11	150	0.20
Gray cast iron	75	0.14	130	0.25

tool point continues to cut after the 76 min, there may be a reduction in surface quality or the machining horsepower may increase.

Tool Changing Cost

The third component cost of the classical method is the tool changing cost per operation. Define it as

$$\text{tool changing cost} = C_p t_c \frac{t_m}{T} \tag{31.6}$$

where

$$t_c = \text{Tool changing time, min}$$

The *tool changing time*, t_c, is the time to remove a worn-out tool, replace or index the tool, reset it for dimension and tolerance, and adjust for cutting. The time depends on whether the tool being changed is a disposable insert or a regrindable tool for which the tool must be removed and a new one reset. In lathe turning and milling there is the option of an indexable or regrindable tool. The drill is only reground. In Figure 31.2C the relationship of tool changing cost to cutting speed is shown.

Define the following as

$$\text{tool cost per operation} = C_t \frac{t_m}{T} \tag{31.7}$$

where

$$C_t = \text{Tool cost, dollars}$$

As machining velocity, V, increases, there is increasing cost for tool changing because the tool is changed more frequently.

Tool Cost

Tool cost, C_t, depends on the tool being a disposable tungsten carbide insert or a regrindable tool. For inserts, tool cost is a function of the insert price, and the number of cutting edges per insert. For regrindable tooling the tool cost is a function of original price, and total number of cutting edges. As the speed increases, the cost for the tool increases, as shown in Figure 31.2D. Table 31.3 provides tool costs and changing times.

The total cost per operation is composed of these four items. Machining cost decreases with increasing cutting speed, whereas tool and tool changing costs increase. Handling costs are independent of cutting speed. Thus, unit cost, C_u, is given as

$$\Sigma = [C_o t_h + \frac{t_m}{T}(C_t + C_o t_c) + C_o t_m] \tag{31.8}$$

Table 31.3 **Tool Changing or Indexing Time and Costs**

Elements	Time
Time to index a turning type of carbide tool	2 min
Time to set a high-speed tool	4 min
Large milling tool replacement	10 min
Remove, regrind, and replace drill	3 min
Cost per tool cutting corner for turning, carbide	$3
Cost for high-speed steel tool point	$5
Cost per milling cutter, 6 in. carbide	$1500
Drill cost	$3

On substitution of t_m and T and after taking the derivative of this equation with respect to velocity (V) and equating the derivative to zero, the minimum cost is found as

$$V_{min} = \left[\frac{K}{\left(\frac{1}{n} - 1\right)\left(\frac{C_o t_c + C_t}{C_o}\right)} \right]^n \tag{31.9}$$

which gives the velocity for the unit cost of a rough-turning operation. In this formula there is no recognition of revenues that are produced by the machine. V_{min} identifies the minimum velocity without revenue considerations. If Equation (31.8) is plotted for a range of V_c, then a minimum cost per unit is found as 200 ft/min and seen in Figure 31.3. This graphical value is the same as that found by Equation (31.9).

Occasionally, to avoid *bottleneck* situations, there is a need to accelerate production at cutting speeds greater than that recommended for minimum cost. In these expedited operations, it is assumed the tool cost to be negligible, or $C_t = 0$. If the costs in the basic model are not considered, the model gives the time to produce a workpiece, and the following is developed

$$T_u = t_h + t_m + t_c \frac{t_m}{T} \tag{31.10}$$

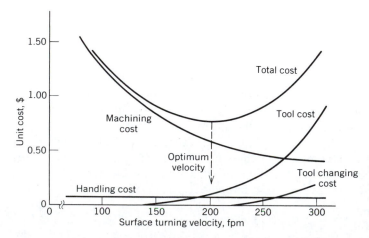

Figure 31.3
Study of the various effects of cutting speed upon handling, machining, tool, and tool changing cost to give optimum.

where

$$T_u = \text{min per unit}$$

The production rate (units per minute) is the reciprocal of T_u. The equation that gives the cutting speed that corresponds to maximum production rate is

$$V_{max} = \frac{K}{\left[\left(\frac{1}{n} - 1\right)t_c\right]^n} \tag{31.11}$$

The tool life that corresponds to maximum production rate is given by

$$T_{max} = \left(\frac{1}{n} - 1\right)t_c \tag{31.12}$$

EXAMPLE 31.1

We want to machine 430 F stainless steel, 1.750-in. OD bar stock. The cutting length is 16.50 in. plus a $\frac{1}{32}$ in. for approach length, giving a total of 16.35 in. The turning operation uses cemented tungsten carbide tooling, insertable and indexable eight-corner tool (about the size of a dime), which costs \$3 per corner. Time to reset the tool is 2 min. Handling of the part is 0.16 min. The operator wage is \$15.20/h. The Taylor tool life equation is $VT^{0.16} = 400$ (in U.S. customary units) for the tool and workpiece material. Feed of the rough-turning element is 0.015 ipr, for a depth of cut of 0.15 in. per side. These data are given by Tables 11.1, 31.1, 31.2, and 31.3. If the four items of the cost equation are plotted with several velocity V values as the x-variable, Figure 31.3 shows the optimum cost velocity at about 200 ft/min.

Preparation of operations sheets and estimating is more than computing (Figure 31.3). Usually, this analysis may not be possible and operations sheet estimates are created using other approaches. The student should appreciate this historical understanding of the Taylor tool life cost.

31.2 INDUSTRIAL COST ESTIMATING PRACTICES

There are thousands of industrial operations. Clearly, the classical metal cutting analysis is a very narrow approach, and is virtually unused in industry. Metal cutting operations, along with the many other types of manufacturing operations, are cost estimated using other principles.

The operations sheet is fundamental to manufacturing planning and cost estimating. Operations planning is discussed in Chapter 23. The operations plan identifies the workstation, operation sequence number, description, and special or standard tools. Cost estimating is an extension of operations planning, and is done during the preparation of the *route sheet*.

Each operation is estimated for setup and cycle time, and then extended for operational cost, one which we have coined as Productive Hour Cost. Figure 31.4 is a picture of the cost elements to estimate full cost and profit, leading to product price.

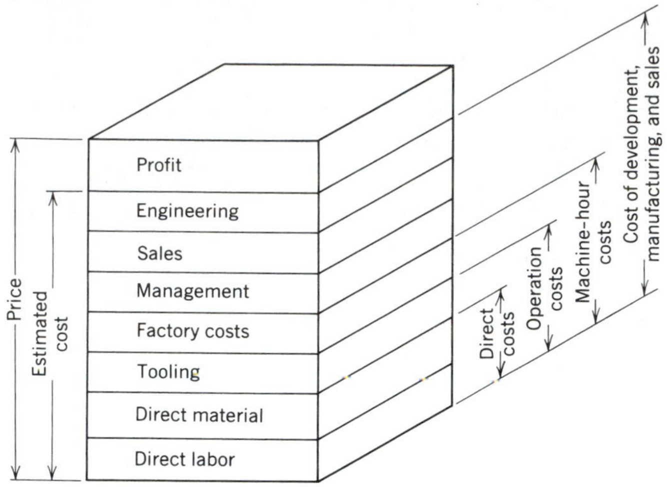

Figure 31.4
Elements in a cost–price model for a manufacturing enterprise.

31.3 ESTIMATING SETUP AND CYCLE TIME

Setup and cycle time are estimated for each manufacturing operation that appears on the operations process plan. If the operation uses an operator (and not all operations require machinists), the worker is called *direct labor* (which is an accounting term), as the worker is operating the machine tool or process in the conversion of the material from one state to another processed level.

The times for setup and cycle time are frequently determined from a *time study*, foreman's job tickets, or historical values for similar operations. As a time study is

Table 31.4   **Setup Elements, Hours**

Elements		Time
Punch in and out, study drawing		0.1
Turret lathe		
First tool		0.5
Each additional tool		0.3
Collet fixture		0.2
Chuck fixture		0.1
Milling machine		
Vise		1.1
Angle plate		1.4
Shoulder cut milling cutter		1.5
Slot cut milling cutter		1.6
Tight tolerance		0.5
Drill press		
Jig or fixture		0.14
Vise		0.05
Numerically controlled turret drill press		
First turret tool		0.25
Additional turret tools	each	0.07

composed of elements of work, so estimating uses these elements in various ways that reflect the effort for a new but similar operation.

Setup time includes the work to prepare the machine for product parts or the cycle. Setup is expressed in hours. Setup work includes punch in/out, complete and study any paperwork, obtain tools, position unprocessed materials next to the work area, adjust the tools for making the part, meet tolerance requirements, and inspection. Setup includes the return of tooling, cleanup of the chips, and teardown of the machine ready for the next job. Setup does not include the time to make parts or perform the repetitive cycle. Setup is often related to the number of tools that are assigned to the job. Table 31.4 is

Table 31.5   **Cycle Elements, Minutes**

Elements		Minutes
All machines		
Start and stop machine		0.08
Change speed of spindle		0.04
Engage spindle or feed		0.05
Air clean part and fixture		0.06
Inspect dimension with micrometer		0.30
Brush chips		0.14
Turret lathe		
Advance turret and feed stock		0.18
Turret advance and return		0.04
Turret return, index, and advance		0.08
Cross slide advance and engage feed		0.07
Index square turret		0.04
Cross slide advance, engage feed, and return		0.14
Place and remove oil guard		0.09
Milling machine		
Pick up part, move, and place; remove and lay aside		
5 lb		0.13
10 lb		0.16
15 lb		0.19
Open and close vise		0.14
Clamp, unclamp vise, ¼ turn		0.05
Quick clamp collet		0.06
Clamp and unclamp hex nut	each	0.21
Numerically controlled turret drill press		
Pick up part, move, and place; pick up and lay aside		
To chuck:		
2.5 lb		0.10
5.0 lb		0.13
10.0 lb		0.16
Clamp and unclamp		
Vise, ¼ turn		0.05
Air cylinder		0.05
C-clamp		0.26
Machine operation		
Change tool		0.06
Start control tape		0.02
Raise tool, position to next x–y location, advance tool to work		0.06/hole
Index turret		0.03/tool

a small example of setup work elements for estimating.

Then a selection of these elements are made for each operation.

$$\text{Setup} = \Sigma \text{ setup elements} \qquad (31.13)$$

Cycle time or *run time* is dimensioned in minutes per unit and is the work to complete each unit after the setup is concluded. It does not include any element involved in setup. Each cycle includes loading the part; machine movements, such as rapid traverse, index of the turret, and start machine; machining time to cut; change the tooling; unloading the part; and inspection. These cycle element minutes vary for the type of machine, size of the part, and the machining conditions. If there are several rough and finishing passes, it includes the time for readjustment of the machine to permit these repeated machine cutting passes. Table 31.5 is a small example of cycle minutes for metal-working estimating. An estimator will choose the elements that are needed for the new job. The student needs to be reminded that each manufacturing process (remember there are thousands) has its own table of time elements.

If the cycle minutes are totaled for each operation, one can find

$$\text{Pieces per hour} = \frac{60}{\Sigma \text{ cycle minutes}} \qquad (31.14)$$

Lot hours for the operation and the batch quantity are found using

$$\text{Lot hours} = \Sigma \text{ setup} + N(\Sigma \text{ cycle min}/60) \qquad (31.15)$$

where

$$N = \text{Number of units in the lot}$$

These lot hours are multiplied by the *productive hour cost rate* (*PHCR*). The PHCR includes direct labor cost, machine costs such as depreciation, power, space and heating, and supervision costs provided for the foreman. Once a share of the engineering and management costs are added to the PHCR, the part or product cost is ready for the inclusion of material cost, which is separate from the processing cost for each operation.

31.4 MATERIAL ESTIMATING

Information about the lot quantity, dimensions of the material, and specifications and grade of the material is necessary in estimating material costs. Much of this information is obtained from engineering drawing, operations process plan, and supplier catalogs. Material costs are important to the cost of the product.

A *material specification* for a steel bar is as follows: "Bar, cold-rolled steel, Type C 1030, 3.00 in. round OD × 12 ft lengths, and unannealed." The amount that is required for the order sets the cost of the purchased material. The company will purchase the material from a steel supplier. As an example

Purchased Weight (lb)	Cost: $/100 lb
97.0 (= 1 bar)	175
500	165
4000	150
10,000 (base amount)	120

The *base quantity* amount is that weight at which price is no longer sensitive to additional amounts of purchase order.

Direct materials are calculated using the engineering drawing and physically become the product. An estimator calculates the weight from these drawings. This computed drawing quantity is increased by losses, such as *waste* and *scrap*. *Indirect materials*, such as cutting oils, supplies, and the like, which do not appear in the product and on the drawing, are included costwise in an overhead cost such as the PHCR. A general formula for material estimating is

$$C_{dm} = W(1 + L_1 + L_2)C_m \qquad (31.16)$$

where

C_{dm} = Direct material cost, \$/unit
W = Theoretical finished weight from drawing, lb
L_1 = Percentage loss resulting from scrap, which is a consequence of errors in manufacturing or engineering. Scrap results in the part being rejected, and having little if any value.
L_2 = Percentage loss resulting from waste, which is caused by the manufacturing process, such as chips, cutoff, overburden, skeleton, runners, sprue, dross, and the like.
C_m = Cost of the material, \$/lb

The value for C_m is received from suppliers that sell the raw material.

EXAMPLE 31.2

A spacer is shown in Figure 31.5. The total annual order is 12,000, with monthly lot sizes of 1000 units for *Just-In-Time* (JIT) deliveries. An operation sheet, Figure 31.6, gives additional details and is expanded for cost estimating. The C 1030 3-in. OD steel bar stock is in initial 12-ft lengths. The operation planner determines that two operations will be adequate. A CNC turret lathe and a turret drill press are the selected workstations for operation numbers 10 and 20.

For operation number 10 the sequence of metal cuts on the lathe is shown in Figure 31.6. The next step is to calculate the time for machining the elements, which is found using Equation (31.2), Table 31.1, and the drawing.

Element	L	D	V_S	f	t_m
Face	1.5	3.00	400	0.007	0.37
Drill	1.2	⅜		0.2 in./min	0.24
Turn	1.1	3.00	400	0.007	0.31
Chamfer	0.12	3.00	400	0.007	0.03
Cutoff	1.5	3.00	325	0.020	0.18
				Σ metal cut =	1.13 min

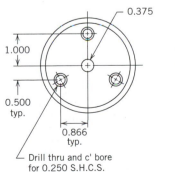
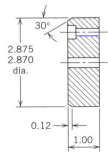

Figure 31.5
Spacer.

Part No. 675V Ordering quantity 12,000 Material 3.00 in. OD bar, 12 ft. long
Part name Spacer Lot requirement 1,000/mo. C1030 Steel
Operation planner JM Cost = $4.10

Workstation	Op. No.	Description	Special or Standard tools	Setup	Cycle Min.	Lot Hr.	PHCR	Op. Cost
CNC lathe	10	Face end Drill	Carbide insert H.S.S.	2.00	1.91	33.8	65	$2199
		Turn 2.875 Chamfer Cutoff	Carbide insert Carbide insert Carbide insert					
CNC turret Drill press	20	Drill 3 holes		0.42	1.63	27.6	40	$1103
		C'bore for socket head cap screw						

Unit cost = $7.40

Figure 31.6
Operation sheet for cost estimating.

The nonmachining time is found using Table 31.5. The number of machining cuts is assumed equal to the number of tools:

Start and stop		0.08
Speed changes, 3 × 0.04		0.12
Spindle engagements		0.05
Air clean		0.06
Turret operation, 3 × 0.08		0.24
Cross slide operation, 2 × 0.07		0.14
Place and aside oil guard		0.09
Σ nonmachining elements	=	0.78 min

The cycle time is 1.91 min, or 31.4 pieces/h. The 1.91 is posted on the operations process sheet in Figure 31.6. Next, the setup time is estimated using Table 31.4.

Punch in and out		0.1
First tool setup		0.5
Four additional tools, 4 by 0.3		1.2
Collet handling		0.2
Σ setup elements	=	2.0 h

The 2.0 hour is posted on the operations process sheet in Figure 31.6. Lot hours are calculated as 33.8 (= 2.0 + 1000 × 1.91/60) and posted on the operations plan. The PHCR is assumed to be \$65/h and the operation cost is \$2199.17 for the lot, or \$2.20 per unit.

The calculation of the unit material cost starts with the raw material, which is 12 ft long. Each piece requires 0.015 in. for facing stock, 1.00 in. for design length and $\frac{3}{16}$ in. for cutoff stock. Then each part will consume 1.203 in. There are 119 pieces in each 12-ft bar. Nine bars will be required for the 1000-lot quantity. Each bar weighs 97 lb, so about 873 lb are required. From the cost table as given above, the purchase cost of the material is $165/100 lb. Material costs for a single piece is $4.10 ($\frac{3}{4}^2 \times \pi \times 1.203 \times 0.293 \times 1.65$), which is entered on the operations process plan. The total cost of the spacer is $7.40 (= 4.10 + 2.199 + 1.103). If the firm has the practice of 40% profit as an add-on, then the price is $10.36 per unit.

QUESTIONS

1. Give an explanation of the following terms:

Cost estimates

Profit

Operation unit cost

Handling time

Handling cost

Tool life

Bottleneck

Time study

Setup time

Cycle time

Lot hours

PHCR

Material specification

Direct materials

Waste

Scrap

2. What are the principal elements of the classical metal cutting analysis?

3. What factors affect the cost for materials?

4. What are the differences between metal cutting feed mode and rapid traverse velocity mode?

5. Define tool life.

6. What is an estimate?

7. Describe the losses in metal working that must be added to the cost of the product.

8. Why does industrial practice require a more diversified cost estimating approach than that given by the classical method?

PROBLEMS

31.1. An operation requires 81 pieces per hour.

a. What are the h/100 and the unit time in min for this operation?

b. Repeat for 161 pieces per hour.

31.2. If the setup and cycle time are 3 h and 2 min, find the lot time for 75 units. Repeat for 750 units. What is the unit cost if the PHCR is $40?

31.3. Solve the following:

a. An operator earns $15/h and handling and other metal cutting elements total 1.35 min. What is the cost for the element?

b. The length of a machining element is 20 in. and the part diameter is 4-in. OD. Velocity and feed for this material are 275 ft/min and 0.020 ipr. What is the time to machine?

c. The Taylor tool life equation is $VT^{0.1} = 372$. What is the expected average tool life for $V = 275$ ft/min?

d. Tool changing time is 4 min, tool life equation is $VT^{0.1} = 372$, $V = 275$ ft/min, $C_o = $0.25/min, $L = 20$ in., $D = 4$ in., and $f = 0.020$ ipr. Determine the tool changing cost.

e. Using the information in item d. and $C_t = $5, find the tool cost per operation. Would you recommend these machining conditions?

31.4. Solve the following:

a. Operator and machine expenses are $60/h and handling is 1.65 min. Find the handling cost for this element.

b. The length and diameter of a gray iron casting are 8½ in. by 8.6 in. Velocity is 300 ft/min and feed is 0.020 ipr. Find the machining time.

c. The Taylor tool life equation is $VT^{0.15} = 500$. Find the expected tool life for 300 ft/min.

d. The time to remove an insert and index to another new corner is 2 min, the tool life equation is $VT^{0.15} = 500$, $V = 300$ ft/min, $C_o = $1/min, $L = 8.6$ in., $D = 8$ in., and $f = 0.020$ ipr. What is the cost to change tools?

e. Using the information in item d., and that an eight-corner insert costs $24, find the tool cost per operation.

31.5. A simple job is estimated to require 2 h for setup and 3 min for cycle time. Lot quantity is 83. Find the pieces/h, h/100 units, and unit estimated time, including a fair share of the setup. The PHCR = $45 per hour. Find the full cost of the job if the material = $5 per unit. The cost plus markup is 10%. What is the price for the unit and the lot?

31.6. The planner estimates that a work order will need several operations. One operation will need elements for the cycle consisting of 0.17, 0.23, 0.81, and 0.17 min. The setup is 2 h and the quantity is 182 units. The cost for the PHCR is $54.50/h. Find the pieces/h, h/100 units, the unit estimate, and cost of labor and overhead for 1 unit. What is the lot cost for this operation?

31.7. If PHCR and material cost $25/unit and the markup rate is 25%, find the price for the lot of 715 units.

31.8. A bar is C 1030 steel and 3-in. OD. The part length is 7.25 in. and a cutoff width and facing stock allowance is ¼ in. and ⅟32 in., respectively. If the safety stock allowance is ⅟16 in. and the bar is 12 ft long and 15 parts are required, find the unit and lot cost for this material.

31.9. The part length = 22.44 in., safety = ⅟32 in., approach = 1.4 in., overtravel = 1.4 in., and the material is C 1030 steel. Find the cost for the following conditions

a. N = 75 units and 12-ft bar length.

b. N = 750 units and 12-ft bar length. Is it cheaper if the bar length is 16 ft long?

c. N = 7 units and 12-ft bar length.

31.10. Stainless steel material is to be rough- and finish-turned. Diameter and length are 4 in. by 30 in. Recommended rough and finish cutting velocity and feed for tungsten carbide tool material are (350 ft/min, 0.015 ipr) and (350 ft/min, 0.007 ipr). Determine rough and finish cutting time.

31.11. Medium-carbon steel is to be rough- and finish-turned using high-speed steel tool material. The part diameter and cutting length are 4 in. and 20 in. Determine the total time to machine.

31.12. Gray cast iron is to be rough- and finish-turned with tungsten carbide tooling. Part diameter and cutting length are 8½ in. and 8.6 in. What is the part rpm for the rough and finish? Find the turning time.

31.13. A stainless steel part is to be drilled ⅟16 in. for a

depth of 1 in. It is followed by a ⅟16-16 tap for a depth of ⅞ in. Find the drilling length and the drilling and tapping time. Repeat for steel.

31.14. Steel is tap drilled ⅜ in. for a depth of 1 in. It is followed with a tapping element of ⅟16-10 hole for ⅞ in. Find the drilling length and the drilling and tapping time. Repeat for stainless steel.

31.15. Find the cutting time for a hard copper shaft 2 by 20 in. long. A surface velocity of 350 ft/min is suggested with a feed of 0.008 in. per revolution. Convert to metric units and repeat.

31.16. An end-facing cut is required for a 10-in. diameter workpiece. The rpm of the lathe is controlled to maintain 400 surface ft/min from the center out to the surface. Feed is 0.009 ipr. Find the time for the cut. Convert to SI and repeat.

31.17. If the tool is K-3H carbide, the material is AISI 4140 steel, depth of cut is 0.050 in., and the feed is 0.010 ipr, what is the turning speed in surface ft/min for a 4-in. bar with a 6 min life if the tool life equation is $VT^{0.3723} = 1022$? Also find the rpm.

31.18. Consider the Taylor tool life model, $VT^n = K$ for the following tool and work materials:

Tool	Work	n	K
High-speed steel	Cast iron	0.14	75
High-speed steel	Steel	0.125	47
Cemented carbide	Steel	0.20	150
Cemented carbide	Cast iron	0.25	130

For a tool life of 10 min for each of these combinations, find the cutting velocity.

31.19. A rough-turning operation is performed on a medium-carbon steel. Tool material is high-speed steel. The part diameter is 4-in. OD and the cutting length is 20 in. The tool point costs $5. Time to reset the tool is 4 min. Part handling is 2 min and the operator wage is $20/h. The Taylor tool life equation $VT^{0.1} = 172$. Feed of the turning operation is 0.020 ipr for a depth of cut = 0.25 in.

a. Plot the four items of cost to find total cost curve and select optimum velocity.

b. Determine the minimum velocity analytically.

31.20. Let the cost of C_t = 0 for Problem 31.19 and plot the elements of time to find optimum time. Compute V_{max}, T_{max}, and T_u.

31.21. Assume that the cost of C_t = 0 for Problem 31.19 and graphically find optimum time, V_{max} and T_{max}.

MORE DIFFICULT PROBLEMS

31.22. Construct individual cost curves similar to Figure 31.3 for the following machining work. A gray iron casting having a diameter of 8½ in. is rough turned to 8.020/8.025 in. for a length of 8.6 in. A renewable square carbide insert is used. The insert has eight corners suitable for turning work and costs $24. Operator and variable expenses less tooling costs are $60/h. The feed for this turning operation is 0.020 ipr. Taylor's tool life equation for part and tool material is $VT^{0.15} = 500$. The time for the operator to remove the insert, install another new corner, and qualify the tool ready to cut is 2 min. Part handling time is 1.65 min for a casting mounted in a fixture.

a. If the y-axis is unit cost and x-axis is ft/min, plot the curves and locate optimum velocity.

b. Determine optimum velocity analytically. (Hints for the following problems: Use Tables 31.4 and 31.5 for the following problems. It will be necessary to determine an operations process plan. Organize the operations around Tables 31.4 and 31.5, assuming that this equipment is available. If there are operations that cannot be manufactured by these tables for these problems, then use your judgment in the setup and cycle estimation. Call a steel service center for prices of raw materials. It may be necessary to make assumptions for quantity and productive hour cost rates that will allow the processing and estimating. Your instructor will give additional hints and boundaries of the problems.)

31.23. Estimate the time for Figure C9.1. Reread the case study for Chapter 9.

31.24. Construct an operations processing plan and estimate for Figure C10.1.

31.25. Construct an operations processing plan and estimate for Figure P11.24.

31.26. Construct an operations processing plan and estimate for Figure P11.28.

PRACTICAL APPLICATION

Call a manufacturing firm and speak to the cost estimator. Discuss the estimator's role in dealing with customers, pricing, full costing, and the manufacturing process.

CASE STUDY: PINION

Find the material cost and setup and cycle time for the operations of a stainless steel pinion similar in all respects to Figure 23.2. (Hints: Assume that stainless steel is three times more costly than C 1030 steel. Complete the operations process sheet and find the unit cost as demonstrated by Figure 31.3. Use Tables 31.4 and 31.5.)

BIBLIOGRAPHY

Alexander, J. M., R. C. Brewer, and G. W. Rowe (1987). *Manufacturing Technology,* 2 vols. Ellis Horwood, Chichester, England.

Amirouche, F. M. L. (1993). *Computer-Aided Design and Manufacturing.* Prentice-Hall, Englewood Cliffs, NJ.

Boothroyd, G., P. Dewhurst, and W. Knight (1994). *Product Design for Manufacture and Assembly.* Marcel Dekker, New York.

Brewer, R. F. (1996). *Design of Experiments for Process Improvement and Quality Assurance.* Engineering & Management Press, Norcross, GA.

Chang, T.-C., R. Wysk, and H.-P. Wang (1991). *Computer-Aided Manufacturing.* Prentice-Hall, Englewood Cliffs, NJ.

Chapanis, A. (1996). *Human Factors in Systems Engineering.* Engineering & Management Press, Norcross, GA.

Degarmo, E. P., J. T. Black, and R. A. Kohser (1996). *Materials and Processes in Manufacturing,* 8th ed. Prentice-Hall, Englewood Cliffs, NJ.

Dieter, G. E. (1991). *Engineering Design: A Materials and Processing Approach,* 2nd ed. McGraw-Hill, New York.

Doyle, L. E., C. E. Keyser, J. L. Leach, G. F. Schrader, and M. S. Singer (1985). *Manufacturing Processes & Materials for Engineers,* 3rd ed. Prentice-Hall, Englewood Cliffs, NJ.

El Wakil, S. D. (1989). *Processes and Design for Manufacturing.* Prentice-Hall, Englewood Cliffs, NJ.

Genevro, G. W. (1990). *Machine Tools: Processes and Applications,* 2nd ed. Prentice-Hall, Englewood Cliffs, NJ.

Gershwin, S. B. (1994). *Manufacturing Systems Engineering.* Prentice-Hall, Englewood Cliffs, NJ.

Groover, M. P. (1996). *Fundamentals of Modern Manufacturing (Materials, Processes and Systems).* Prentice-Hall, Englewood Cliffs, NJ.

Groover, M. P., and E. W. Zimmers (1984). *CAD/CAM: Computer-Aided Design and Manufacturing.* Prentice-Hall, Englewood Cliffs, NJ.

Kalpakjian, S. (1991). *Manufacturing Processes for Engineering Materials.* Addison-Wesley, Reading, MA.

Kalpakjian, S. (1992). *Manufacturing Engineering and Technology,* 3rd ed. Addison-Wesley, Reading, MA.

Koenig, D. T. (1994). *Manufacturing Engineering: Principles for Optimization,* 2nd ed. Taylor & Francis, Washington, DC.

Linbeck, J. R. (1995). *Product Design and Manufacture.* Prentice-Hall, Englewood Cliffs, NJ.

Lindberg, R. A. (1990). *Processes and Materials of Manufacture,* 4th ed. Prentice-Hall, Englewood Cliffs, NJ.

Ludema, K. C., R. M. Caddell, and A. G. Atkins (1987). *Manufacturing Engineering: Economics and Processes.* Prentice-Hall, Englewood Cliffs, NJ.

Mitchell, F. H. (1991). *CIM Systems. An Introduction to Computer-Integrated Manufacturing.* Prentice-Hall, Englewood Cliffs, NJ.

Niebel, B. W., A. B. Draper, and R. A. Wysk (1989). *Modern Manufacturing Process Engineering.* McGraw-Hill, New York.

Oberg, E., et al. (1992). *Machinery's Handbook.* Industrial Press, New York.

Ostwald, P. F. (1988). *AM Cost Estimator,* 4th ed. Penton Publishing, Cleveland, OH.

Ostwald, P. F. (1992). *Engineering Cost Estimating,* 3rd ed. Prentice-Hall, Englewood Cliffs, NJ.

Schey, J. A. (1987). *Introduction to Manufacturing Processes,* 2nd ed. McGraw-Hill, New York.

Society of Manufacturing Engineers (1983–1995). *Tool and Manufacturing Engineers' Handbook,* multivolume set. Society of Manufacturing Engineers, Dearborne, MI.

Wysk, R. (1988). *Computer Integrated Engineering.* Delmar, Albany, NY.

ACKNOWLEDGMENTS AND PHOTO CREDITS

Chapter 1. Fig. 1.1: National Machinery Company. Fig. 1.2: Smithsonian Institution Photo No. 64173. Fig. 1.4: IBM Corporation.

Chapter 4. Fig. 4.1: Aluminum Company of America. Fig. 4.2: Dow Chemical Company.

Chapter 6. Fig. 6.1: Paramount Dire Casting Division, Hayes-Albion Corporation. Fig. 6.3: Titan Metal Manufacturing, Division of Cerro Corporation. Fig. 6.4: HPM Division, Koehring Company. Fig. 6.6: Kaiser Aluminum and Chemical Corporation. Fig. 6.7: Easton Manufacturing Company. Fig. 6.8: American Cast Iron Pipe Company. Fig. 6.9: American Cast Iron Pipe Company. Fig. 6.10: American Cast Iron Pipe Company. Fig. 6.12: Casting Engineers. Fig. 6.13: Misco Precision Casting Company. Fig. 6.14: Universal Casting Corporation. Fig. 6.15: Universal Casting Corporation. Fig. 6.16: The Borden Company. Fig. 6.18: Dow Corning. Fig. 6.19: American Smelting and Refining Company. Fig. 6.20: Aluminum Corporation of America.

Chapter 7. Fig. 7.3: Giddings and Lewis Machine Tool Company. Fig. 7.4: Pope Machinery Company. Fig. 7.6: Setco Industries. Fig. 7.10: South Bend Industries. Fig. 7.15: O.S. Walker Company. Fig. 7.16: Brown and Sharpe Company. Fig. 7.17: Bullard Machine Tool Company. Fig. 7.18: Burgmaster Houidalle.

Chapter 8. Fig. 8.5: Kestler Instrumente A.G. Fig. 8.11: Kennametal Inc.

Chapter 9. Fig. 9.1: Kennametal Inc. Fig. 9.6: George Gorton Machine Company. Fig. 9.10: South Bend Lathe Works. Fig. 9.11: Kennametal Inc. Fig. 9.14*A*, Milacron Inc. Fig. 9.15: Giddings and Lewis Machine Tool Company Fig. 9.17: Quality Machines Inc. Fig. 9.18: Barnes Drill Company.

Chapter 10. Fig. 10.17: Giddings and Lewis Machine Tool Company.

Chapter 12. Fig. 12.10: Mattison Machine Works. Fig. 12.11: The Fosdick Machine Tool Company. Fig. 12.16: Extrude Hone Corporation. Fig. 12.19: Queen Products Division, King-Seeley Thermos Corporation.

Chapter 13. Fig. 13.6: Linde Division, Union Carbide Corporation. Fig. 13.7: Linde Division, Union Carbide Corporation. Fig. 13.10: American Welding Society. Fig. 13.20: Koldweld Corporation. Fig. 13.21: Skiaky Brothers. Fig. 13.24: E. F. Industries. Fig. 13.26: Skiaky Brothers. Fig. 13.27: Skiaky Brothers. Fig. 13.29: Metal and Thermit Corporation. Fig. 13.34: Linde Division, Union Carbide Corporation.

Chapter 14. Fig. 14.4: Chambersburg Engineering Corporation. Fig 14.7: Wyman-Gordon Company. Fig. 14.9: The Ajax Manufacturing Company. Fig. 14.11: Edgewater Steel Company. Fig. 14.6: U.S. Steel. Fig. 14.7: Hydropress. Fig. 14.18: U.S. Steel. Fig. 14.19: The Holokrome Screw Corporation. Fig. 14.21: Reynolds Metal Corporation.

Chapter 15. Fig. 15.2: U.S. Steel. Fig. 15.5: Bethlehem Steel Corporation. Fig. 15.7: Phoenix Products Company. Fig. 15.8: Cincinnati Milacron. Fig. 15.11: National Machinery. Fig. 15.13: National Machinery. Fig. 15.14: Universal Engineering Company. Fig. 15.19: The Yoder Company. Fig. 15.20: The Yoder Company. Fig. 15.24: General Dynamics. Fig. 15.28: USI—Clearing, Division of U.S. Industries, Incorporated. Fig. 15.29: Wheelabrator Corporation.

Chapter 16. Fig. 16.1: Verson Allsteel Press Company. Fig. 16.3: E.W. Bliss Company. Fig. 16.4: E.W. Bliss Company. Fig. 16.5: Verson Allsteel Press Company. Fig. 16.6:

Bethlehem Steel Corporation. Fig. 16.7: Wiedemann Division, The Warner and Swasey Company. Fig. 16.8: The Hydraulic Press Manufacturing Company. Fig. 16.9: E.W. Bliss Company. Fig. 16.13: Niagara Machine and Tool Works. Fig. 16.24: Cincinnati Milacron. Fig. 16.26: Cincinnati Milacron. Fig. C16.1: Coors Container Company.

Chapter 17. Fig. 17.1: Bureau of Standards, U.S. Department of Commerce. Fig. 17.10: Burns, J.L., T.L. Moore, and R.S. Archer, "Quantitative Hardening," Transactions of the ASM, Vol. XXVI, 1938. Fig. 17.11: Bethlehem Steel Works. Fig. 17.16: The Ohio Crankshaft Company. Fig. 17.17: Linde Division, Union Carbide.

Chapter 18. Fig. 18.1: General Electric Company. Fig. 18.2: HPM Division, Koehring Company. Fig. 18.4: F. J. Stokes Company, Division Pennsalt Chemicals Corporation. Fig. 18.5: Owens-Corning Fiberglass Corporation. Fig. 18.6: Owens-Corning Fiberglass Corporation. Fig. 18.7: Owens-Corning Fiberglass Corporation. Fig. 18.9: HPM Division, Koehring Company. Fig. 18.13: Eastman Chemical Products. Fig. 18.14: U.S. Industrial Chemicals Corporation. Fig. 18.15: U.S. Industrial Chemicals Corporation. Fig. 18.16: Phillips Petroleum Company. Fig. 18.17: Centro Inc. Fig. 18.18: Carbide Plastics Company. Fig. 18.19: Carbide Plastics Company. Fig. 18.20: Rohm and Haas Company.

Chapter 19. Fig. 19.3: Universal Instruments. Fig. 19.7: International Rectifier. Fig. 19.9: IBM Product Development Labs. Fig. 19.13: Hi-Tech Manufacturing. Fig. 19.15: Chomerics Inc. Fig. 19.18: Avex Electronics. Fig. 19.19: Martin Marietta, EIM Group. Fig. P19.2: International Rectifier. Fig. P19.8: International Rectifier.

Chapter 20. Fig. 20.3: Cincinnati Milacron. Fig. 20.6: Hammud Machinery Builders. Fig. 20.10: Cincinnati Milacron. Fig. 20.14: Chemcut Corporation. Fig. 20.16: Chemcut Corporation. Fig. 20.17: The International Nickel Company. Fig. 20.18: Bureau of Standards, U.S. Department of Commerce. Fig. 24.19: The International Nickel Company. Fig. 20.20: Metco, Incorporated. Fig. 20.21: Metco, Incorporated. Fig. 20.25: The United States Graphite Company, Division of the Wickes Corporation. Fig. 20.27: National Forge Company. Fig. 20.28: National Aeronautics and Space Administration, NASA SP-5060. Fig. 20.28: Japax Scientific Corporation. Fig. 20.29: Keystone Carbide Corporation. Fig. 20.30: Amplex Division, Chrysler Corporation.

Chapter 21. Fig. 21.24: Fellows Corporation. Fig. 21.30: Gould and Eberhardt, Division of Norton Company.

Chapter 22. Fig. 22.1: Potters Industries, Incorporated. Fig. 22.2: Potters Industries, Incorporated. Fig. 22.3: J. B. Mohler. Fig. 22.4: J. B. Mohler. Fig. 22.5: Myron L Company. Fig. 22.6: Fischer Instrumentation Ltd. Fig. 22.7: Army Materials and Mechanics Research Center. Fig. 22.8: Army Materials and Mechanics Research Center.

Chapter 24. Figs. 24.2 to Fig. 24.19: American National Standards Institute.

Chapter 25. Fig. 25.16: Carbidex Corporation. Fig. 25.21: Verson Allsteel Press Company. Fig. 25.20: Coors Container Company.

Chapter 26. Fig. 26.8: L.S. Starret Company. Fig. 26.9: Pratt and Whitney, Division Miles-Bement-Pond Company. Fig. 26.10: Van Keuren Company. Fig. 26.11: American Society of Mechanical Engineers. Fig. 26.12: Bendix Automation and Measurement Division. Fig. 26.16: Deltronic Corporation. Fig. 26.20: Bendix Automation and Measurement Division. Fig. 26.22: The Esbenson Company. Fig. 26.23: Webster Instruments, Incorporated.

Chapter 28. Fig. 28.1: Makino Machine Tool Company.   Fig. 28.2: Makino Machine Tool Company.   Fig. 28.7: Electronic Industries Association.   Fig. 28.12: *Modern Machine Shop, NC Guide-book and Directory.*   Fig. P28.10: *Modern Machine Shop, NC Guide-book and Directory.*

Chapter 29. Fig. 29.1: Micro Analysis and Design, Incorporated.   Fig. 29.2: Balance Engineering Division.   Fig. 29.5: *Automatic Assembly*, Marcel Dekker, New York.   Fig. 29.6: Gillette Safety Razor Company   Fig. 29.7: Gillette Safety Razor Company   Fig. 29.8: Gillette Safety Razor Company   Fig. 29.13: Westinghouse Corporation.   Fig. 29.14: Cincinnati Milacron.   Fig. 29.24: Westinghouse Corporation.   Fig. 29.32: Westinghouse Corporation.

Chapter 30. Fig. 30.2: North Coast Medical, Incorporated.   Fig. 30.3: Vestil Manufacturing Company.   Fig. 30.5: NASA Anthropometric Source Book, Pub. No. 1024.   Fig. 30.7: Lyon Metal Products, Incorporated.   Figs. 30.11 to 30.13: Ergo-Tech Incorporated.

INDEX